EISENSTEIN SERIES AND AUTOMORPHIC REPRESENTATIONS

This introduction to automorphic forms on adelic groups $G(\mathbb{A})$ emphasises the role of representation theory. The exposition is driven by examples and collects and extends many results scattered throughout the literature, in particular the Langlands constant term formula for Eisenstein series on $G(\mathbb{A})$ as well as the Casselman–Shalika formula for the p-adic spherical Whittaker function. This book also covers more advanced topics such as spherical Hecke algebras and automorphic L-functions.

Many of these mathematical results have natural interpretations in string theory, and so some basic concepts of string theory are introduced with an emphasis on connections with automorphic forms. Throughout the book special attention is paid to small automorphic representations which are of particular importance in string theory, but are also of independent mathematical interest. Numerous open questions and conjectures, partially motivated by physics, are included to prompt the reader's own research.

Philipp Fleig is a Postdoctoral Researcher at the Max Planck Institute for Dynamics and Self-Organization, Germany.

Henrik P. A. Gustafsson is a Postdoctoral Researcher in the Department of Mathematics at Stanford University, California.

Axel Kleinschmidt is a Senior Scientist at the Max Planck Institute for Gravitational Physics (Albert Einstein Institute), Germany, and at the International Solvay Institutes, Brussels.

Daniel Persson is an Associate Professor in the Department of Mathematical Sciences at Chalmers University of Technology, Gothenburg.

CAMBRIDGE STUDIES IN ADVANCED MATHEMATICS

All the titles listed below can be obtained from good booksellers or from Cambridge University Press. For a complete series listing visit: www.cambridge.org/mathematics.

Already published

138 C. Muscalu & W. Schlag *Classical and multilinear harmonic analysis, II*
139 B. Helffer *Spectral theory and its applications*
140 R. Pemantle & M. C. Wilson *Analytic combinatorics in several variables*
141 B. Branner & N. Fagella *Quasiconformal surgery in holomorphic dynamics*
142 R. M. Dudley *Uniform central limit theorems* (2nd Edition)
143 T. Leinster *Basic category theory*
144 I. Arzhantsev, U. Derenthal, J. Hausen & A. Laface *Cox rings*
145 M. Viana *Lectures on Lyapunov exponents*
146 J.-H. Evertse & K. Győry *Unit equations in Diophantine number theory*
147 A. Prasad *Representation theory*
148 S. R. Garcia, J. Mashreghi & W. T. Ross *Introduction to model spaces and their operators*
149 C. Godsil & K. Meagher *Erdős–Ko–Rado theorems: Algebraic approaches*
150 P. Mattila *Fourier analysis and Hausdorff dimension*
151 M. Viana & K. Oliveira *Foundations of ergodic theory*
152 V. I. Paulsen & M. Raghupathi *An introduction to the theory of reproducing kernel Hilbert spaces*
153 R. Beals & R. Wong *Special functions and orthogonal polynomials*
154 V. Jurdjevic *Optimal control and geometry: Integrable systems*
155 G. Pisier *Martingales in Banach spaces*
156 C. T. C. Wall *Differential topology*
157 J. C. Robinson, J. L. Rodrigo & W. Sadowski *The three-dimensional Navier–Stokes equations*
158 D. Huybrechts *Lectures on K3 surfaces*
159 H. Matsumoto & S. Taniguchi *Stochastic analysis*
160 A. Borodin & G. Olshanski *Representations of the infinite symmetric group*
161 P. Webb *Finite group representations for the pure mathematician*
162 C. J. Bishop & Y. Peres *Fractals in probability and analysis*
163 A. Bovier *Gaussian processes on trees*
164 P. Schneider *Galois representations and* (φ, Γ)*-modules*
165 P. Gille & T. Szamuely *Central simple algebras and Galois cohomology* (2nd Edition)
166 D. Li & H. Queffelec *Introduction to Banach spaces, I*
167 D. Li & H. Queffelec *Introduction to Banach spaces, II*
168 J. Carlson, S. Müller-Stach & C. Peters *Period mappings and period domains* (2nd Edition)
169 J. M. Landsberg *Geometry and complexity theory*
170 J. S. Milne *Algebraic groups*
171 J. Gough & J. Kupsch *Quantum fields and processes*
172 T. Ceccherini-Silberstein, F. Scarabotti & F. Tolli *Discrete harmonic analysis*
173 P. Garrett *Modern analysis of automorphic forms by example, I*
174 P. Garrett *Modern analysis of automorphic forms by example, II*
175 G. Navarro *Character theory and the McKay conjecture*
176 P. Fleig, H. P. A. Gustafsson, A. Kleinschmidt & D. Persson *Eisenstein series and automorphic representations*

Eisenstein Series and Automorphic Representations

with Applications in String Theory

PHILIPP FLEIG
Max-Planck-Institut für Dynamik und Selbstorganisation, Germany

HENRIK P. A. GUSTAFSSON
Stanford University, California

AXEL KLEINSCHMIDT
Max-Planck-Institut für Gravitationsphysik, Germany

DANIEL PERSSON
Chalmers University of Technology, Gothenburg

Shaftesbury Road, Cambridge CB2 8EA, United Kingdom

One Liberty Plaza, 20th Floor, New York, NY 10006, USA

477 Williamstown Road, Port Melbourne, VIC 3207, Australia

314–321, 3rd Floor, Plot 3, Splendor Forum, Jasola District Centre, New Delhi – 110025, India

103 Penang Road, #05–06/07, Visioncrest Commercial, Singapore 238467

Cambridge University Press is part of Cambridge University Press & Assessment, a department of the University of Cambridge.

We share the University's mission to contribute to society through the pursuit of education, learning and research at the highest international levels of excellence.

www.cambridge.org
Information on this title: www.cambridge.org/9781107189928

DOI:10.1017/9781316995860

First published 2018

A catalogue record for this publication is available from the British Library

Library of Congress Cataloging-in-Publication data
Names: Fleig, Philipp, author.
Title: Eisenstein series and automorphic representations with applications in string theory / Philipp Fleig, Max-Planck-Institut für Dynamik und Selbstorganisation, Germany [and three others].
Description: Cambridge : Cambridge University Press, 2018. |
Series: Cambridge studies in advanced mathematics ; 176 |
Includes bibliographical references and index.
Identifiers: LCCN 2018018863 | ISBN 9781107189928 (hardback : alk. paper)
Subjects: LCSH: Eisenstein series. | Automorphic functions. | String models.
Classification: LCC QA353.A9 E47 2018 | DDC 512.7/3–dc23
LC record available at https://lccn.loc.gov/2018018863

ISBN 978-1-107-18992-8 Hardback

Meinen ersten Schritten in der Wissenschaft

–P.F.

To my beloved family

–H.P.A.G.

Für meine liebe Mutter

–A.K.

Till min underbara familj

–D.P.

Contents

Definitions and Theorems

Examples

Preface

This book is the result of our endeavour to give a comprehensive presentation of the theory of automorphic representations and the structure of Fourier expansions of automorphic forms with a particular emphasis on adelic methods and Eisenstein series. Our intention is also to open up a channel of communication between mathematicians and physicists, in particular string theorists, interested in these topics. Many of the results presented herein exist already in the literature and we benefited greatly from [447, 116, 291, 292, 569, 164, 290, 598, 599]; our exposition differs, however, from these references in the selection of material and also in highlighting the connection to theoretical physics. Several new results and examples are included as well; in particular, we provide many techniques for working out aspects of the Fourier expansion of Eisenstein series and studying automorphic representations of small functional dimension.

This book is divided into three parts:

PART ONE Automorphic Representations
contains the mathematical backbone of the book, including adeles, Lie theory, Eisenstein series, automorphic representations, Fourier expansions and Hecke theory

PART TWO Applications in String Theory
contains the salient features of string theory to understand the way automorphic forms and representations arise; including scattering amplitudes, perturbative and non-perturbative aspects and supersymmetry

PART THREE Advanced Topics
contains a mix of topics from mathematics and physics, often situated at the interface and highlighting ideas and open problems that can or should be transferred from one field to the other

Many chapters in PART ONE and PART TWO contain sections that are marked with an asterisk and these are more advanced and not strictly necessary for the development of the theory. PART THREE as a whole is advanced by definition and includes many appetisers for current research.

Section 1.4 contains a detailed reader's guide to this book and we refrain from duplicating this summary here. The labelling of theorems, definitions, propositions, remarks and examples is done within each chapter without reference to the (sub)section they are in. Two separate lists, one for examples and one for everything else, are provided at the beginning of the book.

Our understanding of the material presented here was greatly facilitated by numerous discussions with colleagues both from the mathematics community and from the string theory community. We are especially indebted to Guillaume Bossard, David Ginzburg, Joseph Hundley, Stephen D. Miller, Hermann Nicolai, Bengt E.W. Nilsson and Boris Pioline for many clarifying and stimulating discussions and collaborations over many years.

In addition, we gratefully acknowledge the exchanges with Olof Ahlén, Sergei Alexandrov, Marcus Berg, Robert Berman, Benjamin Brubaker, Daniel Bump, Lisa Carbone, Martin Cederwall, Brian Conrad, Thibault Damour, Eric D'Hoker, Dennis Eriksson, Alex J. Feingold, Sol Friedberg, Matthias Gaberdiel, Terry Gannon, Ori Ganor, Howard Garland, Dmitri Gourevitch, Michael B. Green, Murat Günaydin, Stefan Hohenegger, Shamit Kachru, Henry Kim, Ralf Köhl (né Grämlich), Kyu-Hwan Lee, Baiying Liu, Carlos R. Mafra, Gregory W. Moore, Jakob Palmkvist, Manish Patnaik, Christoffer Petersson, Martin Raum, Siddhartha Sahi, Per Salberger, Gordan Savin, Oliver Schlotterer, Philippe Spindel, Stefan Theisen, Pierre Vanhove, Roberto Volpato and Peter West.

Finally, we are indebted to Olof Ahlén, Marcus Berg, Guillaume Bossard, Brian Conrad, Jan Gerken, Stephen D. Miller, Sol Friedberg, Boris Pioline and Freydoon Shahidi for valuable comments on earlier drafts of this book.

We would be very grateful to learn of any omissions and mistakes that we have made unintentionally. A list of errata will be maintained on the website

`www.aei.mpg.de/~axkl/AutomorphicBook`

Philipp Fleig
Henrik Gustafsson
Axel Kleinschmidt
Daniel Persson

1

Motivation and Background

An efficient, but abstract, way to approach the subject of automorphic forms is by the introduction of adeles, rather ungainly objects that nevertheless, once familiar, spare much unnecessary thought and many useless calculations.

— Robert P. Langlands[1]

This first chapter serves as a *tour d'horizon* of the topics covered in this book. Its purpose is to introduce and survey the key concepts relevant for the study of automorphic forms and automorphic representations in the adelic language (PART ONE of the book), their application in string theory (PART TWO) and how they relate to other questions of current research interest in mathematics and physics (PART THREE).

1.1 Automorphic Forms and Eisenstein Series

Automorphic forms are complex functions $f(g)$ on a real Lie group G that

(1) are invariant under the action of a discrete subgroup $\Gamma \subset G$: $f(\gamma \cdot g) = f(g)$ for all $\gamma \in \Gamma$,
(2) satisfy eigenvalue differential equations under the action of the ring of G-invariant differential operators, and
(3) have well-behaved *growth conditions*.

[1] Mostow, G. D. and Caldi, D. G. (eds) 1989. *Representation Theory: Its Rise and Its Role in Number Theory (Proceedings of the Gibbs Symposium)*. American Mathematical Society, Providence, RI.

A more explicit and refined form of these conditions will be given in Chapter 4 when we properly define automorphic forms; here we content ourselves with a qualitative description based on examples. We will mainly be interested in automorphic forms $f(g)$ that are invariant under the action of the *maximal compact subgroup* K of G when acting from the right: $f(gk) = f(g)$ for all $k \in K$; such forms are called *K-spherical.* The automorphic forms are then functions on the coset space G/K.

The prime example of an automorphic form is obtained when considering $G = SL(2,\mathbb{R})$ and $\Gamma = SL(2,\mathbb{Z}) \subset SL(2,\mathbb{R})$. The maximal compact subgroup is $K = SO(2,\mathbb{R})$ and the coset space G/K is a real two-dimensional constant negative curvature space isomorphic to the *Poincaré upper half-plane*,

$$\mathbb{H} = \{z = x + iy \mid x, y \in \mathbb{R} \text{ and } y > 0\}. \tag{1.1}$$

A function satisfying the three criteria above is then given by the non-holomorphic function with $s \in \mathbb{C}$,

$$f_s(z) = \sum_{\substack{(c,d)\in\mathbb{Z}^2\\(c,d)\neq(0,0)}} \frac{y^s}{|cz + d|^{2s}}. \tag{1.2}$$

The sum converges absolutely for $\text{Re}(s) > 1$. The action of an element $\gamma \in SL(2,\mathbb{Z})$ on $z \in \mathbb{H}$ is given by the standard fractional linear form (see Section 4.1)

$$\gamma \cdot z = \frac{az + b}{cz + d} \quad \text{for} \quad \gamma = \begin{pmatrix} a & b \\ c & d \end{pmatrix} \in SL(2,\mathbb{Z}). \tag{1.3}$$

Property (1) is then verified by noting that the integral lattice $(c, d) \in \mathbb{Z}^2$ is preserved by the action of $SL(2,\mathbb{Z})$ and acting with $\gamma \in SL(2,\mathbb{Z})$ on (1.2) merely reorders the terms in the absolutely convergent sum. Property (2) in this case reduces to a single equation since there is only a single primitive $SL(2,\mathbb{R})$-invariant differential operator on $SL(2,\mathbb{R})/SO(2,\mathbb{R})$. This operator is given by

$$\Delta = y^2 \left(\partial_x^2 + \partial_y^2\right) \tag{1.4}$$

and corresponds to the (scalar) *Laplace–Beltrami operator* on the upper half-plane $\mathbb{H}$. In group theoretical terms it is the quadratic Casimir operator. Acting with it on the function (1.2), one finds

$$\Delta f_s(z) = s(s-1) f_s(z) \tag{1.5}$$

and hence $f_s(z)$ is an eigenfunction of Δ (and therefore of the full ring of differential operators generated by Δ). Condition (3) relating to the growth of the function here corresponds to the behaviour of $f_s(z)$ near the boundary of the upper half-plane, more particularly near the so-called *cusp at infinity* when $y \to \infty$.[2] The growth condition requires $f_s(z)$ to grow at most as a power law as $y \to \infty$. To verify this point it is easiest to consider the *Fourier expansion* of $f_s(z)$. This requires a bit more machinery and also paves the way to the general theory. We will introduce it heuristically in Section 1.3 below and in detail in Chapter 6.

The form of the function $f_s(z)$ is very specific to the action of $SL(2,\mathbb{Z})$ on the upper half-plane $\mathbb{H}$. To prepare the ground for the more general theory of automorphic forms for higher-rank Lie groups we shall now rewrite (1.2) in a more suggestive way. In fact, $f_s(z)$ is (almost) an example of a (non-holomorphic) *Eisenstein series* on $G = SL(2,\mathbb{R})$. To see this, we first extract the greatest common divisor of the coordinates of the lattice point $(c,d) \in \mathbb{Z}^2$:

$$f_s(z) = \left(\sum_{k>0} k^{-2s}\right) \sum_{\substack{(c,d)\in\mathbb{Z}^2 \\ \gcd(c,d)=1}} \frac{y^s}{|cz+d|^{2s}} = \zeta(2s) \sum_{\substack{(c,d)\in\mathbb{Z}^2 \\ \gcd(c,d)=1}} \frac{y^s}{|cz+d|^{2s}}, \tag{1.6}$$

where we have evaluated the sum over the common divisor k using the *Riemann zeta function* [541]

$$\zeta(s) = \sum_{n>0} n^{-s}. \tag{1.7}$$

Referring back to (1.3), we can rewrite the summand using an element γ of the group $SL(2,\mathbb{Z})$:

$$\frac{y^s}{|cz+d|^{2s}} = [\mathrm{Im}\,(\gamma \cdot z)]^s \quad \text{for} \quad \gamma = \begin{pmatrix} a & b \\ c & d \end{pmatrix}. \tag{1.8}$$

For interpreting all summands in (1.6) in this way, two things have to occur: (i) for any coprime pair (c,d) such a matrix $\gamma \in SL(2,\mathbb{Z})$ must exist, and (ii) if several matrices exist we must form equivalence classes such that the sum over coprime pairs (c,d) corresponds exactly to the sum over equivalence classes. For (i), we note that the condition that c and d be coprime is necessary since it would otherwise be impossible to satisfy the determinant condition $ad - bc = 1$ over $\mathbb{Z}$. At the same time, co-primality is sufficient to guarantee existence of integers a_0 and b_0 that complete c and d to a matrix $\gamma \in SL(2,\mathbb{Z})$. In fact, there is a one-parameter family of solutions for γ that can be written as

$$\begin{pmatrix} a_0 + mc & b_0 + md \\ c & d \end{pmatrix} = \begin{pmatrix} 1 & m \\ 0 & 1 \end{pmatrix} \begin{pmatrix} a_0 & b_0 \\ c & d \end{pmatrix} \tag{1.9}$$

[2] For $\Gamma = SL(2,\mathbb{Z})$ this is the only cusp up to equivalence. With this, one means that any fundamental domain of the action of Γ on $\mathbb{H}$ only touches the boundary of the upper half-plane at a single point. See Section 4.1 for illustrations and [439, 417, 73] for more details on discrete subgroups of $SL(2,\mathbb{R})$.

for any integer $m \in \mathbb{Z}$. (That these are all solutions to the determinant condition over $\mathbb{Z}$ is an elementary lemma of number theory, sometimes called *Bézout's lemma* [380].) The form (1.9) tells us also how to resolve point (*ii*): we identify matrices that are obtained from each other by left-multiplication by a matrix belonging to the *Borel subgroup*

$$B(\mathbb{Z}) = \left\{ \begin{pmatrix} \pm 1 & m \\ 0 & \pm 1 \end{pmatrix} \middle| m \in \mathbb{Z} \right\} \subset SL(2,\mathbb{Z}). \tag{1.10}$$

The interpretation of this group is that it is the stabiliser of the x-axis (or the cusp at infinity).

Summarising the steps we have performed, we find that we can write the function (1.2) as

$$f_s(z) = 2\zeta(2s) \sum_{\gamma \in B(\mathbb{Z})\backslash SL(2,\mathbb{Z})} [\mathrm{Im}\,(\gamma \cdot z)]^s . \tag{1.11}$$

Since we included the matrix $-\mathbb{1}$ in the definition of $B(\mathbb{Z})$, an extra factor of 2 arises in this formula.

Dropping the normalising zeta factor, we obtain the function

$$E(\chi_s, z) = \sum_{\gamma \in B(\mathbb{Z})\backslash SL(2,\mathbb{Z})} \chi_s(\gamma \cdot z), \tag{1.12}$$

where we have also introduced the notation $\chi_s(z) = [\mathrm{Im}(z)]^s = y^s$. The reason for this notation is that χ_s is actually induced from a *character* on the real Borel subgroup. We will explain this in detail below in Chapter 4. Note that this way of writing the automorphic form makes the invariance under $SL(2,\mathbb{Z})$ completely manifest because it is a *sum over images*.

The form (1.12) is what we will call an *Eisenstein series* on $SL(2,\mathbb{R})$ and it is this form that generalises straightforwardly to Lie groups $G(\mathbb{R})$ other than $SL(2,\mathbb{R})$ (whereas the form with the sum over a lattice does not, as we discuss in more detail in Section 15.3). In complete analogy with (1.12) we define the *(minimal parabolic) Eisenstein series* on $G(\mathbb{R})$ invariant under the discrete group $G(\mathbb{Z})$ by[3]

$$E(\chi, g) = \sum_{\gamma \in B(\mathbb{Z})\backslash G(\mathbb{Z})} \chi(\gamma g), \tag{1.13}$$

where χ is (induced from) a character on the Borel subgroup $B(\mathbb{R})$ and $g \in G(\mathbb{R})$. Eisenstein series are prototypical automorphic forms and the protagonists of the story we shall develop.

[3] For most of this book, we shall take $G(\mathbb{Z})$ as the discrete Chevalley subgroup of an algebraically simply-connected group split over the real numbers; see Section 3.1.4 below for more details.

1.2 Why Eisenstein Series and Automorphic Forms?

Before delving into the further analysis of Eisenstein series and automorphic forms, let us briefly step back and provide some motivation for their study.

1.2.1 A Mathematician's Possible Answer

Automorphic forms are of great importance in many fields of mathematics such as number theory, representation theory and algebraic geometry. The various ways in which automorphic forms enter these seemingly disparate fields are connected by a web of conjectures collectively referred to as the Langlands program [444, 446, 259, 412, 413, 219] that we will discuss more in Chapter 16.

Much of the arithmetic information is contained in the Fourier coefficients of automorphic forms. The standard examples correspond to modular forms on $G(\mathbb{R}) = SL(2,\mathbb{R})$, where these coefficients yield eigenvalues of Hecke operators (covered in Chapter 11) and the counting of points on elliptic curves; see Section 16.4.

For arbitrary Lie groups $G(\mathbb{R})$ one considers the Hilbert space $L^2(\Gamma\backslash G(\mathbb{R}))$ of square-integrable functions that are invariant under a left-action by a discrete subgroup $\Gamma \subset G(\mathbb{R})$. This space carries a natural action of $g \in G(\mathbb{R})$, called the *right-regular action*, through

$$[\pi(g)f](x) = f(xg), \tag{1.14}$$

where $f \in L^2(\Gamma\backslash G(\mathbb{R}))$, $g, x \in G(\mathbb{R})$ and $\pi\colon G(\mathbb{R}) \to \mathrm{Aut}(L^2(\Gamma\backslash G(\mathbb{R})))$ is the right-regular representation map. Since the functions are square-integrable the representation is unitary. This representation-theoretic viewpoint on automorphic forms was first proposed by Gelfand, Graev and Piatetski-Shapiro [267] and later developed considerably by Jacquet and Langlands [376]. This perspective provides the key to generalising the classical theory of modular forms on the complex upper half-plane to higher-rank Lie groups.

It is an immediate, important and difficult question as to what the decomposition of the space $L^2(\Gamma\backslash G(\mathbb{R}))$ into irreducible representations of $G(\mathbb{R})$ looks like. The irreducible constituents in this decomposition are called *automorphic representations*. This spectral problem was tackled and solved by Langlands [447]. The Eisenstein series (and their analytic continuations) form an integral part in the resolution although they themselves are not square-integrable.[4] We shall discuss aspects of this in Chapter 5.

[4] A passing physicist might note that this is very similar to using non-normalisable plane waves as a 'basis' for wave functions in quantum mechanics. Indeed the piece $\chi(\gamma g)$ in (1.13) is exactly like a plane wave; the γ-sum is there to make it invariant under the discrete group by the method of images so that $E(\chi, g)$ are the simplest Γ-invariant plane waves. The decomposition

1.2.2 A Physicist's Possible Answer

Many problems in quantum mechanics are characterised by discrete symmetries. At the heart of many of them lies *Dirac quantisation*, where charges (e.g., electric or magnetic) of physical states are restricted to lie in certain lattices rather than in continuous spaces. The discrete symmetries preserving the lattice are often called *dualities* and can give very interesting different angles on a physical problem. This happens in particular in string theory (see PART TWO), where such dualities can mix perturbative and non-perturbative effects.

For the discrete symmetry to be a true symmetry of a physical theory, all observable quantities must be given by functions that are invariant under the discrete symmetry, corresponding to property (1) discussed at the beginning of Section 1.1. Similarly, the dynamics or other symmetries of the theory impose differential equations on the observables, corresponding to property (2), and the growth condition (3) is typically associated with having well-defined perturbative regimes of the theory. The main example we have in mind here comes from string theory and the construction of scattering amplitudes of type II strings in various maximally supersymmetric backgrounds, as is discussed in PART TWO of this book. However, the logic is not necessarily restricted to this; see also [592, 515, 65, 33] for some other uses of automorphic forms in physics.

For these reasons one is often led naturally to the study of automorphic forms in physical systems with discrete symmetries. Via this route one also encounters a spectral problem similar to the one posed by mathematicians when one needs to determine to which automorphic representation a given physical observable belongs. Again, the Eisenstein series and their properties are important building blocks of such spaces and it is important to understand them well. Furthermore, in a number of examples from string theory it was actually possible to show that the observable is given by an Eisenstein series itself [308, 311, 321, 407, 503, 318, 313, 520, 315].[5]

1.3 Analysing Automorphic Forms and Adelisation

We now return to the study of Eisenstein series defined by (1.13) and their properties, starting again with the very explicit example (1.2) for $SL(2,\mathbb{R})$.

of an automorphic function ('wave-packet') in this basis (extended by the discrete spectrum of cusp forms and residues) is the content of various trace formulas discussed in Section 5.5.

[5] That Eisenstein series are mostly not square-integrable is no problem in these cases since the object computed (part of a scattering amplitude) is not a wavefunction and not required to be normalisable.

1.3.1 Fourier Expansion of the $SL(2,\mathbb{R})$ Series

The discrete Borel subgroup $B(\mathbb{Z})$ of (1.10) acts on the variable $z = x + iy$ through translations by

$$\begin{pmatrix} \pm 1 & m \\ 0 & \pm 1 \end{pmatrix} \cdot z = z \pm m \quad \text{for } m \in \mathbb{Z} \tag{1.15}$$

and therefore any $SL(2,\mathbb{Z})$-invariant function is periodic in the x direction with period equal to 1 corresponding to the smallest non-trivial $m = 1$. This means that we can Fourier expand it in modes $e^{2\pi i n x}$. Applying this to (1.12) leads to

$$E(\chi_s, z) = \underbrace{C(y)}_{\substack{\text{constant term} \\ \text{zero mode}}} + \underbrace{\sum_{n\neq 0} a_n(y) e^{2\pi i n x}}_{\text{non-zero mode}}. \tag{1.16}$$

As we indicated, it is natural to divide the *Fourier expansion* into two parts depending on whether one deals with the zero Fourier mode (a.k.a. *constant term*) or with a non-zero Fourier mode. Since the Fourier expansion was only in the x-direction, the Fourier coefficients still depend on the second variable y.[6]

Determining the explicit form of the Fourier coefficients is one of the key problems in the study of Eisenstein series. In the example of $SL(2,\mathbb{R})$, this can for instance be done by making recourse to the formulation in terms of a lattice sum that was given in (1.2) and using the technique of *Poisson resummation.* The calculation is reviewed in Appendix A and leads to the following explicit expression:

$$\begin{aligned} E(\chi_s, z) = {} & y^s + \frac{\xi(2s-1)}{\xi(2s)} y^{1-s} \\ & + \frac{2y^{1/2}}{\xi(2s)} \sum_{m\neq 0} |m|^{s-1/2} \sigma_{1-2s}(m) K_{s-1/2}(2\pi |m| y) e^{2\pi i m x}, \end{aligned} \tag{1.17}$$

where

$$\xi(s) = \pi^{-s/2}\Gamma(s/2)\zeta(s) \tag{1.18}$$

is the *completion of the Riemann zeta function* (1.7) with the standard Γ-function, $K_s(y)$ is the *modified Bessel function of the second kind* (which decreases exponentially for $y \to \infty$ in accordance with the growth condition) and

$$\sigma_{1-2s}(m) = \sum_{d|m} d^{1-2s} \tag{1.19}$$

[6] If one dealt with an automorphic form holomorphic in z (called modular form in Chapter 4 below) this would not be true since the holomorphicity condition links the x and y dependence. The Fourier coefficients in an expansion in $q = e^{2\pi i(x+iy)} = e^{2\pi i z}$ would be pure numbers. This is the origin of the name *constant term* for the zero mode in (1.16).

is called a *divisor sum* (or the *instanton measure* in physics; see Section 13.6.3) and given by a sum over the positive divisors of $m \neq 0$.

As is evident from (1.17), the explicit form of the Fourier expansion can appear quite complicated and involves special functions as well as number-theoretic objects. For the case of more general groups $G(\mathbb{R})$, the method of Poisson resummation is not necessarily available as there is not always a form of the Eisenstein series as a lattice sum. *It is therefore desirable to develop alternative techniques for obtaining (parts of) the Fourier expansion under more general assumptions.*[7] This is achieved by lifting the Eisenstein series into an adelic context which we now sketch and explain in more detail in Section 4.4.

1.3.2 Adelisation of Eisenstein Series

A standard elementary technique in number theory for analysing equations over $\mathbb{Z}$ is by analysing them instead as congruences for every prime (and its powers) separately (sometimes known as the *Hasse principle* or the *local-global principle* based on the *Chinese remainder theorem*) [23, 500]. One way of writing all the terms together is to use the *ring of adeles* $\mathbb{A}$. The adeles can formally be thought of as an infinite tuple

$$a = (a_\infty; a_2, a_3, a_5, a_7, \ldots) \in \mathbb{A} = \mathbb{R} \times \prod_{p<\infty}{}' \mathbb{Q}_p, \tag{1.20}$$

where $\mathbb{Q}_p$ denotes the *p-adic numbers*. The p-adic numbers are a completion of the rational number $\mathbb{Q}$ that is inequivalent to the standard one (leading to $\mathbb{R}$) and that is parametrised by a prime number p and defined properly in Section 2.1. The product is over all prime numbers and the prime on the product symbol indicates that the entries a_p in the tuple are restricted in a certain way (see (2.59) below for the exact statement). The real numbers $\mathbb{R}$ can be written as $\mathbb{Q}_\infty$ in this context and interpreted as the completion of $\mathbb{Q}$ at the 'prime' $p = \infty$. *Very* crudely, an adele can be thought of as summarising the information of an object modulo all primes.

Strong approximation is a similar method that lifts a general automorphic form from being defined on the space $G(\mathbb{Z})\backslash G(\mathbb{R})/K(\mathbb{R})$ to the space $G(\mathbb{Q})\backslash G(\mathbb{A})/K(\mathbb{A})$ so that $G(\mathbb{Q})$ plays the rôle of the discrete subgroup that was played by $G(\mathbb{Z})$ before. However, $G(\mathbb{Q})$ is a nicer group than $G(\mathbb{Z})$ since $\mathbb{Q}$ is a field whereas $\mathbb{Z}$ is only a ring. This facilitates the analysis and allows the application of many theorems for algebraic groups.

[7] Additional care has to be taken for the Fourier expansion for general $G(\mathbb{R})$ also because the translation group $B(\mathbb{Z})$ is in general not abelian. One can still define (abelian) Fourier coefficients, as we will see; however, they fail to capture the full Eisenstein series. There are also non-abelian parts to the Fourier expansion, as we discuss in Chapter 6.

A consequence of using strong approximation and adeles is that the result of many calculations factorises according to (1.20) and one can do the calculation for all primes and $p = \infty$ separately. Indeed, the explicit form (1.17) for the Fourier expansion of the $SL(2,\mathbb{R})$ Eisenstein series already secretly had this form. This can be seen for example in the constant term, since

$$\frac{\xi(2s-1)}{\xi(2s)} = \pi^{1/2}\frac{\Gamma(s-1/2)}{\Gamma(s)}\prod_{p<\infty}\frac{1-p^{-2s}}{1-p^{1-2s}}, \tag{1.21}$$

where we have used the definition of the completed zeta function from (1.18) and the *Euler product formula* for the Riemann zeta function [541]

$$\zeta(s) = \sum_{n>0} n^{-s} = \prod_{p<\infty}\frac{1}{1-p^{-s}}. \tag{1.22}$$

In (1.21), we clearly recognise a factorised form that is very similar to (1.20). That this is not an accident will be demonstrated in Section 7.2 for $SL(2,\mathbb{R})$. The factorisation of the (completed) Riemann zeta function itself can also be understood in this way; see Section 2.7. For the other Fourier modes in (1.17) we get a similar factorisation with the modified Bessel function belonging to the $p = \infty$ factor and

$$\sigma_{1-2s}(m) = \prod_{p<\infty}\gamma_p(m)\frac{1-p^{-(2s-1)}\,|m|_p^{2s-1}}{1-p^{-(2s-1)}}, \tag{1.23}$$

where $|m|_p$ is the p-adic norm of m defined in Section 2.1 and $\gamma_p(m)$ selects all factors with $|m|_p \leq 1$ as shown in Section 2.4. The complete derivation for the non-constant terms can be found in Section 7.3 for the $SL(2,\mathbb{R})$ Eisenstein series.

The adelic methods are so powerful that one can obtain a closed, simple and group-theoretic formula for the constant term of Eisenstein series on any (split real) Lie group $G(\mathbb{R})$. This formula, known as the *Langlands constant term formula*, will be the topic of Chapter 8.

For the (abelian) Fourier coefficients, the adelic methods also help to obtain fairly general results, in particular for the part that involves the finite primes $p < \infty$. For the contribution coming from the $\mathbb{R}$ in (1.20) the results are not quite as general; already for $SL(2,\mathbb{R})$ this is what gives the modified Bessel function. We discuss the computation of Fourier coefficients in Chapter 9.

1.4 Reader's Guide and Main Theorems

The following is a brief outline of the contents and results of this book that is divided into three parts. Part One deals with automorphic forms and automorphic representations, PART TWO addresses their applications in string theory and PART THREE discusses advanced topics of current research in mathematics and physics.

As a general rule, we have placed a lot of emphasis on examples and explicit calculations where possible to illustrate the key concepts. The main theorems are proved in full detail and many advanced ideas are illustrated with the main steps, and references are given for additional information. Topics that are somewhat supplementary to the main thread of the exposition are contained in sections marked with an asterisk, and they are not essential to follow the development of the main theory.

We also would like to give some disclaimers: unless otherwise specified, all groups $G(\mathbb{R})$ that will be considered here are associated with split real forms and we also restrict to algebraically simply-connected groups. Except for certain sections in PART THREE, the base field for the ring of adeles will be the rational numbers $\mathbb{Q}$. Often we will perform formal manipulations of infinite sums and integrals without paying attention to whether the expressions are (absolutely) convergent or not. The expressions typically depend on a set of parameters and for some range of parameters convergence can be established, and we assume them to be in this range. In many cases, the results can be extended by analytic continuation.

Contents of PART ONE: AUTOMORPHIC REPRESENTATIONS

Chapters 2 and 3 are introductory and provide some background material and set the notation for the rest of the book. More precisely, Chapter 2 introduces the basic machinery of p-adic and adelic analysis which will be crucial for almost everything we do later. The main thrust of the chapter is provided by the numerous examples of computing p-adic integrals that will be used extensively in proving Langlands' constant term formula, and computing Fourier coefficients of Einstein series. In Chapter 3, we introduce some basic features of Lie algebras and Lie groups that will be used throughout the book. We first discuss Lie groups and Lie algebras over $\mathbb{R}$, their discrete subgroups and then move on to algebraic groups over $\mathbb{Q}_p$ as well as adelic groups.

The proper discussion of automorphic forms and automorphic representations begins after these preliminaries, and we now summarise the structure of the remainder of PART ONE in a little more detail, with emphasis on the central results in each chapter.

- In Chapters 4 and 5, we introduce the general theory of automorphic forms and automorphic representations. We start out gently by discussing how to pass from modular forms on the upper half-plane to automorphic forms on the adelic group $SL(2, \mathbb{A})$. We then move on to the general case of arbitrary Lie groups. We define *Eisenstein series* $E(\chi, g)$ for general split real Lie groups $G(\mathbb{R})$ that are invariant under the discrete Chevalley subgroup $G(\mathbb{Z})$. The definition (1.13) requires the choice of a character χ of a parabolic subgroup P of G; alternatively, we can think of χ as being defined by a choice of weight vector λ of the (split real) Lie algebra of $G(\mathbb{R})$. We explain how this can be understood from the point of view of the representation theory of $G(\mathbb{R})$, and we show how to lift the function from being defined on $G(\mathbb{R})$ to a function defined on $G(\mathbb{A})$ where $\mathbb{A}$ are the *adeles* of the rational number field $\mathbb{Q}$.

- A major part of this book is devoted to analysing Fourier expansions of automorphic forms. This is a highly non-trivial subject with many interesting connections to representation theory as well as to physics. In Chapter 6, we introduce the general theory of Fourier coefficients and Whittaker coefficients, with emphasis on the representation-theoretic viewpoint. Of particular interest will be so-called small automorphic representations. Toward the end of the chapter we also introduce some more advanced topics, such as nilpotent orbits and wavefront sets, as well as the Piatetski-Shapiro–Shalika method.
- In Chapter 7, we illustrate all the general techniques in the context of Eisenstein series on $SL(2)$. Specifically, using adelic techniques, we provide a detailed proof of the following classic theorem:

Theorem 1.1 ($SL(2)$ **Eisenstein series**) *The complete Fourier expansion of the Eisenstein series* $E(\chi_s, z)$ *for* $z = x + iy \in SL(2,\mathbb{R})/SO(2,\mathbb{R})$ *is given by*

$$E(\chi_s, z) = y^s + \frac{\xi(2s-1)}{\xi(2s)} y^{1-s} + \sum_{m \neq 0} \frac{2y^{1/2}}{\xi(2s)} |m|^{s-1/2} \sigma_{1-2s}(m) K_{s-1/2}(2\pi|m|y) e^{2\pi i m x}. \quad (1.24)$$

Furthermore, the Eisenstein series satisfies the functional relation

$$E(\chi_s, z) = \frac{\xi(2s-1)}{\xi(2s)} E(\chi_{1-s}, z), \quad (1.25)$$

where ξ *is the (completed) Riemann zeta function. The rest of the notation is explained in Chapter 7.*

- The first two terms in the Fourier expansion above correspond to the zeroth Fourier coefficients. These are often collectively referred to as the *constant term* of the Eisenstein series. A very important and general result in this context is provided by the so-called Langlands constant term formula, which yields a remarkably simple expression for the complete constant term of Eisenstein series on arbitrary semi-simple Lie groups. In Chapter 8, we give a complete proof of the following theorem of Langlands:

Theorem 1.2 (Langlands constant term formula) *Let G be a $\mathbb{Q}$-split semi-simple simply-connected algebraic group with $G(\mathbb{R})$ a real semi-simple Lie group and $G(\mathbb{A})$ the corresponding adelic group. Let λ be a weight of the Lie algebra $\mathfrak{g}$, $\mathcal{W}$ the associated Weyl group and N a maximal unipotent radical of G. We then have*

$$\int\limits_{N(\mathbb{Z})\backslash N(\mathbb{R})} E(\lambda, ng)dn = \sum_{w\in\mathcal{W}} a^{w\lambda+\rho} \prod_{\alpha>0:\, w\alpha<0} \frac{\xi(\langle\lambda|\alpha\rangle)}{\xi(1+\langle\lambda|\alpha\rangle)}, \tag{1.26}$$

where a belongs to the Cartan torus $A \subset G$, and the product runs over positive roots α of $\mathfrak{g}$ that are mapped to negative roots by the Weyl word $w \in \mathcal{W}$.

- The infinite sum in the Fourier expansion (1.24) corresponds to the non-zero coefficients and this is generally referred to as the *non-constant term*. In Chapter 9, we discuss the general structure of Fourier coefficients of Eisenstein series on reductive groups G. For this part of the expansion much less is known explicitly. However, there exists a beautiful formula due to Kato–Shintani–Casselman–Shalika, commonly known as the *Casselman–Shalika formula*, which gives an explicit expression for the so-called p-adic *Whittaker coefficient*. This corresponds to a local version of the Fourier coefficient of the Eisenstein series, which enters in assembling the full (global) coefficient. A large part of Chapter 9 is therefore devoted to proving the following theorem:

Theorem 1.3 (Casselman–Shalika formula) *Let $G(\mathbb{Q}_p)$ $(p < \infty)$ be a p-adic split semi-simple Lie group, $N(\mathbb{Q}_p)$ a maximal unipotent radical of $G(\mathbb{Q}_p)$ and ψ an unramified unitary character on $N(\mathbb{Q}_p)$. The Casselman–Shalika formula is given by*

$$\int_{N(\mathbb{Z}_p)\backslash N(\mathbb{Q}_p)} E(\lambda, na)\overline{\psi(n)}dn = \frac{\epsilon(\lambda)}{\xi(\lambda)} \sum_{w\in\mathcal{W}} (\det(w))|a^{w\lambda+\rho}| \prod_{\substack{\alpha>0\\ w\alpha<0}} p^{\langle\lambda|\alpha\rangle}, \tag{1.27}$$

where a belongs to a Cartan subgroup $A \subset G$ and $\epsilon(\lambda)$ and $\xi(\lambda)$ are defined in terms of the algebraic data of G in Chapter 9.

- For certain special types of Fourier coefficients, so-called *degenerate Whittaker coefficients*, one can take this one step further and compute the full global coefficient (and not just the p-adic version). In Chapter 9 we also prove the following theorem which gives such a formula:

Theorem 1.4 (Reduction for degenerate Whittaker coefficients) *Let $\psi\colon N(\mathbb{Q})\backslash N(\mathbb{A}) \to U(1)$ be a degenerate character with associated* *subgroup $G'(\mathbb{A}) \subset G(\mathbb{A})$ on which it is supported. Then the degenerate Whittaker function on $G(\mathbb{A})$ is given by*

$$W_\psi^\circ(\chi, a) = \sum_{w_c w'_{\text{long}} \in \mathcal{W}/\mathcal{W}'} a^{(w_c w'_{\text{long}})^{-1}\lambda+\rho} M(w_c^{-1}, \lambda) W'^\circ_{\psi^a}(w_c^{-1}\lambda, \mathbb{1}), \tag{1.28}$$

where W'°_ψ denotes a Whittaker coefficient on the $G'(\mathbb{A})$ subgroup of $G(\mathbb{A})$. The weight $w_c^{-1}\lambda$ is given as a weight of $G'(\mathbb{A})$ by orthogonal projection.

A more complete formulation is provided in Section 9.5. In Section 9.8 we also provide an extensive example of how to calculate Whittaker coefficients for Eisenstein series on $SL(3, \mathbb{A})$.

- In Chapter 10 we illustrate how to perform calculations with Eisenstein series in practice. More specifically we explain how to evaluate the Langlands constant term formula in concrete examples, which, in particular, involves a detailed analysis of the functional equation. We also show how to perform similar evaluations of the Whittaker coefficients that appear in the non-constant Fourier coefficients. We provide some explicit examples for exceptional Lie groups and discuss their relation to automorphic representation of small functional dimension.
- It is interesting to note that both the Langlands constant term formula (1.26) and the Casselman–Shalika formula (1.27) have cunning similarities to the Weyl character formula. That this is not a coincidence is a central insight of Langlands. To understand this requires the additional machinery of *Hecke theory*, which is the topic of Chapter 11. Here we explain how to pass from the classical notion of Hecke operators acting on modular forms to the general notion of *spherical Hecke algebras* on adelic groups. This analysis leads us to a reformulation of the Casselman–Shalika formula that clearly illustrates the intimate connection with the Weyl character formula. In this context we are naturally led to the notion of the *Langlands dual group* ^LG (also called L-group) and to the notion of *automorphic L-functions*, which form a central ingredient in the Langlands program. The chapter concludes with a discussion of the *Langlands–Shahidi method*, which is a powerful way to construct L-functions from automorphic representations.
- One useful tool for constructing automorphic forms beyond Eisenstein series is the so-called *theta correspondence*. We begin Chapter 12 by looking at classical theta functions before introducing theta correspondences through dual pairs more generally. Examples of how one can use this technique, for instance to study minimal representations of exceptional groups, are provided.

Contents of Part Two: Applications in String Theory

The second part of the book begins with an introduction to string theory. This is done with an emphasis on the elements that are most important for the relation to automorphic forms and representation theory.

- Chapter 13 describes the basic formulation of perturbative string theory through world-sheet actions and discusses also the continuous and discrete symmetries. We then focus on the calculation of scattering amplitudes of string theory states. Detailed examples of the four-graviton scattering amplitude for the first two orders in the perturbative string expansion are given and we discuss a connection of the resulting expression to theta lifts and automorphic forms, a topic already discussed in Chapter 12.
- As more advanced items in Chapter 13, we give an introduction to aspects of string theory that lie beyond the standard perturbative expansion. The associated non-perturbative objects can for example be branes that are needed for the consistency of the theory. In the calculation of scattering amplitudes, non-perturbative effects also arise in the form of instanton backgrounds and we demonstrate the basic steps in finding such contributions.
- In Chapter 14, we then properly identify the way automorphic forms arise in string scattering amplitudes. An important notion in this context is the low-energy or so-called α'-expansion of the scattering amplitude. The terms appearing at various orders in this expansion can be shown to satisfy discrete duality invariances, obey differential equations and have nice growth conditions. They therefore fit the framework of automorphic forms nicely if the differential conditions can be understood as eigenvalue equations. We review the cases where this can be done and identify the corresponding automorphic forms that turn out to be Eisenstein series on split real Lie groups. We discuss how the Fourier expansion techniques developed in Part One can be applied to them and what the physical interpretation of the various pieces is. The constant term will be seen to correspond to certain perturbative contributions whereas the non-zero Fourier coefficients very naturally are associated with non-perturbative contributions. As the direct determination of non-perturbative physics can be very involved, the technique of Fourier expansion gives very precise and valuable predictions for these effects. It is also possible to relate the automorphic representation constraints to constraints arising from supersymmetry in string theory.
- The last Chapter 15 discusses generalisations of the notion of automorphic forms that string theory seems to also call for and other places in string theory where automorphic forms occur. The differential conditions on the various pieces in the low-energy expansion of a string scattering amplitude need not be eigenvalue equations and indeed typically are not. We discuss in some

detail the simplest such case, called the $\nabla^6 R^4$ correction, and what is known about the more general automorphic form that it represents, with particular emphasis on the possible interpretation as an automorphic representation.

- String theory is also a rich source of $SL(2,\mathbb{Z})$-invariant functions that are generated as so-called *modular graph functions* in the string perturbative expansion at the one-loop order. We review this construction and the known theorems and conjectures on these functions that are also related to (single-valued) elliptic multi-zeta values in Section 15.2.
- In string theory, automorphic functions often arise as sums over lattices, where the sum can be constrained or unconstrained. We discuss the relation of this description of duality invariant functions to the orbit description as it arises for Eisenstein series in Section 15.3.
- Section 15.4 discusses a slightly different aspect of automorphic forms in string theory, namely how they relate to counting problems, in particular understanding the degeneracy of black hole states in different string theory models with different amounts of supersymmetry. So-called one-loop partition functions of string theory also count specific states, and if the string theory model is set up correctly, these partition functions are characters of modules that also appear in the discussion of the moonshine phenomenon, where representations and characters of finite sporadic simple groups are related to automorphic forms. These developments are discussed in Section 15.5.

Contents of Part Three: Advanced Topics

The final third part of the book is devoted to many different developments in mathematics and physics that build on the ideas and concepts of the first two parts. In the discussion of these developments, we emphasise open questions and conjectures and provide references for readers wishing to submerge themselves in one of these exciting subjects.

- The first topic to be explored in Chapter 16 is the Langlands program that comes in two flavours. The original and classical version deals with a conjectured relation between automorphic representations and Galois representations or, more generally, the functoriality idea that transfers automorphic representations from one group to another in such a way that certain L-functions are related. The more recent geometric version of Langlands' conjectures places them in the world of algebraic geometry where all the objects from the classical Langlands program have geometric counterparts. We exemplify this by presenting the geometric version of Eisenstein series. A celebrated theorem that is connected to the classical Langlands program is the *modularity theorem* that relates rational elliptic curves to modular properties. These concepts were crucial in the proof of Fermat's last theorem as we also review.

- Another recent development is the connection between automorphic forms on metaplectic covers of real Lie groups and integrable statistical systems. This connection proceeds through the so-called multiple Dirichlet series and Yang–Baxter equations. We focus in Chapter 17 on the properties of the Eisenstein series and Whittaker functions on the metaplectic covers and how their combinatorics works in the p-adic case.
- Most of the book deals with split real Lie groups. Other cases are of interest not only in mathematics but also in string theory and therefore in Chapter 18 we treat an explicit example based on the real form $SU(2,1)$ of $SL(3,\mathbb{C})$. We show how to construct Eisenstein series invariant under certain Picard groups and what the changes in the Langlands constant formula are. We also discuss how this relates conjecturally to properties of string theory compactifications on rigid Calabi–Yau spaces where certain moduli spaces can be expressed in terms of twistor spaces on which the Picard groups should act.
- As the last topic of this book, we discuss in Chapter 19 the way in which many of the notions that pervade this text are expected to generalise to infinite-dimensional Lie groups, especially Kac–Moody groups. The case of loop groups (or affine groups) has been studied in most detail but the number-theoretic implications encoded in the Fourier expansions have not been unveiled. Moreover, we look at the case where the Kac–Moody group is not affine but hyperbolic or more generally indefinite. The theory of automorphic forms in this case is in its infancy and we review what is known. Remarkably, these groups also arise in string theory and maximal supergravity and we discuss what conjectures for the theory of automorphic forms and representations can be deduced from physics.

Four appendices contain complementary details to the material of this book.

PART ONE

AUTOMORPHIC REPRESENTATIONS

2

Preliminaries on p-adic and Adelic Technology

As seen in Section 1.3.2, the Fourier expansion of the $SL(2,\mathbb{R})$ Eisenstein series factorises into an Euler product over all primes p. This number-theoretic information is best captured by introducing the p-adic numbers $\mathbb{Q}_p$ which, for any prime p, are an extension of the rational numbers $\mathbb{Q}$, and furthermore the ring of adeles $\mathbb{A} \equiv \mathbb{A}_{\mathbb{Q}}$ that encapsulates all the different p-adic extensions in a single object.

This chapter is intended as an introduction to these objects as well as providing sample calculations that will be used throughout the remaining text. Additional reading can for example be found in [500, 164, 23, 263]. Readers familiar with the subject are welcome to proceed to the next chapter and come back to the explicit examples when needed later on in the text. Further reading can be found in [92] and [164].

Throughout this book, p will denote a prime number in the integers $\mathbb{Z}$, unless stated otherwise.

2.1 p-adic Numbers

We start by providing the basic definitions and discussing some of the properties of p-adic numbers.

Definition 2.1 (p-adic integers $\mathbb{Z}_p$) The *p-adic integers* $\mathbb{Z}_p$ are formal power series in p with coefficients between 0 and $p-1$

$$x \in \mathbb{Z}_p \Leftrightarrow x = x_0 p^0 + x_1 p^1 + \cdots \qquad \text{with } x_i \in \mathbb{Z}/p\mathbb{Z} \cong \{0, 1, \ldots, p-1\}\ . \tag{2.1}$$

The p-adic integers form a ring.

Arithmetic operations on the p-adic integers work in the usual manner. However, since all coefficients in the expansion are positive it may not be immediately obvious how the additive inverse (i.e., subtraction) works. As an example, consider the equation $x + 1 = 0$ that should have a solution over the ring $\mathbb{Z}_p$. The inverse is given by the infinite power series in p with all coefficients equal to $p - 1$:

$$x = \sum_{i=0}^{\infty} (p-1)p^i . \tag{2.2}$$

This is a bit like evaluating the (non-converging) sum $x = \sum_{k=0}^{\infty} 10^k = 1 + 10 + 100 + \cdots$ in decimal notation to be an infinite string of 1s. Multiplying by 9 and then adding 1 creates a zero for every decimal place. Hence $9x + 1 = 0 \Leftrightarrow x = -1/9 = 1/(1-10)$ in agreement with a naive application of the geometric series definition.

Next we define the p-adic number field.

Definition 2.2 (p-adic number field $\mathbb{Q}_p$) The associated number field is given by the *p-adic numbers* $\mathbb{Q}_p$ that are formal Laurent series in p with a finite number of terms of degree less than zero, i.e., finite polar part

$$x = x_k p^k + x_{k+1} p^{k+1} + \cdots \quad \text{with } x_k \neq 0 , \tag{2.3}$$

where k is some integer not necessarily positive.

The uncountable field $\mathbb{Q}_p$ has characteristic zero and cannot be ordered. As discussed in more detail below in Remark 2.7, the rationals $\mathbb{Q}$ are a subfield of $\mathbb{Q}_p$ for any p and the p-adic numbers $\mathbb{Q}_p$ can be thought of as the completion of $\mathbb{Q}$ with respect to the following norm.

Definition 2.3 (p-adic norm $|\cdot|_p$) The *p-adic norm* on $\mathbb{Q}_p$ is given by

$$|x|_p = p^{-k} \quad \Leftrightarrow \quad \text{with } x_k \neq 0 . \tag{2.4}$$

The p-adic norm is multiplicative

$$|x \cdot y|_p = |x|_p |y|_p \tag{2.5}$$

and satisfies a stronger triangle inequality than generic norms, namely

$$|x + y|_p \leq \max(|x|_p, |y|_p) , \tag{2.6}$$

for $x, y \in \mathbb{Q}_p$. This second property is called an *ultrametric property* and a space with a norm of this type is called *non-archimedean* in contrast with *archimedean spaces* satisfying the usual archimedean triangle inequality. The p-adic norm of 0 is $|0|_p = 0$.

The integer k in (2.4) is called the p-adic *valuation* of $\mathbb{Q}$ or $\mathbb{Q}_p$ and is often also denoted by $v_p(x)$. Two properties of the p-adic valuation, equivalent to the ones above for the p-adic norm, are

$$\nu_p(x \cdot y) = \nu_p(x) + \nu_p(y) \tag{2.7}$$

and

$$\nu_p(x + y) \geq \min(\nu_p(x), \nu_p(y)), \tag{2.8}$$

where in the last property equality is achieved if $\nu_p(x) \neq \nu_p(y)$.

The integers in the normed space $\mathbb{Q}_p$ can then be expressed as

$$\mathbb{Z}_p = \left\{ x \in \mathbb{Q}_p \,\middle|\, |x|_p \leq 1 \right\}, \tag{2.9}$$

i.e., they have an exponent $k \geq 0$ of p. This shows that the p-adic integers are compactly embedded in $\mathbb{Q}_p$. The complementary set to $\mathbb{Z}_p$ in $\mathbb{Q}_p$ is given by

$$\mathbb{Q}_p \setminus \mathbb{Z}_p = \{x \in \mathbb{Q}_p \mid |x|_p > 1\}. \tag{2.10}$$

Let us provide two simple examples illustrating the p-adic expansion of a rational number.

Example 2.4: p-adic expansions

We consider the p-adic expansion of the rational number $x = \frac{1}{2} \in \mathbb{Q}$ for $p = 2$ and $p = 3$.

For $p = 2$ one has $|x|_2 = 2^1 = 2$ or $\nu_2(x) = -1$ and hence $\frac{1}{2}$ is not a 2-adic integer. As an element of the 2-adic numbers $\mathbb{Q}_2$ one finds $\frac{1}{2} = 1 \cdot 2^{-1}$ as the expansion of the form (2.3).

For $p = 3$ one has $|x|_3 = 3^0 = 1$ or $\nu_3(x) = 0$ and hence $\frac{1}{2}$ is a 3-adic integer. Its expansion of the form (2.3) is $\frac{1}{2} = 2 \cdot 3^0 + \sum_{k>0} 3^k$.

Another useful property for the p-adic norm of the greatest common divisor of two integers, which will be used in Section 9.8, is introduced in the following example.

Example 2.5: Norm of a greatest common divisor

Let m and n be two integers, $d = \gcd(m, n)$, $m' = m/d$ and $n' = n/d$. Then $1 = \gcd(m', n')$ which, for a prime p, means that if $|m'|_p < 1$ (that is, $p \mid m'$) then $|n'|_p = 1$ (that is, $p \nmid n'$) and vice versa. Thus, $1 = |\gcd(m', n')|_p = \max(|m'|_p, |n'|_p)$. Hence,

$$|d|_p = |d|_p \max\left(\left|m'\right|_p, \left|n'\right|_p\right) = \max\left(\left|m'd\right|_p, \left|n'd\right|_p\right) = \max(|m|_p, |n|_p). \tag{2.11}$$

We also define the multiplicatively invertible p-adic numbers in $\mathbb{Z}_p$ and $\mathbb{Q}_p$.

Definition 2.6 (Multiplicatively invertible numbers $\mathbb{Z}_p^\times$ and $\mathbb{Q}_p^\times$) The set of *multiplicatively invertible elements* in $\mathbb{Z}_p$ will be denoted by

$$\begin{aligned} \mathbb{Z}_p^\times &= \left\{ x \in \mathbb{Z}_p \,\middle|\, x^{-1} \text{ exists in } \mathbb{Z}_p \right\} = \left\{ x \in \mathbb{Z}_p \,\middle|\, |x|_p = 1 \right\} \\ &= \left\{ x \in \mathbb{Q}_p \,\middle|\, |x|_p = 1 \right\}. \end{aligned} \tag{2.12}$$

They correspond to those x in (2.1) for which $x_0 \neq 0$. The set of multiplicatively invertible elements $\mathbb{Q}_p^\times$ in $\mathbb{Q}_p$ is defined as

$$\mathbb{Q}_p^\times = \left\{ x \in \mathbb{Q}_p \,\middle|\, |x|_p \neq 0 \right\}. \tag{2.13}$$

By definition, we formally allow p to also take the value $p = \infty$ and associate it with standard calculus via

$$\mathbb{Q}_\infty = \mathbb{R}. \tag{2.14}$$

In accord with the terminology used for more general number fields, the case of a finite prime, i.e., $p < \infty$, is sometimes referred to as the *non-archimedean place*, while $p = \infty$ is called the *archimedean place*.

The p-adic numbers were introduced in number theory by Hensel with the intention of transferring the powerful tools of complex analysis to power and Laurent series. A *theorem by Ostrowski* [416] states that, up to isomorphism, any non-trivial norm on $\mathbb{Q}$ is either the standard Euclidean norm (leading to the real numbers upon completion) or one of the p-adic norms.

Remark 2.7 (Alternative construction of $\mathbb{Q}_p$) Another way of defining the p-adic numbers is through the following equivalent definition of the p-adic norm of an ordinary rational number $x \in \mathbb{Q}$:

$$|x|_p = p^{-k}, \tag{2.15}$$

where $k \in \mathbb{Z}$ is the largest integer such that $x = p^k y$ with $y \in \mathbb{Q}$ not containing any powers of p in its numerator or denominator (in cancelled form); this is often stated as p^k divides x. It is from this construction that one obtains $\mathbb{Q}_p$ as the completion of $\mathbb{Q}$ and one obtains an embedding of $\mathbb{Q}$ into $\mathbb{Q}_p$. The definition implies that, for a prime q and $k \in \mathbb{Z}$,

$$|q^k|_p = \begin{cases} p^{-k} & \text{if } p = q \\ 1 & \text{otherwise}. \end{cases} \tag{2.16}$$

2.2 p-adic Integration

Integration on $\mathbb{Q}_p$ can be defined with respect to the *additive measure* dx that is invariant under translation and has a simple scaling transformation

$$d(x+a) = dx\,, \qquad d(ax) = |a|_p dx\,. \tag{2.17}$$

The measure is by convention normalised so as to give the p-adic integers unit volume:

$$\int_{\mathbb{Z}_p} dx = 1\,. \tag{2.18}$$

We will now provide a series of examples of basic p-adic integrals. When evaluating such integrals it is often useful to employ different decompositions of $\mathbb{Z}_p$. One such decomposition is to write $\mathbb{Z}_p$ as a disjoint union

$$\mathbb{Z}_p = \bigsqcup_{k=0}^{p-1} C_k\,, \tag{2.19}$$

where C_k denotes the set of those p-adic integers with 'constant' coefficient (the coefficient of p^0 in (2.1)) equal to k. Another decomposition of $\mathbb{Z}_p$ employed is to write it as

$$\mathbb{Z}_p = \bigsqcup_{k=0}^{\infty} p^k \mathbb{Z}_p^\times\,. \tag{2.20}$$

This decomposition corresponds to taking shells of constant p-adic norm within the p-adic integers $\mathbb{Z}_p$.

Example 2.8: Volume of invertible integers $\mathbb{Z}_p$

$$\int_{\mathbb{Z}_p^\times} dx = \frac{p-1}{p}. \tag{2.21}$$

The integral is a simple consequence of the definition (2.18) and can be understood intuitively by noting that only $p-1$ out of the p choices for the constant coefficient of $x \in \mathbb{Z}_p$ correspond to elements in $\mathbb{Z}_p^\times$. For a more formal derivation we use decomposition (2.19) of $\mathbb{Z}_p$ and integrate over each C_k separately. By translation invariance of the measure (2.17) all C_k have the same volume $1/p$. Integrating over all C_k except for the one with $k = 0$ one thus obtains the above formula.

The following two examples explore the integration of the p-adic norm $|\cdot|_p$ defined in (2.4).

Example 2.9: Integration of the norm over $\mathbb{Z}_p$

$$\int_{\mathbb{Z}_p} |x|_p^s dx = \frac{p-1}{p}\frac{1}{1-p^{-s-1}}, \tag{2.22}$$

with convergence for $\mathrm{Re}(s) > -1$. This is derived in a few steps:

$$\int_{\mathbb{Z}_p} |x|_p^s dx = \sum_{k=0}^{\infty} \int_{p^k\mathbb{Z}_p^\times} |x|_p^s dx = \sum_{k=0}^{\infty} p^{-ks} \int_{p^k\mathbb{Z}_p^\times} dx = \sum_{k=0}^{\infty} p^{-ks} \int_{\mathbb{Z}_p^\times} p^{-k} dy$$
$$= \frac{p-1}{p}\sum_{k=0}^{\infty} p^{-k(s+1)} = \frac{p-1}{p}\frac{1}{1-p^{-s-1}}. \tag{2.23}$$

In the first step we have used the decomposition (2.20) of the p-adic integers. Then we have used the fact that for $x \in p^k\mathbb{Z}_p^\times$ the norm is $|x|_p = p^{-k}$. After that we have changed variables to $x = p^k y$ with $y \in \mathbb{Z}_p^\times$, used the resulting volume of $\mathbb{Z}_p^\times$ computed in Example 2.8 and carried out the geometric sum.

Using the identity from the previous example, we can also evaluate the following integral which will be used in Chapter 7 and Section 8.6.2.

Example 2.10: Integration of the norm over $\mathbb{Q}_p \setminus \mathbb{Z}_p$

$$\int_{\mathbb{Q}_p\setminus\mathbb{Z}_p} |x|_p^s dx = \frac{p-1}{p}\frac{p^{s+1}}{1-p^{s+1}}, \tag{2.24}$$

with $\mathrm{Re}(s) < -1$ and the domain of integration as defined in (2.10). The integral is then evaluated to be

$$\int_{\mathbb{Q}_p\setminus\mathbb{Z}_p} |x|_p^s dx = \int_{|x|_p>1} |x|_p^s dx = \sum_{k=1}^{\infty} p^{ks} \int_{p^{-k}\mathbb{Z}_p^\times} dx = \sum_{k=1}^{\infty} p^{k(s+1)} \int_{\mathbb{Z}_p^\times} dx$$
$$= \frac{p-1}{p}\sum_{k=1}^{\infty} p^{k(s+1)} = \frac{p-1}{p}\frac{p^{s+1}}{1-p^{s+1}}. \tag{2.25}$$

The integral converges for $\mathrm{Re}(s) < -1$. Note that the same integral over all of $\mathbb{Q}_p$ does not exist.

Remark 2.11 (Multiplicative measure $dx^\times$) We denote the *multiplicative measure* on $\mathbb{Q}_p^\times$ by $d^\times x$ with its defining relation

$$d^\times x = \frac{p}{p-1}\frac{dx}{|x|_p}. \tag{2.26}$$

It satisfies $\int_{\mathbb{Z}_p^\times} d^\times x = 1$. It transforms as $d^\times(ax) = d^\times x$. Integrating the function $|x|_p^s$ against the multiplicative measure $d^\times x$ the result (2.22) simplifies to

$$\int_{\mathbb{Z}_p} |x|_p^s \, d^\times x = \sum_{k=0}^{\infty} p^{-ks} \int_{\mathbb{Z}_p^\times} d^\times x = \sum_{k=0}^{\infty} p^{-ks} = \frac{1}{1 - p^{-s}}, \tag{2.27}$$

where in the first step we used the property (2.20). Note that the same result is obtained if we restrict the integration domain to $\mathbb{Z}_p \setminus \{0\}$ (different from $\mathbb{Z}_p^\times$), which will be useful in the proof of Proposition 2.26.

2.3 p-adic Characters and the Fourier Transform

In this section, we introduce the concept of a character, which is then used to define the p-adic Fourier transform. As before we provide explicit computations of various integrals serving as prototypical examples for later calculations.

Definition 2.12 (Fractional part of a p-adic number) The *fractional part* $[y]_p$ of a p-adic number $y \in \mathbb{Q}_p$ is given by its class in $\mathbb{Q}_p/\mathbb{Z}_p$, or more concretely by the terms in its series expansion with negative powers of p:

$$\left[x_k p^k + \cdots + x_{-1} p^{-1} + x_0 p^0 + x_1 p^1 + \cdots\right]_p = \begin{cases} x_k p^k + \cdots + x_{-1} p^{-1} & \text{if } k < 0, \\ 0 & \text{otherwise.} \end{cases} \tag{2.28}$$

Note that we will often suppress the subscript p when there is no risk of confusion.

We will now show that, given a rational number x, subtracting all the fractional parts of x with respect to all $\mathbb{Z}_p$ from x leaves a normal integer. This will, for example, be used in Sections 2.5 and 9.8.

Proposition 2.13 (Normal integer of a rational number) *Let $x \in \mathbb{Q}$. Then,*

$$x - \sum_{p<\infty} [x]_p \in \mathbb{Z}. \tag{2.29}$$

Proof By design, $x - [x]_p \in \mathbb{Z}_p$ and for any prime $q \neq p$

$$\left|[x]_q\right|_p \leq \max\left(\left|x_k q^k\right|_p, \ldots, \left|x_{-1} q^{-1}\right|_p\right) \leq 1 \tag{2.30}$$

if $k < 0$ (since $x_i \in \mathbb{Z}$) and otherwise $[x]_q = 0$, which means that $[x]_q \in \mathbb{Z}_p$. Hence, for any prime p

$$x - \sum_{q<\infty} [x]_q = (x - [x]_p) - \sum_{q \neq p} [x]_q \in \mathbb{Z}_p, \tag{2.31}$$

which proves the statement. □

With the definition of the fractional part of a p-adic number introduced we can now provide the definition of an additive character.

Definition 2.14 (Additive characters) *Additive characters* on $\mathbb{Q}_p$ are defined for any $u \in \mathbb{Q}_p$ by

$$\psi_p \equiv \psi_{p,u} : \mathbb{Q}_p \to U(1), \quad \psi_{p,u}(x) = e^{-2\pi i [ux]_p} \qquad x, u \in \mathbb{Q}_p . \tag{2.32}$$

The additive characters of (2.32) satisfy the relations $\psi_{p,u}(x)\psi_{p,u}(y) = \psi_{p,u}(x+y)$ and $\psi_{p,u}\psi_{p,v} = \psi_{p,u+v}$, as well as $\overline{\psi_{p,u}(x)} = \psi_{p,-u}(x) = \psi_{p,u}(-x)$. The *conductor* of the character is its kernel $|u|_p\mathbb{Z}_p$, but often we simply call $|u|_p$ the conductor.

Note that in the following we shorten the notation to $\psi_p \equiv \psi_{p,u}$ since it will be more important to keep track of the dependence on the prime p; the 'mode number' u will be given explicitly where needed. Also in the interest of simplicity of notation we will often drop the prime p script on the symbol for the fractional part $[\cdot]_p$ when writing out characters explicitly.

Example 2.15: Integration of a character over $p^k\mathbb{Z}_p$

For $k \in \mathbb{Z}$ one has

$$\int_{p^k\mathbb{Z}_p} e^{-2\pi i [ux]} dx = p^{-k} \gamma_p(up^k), \tag{2.33}$$

where the *characteristic function* $\gamma_p(u)$ of $\mathbb{Z}_p$ in $\mathbb{Q}_p$ is defined as

$$\gamma_p(u) := \int_{\mathbb{Z}_p} e^{-2\pi i [ux]} dx = \begin{cases} 1 & \text{if } u \in \mathbb{Z}_p , \\ 0 & \text{otherwise} . \end{cases} \tag{2.34}$$

The function $\gamma_p(u)$ is also called the p-adic *Gaussian* and will be discussed in more detail in Section 2.4. In order to derive this result we start with the case when $k = 0$:

$$\int_{\mathbb{Z}_p} \psi_u(x) dx = \int_{\mathbb{Z}_p} e^{-2\pi i [ux]} dx \tag{2.35}$$

and the integral only depends on the conductor $|u|_p$. We distinguish two cases: (i) $u \in \mathbb{Z}_p$ and (ii) $u \notin \mathbb{Z}_p$:

(i) If $u \in \mathbb{Z}_p$ then $[ux]_p = 0$ for $x \in \mathbb{Z}_p$ and hence the integral equals $\int_{\mathbb{Z}_p} dx = 1$.

(ii) If $u \notin \mathbb{Z}_p$ then we are effectively integrating a periodic function over a full period and hence the integral gives zero. More concretely, consider the example when $u = p^{-1}$; then

$$\int_{\mathbb{Z}_p} e^{-2\pi i [p^{-1}x]} dx = \sum_{k=0}^{p-1} e^{-2\pi i k/p} \int_{C_k} dx = \frac{1}{p} \sum_{k=0}^{p-1} e^{-2\pi i k/p} = 0, \tag{2.36}$$

with C_k defined as in (2.19) and where we have used the fact that $\int_{C_k} dx = 1/p$; see also Example 2.8. If u is 'more rational' one has to refine the summation region more but will always encounter sums that average to zero. We have thus derived (2.33) for the case of $k = 0$.

The result for the integral in the case when $k \neq 0$ then follows by a simple change of variables:

$$\int_{p^k \mathbb{Z}_p} e^{-2\pi i [ux]} dx = p^{-k} \int_{\mathbb{Z}_p} e^{-2\pi i [up^k x]} dx = p^{-k} \gamma_p(up^k). \tag{2.37}$$

We will also require the integral over 'shells' of p-adic numbers.

Example 2.16: Integration of a character over $p^k \mathbb{Z}_p^\times$

For $k \in \mathbb{Z}$ we have

$$\int_{p^k \mathbb{Z}_p^\times} e^{-2\pi i [ux]} dx = \begin{cases} \frac{p-1}{p} p^{-k} & \text{for } |u|_p \leq p^k \\ -p^{-(k+1)} & \text{for } |u|_p = p^{k+1} \\ 0 & \text{for } |u|_p > p^{k+1} \end{cases}. \tag{2.38}$$

Starting as before with the case $k = 0$, this can be related to the preceding example by noting that $\mathbb{Z}_p^\times = \mathbb{Z}_p \setminus (p\mathbb{Z}_p)$:

$$\begin{aligned} \int_{\mathbb{Z}_p^\times} e^{-2\pi i [ux]} dx &= \int_{\mathbb{Z}_p} e^{-2\pi i [ux]} dx - \int_{p\mathbb{Z}_p} e^{-2\pi i [ux]} dx \\ &= \gamma_p(u) - p^{-1} \int_{\mathbb{Z}_p} e^{-2\pi i [upx]} dx \\ &= \gamma_p(u) - p^{-1} \gamma_p(pu) \\ &= \begin{cases} \frac{p-1}{p} & \text{for } |u|_p \leq 1, \text{ i.e., } u \in \mathbb{Z}_p \\ -p^{-1} & \text{for } |u|_p = p \\ 0 & \text{for } |u|_p > p \end{cases}. \end{aligned} \tag{2.39}$$

The result of $p^k \mathbb{Z}_p^\times$ for $k \neq 0$ then follows by a change of variables:

$$\begin{aligned} \int_{p^k \mathbb{Z}_p^\times} e^{-2\pi i [ux]} dx &= p^{-k} \int_{\mathbb{Z}_p^\times} e^{-2\pi i [up^k x]} dx = p^{-k} \gamma_p(up^k) - p^{-(k+1)} \gamma_p(up^{k+1}) \\ &= \begin{cases} \frac{p-1}{p} p^{-k} & \text{for } |u|_p \leq p^k \\ -p^{-(k+1)} & \text{for } |u|_p = p^{k+1} \\ 0 & \text{for } |u|_p > p^{k+1} \end{cases}. \end{aligned} \tag{2.40}$$

Note also that this implies that

$$\int_{\mathbb{Q}_p \setminus \mathbb{Z}_p} e^{-2\pi i [ux]} dx = \sum_{k=-1}^{-\infty} \int_{p^k \mathbb{Z}_p^\times} e^{-2\pi i [ux]} dx = -\gamma_p(u) \,. \tag{2.41}$$

An important comment here concerns the integral of a character over all of $\mathbb{Q}_p$: since $\mathbb{Q}_p$ is formally the sum of $p^k \mathbb{Z}_p^\times$ over all $k \in \mathbb{Z}$, we see from the above result that for any $u \in \mathbb{Q}_p$ we obtain formally

$$\int_{\mathbb{Q}_p} e^{-2\pi i [ux]} dx = 0 \qquad \text{(not well-defined!)}, \tag{2.42}$$

which is the analogue of the incorrect equation $\int_{\mathbb{R}} e^{2\pi i u x} dx = 0$ which could be derived by splitting up $\mathbb{R}$ into an infinite number of intervals of length $1/u$ on each of which the integral vanishes. As is well known, the integral over the whole real line of $e^{2\pi i u x}$ is not well-defined but rather yields a δ-distribution. We will now see that something similar is true for the p-adic character $e^{-2\pi i [ux]}$ integrated over $\mathbb{Q}_p$.

Before introducing the concept of a p-adic Fourier transform let us make a short comment about function spaces used. The functions which we will be integrating are elements of $\mathcal{S}(\mathbb{Q}_p)$, which is the *Schwartz–Bruhat space* that will be characterised below. These functions generalise the *Schwartz functions*, which are infinitely differentiable, with rapidly decreasing derivatives.

Definition 2.17 (p-adic Fourier transform) One defines the *Fourier transform* over $\mathbb{Q}_p$, by integrating a function f_p on $\mathbb{Q}_p$ against the additive character $\psi_p(x) \equiv \psi_{p,u}(x)$:

$$\tilde{f}_p(u) = \int_{\mathbb{Q}_p} f_p(x) \psi_p(x) dx = \int_{\mathbb{Q}_p} f_p(x) e^{-2\pi i [ux]} dx \,. \tag{2.43}$$

The inverse transform uses the conjugate character

$$f_p(x) = \int_{\mathbb{Q}_p} \tilde{f}_p(u) \overline{\psi_p(x)} du = \int_{\mathbb{Q}_p} \tilde{f}_p(u) e^{2\pi i [ux]} du \,. \tag{2.44}$$

One can now ask for which functions f_p the transform is well-defined and can actually be inverted. As a first step we calculate the composition of the transforms of the characteristic function of a ball $p^k\mathbb{Z}_p \subset \mathbb{Q}_p$, i.e., $f_p(x) = \gamma_p(p^{-k}x)$:

$$\begin{aligned}\int_{\mathbb{Q}_p} \overline{\psi_p(x)} \int_{\mathbb{Q}_p} \psi_p(y)\gamma_p(p^{-k}y)dydu &= \int_{\mathbb{Q}_p} \overline{\psi_p(x)} \int_{p^k\mathbb{Z}_p} \psi_p(y)dydu \\ &= \int_{\mathbb{Q}_p} e^{2\pi i[ux]}p^{-k}\gamma_p(up^k)du = \int_{\mathbb{Z}_p} e^{2\pi i[up^{-k}x]}du = \gamma_p(p^{-k}x). \end{aligned} \tag{2.45}$$

From this calculation we see that restricting to compactly supported (and bounded) functions makes the integrals well-defined. We can relax the assumption of compact support if the function decreases sufficiently fast for larger and larger balls. This is for instance the case when $f_p(x) = |x|_p^s$ with $\mathrm{Re}(s)$ sufficiently negative. However, since in this case $|x|_p^s$ blows up for $|x|_p \to 0$ one has to cut out that region or replace $f_p(x)$ by a different function there. In summary, the p-adic Fourier transform is only well-defined on functions that are locally constant (i.e., constant on each $p^k\mathbb{Z}_p^\times$) and have compact support or a sufficiently fast decrease when $|x|_p \to \infty$. This is the Schwartz–Bruhat space mentioned above.

In the later development of automorphic forms, we will often encounter Fourier integrals that are not over all of $\mathbb{Q}_p$ but only over a subset of it. The following two examples will be particularly useful and also illustrate general techniques for performing such integrals.

Example 2.18: Fourier integral of $|x|_p^s$ over $\mathbb{Q}_p \setminus \mathbb{Z}_p$

Here, we cut out the compact region of the integers and consider the effect of the damping function $|x|_p^s$ with $\mathrm{Re}(s) < -1$. The result is

$$\int_{\mathbb{Q}_p\setminus\mathbb{Z}_p} |x|_p^s\psi_p(x)dx = \gamma_p(u)\left((1-p^s)\frac{1-p^{s+1}|u|_p^{-s-1}}{1-p^{s+1}} - 1\right). \tag{2.46}$$

To show this, we denote the integral by I and distinguish two cases: (i) u integral and (ii) u non-integral:

(i) If $u \in \mathbb{Z}_p$ and has conductor p^k with $k \geq 0$, then we evaluate the integral as

$$\begin{aligned} I &= \int_{\mathbb{Q}_p\setminus\mathbb{Z}_p} |x|_p^s e^{-2\pi i[p^kx]}dx = \sum_{\ell=1}^{\infty} p^{s\ell}\int_{p^{-\ell}\mathbb{Z}_p^\times} e^{-2\pi i[p^kx]}dx \\ &= \sum_{\ell=1}^{\infty} p^{(s+1)\ell}\int_{\mathbb{Z}_p^\times} e^{-2\pi i[p^{k-\ell}x]}dx \underset{\text{Ex. 2.16}}{=} \frac{p-1}{p}\sum_{\ell=1}^{k} p^{(s+1)\ell} - \frac{1}{p}p^{(k+1)(s+1)} \\ &= (1-p^s)\frac{1-p^{s+1}|u|_p^{-s-1}}{1-p^{s+1}} - 1. \end{aligned} \tag{2.47}$$

(*ii*) If $u \notin \mathbb{Z}_p$, so that the character has conductor p^k with $k < 0$, we find

$$I = \int_{\mathbb{Q}_p \setminus \mathbb{Z}_p} |x|_p^s e^{-2\pi i [p^k x]} dx = \sum_{\ell=1}^{\infty} p^{(s+1)\ell} \int_{\mathbb{Z}_p^\times} e^{-2\pi i [p^{k-\ell} x]} dx = 0 \quad (2.48)$$

by Example 2.16 since $k - \ell < -1$ for all $\ell \geq 1$ and $k < 0$.

This Fourier integral over $\mathbb{Q}_p \setminus \mathbb{Z}_p$ converged for $\mathrm{Re}(s) < -1$. The Fourier integral over $\mathbb{Z}_p$ converges in the complementary region $\mathrm{Re}(s) > -1$ and the result is slightly more complicated.

Example 2.19: Fourier integral of $|x|_p^s$ over $\mathbb{Z}_p$

Here we cut out the region $\mathbb{Q}_p \setminus \mathbb{Z}_p$. In order to have a bounded function for $|x|_p \to 0$ we now require $\mathrm{Re}(s) > -1$. The Fourier transform now evaluates to

$$\int_{\mathbb{Z}_p} |x|_p^s e^{-2\pi i [xu]} dx = \gamma_p(u) \frac{p-1}{p} \frac{1}{1-p^{-s-1}} + (1-\gamma_p(u)) |u|_p^{-s-1} \frac{1-p^s}{1-p^{-s-1}} . \quad (2.49)$$

This can be derived in a few steps:

$$\begin{aligned} \int_{\mathbb{Z}_p} |x|_p^s e^{-2\pi i [xu]} dx &= \sum_{\ell=0}^{\infty} p^{-s\ell} \int_{p^\ell \mathbb{Z}_p^\times} e^{-2\pi i [ux]} dx \\ &= \sum_{\ell=0}^{\infty} p^{-(s+1)\ell} \int_{\mathbb{Z}_p^\times} e^{-2\pi i [up^\ell x]} dx \\ &= \gamma_p(u) \frac{p-1}{p} \frac{1}{1-p^{-s-1}} + (1-\gamma_p(u)) |u|_p^{-s-1} \frac{1-p^s}{1-p^{-s-1}}, \end{aligned} \quad (2.50)$$

where we have treated the cases $u \in \mathbb{Z}_p$ and $u \notin \mathbb{Z}_p$ separately and used Equation (2.22) and Example 2.16.

2.4 *p*-adic Gaussian and Bessel Functions

In this section, we will discuss two special functions: the *p-adic Gaussian* and the *p-adic Bessel function*, which will play a rôle later on in the text.

The p-adic analogue of the real Gaussian $e^{-\pi x^2}$ is given by the function

$$\gamma_p(x) = \begin{cases} 1 & \text{if } x \in \mathbb{Z}_p, \text{ i.e., } |x|_p \leq 1, \\ 0 & \text{if } x \notin \mathbb{Z}_p, \text{ i.e., } |x|_p > 1, \end{cases} \quad (2.51)$$

which we have already encountered in Example 2.15 of the previous section. In order to see why it is the generalisation of the real Gaussian $e^{-\pi x^2}$ we recall that the real Gaussian is invariant under Fourier transformation. Using (2.33) this property is then also easily checked for the p-adic version:

$$\tilde{\gamma}_p(u) = \int_{\mathbb{Q}_p} \psi_u(x)\gamma_p(x)dx = \int_{\mathbb{Z}_p} e^{-2\pi i[ux]}dx = \gamma_p(u)\,. \tag{2.52}$$

Let us also note the following useful property of the Euler product of the p-adic Gaussian which will be used in Chapter 7 in the computation of the Fourier coefficients of Eisenstein series on $SL(2,\mathbb{A})$; see Equation (7.78):

$$\prod_{p<\infty} \gamma_p(m) = \begin{cases} 1 & \text{if } m \in \mathbb{Z}, \\ 0 & \text{otherwise}\,. \end{cases} \tag{2.53}$$

In order to introduce the p-adic version of the Bessel function, recall that the real (modified) *Bessel function* K_s can be written as the (inverse) Fourier transform of the vector norm function $||(1,u)||^{-2s} = (1+u^2)^{-s}$ via

$$\int_{\mathbb{R}} (1+u^2)^{-s} e^{-2\pi i m u} du = \frac{2\pi^s}{\Gamma(s)} |m|^{s-1/2} K_{s-1/2}(2\pi|m|)\,, \tag{2.54}$$

where $\Gamma(s)$ is the standard gamma function. The p-adic generalisation of the integrand is through $||(1,u)||_p^{-2s} = (\max(1,|u|_p))^{-2s}$. The normalisation to be chosen is [393, 394]

$$\tilde{f}(u) = \frac{1}{1-p^{-2s}} (\max(1,|u|_p))^{-2s}\,. \tag{2.55}$$

The Fourier transform of this function is

$$f(x) = \gamma_p(x) \frac{1-p^{-2s+1}|x|_p^{2s-1}}{1-p^{-2s+1}}\,, \tag{2.56}$$

which we call the p-adic Bessel function, which will be used, for example, in Section 7.3. To demonstrate these properties of the p-adic Bessel function we perform the Fourier transform in the following example.

Example 2.20: Fourier transform of p-adic Bessel function

Consider the following calculation that converges for $\mathrm{Re}(s) > 1/2$:

$$\begin{aligned} &\int_{\mathbb{Q}_p} e^{2\pi i[ux]} \frac{1}{1-p^{-2s}} (\max(1,|u|_p))^{-2s} du \\ &= \frac{1}{1-p^{-2s}} \int_{\mathbb{Z}_p} e^{2\pi i[ux]} du + \frac{1}{1-p^{-2s}} \int_{\mathbb{Q}_p\setminus\mathbb{Z}_p} |u|_p^{-2s} e^{2\pi i[ux]} du\,. \end{aligned} \tag{2.57}$$

We have separated the integral according to the two possible cases of the max function. The first integral is given in Example 2.15 and the second one in Example 2.18.

Combining the results we obtain

$$\int_{\mathbb{Q}_p} e^{2\pi i[ux]} \frac{1}{1-p^{-2s}} (\max(1,|u|_p))^{-2s} du = \gamma_p(x) \frac{1-p^{-2s+1}|x|_p^{2s-1}}{1-p^{-2s+1}}. \tag{2.58}$$

2.5 Adeles

In the previous sections, we have introduced the concept of the p-adic completions $\mathbb{Q}_p$ of $\mathbb{Q}$ and we have shown in a number of examples how integration can be carried out locally and also that the real Gaussian and Bessel function have p-adic counterparts. The next step will be to organise the completions $\mathbb{Q}_p$ of the rational numbers $\mathbb{Q}$ into the *ring of adeles* of $\mathbb{Q}$, denoted by $\mathbb{A}$, which comprises the p-adic completions at all primes, including the prime at infinity, at the same time. The introduction of the adeles as a global number field is in line with the so-called *local-global principle*.

Definition 2.21 (Adeles $\mathbb{A}$) The *adeles* $\mathbb{A} = \mathbb{A}_{\mathbb{Q}}$ of $\mathbb{Q}$ are defined as a *restricted direct product*

$$\mathbb{A} = \mathbb{R} \times \prod_{p<\infty}{}' \mathbb{Q}_p, \tag{2.59}$$

where the restriction on the product (signified by the prime) means that $\mathbb{A}$ consists of those elements

$$a = (a_p) = (a_\infty; a_2, a_3, a_5, a_7, \dots) \tag{2.60}$$

such that for almost all finite primes p one has $a_p \in \mathbb{Z}_p$.

The restriction on the direct product in the definition of the adeles makes them locally compact, which is needed for the existence of a *Haar measure*. Also, as a consequence of the definition, the adeles are endowed with a natural topology, and they are in fact a locally compact ring. We refer the reader to [263] for more details on these issues. It will sometimes be useful to talk about the *finite* adeles $\mathbb{A}_f$, which are defined as the restricted direct product over the finite primes:

$$\mathbb{A}_f = \prod_{p<\infty}{}' \mathbb{Q}_p. \tag{2.61}$$

We also define the set of invertible elements of the adeles.

Definition 2.22 (Ideles $\mathbb{A}^\times$) The *ideles* $\mathbb{A}^\times$ are the set of invertible elements in $\mathbb{A}$. They are defined as

$$\mathbb{A}^\times = \mathbb{R}^\times \times \prod_{p<\infty}{}' \mathbb{Q}_p^\times. \tag{2.62}$$

The norm for the adeles is induced directly from the local norms.

Definition 2.23 (Global norm $|\cdot|_\mathbb{A}$) The *global norm* $|\cdot|_\mathbb{A}$ on $\mathbb{A}$ is induced from the norm $|\cdot|_p$ on the local factors $\mathbb{Q}_p$ according to the formula

$$|a|_\mathbb{A} = \prod_{p\le\infty} |a_p|_p. \tag{2.63}$$

This is in fact a convergent product since almost all $a_p \in \mathbb{Z}_p$ and hence satisfy $|a_p|_p \le 1$. Furthermore, the adelic norm $|a|_\mathbb{A}$ is only non-zero if $a \in \mathbb{A}^\times$ and then is a finite product.

The *strong approximation principle* [164] states that the set

$$J = \mathbb{R}_+ \times \prod_{p<\infty} \mathbb{Z}_p^\times \tag{2.64}$$

is a fundamental domain for $\mathbb{Q}^\times \backslash \mathbb{A}^\times$. Hence we can write the ideles as the (disjoint) union

$$\mathbb{A}^\times = \bigcup_{k\in\mathbb{Q}^\times} k \cdot J. \tag{2.65}$$

(A higher-rank version of strong approximation is proven in Section 3.2.2.)

The rational numbers *embed diagonally* into the adeles, i.e., $\mathbb{Q} \hookrightarrow \mathbb{A}$, by simply taking

$$\mathbb{Q} \ni x \longmapsto (x; x, x, x, \dots) \in \mathbb{A}. \tag{2.66}$$

One can see that this is indeed an element of the adeles since for $x \in \mathbb{Q}$ the norm $|x|_p$ is non-trivial only for the finite number of p's which divide x in the sense of (2.15). In other words, the prime factorisations of the coprime numerator and denominator of x contain only a finite number of primes. By factorising $x \in \mathbb{Q}^\times$ into its prime factors we see that

$$|x|_\mathbb{A} = |x|_\infty \prod_{p<\infty} |x|_p = |x|_\infty |x|_\infty^{-1} = 1. \tag{2.67}$$

Following [164], we will now show that with this embedding $\mathbb{Q}$ sits discretely inside $\mathbb{A}$, mimicking the way the integers $\mathbb{Z}$ are embedded as a lattice inside $\mathbb{R}$. As we will see, this fact lies at the heart of the analysis in subsequent sections.

Proposition 2.24 (Discrete embedding of $\mathbb{Q}$ in $\mathbb{A}$) *$\mathbb{Q}$ sits discretely inside $\mathbb{A}$.*

Proof Let us first consider $0 \in \mathbb{Q}$ and construct

$$V = \left(-\frac{1}{2}, \frac{1}{2}\right) \times \prod_{p<\infty} \mathbb{Z}_p \subset \mathbb{A}\,. \tag{2.68}$$

The subgroup $\mathbb{Z}_p$ is an open ball in $\mathbb{Q}_p$ since $|x|_p$ takes only a discrete set of values, that is, $\mathbb{Z}_p = \{x \in \mathbb{Q}_p \mid |x|_p \leq 1\} = \{x \in \mathbb{Q}_p \mid |x|_p < \alpha\}$ for any $1 < \alpha < p$.

Thus, V is an open neighbourhood of 0 in $\mathbb{A}$ and for any $x \in V \cap \mathbb{Q}$ we have that $|x|_p \leq 1$ for all $p < \infty$, which means that $x \in \mathbb{Z}$, and $|x|_\infty < 1/2$ which then gives that $x = 0$. Hence, we have found an open neighbourhood V to 0 in $\mathbb{A}$ such that $V \cap \mathbb{Q} = \{0\}$. For a general point $r \in \mathbb{Q}$ these arguments generalise by instead considering $r + V$, which makes $\mathbb{Q}$ discrete in $\mathbb{A}$. □

With the definition of the adeles as the collection of all local factors at hand, we will now see how to turn a set of local functions into a global one.

2.6 Adelisation

One can extend a collection of local functions f_p on $\mathbb{Q}_p$ to a global function $f_\mathbb{A}$ on $\mathbb{A}$:

$$f_\mathbb{A}(a) = f_\mathbb{A}(a_\infty; a_f)\,, \qquad a_f = (a_2, a_3, a_5, \dots) \in \mathbb{A}_f\,, \tag{2.69}$$

via an Euler product

$$f_\mathbb{A}(a) = \prod_{p\leq\infty} f_p(a_p)\,. \tag{2.70}$$

Starting from $f_\mathbb{A}$ we can recover a function on $\mathbb{R}$ by setting

$$f_\infty(a_\infty) = f_\mathbb{A}(a_\infty; 1, 1, 1, \dots)\,. \tag{2.71}$$

One says that $f_\mathbb{A}$ is the *adelisation* of $f_\mathbb{R}$. Similarly we can extend to $\mathbb{A}$ the notion of local additive characters ψ on $\mathbb{Q}_p$.

Let $u = (u_\infty, u_2, u_3, \dots) \in \mathbb{A}$ and $\psi_p : \mathbb{Q}_p \to U(1)$ be an additive character, such that for finite p this coincides with the character

$$\psi_p(x_p) = e^{-2\pi i [u_p x_p]_p}\,, \qquad u_p, x_p \in \mathbb{Q}_p\,, \tag{2.72}$$

defined in Section 2.3, while for $p = \infty$ this is the standard character on $\mathbb{R}$:

$$\psi_\infty(x_\infty) = e^{2\pi i u_\infty x_\infty}\,, \qquad u_\infty, x_\infty \in \mathbb{R}. \tag{2.73}$$

We can then consider a *global character*

$$\psi_\mathbb{A} : \mathbb{A} \to U(1) \tag{2.74}$$

as the adelisation of $\psi_{\mathbb{R}}$, i.e., as the Euler product

$$\psi_{\mathbb{A}}(x) = \prod_{p \leq \infty} \psi_p(x_p) = e^{2\pi i u_\infty x_\infty} \prod_{p<\infty} e^{-2\pi i [u_p x_p]_p}, \tag{2.75}$$

which we will denote as $\psi_{\mathbb{A}}(x) = e^{2\pi i u x}$ for short.

The sign difference in the exponentials of the characters at the archimedean and non-archimedean places has been introduced for the following reason. For $u = m \in \mathbb{Q}$ diagonally in $\mathbb{A}$, the character $\psi_{\mathbb{A}}$ is periodic in $\mathbb{Q}$ since, for $x \in \mathbb{A}$ and $r \in \mathbb{Q}$,

$$\psi_{\mathbb{A}}(x + r) = \psi_{\mathbb{A}}(x)\psi_{\mathbb{A}}(r), \tag{2.76}$$

with

$$\psi_{\mathbb{A}}(r) = \prod_{p \leq \infty} \psi_p(r) = \exp\Big(2\pi i (mr - \sum_{p<\infty} [mr]_p)\Big) = 1, \tag{2.77}$$

using Proposition 2.13. Thus, for rational u, $\psi_{\mathbb{A}}$ is a character on $\mathbb{Q}\backslash\mathbb{A}$. That these are all the characters on $\mathbb{Q}\backslash\mathbb{A}$ is shown for example in [164].

Integration over the adeles is similarly defined using Euler products [443]. For instance, the integral over an adelic function $f_{\mathbb{A}}(x)$ can be written as

$$\int_{\mathbb{A}} f_{\mathbb{A}}(x)\, dx = \left(\int_{\mathbb{R}} f_{\mathbb{R}}(x)\, dx\right)\left(\prod_{p<\infty} \int_{\mathbb{Q}_p} f_p(x)\, dx\right). \tag{2.78}$$

Definition 2.25 (Adelic Fourier transform) The adelic Fourier transform is defined using the global character $\psi_{\mathbb{A}}$ as follows:

$$\tilde{f}_{\mathbb{A}}(u) = \int_{\mathbb{A}} f_{\mathbb{A}}(x)\overline{\psi}_{\mathbb{A}}(x) dx. \tag{2.79}$$

We will perform several integrals of this type in subsequent sections. In the following section, we will illustrate the usefulness of the adelic framework in the context of the Riemann zeta function.

2.7 Adelic Analysis of the Riemann Zeta Function

In this section, we will illustrate the power of the adelic formalism by analysing the Riemann zeta function from this point of view. This was one of the main points of the celebrated thesis of Tate [594], which first introduced the notion of Fourier analysis over the adeles.

2.7.1 The completed Riemann zeta function

The first task will be to illustrate how the *completed Riemann zeta function* is a much more natural object, from an adelic perspective, than the ordinary zeta function. Recall first that the completed Riemann zeta function takes the form

$$\xi(s) = \pi^{-s/2}\Gamma(s/2)\zeta(s). \tag{2.80}$$

We now have:

Proposition 2.26 (Tate's global Riemann integral) *The completed Riemann zeta function $\xi(s)$ can be written in the following global form* [594]*:*

$$\xi(s) = \int_{\mathbb{A}^\times} \gamma_{\mathbb{A}}(x)|x|_{\mathbb{A}}^s d^\times x, \tag{2.81}$$

where $s \in \mathbb{C}$ and $\gamma_{\mathbb{A}} = \prod_{p\leq\infty} \gamma_p$ with γ_p the p-adic Gaussian (2.51) and $\gamma_\infty = e^{-\pi x^2}$ the real Gaussian.

Proof Splitting the integral into an Euler product yields

$$\int_{\mathbb{A}^\times} \gamma_{\mathbb{A}}(x)|x|_{\mathbb{A}}^s d^\times x = \left(\int_{\mathbb{R}^\times} e^{-\pi x^2}|x|_\infty^s d^\times x\right) \prod_{p<\infty} \int_{\mathbb{Q}_p^\times} \gamma_p(x)|x|_p^s d^\times x. \tag{2.82}$$

The archimedean integral can be evaluated in terms of a gamma function:

$$\int_{\mathbb{R}^\times} e^{-\pi x^2}|x|_\infty^s d^\times x = \int_{\mathbb{R}} e^{-\pi x^2}|x|_\infty^{s-1} dx = \pi^{-s/2}\Gamma(s/2), \tag{2.83}$$

where we have made use of (2.26).

Due to the γ_p-factor, the p-adic integrals localise on the p-adic integers

$$\int_{\mathbb{Q}_p^\times} \gamma_p(x)|x|_p^s d^\times x = \int_{\mathbb{Z}_p\setminus\{0\}} |x|_p^s d^\times x. \tag{2.84}$$

By Remark 2.11, this yields

$$\int_{\mathbb{Z}_p\setminus\{0\}} |x|_p^s d^\times x = \int_{\mathbb{Z}_p} |x|_p^s d^\times x = \frac{1}{1-p^{-s}}. \tag{2.85}$$

Combining everything and performing the product over primes we obtain

$$\int_{\mathbb{A}^\times} \gamma_{\mathbb{A}}(x)|x|_{\mathbb{A}}^s d^\times x = \pi^{-s/2}\Gamma(s/2) \prod_{p<\infty} \frac{1}{1-p^{-s}} = \pi^{-s/2}\Gamma(s/2)\zeta(s) = \xi(s). \tag{2.86}$$

□

Remark 2.27 The above result illustrates that the adelic approach gives an elegant integral representation of the completed zeta function, where the normalisation factor corresponds to the contribution from the archimedean place $p = \infty$. Such integrals were first considered in the thesis of Tate [594], and then developed further by Jacquet and Langlands [376].

Remark 2.28 It is common to define the archimedean zeta factor by $\zeta_\infty(s) = \pi^{-s/2}\Gamma(s/2)$ and write the global Euler product form of the completed Riemann zeta function as

$$\xi(s) = \prod_{p\leq\infty} \zeta_p(s). \tag{2.87}$$

Anticipating later notions, the completed Riemann zeta function can be thought of as an automorphic form on the group $GL(1, \mathbb{A})$.

2.7.2 The Functional Relation

We shall now take the analysis one step further and prove the following famous theorem by Riemann using the adelic framework.

Theorem 2.29 (Functional relation for $\xi(s)$) *The completed Riemann zeta function $\xi(s)$ satisfies the functional relation*

$$\xi(s) = \xi(1 - s). \tag{2.88}$$

To prove the theorem using the approach of Tate [594] we first need the following lemma.

Lemma 2.30 (Adelic Poisson resummation) *For any (sufficiently nice) function $f_\mathbb{A}$ with adelic Fourier transform $\tilde{f}_\mathbb{A}$ we have the Poisson summation formula*

$$\sum_{\gamma\in\mathbb{Q}} f_\mathbb{A}(\gamma) = \sum_{\gamma\in\mathbb{Q}} \tilde{f}_\mathbb{A}(\gamma). \tag{2.89}$$

Proof The proof is similar to the proof of the ordinary Poisson summation formula so we will be brief. Define

$$F_\mathbb{A}(x) = \sum_{\gamma\in\mathbb{Q}} f_\mathbb{A}(x + \gamma). \tag{2.90}$$

This function is periodic under the *diagonally embedded* rationals $\mathbb{Q}$ by construction and so has a Fourier expansion. The Fourier coefficients F_{ψ_γ} of $F_\mathbb{A}$ with respect to a unitary character ψ_γ precisely equal the Fourier transform $\tilde{f}_\mathbb{A}(\gamma)$ of the seed function $f_\mathbb{A}(\gamma)$ and so we can write

$$F_\mathbb{A}(x) = \sum_{\gamma\in\mathbb{Q}} \tilde{f}_\mathbb{A}(\gamma)\psi_\gamma(x). \tag{2.91}$$

Putting $x = 0$ in this formula and equating it with $F_\mathbb{A}(0)$ from the definition (2.90) then establishes the result. □

To complete the proof of the theorem we need also the following lemma.

Lemma 2.31 (Functional equation of adelic theta function) *The* global theta function

$$\Theta_{\mathbb{A}}(x) = \sum_{k \in \mathbb{Q}} \gamma_{\mathbb{A}}(kx) \tag{2.92}$$

satisfies the functional relation

$$\Theta_{\mathbb{A}}(x) = \frac{1}{|x|_{\mathbb{A}}} \Theta_{\mathbb{A}}(1/x), \qquad \textit{for all } x \in \mathbb{A}^{\times}. \tag{2.93}$$

Proof This follows from applying the Poisson summation formula and the fact that the global Gaussian $\gamma_{\mathbb{A}}(x)$ is invariant under Fourier transformation. □

Remark 2.32 The function $\Theta_{\mathbb{A}}(x)$ defined in (2.92) is called a theta function because at the real place it is a sum over terms of the type $e^{-\pi k^2 x^2}$ which is a hallmark of classical theta functions, which we shall discuss in much more detail in Chapter 12.

We now return to the

Proof of Theorem 2.29 Now let J be the fundamental domain defined in (2.64) for $\mathbb{Q}^{\times} \backslash \mathbb{A}^{\times}$. By Lemma 2.31 we then have

$$\int_J \Theta_{\mathbb{A}}(x)|x|_{\mathbb{A}}^s d^{\times}x = \int_J \Theta_{\mathbb{A}}(1/x)|x|_{\mathbb{A}}^{s-1} d^{\times}x = \int_J \Theta_{\mathbb{A}}(x)|x|_{\mathbb{A}}^{1-s} d^{\times}x, \tag{2.94}$$

where in the last step we used the fact that the multiplicative measure is invariant under $x \to x^{-1}$. Finally, using the factorisation $\mathbb{A}^{\times} = \bigcup_{k \in \mathbb{Q}^{\times}} k \cdot J$ (see (2.65)) and the fact that $|x|_{\mathbb{A}} = 1$ for $x \in \mathbb{Q}$, we can rewrite (2.94) as

$$\int_{\mathbb{A}^{\times}} \gamma_{\mathbb{A}}(x)|x|_{\mathbb{A}}^s d^{\times}x = \int_{\mathbb{A}^{\times}} \gamma_{\mathbb{A}}(x)|x|_{\mathbb{A}}^{1-s} d^{\times}x, \tag{2.95}$$

thus establishing the functional relation (2.88) for the completed Riemann zeta function using Proposition 2.26. □

3

Basic Notions from Lie Algebras and Lie Groups

We will make use of some standard terminology from the theory of Lie groups and Lie algebras, which we summarise for definiteness. We first address real Lie algebras and groups and their complexifications before we turn to the adelic setting with emphasis on the strong approximation theorem.

3.1 Real Lie Algebras and Real Lie Groups

The material reviewed in this section can be found for example in [368, 232, 353, 386].

3.1.1 Split Real Simple Lie Algebras and Root Systems

Let $\mathfrak{g}(\mathbb{C})$ be a finite-dimensional and simple complex Lie algebra from the Cartan–Killing classification. We will consider here only the *split real* form $\mathfrak{g} \equiv \mathfrak{g}(\mathbb{R})$ of the Lie algebra. We choose a *Cartan subalgebra* $\mathfrak{h} \subset \mathfrak{g}$, that is, a maximal abelian subalgebra of semi-simple elements. This means that we can decompose $\mathfrak{g}$ into eigenspaces of $\mathfrak{h}$ in what is called the *root space decomposition*:

$$\mathfrak{g} = \mathfrak{h} \oplus \bigoplus_{\alpha \in \Delta} \mathfrak{g}_\alpha, \tag{3.1}$$

where the root space $\mathfrak{g}_\alpha$ for a generalised eigenvalue $\alpha \colon \mathfrak{h} \to \mathbb{R}$ is given by

$$\mathfrak{g}_\alpha = \{x \in \mathfrak{g} \mid [h, x] = \alpha(h)x \quad \text{for all } h \in \mathfrak{h}\}\,. \tag{3.2}$$

The set of $\alpha \neq 0$ for which $\mathfrak{g}_\alpha \neq \{0\}$ is called the set of *roots* Δ. By our assumption on the Lie algebra $\mathfrak{g}$ we have that $\dim(\mathfrak{g}_\alpha) = 1$ for all $\alpha \in \Delta$. Since $\mathfrak{g}_\alpha$ is one-dimensional there is, for each root $\alpha \in \Delta$, a unique element $H_\alpha \in [\mathfrak{g}_\alpha, \mathfrak{g}_{-\alpha}] \subset \mathfrak{h}$ such that $\alpha(H_\alpha) = 2$.

One can prove that for any root α one necessarily has that $-\alpha$ is also a root. Thus, the set of roots $\Delta \subset \mathfrak{h}^*$ can be consistently divided into two halves by picking one out of the pair $(\alpha, -\alpha)$ to be positive and the other one to be negative. This yields a disjoint decomposition

$$\Delta = \Delta_- \cup \Delta_+, \quad \Delta_- = -\Delta_+ . \tag{3.3}$$

An element $\alpha \in \Delta_+$ ($\alpha \in \Delta_-$) is called a *positive root* (*negative root*) and we use the notation $\alpha > 0$ to denote positive roots as well as $\alpha < 0$ for negative roots. With respect to a set of positive roots Δ_+ we have a system of *simple roots*

$$\Pi = \{\alpha_1, \dots, \alpha_r\}, \tag{3.4}$$

where $r = \dim(\mathfrak{h})$ is the *rank* of the Lie algebra. Simple roots are distinguished as those positive roots that cannot be written as *sums* of other positive roots. Then any root $\alpha \in \Delta$ can be written as an integral linear combination of the simple roots

$$\alpha = \sum_{i=1}^{r} m_i \alpha_i, \tag{3.5}$$

where either all $m_i \geq 0$ for positive roots or all $m_i \leq 0$ for negative roots. There is a unique *highest root* $\theta \in \Delta$ for which the *height* $\mathrm{ht}(\alpha) = \sum_i m_i$ is maximal. Another important element is the *Weyl vector* (which is not necessarily an element of Δ)

$$\rho = \frac{1}{2} \sum_{\alpha \in \Delta_+} \alpha. \tag{3.6}$$

We define the spaces of *positive/negative step operators* by

$$\mathfrak{n} \equiv \mathfrak{n}_+ = \bigoplus_{\alpha \in \Delta_+} \mathfrak{g}_\alpha \quad \text{and} \quad \mathfrak{n}_- = \bigoplus_{\alpha \in \Delta_-} \mathfrak{g}_\alpha, \tag{3.7}$$

as well as the *(upper) Borel subalgebra*

$$\mathfrak{b} = \mathfrak{h} \oplus \mathfrak{n}. \tag{3.8}$$

The spaces $\mathfrak{n}_\pm$ are nilpotent subalgebras of $\mathfrak{g}$; the Borel subalgebra $\mathfrak{b}$ is solvable. One can think of $\mathfrak{n}_\pm$ as strictly upper/lower triangular matrices and $\mathfrak{h}$ as diagonal matrices, so that $\mathfrak{b}$ consists of upper triangular matrices.

More formally, the notions of nilpotency and solvability for Lie algebras are defined as follows. A *nilpotent Lie algebra* is one whose *lower central series* $D_k(\mathfrak{g}) := [\mathfrak{g}, D_{k-1}(\mathfrak{g})]$ vanishes for some finite k. A *solvable Lie algebra* is one whose *derived series* $D^k(\mathfrak{g}) := \left[D^{k-1}(\mathfrak{g}), D^{k-1}(\mathfrak{g})\right]$ vanishes for some finite k. In both cases we set $D_0(\mathfrak{g}) = D^0(\mathfrak{g}) = \mathfrak{g}$. The Borel subalgebra $\mathfrak{b}$ includes semi-simple elements whence the lower central series does not vanish.

The semi-simple elements disappear in $D^1(\mathfrak{b}) = [\mathfrak{b}, \mathfrak{b}]$ and thus the derived series vanishes, rendering $\mathfrak{b}$ solvable. The derived series will play a rôle when discussing Fourier expansions of automorphic forms in Chapter 6.

On $\mathfrak{g}$ one can define an *invariant bilinear form* that we will write as $\langle x|y\rangle$ for $x, y \in \mathfrak{g}$. Invariance means compatibility with the Lie bracket:

$$\langle [x, y]\,|z\rangle = \langle x|\,[y, z]\rangle. \tag{3.9}$$

This form is proportional to the *Killing metric*. We have that $H_\alpha \in [\mathfrak{g}_\alpha, \mathfrak{g}_{-\alpha}]$ implies that $H_\alpha = [X_\alpha, Y_\alpha]$ for some $X_\alpha \in \mathfrak{g}_\alpha$ and $Y_\alpha \in \mathfrak{g}_{-\alpha}$.

Then, for any $h \in \mathfrak{h}$,

$$\langle H_\alpha|h\rangle = \langle [X_\alpha, Y_\alpha]|h\rangle = \langle X_\alpha|[Y_\alpha, h]\rangle = \alpha(h)\langle X_\alpha|Y_\alpha\rangle\,. \tag{3.10}$$

With $h = H_\alpha$ this becomes

$$\langle H_\alpha|H_\alpha\rangle = \alpha(H_\alpha)\langle X_\alpha|Y_\alpha\rangle = 2\langle X_\alpha|Y_\alpha\rangle. \tag{3.11}$$

Thus, by insertion into (3.10),

$$\alpha(h) = \frac{2\langle H_\alpha|h\rangle}{\langle H_\alpha|H_\alpha\rangle}. \tag{3.12}$$

Sometimes we will also use the notation $\langle\alpha|h\rangle$ for $\alpha(h)$.

The Cartan element $T_\alpha = 2H_\alpha/\langle H_\alpha|H_\alpha\rangle$ can then be used to define an inner product on $\mathfrak{h}^*$ by

$$\langle\alpha|\beta\rangle = \langle T_\alpha|T_\beta\rangle = \alpha(T_\beta) = \beta(T_\alpha)\,. \tag{3.13}$$

Since $\mathfrak{g}$ is finite-dimensional and simple this bilinear form on $\mathfrak{h}^*$ is positive definite and can be used to define the lengths of root vectors α. We normalise it such that the highest root θ has length $\theta^2 := \langle\theta|\theta\rangle = 2$.

The bilinear form on $\mathfrak{h}^*$ (spanned over $\mathbb{R}$ by the simple roots) can be used to define a basis of $\mathfrak{h}^*$ dual to the simple roots. The corresponding basis elements are called the *fundamental weights* Λ_i and satisfy

$$\langle\Lambda_i|\alpha_j\rangle = \frac{1}{2}\langle\alpha_i|\alpha_i\rangle\delta_{ij} \quad \text{for } i, j = 1, \ldots, r. \tag{3.14}$$

In terms of the fundamental weights one can re-express the Weyl vector of Equation (3.6) as $\rho = \sum_{i=1}^r \Lambda_i$. A general element of $\mathfrak{h}^*$ will be called a *weight* and denoted by λ.

Associated with the choice of simple roots α_i is also a realisation of the *Weyl group* of $\mathfrak{g}$. This is a finite *Coxeter group* that is generated by the *fundamental reflections* w_i $(i = 1, \ldots, r)$ that are defined through their action on weights $\lambda \in \mathfrak{h}^*$ by

$$w_i(\lambda) = \lambda - \frac{2\langle\lambda|\alpha_i\rangle}{\langle\alpha_i|\alpha_i\rangle}\alpha_i, \tag{3.15}$$

so that in particular $w_i(\rho) = \rho - \alpha_i$. A general *word* of the Weyl group is given by a succession of fundamental reflections $w = w_{i_1} \cdots w_{i_\ell}$ and we call $\ell = \ell(w)$ the *length* of the Weyl word w. This assumes that the expression is in reduced form, i.e., that the relations between the generating fundamental w_i have been used to make the word as short as possible. We denote the Weyl group by $\mathcal{W} \equiv \mathcal{W}(\mathfrak{g})$ and its distinguished longest element by w_{long}. The *longest Weyl word* has the property $w_{\text{long}}(\Delta_+) = \Delta_-$; all other Weyl words map some positive roots to other positive roots.

Since $\dim(\mathfrak{g}_\alpha) = 1$ for all roots $\alpha \in \Delta$ and Δ_- is opposite to Δ_+ we can define for any $\alpha > 0$ a triplet

$$(E_\alpha, H_\alpha, F_\alpha) \in \mathfrak{g}_\alpha \oplus \mathfrak{h} \oplus \mathfrak{g}_{-\alpha}, \tag{3.16}$$

such that the triplet spans a standard $\mathfrak{sl}(2, \mathbb{R})$ subalgebra of $\mathfrak{g}$. The relations of one such $\mathfrak{sl}(2, \mathbb{R})$ algebra are

$$[H_\alpha, E_\alpha] = 2E_\alpha, \quad [H_\alpha, F_\alpha] = -2F_\alpha, \quad [E_\alpha, F_\alpha] = H_\alpha. \tag{3.17}$$

We also use the notation $E_{-\alpha} = F_\alpha$ and use the normalisation $\langle E_\alpha | F_\alpha \rangle = 1$.

Furthermore, we introduce the following notation for the $\mathfrak{sl}(2, \mathbb{R})$ triples associated with the simple roots α_i for $i = 1, \ldots, r$:

$$e_i \equiv E_{\alpha_i}, \quad f_i \equiv F_{\alpha_i}, \quad h_i \equiv H_{\alpha_i}. \tag{3.18}$$

The h_i form a basis of the Cartan subalgebra $\mathfrak{h}$. The r triples (e_i, h_i, f_i) are sometimes referred to as the *simple Chevalley generators*.

The *Cartan matrix* A is an $r \times r$ matrix defined by the elements

$$A_{ij} = \frac{2\langle \alpha_i | \alpha_j \rangle}{\langle \alpha_i | \alpha_i \rangle} = \frac{2\alpha_j(h_i)}{\alpha_i(h_i)} = \alpha_j(h_i). \tag{3.19}$$

We then have that

$$\begin{aligned} [h_i, e_j] &= \alpha_j(h_i) e_j = A_{ij} e_j \\ [h_i, f_j] &= -\alpha_j(h_i) f_j = -A_{ij} f_j, \end{aligned} \tag{3.20}$$

as well as the *Serre relations*

$$(\operatorname{ad} e_i)^{1-A_{ij}}\, e_j = 0, \quad (\operatorname{ad} f_i)^{1-A_{ij}}\, f_j = 0 \tag{3.21}$$

for $i \neq j$ and where the adjoint action of $\mathfrak{g}$ on itself is $(\operatorname{ad} x)y = [x, y]$. *Simplicity* of the algebra $\mathfrak{g}$ is equivalent to the indecomposability of the Cartan matrix A.

The complex Lie algebra $\mathfrak{g}(\mathbb{C})$ has a *split real form* $\mathfrak{g}(\mathbb{R})$ that is obtained by taking only real linear combinations of the Chevalley generators and their Lie brackets. This split real form $\mathfrak{g}$ was considered above and simply called $\mathfrak{g}$. The invariant bilinear form on $\mathfrak{g}$ is of indefinite signature and the $\mathfrak{g}$ elements with negative 'norm' in this bilinear form span a subalgebra of $\mathfrak{g}$ that is called the *maximal compact subalgebra* $\mathfrak{k}$. It can also be seen as the fixed point set of the so-called *Chevalley involution* on $\mathfrak{g}$ and $\mathfrak{k}$ is spanned by the combinations $E_\alpha - E_{-\alpha}$. It is of dimension equal to the number of positive roots.

3.1.2 Split Real Lie Groups and Highest Weight Representations

Many of the notions just introduced carry over to the group level. Let $G(\mathbb{R})$ be a real Lie group with Lie algebra $\mathfrak{g}(\mathbb{R})$. We only consider the case when $G(\mathbb{R})$ corresponds to the real points of an (algebraically) simply-connected Chevalley group that is split over $\mathbb{R}$ in Part One and Part Two; some more general cases are treated in Part Three.

The link between the Lie algebra and Lie group is given by the standard *exponential map* (in the identity component of $G(\mathbb{R})$) and we shall make this more precise by employing the *Steinberg presentation* below in Section 3.1.4. Our assumptions mean that we associate to the split real Lie algebra $\mathfrak{sl}(n,\mathbb{R})$ the Lie group $SL(n,\mathbb{R})$, to the split real Lie algebra $\mathfrak{so}(n,n;\mathbb{R})$ the Lie group $Spin(n,n;\mathbb{R})$ and so forth.

The Cartan subalgebra $\mathfrak{h}$ of commuting elements is the Lie algebra of an abelian subgroup $A(\mathbb{R}) \subset G(\mathbb{R})$ that we take to be the exponential of $\mathfrak{h}$. Topologically, $A(\mathbb{R}) \cong (GL(1,\mathbb{R})_+)^r$, where the $+$ subscript indicates that we restrict to positive elements. An important remark here is that there is a larger abelian subgroup, sometimes called the (split) *Cartan torus*, that is of the form $(GL(1,\mathbb{R}))^r$ and covers $A(\mathbb{R})$. We will sometimes abuse terminology and refer to $A(\mathbb{R})$ as the Cartan torus or even refer to the Cartan torus as $A(\mathbb{R})$, as it should always be clear from the context which abelian subgroup is meant.

The space of nilpotent elements $\mathfrak{n} \equiv \mathfrak{n}_+$ is the Lie algebra of a *unipotent* subgroup $N(\mathbb{R}) \subset G(\mathbb{R})$. The compact subalgebra $\mathfrak{k} \subset \mathfrak{g}$ is the Lie algebra of a (maximal) *compact* subgroup $K(\mathbb{R}) \subset G(\mathbb{R})$.

The *Iwasawa decomposition* states that one can write any element $g \in G(\mathbb{R})$ uniquely as the product of elements of the three subgroups just introduced, i.e.,

$$G(\mathbb{R}) = N(\mathbb{R})A(\mathbb{R})K(\mathbb{R}) \tag{3.22}$$

with uniqueness of decomposition [353].

The split real Lie algebras $\mathfrak{g}(\mathbb{R})$ have (non-unitary) irreducible finite-dimensional representations labelled by a *dominant highest weight* Λ. This is an element of $\mathfrak{h}^*$ that has integral non-negative coefficients when expanded in the basis of fundamental weights Λ_i that was introduced in (3.14). In other words, a dominant highest weight Λ satisfies

$$\langle \Lambda | \alpha_i \rangle \in \mathbb{N}_0 \quad \text{for all } i = 1, \ldots, r. \tag{3.23}$$

We denote the *highest weight representation* of a dominant highest weight Λ by V_Λ. The notion of highest weight implies that there is a *highest weight vector* $v_\Lambda \in V_\Lambda$ that satisfies

$$h \cdot v_\Lambda = \Lambda(h) v_\Lambda \qquad \text{for all } h \in \mathfrak{h}, \tag{3.24a}$$

$$E_\alpha \cdot v_\Lambda = 0 \qquad \text{for all positive roots } \alpha \in \Delta_+. \tag{3.24b}$$

The first condition reflects that the vector v_Λ is in the Λ-eigenspace of the action of $\mathfrak{g}$ (hence it is a weight vector) and the second condition shows that it is annihilated by all raising operators (hence at highest weight). Here, we have denoted the action of $\mathfrak{g}$ on the representation space V_Λ by $\cdot$ for brevity.

The structure of highest weight representations V_Λ can be conveniently summarised in terms of its (formal) *character*

$$\mathrm{ch}_\Lambda = \sum_{\mu \in \mathfrak{h}^*} \mathrm{mult}_{V_\Lambda}(\mu) e^\mu, \tag{3.25}$$

where $\mathrm{mult}_{V_\Lambda}(\mu)$ denotes the *weight multiplicity* of a weight $\mu \in \mathfrak{h}^*$ in the representation V_Λ, i.e., the dimension of the μ-eigenspace of the action of $\mathfrak{g}$ on V_Λ. The expression e^μ denotes an element of the *group algebra* of $\mathfrak{h}^*$ and satisfies $e^{\mu_1} e^{\mu_2} = e^{\mu_1+\mu_2}$ for two weights μ_1 and μ_2. Any representation has a character but the advantage of highest weight representations V_Λ is that there is a nice compact formula that determines the character ch_Λ in terms of Λ, the root structure of $\mathfrak{g}$ and its Weyl group. This formula is the *Weyl character formula* [368, 232]:

$$\mathrm{ch}_\Lambda = \frac{\sum_{w \in \mathcal{W}} \epsilon(w) e^{w(\Lambda+\rho)-\rho}}{\prod_{\alpha>0}(1 - e^{-\alpha})}. \tag{3.26}$$

The product in the denominator is over all positive roots $\alpha \in \Delta_+$ of the algebra $\mathfrak{g}$ and ρ is the Weyl vector defined in (3.6). The sign $\epsilon(w) = (-1)^{\ell(w)}$ gives the signature of w as an even or odd element in $\mathcal{W}$. As a special case for $\Lambda = 0$ one obtains the one-dimensional *trivial representation* with $\mathrm{ch}_{\Lambda=0} = 1$. This implies the *denominator formula*

$$\sum_{w \in \mathcal{W}} \epsilon(w) e^{w(\rho)-\rho} = \prod_{\alpha>0}(1 - e^{-\alpha}) \tag{3.27}$$

that ties the structure of the Weyl group to the structure of the root system. There is an alternative form of the character formula that will play a rôle later on. This is based on observing that

$$w\left(e^\rho \prod_{\alpha>0}(1 - e^{-\alpha})\right) = \epsilon(w) e^\rho \prod_{\alpha>0}(1 - e^{-\alpha}) \tag{3.28}$$

is $\mathcal{W}$ skew-invariant, as follows for example from the denominator identity. This implies that one can write the character ch_Λ alternatively as

$$\mathrm{ch}_\Lambda = \sum_{w\in\mathcal{W}} w\left(\frac{e^{\Lambda+\rho}}{e^\rho\prod_{\alpha>0}(1-e^{-\alpha})}\right) = \sum_{w\in\mathcal{W}} w\left(\frac{e^{\Lambda}}{\prod_{\alpha>0}(1-e^{-\alpha})}\right). \tag{3.29}$$

The character ch_Λ is not only a formal object but can actually be interpreted as a function $\mathrm{ch}_\Lambda : \mathfrak{h}(\mathbb{C}) \to \mathbb{C}$ on the complexified Cartan subalgebra $\mathfrak{h}(\mathbb{C})$ by replacing $e^\Lambda(h) = e^{\Lambda(h)}$ etc. everywhere. Then one has $\mathrm{ch}_\Lambda(h) = \mathrm{Tr}_{V_\Lambda}\exp(h)$ for $h \in \mathfrak{h}$, the trace in the highest weight representation V_Λ. The resulting expression converges everywhere on the complexified Cartan subalgebra. We can also evaluate the character on elements of the maximal torus by the exponential map. Let $a \in A$; then

$$\mathrm{ch}_\Lambda(a) = \frac{\sum_{w\in\mathcal{W}}\epsilon(w)a^{w(\Lambda+\rho)}a^{-\rho}}{\prod_{\alpha>0}(1-a^{-\alpha})} = \sum_{w\in\mathcal{W}} w\left(\frac{a^{\Lambda}}{\prod_{\alpha>0}(1-a^{-\alpha})}\right). \tag{3.30}$$

Remark 3.1 Any representation of the Lie algebra $\mathfrak{g}(\mathbb{R})$ can be promoted to a representation of the topologically simply-connected group obtained from $\mathfrak{g}(\mathbb{R})$ by the exponential map. The highest weight representations V_Λ for split real $G(\mathbb{R})$ are finite-dimensional, but not unitary. For complex $G(\mathbb{C})$ the representation V_Λ is irreducible and *unitarisable* for dominant highest weights in the sense of Kac [386, Thm. 11.7].

Remark 3.2 For Kac–Moody algebras with symmetrizable Cartan matrix, convergence is restricted to the interior of the complexified *Tits cone* [386, Sec. 10.6]. We shall discuss some aspects of Kac–Moody algebras in Chapter 19.

3.1.3 Borel and Parabolic Subgroups

An important notion for the development of automorphic representations will be that of Borel and parabolic subgroups. The *(upper) Borel subgroup* is given by

$$B(\mathbb{R}) = A(\mathbb{R})N(\mathbb{R}) = N(\mathbb{R})A(\mathbb{R}). \tag{3.31}$$

Here, the abelian group $A(\mathbb{R})$ denotes the *full* Cartan torus that covers the exponential of the Cartan subalgebra $\mathfrak{h}$, where we have chosen a set of simple roots.

A *(standard) parabolic subgroup* $P(\mathbb{R})$ of $G(\mathbb{R})$ is a proper subgroup that contains the standard Borel subgroup $B(\mathbb{R})$. If we think of $B(\mathbb{R})$ as consisting of upper triangular matrices (in $G(\mathbb{R})$), then a parabolic subgroup $P(\mathbb{R})$ contains all upper triangular matrices as well as some lower triangular ones. The discussion of this section is valid for both $\mathbb{R}$ and $\mathbb{C}$, and from here on we will suppress the notation of the underlying field.

Standard parabolic subgroups can be described by choosing a subset Σ of the simple roots Π of $\mathfrak{g}$ [146]. The subset $\Sigma \subset \Pi$ generates a root system $\langle\Sigma\rangle$ which defines a *parabolic subalgebra* as follows:

$$\mathfrak{p} = \mathfrak{h} \oplus \bigoplus_{\alpha \in \Delta(\mathfrak{p})} \mathfrak{g}_\alpha \qquad \text{where } \Delta(\mathfrak{p}) = \Delta_+ \cup \langle\Sigma\rangle\,. \tag{3.32}$$

For clarity of notation, we suppress typically the dependence on the subset Σ.

The parabolic subalgebra can be decomposed into a semi-simple *Levi subalgebra* $\mathfrak{l}$ and a *nilpotent subalgebra* $\mathfrak{u}$:

$$\mathfrak{p} = \mathfrak{l} \oplus \mathfrak{u}, \tag{3.33}$$

which is called a *Levi decomposition*. Explicitly,

$$\mathfrak{l} = \mathfrak{h} \oplus \bigoplus_{\alpha \in \langle\Sigma\rangle} \mathfrak{g}_\alpha \qquad \mathfrak{u} = \bigoplus_{\alpha \in \Delta_+ \setminus \langle\Sigma\rangle_+} \mathfrak{g}_\alpha\,, \tag{3.34}$$

where $\langle\Sigma\rangle_+ = \Delta_+ \cap \langle\Sigma\rangle$. Henceforth we will often denote the set difference $\Delta_+ \setminus \langle\Sigma\rangle_+$ as $\Delta(\mathfrak{u})$. We also note that $\mathfrak{l}$ has the same rank as $\mathfrak{g}$.

The reductive Levi subalgebra is often decomposed further into

$$\mathfrak{l} = \mathfrak{m} \oplus \mathfrak{a}_P, \tag{3.35}$$

with $\mathfrak{m} = [\mathfrak{l}, \mathfrak{l}]$ being semi-simple and $\mathfrak{a}_P \subset \mathfrak{h}$ being abelian. The decomposition

$$\mathfrak{p} = \mathfrak{m} \oplus \mathfrak{a}_P \oplus \mathfrak{u} \tag{3.36}$$

of the parabolic subalgebra is referred to as the *Langlands decomposition*. Note that we have decorated $\mathfrak{a}_P$ with a subscript P to distinguish its corresponding group A_P from the A in the Iwasawa decomposition. Recall that we use $\mathfrak{h}$ (and not $\mathfrak{a}$) for the Cartan subalgebra of $\mathfrak{g}$. Explicitly we have that

$$\begin{aligned} \mathfrak{a}_P &= \{h \in \mathfrak{h} \mid \alpha(h) = 0 \ \text{ for all } \alpha \in \Sigma\} \\ \mathfrak{m} &= [\mathfrak{l}, \mathfrak{l}] = \mathfrak{a}_P^\perp \oplus \bigoplus_{\alpha \in \langle\Sigma\rangle} \mathfrak{g}_\alpha\,, \end{aligned} \tag{3.37}$$

where the orthogonal complement $\mathfrak{a}_P^\perp$ is taken within $\mathfrak{h}$ with respect to the invariant bilinear form $\langle\cdot|\cdot\rangle$.

Example 3.3: Parabolic subalgebras of $\mathfrak{sl}(3,\mathbb{R})$

As an example we consider the Lie algebra $\mathfrak{g}(\mathbb{R}) = \mathfrak{sl}(3,\mathbb{R})$ of type A_2. It has two simple roots $\Pi(\mathfrak{g}) = \{\alpha_1, \alpha_2\}$ and positive roots given by

$$\Delta_+(\mathfrak{g}) = \{\alpha_1, \alpha_2, \alpha_1 + \alpha_2\}\,. \tag{3.38}$$

Choosing the subset $\Sigma = \{\alpha_1\}$ defines a parabolic subalgebra $\mathfrak{p}(\mathbb{R}) \subset \mathfrak{sl}(3, \mathbb{R})$ with root system

$$\Delta(\mathfrak{p}) = \underbrace{\{\alpha_1, -\alpha_1\}}_{\Delta(\mathfrak{l})} \cup \underbrace{\{\alpha_2, \alpha_1 + \alpha_2\}}_{\Delta(\mathfrak{u})} . \tag{3.39}$$

The Levi subalgebra $\mathfrak{l}(\mathbb{R})$ consists of the embedded $\mathfrak{sl}(2, \mathbb{R})$ associated with the simple root α_1, together with an additional abelian element:

$$\mathfrak{l}(\mathbb{R}) = \underbrace{\mathfrak{sl}(2, \mathbb{R})}_{\mathfrak{m}(\mathbb{R})} \oplus \underbrace{\mathbb{R}}_{\mathfrak{a}(\mathbb{R})} . \tag{3.40}$$

The nilpotent part $\mathfrak{u}(\mathbb{R})$ is a two-dimensional abelian Lie algebra and transforms in the two-dimensional representation of $\mathfrak{l}(\mathbb{R})$.

As (traceless) (3×3) matrices the elements of $\mathfrak{p}(\mathbb{R})$, $\mathfrak{l}(\mathbb{R})$ and $\mathfrak{u}(\mathbb{R})$ take the forms

$$\mathfrak{p}(\mathbb{R}): \begin{pmatrix} * & * & * \\ * & * & * \\ 0 & 0 & * \end{pmatrix}, \quad \mathfrak{l}(\mathbb{R}): \begin{pmatrix} * & * & 0 \\ * & * & 0 \\ 0 & 0 & * \end{pmatrix}, \quad \mathfrak{u}(\mathbb{R}): \begin{pmatrix} 0 & 0 & * \\ 0 & 0 & * \\ 0 & 0 & 0 \end{pmatrix}. \tag{3.41}$$

At the level of Lie groups there are corresponding notions. Let P be a connected group having $\mathfrak{p}$ as its Lie algebra. Then there are (unique) decompositions

$$P = LU = MA_P U , \tag{3.42}$$

also called the *Levi decomposition* and the *Langlands decomposition*. The subgroup L is called the *Levi subgroup* and U the *unipotent subgroup* or *unipotent radical* of the parabolic subgroup $P \subset G$.

A particularly important class of parabolic subgroups is furnished by the so-called *maximal parabolic subgroups*. These are in a sense the largest (proper) parabolic subgroups and are characterised by choosing as a defining set Σ of all simple roots of G but one: $\Sigma = \Pi \setminus \{\alpha_{i_*}\}$, where we denoted the simple root that is left out by α_{i_*}. We will use the notation P_{i_*} to denote the maximal parabolic subgroup associated with such a choice. For maximal parabolic subgroups one has that

$$L = GL(1) \times M, \tag{3.43}$$

where M is a semi-simple Lie group. The Dynkin diagram of its Lie algebra is obtained by removing the node i_* from the Dynkin diagram of $\mathfrak{g}$. The parabolic subgroup of Example 3.3 is maximal and corresponds to the choice $i_* = 2$. We note that our writing of the product (3.43) does not signify that the L is a direct product but rather that every element in L can be written as a product of elements of the two groups.

3.1.4 Chevalley Group Notation and Discrete Subgroups

Using the exponential map, we will often parametrise group elements in terms of some basic elements. This is often referred to as the *Steinberg presentation*. Concretely, we define for roots $\alpha \in \Delta$ and $u \in \mathbb{R}$ (or another base field of the split Lie algebra)

$$x_\alpha(u) = \exp(uE_\alpha), \tag{3.44}$$

where E_α is the distinguished element of the root space $\mathfrak{g}_\alpha$ that appears in the *Chevalley basis* constructed in (3.16). The *one-parameter subgroup* generated by $x_\alpha(u)$ for $u \in \mathbb{R}$ will be denoted by $N_\alpha(\mathbb{R})$. The one-parameter subgroups $N_\alpha(\mathbb{R})$ belong to the simply-connected (central) cover of the group $G(\mathbb{R})$, which equals $G(\mathbb{R})$ due to our assumptions on the real Lie group stated at the beginning of Section 3.1.2. These assumptions can also be restated by saying that we consider an evaluation of the exponential in (3.44) as a matrix exponential in a *fundamental representation* of $\mathfrak{g}(\mathbb{R})$. We define a fundamental representation to be one whose tensor power series covers all finite-dimensional representations of $\mathfrak{g}(\mathbb{R})$. In other words, the weights of the fundamental representation generate the whole weight lattice of $\mathfrak{g}(\mathbb{R})$. For $\mathfrak{sl}(n,\mathbb{R})$ an example would be the n-dimensional standard representation; for $\mathfrak{so}(n,n;\mathbb{R})$ one can take a spinor representation.

Besides the one-parameter subgroups defined in (3.44) we also let

$$w_\alpha(u) = x_\alpha(u)x_{-\alpha}(-u^{-1})x_\alpha(u) \quad \text{and} \quad h_\alpha(u) = w_\alpha(u)w_\alpha(1)^{-1}. \tag{3.45}$$

The notation $w_\alpha(u)$ is connected to the Weyl group defined above by noting that the $w_{\alpha_i}(1)$ (for simple α_i) generate a cover of the Weyl group [386]. For $u \approx 1$, the element $h_\alpha(u)$ yields H_α at linear order.

These elements so defined satisfy

$$x_\alpha(u)x_\alpha(v) = x_\alpha(u+v) \quad \text{and} \quad h_\alpha(u)h_\alpha(v) = h_\alpha(uv) \tag{3.46}$$

and for $\alpha \neq -\beta$

$$x_\alpha(u)x_\beta(v)x_\alpha(u)^{-1}x_\beta(v)^{-1} = \prod_{\substack{m,n>0\\ m\alpha+n\beta\in\Delta}} x_{m\alpha+n\beta}(c_{mn}^{\alpha\beta}u^m v^n), \tag{3.47}$$

which is the exponentiation of the relation $\left[E_\alpha, E_\beta\right] \propto E_{\alpha+\beta}$. The constants $c_{mn}^{\alpha\beta}$ depend on the chosen order in the product and the structure constants of the basis $\{E_\alpha\}$. If $\alpha = -\beta$ we obtain instead

$$w_\alpha(u)x_\alpha(v)w_\alpha(-u) = x_{-\alpha}(-u^{-2}v). \tag{3.48}$$

This is related to the commutator $[E_\alpha, E_{-\alpha}] = H_\alpha$. We also have that

$$h_\alpha(t)x_\beta(u)h_\alpha(t)^{-1} = x_\beta(t^{\beta(H_\alpha)}u). \tag{3.49}$$

Definition 3.4 (Discrete Chevalley group) Let $G(\mathbb{R})$ be the real points of an algebraically simply-connected group split over $\mathbb{Q}$ (discussed below) and consider a Chevalley basis as defined in Section 3.1.1. The group $G(\mathbb{Z})$ generated by the $x_\alpha(u)$ for integer $u \in \mathbb{Z}$ is a discrete subgroup of $G(\mathbb{R})$ that we call the discrete *Chevalley group*.

The group $G(\mathbb{Z})$ contains in particular representatives of (a covering of) the Weyl group.

Remark 3.5 Within the Lie algebra $\mathfrak{g}(\mathbb{R})$ one can define the *Chevalley lattice* as the lattice spanned by a Chevalley basis like the one we introduced in Section 3.1.1. The group $G(\mathbb{Q})$ acts on this lattice by the adjoint action. The stabiliser of the lattice inside $G(\mathbb{Q})$ is $G(\mathbb{Z})$. Note that $G(\mathbb{Q})$ (and $G(\mathbb{Z})$) can have a non-trivial centre that in particular acts trivially on the Chevalley lattice.

More generally, one can consider the case when $G_0(\mathbb{R})$ is a quotient of $G(\mathbb{R})$ by a central subgroup. If G_0 is an algebraic group one can define independently the discrete group $G_0(\mathbb{Z})$ and the projection of $G(\mathbb{Z})$ to the quotient. However, it can happen that the projection of $G(\mathbb{Z})$ to the quotient $G_0(\mathbb{R})$ does not surject onto $G_0(\mathbb{Z})$.

In most of this book, we will deal with the case where $G(\mathbb{R})$ is the simply-connected group and $G(\mathbb{Z})$ is given as in the above definition. In some rare exceptions, we shall encounter also algebraic groups $G_0(\mathbb{R})$ that arise as quotients of $G(\mathbb{R})$. $G_0(\mathbb{Z})$ will then be defined from the algebraic group conditions of $G_0(\mathbb{R})$ and not from the quotient construction. In these cases we will also have that, if $G_0(\mathbb{R})$ is disconnected, the discrete group $G_0(\mathbb{Z})$ meets all the connected components. Under these conditions there is a bijection between functions on $G(\mathbb{R})$ invariant under $G(\mathbb{Z})$ and functions on $G_0(\mathbb{R})$ invariant under $G_0(\mathbb{Z})$. For instance, we shall interchangeably talk about Eisenstein series for $Spin(d,d)$ and $SO(d,d)$.

3.2 p-adic and Adelic Groups

In this section, we introduce some basic properties of Lie groups for other fields and rings. In our treatment we shall mostly consider the fields $\mathbb{Q}$, $\mathbb{Q}_p$ and $\mathbb{R}$ or the ring of adeles $\mathbb{A}$. For more details and proofs, see for instance [72, 267, 263, 116].

Recall that a complex Lie group $G(\mathbb{C})$ is a differentiable manifold with a compatible group structure. More generally, one can consider an *algebraic group* G over $\mathbb{Q}$, which, formally, is an (affine) algebraic variety over $\mathbb{Q}$ equipped with a group structure. We will be interested in *linear algebraic groups* over $\mathbb{Q}$, which are subgroups G of the group $GL(n)$ of invertible $n \times n$ matrices restricted by polynomial equations of the matrix elements with coefficients in $\mathbb{Q}$. For a $\mathbb{Q}$-algebra R, the R-points $G(R)$ of G are then the solutions to these equations for values in R and form a subgroup of $GL(n,R)$. The typical example of a linear algebraic group is $SL(n)$, which is defined as the subgroup of $GL(n)$ such that the polynomial equation $\det(g) = 1$ is satisfied. We will mostly consider algebraic groups that are $\mathbb{Q}$-split (see [366]) and study the $\mathbb{Q}$-algebras $R = \mathbb{Q}_p$, $\mathbb{R}$, $\mathbb{C}$ or $\mathbb{A}$ with the corresponding points $G(R)$.

Remark 3.6 As previously mentioned, we restrict to algebraically simply-connected groups G [366, 480] which, in the case of the Chevalley groups described in Section 3.1.4, amounts to evaluating the matrix exponential for $x_\alpha(u)$ in a fundamental representation. This does not, in general, imply that the group $G(R)$ for a $\mathbb{Q}$-algebra R is *topologically* simply-connected. We have, however, in the case of $R = \mathbb{R}$, that if G is algebraically simply-connected, then $G(\mathbb{R})$ is connected [74].

3.2.1 p-adic Groups

We shall now take a closer look at the groups $G(\mathbb{Q}_p)$. At the infinite place, $\mathbb{Q}_\infty = \mathbb{R}$, this is just the split real Lie group $G(\mathbb{R})$ of the complex Lie group $G(\mathbb{C})$.

Let us focus on the non-archimedean completions $\mathbb{Q}_p$ of $\mathbb{Q}$ for $p < \infty$, comparing with the more familiar situation of real Lie groups $G(\mathbb{R})$ where it is appropriate. An important fact is that the notion of maximal compact subgroup carries over to the p-adic setting. Recall that for a real Lie group $G(\mathbb{R})$ in its split real form the *maximal compact subgroup* $K(G)$ is defined as the fixed point set of G under the Chevalley involution. For example, in the case of $G(\mathbb{R}) = SL(n,\mathbb{R})$ we have $K(G) = SO(n)$. For real Lie groups, the maximal compact subgroup is unique up to conjugacy. To understand the analogous notion of maximal compact subgroup of $G(\mathbb{Q}_p)$, recall that the p-adic integers $\mathbb{Z}_p$ form a compact ring inside $\mathbb{Q}_p$. It follows that the subgroup of integer points

$$G(\mathbb{Z}_p) = G \cap GL(n,\mathbb{Z}_p) \tag{3.50}$$

sits compactly inside $G(\mathbb{Q}_p)$. For finite primes $p < \infty$ a maximal compact subgroup of $G(\mathbb{Q}_p)$ is $K_p = G(\mathbb{Z}_p)$.

Remark 3.7 In contrast to the archimedean case, maximal compact subgroups of $G(\mathbb{Q}_p)$ for $p < \infty$ are not all conjugate. The different conjugacy classes of maximal compact subgroups of $G(\mathbb{Q}_p)$ are most easily classified by *Bruhat–Tits theory* [112, 111, 367]. For example, for $SL(2,\mathbb{Q}_p)$ there are two conjugacy classes of maximal subgroups. One is represented by $SL(2,\mathbb{Z}_p)$ and the other is represented by $\begin{pmatrix} p & \\ & 1 \end{pmatrix} SL(2,\mathbb{Z}_p) \begin{pmatrix} p^{-1} & \\ & 1 \end{pmatrix}$ [367]. For $SL(n,\mathbb{Q}_p)$ there are n

classes of maximal compact subgroups in general. (However, there is a unique *Iwahori subgroup* that generalises the notion of Borel subgroup to local fields.)

For real Lie groups $G(\mathbb{R})$ we always have a unique *Iwasawa decomposition*

$$G(\mathbb{R}) = N(\mathbb{R})A(\mathbb{R})K(\mathbb{R}), \tag{3.51}$$

where $K(\mathbb{R})$ is the maximal compact subgroup, $A(\mathbb{R})$ is the Cartan torus and $N(\mathbb{R})$ is the nilpotent subgroup generated by the positive Chevalley generators of the Lie algebra of G. The notion of Iwasawa decomposition carries over to the p-adic situation, where we have a decomposition of the form

$$G(\mathbb{Q}_p) = N(\mathbb{Q}_p)A(\mathbb{Q}_p)G(\mathbb{Z}_p). \tag{3.52}$$

In contrast to the case of real groups, the p-adic Iwasawa decomposition is *not* unique; however its restriction to the norm on A is, and this fact will play a crucial rôle later. Many details on Iwasawa decompositions for $SL(n, \mathbb{Q}_p)$ can be found for example in [291, 1].

Example 3.8: Iwasawa decompositions in $SL(2, \mathbb{Q}_p)$ for $p \leq \infty$

We now consider in more detail the example of $G(\mathbb{Q}_p) = SL(2, \mathbb{Q}_p)$. The maximal compact subgroup is $K_p = SL(2, \mathbb{Z}_p)$ and the Iwasawa decomposition reads

$$SL(2, \mathbb{Q}_p) = N(\mathbb{Q}_p)A(\mathbb{Q}_p)SL(2, \mathbb{Z}_p), \tag{3.53}$$

where

$$N(\mathbb{Q}_p) = \left\{ \begin{pmatrix} 1 & x \\ & 1 \end{pmatrix} \middle| x \in \mathbb{Q}_p \right\}, \qquad A(\mathbb{Q}_p) = \left\{ \begin{pmatrix} a & \\ & a^{-1} \end{pmatrix} \middle| a \in \mathbb{Q}_p^\times \right\}. \tag{3.54}$$

To illustrate this further, let us consider the explicit Iwasawa decomposition of a specific element

$$g = \begin{pmatrix} 1 & \\ u & 1 \end{pmatrix} \in SL(2, \mathbb{Q}_p), \tag{3.55}$$

which will be of relevance for the analysis in subsequent sections. First notice that if $u \in \mathbb{Z}_p$ then g is already in $SL(2, \mathbb{Z}_p)$ and the decomposition is trivial. Consider therefore the case when $u \in \mathbb{Q}_p \backslash \mathbb{Z}_p$, for which one could write a $g = nak$ decomposition as follows:

$$\begin{pmatrix} 1 & \\ u & 1 \end{pmatrix} = \begin{pmatrix} 1 & u^{-1} \\ & 1 \end{pmatrix} \begin{pmatrix} u^{-1} & \\ & u \end{pmatrix} \begin{pmatrix} & -1 \\ 1 & u^{-1} \end{pmatrix}. \tag{3.56}$$

Notice that, for $u \in \mathbb{Q}_p \backslash \mathbb{Z}_p$, the element $u^{-1} \in \mathbb{Z}_p$ and therefore the matrix on the right is in $K_p = SL(2, \mathbb{Z}_p)$ such that this represents a valid Iwasawa decomposition.

As emphasised above, the Iwasawa decomposition for groups over non-archimedean fields is not unique. In the present example, all possible Iwasawa decompositions are of the form

$$\begin{pmatrix} 1 & \\ u & 1 \end{pmatrix} = \begin{pmatrix} 1 & u^{-1} - ke^{-1}u^{-2} \\ & 1 \end{pmatrix} \begin{pmatrix} (eu)^{-1} & \\ & eu \end{pmatrix} \begin{pmatrix} k & ku^{-1} - e \\ e^{-1} & e^{-1}u^{-1} \end{pmatrix} \tag{3.57}$$

for arbitrary $k \in \mathbb{Z}_p$ and $e \in \mathbb{Z}_p^\times$. We note that, since $|e|_p = 1$, the norms of the entries of the element $a \in A(\mathbb{Q}_p)$ are unambiguously defined even though the full Iwasawa decomposition is not unique. The relation (3.56) corresponds to $k = 0$ and $e = 1$. (One can render the p-adic Iwasawa decomposition unique by imposing further restrictions on the individual elements [291]. We will not use this here.)

It is illuminating to compare (3.56) with the decomposition of the analogous element in $SL(2, \mathbb{R})$. Thus, take

$$\begin{pmatrix} 1 & \\ x & 1 \end{pmatrix} \in SL(2, \mathbb{R}), \tag{3.58}$$

so that in this case $x \in \mathbb{R}$. The *unique* Iwasawa decomposition of this element is

$$\begin{pmatrix} 1 & \\ x & 1 \end{pmatrix} = \begin{pmatrix} 1 & \frac{x}{1+x^2} \\ & 1 \end{pmatrix} \begin{pmatrix} 1/\sqrt{1+x^2} & \\ & \sqrt{1+x^2} \end{pmatrix} k, \tag{3.59}$$

with

$$k = \frac{1}{\sqrt{1+x^2}} \begin{pmatrix} 1 & -x \\ x & 1 \end{pmatrix} \in SO(2, \mathbb{R}). \tag{3.60}$$

Hence, the component along the Cartan torus in (3.56) is in fact simpler in the Iwasawa decomposition of $SL(2, \mathbb{Q}_p)$ compared with that of $SL(2, \mathbb{R})$.

3.2.2 Adelisation and Strong Approximation

We now discuss the central notion of strong approximation, which allows the reformulation of many questions concerning $G(\mathbb{R})$ and its automorphic forms in terms of questions on the adelic group $G(\mathbb{A})$. The description in this section is general; the following Section 3.2.3 gives more details for the case of $G = SL(2)$.

Given an algebraic group G defined over $\mathbb{Q}$ we can consider the adelic group $G(\mathbb{A})$ as the restricted direct product

$$G(\mathbb{A}) = G(\mathbb{R}) \times G_f, \tag{3.61}$$

where

$$G_f = \prod_{p<\infty}{}' G(\mathbb{Q}_p), \tag{3.62}$$

consisting of elements $g = (g_p) = (g_\infty; g_2, g_3, g_5, \dots)$ such that all but finitely many $g_p \in G(\mathbb{Z}_p)$.

Remark 3.9 The adelic group $G(\mathbb{A})$ is attached to the algebraic group G over $\mathbb{Q}$ (or more precisely over $\mathrm{Spec}(\mathbb{Q})$). We typically start from a Chevalley group as in the preceding Section 3.1.4, which provides a clear basis for the construction of the adelic group $G(\mathbb{A})$.

We further set

$$K_f = \prod_{p<\infty} G(\mathbb{Z}_p) \tag{3.63}$$

and we then have the notion of maximal compact subgroup $K_{\mathbb{A}}$ of $G(\mathbb{A})$ defined as

$$K_{\mathbb{A}} = K_\infty \times K_f, \tag{3.64}$$

where K_∞ is the maximal compact subgroup of $G(\mathbb{R})$. The adelic version of the Iwasawa decomposition thus reads

$$G(\mathbb{A}) = N(\mathbb{A})A(\mathbb{A})K_{\mathbb{A}}. \tag{3.65}$$

When G is split of rank r, the adelic Cartan torus is given by

$$A(\mathbb{A}) = GL(1, \mathbb{A}) \times \cdots \times GL(1, \mathbb{A}) \cong (\mathbb{A}^\times)^r. \tag{3.66}$$

Since $\mathbb{Q}$ is discrete in $\mathbb{A}$ according to Proposition 2.24, it follows that $G(\mathbb{Q})$ is a discrete subgroup of $G(\mathbb{A})$. This implies that the arithmetic coset space $G(\mathbb{Q})\backslash G(\mathbb{A})$ corresponds to the adelisation of $G(\mathbb{Z})\backslash G(\mathbb{R})$. In fact, topologically $G(\mathbb{Q})\backslash G(\mathbb{A})$ is the total space of a fiber bundle over $G(\mathbb{Z})\backslash G(\mathbb{R})$ [267]:

$$\begin{array}{ccc} K_f & \hookrightarrow & G(\mathbb{Q})\backslash G(\mathbb{A}) \\ & & \downarrow \\ & & G(\mathbb{Z})\backslash G(\mathbb{R}). \end{array} \tag{3.67}$$

One way of stating *strong approximation* then asserts that

$$G(\mathbb{Z})\backslash G(\mathbb{R}) \cong G(\mathbb{Q})\backslash G(\mathbb{A})/K_f. \tag{3.68}$$

This has the very useful consequence that any function $\phi_{\mathbb{R}}$ on $G(\mathbb{Z})\backslash G(\mathbb{R})$ can be lifted to a function $\phi_{\mathbb{A}}$ on the adelisation $G(\mathbb{Q})\backslash G(\mathbb{A})$ where $\phi_{\mathbb{A}}$ is characterised by being right-invariant under K_f. The consequences of this for automorphic forms will be discussed in Chapter 4.

The strong approximation theorem (3.68) can be stated even more generally for open subgroups K_Γ of K_f according to [164] (see also [531, 537]):

Theorem 3.10 (Strong approximation theorem) *Let G be a topological group with $G(\mathbb{Q})$ dense in G_f, and let K_Γ be an open subgroup of K_f and $\Gamma = K_\Gamma \cap G(\mathbb{Q})$. Then*

$$\begin{aligned} \phi\colon \Gamma\backslash G(\mathbb{R}) &\to G(\mathbb{Q})\backslash G(\mathbb{A})/K_\Gamma \\ \Gamma x_\infty &\mapsto G(\mathbb{Q})(x_\infty; \mathbb{1})K_\Gamma \end{aligned} \tag{3.69}$$

is a homeomorphism. Here, $G(\mathbb{Q})$ is diagonally embedded in $G(\mathbb{A})$; $G(\mathbb{R})$ is embedded as $(x_\infty; \mathbb{1})$ and K_Γ as $(\mathbb{1}; k_p)$.

Remark 3.11 An assumption in the theorem is that $G(\mathbb{Q})$ should be dense in G_f, and this is equivalent to the statement that for all open subsets U of G_f we have that $U \cap G(\mathbb{Q}) \neq \emptyset$. An example of such a group G that will be useful for us is $SL(n)$ [116]. The assumption of denseness can also be reformulated in terms of the semi-simple group G being algebraically simply-connected [531].

Proof (of Theorem 3.10)

- ϕ is well-defined (independent of coset representative)

  Let $x_\infty, y_\infty \in G(\mathbb{R})$ such that $\Gamma x_\infty = \Gamma y_\infty$, that is, there exists a $\gamma \in \Gamma$ such that $x_\infty = \gamma y_\infty$.

  We have that $\Gamma = K_\Gamma \cap G(\mathbb{Q})$ and thus,

$$\begin{aligned}\phi(\Gamma x_\infty) &= G(\mathbb{Q})(x_\infty; \mathbb{1})K_\Gamma = G(\mathbb{Q})(\gamma y_\infty; \mathbb{1})K_\Gamma \\ &= G(\mathbb{Q})(\gamma y_\infty; \gamma)K_\Gamma = G(\mathbb{Q})(y_\infty; \mathbb{1})K_\Gamma = \phi(\Gamma y_\infty)\,. \end{aligned} \tag{3.70}$$

- ϕ is injective

  Assume $\phi(\Gamma x_\infty) = \phi(\Gamma y_\infty)$. Then $G(\mathbb{Q})(x_\infty; \mathbb{1})K_\Gamma = G(\mathbb{Q})(y_\infty; \mathbb{1})K_\Gamma$, that is, there exists a $\gamma \in G(\mathbb{Q})$ and $k \in K_\Gamma$ such that $(x_\infty; \mathbb{1}) = \gamma(y_\infty; \mathbb{1})k = (\gamma y_\infty; \gamma k)$.

  This means that $x_\infty = \gamma y_\infty$ and $\gamma = k^{-1}$. Since $\gamma \in G(\mathbb{Q})$ and $k \in K_\Gamma$ we then have that $\gamma = k^{-1} \in K_\Gamma \cap G(\mathbb{Q}) = \Gamma$. Thus, $x_\infty = \gamma y_\infty$ implies that $\Gamma x_\infty = \Gamma y_\infty$.
- ϕ is surjective ($G(\mathbb{A}) = G(\mathbb{Q})G(\mathbb{R})K_\Gamma$)

  Let $x = (x_\infty; x_f)$ be an arbitrary element in $G(\mathbb{A}) = G(\mathbb{R}) \times G_f$. We will now show that since $G(\mathbb{Q})$ is dense in G_f, there exists a $\gamma \in G(\mathbb{Q})$ such that $\gamma x_f \in K_\Gamma$.

  Consider the continuous map $f\colon G_f \to G_f, g \mapsto g x_f$. Since K_Γ is an open set in G_f and f is continuous, $V = f^{-1}(K_\Gamma)$ is an open set in G_f. Since $G(\mathbb{Q})$ is dense in G_f this means that $V \cap G(\mathbb{Q})$ is non-empty. Let $\gamma \in V \cap G(\mathbb{Q})$; then $\gamma x_f = f(\gamma) \in f(V) = K_\Gamma$.

  Thus, for any element $(x_\infty; x_f) \in G(\mathbb{A})$ with $\gamma \in V \cap G(\mathbb{Q})$ as above,

$$G(\mathbb{Q})(x_\infty; x_f)K_\Gamma = G(\mathbb{Q})(\gamma x_\infty; \gamma x_f)K_\Gamma = G(\mathbb{Q})(\gamma x_\infty; \mathbb{1})K_\Gamma = \phi(\Gamma \gamma x_\infty)\,. \tag{3.71}$$

□

Remark 3.12 The generalisation to open subgroups K_Γ is important since it allows us to treat different discrete subgroups Γ in a uniform way. Typically these subgroups are associated with arithmetically defined congruence subgroups.

Remark 3.13 Let U be a subgroup of G such that $U(\hat{\mathbb{Z}}) = \prod_{p<\infty} U(\mathbb{Z}_p)$ is an open set in G_f and let $x_f \in U_f = \prod'_{p<\infty} U(\mathbb{Q}_p)$. With $f\colon G_f \to G_f$, $g \mapsto g x_f$ as above, we have that $V = f^{-1}(U(\hat{\mathbb{Z}})) = U(\hat{\mathbb{Z}})x_f^{-1}$ is an open subset of $U_f \subset G_f$. If $G(\mathbb{Q})$ is dense in G_f, this means that $V \cap G(\mathbb{Q})$ is non-empty and we thus have a $\gamma \in V \cap G(\mathbb{Q}) \subseteq U_f \cap G(\mathbb{Q}) = U(\mathbb{Q})$ such that $\gamma x_f \in U(\hat{\mathbb{Z}})$.

Analogously to the proof of Theorem 3.10, we can then show that

$$\begin{aligned} \Phi\colon U(\mathbb{Z})\backslash U(\mathbb{R}) &\to U(\mathbb{Q})\backslash U(\mathbb{A})/U(\hat{\mathbb{Z}}) \\ U(\mathbb{Z})x_\infty &\mapsto U(\mathbb{Q})(x_\infty; \mathbb{1})U(\hat{\mathbb{Z}}) \end{aligned} \tag{3.72}$$

is also a homeomorphism.

3.2.3 Strong Approximation for $SL(2,\mathbb{R})$

In this section, we illustrate the concepts of the preceding section in some examples involving $G = GL(2)$ and $G = SL(2)$.

Example 3.14: Discreteness of $GL(2,\mathbb{Q})$ in $GL(2,\mathbb{A})$

We will now show that $GL(2,\mathbb{Q})$ is discrete in $GL(2,\mathbb{A})$ by first considering the identity element. The line of reasoning is analogous to the case of $\mathbb{Q}$ being discretely embedded in $\mathbb{A}$ that was treated in Proposition 2.24.

Let $U \subset GL(2,\mathbb{R})$ be an open neighbourhood of $\mathbb{1}$ such that $U \cap GL(2,\mathbb{Z}) = \{\mathbb{1}\}$. Then

$$V = U \times \prod_{p<\infty} GL(2,\mathbb{Z}_p) \subset GL(2,\mathbb{A}) \tag{3.73}$$

is an open neighbourhood of $\mathbb{1}$ in $GL(2,\mathbb{A})$. Since, with the diagonal embedding,

$$GL(2,\mathbb{Q}) \cap \prod_{p<\infty} GL(2,\mathbb{Z}_p) = GL(2,\mathbb{Z}), \tag{3.74}$$

we then have that $GL(2,\mathbb{Q}) \cap V = \{\mathbb{1}\}$. For an arbitrary element $g \in GL(2,\mathbb{Q})$ these arguments generalise directly by instead considering gV. We have then that $GL(2,\mathbb{Q})$ is discrete in $GL(2,\mathbb{A})$.

Perhaps more importantly for our further calculations, it can similarly be shown that $SL(2,\mathbb{Q})$ is discrete in $SL(2,\mathbb{A})$.

The next example discusses how the strong approximation theorem, Theorem 3.10, works for $SL(2)$ and different choices of subgroup Γ.

Example 3.15: Strong approximation for $SL(2)$

Let $G = SL(2)$. First, let $K_\Gamma = K_f = \prod_{p<\infty} G(\mathbb{Z}_p)$. Then $\Gamma = K_\Gamma \cap G(\mathbb{Q}) = G(\mathbb{Z}) = SL(2,\mathbb{Z})$, the standard modular group. From the above theorem we then get that

$$G(\mathbb{Z})\backslash G(\mathbb{R}) \cong G(\mathbb{Q})\backslash G(\mathbb{A})/K_f. \tag{3.75}$$

The second example addresses the principal congruence subgroup $\Gamma_0(N)$. Let locally

$$\Gamma_0(N)_p = \left\{ \begin{pmatrix} a & b \\ c & d \end{pmatrix} \in SL(2,\mathbb{Z}_p) : c \equiv 0 \mod N\mathbb{Z}_p \right\} \tag{3.76}$$

and $K_\Gamma(N) = K_0(N) := \prod_{p<\infty} K_\Gamma(N)_p$, where

$$K_\Gamma(N)_p = \begin{cases} SL(2,\mathbb{Z}_p) & p \nmid N, \\ \Gamma_0(N)_p & p \mid N. \end{cases} \tag{3.77}$$

Since $K_\Gamma \subset K_f = \prod_{p<\infty} SL(2,\mathbb{Z}_p)$ we know that $\Gamma = K_\Gamma \cap SL(2,\mathbb{Q}) \subset SL(2,\mathbb{Z})$. That $c \equiv 0 \mod N\mathbb{Z}_p$ for all divisors p of N means that (with $c \in \mathbb{Z}$)

$$\begin{aligned} c \in N\mathbb{Z}_p \quad \forall p \mid N &\iff \left|\frac{c}{N}\right|_p \leq 1 \quad \forall p \mid N \\ &\iff \frac{c}{N} \text{ has no } p \text{ in the denominator} \quad \forall p \mid N \\ &\iff c \equiv 0 \mod N. \end{aligned} \tag{3.78}$$

Thus $\Gamma = \Gamma_0(N)$ and, from the above theorem,

$$\Gamma_0(N)\backslash SL(2,\mathbb{R}) \cong SL(2,\mathbb{Q})\backslash SL(2,\mathbb{A})/K_\Gamma(N). \tag{3.79}$$

We finally exhibit an isomorphism of cosets of the discrete subgroups in $G(\mathbb{A})$ with cosets of discrete subgroups in $G(\mathbb{R})$ that will be central in Section 4.4.1.

Example 3.16: Bijection of Borel cosets

In this example we will (based on the notes [255]) show that

$$\begin{aligned} \phi \colon B(\mathbb{Z})\backslash SL(2,\mathbb{Z}) &\to B(\mathbb{Q})\backslash SL(2,\mathbb{Q}) \\ B(\mathbb{Z})\gamma &\mapsto B(\mathbb{Q})\gamma \end{aligned} \tag{3.80}$$

is an isomorphism, where

$$B(\mathbb{X}) = \left\{ \begin{pmatrix} * & * \\ 0 & * \end{pmatrix} \right\} \cap SL(2,\mathbb{X}), \tag{3.81}$$

with $\mathbb{X} = \mathbb{Q}$ or $\mathbb{Z}$. The mapping is well-defined since if $B(\mathbb{Z})\gamma' = B(\mathbb{Z})\gamma$ then $B(\mathbb{Q})\gamma' = B(\mathbb{Q})\gamma$ as $B(\mathbb{Z}) \subset B(\mathbb{Q})$.

It is injective because if $B(\mathbb{Q})\gamma' = B(\mathbb{Q})\gamma$ then there exists a b in $B(\mathbb{Q})$ such that $\gamma' = b\gamma$, but then $b = \gamma'\gamma^{-1} \in SL(2,\mathbb{Z})$ which means that $b \in B(\mathbb{Q}) \cap SL(2,\mathbb{Z}) = B(\mathbb{Z})$. Thus, $B(\mathbb{Z})\gamma' = B(\mathbb{Z})\gamma$.

For the surjectivity we need to show that every $B(\mathbb{Q})g$ with $g \in SL(2,\mathbb{Q})$ has a representative in $SL(2,\mathbb{Z})$. Let

$$g = \begin{pmatrix} a & b \\ c & d \end{pmatrix} \in SL(2,\mathbb{Q}), \qquad b = \begin{pmatrix} q & m \\ 0 & q^{-1} \end{pmatrix} \in SL(2,\mathbb{Q}),$$
$$bg = \begin{pmatrix} qa + mc & qb + md \\ q^{-1}c & q^{-1}d \end{pmatrix}, \tag{3.82}$$

where $c = c_1/c_2$ and $d = d_1/d_2$ with $c_i, d_i \in \mathbb{Z}$ in shortened form with positive denominators. Now set $q = \gcd(c_1d_2, c_2d_1)/(c_2d_2)$, which makes $q^{-1}c$ and $q^{-1}d$ coprime integers, and thus there exist integers α and β such that $\alpha q^{-1}d - \beta q^{-1}c = 1$ by Bézout's lemma.

If $c = 0$ then $d \neq 0$, $a = 1/d$ and $q = \gcd(0, c_2d_1)/(c_2d_2) = |c_2d_1|/(c_2d_2) = |d|$, meaning that $qa = q^{-1}d = \pm 1$ and we may choose m such that $qb + md$ is integer. On the other hand, if $c \neq 0$ we may choose $m = (\alpha - qa)/c$ giving $qa + mc = \alpha$ and $qb + md = \beta$ which are both integers. This shows that ϕ from Equation (3.80) is an isomorphism.

4

Automorphic Forms

In this chapter, we explain how to think about automorphic forms as functions on $G(\mathbb{Q})\backslash G(\mathbb{A})$, as opposed to the more familiar concept of $G(\mathbb{Z})$-invariant functions on a real Lie group $G(\mathbb{R})$. After introducing some standard terminology for $SL(2,\mathbb{R})$ in Section 4.1, and reviewing some standard facts about classical modular forms in Section 4.2, we discuss in Section 4.3 the passage from classical modular forms on the upper half-plane $\mathbb{H}$ to automorphic forms on the real group $SL(2,\mathbb{R})$.

The next step is the transition to the adelic language, which will be carried out in Section 4.4, beginning with the example of $SL(2,\mathbb{A})$. The general adelic picture of automorphic forms on $G(\mathbb{A})$ is then introduced, and Eisenstein series on $G(\mathbb{A})$ are defined in the following Section 4.5. These notions will lead naturally to automorphic representations and the close connection with studying the unitary action of $G(\mathbb{A})$ on the Hilbert space $L^2(G(\mathbb{Q})\backslash G(\mathbb{A}))$, which will be discussed in Chapter 5.

4.1 Preliminaries on $SL(2,\mathbb{R})$

In this section, we introduce our conventions on notation related to $SL(2,\mathbb{R})$ that will be used throughout this book.

4.1.1 $SL(2,\mathbb{R})$ Lie Group and $\mathfrak{sl}(2,\mathbb{R})$ Lie Algebra

We take $SL(2,\mathbb{R})$ to be the real Lie group defined (in its fundamental representation) by

$$SL(2,\mathbb{R}) = \left\{ g = \begin{pmatrix} a & b \\ c & d \end{pmatrix} \middle| \, a,b,c,d \in \mathbb{R} \text{ and } \det(g) = ad - bc = 1 \right\}. \quad (4.1)$$

The *maximal compact subgroup* is $K = SO(2,\mathbb{R})$ corresponding to the orthogonal matrices within $SL(2,\mathbb{R})$.

The Lie algebra $\mathfrak{sl}(2,\mathbb{R})$ has the standard Chevalley basis

$$e = \begin{pmatrix} 0 & 1 \\ 0 & 0 \end{pmatrix}, \quad h = \begin{pmatrix} 1 & 0 \\ 0 & -1 \end{pmatrix}, \quad f = \begin{pmatrix} 0 & 0 \\ 1 & 0 \end{pmatrix} \tag{4.2}$$

with commutation relations

$$[h,e] = 2e, \quad [h,f] = -2f, \quad [e,f] = h. \tag{4.3}$$

The generator h acts diagonally and is called the Cartan generator; e is a positive step operator and f a negative step operator. The compact subgroup $SO(2,\mathbb{R})$ is generated by the combination $e - f$.

The *universal enveloping algebra* $\mathcal{U}(\mathfrak{sl}(2,\mathbb{C}))$ has a distinguished second-order element, called the *Casimir operator*, that we define by

$$\Omega = \frac{1}{4}h^2 + \frac{1}{2}ef + \frac{1}{2}fe = \frac{1}{4}h^2 - \frac{1}{2}h + ef. \tag{4.4}$$

This definition is unique up to normalisation. The Casimir operator commutes with all $\mathfrak{sl}(2,\mathbb{C})$ Lie algebra elements.

The *Iwasawa decomposition* of $SL(2,\mathbb{R})$ can be chosen in the form $SL(2,\mathbb{R}) = NAK$, where N is in the image of the exponential map exp applied to e, the maximal torus is in the image of exp applied to h and K is the compact subgroup $SO(2,\mathbb{R})$ that is the exponential of $e-f$; see Example 3.8. Concretely this means that we can write any element g of $SL(2,\mathbb{R})$ as

$$\begin{aligned} g = nak &= e^{xe} e^{\log(y)h/2} e^{\theta(e-f)} \\ &= \begin{pmatrix} 1 & x \\ 0 & 1 \end{pmatrix} \begin{pmatrix} y^{1/2} & 0 \\ 0 & y^{-1/2} \end{pmatrix} \begin{pmatrix} \cos\theta & \sin\theta \\ -\sin\theta & \cos\theta \end{pmatrix} \end{aligned} \tag{4.5}$$

with $k \in K = SO(2,\mathbb{R})$ and $y > 0$.

4.1.2 The Upper Half-Plane $\mathbb{H}$ and $SL(2,\mathbb{Z})$

A main object of interest to us is the two-dimensional coset space $G/K = SL(2,\mathbb{R})/SO(2,\mathbb{R})$; a representative for any point of this space is given by the first two factors in (4.5). The coset space can therefore be parametrised by elements of the *upper half-plane*

$$\mathbb{H} = \{z = x + iy \mid x, y \in \mathbb{R} \text{ and } y > 0\} \cong G/K. \tag{4.6}$$

The coset space G/K (or, equivalently, the upper half-plane $\mathbb{H}$) carries an action of $SL(2,\mathbb{R})$ by left multiplication: an element $\gamma \in G$ transforms a g

into $g' = \gamma g$. The action on the explicit parameters $z \in \mathbb{H}$ can be read off from writing the new element in Iwasawa form $g' = n'a'k'$. Performing this calculation one finds

$$z' = \gamma \cdot z = \frac{az+b}{cz+d} \quad \text{for} \quad \gamma = \begin{pmatrix} a & b \\ c & d \end{pmatrix} \in SL(2,\mathbb{R}). \tag{4.7}$$

Using the Iwasawa decomposition (4.5), we see that the point i is left-invariant by the maximal compact subgroup $K = SO(2,\mathbb{R})$ and that

$$g \cdot i = x + iy = z. \tag{4.8}$$

For $SL(2,\mathbb{R})$, automorphic forms are functions $f(g)$ that are invariant under the action of a discrete subgroup $\Gamma \subset SL(2,\mathbb{R})$. The choice $\Gamma = SL(2,\mathbb{Z})$ is called the *modular group* and consists of the $SL(2,\mathbb{R})$ matrices with integral entries. The invariance $f(\gamma g) = f(g)$ for all $\gamma \in SL(2,\mathbb{Z})$ means that f is, in fact, a function on the double quotient $SL(2,\mathbb{Z})\backslash SL(2,\mathbb{R})/SO(2,\mathbb{R})$ where $SL(2,\mathbb{Z})$-equivalent points are identified. Using the upper half-plane $\mathbb{H}$ presentation of $SL(2,\mathbb{R})/SO(2,\mathbb{R})$ one can give a very explicit description of the double quotient.

The group $SL(2,\mathbb{Z})$ is well known to be generated by [179]:

$$T = \begin{pmatrix} 1 & 1 \\ 0 & 1 \end{pmatrix} \quad \text{and} \quad S = \begin{pmatrix} 0 & -1 \\ 1 & 0 \end{pmatrix}. \tag{4.9}$$

When acting on $z \in \mathbb{H}$, they generate

$$T \cdot z = z + 1, \quad S \cdot z = -\frac{1}{z}. \tag{4.10}$$

Therefore, T is a translation by one unit and S is inversion in the unit circle combined with a reflection in the y-axis. A fundamental domain for the action of $SL(2,\mathbb{Z})$ on the upper half-plane (parametrising the double quotient above) is depicted in Figure 4.1. The fundamental domain clearly displays a single *cusp* where it touches the boundary of the space. This cusp corresponds to the limit $y \to \infty$. For a congruence subgroup $\Gamma \subset SL(2,\mathbb{Z})$ (discussed more in Section 4.2.2) the cusps are the Γ-equivalence classes of points in $\mathbb{Q} \cup \{\infty\}$ and they can all be reached by $SL(2,\mathbb{Z})$-transformations of the $y \to \infty$ cusp [179]. The corresponding fundamental domain, similar to Figure 4.1, touches the real axis at these points forming additional 'cusps'.

Remark 4.1 What we are dealing with is effectively $PSL(2,\mathbb{Z})$ rather than $SL(2,\mathbb{Z})$, where the 'P' indicates that a matrix has to be identified with minus itself. The reason is that the two matrices have identical action on the upper half-plane as easily verified from (4.7).

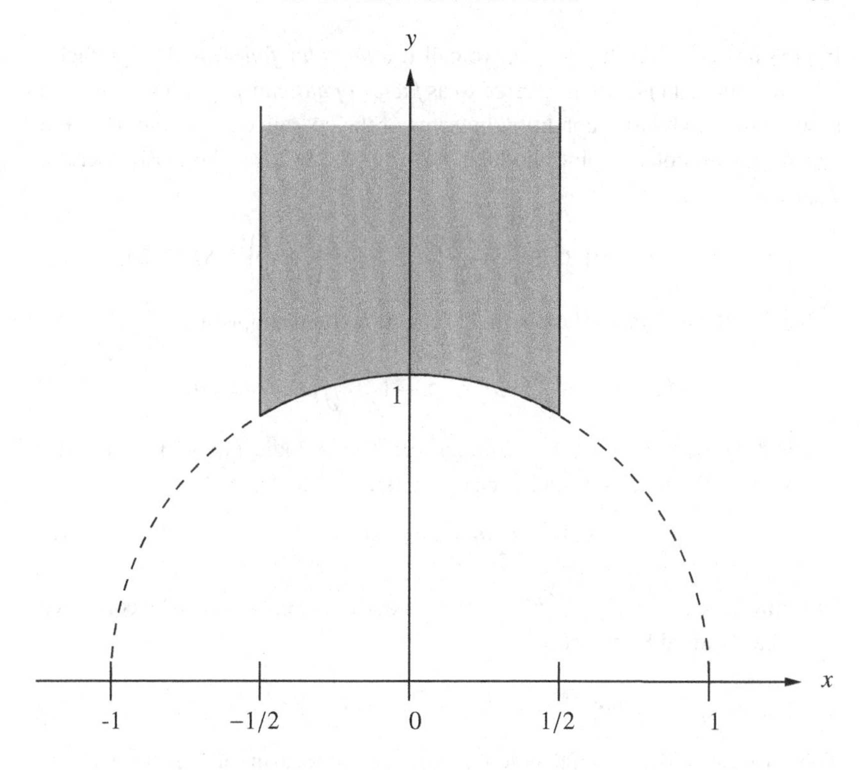

Figure 4.1 A fundamental domain for the action of $SL(2,\mathbb{Z})$ acting on the upper half-plane (grey region). The cusp is at $y \to \infty$.

4.2 Classical Modular Forms

In this section, we discuss some details of the theory of classical modular forms on the complex upper half-plane $\mathbb{H}$. In the next section, we will then see how these modular forms fit into the more general notion of an automorphic form on the Lie group $SL(2,\mathbb{R})$.

4.2.1 Holomorphic Modular Forms

A *holomorphic modular form* of weight $w \geq 0$ is a holomorphic function $f\colon \mathbb{H} \to \mathbb{C}$ which transforms according to

$$f\left(\frac{az+b}{cz+d}\right) = (cz+d)^w f(z) \tag{4.11}$$

under the discrete action of

$$\begin{pmatrix} a & b \\ c & d \end{pmatrix} \in SL(2,\mathbb{Z}). \tag{4.12}$$

If $f(z)$ has zero weight, $w = 0$, we call it a *modular function*. The prefactor $(cz+d)^w$ in (4.11) is often referred to as *factor of automorphy*. It is convenient to write the modularity condition in terms of the co-called *slash operator*. For f a weight w holomorphic modular form, $g \in SL(2,\mathbb{R})$, the slash operator $f|_w g : \mathbb{H} \to \mathbb{C}$ is defined by

$$(f|_w g)(z) := (cz+d)^{-w} f\left(\frac{az+b}{cz+d}\right), \qquad g = \begin{pmatrix} a & b \\ c & d \end{pmatrix} \in SL(2,\mathbb{R}). \tag{4.13}$$

Using this the defining relation, (4.11) can be written simply as

$$(f|_w \gamma)(z) = f(z), \qquad \gamma = \begin{pmatrix} a & b \\ c & d \end{pmatrix} \in SL(2,\mathbb{Z}). \tag{4.14}$$

The defining Equation (4.11) implies that f is periodic $f(z+1) = f(z)$ (for any weight w) and thus has a Fourier expansion of the form

$$f(z) = \sum_{n\in\mathbb{Z}} a_n q^n, \qquad q := e^{2\pi i z}. \tag{4.15}$$

Decomposing $q = e^{2\pi i z} = e^{2\pi i x} e^{-2\pi y}$ the Fourier coefficients can be computed from the standard Fourier transform

$$a_n e^{-2\pi n y} = \int_0^1 e^{-2\pi i n x} f(x+iy) dx. \tag{4.16}$$

This formula (and its generalisations) will play a key rôle in subsequent chapters.

The moderate growth condition mentioned in Section 1.1 can be formulated as the statement that

$$|f(x+iy)| \leq C \cdot y^N \tag{4.17}$$

for some constants C, N as $y \to \infty$ for any $x \in \mathbb{R}$. For holomorphic modular forms this is in fact equivalent to the statement that all negative Fourier coefficients a_n with $n < 0$, in (4.15) vanish. To see this we simply use the integral representation (4.16) for the Fourier coefficient and calculate its norm:

$$|a_n e^{-2\pi n y}| = \left| \int_0^1 e^{-2\pi i n x} f(x+iy) dx \right|. \tag{4.18}$$

Removing the oscillating exponential, we obtain a sequence of inequalities

$$\left| \int_0^1 e^{-2\pi i n x} f(x+iy) dx \right| \leq \int_0^1 |f(x+iy)|\, dx \leq \int_0^1 C \cdot y^N dx = C \cdot y^N. \tag{4.19}$$

Thus we arrive at the inequality

$$|a_n e^{-2\pi n y}| \leq C \cdot y^N, \tag{4.20}$$

and when $n < 0$ the exponential $e^{-2\pi n y}$ blows up as $y \to \infty$ so therefore we must have $a_n = 0$ for $n < 0$ as claimed.

Example 4.2: Classical holomorphic Eisenstein series

Classic examples of holomorphic modular forms on $\mathbb{H}$ are provided by the *holomorphic Eisenstein series* defined by

$$E_{2w}(z) = \frac{1}{2} \sum_{\substack{(c,d)\in\mathbb{Z}^2 \\ (c,d)=1}} \frac{1}{(cz+d)^{2w}}. \tag{4.21}$$

One can check that this satisfies all the criteria stated above for integral $w \geq 2$. An alternative form of this can be found using the slash operator (4.13). To this end, define the subgroup

$$\Gamma_\infty = \left\{ \pm \begin{pmatrix} 1 & n \\ & 1 \end{pmatrix} \mid n \in \mathbb{Z} \right\}. \tag{4.22}$$

We can then write the Eisenstein series as

$$E_{2w}(z) = \sum_{\gamma\in\Gamma_\infty\backslash SL(2,\mathbb{Z})} 1|_{2w}\gamma. \tag{4.23}$$

The (finite-dimensional) space $\mathcal{M}_{2w}(SL(2,\mathbb{Z}))$ of weight $2w$ holomorphic modular forms is a ring, famously generated by the Eisenstein series $E_4(z)$ and $E_6(z)$ (see, e.g., [630] for a proof). The Fourier expansions of $E_4(z)$ and $E_6(z)$ are given by

$$\begin{aligned} E_4(z) &= 1 + 240 \sum_{n=1}^{\infty} \sigma_3(n) q^n = 1 + 240q + 2160q^2 + \cdots, \\ E_6(z) &= 1 - 504 \sum_{n=1}^{\infty} \sigma_5(n) q^n = 1 - 504q - 16632q^2 + \cdots, \end{aligned} \tag{4.24}$$

where

$$\sigma_s(n) = \sum_{d|n} d^s \tag{4.25}$$

is the divisor function. For proofs, see for example the classic book by Serre [559].

Besides the space $\mathcal{M}_k$ of modular forms of weight k one can also consider the subspace $\mathcal{S}_k$ of *cusp forms* of weight k. These holomorphic modular forms satisfy $a_0 = 0$ in their Fourier expansion. The holomorphic Eisenstein series $E_4(z)$ and $E_6(z)$ in Example 4.21 are clearly not cusp forms. As $\mathcal{S}_k \subset \mathcal{M}_k$ and all holomorphic modular forms are generated by $E_4(z)$ and $E_6(z)$, the first non-trivial cusp form arises for weight $k = 12$ and is given by a specific linear combination of E_4^3 and E_6^2 such that the constant Fourier coefficients cancel. It is customary to take

$$\Delta(z) = \frac{1}{1728} \left(E_4(z)^3 - E_6(z)^2 \right) = \sum_{n=1}^{\infty} \tau_n q^n = q - 24q^2 + \cdots. \tag{4.26}$$

This cusp form of weight $k = 12$ is called the *discriminant modular form* and was studied in detail by Ramanujan. Its Fourier coefficients are the values of the *Ramanujan tau function*. One can also write $\Delta(z) = \eta(z)^{24}$, where $\eta(z)$ is the *Dedekind eta function*, which will be defined in (10.7) later.

Closely associated with the discriminant is the meromorphic *modular invariant j-function*

$$j(z) = \frac{E_4(z)^3}{\Delta(z)} = \frac{1}{q} + 744 + 196884q + \cdots . \tag{4.27}$$

Note that $j(z)$ has a simple pole at the cusp and therefore it is not holomorphic. However, j is modular invariant: $j(\gamma z) = j(z)$ for all $\gamma \in SL(2,\mathbb{Z})$ and in fact any meromorphic modular invariant can be expressed as a modular function of j. We shall come back to these functions in a different context in Sections 15.5 and 16.4.

4.2.2 Modular Forms for Congruence Subgroups*

It is often of interest in number theory to consider holomorphic modular forms for *congruence subgroups* $\Gamma \subset SL(2,\mathbb{Z})$ (a good reference is the book by Diamond and Shurman [179]). These satisfy an analogous relation to (4.11) but with extra restrictions on the transformation matrix and possibly with a character appearing on the right-hand side. We recall that congruence subgroups are subgroups of $SL(2,\mathbb{Z})$ that contain a *principal congruence subgroup*

$$\Gamma(N) = \left\{ \begin{pmatrix} a & b \\ c & d \end{pmatrix} \in SL(2,\mathbb{Z}) \,\middle|\, N|b, c, a-1, d-1 \right\} \tag{4.28}$$

for some N. Examples are the congruence subgroups

$$\Gamma_1(N) = \left\{ \begin{pmatrix} a & b \\ c & d \end{pmatrix} \in SL(2,\mathbb{Z}) \,\middle|\, N|c, a-1, d-1 \right\},$$

$$\Gamma_0(N) = \left\{ \begin{pmatrix} a & b \\ c & d \end{pmatrix} \in SL(2,\mathbb{Z}) \,\middle|\, N|c \right\}, \tag{4.29}$$

where $\Gamma_0(N)$ contains $\Gamma_1(N)$ as a normal subgroup of finite index $\phi(N)$. Here, $\phi(N)$ is the *Euler totient function* given by the number of integers up to N that are coprime to N. The space of weight w modular forms for $\Gamma_1(N)$ (resp. $\Gamma_0(N)$) is then denoted by $\mathcal{M}_w(\Gamma_1(N))$ (resp. $\mathcal{M}_w(\Gamma_0(N))$). Since the quotient $\Gamma_0(N)/\Gamma_1(N)$ is isomorphic to the multiplicative group $(\mathbb{Z}/N\mathbb{Z})^\times$ of order $\phi(N)$, one can relate modular forms on these different congruence subgroups through the introduction of Dirichlet characters. A *Dirichlet character* χ is a group homomorphism

$$\chi\colon (\mathbb{Z}/N\mathbb{Z})^\times \to \mathbb{C}^\times, \tag{4.30}$$

where the product between two Dirichlet characters χ_1 and χ_2 is defined by $(\chi_1\chi_2)(g) = \chi_1(g)\chi_2(g)$ for $g \in (\mathbb{Z}/N\mathbb{Z})^\times$. One can then decompose the space $\mathcal{M}_w(\Gamma_1(N))$ in terms of modular forms for the larger group $\Gamma_0(N)$, at the expense of introducing Dirichlet characters:

$$\mathcal{M}_w(\Gamma_1(N)) = \bigoplus_{\chi} \mathcal{M}_w(\Gamma_0(N), \chi), \tag{4.31}$$

where functions in the χ-eigenspace $\mathcal{M}_w(\Gamma_0(N), \chi)$ obey a generalisation of (4.11):

$$f\left(\frac{az+d}{cz+d}\right) = \chi(d)(cz+d)^w f(z), \qquad f \in \mathcal{M}_w(\Gamma_0(N), \chi). \tag{4.32}$$

Functions in $\mathcal{M}_w(\Gamma_0(N), \chi)$ are said to be of *level* N.

Example 4.3: Classical level N Eisenstein series

An example of a modular form for $\Gamma_0(N)$ is provided by the level N, weight $2w$ Eisenstein series

$$E_{2w}(z;\chi) = \sum_{\substack{(m,n)\in\mathbb{Z}^2 \\ (m,n)=1}} \frac{\chi(n)}{(mz+n)^{2w}}, \tag{4.33}$$

generalising the classical series (4.21).

Remark 4.4 An additional notion when discussing the modular forms for congruence subgroups $\Gamma_0(N)$ is that of *newforms* and *oldforms*. A oldform is a cusp form $f(z)$ for $\Gamma_0(N)$ that is of the form $f(z) = g(dz)$ where $g(z)$ is a cusp form for $\Gamma_0(M)$ where M is a non-trivial divisor of N and d a divisor of N/M. It thus comes from a cusp form on a congruence subgroup of smaller level. The orthogonal complement (with respect to the *Petersson inner product* [24]) to the space of oldforms of weight k for $\Gamma_0(N)$ is the space of newforms.

4.2.3 Jacobi Forms*

There exist certain two-variable generalisations of modular forms, originally developed by Eichler and Zagier [202]. Let $SL(2,\mathbb{Z})^J = SL(2,\mathbb{Z}) \ltimes \mathbb{Z}^2$ be the *(Fourier–)Jacobi group*, defined such that for two elements $[M, X]$ and $[M', X']$ in $SL(2,\mathbb{Z})^J$ we have the group law

$$[M, X] \cdot [M', X'] = [MM', XM' + X']. \tag{4.34}$$

The Jacobi group acts on functions $\phi\colon \mathbb{H}\times\mathbb{C}\to\mathbb{C}$ via a generalisation of the slash operator in (4.13). For integers k and m we define

$$(\phi|_{k,m}\gamma)(\tau,z) = (c\tau+d)^{-k} e^{2\pi i m\left[-\frac{c(z+\lambda\tau+\mu)^2}{c\tau+d}+\lambda^2\tau+2\lambda z+\lambda\nu\right]}\phi\left(\frac{az+b}{cz+d},\frac{z}{c\tau+d}\right) \tag{4.35}$$

for

$$\gamma = \left[\begin{pmatrix} a & b\\ c & d\end{pmatrix},(\lambda,\mu)\right] \in SL(2,\mathbb{Z})^J. \tag{4.36}$$

Remark 4.5 We here follow the standard literature convention for Jacobi forms that τ is the variable on the upper half-plane $\mathbb{H}$ while $z\in\mathbb{C}$. This should hopefully not lead to any confusion with the cases when we only discuss the upper half-plane for $SL(2,\mathbb{R})$, in which case z denotes a parameter in $\mathbb{H}$.

A *Jacobi form* of weight k and index m is a meromorphic function

$$\phi\colon \mathbb{H}\times\mathbb{C}\longrightarrow\mathbb{C} \tag{4.37}$$

that satisfies the following invariance property:

$$(\phi|_{k,m}\gamma)(\tau,z) = \phi(\tau,z), \qquad \text{for all} \quad \gamma\in SL(2,\mathbb{Z})^J. \tag{4.38}$$

In the two special cases $\gamma = [\left(\begin{smallmatrix} a & b\\ c & d\end{smallmatrix}\right),(0,0)]$ and $\gamma = [\left(\begin{smallmatrix} 0 & 0\\ 0 & 0\end{smallmatrix}\right),(\lambda,\mu)]$ this condition can be decomposed into a *modular* property and an *elliptic* property, respectively, given by

$$\begin{aligned} \phi\left(\begin{pmatrix} a & b\\ c & d\end{pmatrix},\frac{z}{c\tau+d}\right) &= (c\tau+d)^k e^{2\pi i m\frac{cz^2}{c\tau+d}}\phi(\tau,z),\\ \phi(\tau,z+\lambda\tau+\mu) &= e^{-2\pi i m(\lambda^2\tau+2\lambda z)}\phi(\tau,z). \end{aligned} \tag{4.39}$$

Any Jacobi form admits a Fourier expansion of the form

$$\phi(\tau,z) = \sum_{r\in\mathbb{Z}}\sum_{n\in\mathbb{Z}} c(n,r)q^n\zeta^r, \qquad q:=e^{2\pi i\tau},\ \zeta:=e^{2\pi i z}. \tag{4.40}$$

Following [157] we call this function a *holomorphic Jacobi form* if all coefficients vanish unless $4mn\geq r^2$, and a *Jacobi cusp form* if the stronger condition $4mn>r^2$ holds. One also defines *weak Jacobi forms* as those satisfying the weaker condition $c(n,r)=0$ unless $n\geq 0$.

Let us now consider a classic example of a Jacobi form, namely the *Jacobi–Eisenstein series*.

Example 4.6: Jacobi–Eisenstein series

Define the subgroup $\Gamma_\infty^J \subset SL(2,\mathbb{Z})^J$ by

$$\Gamma_\infty^J := \left\{ \left[\pm \begin{pmatrix} 1 & n \\ & 1 \end{pmatrix}, (0,\mu) \right] \mid n, \mu \in \mathbb{Z} \right\}. \tag{4.41}$$

Analogously to the classical holomorphic Eisenstein series E_{2w} in (4.23), a Jacobi–Eisenstein series $E_{k,m}(\tau, z)$ of weight k and index m is defined by

$$E_{k,m}(\tau, z) = \sum_{\gamma \in \Gamma_\infty^J \backslash SL(2,\mathbb{Z})^J} 1|_{k,m}\gamma. \tag{4.42}$$

Remark 4.7 We will return to Jacobi forms in Section 12.1 in the form of the classical Jacobi theta functions, and in Section 15.5 in the context of Mathieu moonshine and black hole counting in string theory.

4.3 From Classical Modular Forms to Automorphic Forms

We shall now see how to adapt the theory of holomorphic modular forms on $\mathbb{H} = SL(2,\mathbb{R})/SO(2,\mathbb{R})$ to the more general framework of automorphic forms defined on $G = SL(2,\mathbb{R})$, invariant under the left-action of $SL(2,\mathbb{Z})$.

4.3.1 Lifting Modular Forms to $SL(2,\mathbb{R})$

Given a weight w holomorphic modular form $f\colon \mathbb{H} \to \mathbb{C}$ as in (4.11) we define a new (complex) function φ_f on $SL(2,\mathbb{R})$ through the assignment

$$f \longmapsto \varphi_f(g) = (ci + d)^{-w} f(g \cdot i), \tag{4.43}$$

where $g = \left(\begin{smallmatrix} a & b \\ c & d \end{smallmatrix}\right) \in SL(2,\mathbb{R})$ and $g \cdot i$ was given in (4.8). The prefactor here is chosen in such away as to cancel the factor of automorphy in (4.11) in order for the function φ_f to be *invariant* under $SL(2,\mathbb{Z})$:

$$\varphi_f(\gamma g) = \varphi_f(g), \qquad \gamma \in SL(2,\mathbb{Z}). \tag{4.44}$$

According to our definition in Section 1.1, φ_f is thus an automorphic function on $SL(2,\mathbb{R})$. Note that the condition of moderate growth is satisfied automatically since the seed function f is holomorphic.

We can make the *lift* (4.43) *from* $\mathbb{H}$ *to* $SL(2,\mathbb{R})$ more explicit by making use of the Iwasawa decomposition (4.5) of an element $g \in SL(2,\mathbb{R})$:

$$g = nak = \begin{pmatrix} 1 & x \\ & 1 \end{pmatrix} \begin{pmatrix} y^{1/2} & \\ & y^{-1/2} \end{pmatrix} \begin{pmatrix} \cos\theta & \sin\theta \\ -\sin\theta & \cos\theta \end{pmatrix}, \tag{4.45}$$

with $n \in N(\mathbb{R})$, $a \in A(\mathbb{R})$, $k \in SO(2,\mathbb{R})$. We also recall from (4.8) that

$$g \cdot i = x + iy \equiv z. \tag{4.46}$$

We recall that $K = SO(2,\mathbb{R})$ leaves the point i invariant. Plugging the Iwasawa decomposition of g into the right-hand side of (4.43) we can write φ_f as a function of the three variables (x, y, θ):

$$\varphi_f(g) = \varphi_f(x, y, \theta) = e^{iw\theta} y^{w/2} f(x + iy). \tag{4.47}$$

Moreover, under the right-action of

$$k = \begin{pmatrix} \cos\vartheta & \sin\vartheta \\ -\sin\vartheta & \cos\vartheta \end{pmatrix} \in SO(2,\mathbb{R}) \tag{4.48}$$

it transforms by a phase:

$$\varphi_f(gk) = e^{iw\vartheta}\varphi_f(g). \tag{4.49}$$

This implies that the original transformation property (4.11) of f under $SL(2,\mathbb{Z})$ has been traded for the above phase transformation of $\varphi_f(g)$ under $K = SO(2,\mathbb{R})$. While f itself was invariant under $SO(2,\mathbb{R})$, one instead says that φ_f is *K-finite*, implying that the action of K on f generates a finite-dimensional vector space; in the present example this is represented by the one-dimensional space of characters $\sigma : k \mapsto e^{iw\vartheta}$ through

$$\varphi_f(gk) = \sigma(k)\varphi_f(g). \tag{4.50}$$

Next, we address the question of how the automorphic form φ_f incorporates the holomorphy of f on $\mathbb{H}$:

$$\frac{\partial}{\partial \bar{z}} f = \frac{1}{2}\left(\frac{\partial}{\partial x} + i\frac{\partial}{\partial y}\right) f = 0, \tag{4.51}$$

where $z = x + iy$. The corresponding statement for φ_f is that it satisfies

$$F\varphi_f = -2ie^{-2i\theta}\left(y\frac{\partial}{\partial \bar{z}} - \frac{1}{4}\frac{\partial}{\partial\theta}\right)\varphi_f = 0. \tag{4.52}$$

We will now give a group-theoretic interpretation to this differential condition.

The group $SL(2,\mathbb{R})$ acts on smooth functions on $SL(2,\mathbb{R})$ via the *right-regular action*, where we recall that 'regular' here refers to the fact that the so-called *Radon–Nikodym derivative* is trivial. Let g' be an element of $SL(2,\mathbb{R})$ and $\varphi(g)$ a function on $SL(2,\mathbb{R})$. The right-regular action is defined by

$$\big(\pi(g')\varphi\big)(g) = \varphi(gg'). \tag{4.53}$$

The action of the Lie algebra $\mathfrak{sl}(2,\mathbb{R})$ is then given by differential operators acting on *smooth* functions. Using (4.53) one finds the following differential operators corresponding to the Chevalley basis generators:

$$h = -2\sin(2\theta)y\partial_x + 2\cos(2\theta)y\partial_y + \sin(2\theta)\partial_\theta, \tag{4.54a}$$

$$e = \cos(2\theta)y\partial_x + \sin(2\theta)y\partial_y + \sin^2\theta\partial_\theta, \tag{4.54b}$$

$$f = \cos(2\theta)y\partial_x + \sin(2\theta)y\partial_y - \cos^2\theta\partial_\theta. \tag{4.54c}$$

The compact generator $e-f$ of $SO(2,\mathbb{R})$ acts by ∂_θ. We record also the inverse relations

$$y\partial_x = \frac{1}{2}((e+f)\cos(2\theta) + e - f - h\sin(2\theta)), \tag{4.55a}$$

$$y\partial_y = \frac{1}{2}((e+f)\sin(2\theta) + h\cos(2\theta)), \tag{4.55b}$$

$$\partial_\theta = e - f. \tag{4.55c}$$

The Casimir operator (4.4) then becomes a second-order differential operator, namely the *Laplacian*

$$\Delta_{SL(2,\mathbb{R})} = y^2\left(\partial_x^2 + \partial_y^2\right) - y\partial_x\partial_\theta. \tag{4.56}$$

We shall also make use of the so-called *compact basis*. This is a representation of $\mathfrak{sl}(2,\mathbb{R})$ in terms of (2×2) matrices different from (4.2) and given explicitly by

$$H = -i(e-f), \quad E = \frac{1}{2}\left(h + i(e+f)\right), \quad F = \frac{1}{2}\left(h - i(e+f)\right), \tag{4.57}$$

that is,

$$H = i\begin{pmatrix} 0 & -1 \\ 1 & 0 \end{pmatrix}, \quad E = \frac{1}{2}\begin{pmatrix} 1 & i \\ i & -1 \end{pmatrix}, \quad F = \frac{1}{2}\begin{pmatrix} 1 & -i \\ -i & -1 \end{pmatrix}. \tag{4.58}$$

The generators satisfy the standard $\mathfrak{sl}(2,\mathbb{R})$ commutation relations

$$[H,E] = 2E, \quad [H,F] = -2F, \quad [E,F] = H. \tag{4.59}$$

The Cartan generator H is Hermitian in this basis, and this is the reason for the name compact basis.

The representation (4.58) of $\mathfrak{sl}(2,\mathbb{R})$ is unitarily equivalent to the standard representation (4.2) through the transformation

$$UHU^\dagger = h, \text{etc.} \qquad \text{for} \quad U = \frac{1}{2}\begin{pmatrix} -1+i & 1+i \\ -1+i & -1-i \end{pmatrix}. \tag{4.60}$$

The differential operators associated with this basis are then given by

$$H = -i\partial_\theta, \tag{4.61a}$$

$$E = 2ie^{2i\theta}\left(y\partial_z - \frac{1}{4}\partial_\theta\right), \tag{4.61b}$$

$$F = -2ie^{-2i\theta}\left(y\partial_{\bar{z}} - \frac{1}{4}\partial_\theta\right), \tag{4.61c}$$

where we have used standard holomorphic and anti-holormorphic derivatives:

$$\partial_z = \frac{1}{2}\left(\partial_x - i\partial_y\right), \quad \partial_{\bar{z}} = \frac{1}{2}\left(\partial_x + i\partial_y\right). \tag{4.62}$$

Because the compact basis is unitarily equivalent, the Casimir operator does not change.

Remark 4.8 The change of basis is basically that induced by the *Sekiguchi isomorphism* [553, 423] that enters in the description of real nilpotent orbits.

The above discussion implies that the differential operator F in (4.52) may in fact be identified with the lowering operator in the basis (E, F, H) of the Lie algebra $\mathfrak{sl}(2,\mathbb{R})$. This implies that φ_f may be viewed as a lowest weight state of a representation of $\mathfrak{sl}(2,\mathbb{R})$. Furthermore, we note that the (Hermitian) generator H in this basis corresponds to

$$e^{i\theta H} = \begin{pmatrix} \cos\theta & \sin\theta \\ -\sin\theta & \cos\theta \end{pmatrix} \in SO(2,\mathbb{R}), \tag{4.63}$$

and hence corresponds to the differential operator $H = -i\partial_\theta$. Therefore, H is diagonal on φ_f with eigenvalue w:

$$H\varphi_f = w\varphi_f. \tag{4.64}$$

From the commutation relations we further deduce that E raises the H-eigenvalue w by $+2$, while F lowers it by the same amount. This implies that the holomorphic Eisenstein series E_{2w} can be viewed as lowest weight vectors in the holomorphic discrete series of $SL(2,\mathbb{R})$, providing our first glimpse of the general connection between automorphic forms and representation theory, a topic that will be discussed in more generality in Chapter 5 and onwards. See also Section 5.6 for some more details on the specific case of $SL(2)$ treated above.

Before we proceed we shall mention one final important property of φ_f, namely that it is an eigenfunction of the Laplacian (4.56) on $SL(2,\mathbb{R})$. Acting on φ_f one obtains that

$$\Delta_{SL(2,\mathbb{R})}\varphi_f = \frac{w}{2}\left(\frac{w}{2} - 1\right)\varphi_f. \tag{4.65}$$

As we will see, all the properties of φ_f discussed above will have counterparts in the general theory of automorphic forms.

The automorphic lift of weight w, level N modular forms for $\Gamma_0(N)$ will be treated in Section 4.4.3.

Example 4.9: Lift of a holomorphic Eisenstein series and slash operator

The lift of the holomorphic Eisenstein series $f(z) = E_{2w}(z)$ to an automorphic form on $SL(2,\mathbb{R})$ can be written using (4.47) as

$$\varphi_f(g) = e^{2iw\theta} y^w E_{2w}(x+iy). \tag{4.66}$$

We use the slash operator defined in (4.13) to write the lift. By a calculation similar to the one used in proving Equation (1.8) we can now rewrite the Eisenstein series $E_{2w}(z)$ in (4.21) directly as a function on $SL(2,\mathbb{R})$. Parametrising an element $g \in SL(2,\mathbb{R})$ in Iwasawa form as $g = nak$ (see 4.45) we obtain

$$\varphi_f(nak) = (f|_w nak)(i) = e^{iw\theta} y^{w/2} \sum_{\gamma \in N(\mathbb{Z})\backslash SL(2,\mathbb{Z})} (1|_w\gamma)(nak \cdot i). \tag{4.67}$$

Remark 4.10 We would like to make a cautionary remark regarding the generalisation of the above discussion to arbitrary groups $G(\mathbb{R})$. Modular forms are holomorphic functions on $\mathbb{H} \cong SL(2,\mathbb{R})/SO(2,\mathbb{R})$ with simple transformation properties under $SL(2,\mathbb{Z})$, and it seems natural to try to generalise this construction to higher-rank real Lie groups $G(\mathbb{R})$. One might suspect a generalisation to 'holomorphic functions' $f\colon G(\mathbb{R})/K \to \mathbb{C}$, where K is the maximal compact subgroup of $G(\mathbb{R})$, transforming with some weight under the action of a discrete subgroup $G(\mathbb{Z}) \subset G(\mathbb{R})$. However, this only works whenever the coset $G(\mathbb{R})/K$ carries a complex structure. In the case above this complex structure is provided by the fact that the maximal compact subgroup $K = SO(2,\mathbb{R}) \cong U(1)$. In general, the maximal subgroup K of some $G(\mathbb{R})$ does not have an isolated $U(1)$ factor that can provide a complex structure on $G(\mathbb{R})/K$ and therefore we could not expect to have a general theory of holomorphic modular forms on G/K. A standard example with a complex structure is provided by $G = Sp(2n,\mathbb{R})$, $K = U(n)$, in which case $Sp(2n;\mathbb{R})/U(n)$ is a Hermitian symmetric domain known as the *Siegel upper half-plane*. This leads to the theory of *holomorphic Siegel modular forms* (see, e.g., [113] for a review).

4.3.2 Maaß Forms and Non-holomorphic Eisenstein Series

As just discussed, it is in general too restrictive (and often impossible) to consider holomorphic modular forms. It is therefore necessary to look for a theory of arbitrary (*non-holomorphic*) functions $f\colon G(\mathbb{R})/K \to \mathbb{C}$ which transform nicely under the action of some discrete subgroup $G(\mathbb{Z}) \subset G(\mathbb{R})$. This leads to the notion of an *automorphic form*, which we will now discuss for $SL(2,\mathbb{R})$.

In addition to the holomorphic modular forms, the classical theory also contains an interesting class of *non-holomorphic functions* $f\colon SL(2,\mathbb{Z})\backslash\mathbb{H} \to \mathbb{R}$. These non-holomorphic functions are eigenfunctions of the Laplacian $\Delta_{\mathbb{H}}$ on $\mathbb{H} = SL(2,\mathbb{R})/SO(2,\mathbb{R})$ (which is simply obtained from (4.56) since $\partial_\theta = 0$ on $\mathbb{H}$):

$$\Delta_{\mathbb{H}} f = y^2\left(\frac{\partial^2}{\partial x^2} + \frac{\partial^2}{\partial y^2}\right) f = \lambda f \tag{4.68}$$

and by definition are *invariant* under $SL(2,\mathbb{Z})$:

$$f(\gamma\cdot z) = f(z); \tag{4.69}$$

there is no non-trivial weight w compared to (4.11). Similarly to the holomorphic case, we require that $f(z)$ is of moderate growth, i.e., that it grows at most polynomially for $y \to \infty$ (see (4.17)).

Functions on $SL(2,\mathbb{R})$ satisfying these conditions are called *Maaß (wave) forms*, and they can also be fit into the general framework of automorphic forms, with even less effort than for the holomorphic modular forms that we discussed in Section 4.3 above. Given a Maaß form f on $\mathbb{H}$ we lift this to a function φ_f on $SL(2,\mathbb{R})$ according to (4.47):

$$f \longmapsto \varphi_f(g) = \varphi_f\left(\begin{pmatrix}1 & x\\ & 1\end{pmatrix}\begin{pmatrix}y^{1/2} & \\ & y^{-1/2}\end{pmatrix}k\right) = f(x+iy), \tag{4.70}$$

where we have used the Iwasawa decomposition $g = nak \in SL(2,\mathbb{R})$ given in Equation (5.56). The lift in this case is trivial since $w = 0$. This property is called *K-spherical* or *K-trivial.*

The associated function $\varphi_f(g)$ then satisfies

$$\varphi_f(\gamma g k) = \varphi_f(g), \qquad \gamma \in SL(2,\mathbb{Z}), \quad k \in SO(2,\mathbb{R}), \tag{4.71}$$

and so is indeed an automorphic form on $SL(2,\mathbb{R})$. It is also an eigenfunction of the $SL(2,\mathbb{R})$ Laplacian

$$\Delta_{SL(2,\mathbb{R})}\varphi_f = \lambda f\,. \tag{4.72}$$

Important examples of Maaß forms are provided by the *non-holomorphic Eisenstein series* with parameter $s \in \mathbb{C}$ that we encountered in the introduction:

$$E(s,z) = \frac{1}{2} \sum_{\substack{(c,d)\in\mathbb{Z}^2 \\ \gcd(c,d)=1}} \frac{y^s}{|cz+d|^{2s}}. \tag{4.73}$$

This converges absolutely for $\mathrm{Re}(s) > 1$, but according to Langlands' general theory [447] it can be analytically continued to a meromorphic function of $s \in \mathbb{C}\backslash\{0,1\}$. This crucial fact relies on the *functional relation*

$$\xi(s)E(s,z) = \xi(1-s)E(1-s,z), \tag{4.74}$$

where $\xi(s)$ is the completed Riemann zeta function (1.18). We shall discuss this relation in much more detail in Chapter 7 and Section 8.8.

One can verify that the Eisenstein series $E(s,z)$ indeed defines an $SL(2,\mathbb{Z})$-invariant eigenfunction of the Laplacian $\Delta_{\mathbb{H}}$ with eigenvalue $\lambda = s(s-1)$. The non-holomorphic Eisenstein series $E(s,z)$ provides the simplest example of the class of Eisenstein series on a group $G(\mathbb{R})$ that will be our main concern in the following.

It is instructive to rewrite $E(s,z)$ as defined in (4.73). We parametrise an arbitrary group element $g \in SL(2,\mathbb{R})$ according to the same Iwasawa decomposition $g = nak$ as in (4.5). Then introduce a character $\chi_s : B = NA \to \mathbb{C}^\times$ defined by

$$\chi_s(na) = y^s, \qquad n \in N,\ a \in A, \tag{4.75}$$

and extend it to all of $SL(2,\mathbb{R})$ by requiring it to be trivial on $SO(2,\mathbb{R})$: $\chi_s(nak) = \chi_s(na)$. The Eisenstein series $E(s,z)$ can now be equivalently written as

$$E(s,g) = \sum_{\gamma\in B(\mathbb{Z})\backslash SL(2,\mathbb{Z})} \chi_s(\gamma g), \tag{4.76}$$

where the quotient by the discrete Borel subgroup $B(\mathbb{Z}) = \{\begin{pmatrix} \pm 1 & m \\ & \pm 1 \end{pmatrix} \mid m \in \mathbb{Z}\}$ is needed since it leaves χ_s invariant. (This was also explained in the introduction.) It should be apparent that this reformulation of the Eisenstein series is well suited for generalisations to higher-rank Lie groups $G(\mathbb{R})$. This will be discussed in Section 4.5 below.

4.3.3 Maaß Forms of Non-zero Weight*

One can generalise the definition of the Maaß form given above to include non-holomorphic functions that transform with a weight. We define a *weight w Maaß form* to be a non-holomorphic function $f : \mathbb{H} \to \mathbb{C}$ satisfying

$$f\left(\frac{az+d}{cz+d}\right)=\left(\frac{cz+d}{|cz+d|}\right)^w f(z), \qquad w\in\mathbb{Z}. \tag{4.77}$$

A weight w Maaß form is furthermore an eigenfunction of the weight w Laplacian Δ_w which is a modification of (4.68):

$$\Delta_{\mathbb{H}}^{w} := y^2\left(\frac{\partial^2}{\partial x^2}+\frac{\partial^2}{\partial y^2}\right)-iwy\frac{\partial}{\partial x}. \tag{4.78}$$

We can elucidate the meaning of this differential operator by lifting the weight w Maaß form f to an automorphic form φ_f on $SL(2,\mathbb{R})$ through a straightforward generalisation of (4.70):

$$f\longmapsto \varphi_f(g) := \left(\frac{ci+d}{|ci+d|}\right)^{-w} f(g\cdot i), \qquad g=\begin{pmatrix}a & b\\ c & d\end{pmatrix}\in SL(2,\mathbb{R}). \tag{4.79}$$

Rewriting this in Iwasawa form (4.5) yields

$$\varphi_f(g)=\varphi_f\left(\begin{pmatrix}1 & x\\ & 1\end{pmatrix}\begin{pmatrix}y^{1/2} & \\ & y^{-1/2}\end{pmatrix}k\right)=e^{iw\theta}f(x+iy). \tag{4.80}$$

We then recognise the weight w Laplacian $\Delta_{\mathbb{H}}^{w}$ in (4.78) as nothing but the full Laplacian on $SL(2,\mathbb{R})$ (4.56) after evaluating the derivative on θ:

$$\Delta_{SL(2,\mathbb{R})}\varphi_f(g)=e^{iw\theta}\left[y^2\left(\frac{\partial^2}{\partial x^2}+\frac{\partial^2}{\partial y^2}\right)-iwy\frac{\partial}{\partial x}\right]f(x+iy)=e^{iw\theta}\Delta_{\mathbb{H}}^{w}f(z). \tag{4.81}$$

Example 4.11: Non-holomorphic Eisenstein series of weight w

A classic example of a weight w Maaß form is the following generalisation of the non-holomorphic Eisenstein series (4.73):

$$E_w(s,z)=\sum_{\substack{(c,d)\in\mathbb{Z}^2\\(c,d)=1}}\frac{y^s}{|cz+d|^{2s}}\left(\frac{cz+d}{c\bar{z}+d}\right)^w, \tag{4.82}$$

which transforms as

$$E_w\left(s,\frac{az+b}{cz+d}\right)=\left(\frac{cz+d}{c\bar{z}+d}\right)^{w/2}E_w(s,z)=\left(\frac{cz+d}{|cz+d|}\right)^w E_w(s,z). \tag{4.83}$$

Remark 4.12 The generalisation of Maaß forms discussed in the example above will be analysed from a representation-theoretic perspective in Section 5.6. They also play a rôle in string theory as will be seen in Section 14.4.

4.4 Adelic Automorphic Forms

We shall now treat automorphic forms in the adelic framework. There are various degrees of generality here; in particular, one can define the theory of automorphic forms over the adeles $\mathbb{A}_{\mathbb{F}}$ of an arbitrary number field $\mathbb{F}$ (or even a function field). Although most of what follows can be straightforwardly extended to general number fields, we shall for the sake of exposition restrict to the case $\mathbb{F} = \mathbb{Q}$. More general number fields will however make an appearance in PART THREE of this book.

The framework of adelic automorphic forms was originally developed in the books by Gelfand–Graev–Piatetski-Shapiro [267], and Jacquet–Langlands [376]. Good introductions can be found in the books by Gelbart [263], Bump [116] and Goldfeld–Hundley [291, 292].

Even though our main interest often lies with automorphic forms on real Lie groups $G(\mathbb{R})$, the adelic reformulation turns out to be extremely convenient for many purposes, not the least of which being the calculational advantages that it brings when computing Fourier expansions of automorphic forms.

4.4.1 Adelisation of Non-holomorphic Eisenstein Series

As a motivational example, we explain how to view non-holomorphic $SL(2,\mathbb{R})$ Eisenstein series as functions on the adelic group $SL(2,\mathbb{A})$. In Section 3.2.2 we saw that strong approximation ensures that we can always lift a function on $SL(2,\mathbb{Z})\backslash SL(2,\mathbb{R})$ to an adelic function on $SL(2,\mathbb{Q})\backslash SL(2,\mathbb{A})$, where the rôle of the discrete subgroup is now played by $SL(2,\mathbb{Q})$. Recall that this lift also requires that the resulting function is right-invariant under $K_f = \prod_{p<\infty} SL(2,\mathbb{Z}_p)$. It is now a simple matter to generalise the Eisenstein series $E(s,g)$ to such an adelic function. First extend the definition of χ_s to a function $\chi_s\colon B(\mathbb{A}) \to \mathbb{C}$, which is invariant under the left-action of $B(\mathbb{Q})$. We extend it to all of $SL(2,\mathbb{A})$ using the global Iwasawa decomposition $SL(2,\mathbb{A}) = B(\mathbb{A})K_{\mathbb{A}}$ and demanding that it be trivial on $K_{\mathbb{A}} = SO(2,\mathbb{R}) \times K_f$. Note that this automatically takes care of the required condition of K_f-invariance on the right. The adelic Eisenstein series then takes the form

$$E(s,g_{\mathbb{A}}) = \sum_{\gamma\in B(\mathbb{Q})\backslash SL(2,\mathbb{Q})} \chi_s(\gamma g_{\mathbb{A}}), \tag{4.84}$$

which is a function on $SL(2,\mathbb{A})$ satisfying

$$E(s,\gamma g_{\mathbb{A}} k_{\mathbb{A}}) = E(s,g_{\mathbb{A}}), \qquad \gamma \in SL(2,\mathbb{Q}), \quad k_{\mathbb{A}} \in K_{\mathbb{A}}. \tag{4.85}$$

As shown in Example 3.16, the range of the sum in (4.84) is in fact in bijection with the range of summation in (4.76):

$$B(\mathbb{Q})\backslash SL(2,\mathbb{Q}) \cong B(\mathbb{Z})\backslash SL(2,\mathbb{Z}). \tag{4.86}$$

Therefore, if we restrict to elements $g_{\mathbb{A}} = (g_\infty; 1, 1, \dots) \in SL(2,\mathbb{A})$, with $g_\infty \in SL(2,\mathbb{R})$, then the adelic Eisenstein series reduces to the real Eisenstein series (4.76). More details of this procedure can be found in Example 5.6.

With a little more effort one can also obtain the adelisation of the function $\varphi_f(g)$ in (4.43), for f a weight w holomorphic modular form on $\mathbb{H}$. This analysis is done in Section 4.4.3.

The adelic Eisenstein series (4.84) on $SL(2,\mathbb{A})$ has the following properties. (1) It is invariant under the left-action of $SL(2,\mathbb{Q})$ since it is formed as a sum over images. (2) It is invariant under the right-action of $SO(2) \times \prod_{p<\infty} SL(2,\mathbb{Z}_p)$ since the character χ_s is invariant. (3) It is an eigenfunction of the scalar Laplacian by (4.72). (4) It grows at most polynomially when approaching the cusp (and elsewhere) as will follow from the Fourier expansion given in Chapter 7. These characteristic properties will now be formulated more precisely for more general groups.

4.4.2 Main Definition

In what follows we let G be a split algebraic group defined over $\mathbb{Q}$, and $G(\mathbb{A})$ its adelisation as in Section 3.2.2. The typical example we have in mind is $G(\mathbb{A}) = SL(n,\mathbb{A})$. Let us now state our definition of an automorphic form.

Definition 4.13 (Automorphic form) An *automorphic form* is a smooth function $\varphi\colon G(\mathbb{Q})\backslash G(\mathbb{A}) \to \mathbb{C}$ satisfying the following conditions:

(1) *left $G(\mathbb{Q})$-invariance:* $\varphi(\gamma g) = \varphi(g), \qquad \gamma \in G(\mathbb{Q})$
(2) *right K-finiteness:* $\dim_{\mathbb{C}}\langle \varphi(gk) | k \in K_{\mathbb{A}}\rangle < \infty$
(3) *$\mathcal{Z}(\mathfrak{g}_{\mathbb{C}})$-finiteness:* $\dim_{\mathbb{C}}\langle X\varphi(g) | X \in \mathcal{Z}(\mathfrak{g}_{\mathbb{C}})\rangle < \infty$
(4) *φ is of moderate growth:* for any norm $||\cdot||$ on $G(\mathbb{A})$ there exists a positive integer n and a constant C such that $|\varphi(g)| \le C||g||^n$.

The angled brackets denote the linear span.

Remark 4.14 We denote by $\mathcal{A}(G(\mathbb{Q})\backslash G(\mathbb{A}))$ the space of automorphic forms satisfying Definition 4.13. This is a subspace of the space $C^\infty(G(\mathbb{Q})\backslash G(\mathbb{A}))$ of *smooth* functions on $G(\mathbb{Q})\backslash G(\mathbb{A})$. An adelic function $\varphi(g)$, with $g = (g_\infty; g_f) \in G(\mathbb{A}) = G(\mathbb{R}) \times G(\mathbb{A}_f)$, is said to be smooth if it is C^∞ with respect to the archimedean variables $g_\infty \in G(\mathbb{R})$ and locally constant with respect to the non-archimedean variables $g_f \in G(\mathbb{A}_f)$.

Let us now elaborate a little on Definition 4.13 of an adelic automorphic form:

- Condition (1) ensues as a straightforward generalisation of invariance of the function under a discrete subgroup of $G(\mathbb{A})$.
- The condition of right K-*finiteness* means that the complex vector space V spanned by the functions $k \mapsto \varphi(gk)$, $k \in K_{\mathbb{A}}$, is finite-dimensional. We have already seen an example of a non-trivial K-representation in (4.50). When φ is K-invariant on the right, it is called K-*spherical.*
- In condition (3), $\mathfrak{g}$ is the Lie algebra associated with the group G and $\mathcal{Z}(\mathfrak{g})$ is the centre of its universal enveloping algebra $\mathcal{U}(\mathfrak{g}_{\mathbb{C}})$. The centre $\mathcal{Z}(\mathfrak{g})$ is the space of bi-invariant differential operators on G, i.e., the quadratic Casimir and higher-order operators. The condition of $\mathcal{Z}(\mathfrak{g})$-finiteness then implies that φ is contained in a $\mathcal{Z}(\mathfrak{g})$-invariant finite-dimensional subspace of $C^\infty(G(\mathbb{Q})\backslash G(\mathbb{A}))$. Equivalently, if $X \in \mathcal{Z}(\mathfrak{g})$ then $\mathcal{Z}(\mathfrak{g})$-finiteness implies that there exists a polynomial $R(X)$ such that $R(X)\varphi = 0$.

Remark 4.15 It is sometimes useful to specify the transformation properties of an automorphic form with respect to the centre $Z(\mathbb{A})$ of $G(\mathbb{A})$. To this end, let ω be a *central character*, i.e., a homomorphism $\omega\colon Z(\mathbb{A}) \to \mathbb{C}^\times$, which is trivial on $Z(\mathbb{Q})$. An automorphic form f is then said to have central character ω if it satisfies conditions (1)–(4) along with the additional condition

(5) $f(zg) = \omega(z)f(g)$.

We shall now illustrate these defining properties of an automorphic form by considering the example of the non-holomorphic Eisenstein series (4.84), and we will verify its properties according to the above definition.

Example 4.16: Automorphic properties of an Eisenstein series on $SL(2,\mathbb{A})$

Consider the case when $G(\mathbb{A}) = SL(2,\mathbb{A})$. We now verify the conditions (1)–(4) in Definition 4.13 for the non-holomorphic Eisenstein series defined in (4.84):

$$E(s,g) = \sum_{\gamma \in B(\mathbb{Q})\backslash SL(2,\mathbb{Q})} \chi_s(\gamma g), \qquad g \in SL(2,\mathbb{A}). \tag{4.87}$$

- By construction, $E(s,g)$ is left $SL(2,\mathbb{Q})$-invariant and so satisfies condition (1).
- Moreover, by definition the function χ_s is invariant under any $k_p \in SL(2,\mathbb{Z}_p)$, $\chi_s(gk_p) = \chi_s(g)$, and hence condition (2) is also satisfied.
- To understand condition (3) we recall that $E(s,g)$ is an eigenfunction of the $\mathfrak{g}$-invariant Laplacian $\Delta_{\mathbb{H}}$ on $SL(2,\mathbb{R})/SO(2,\mathbb{R})$ with eigenvalue $\lambda = s(s-1)$. Hence, $E(s,g)$ is in the kernel of the operator $(\Delta_{\mathbb{H}} - \lambda) \in \mathcal{Z}(\mathfrak{g})$, and since for $\mathfrak{g} = \mathfrak{sl}(2,\mathbb{A})$, $\mathcal{Z}(\mathfrak{g}) = \mathbb{C}[\Delta_{\mathbb{H}}]$, we have that condition (3) is satisfied.

- The final part consists in verifying the moderate growth condition (4). To this end one must translate the classical moderate growth condition (4.17) to the adelic picture. A norm $|| \; ||$ on $SL(2,\mathbb{A})$ can be defined as follows (see, e.g., [291]):

$$||g|| := \prod_{p\leq\infty} \max\left\{|a|_p, |b|_p, |c|_p, |d|_p, |ad-bc|_p^{-1}\right\}, \qquad g = \begin{pmatrix} a & b \\ c & d \end{pmatrix}, \tag{4.88}$$

where it is understood that $|a|_p = |a_p|_p$ etc., with a_p the pth component of the adele $a = (a_\infty, a_2, a_3, \dots) \in \mathbb{A}$. For a proof that the moderate growth condition of $E(s,g)$ with respect to this norm follows from the classical moderate growth on $SL(2,\mathbb{R})$, see pp. 122–123 of [291].

Before we move on to analysing Eisenstein series on arbitrary reductive groups we shall give an additional important definition.

Definition 4.17 (Cusp form) An automorphic form $f \in \mathcal{A}(G(\mathbb{Q})\backslash G(\mathbb{A}))$ is a *cusp form* if for all parabolic subgroups $P(\mathbb{A}) \subset G(\mathbb{A})$, defined in Section 3.1.3, it satisfies

$$\int_{U(\mathbb{Q})\backslash U(\mathbb{A})} f(ug)du = 0, \tag{4.89}$$

where U is the unipotent radical in the Levi decomposition $P(\mathbb{A}) = L(\mathbb{A})U(\mathbb{A})$, and du is the left-invariant Haar measure on U. The subspace of cusp forms will be denoted by $\mathcal{A}_0(G(\mathbb{Q})\backslash G(\mathbb{A})) \subset \mathcal{A}(G(\mathbb{Q})\backslash G(\mathbb{A}))$.

Remark 4.18 This definition generalises the notion of cusp form found in the classical theory, namely holomorphic modular forms $f(z)$ whose Fourier expansion in $q = e^{2\pi i z}$ contains no term of order q^0. An example is provided by Ramanujan's discriminant $\Delta(z)$ of weight $w = 12$ that was discussed in Section 4.2.1.

Remark 4.19 The integral in (4.89) can be thought of as the zeroth Fourier coefficient of $f(g)$ with respect to U; by analogy with the classical theory it is called the 'constant term' of $f(g)$, although in general it is by no means constant. From this perspective a cusp form is simply an automorphic form with vanishing constant term. Constant terms are analysed in detail for $SL(2,\mathbb{A})$ in Chapter 7 and in full generality in Chapter 8.

4.4.3 Adelic Lift with Hecke Character*

We shall now construct the adelic lift of a holomorphic modular form f. To illustrate the power of the adelic formalism we will consider the general case addressed in Section 4.2.2, namely let $f \in M_w(\Gamma_0(N), \chi)$, i.e a level N holomorphic modular form for $\Gamma_0(N)$ with Dirichlet character χ. We can now use strong approximation (see Section 3.2.2) to lift f to a function on

$SL(2,\mathbb{A})$. Recall from Section 3.2.2 (see in particular Example 3.15) that strong approximation implies that any $g \in SL(2,\mathbb{A})$ can be (non-uniquely) written as

$$g = \gamma g_\infty k_f, \qquad \gamma \in SL(2,\mathbb{Q}),\ g_\infty \in SL(2,\mathbb{R}),\ k_f \in K_0(N), \tag{4.90}$$

where $K_0(N) \subset K_f = \prod_{p<\infty} SL(2,\mathbb{Z}_p)$ was defined in Example 3.15. In order to define a lift to $SL(2,\mathbb{A})$ we must first lift the Dirichlet character χ to the adelic setting. This can be done using the $GL(1,\mathbb{A}) = \mathbb{A}^\times$ version of strong approximation from (2.65):

$$\mathbb{A}^\times = \mathbb{Q}^\times \mathbb{R}_+ \prod_{p<\infty} \mathbb{Z}_p^\times. \tag{4.91}$$

This implies that any Dirichlet character $\chi\colon (\mathbb{Z}/N\mathbb{Z})^* \to \mathbb{C}^\times$ has a canonical lift to an *adelic (Hecke) character*, sometimes also called a *Größencharakter*:

$$\omega_\chi\colon \mathbb{Q}^\times\backslash\mathbb{A}^\times \to \mathbb{C}^\times. \tag{4.92}$$

Indeed, such a character can be decomposed as

$$\omega_\chi = \omega_{\chi,\infty} \prod_{p<\infty} \omega_{\chi,p}, \tag{4.93}$$

where the archimedean factor $\omega_{\chi,\infty}$ is taken to be trivial, and each local factor $\omega_{\chi,p}$ corresponds to the lift of a Dirichlet character χ (see, e.g., [291] for details).

Next, we lift the local character $\omega_{\chi,p}$ to a character on $SL(2,\mathbb{Z}_p)$ via the map $\left(\begin{smallmatrix} a & b \\ c & d \end{smallmatrix}\right) \mapsto \omega_{\chi,p}(d)$. The adelic lift of the holomorphic modular form f is then defined by

$$\varphi_f(g) := (ci+d)^{-w} f(g_\infty \cdot i)\omega_\chi(k_f), \tag{4.94}$$

where $g_\infty = \left(\begin{smallmatrix} a & b \\ c & d \end{smallmatrix}\right) \in SL(2,\mathbb{R})$. We can also write this in terms of the slash operator defined in (4.13):

$$\varphi_f(g) = (f|_w g_\infty)(i)\, \omega_\chi(k_f). \tag{4.95}$$

Having extended the definition of φ_f to an adelic automorphic form, we wish to verify the conditions (1)–(4) of Definition 4.13:

- Condition (1) is satisfied by construction: $\varphi_f(\gamma g) = \varphi_f(g)$, for any $g \in SL(2,\mathbb{A})$ and $\gamma \in SL(2,\mathbb{Q})$.
- Condition (2), concerning right K-finiteness, can be seen as follows. Finiteness under the non-archimedean K_f is a consequence of the relation

$$\varphi_f(gk_f) = \varphi_f(g)\omega_\chi(k_f), \tag{4.96}$$

while at the archimedean place we have

$$\varphi_f(gk_\infty) = \varphi_f(g)e^{iw\theta}, \tag{4.97}$$

where $k_\infty = k_\infty(\theta) \in SO(2,\mathbb{R})$ as in (4.63).

- $\mathcal{Z}(\mathfrak{g})$-finiteness (condition 3) again follows from the fact that φ_f is an eigenfunction of the Laplacian:

$$\Delta\varphi_f = \frac{w}{2}\left(\frac{w}{2}-1\right)\varphi_f. \tag{4.98}$$

- Finally the condition of moderate growth (condition 4) is satisfied if the coefficients a_n in the q-expansion of $f(z)$ satisfy $a_n = 0$, whenever $n < 0$ which holds since f is holomorphic.

Finally, we shall see that φ_f is in fact an example of an automorphic form *with central character*, as in the supplementary condition (5) mentioned in Remark 4.15. To this end we must first view φ_f as a function on $GL(2,\mathbb{A})$ as opposed to $SL(2,\mathbb{A})$. The defining relation (4.94) is still valid for $g \in GL(2,\mathbb{A})$ and conditions (1)–(4) go through without change. Our aim is now to check how $\varphi_f(g)$ transforms under the non-trivial centre $Z(GL(2,\mathbb{A})) = \mathbb{A}^\times$. An element $z \in Z(GL(2,\mathbb{A}))$ can be represented by the diagonal matrix

$$z = \begin{pmatrix} r & \\ & r\end{pmatrix}, \qquad r \in \mathbb{A}^\times. \tag{4.99}$$

Strong approximation then yields the decompositions

$$\begin{aligned} g &= \gamma g_\infty k_f, \qquad && \gamma \in GL(2,\mathbb{Q}),\ g_\infty \in GL(2,\mathbb{R})^+,\ k_f \in K_0(N),\\ r &= \alpha\, r_\infty r_f, && \alpha \in \mathbb{Q}^\times,\ r_\infty \in \mathbb{R}_+,\ r_f \in \prod_{p<\infty}\mathbb{Z}_p^\times, \end{aligned} \tag{4.100}$$

and consequently

$$zg = \begin{pmatrix}\alpha & \\ & \alpha\end{pmatrix}\gamma\begin{pmatrix} r_\infty & \\ & r_\infty\end{pmatrix} g_\infty \begin{pmatrix} r_f & \\ & r_f\end{pmatrix} k_f \in GL(2,\mathbb{A}). \tag{4.101}$$

We can now proceed to calculate the action of Z on the automorphic form φ_f:

$$\varphi_f(zg) = \left(f|_w \begin{pmatrix} r_\infty & \\ & r_\infty\end{pmatrix} g_\infty\right)(i)\,\omega_\chi\left(\begin{pmatrix} r_f & \\ & r_f\end{pmatrix} k_f\right). \tag{4.102}$$

To evaluate this we first notice that

$$\left(f|_w \begin{pmatrix} r_\infty & \\ & r_\infty\end{pmatrix}\right) = f, \tag{4.103}$$

and hence

$$\left(f|_w \begin{pmatrix} r_\infty & \\ & r_\infty\end{pmatrix} g_\infty\right)(i) = (f|_w g_\infty)(i). \tag{4.104}$$

Using the multiplicative property of the Hecke character we further have

$$\omega_\chi\left(\begin{pmatrix} r_f & \\ & r_f \end{pmatrix} k_f\right) = \omega_\chi\left(\begin{pmatrix} r_f & \\ & r_f \end{pmatrix}\right) \omega_\chi(k_f). \tag{4.105}$$

By definition the Hecke character ω_χ is trivial on $\mathbb{Q}^\times$ and at the archimedean place. Thus, using strong approximation we can write

$$\omega_\chi(r_f) = \omega_\chi(\alpha r_\infty r_f) = \omega_\chi(z), \qquad z \in Z(GL(2, \mathbb{A}). \tag{4.106}$$

Combining everything we then find

$$\varphi_f(zg) = \omega_\chi(z)\varphi_f(g), \tag{4.107}$$

verifying that φ_f is an automorphic form with central character $\omega = \omega_\chi$ as in Remark 4.15.

Remark 4.20 Automorphic forms on $GL(1, \mathbb{A})$ are completely classified by the Hecke characters ω_χ. Indeed any automorphic form $\phi\colon GL(1, \mathbb{A}) \to \mathbb{C}$ can be written in the form

$$\phi(g) = c \cdot \omega_\chi(g) \cdot |g|_{\mathbb{A}}^{it}, \tag{4.108}$$

where $c \in \mathbb{C}$ is a constant, χ is a Dirichlet character and $t \in \mathbb{R}$. See, e.g., [291] for more details on the $GL(1, \mathbb{A})$ case.

4.5 Eisenstein Series

We now want to generalise the construction of adelic Eisenstein series given in Section 4.4.1 from $SL(2, \mathbb{A})$ to arbitrary reductive groups $G(\mathbb{A})$. To this end we must first recall the process of constructing representations of G via induction from a standard parabolic subgroup $P \supset B$. In this section we shall take $P = B$, the Borel subgroup which is the minimal parabolic subgroup. The case of arbitrary (standard) parabolic subgroups will be treated in Section 5.7.

4.5.1 Adelic Multiplicative Characters

Fix a Borel subgroup $B(\mathbb{A}) \subset G(\mathbb{A})$ with Levi decomposition $B(\mathbb{A}) = N(\mathbb{A})A(\mathbb{A})$. Recall that since $G(\mathbb{A})$ is split, $A(\mathbb{A}) \cong (\mathbb{A}^\times)^{\operatorname{rank}\mathfrak{g}}$. Introduce a multiplicative character on $B(\mathbb{A})$:

$$\chi\colon B(\mathbb{Q})\backslash B(\mathbb{A}) \to \mathbb{C}^\times, \tag{4.109}$$

determined by its restriction to $A(\mathbb{A})$:

$$\chi(na) = \chi(a), \qquad n \in N(\mathbb{A}),\ a \in A(\mathbb{A}). \tag{4.110}$$

Using the Iwasawa decomposition we can extend χ to all of $G(\mathbb{A})$ by demanding that it be trivial on $K_{\mathbb{A}}$:

$$\chi(g) = \chi(nak) = \chi(na) = \chi(an) = \chi(a), \qquad k \in K_{\mathbb{A}}. \tag{4.111}$$

Although we extend the character to all of $G(\mathbb{A})$ it is only multiplicative on $B(\mathbb{A})$:

$$\chi(bb') = \chi(b)\chi(b') = \chi(a)\chi(a'), \qquad b, b' \in B(\mathbb{A}). \tag{4.112}$$

On the other hand, to evaluate it on a product of two elements $g, g' \in G(\mathbb{A})$ we have

$$\chi(gg') = \chi(bkb'k') = \chi(bkb') = \chi(b\tilde{b}\tilde{k}) = \chi(b\tilde{b}) = \chi(b)\chi(\tilde{b}), \tag{4.113}$$

where $\tilde{b}\tilde{k}$ is the Iwasawa decomposition of kb'. From this we see also

$$\chi(bg) = \chi(b)\chi(g), \qquad b \in B(\mathbb{A}), g \in G(\mathbb{A}). \tag{4.114}$$

The global character splits into an Euler product over local factors:

$$\chi(g) = \prod_{p\leq\infty} \chi_p(g_p), \qquad g_p \in G(\mathbb{Q}_p). \tag{4.115}$$

There is a one-to-one correspondence between such characters and weights of the Lie algebra $\mathfrak{g}(\mathbb{R})$, or, more precisely, complex linear functionals $\lambda \in \mathfrak{h}^*_{\mathbb{C}} = \mathfrak{h}(\mathbb{R})^* \otimes_{\mathbb{R}} \mathbb{C}$, where $\mathfrak{h}(\mathbb{R})$ is the Cartan subalgebra of $\mathfrak{g}(\mathbb{R})$.

We define a logarithm map H as follows:

$$H\colon G(\mathbb{A}) \to \mathfrak{h}(\mathbb{R}), \tag{4.116}$$

defined by

$$H(g) = H(nak) = H(a) = \log|a|. \tag{4.117}$$

The absolute value is defined as follows. Let $a \in A(\mathbb{A})$ be parametrised by

$$a = \prod_{\alpha\in\Pi} h_\alpha(t_\alpha), \qquad t_\alpha \in \mathbb{A}^\times, \tag{4.118}$$

where h_α was defined in Section 3.1.4.

With the above parametrisation for $a \in A(\mathbb{A})$ we then define the map

$$\begin{aligned} \mathfrak{h}^*_{\mathbb{C}} \times A(\mathbb{A}) &\to \mathbb{C}^\times \\ (\lambda, a) &\mapsto \prod_{\alpha\in\Pi} |t_\alpha|_{\mathbb{A}}^{\langle\lambda, H_\alpha\rangle}, \end{aligned} \tag{4.119}$$

which we will denote by $|a^\lambda|$. Let $|a^\lambda|_p$ denote the analogous definition using the p-adic norm.

Then we define

$$\log|a| := \sum_{\alpha\in\Pi} H_\alpha \log|a^{\Lambda_\alpha}| \in \mathfrak{h}(\mathbb{R}), \tag{4.120}$$

where H_α is a Chevalley generator and Λ_α a fundamental weight. We have that

$$\log\left|\prod_{\alpha\in\Pi} h_\alpha(t_\alpha)\right| = \sum_{\alpha\in\Pi} H_\alpha \log|t_\alpha|_{\mathbb{A}}\,. \tag{4.121}$$

The choice of character χ can now be parametrised by the choice of linear functional λ according to the formula

$$\chi(g) = e^{\langle\lambda+\rho|H(g)\rangle} = |a^{\lambda+\rho}|. \tag{4.122}$$

The translation by the Weyl vector ρ, which was defined in (3.6), constitutes a convenient choice of normalisation.

Remark 4.21 (Infinitesimal character) The weight $-\lambda$ is often referred to as the *infinitesimal character*. We shall see later in Section 5.3 that it encodes the values of all the invariant differential operators in a representation induced from χ.

Remark 4.22 (Modulus character) The map

$$b \mapsto e^{\langle 2\rho|H(b)\rangle} \equiv \delta_B(b), \qquad b \in B(\mathbb{A}), \tag{4.123}$$

is often called the *modulus character* (or '*modular function*') of B. It is defined by

$$\delta_B(b) = \left|\det \mathrm{ad}(b)|_{\mathfrak{n}}\right|. \tag{4.124}$$

In words, it is the modulus of the determinant of the adjoint representation of $b \in B(\mathbb{A})$, restricted to the Lie algebra $\mathfrak{n}$ of the unipotent radical N. By virtue of the properties (4.117) of the logarithm map we have

$$\delta_B(b) = \delta_B(na) = \delta_B(a). \tag{4.125}$$

The modulus character corresponds to the Jacobian that relates the left- and right-invariant Haar measures on B. This implies in particular that under conjugation by $a \in A(\mathbb{A})$, i.e.,

$$n \mapsto ana^{-1}, \tag{4.126}$$

the Haar measure dn on $N(\mathbb{A})$ transforms as

$$dn \mapsto \delta_B(a)dn. \tag{4.127}$$

This fact will play a crucial rôle in our calculations in Chapter 7 and onwards.

See Example 4.23 for an explicit description for $SL(2,\mathbb{A})$. Using the modulus character we can write χ in the alternative form

$$\chi(g) = e^{\langle\lambda|H(g)\rangle}\delta_B^{1/2}(g). \tag{4.128}$$

This form of the character is also common in the literature. We will typically drop the subscript B on the modulus character of the Borel. In Section 5.7 we shall encounter similar modulus characters for other parabolic subgroups P of G.

Example 4.23: Haar measure and modulus character for the Borel of $SL(2,\mathbb{A})$

For $SL(2,\mathbb{A})$ we can take the standard Borel

$$b = na = \begin{pmatrix} 1 & u \\ & 1 \end{pmatrix}\begin{pmatrix} v & \\ & v^{-1} \end{pmatrix}, \tag{4.129}$$

in which case the right-invariant Haar measure is

$$dnda = \frac{dudv}{v} \tag{4.130}$$

and the modulus character is given by

$$\delta_B(na) = \delta_B\left(\begin{pmatrix} 1 & u \\ & 1 \end{pmatrix}\begin{pmatrix} v & \\ & v^{-1} \end{pmatrix}\right) = |v|^2. \tag{4.131}$$

4.5.2 Eisenstein Series

With the definition of the character χ on the Borel subgroup, we are now in a position to state Langlands' definition of an Eisenstein series for an arbitrary reductive group $G(\mathbb{A})$.

Definition 4.24 (Eisenstein series) Let $G(\mathbb{A})$ be an adelic group with discrete subgroup $G(\mathbb{Q})$ and Borel subgroup $B(\mathbb{Q})$. Let χ be a character on $B(\mathbb{A})$ that is trivial on $B(\mathbb{Q})$. The *(minimal) Eisenstein series* is defined as the sum over images of the coset $B(\mathbb{Q})\backslash G(\mathbb{Q})$ by

$$E(\chi, g) = \sum_{\gamma\in B(\mathbb{Q})\backslash G(\mathbb{Q})} \chi(\gamma g), \tag{4.132}$$

and using the explicit parametrisation (4.122), the definition reads

$$E(\lambda, g) = \sum_{\gamma\in B(\mathbb{Q})\backslash G(\mathbb{Q})} e^{\langle\lambda+\rho|H(\gamma g)\rangle}. \tag{4.133}$$

Remark 4.25 The series defined here is not the only possible type of Eisenstein series that one can define, although it is the one that we will be most interested in. In Section 5.4, we will treat Eisenstein series in the context of automorphic representations, and there we shall encounter generalisations where the minimal parabolic subgroup $B(\mathbb{Q})$ is replaced by other parabolic subgroups. This will then provide us with a way of deriving different types of Eisenstein series, including the one just defined.

For the above series *Godement* [288, 69] proved that the sum converges absolutely whenever λ lies in the open range

$$\{\lambda \in \mathfrak{h}^*_{\mathbb{C}} \mid \mathrm{Re}(\lambda) \in \rho + (\mathfrak{h}^*)^+\}, \tag{4.134}$$

where the positive chamber $(\mathfrak{h}^*)^+$ in $\mathfrak{h}^* \equiv \mathfrak{h}^*_{\mathbb{R}}$ is defined by

$$(\mathfrak{h}^*)^+ = \{\Lambda \in \mathfrak{h}^* \mid \langle \Lambda | H_\alpha \rangle > 0, \forall\, \alpha \in \Pi\}, \tag{4.135}$$

so that we require $\langle \lambda | H_\alpha \rangle > 1$ for all simple roots α. The origin of the Godement range will become clearer when we discuss the so-called constant term formula in Section 8.8.

For discussing the spectral decomposition of $\mathcal{A}(G(\mathbb{Q})\backslash G(\mathbb{A}))$ (see also Section 5.5), one is mainly interested in the case when λ lies on certain critical lines or planes in the complex space $\mathfrak{h}^*_{\mathbb{C}}$. This choice for λ is motivated in Section 5.3 by the fact that the inducing representation appears in the continuous part of the spectral decomposition. However, such values lie outside the domain of absolute convergence (4.134) of $E(\lambda, g)$, which seems worrisome for the spectral decomposition. This puzzle is resolved by the remarkable result of Langlands that the Eisenstein series $E(\lambda, g)$ can in fact be analytically continued outside of the domain (4.134) to a meromorphic function on all of $\mathfrak{h}^*_{\mathbb{C}}$. To establish the analytic continuation a crucial property of $E(\lambda, g)$ is its *functional relation*, which relates its value at λ to its value at the Weyl-transform of λ:

$$E(\lambda, g) = M(w, \lambda) E(w\lambda, g), \qquad w \in \mathcal{W}(\mathfrak{g}), \tag{4.136}$$

where $M(w, \lambda)$ is defined later in (8.42). The functional relation will be discussed in more detail in Chapter 8. Another important property is that $E(\lambda, g)$ is an eigenfunction of the Laplace operator $\Delta_{G/K}$:

$$\Delta_{G/K} E(\lambda, g) = \frac{1}{2} \left(\langle \lambda | \lambda \rangle - \langle \rho | \rho \rangle \right) E(\lambda, g). \tag{4.137}$$

This formula is derived in Appendix B; the action here coincides with that of the quadratic Casimir operator. In fact, the Eisenstein series $E(\lambda, g)$ is a common eigenfunction of all $G(\mathbb{A})$-invariant differential operators, which is a reflection of its $\mathcal{Z}(\mathfrak{g})$-finiteness. The following is a useful property of Eisenstein series:

Proposition 4.26 (Eisenstein series at $\lambda = -\rho$) *In the special case when $\lambda = -\rho$ we have*

$$E(-\rho, g) = 1. \tag{4.138}$$

Proof We note first that the value $\lambda = -\rho$ is outside the region of absolute convergence of the Eisenstein and the function must therefore be defined by analytic continuation using the functional Equation (4.136). One also observes that, by (4.137), $E(-\rho, g)$ is an eigenfunction of the Laplacian $\Delta_{G/K}$ with eigenvalue zero; hence it must be a constant function. To fix the constant to unity, we note that by Langlands' constant term formula (see Chapter 8), the constant term of $E(-\rho, g)$ with respect to the maximal unipotent radical N is

$$\int_{N(\mathbb{Q})\backslash N(\mathbb{A})} E(-\rho, ng)dn = 1, \tag{4.139}$$

from which the claim follows. □

5

Automorphic Representations and Eisenstein Series

In this chapter, we introduce the concept of an automorphic representation associated with an adelic group $G(\mathbb{A})$. Since this is a rather difficult concept to grasp at first sight, we shall begin with a heuristic discussion before we delve into the technical definition. Our main focus will then lie with the so-called principal series, which is the representation relevant for general Eisenstein series. We provide some remarks on the problem of classifying all automorphic representations of $\mathcal{A}(G(\mathbb{Q})\backslash G(\mathbb{A}))$. The theory of automorphic representations is illustrated through a detailed discussion of Eisenstein series from the point of view of representations induced from parabolic subgroups $P(\mathbb{A}) \subset G(\mathbb{A})$. We shall see from a representation-theoretic point of view how the holomorphic and non-holomorphic Eisenstein series are unified via the embedding of the holomorphic discrete series inside the principal series of the non-compact group $SL(2,\mathbb{R})$. We will also discuss Eisenstein series for non-minimal parabolic subgroups and the theory of induced representations in relation to spherical vectors.

5.1 A First Glimpse at Automorphic Representations

We have already seen hints in Section 4.3 that automorphic forms on $SL(2,\mathbb{Z})\backslash SL(2,\mathbb{R})$ are intimately related to the representation theory of $SL(2,\mathbb{R})$. Here we will further develop this point of view and also generalise it to the adelic framework.

The main idea is that the space of smooth functions $\varphi\colon SL(2,\mathbb{Z})\backslash SL(2,\mathbb{R}) \to \mathbb{C}$ carries several actions:

- First, we have the action π of $SL(2,\mathbb{R})$ by *right-translation*:

$$[\pi(h)\varphi](g) := \varphi(gh), \qquad g, h \in SL(2,\mathbb{R}). \tag{5.1}$$

- Second, we have the action of the universal enveloping algebra $\mathcal{U}(\mathfrak{sl}(2,\mathbb{C}))$ by differential operators:

$$(D_X \cdot \varphi)(g) := \frac{d}{dt}\varphi(g \cdot e^{tX})\Big|_{t=0}, \qquad X \in \mathcal{U}(\mathfrak{sl}(2,\mathbb{C})). \tag{5.2}$$

Whenever one has a group action on a space it is natural to look for a decomposition into irreducible representations of the group. Moreover, since the centre $\mathcal{Z}(\mathfrak{g})$ of the universal enveloping algebra $\mathcal{U}(\mathfrak{g}_\mathbb{C})$ commutes with $SL(2,\mathbb{R})$ it is also natural to distinguish the irreducible components in terms of their eigenvalues with respect to differential operators in $\mathcal{Z}(\mathfrak{g})$. For these reasons the theory of automorphic forms on $SL(2,\mathbb{R})$ is closely related to the decomposition of the space $\mathcal{A}(SL(2,\mathbb{Z})\backslash SL(2,\mathbb{R}))$ into irreducible representations with respect to the right-regular action of $SL(2,\mathbb{R})$, compatible with the action by $\mathcal{U}$. To get an idea of what this entails, let us now look at an extremely simplified, though still enlightening, example.

Example 5.1: Fourier analysis on $\mathbb{Z}\backslash\mathbb{R}$

In this example, we will look at the abelian situation where the space $SL(2,\mathbb{Z})\backslash SL(2,\mathbb{R})$ is replaced by $\mathbb{Z}\backslash\mathbb{R}$, the circle group. This is the setting of classical Fourier analysis. The space of smooth functions $C^\infty(\mathbb{Z}\backslash\mathbb{R})$ is then just Fourier series where the coefficients are constrained to decay rapidly with increasing Fourier number. Let us now formalise this a little and try to analyse it in the spirit of automorphic forms. Consider the unitary character $\psi_k : \mathbb{Z}\backslash\mathbb{R} \to U(1)$, defined by $\psi_k(x) = e^{2\pi i k x}$ for $x \in \mathbb{R}$, $k \in \mathbb{Z}$. Any function $f \in C^\infty(\mathbb{Z}\backslash\mathbb{R})$ can then be expanded in a Fourier series in terms of these characters:

$$f(x) = \sum_{k\in\mathbb{Z}} c_k \psi_k(x). \tag{5.3}$$

Recall that an automorphic form is also required to satisfy a moderate growth condition. In the present setting we can choose square-integrability of $f(x)$ as a suitable condition for moderate growth. Thus the space of automorphic forms on $\mathbb{Z}\backslash\mathbb{R}$ can be taken to be the Hilbert space $L^2(\mathbb{Z}\backslash\mathbb{R}) \subset C^\infty(\mathbb{Z}\backslash\mathbb{R})$ where the Fourier coefficients satisfy the square-integrability condition

$$\sum_{k\in\mathbb{Z}} |c_k|^2 < \infty. \tag{5.4}$$

Each character ψ_k generates a one-dimensional irreducible subspace $V_k = \mathbb{C}\psi_k \subset L^2(\mathbb{Z}\backslash\mathbb{R})$ and the *regular representation* π of $\mathbb{R}$ defined by

$$(\pi(y)f)(x) := f(x+y), \qquad x, y \in \mathbb{R}, \tag{5.5}$$

is diagonalised by the subspaces V_k:

$$\pi(y) \cdot v = \psi_k(y)v, \qquad v \in V_k,\ y \in \mathbb{R}. \tag{5.6}$$

The set of equivalence classes of unitary representations of a group G is called the *unitary dual*, usually denoted $\widehat{G}$. In our example, the unitary dual $\widehat{\mathbb{R}}$ is simply the space of Fourier coefficients subject to the condition (5.4):

$$\widehat{\mathbb{R}} = L^2(\mathbb{Z}) = \left\{ (c_k) \mid \sum_k |c_k|^2 < \infty \right\}. \tag{5.7}$$

This gives the spectral decomposition of the Hilbert space $L^2(\mathbb{Z}\backslash\mathbb{R})$. The fact that the spectrum is discrete is a general feature of spectral theory on compact spaces like $S^1 \cong \mathbb{Z}\backslash\mathbb{R}$.

Before we proceed with the adelic perspective we shall consider one more simple example that illustrates another feature that has a counterpart on the general theory of automorphic forms.

Example 5.2: Fourier analysis on $\mathbb{R}$

Consider the same setting as in the previous example, namely $G = \mathbb{R}$, but we now take the discrete subgroup $G(\mathbb{Z})$ to be trivial. In other words, we are interested in the regular representation of $\mathbb{R}$ on the Hilbert space $L^2(\mathbb{R})$. The regular action π is defined in the same way as in (5.5), but now this action is diagonalised on a continuous family of characters $\chi_\zeta : \mathbb{R} \to U(1)$ defined by $\chi_\zeta(x) = e^{2\pi i \zeta x}$ with $\zeta, x \in \mathbb{R}$. On the irreducible subspaces $V_\zeta = \mathbb{C}\chi_\zeta$ we then have

$$\pi(x) \cdot v = \chi_\zeta(x) v, \qquad \zeta, x \in \mathbb{R},\ v \in V_\zeta. \tag{5.8}$$

The unitary dual in this case is a 'continuous direct sum', or *direct integral* of irreducible representations, meaning that any function $f \in L^2(\mathbb{R})$ can be written as a continuous version of a Fourier series:

$$f(x) = \int_{\mathbb{R}} \hat{f}(\zeta)\chi_\zeta(x) d\zeta, \tag{5.9}$$

where $\hat{f}(\zeta)$ is the standard Fourier transform of f with respect to the character χ_ζ.

From the above analysis we conclude that the spectral decomposition of $L^2(\mathbb{R})$ with respect to the regular action of $\mathbb{R}$ has only a continuous part, in stark contrast with the situation in Example 5.1 above. The appearance of a continuous spectrum is a general feature of spectral analysis on non-compact spaces, just like the discrete spectrum always appears for compact spaces. We also note a curious feature, namely that although the characters χ_k are used to decompose the spectrum $L^2(\mathbb{R})$ they are in fact *not* square-integrable. This is not a problem since the Fourier transform always preserves square-integrability. An elaborate version of this phenomenon will reappear later in this chapter.

In the previous examples, we have illustrated how the spectral analysis on compact and non-compact spaces reveals very different properties. When we generalise this to the non-abelian setting of $SL(2,\mathbb{Z})\backslash SL(2,\mathbb{R})$ we actually combine these properties in the following sense. Consider for a moment the space of square-integrable automorphic forms $L^2(SL(2,\mathbb{Z})\backslash SL(2,\mathbb{R}))$ which is a subspace of all automorphic forms where the moderate growth condition is replaced by the square-integrability condition. In contrast to the abelian case $\mathbb{Z}\backslash\mathbb{R}$, the space $SL(2,\mathbb{Z})\backslash SL(2,\mathbb{R})$ is certainly non-compact and we therefore expect that a spectral analysis would give rise to a continuous spectrum. In addition, the quotient $SL(2,\mathbb{Z})\backslash SL(2,\mathbb{R})$ has *finite volume* and therefore also gives rise to a discrete spectrum; $SL(2,\mathbb{Z})$ is what is known as a *co-compact subgroup*. Indeed, it was proven by Selberg that

$$\begin{aligned} L^2(SL(2,\mathbb{Z})\backslash SL(2,\mathbb{R})) = \mathbb{C} \oplus L^2_{\text{cusp}}(SL(2,\mathbb{Z})\backslash SL(2,\mathbb{R})) \\ \oplus L^2_{\text{cont}}(SL(2,\mathbb{Z})\backslash SL(2,\mathbb{R})), \end{aligned} \tag{5.10}$$

where:

- the first factor $\mathbb{C}$ represents the constant functions (these are considered to be part of the discrete spectrum and arise also as the residue of the non-holomorphic Eisenstein series $E(s,z)$ at $s = 1$; see also Section 10.1.1)
- the remainder of the discrete spectrum is $L^2_{\text{cusp}}(SL(2,\mathbb{Z})\backslash SL(2,\mathbb{R}))$ which is spanned by *Maaß cusp forms* φ with a discrete set of eigenvalues $\lambda_n, n = 1, 2, 3, \ldots$, with respect to the Laplacian $\Delta_{\mathbb{H}}$. Maaß cusp forms are only known numerically and examples can be found in [598]
- the continuous spectrum $L^2_{\text{cont}}(SL(2,\mathbb{Z})\backslash SL(2,\mathbb{R}))$ is a direct integral of non-holomorphic Eisenstein series $E(s,z)$ (4.73).

Remark 5.3 Note that the non-holomorphic Eisenstein series $E(s,z)$ are not square-integrable and so are not themselves part of $L^2(SL(2,\mathbb{Z})\backslash SL(2,\mathbb{R}))$. They nevertheless play a key rôle in parametrising the unitary dual, in a very similar vein as the non-square-integrable, continuous characters χ_ζ occurred in the spectral decomposition of $L^2(\mathbb{R})$ in Example 5.2. The spectral decomposition (5.10) forms a crucial ingredient in the *Selberg trace formula* (see [29] for a nice introduction).

Although Eisenstein series are not square-integrable they are still important for the representation-theoretic aspects of automorphic forms and therefore it is natural to enhance the space of automorphic forms to include non-square-integrable objects. One then replaces the square-integrable condition with a more general moderate growth condition, leading to the full space of automorphic forms $\mathcal{A}(SL(2,\mathbb{Z})\backslash SL(2,\mathbb{R}))$. We thus have the inclusions of function spaces

$$\begin{aligned} L^2_{\text{cusp}}(SL(2,\mathbb{Z})\backslash SL(2,\mathbb{R})) \subset L^2(SL(2,\mathbb{Z})\backslash SL(2,\mathbb{R})) \\ \subset \mathcal{A}(SL(2,\mathbb{Z})\backslash SL(2,\mathbb{R})). \end{aligned} \tag{5.11}$$

The representation-theoretic aspects of automorphic forms are however not yet complete, as the above treatment is missing an important ingredient. The space $\mathcal{A}(SL(2,\mathbb{Z})\backslash SL(2,\mathbb{R}))$ carries an additional action by *Hecke operators*, which has not yet been taken into account. We will treat Hecke operators in detail in Chapter 11 so here we shall merely offer some qualitative remarks. A Hecke operator is an operation $T_p\colon \mathcal{A} \to \mathcal{A}$, parametrised by a prime number $p < \infty$. The set of all Hecke operators $\{T_p\}_{p<\infty}$ forms a commutative ring, called the *Hecke algebra*. An element $\varphi \in \mathcal{A}(SL(2,\mathbb{Z})\backslash SL(2,\mathbb{R}))$ which is an eigenvector for all Hecke operators

$$T_p\varphi = \lambda_p\varphi \tag{5.12}$$

is called a *Hecke eigenform*. Here the eigenvalues λ_p carry the arithmetic information contained in φ. However, the right-regular action of $SL(2,\mathbb{R})$ on $\mathcal{A}$ cannot be used to accommodate the action of the Hecke algebra and so the analysis of automorphic forms in terms of the representation theory of $SL(2,\mathbb{R})$ is incomplete. One of the reasons for passing to the adelic picture is precisely to remedy this problem. The basic idea is this: if we consider the right-regular action of $SL(2,\mathbb{A})$ on the space $\mathcal{A}(SL(2,\mathbb{Q})\backslash SL(2,\mathbb{A}))$, then the Hecke eigenvalues $\{\lambda_p\}$ parametrise the irreducible representations of the right-regular action of the local subgroups $SL(2,\mathbb{Q}_p)$ on $\mathcal{A}(SL(2,\mathbb{Q})\backslash SL(2,\mathbb{A}))$. Thus, from the adelic perspective the Hecke algebra plays the same rôle at the non-archimedean places of $SL(2,\mathbb{A})$ as the universal enveloping algebra $\mathcal{U}(\mathfrak{sl}(2,\mathbb{C}))$ does at the archimedean place. The Hecke action is thus implicitly already taken into account in the general Definition 4.13 of an adelic automorphic form.

Example 5.4: The action of Hecke operators on non-holomorphic Eisenstein series

For illustration we consider here a simple example of how the Hecke operators act on the non-holomorphic Eisenstein series $E(s,z)$. For $SL(2,\mathbb{Z})$ it is customary to define Hecke operators T_n for any natural number $n > 0$ and not only primes p. They are given by

$$(T_nE)(s,z) := \frac{1}{n}\sum_{d|n}\sum_{b=0}^{d-1} E\left(s, \frac{nz+bd}{d^2}\right). \tag{5.13}$$

In Chapter 11 we will show that

$$(T_nE)(s,z) = \lambda_n E(s,z), \tag{5.14}$$

with

$$\lambda_n = n^{s-1/2}\sigma_{1-2s}(n). \tag{5.15}$$

This is precisely the numerical Fourier coefficient in the Fourier expansion of $E(s,z)$; see Equation (1.17). The operators T_n further satisfy the basic relation

$$T_m T_n = \sum_{d|(m,n)} \frac{1}{d} T_{mn/d^2} \tag{5.16}$$

characterising the Hecke algebra. This algebra is generated by the subset of Hecke operators T_p for p a (finite) prime, hence the fundamental information is contained in the prime eigenvalues λ_p, as claimed in the main text. In Chapter 11, we provide much more detail on Hecke operators and explain their link with the representation theory of $SL(2, \mathbb{Q}_p)$.

We now want to make sense of the combined action on $\mathcal{A}(SL(2,\mathbb{Q})\backslash SL(2,\mathbb{A}))$ of $SL(2,\mathbb{A})$ by right-translation as well as the action of the universal enveloping algebra $\mathcal{U}(\mathfrak{sl}(2,\mathbb{C}))$ by differential operators mentioned at the beginning of this Section 5.1. To this end it is useful to distinguish between the actions of the finite part $SL(2,\mathbb{A}_f) = \prod_{p<\infty} SL(2,\mathbb{Q}_p)$ and the archimedean part $SL(2,\mathbb{R})$. For any $\varphi \in \mathcal{A}(SL(2,\mathbb{Q})\backslash SL(2,\mathbb{A}))$ we then have

$$\begin{aligned}
(\pi(h_f)\varphi)(g) &= \varphi(g h_f), \qquad g \in SL(2,\mathbb{A}),\ h_f \in SL(2,\mathbb{A}_f),\\
(\pi(h_\infty)\varphi)(g) &= \varphi(g h_\infty), \qquad g \in SL(2,\mathbb{A}),\ h_\infty \in SL(2,\mathbb{R}).
\end{aligned} \tag{5.17}$$

Here it is understood that the elements h_f and h_∞ are embedded in the canonical way into the adelic group. For instance, when we write $g h_\infty$ we really mean

$$g \cdot \left(h_\infty, \begin{pmatrix} 1 & \\ & 1 \end{pmatrix}, \begin{pmatrix} 1 & \\ & 1 \end{pmatrix}, \ldots, \begin{pmatrix} 1 & \\ & 1 \end{pmatrix}\right), \qquad g \in SL(2,\mathbb{A}),\ h_\infty \in SL(2,\mathbb{R}). \tag{5.18}$$

These two actions of course commute with each other. The action by $SL(2,\mathbb{A}_f)$ at the non-archimedean places also commutes with the $\mathcal{U}(\mathfrak{sl}(2,\mathbb{C}))$ action at the archimedean place, and so this gives a well-defined representation. On the other hand, the right-regular action of $K_\infty = SO(2,\mathbb{R}) \subset SL(2,\mathbb{R})$ does *not* commute with $\mathcal{U}(\mathfrak{sl}(2,\mathbb{C}))$. Rather, for $X \in \mathcal{U}(\mathfrak{sl}(2,\mathbb{C}))$ and $k_\infty \in SO(2,\mathbb{R})$ one has

$$D_X \cdot \pi(k_\infty) = \pi(k_\infty) \cdot D_{k_\infty^{-1} X k_\infty}, \tag{5.19}$$

where D_X is the differential operator (5.2). One can check this by a direct calculation:

$$\begin{aligned}
D_X \cdot (\pi(k_\infty)\varphi)(g) &= D_X \cdot \varphi(g k_\infty)\\
&= D_X \cdot \varphi\left(g \cdot \left(k_\infty, \begin{pmatrix} 1 & \\ & 1 \end{pmatrix}, \begin{pmatrix} 1 & \\ & 1 \end{pmatrix}, \ldots, \begin{pmatrix} 1 & \\ & 1 \end{pmatrix}\right)\right),
\end{aligned} \tag{5.20}$$

where $k_\infty \in SO(2,\mathbb{R})$. Now using the definition of D_X we find that the right-hand side can be written as

$$\frac{d}{dt}\varphi\left(g\cdot\left(e^{Xt}k_\infty,\begin{pmatrix}1 & \\ & 1\end{pmatrix},\begin{pmatrix}1 & \\ & 1\end{pmatrix},\dots,\begin{pmatrix}1 & \\ & 1\end{pmatrix}\right)\right)\bigg|_{t=0}. \tag{5.21}$$

Inserting the identity $k_\infty k_\infty^{-1}$ and using the following property of the matrix exponential

$$k_\infty^{-1}e^{Xt}k_\infty = e^{k_\infty^{-1}Xk_\infty t}, \tag{5.22}$$

we can rewrite Equation (5.21) as

$$\begin{aligned}\frac{d}{dt}\varphi&\left(gk_\infty\cdot\left(k_\infty^{-1}e^{Xt}k_\infty,\begin{pmatrix}1 & \\ & 1\end{pmatrix},\dots,\begin{pmatrix}1 & \\ & 1\end{pmatrix}\right)\right)\bigg|_{t=0}\\ &= \pi(k_\infty)\cdot\left(D_{k_\infty^{-1}Xk_\infty}\cdot\varphi(g)\right),\end{aligned} \tag{5.23}$$

which is the right-hand side of (5.19). This turns out to be the characteristic property of a so-called $(\mathfrak{g}, K_\infty)$*-module*, a notion which will be properly defined in the next section.

Remark 5.5 To ensure that the space $\mathcal{A}(SL(2,\mathbb{Q})\backslash SL(2,\mathbb{A}))$ is preserved under all three actions defined above, one must of course verify that they are compatible with Definition 4.13. In other words one should check that the three functions

$$D_X\cdot\varphi(g), \qquad (\pi(h_\infty)\varphi)(g), \qquad (\pi(h_f)\varphi)(g), \tag{5.24}$$

all satisfy conditions (1)–(4) in Definition 4.13. See, e.g., Section 5.1 of [291] for a detailed check of this.

Remark 5.6 When speaking about an *automorphic representation* of $SL(2,\mathbb{A})$ one really refers to a structure that carries a standard group representation with respect to the finite part $SL(2,\mathbb{A}_f)$, and a $(\mathfrak{g}, K_\infty)$-module structure at the archimedean place. In the following section we will give the precise definition for an arbitrary reductive group $G(\mathbb{A})$ and discuss some central features of automorphic representations.

5.2 Automorphic Representations

In this section, we shall give the precise definition of an automorphic representation of an adelic group $G(\mathbb{A})$ and present some of the key features that will be important in subsequent chapters. Just as in the $SL(2,\mathbb{A})$ discussion of the previous section, we are interested in the combined actions of $G(\mathbb{A}_f)$ and K_∞ by right-translation and the action of $\mathcal{U}(\mathfrak{g}_\mathbb{C})$ by differential operators.

The general analysis goes through in a similar vein and the conclusion is that the space $\mathcal{A}(G(\mathbb{Q})\backslash G(\mathbb{A}))$ does not carry a group representation with respect to the whole group $G(\mathbb{A})$, but only with respect to the finite part $G(\mathbb{A}_f)$. At the real place one has instead that $\mathcal{A}(G(\mathbb{Q})\backslash G(\mathbb{A}))$ carries the structure of a $(\mathfrak{g}, K_\infty)$-module, whose definition we will now recall.

Definition 5.7 (($\mathfrak{g}$, K)-module) A $(\mathfrak{g}, K_\infty)$*-module* is a complex vector space V which carries an action of both the Lie algebra $\mathfrak{g}$ and K_∞, such that all vectors $v \in V$ are K-finite, i.e., $\dim \langle k_\infty \cdot v \mid k_\infty \in K_\infty \rangle < \infty$. The actions of $\mathfrak{g}$ and K_∞ are furthermore required to be compatible in the following sense:

$$X \cdot k_\infty \cdot v = k_\infty \cdot \mathrm{Ad}_{k_\infty^{-1}}(X) \cdot v, \qquad k_\infty \in K_\infty,\ X \in \mathfrak{g}. \tag{5.25}$$

Remark 5.8 In our context, the complex vector space V is $\mathcal{A}(G(\mathbb{Q})\backslash G(\mathbb{A}))$, the action by $X \in \mathfrak{g}$ is by the differential operator D_X and the action by $k_\infty \in K_\infty$ is by right-translation. In this setting $k_\infty \cdot \mathrm{Ad}_{k_\infty^{-1}}(X)$ means $\pi(k_\infty) \cdot D_{k_\infty^{-1} X k_\infty}$ and hence Equation (5.19) is precisely the compatibility condition (5.25) for a $(\mathfrak{g}, K_\infty)$-module.

Remark 5.9 Let us offer some remarks on the usefulness of $(\mathfrak{g}, K_\infty)$-modules. The notion of $(\mathfrak{g}, K_\infty)$-module was introduced by Harish-Chandra in his efforts on 'algebraisation' of representations. Function spaces on groups are themselves typically not specific enough and there can be many function spaces that share the same algebraic features. For example, one can consider continuous functions on a group manifold and they are a perfectly nice representation of G. However, unless the functions are differentiable, this representation does not give rise to a representation of the Lie algebra $\mathfrak{g}$ that would be represented by an algebra of differential operators. There are many different types of differentiable functions on G and the notion of $(\mathfrak{g}, K_\infty)$-module mainly serves to eliminate the ambiguities related to choosing a type of differentiable function [342, 70, 116].

With the above concepts introduced, we are now ready to state the definition of an automorphic representation.

Definition 5.10 (Automorphic representation) A representation π of $G(\mathbb{A})$ is called an *automorphic representation* if it occurs as an irreducible constituent in the decomposition of $\mathcal{A}(G(\mathbb{Q})\backslash G(\mathbb{A}))$ with respect to the simultaneous action by

$$(\mathfrak{g}_\infty, K_\infty) \times G(\mathbb{A}_f), \tag{5.26}$$

where K_∞ and $G(\mathbb{A}_f)$ act by right-translation and $\mathfrak{g}_\infty$ by differential operators at the archimedean place.

We shall for short denote by V the complex vector space on which $(\mathfrak{g}_\infty, K_\infty) \times G(\mathbb{A}_f)$ acts. Then V is simultaneously a $(\mathfrak{g}_\infty, K_\infty)$-module and a $G(\mathbb{A}_f)$-module and the automorphic representation is also often denoted by the pair (π, V).

Let $K_f \subset G(\mathbb{A}_f)$ be a compact open subgroup (not necessarily maximal) and σ an irreducible representation of $K_\infty \times K_f$. Denote by $V[\sigma]$ the space of vectors in V that transform according to σ under the action of $K_\infty \times K_f$. We then have the following important definitions.

Definition 5.11 (Admissible representation) A $(\mathfrak{g}_\infty, K_\infty) \times G(\mathbb{A}_f)$-module V is called *admissible* if the subspace $V[\sigma] \subset V$ is finite-dimensional for all σ.

Definition 5.12 (K-type decomposition) The space of K-finite vectors in V is the algebraic direct sum

$$V = \bigoplus_{\sigma \in \widehat{K}} V[\sigma], \tag{5.27}$$

where $\widehat{K}$ is the unitary dual, i.e., the equivalence classes of irreducible representations of K.

One then has the following central result due to Flath.

Theorem 5.13 (Tensor decomposition theorem) *Let (π, V) be an admissible automorphic representation. Then there exists an Euler product decomposition into local factors*

$$(\pi, V) = \bigotimes_{p \leq \infty} (\pi_p, V_p), \tag{5.28}$$

where the archimedean component (π_∞, V_∞) is a $(\mathfrak{g}_\infty, K_\infty)$-module according to the discussion above, while the non-archimedean components (π_p, V_p) furnish representations of $G(\mathbb{Q}_p)$.

Proof See [211]. □

Let us finally also introduce the notion of an unramified (or spherical) representation and associated spherical vector.

Definition 5.14 (Unramified representation) An admissible representation π_p is called *unramified* (or *spherical*) if V_p contains a non-zero vector $\mathfrak{f}_p$ which is invariant under $K_p = G(\mathbb{Z}_p)$. We then call such an $\mathfrak{f}_p$ a *spherical vector*. Globally, one has the important notion that (π, V) is a spherical automorphic representation if π_p is spherical for almost all p.

5.3 Principal Series Representations

Fix a Borel subgroup B and a quasi-character $\chi\colon B \to \mathbb{C}^\times$ defined as in (4.122):

$$\chi = e^{\langle \lambda+\rho|H\rangle}. \tag{5.29}$$

Consider now the following space of smooth functions on $G(\mathbb{A})$:

$$I(\chi) = \{f\colon G(\mathbb{A}) \to \mathbb{C} \,|\, f(bg) = \chi(b)f(g),\ b \in B(\mathbb{A})\}. \tag{5.30}$$

This is the function space of an induced representation of $G(\mathbb{A})$ called the *principal series representation*; it is also often denoted by $\mathrm{Ind}_{B(\mathbb{A})}^{G(\mathbb{A})}\chi$. The principal series $I(\chi)$ provides an important example of an automorphic representation thanks to the theory of Eisenstein series which will be discussed in Section 5.4 below. In general $I(\chi)$ is a reducible representation. Moreover, when $\chi = \otimes_p \chi_p$ is an unramified character (i.e., spherical for almost all p; see Definition 5.14), $I(\chi)$ is an irreducible, admissible representation and so affords a decomposition into local factors:

$$I(\chi) = \bigotimes_{p\le\infty} I_p(\chi) = \bigotimes_{p\le\infty} \mathrm{Ind}_{B(\mathbb{Q}_p)}^{G(\mathbb{Q}_p)}\chi_p. \tag{5.31}$$

Remark 5.15 Since we use χ and λ interchangeably for characterising the quasi-character, we also write $I(\lambda)$. The space $I(\lambda)$ can be viewed as the total space of a fiber bundle $I(\lambda) \to \mathfrak{h}_\mathbb{C}^*$, with the fiber over each point $\lambda \in \mathfrak{h}_\mathbb{C}^*$ consisting of the space of functions on $G(\mathbb{A})$ which transform by the character $e^{\langle\lambda+\rho|H\rangle}$ under the left-action of $B(\mathbb{A})$.

Definition 5.16 (Standard section) An element $\mathfrak{f}_\lambda \in I(\lambda)$ is called a *standard section* if it is $K_\mathbb{A}$-finite and its restriction to $K_\mathbb{A}$ is independent of λ.

By virtue of (5.31), any standard section $\mathfrak{f}_\lambda \in I(\lambda)$ splits into a product of local factors

$$\mathfrak{f}_\lambda = \bigotimes_{p\le\infty} \mathfrak{f}_{\lambda,p}. \tag{5.32}$$

Remark 5.17 For $\chi = \otimes_{p\le\infty}\chi_p$ an unramified character of $B(\mathbb{A})$, the induced representation $I(\chi) = \otimes_{p\le\infty} I(\chi_p)$ is an unramified principal series (see Definition 5.14). That is, for almost all p, there is a unique (up to scaling) spherical vector $\mathfrak{f}_p^\circ \in I(\chi_p)$ defined by [116]

$$\mathfrak{f}_p^\circ(g) = \mathfrak{f}_p^\circ(nak) = \chi_p(a), \qquad \mathfrak{f}_p^\circ(k) = \mathfrak{f}_p^\circ(1) = 1. \tag{5.33}$$

Definition 5.18 (Gelfand–Kirillov dimension) Although the principal series representations $I_p(\lambda)$ are infinite-dimensional, one can still attach to them a notion of 'size', which is called the *functional*, or *Gelfand–Kirillov, dimension* and is denoted by GKdim. This is defined as the smallest number of variables on which we can realise the functions in $I_p(\lambda)$. A more formal algebraic definition can be found in [75].

Example 5.19: Gelfand–Kirillov dimension

The Gelfand–Kirillov dimension of the Hilbert space $L^2(\mathbb{R}^n)$ is n. Similarly, by the Iwasawa decomposition $G(\mathbb{R}) = B(\mathbb{R})K(\mathbb{R})$, the representation induced from a spherical function at the archimedean place has functional dimension (for generic χ_∞)

$$\mathrm{GKdim}\,\mathrm{Ind}_{B(\mathbb{R})}^{G(\mathbb{R})}\chi_\infty = \dim_{\mathbb{R}}\left(B(\mathbb{R})\backslash G(\mathbb{R})\right) = \dim_{\mathbb{R}} G(\mathbb{R}) - \dim_{\mathbb{R}} B(\mathbb{R})\,. \tag{5.34}$$

5.4 Eisenstein Series and Induced Representations

Let us now discuss the definition of Eisenstein series from the perspective of induced representations. One can think of an Eisenstein series as providing a $G(\mathbb{A})$-equivariant map from the induced representation $I(\lambda)$ of (5.30) into the space of automorphic forms:

$$E:\ I(\lambda)\ \to\ \mathcal{A}(G(\mathbb{Q})\backslash G(\mathbb{A})). \tag{5.35}$$

For any standard section $\mathfrak{f}_\lambda \in I(\lambda)$ the construction of the corresponding Eisenstein series is given by

$$E(\mathfrak{f}_\lambda, g) = \sum_{\gamma\in B(\mathbb{Q})\backslash G(\mathbb{Q})} \mathfrak{f}_\lambda(\gamma g). \tag{5.36}$$

As $\mathfrak{f}_\lambda$ varies in the fiber of $I(\lambda) \to \mathfrak{h}_{\mathbb{C}}^*$ we thus obtain a family of Eisenstein series $E(\mathfrak{f}_\lambda, g)$ that satisfy all the conditions of Definition 4.13 for an automorphic form; in particular, $K_{\mathbb{A}}$-finiteness follows from the fact that $\mathfrak{f}_\lambda$ is a standard section. By virtue of the decomposition $\mathfrak{f}_\lambda = \otimes_{p\le\infty}\mathfrak{f}_{\lambda,p}$, the Eisenstein series $E(\mathfrak{f}_\lambda, g)$ can be defined by choosing all the local factors $\mathfrak{f}_{\lambda,p}$ separately. This gives a lot of freedom in defining the Eisenstein series and is one of the main reasons why the adelic formalism is so powerful.

Remark 5.20 In order to recover the particular Eisenstein series of definition (4.24), one chooses the standard section $\mathfrak{f}_\lambda$ to be equal to the inducing global character, $\mathfrak{f}_\lambda = e^{\langle\lambda+\rho|H\rangle} = \chi$.

Remark 5.21 In Section 5.6 we will illustrate for $G(\mathbb{A}) = SL(2,\mathbb{A})$ how the more general construction in (5.36) interpolates between holomorphic and non-holomorphic Eisenstein series on the upper half-plane $\mathbb{H}$.

5.5 Classifying Automorphic Representations

It is one of the central unsolved problems in the theory of automorphic forms to classify all automorphic representations. As an important intermediate result, all admissible automorphic representations have been classified (see, e.g., [116]). This includes in particular the spherical, or unramified, representations.

The task of decomposing $\mathcal{A}(G(\mathbb{Q})\backslash G(\mathbb{A}))$ into irreducible representations is closely connected to the problem of decomposing the Hilbert space $L^2(G(\mathbb{Q})\backslash G(\mathbb{A}))$ under the unitary action of G. A priori this might seem a little surprising since an automorphic form need not be square-integrable; indeed the Eisenstein series $E(s, g)$ considered in Section 4.4.1 provides an example of a non-square-integrable automorphic form. The decomposition of $L^2(G(\mathbb{Q})\backslash G(\mathbb{A}))$ splits into two orthogonal spaces

$$L^2(G(\mathbb{Q})\backslash G(\mathbb{A})) = L^2_{\text{discrete}}(G(\mathbb{Q})\backslash G(\mathbb{A})) \oplus L^2_{\text{continuous}}(G(\mathbb{Q})\backslash G(\mathbb{A})), \quad (5.37)$$

corresponding respectively to the *discrete* and *continuous parts of the spectrum*. It turns out that the discrete spectrum is spanned by cusp forms and *residues of Eisenstein series* [447, 488].

It is a fundamental result in the spectral theory of automorphic forms that the space $\mathcal{A}_0(G(\mathbb{Q})\backslash G(\mathbb{A}))$ is the subspace of $L^2_{\text{discrete}}(G(\mathbb{Q})\backslash G(\mathbb{A}))$ corresponding to smooth, cuspidal, K-finite and $\mathcal{Z}(\mathfrak{g})$-finite vectors occurring in the decomposition of the unitary representation $L^2(G(\mathbb{Q})\backslash G(\mathbb{A}))$. This is the reason that cusp forms constitute an essential part in the theory of automorphic forms.

While the discrete spectrum can be understood in this way as a direct sum of invariant subspaces (part of which is spanned by cusp forms; for the remaining part, see below), the space $L^2_{\text{continuous}}(G(\mathbb{Q})\backslash G(\mathbb{A}))$ rather decomposes into a *direct integral* over principal series representations of $G(\mathbb{R})$. Such integrals turn out to be parametrised by Eisenstein series, even though these by themselves are not square-integrable (see the following section and also [289, 263] for more on the continuous spectrum and the relation with Eisenstein series). This situation is a generalisation of the problem of decomposing the Hilbert space $L^2(\mathbb{R})$ via Fourier analysis in terms of the non-square-integrable characters (Fourier modes) $e^{2\pi ixy}$, $x, y \in \mathbb{R}$, as discussed in Example 5.2. The construction of Eisenstein series on $G(\mathbb{Q})\backslash G(\mathbb{A})$, generalising the function $E(g, s)$ of Section 4.4.1, therefore constitutes an equally important part of the theory of automorphic forms as that of analysing the space of cusp forms. Moreover, the complement of $\mathcal{A}_0(G(\mathbb{Q})\backslash G(\mathbb{A}))$ inside the discrete spectrum $L^2_{\text{discrete}}(G(\mathbb{Q})\backslash G(\mathbb{A}))$ is conjecturally spanned by *residues* of Eisenstein series $E(\lambda, g)$ for special values of the weight λ. Thus, one expects that the discrete spectrum decomposes according to

$$L^2_{\text{discrete}}(G(\mathbb{Q})\backslash G(\mathbb{A})) = L^2_{\text{cusp}}(G(\mathbb{Q})\backslash G(\mathbb{A})) \oplus L^2_{\text{res}}(G(\mathbb{Q})\backslash G(\mathbb{A})). \quad (5.38)$$

Remark 5.22 Arthur has outlined a set of conjectures that characterise precisely the weights λ for which the representation becomes square-integrable [27] (for proofs of Arthur's conjectures in some cases, see [375, 396, 483, 487, 474]). See also Section 10.2.3 for further discussion of square-integrability of Eisenstein series.

Example 5.23: Representation-theoretic viewpoint on Eisenstein series

We now analyse the general Eisenstein series $E(\lambda, g)$ more explicitly for $G(\mathbb{A}) = SL(2,\mathbb{A})$. In this case the space of (complex) weights $\mathfrak{h}^*_{\mathbb{C}}$ is one-dimensional and spanned by the fundamental weight Λ dual to the unique simple root α of the Lie algebra $\mathfrak{sl}(2)$. The Weyl vector ρ is also identical to $\Lambda = \alpha/2$. Therefore, we can parametrise the weight appearing in (4.133) with a single parameter $s \in \mathbb{C}$ according to

$$\lambda = 2s\Lambda - \rho = (2s-1)\Lambda \quad \Longrightarrow \quad \lambda + \rho = 2s\Lambda\,. \quad (5.39)$$

The character $\chi\colon B(\mathbb{Q})\backslash B(\mathbb{A}) \to \mathbb{C}^\times$ in (4.122) can now be written as

$$\chi_s(g) \equiv e^{\langle \lambda+\rho | H(g)\rangle} = e^{\langle 2s\Lambda | H(a)\rangle}\,. \quad (5.40)$$

We can write all these objects explicitly in the fundamental representation of $SL(2,\mathbb{A})$:

$$g = nak = \begin{pmatrix} 1 & u \\ & 1 \end{pmatrix}\begin{pmatrix} v & \\ & v^{-1} \end{pmatrix} k, \quad (5.41)$$

with $k \in K_{\mathbb{A}}$. Evaluated on the group element (5.41) we then find

$$\chi_s(g) = e^{2s\langle \Lambda | H(a)\rangle} = |v|^{2s}, \quad (5.42)$$

since $H(a) = \log|v| \cdot H_\alpha$, where H_α is the Cartan generator of $\mathfrak{sl}(2)$; see (4.122) above. The general Eisenstein series $E(\lambda, g)$ in (4.133) now becomes

$$E(s, g) = \sum_{\gamma \in B(\mathbb{Q})\backslash SL(2,\mathbb{Q})} \chi_s(\gamma g), \quad (5.43)$$

which is indeed equivalent to (4.84). This Eisenstein series is attached to the induced representation

$$I(s) = \text{Ind}_{B(\mathbb{A})}^{SL(2,\mathbb{A})} \chi_s. \quad (5.44)$$

This representation is unitary when $s = 1/2 + it \in 1/2 + i\mathbb{R}_+$. In this simple example one can also give a more explicit description of the spectral problem of decomposing the space $L^2(SL(2,\mathbb{Q})\backslash SL(2,\mathbb{A}))$ with respect to the right-regular action of $SL(2,\mathbb{A})$. The decomposition (5.37) becomes in this case

$$L^2(SL(2,\mathbb{Q})\backslash SL(2,\mathbb{A})) = L^2_{\text{cusp}}(SL(2,\mathbb{Q})\backslash SL(2,\mathbb{A})) \oplus \mathbb{C} \oplus \int_0^\infty I(1/2+it)\,dt, \tag{5.45}$$

where the discrete spectrum L^2_{discrete} is represented by the space of cusp forms L^2_{cusp} together with the space $\mathbb{C}$ of constant functions that correspond to the residual spectrum from the pole of $E(s,g)$ at $s=1$. The continuous spectrum $L^2_{\text{continuous}}$ corresponds to the integral over the principal series $I(s)$, restricted to the unitary domain $s \in 1/2 + i\mathbb{R}_+$. For a proof of this statement, see the book by Gelbart [263].

5.6 Embedding of the Discrete Series in the Principal Series

Our aim in this section is to illustrate in greater detail the construction of the general Eisenstein series $E(\mathfrak{f}_\lambda, g)$ in (5.36) for the special case of $SL(2,\mathbb{A})$. We will in particular demonstrate that when restricted to a function on $SL(2,\mathbb{Z})\backslash SL(2,\mathbb{R})$ this yields a generalisation of the classical non-holomorphic Eisenstein series, which in fact interpolates between the non-holomorphic function $E(s,z)$, $z \in \mathbb{H}$, and the weight w holomorphic Eisenstein series $E_w(z)$. We explain how to understand this representation-theoretically in terms of the embedding of the holomorphic discrete series of $SL(2,\mathbb{R})$ into the principal series.

5.6.1 Eisenstein Series for Arbitrary Standard Sections

Let $I(\lambda) = \text{Ind}_{B(\mathbb{A})}^{SL(2,\mathbb{A})} \chi_s$ be the induced representation (5.30) for $SL(2,\mathbb{A})$. As in Example 5.23, we take the inducing character $\chi_s \colon B(\mathbb{Q})\backslash B(\mathbb{A}) \to \mathbb{C}^\times$ (extended to all of $SL(2,\mathbb{A})$) to be defined by

$$\chi_s(bk) = \chi_s(b) = \chi_s\begin{pmatrix} v & * \\ & v^{-1} \end{pmatrix} = |v|^{2s}, \qquad s \in \mathbb{C}, \tag{5.46}$$

where $b \in B(\mathbb{A})$ and $k \in K_{\mathbb{A}} = SO(2,\mathbb{R}) \times \prod_{p<\infty} SL(2,\mathbb{Z}_p)$.

Let $\mathfrak{f}_\lambda = \otimes_p \mathfrak{f}_{\lambda,p} \in I(\lambda)$ with each local factor

$$\mathfrak{f}_{\lambda,p} \in I_p(\lambda) = \text{Ind}_{B(\mathbb{Q}_p)}^{SL(2,\mathbb{Q}_p)} \chi_p, \tag{5.47}$$

determined by its restriction to $SL(2,\mathbb{Z}_p) = B(\mathbb{Q}_p)\backslash SL(2,\mathbb{Q}_p)$. For the purposes of this section we shall now fix these local sections as follows:

- For the non-archimedean places $p < \infty$ we choose the section $\mathfrak{f}_{\lambda,p}$ to be the unique (normalised) spherical vector $\mathfrak{f}^\circ_{\lambda,p}$ in $I_p(\lambda)$ defined by (see also

Remark 5.17)

$$\mathsf{f}^{\circ}_{\lambda,p}(g_p) = \mathsf{f}^{\circ}_{\lambda,p}(b_p k_p) = \chi_{s,p}(b_p), \qquad \mathsf{f}^{\circ}_{\lambda,p}(k_p) = \mathsf{f}^{\circ}_{\lambda,p}(1) = 1, \tag{5.48}$$

where $b_p \in B(\mathbb{Q}_p)$ and $k_p \in SL(2,\mathbb{Z}_p)$.

- For the archimedean place $p = \infty$ we define $\mathsf{f}_{\lambda,\infty} \in I_\infty(\lambda)$ according to

$$\mathsf{f}_{\lambda,\infty}(b_\infty) = \chi_{s,\infty}(b_\infty), \qquad \mathsf{f}_{\lambda,\infty}(g_\infty k_\infty) = e^{iw\theta}\mathsf{f}_{\lambda,\infty}(g_\infty), \qquad \mathsf{f}_{\lambda,\infty}(1) = 1, \tag{5.49}$$

where $w \in \mathbb{Z}$, $g_\infty \in SL(2,\mathbb{R})$, $b_\infty \in B(\mathbb{R})$ and

$$k_\infty = \begin{pmatrix} \cos\theta & \sin\theta \\ -\sin\theta & \cos\theta \end{pmatrix} \in SO(2,\mathbb{R}). \tag{5.50}$$

Notice that with this definition $\mathsf{f}_{\lambda,\infty}$ is a $K_\infty = SO(2,\mathbb{R})$-finite but *non-spherical* section of $I_\infty(\lambda)$.

With these definitions of the local factors, the product

$$\mathsf{f}_\lambda = \mathsf{f}_{\lambda,\infty} \otimes \bigotimes_{p<\infty} \mathsf{f}^{\circ}_{\lambda,p} \tag{5.51}$$

becomes a standard section (because λ and w are independent parameters) of the global representation $I(\lambda)$.

With this choice of section we now construct the Eisenstein series

$$E(\mathsf{f}_\lambda, g) = \sum_{\gamma \in B(\mathbb{Q})\backslash G(\mathbb{Q})} \left(\mathsf{f}_{\lambda,\infty}(\gamma g_\infty) \times \prod_{p<\infty} \mathsf{f}^{\circ}_{\lambda,p}(\gamma g_p) \right), \tag{5.52}$$

with $g = (g_\infty; g_2, g_3, \dots) \in SL(2,\mathbb{A}) = SL(2,\mathbb{R}) \times \prod'_{p<\infty} SL(2,\mathbb{Q}_p)$. We now want to analyse the restriction of this adelic Eisenstein series to a function on $SL(2,\mathbb{R})$. To this end we fix the adelic group element to be the identity at all finite places:

$$g = (g_\infty; 1, 1, \cdots) \in SL(2,\mathbb{A}). \tag{5.53}$$

In Example 3.16, we showed the bijection of cosets $B(\mathbb{Q})\backslash SL(2,\mathbb{Q}) \cong B(\mathbb{Z})\backslash SL(2,\mathbb{Z})$ with each $B(\mathbb{Q})g$ coset having a representative in $SL(2,\mathbb{Z})$ for $g \in SL(2,\mathbb{Q})$. We will now use this to write the Eisenstein series as a sum over $\gamma \in SL(2,\mathbb{Z})$. At the finite places $SL(2,\mathbb{Z})$ embeds into $SL(2,\mathbb{Z}_p)$, and hence, by (5.48), we have

$$\mathsf{f}^{\circ}_{\lambda,p}(\gamma) = 1 \qquad (\text{for } \gamma \in SL(2,\mathbb{Z}) \text{ and } p < \infty). \tag{5.54}$$

The Eisenstein series $E(\mathfrak{f}_\lambda, g)$ is therefore restricted to

$$\begin{aligned} E(\mathfrak{f}_{\lambda,\infty}, g_\infty) &= \sum_{B(\mathbb{Q})\backslash SL(2,\mathbb{Q})} \mathfrak{f}_{\lambda,\infty}(\gamma g_\infty) \times \prod_{p<\infty} \mathfrak{f}^\circ_{\lambda,p}(\gamma) \\ &= \sum_{B(\mathbb{Z})\backslash SL(2,\mathbb{Z})} \mathfrak{f}_{\lambda,\infty}(\gamma g_\infty) \times \prod_{p<\infty} \mathfrak{f}^\circ_{\lambda,p}(\gamma) \\ &= \sum_{B(\mathbb{Z})\backslash SL(2,\mathbb{Z})} \mathfrak{f}_{\lambda,\infty}(\gamma g_\infty). \end{aligned} \tag{5.55}$$

To relate this to a function on the upper half-plane $\mathbb{H}$, we use the Iwasawa decomposition,

$$g_\infty = b_\infty k_\infty = n_\infty a_\infty k_\infty = \begin{pmatrix} 1 & x \\ & 1 \end{pmatrix} \begin{pmatrix} y^{1/2} & \\ & y^{-1/2} \end{pmatrix} \begin{pmatrix} \cos\theta & \sin\theta \\ -\sin\theta & \cos\theta \end{pmatrix}, \tag{5.56}$$

which yields

$$E(\mathfrak{f}_{\lambda,\infty}, g_\infty) = e^{iw\theta} \sum_{B(\mathbb{Z})\backslash SL(2,\mathbb{Z})} \mathfrak{f}_{\lambda,\infty}(\gamma b_\infty). \tag{5.57}$$

It remains to evaluate $\mathfrak{f}_{\lambda,\infty}(\gamma b_\infty)$. To this end we perform an additional Iwasawa decomposition of γb_∞ with the result

$$\gamma b_\infty = b'_\infty k'_\infty, \qquad \gamma = \begin{pmatrix} a & b \\ c & d \end{pmatrix} \in SL(2,\mathbb{Z}), \tag{5.58}$$

with

$$\begin{aligned} b'_\infty &= \begin{pmatrix} \frac{y^{1/2}}{|cz+d|} & * \\ 0 & \frac{|cz+d|}{y^{1/2}} \end{pmatrix}, \\ k'_\infty &= \begin{pmatrix} \cos\theta' & \sin\theta' \\ -\sin\theta' & \cos\theta' \end{pmatrix} = \frac{1}{|cz+d|} \begin{pmatrix} cx+d & -cy \\ cy & cx+d \end{pmatrix}, \end{aligned} \tag{5.59}$$

and $z = x + iy = b_\infty \cdot i$. Further using that $\exp(i\theta) = \cos\theta + i\sin\theta$ we find

$$e^{i\theta'} = \frac{|cz+d|}{cz+d}, \tag{5.60}$$

and hence, by (5.46) and (5.49), the section in (5.57) evaluates to

$$\begin{aligned} \mathfrak{f}_{\lambda,\infty}(\gamma b_\infty) = \mathfrak{f}_{\lambda,\infty}(b'_\infty k'_\infty) &= \left(\frac{|cz+d|}{cz+d}\right)^w \chi_{s,\infty}(b'_\infty) \\ &= \left(\frac{|cz+d|}{cz+d}\right)^w \frac{y^s}{|cz+d|^{2s}}. \end{aligned} \tag{5.61}$$

We thereby arrive at the following explicit expression for the Eisenstein series:

$$E(\mathfrak{f}_{(s,w),\infty}, g_\infty) = e^{iw\theta} \sum_{(c,d)=1} \frac{y^s}{(cz+d)^w |cz+d|^{2s-w}}. \tag{5.62}$$

This is a non-holomorphic function on $SL(2,\mathbb{Z})\backslash SL(2,\mathbb{R})$ with weight $\exp(iw\theta)$ under the right-action of $k_\infty \in SO(2,\mathbb{R})$.

5.6.2 Representation-Theoretic Interpretation

Let us now analyse the Eisenstein series (5.62) a little more closely. First observe that $E(\mathfrak{f}_{(s,w),\infty}, g_\infty)$ interpolates between the classical holomorphic and non-holomorphic Eisenstein series on the upper half-plane. Indeed, restricting the value of s to $s = w/2$ we obtain

$$E(\mathfrak{f}_{(w/2,w),\infty}, g_\infty) = e^{iw\theta} y^{w/2} \sum_{(c,d)=1} \frac{1}{(cz+d)^w} = e^{iw\theta} y^{w/2} E_w(z), \tag{5.63}$$

which we recognise as $\varphi_f(g_\infty)$ in the terminology of Section 4.3 (see Equation (4.47)), with $f = E_w(z)$ being the classical weight w holomorphic Eisenstein series on $\mathbb{H}$. Similarly, restricting to $w = 0$ in (5.62) we obtain the classical non-holomorphic Eisenstein series

$$E(\mathfrak{f}_{(s,0),\infty}, g_\infty) = \sum_{(c,d)=1} \frac{y^s}{|cz+d|^{2s}} = E(s,z). \tag{5.64}$$

Note that this is compatible with the fact that fixing $w = 0$ is equivalent to choosing the local section $\mathfrak{f}_{\lambda,p}$ to be spherical also at the archimedean place, $\mathfrak{f}_{\lambda,\infty} = \mathfrak{f}^\circ_{\lambda,\infty}$. The Eisenstein series $E(\mathfrak{f}_\lambda, g)$ in (5.36) then reduces to (4.84) which is indeed the adelisation of $E(s,z)$.

While the non-holomorphic Eisenstein series $E(s,z)$ is naturally associated with the principal series $\mathrm{Ind}_{B(\mathbb{R})}^{SL(2,\mathbb{R})} \chi_s$, the holomorphic Eisenstein series $E_w(z)$ is rather associated with the so-called *holomorphic discrete series* $\mathcal{D}(w)$ of $SL(2,\mathbb{R})$ for $w \in 2\mathbb{Z}_+$.

In order to exhibit this, we make use of some of the notions of Section 4.3.3 on Maaß forms of non-zero weight. To this end, let $\mathcal{H}(w)$ be the Hilbert space of smooth, complex square-integrable functions $g(z)$ on the upper half-plane $\mathbb{H} \cong SL(2,\mathbb{R})/SO(2,\mathbb{R})$ that are Maaß forms of weight w. Specifically, they transform under $SL(2,\mathbb{Z})$ as

$$g(z) \longmapsto \left(\frac{cz+d}{|cz+d|}\right)^w g\left(\frac{az+b}{cz+d}\right). \tag{5.65}$$

If $f(z)$ is a *holomorphic* modular form of weight w, then $g(z) = y^{w/2} f(z) \in \mathcal{H}(w)$. This is in particular true for $g(z) = y^{w/2} E_w(z)$ with the holomorphic Eisenstein series appearing in (5.63). This transition from modular holomorphic forms to Maaß forms is closely related to the lift (4.47) to $SL(2,\mathbb{R})$.

For Maaß forms of weight w one can define the following differential operators, called *Maaß lowering and raising operators*:

$$L_w := -(z-\bar{z})\frac{\partial}{\partial \bar{z}} - \frac{w}{2}, \tag{5.66a}$$

$$R_w := (z-\bar{z})\frac{\partial}{\partial z} + \frac{w}{2}. \tag{5.66b}$$

These have the property that they lower and raise the weights of a Maaß form by two units:

$$L_w : \mathcal{H}_w \to \mathcal{H}_{w-2}, \qquad R_w : \mathcal{H}_w \to \mathcal{H}_{w+2}. \tag{5.67}$$

If $f(z)$ is a holomorphic modular form of weight w, then one finds that the associated Maaß form $g(z) = y^{w/2} f(z)$ is annihilated by the lowering operator:

$$L_w g(z) = L_w \left(y^{w/2} f(z)\right) = 0. \tag{5.68}$$

Similar operators also arise in the string theory context; see Section 14.4.

The Maaß operators can be understood representation-theoretically as follows. Let

$$\varphi_f(x,y,\theta) = e^{iw\theta} y^{w/2} f(x+iy) = e^{iw\theta} y^{w/2} f(z) \tag{5.69}$$

be the lift of $f \in \mathcal{M}_w \equiv \mathcal{M}_w(SL(2,\mathbb{Z}))$ to $SL(2,\mathbb{R})$ as in Section 4.3. (We see that the Maaß form $g(z) = y^{w/2} f(z)$ on the upper half-plane is in some sense half of the lift.) Recall from (4.52) that φ_f satisfies the differential equation

$$F\varphi_f = -2ie^{-2i\theta}\left(y\frac{\partial}{\partial \bar{z}} - \frac{1}{4}\frac{\partial}{\partial \theta}\right)\varphi_f = 0, \tag{5.70}$$

where F is the differential operator-realisation of the negative Chevalley generator F of $SL(2,\mathbb{R})$ in the compact basis (4.58) (see also Section 4.1 for more details). Using (5.66) we can rewrite this as

$$F\varphi_f = e^{i(w-2)\theta} L_w \left(y^{w/2} f(z)\right) = 0, \tag{5.71}$$

revealing that the Maaß operator L_w is nothing but the Chevalley generator F after evaluating the derivative on θ. Similarly, the Maaß operator R_w is the positive Chevalley operator E. We conclude from this that the holomorphic Eisenstein series $f(z) = E_w(z)$ gives rise to the lowest weight vector in the holomorphic discrete series of $SL(2,\mathbb{R})$ via (5.63). The weight w is measured by the Cartan generator H (see (4.58)):

$$H\varphi_f = w\varphi_f. \tag{5.72}$$

One can also check that

$$H(E\varphi_f) = [H,E]\varphi_f + E(H\varphi_f) = 2E\varphi_f + wE\varphi_f = (w+2)E\varphi_f. \tag{5.73}$$

This means that φ_f for $f(z) = E_w(z)$ is the lowest weight state in a representation of $\mathfrak{sl}(2,\mathbb{R})$, whose states are obtained by acting successively with the raising operator E:

$$\{\varphi_f, E\varphi_f, E^2\varphi_f, \dots\}. \tag{5.74}$$

Here each vector $E^n\varphi_f$, $n \geq 0$, is an automorphic form on $SL(2,\mathbb{A})$, and hence belongs to the space $\mathcal{A}(SL(2,\mathbb{Q})\backslash SL(2,\mathbb{A}))$. The span of the states (5.74) is a subspace V of $\mathcal{A}(SL(2,\mathbb{Q})\backslash SL(2,\mathbb{A}))$ that is clearly invariant under the $\mathfrak{sl}(2,\mathbb{R})$ action. It is furthermore preserved by $K = SO(2,\mathbb{R})$, since each vector $E^n\varphi_f \in \mathcal{A}(SL(2,\mathbb{Q})\backslash SL(2,\mathbb{A}))$ is K-finite by Definition 4.13. Thus, the vector space V spanned by (5.74) is a $(\mathfrak{g}, K)$-module. This is the $(\mathfrak{g}, K)$-module underlying the holomorphic discrete series $\mathcal{D}(w)$. It is conventional to think of the holomorphic Eisenstein series $f = E_w$ as the explicit lowest weight state.

In general, the principal series $\mathrm{Ind}_{B(\mathbb{R})}^{SL(2,\mathbb{R})}\chi_s$ is not a lowest (or highest) weight representation; indeed the general Eisenstein series (5.62) is not annihilated by either of E or F. However, as we restrict to the integer points $s = w/2$ of the complex weight space where χ_s lives, we land on an irreducible submodule of $\mathrm{Ind}_{B(\mathbb{R})}^{SL(2,\mathbb{R})}\chi_s$ which can be identified with the holomorphic discrete series $\mathcal{D}(w)$. In other words, we have discovered the well-known fact that the holomorphic discrete series can be embedded in the principal series for special values of the inducing character:

$$\mathcal{D}(w) \subset \mathrm{Ind}_{B(\mathbb{R})}^{SL(2,\mathbb{R})}\chi_s\big|_{s=w/2}. \tag{5.75}$$

We should in fact be a little more careful. In (5.63) it is understood that the weight is restricted to be a positive integer $w > 0$. We should therefore distinguish between positive and negative weights in the spherical vector (5.49). The case $w > 0$ leads to (5.63) as we just discussed. The negative weight case $w < 0$ leads to the same conclusion, except that the restriction (5.63) now corresponds to the *anti-holomorphic* Eisenstein series $\overline{E}_w(\bar{z})$. This Eisenstein series is then naturally associated with the anti-holomorphic discrete series $\overline{\mathcal{D}}(w)$ of $SL(2,\mathbb{R})$ which is defined analogously to (5.65) for anti-holormorphic functions $\overline{f}(\bar{z})$. The anti-holomorphic Eisenstein series $\overline{E}_w(\bar{z})$ lifts to a function $\varphi_{\overline{f}}$ which is annihilated by E, rather than F, and can therefore be interpreted as a *highest weight vector* of $\overline{\mathcal{D}}(w)$ with weight $-w$. The negative Chevalley generator F then lowers the weight by 2.

The above discussion shows that both the holomorphic and anti-holomorphic discrete series can be embedded in the principal series. The complement is a finite-dimensional representation of $SL(2,\mathbb{R})$, known as Sym^{w-1}; this is the w-dimensional symmetric power representation of $SL(2,\mathbb{R})$ acting on homogeneous degree $w-1$ polynomials in two real variables. Indeed, it is easy to see from the weight diagram in Figure 5.1 that the number of weights that

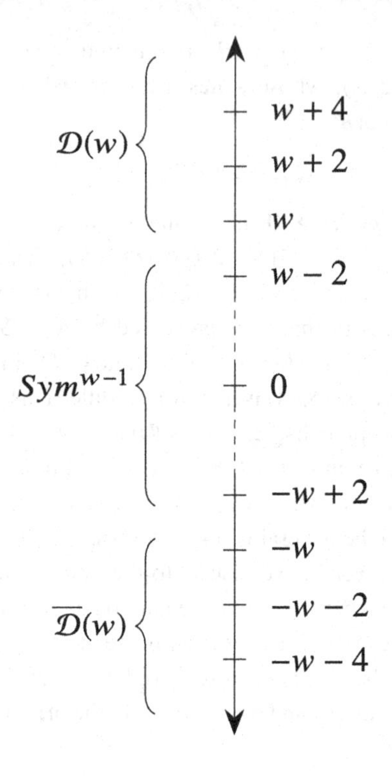

Figure 5.1 Weight diagram for $SL(2,\mathbb{R})$.

are excluded from the holomorphic and anti-holomorphic discrete series are precisely equal to w. This implies that Sym^{w-1} is the following quotient of the principal series by the discrete series for $w \in \mathbb{Z}_+$:

$$\mathrm{Ind}_{B(\mathbb{R})}^{SL(2,\mathbb{R})} \chi_{w/2} \Big/ \left(\mathcal{D}(w) \oplus \overline{\mathcal{D}}(w)\right) = Sym^{w-1}. \tag{5.76}$$

The symmetric power representation is non-unitary.

5.7 Eisenstein Series for Non-minimal Parabolics

With a little effort one can generalise the construction of (Borel–) Eisenstein series to any parabolic subgroup. In what follows we restrict to standard parabolics P, that is, those that contain the Borel subgroup $B(\mathbb{A}) = A(\mathbb{A})N(\mathbb{A})$ as discussed in Section 3.1.3. We recall from that section that standard parabolic subgroups can be defined by a choice of subset Σ of the simple roots Π of $\mathfrak{g}$, where we avoid writing the dependence on Σ on P in order to keep the notation light. Maximal parabolic subgroups are given by removing only a single simple root α_{i_*} from Π such that $\Sigma = \Pi \setminus \{\alpha_{i_*}\}$ and we sometimes write P_{i_*} for such maximal parabolics.

Any standard parabolic subgroup $P(\mathbb{A}) \subset G(\mathbb{A})$ has a Langlands decomposition as in (3.42):

$$P(\mathbb{A}) = L(\mathbb{A})U(\mathbb{A}) = M(\mathbb{A})A_P(\mathbb{A})U(\mathbb{A}). \tag{5.77}$$

The full group $G(\mathbb{A})$ factorises (non-uniquely) as

$$G(\mathbb{A}) = M(\mathbb{A})A_P(\mathbb{A})U(\mathbb{A})K_{\mathbb{A}}. \tag{5.78}$$

For an arbitrary element of $G(\mathbb{A})$ we write

$$g = ulk = uma_Pk, \tag{5.79}$$

with $l \in L(\mathbb{A})$, $m \in M(\mathbb{A})$, $a_P \in A_P(\mathbb{A})$, $u \in U(\mathbb{A})$, $k \in K_{\mathbb{A}}$. See Example 5.24 for some details of these decompositions in the cases $G(\mathbb{A}) = GL(n, \mathbb{A})$ and $G(\mathbb{A}) = SL(n, \mathbb{A})$.

5.7.1 Multiplicative Characters

We now want to define multiplicative characters on $P(\mathbb{A})$ analogously to what we did for the Borel subgroup in Section 4.5.1. These will be homomorphisms

$$\chi_P : P(\mathbb{Q})\backslash P(\mathbb{A}) \to \mathbb{C}^\times, \tag{5.80}$$

determined by their restriction to the Levi subgroup

$$\chi_P(lu) = \chi_P(ul) = \chi_P(l), \qquad l \in L(\mathbb{A}), u \in U(\mathbb{A}). \tag{5.81}$$

As in the minimal parabolic case the characters can be described by roots, but now the image in root space will be $\mathfrak{a}_P^*$, where $\mathfrak{a}_P$ is the Lie algebra of A_P, instead of $\mathfrak{h}^*$, and we will use a generalisation of the logarithm map H from (4.117).

Let $H_P : P(\mathbb{A}) \to \mathfrak{a}_P(\mathbb{R})$ be defined by

$$H_P(p) = H_P(lu) = H_P(ma_Pu) = H_P(a_P) = \log|a_P|, \qquad a_P \in A_P \subseteq A, \tag{5.82}$$

where the absolute value is defined as in (4.120). As is evident from the definition, H_P only depends on the centre A_P of P. We extend H_P to all of G by right K-invariance as we did for H.

In order to define a character on P in analogy with (4.122), we need to pair $H_P(p) \in \mathfrak{a}_P$ with an element of $\mathfrak{a}_P^*$. We can think of $\mathfrak{a}_P^*$ as being embedded in $\mathfrak{h}^*$ by

$$\mathfrak{a}_P^* = \{\lambda \in \mathfrak{h}^* \mid \langle\lambda|\alpha\rangle = 0 \quad \text{for all } \alpha \in \Sigma\} \tag{5.83}$$

in terms of the inner product (3.13) on $\mathfrak{h}^*$ given by the Cartan matrix of $\mathfrak{g}$. This implies that any element $\lambda_P \in \mathfrak{a}_P^*$ can be written as a linear combination of the fundamental weights associated with the complement of Σ in Π.

A character on P can then be defined using a weight $\lambda_P \in \mathfrak{a}_P^*$ as

$$\chi_P(p) = e^{\langle \lambda_P + \rho_P | H_P(p) \rangle}, \tag{5.84}$$

where ρ_P is now the restriction of the full Weyl vector to the positive roots $\Delta(\mathfrak{u}) = \Delta_+ \setminus \langle \Sigma \rangle_+$ of $\mathfrak{g}$ from Section 3.1.3:

$$\rho_P = \frac{1}{2} \sum_{\alpha \in \Delta(\mathfrak{u})} \alpha. \tag{5.85}$$

We have to check that ρ_P lies in $\mathfrak{a}_P^*$. From the definition of the full vector ρ in (3.6) we see that

$$\rho_P = \rho - \frac{1}{2} \sum_{\alpha \in \langle \Sigma \rangle_+} \alpha \equiv \rho - \rho_M, \tag{5.86}$$

where half the sum over the roots in $\langle \Sigma \rangle_+$ represents the Weyl vector of the semi-simple M factor in the Langlands decomposition (5.77) of P. Therefore we deduce that

$$\langle \rho_P | \beta \rangle = \frac{1}{2} \langle \beta | \beta \rangle - \frac{1}{2} \langle \beta | \beta \rangle = 0 \quad \text{for any } \beta \in \langle \Sigma \rangle_+ \tag{5.87}$$

from the standard properties of Weyl vectors; see (3.14). Thus $\rho_P \in \mathfrak{a}_P^*$. We note also that ρ_P is related to the *modulus character* δ_P on P in the same way that ρ is related to the modulus character δ on B; see Remark 4.22.

The character (5.84) can be extended from P to a function on G by right K-invariance as usual. We note that from the definitions of H_P and H it follows that, for any $\lambda_P \in \mathfrak{a}_P^*$,

$$\langle \lambda_P | H_P(g) \rangle = \langle \lambda_P | H(g) \rangle \tag{5.88}$$

when thinking of the spaces $\mathfrak{a}_P$ and $\mathfrak{a}_P^*$ as being embedded in $\mathfrak{h}$ and $\mathfrak{h}^*$, respectively. One also has

$$\langle \rho | H_P(g) \rangle = \langle \rho_P | H_P(g) \rangle \tag{5.89}$$

as only the $\mathfrak{a}_P^*$ component of ρ has a non-vanishing pairing with $\mathfrak{a}_P$.

Example 5.24: Parabolic subgroups and characters for $GL(n, \mathbb{A})$

For $G(\mathbb{A}) = GL(n, \mathbb{A})$ there is a bijection between standard parabolic subgroups $P(\mathbb{A})$ and ordered partitions $(n_1, \ldots, n_q)$ of n. It is then sometimes useful to start from this point of view when parametrising the subgroup P instead of the one based on subsets $\Sigma \subset \Pi$ from Section 3.1.3.

For a given partition we then have that $P(\mathbb{A}) = L(\mathbb{A})U(\mathbb{A}) = M(\mathbb{A})A_P(\mathbb{A})U(\mathbb{A})$ can be expressed explicitly as

$$L(\mathbb{A}) = \left\{ \begin{pmatrix} l_1 & & 0 \\ & \ddots & \\ 0 & & l_q \end{pmatrix} \middle| \, l_i \in GL(n_i, \mathbb{A}) \right\},$$

$$U(\mathbb{A}) = \left\{ \begin{pmatrix} \mathbb{1}_{n_1} & \star & \star \\ & \ddots & \star \\ 0 & & \mathbb{1}_{n_q} \end{pmatrix} \right\},$$

$$M(\mathbb{A}) = \left\{ \begin{pmatrix} m_1 & & 0 \\ & \ddots & \\ 0 & & m_q \end{pmatrix} \middle| \, m_i \in SL(n_i, \mathbb{A}) \right\},$$

$$A_P(\mathbb{A}) = \left\{ \begin{pmatrix} a_1 \mathbb{1}_{n_1} & & 0 \\ & \ddots & \\ 0 & & a_q \mathbb{1}_{n_q} \end{pmatrix} \middle| \, a_i \in GL(1, \mathbb{A}) \right\}, \tag{5.90}$$

where $\mathbb{1}_n$ denotes the $n \times n$ identity matrix.

Similarly, instead of working with the Chevalley basis for $\mathfrak{a}_P$ it is useful to choose a basis that reflects the block form in the parametrisation above.

We choose a basis $\tilde{H}_i$ for $\mathfrak{a}_P$ such that H_P from (5.82) becomes

$$H_P \begin{pmatrix} a_1 \mathbb{1}_{n_1} & & 0 \\ & \ddots & \\ 0 & & a_q \mathbb{1}_{n_q} \end{pmatrix} = \sum_{i=1}^{q} n_i \log |a_i| \, \tilde{H}_i \,. \tag{5.91}$$

For

$$l = \begin{pmatrix} l_1 & & 0 \\ & \ddots & \\ 0 & & l_q \end{pmatrix} \tag{5.92}$$

we then obtain that

$$H_P(l) = \sum_{i=1}^{q} \log |\det l_i| \, \tilde{H}_i \,. \tag{5.93}$$

Now introduce a basis $\tilde{\Lambda}_i$ for $\mathfrak{a}_P^*$ dual to $\tilde{H}_i$, that is, $\langle \tilde{\Lambda}_i | \tilde{H}_j \rangle = \delta_{ij}$, and let λ_P and ρ_P in $\mathfrak{a}_P^*$ be parametrised by $\lambda_P = \sum_{i=1}^{q} s_i \tilde{\Lambda}_i$ and $\rho_P = \sum_{i=1}^{q} \rho_i \tilde{\Lambda}_i$ with $s_i, \rho_i \in \mathbb{C}$. Note that since $\tilde{H}_i$ is not the Chevalley basis, the $\tilde{\Lambda}_i$ are not the standard fundamental weights.

Any character on $P(\mathbb{Q})\backslash P(\mathbb{A})$ can then be constructed by

$$\chi_P(lu) = \chi_P(l) = e^{\langle \lambda_P + \rho_P | H_P(l) \rangle} = \prod_{i=1}^{q} |\det l_i|^{s_i + \rho_i} \,. \tag{5.94}$$

For $G = SL(n, \mathbb{A})$ we have the restriction $\prod_{i=1}^{q} a_i = 1$ which reduces the number of independent elements in the sum (5.91) spanning $\mathfrak{a}_P$. In the same way, the parameters in $\lambda_P = \sum_{i=1}^{q} s_i \tilde{\Lambda}_i$ are also restricted. A general character on $P(\mathbb{Q})\backslash P(\mathbb{A})$ for $SL(n,\mathbb{A})$ can thus be seen as a special case of (5.94). Explicitly, we require that $\langle \lambda_P | \mathbb{1}_n \rangle = 0$ and, since $\mathbb{1}_n = \sum_i n_i \tilde{H}_i$, we get the restriction that $\sum_i n_i s_i = 0$.

5.7.2 Parabolically Induced Representations

Associated with the parabolic subgroup $P(\mathbb{A})$ we now consider the following space of functions:

$$I_P(\lambda) = \left\{ f\colon G(\mathbb{A}) \to \mathbb{C} \;\middle|\; f(pg) = e^{\langle \lambda + \rho_P | H_P(p) \rangle} f(g),\ g \in G(\mathbb{A}),\ p \in P(\mathbb{A}) \right\}, \tag{5.95}$$

where $\lambda \in \mathfrak{a}_P^*$. Note that we have suppressed the subscript P on λ for brevity, thinking of $\mathfrak{a}_P^*$ as being embedded in $\mathfrak{h}^*$.

This is the function space of the induced representation $\mathrm{Ind}_{P(\mathbb{A})}^{G(\mathbb{A})} \chi_P = \mathrm{Ind}_{P(\mathbb{A})}^{G(\mathbb{A})} \exp\left(\langle \lambda + \rho_P | H_P \rangle\right)$. We will also refer to this as the *degenerate principal series*. The generic functional dimension of the representation $\mathrm{Ind}_{P(\mathbb{A})}^{G(\mathbb{A})} \chi_P$ is, similarly to (5.34), given by

$$\mathrm{GKdim}(I_P(\lambda)) = \dim(G) - \dim(P). \tag{5.96}$$

It is now straightforward to construct a *parabolic Eisenstein series* associated with the induced representation $I_P(\lambda)$. For a weight $\lambda_P \in \mathfrak{a}^*$, it takes the form

$$E(\lambda_P, P, g) = \sum_{\gamma \in P(\mathbb{Q})\backslash G(\mathbb{Q})} e^{\langle \lambda_P + \rho_P | H_P(\gamma g) \rangle}. \tag{5.97}$$

In case P is the Borel subgroup we write $E(\lambda, B, g) \equiv E(\lambda, g)$ and we recover the Eisenstein series in (5.36). In view of (5.88), we could also write this as

$$E(\lambda_P, P, g) = \sum_{\gamma \in P(\mathbb{Q})\backslash G(\mathbb{Q})} e^{\langle \lambda_P + \rho_P | H(\gamma g) \rangle}, \tag{5.98}$$

using the 'standard' logarithm map H defined in (4.117). Alternatively, taking any $\lambda \in \mathfrak{h}^*$ and projecting it to $\lambda_P \in \mathfrak{a}_P^*$, we have from (5.89)

$$E(\lambda, P, g) = E(\lambda_P, P, g) = \sum_{\gamma \in P(\mathbb{Q})\backslash G(\mathbb{Q})} e^{\langle \lambda + \rho | H_P(\gamma g) \rangle}. \tag{5.99}$$

Just as for the Borel subgroup, we can of course also start from any standard section $\mathfrak{f}_\lambda \in I_P(\lambda)$ and obtain another Eisenstein series,

$$E(\mathfrak{f}_\lambda, P, g) = \sum_{\gamma \in P(\mathbb{Q})\backslash G(\mathbb{Q})} \mathfrak{f}_\lambda(\gamma g). \tag{5.100}$$

One can generalise the Eisenstein series $E(\lambda, P, g)$ even further by modifying the induced representation $I_P(\lambda)$ as follows. For $P(\mathbb{A}) = L(\mathbb{A})U(\mathbb{A})$ consider the space V_P of functions

$$\phi : (L(\mathbb{Q})U(\mathbb{A}))\backslash G(\mathbb{A}) \to \mathbb{C} \tag{5.101}$$

such that for each fixed $g \in G(\mathbb{A})$ the function

$$\phi_g : l \to \phi(lg) \tag{5.102}$$

and any $l \in L(\mathbb{Q})\backslash L(\mathbb{A})$ belongs to the discrete spectrum of $L(\mathbb{A})$, i.e., $\phi_g \in L^2_{\text{disc}}(L(\mathbb{Q})\backslash L(\mathbb{A}))$. Moreover, we require $\phi \in V_P$ to have finite square norm when integrated over $M(\mathbb{Q})\backslash M(\mathbb{A})K(\mathbb{A})$. Strictly speaking, one should be careful with the real Cartan torus in these definitions (see [29, 569]) but for our exposition these subtleties will not be crucial.

Let $\lambda \in \mathfrak{a}_P^*(\mathbb{C})$. For any such λ and $g \in G(\mathbb{A})$, we define an action $\iota_P(\lambda, g)$ on the space V_P by

$$(\iota_P(\lambda, g)\phi)\,(h) = \phi(hg)e^{\langle\lambda+\rho_P|H_P(hg)\rangle}e^{-\langle\lambda+\rho_P|H_P(h)\rangle}. \tag{5.103}$$

The second unusual term in this modified induced action serves to map V_P to itself and ensures that the space is independent of λ.

Definition 5.25 (Eisenstein series induced from automorphic form on Levi) For each $\phi \in V_P$ defined above and $\lambda \in \mathfrak{a}_P^*(\mathbb{C})$ we define the Eisenstein series

$$E(\lambda, P, \phi, g) = \sum_{\gamma\in P(\mathbb{Q})\backslash G(\mathbb{Q})} \phi(\gamma g)e^{\langle\lambda+\rho_P|H_P(\gamma g)\rangle}. \tag{5.104}$$

If $\phi = 1$, the function $E(\lambda, P, \phi, g)$ reduces to the parabolic Eisenstein series defined in (5.97).

Remark 5.26 Under the assumptions on V_P, we can think of ϕ as an automorphic function on $L(\mathbb{A})$ extended to $G(\mathbb{A})$. In the literature, one often typically has $\phi \in \mathcal{A}_0(L(\mathbb{Q})\backslash L(\mathbb{A}))$, the space of cusp forms on $L(\mathbb{A})$. In this case, Langlands has proven the analytic continuation and functional relation for $E(\lambda, P, \phi, g)$ [447]. The representation of $G(\mathbb{A})$ that the function $E(\lambda, P, \phi, g)$ belongs to is then obtained from a parabolic subgroup by cuspidal induction.

Remark 5.27 The Eisenstein series $E(\lambda, P, \phi, g)$ can be viewed as a vector in the induced representation $I(\lambda, V_P) = \text{Ind}_{P(\mathbb{A})}^{G(\mathbb{A})}(V_P \otimes \exp(\langle\lambda + \rho_P|H_P\rangle))$. Ginzburg noticed [278] that for $GL(n)$ and generic values of $\lambda \in \mathfrak{a}_P^*(\mathbb{C})$ one has the following simple formula for the Gelfand–Kirillov dimension:

$$\text{GKdim}(I(\lambda, V_P)) = \text{GKdim}(V_P) + \dim U, \tag{5.105}$$

where U is the unipotent radical of P. This formula also holds conjecturally for other groups.

The generalisation (5.104) is useful as it connects to the minimal Eisenstein series $E(\lambda, g)$ defined in (4.133), as the following proposition shows.

Proposition 5.28 (Borel–Eisenstein series as a specialisation) *The Eisenstein series $E(\lambda, g)$, induced from the Borel subgroup B, is a special case of the Eisenstein series $E(\lambda, P_{i_*}, \phi, g)$, where P_{i_*} is a maximal parabolic subgroup associated with the simple root α_{i_*} (see Section 3.1.3).*

Proof To see this, we follow the argument in [313]. First note that $B_{L_{i_*}} = L_{i_*} \cap B$ is a Borel subgroup of the Levi $L_{i_*} \subset P_{i_*}$. This implies that any $\gamma \in B(\mathbb{Q})\backslash G(\mathbb{Q})$ can be uniquely decomposed as $\gamma = \gamma_1\gamma_2$, with

$$\gamma_1 \in B_{L_{i_*}}(\mathbb{Q})\backslash L_{i_*}(\mathbb{Q}), \qquad \gamma_2 \in P_{i_*}(\mathbb{Q})\backslash G(\mathbb{Q}). \tag{5.106}$$

We can thus rewrite the Eisenstein series $E(\lambda, g)$ as follows:

$$\begin{aligned} E(\lambda, g) &= \sum_{\gamma \in B(\mathbb{Q})\backslash G(\mathbb{Q})} e^{\langle \lambda+\rho | H(\gamma g)\rangle} \\ &= \sum_{\gamma_1 \in B_{L_{i_*}}(\mathbb{Q})\backslash L_{i_*}(\mathbb{Q})} \sum_{\gamma_2 \in P_{i_*}(\mathbb{Q})\backslash G(\mathbb{Q})} e^{\langle \lambda+\rho | H(\gamma_1\gamma_2 g)\rangle}. \end{aligned} \tag{5.107}$$

Any complex weight $\lambda \in \mathfrak{a}^*(\mathbb{C})$ can be decomposed by orthogonal projection on the plane with normal vector Λ_{i_*}, the fundamental weight associated with node i_*. Explicitly,

$$\lambda = \lambda_{\parallel i_*} + \lambda_{\perp i_*}, \quad \text{with} \quad \lambda_{\parallel i_*} = \frac{\langle \lambda | \Lambda_{i_*}\rangle}{\langle \Lambda_{i_*} | \Lambda_{i_*}\rangle}\Lambda_{i_*}, \quad \lambda_{\perp i_*} = \lambda - \lambda_{\parallel i_*}. \tag{5.108}$$

By the defining properties of Λ_{i_*}, we find that $\lambda_{\perp i_*}$ is a complex linear combination of all simple roots different from α_{i_*}, and $\lambda_{\parallel i_*}$ is proportional to Λ_{i_*}. Furthermore, for any $l \in L_{i_*}$ generated by roots in the Lie algebra $\mathfrak{l}_{i_*}$ we have, for any $g \in G$,

$$\langle \lambda_{\parallel i_*} + \rho_{\parallel i_*} | H(lg)\rangle = \langle \lambda_{\parallel i_*} + \rho_{\parallel i_*} | H(g)\rangle. \tag{5.109}$$

This identity is moreover true for any $l \in L_{i_*}(\mathbb{Q})$ as the global character is trivial on $A_{P_{i_*}}(\mathbb{Q})$. Hence, in the domain of absolute convergence, we may decompose the summation as

$$
\begin{aligned}
E(\lambda, g) &= \sum_{\gamma_1 \in B_{L_{i_*}}(\mathbb{Q})\backslash L_{i_*}(\mathbb{Q})} \sum_{\gamma_2 \in P_{i_*}(\mathbb{Q})\backslash G(\mathbb{Q})} e^{\langle \lambda+\rho | H(\gamma_1\gamma_2 g)\rangle} \\
&= \sum_{\gamma_2 \in P_{i_*}(\mathbb{Q})\backslash G(\mathbb{Q})} \Bigg[e^{\langle \lambda_{\parallel i_*} + \rho_{\parallel i_*} | H(\gamma_2 g)\rangle} \\
&\qquad \times \sum_{\gamma_1 \in B_{L_{i_*}}(\mathbb{Q})\backslash L_{i_*}(\mathbb{Q})} e^{\langle \lambda_{\perp i_*} + \rho_{\perp i_*} | H(\gamma_1\gamma_2 g)\rangle} \Bigg] \\
&= \sum_{\gamma_2 \in P_{i_*}(\mathbb{Q})\backslash G(\mathbb{Q})} e^{\langle \lambda_{\parallel i_*} + \rho_{\parallel i_*} | H(\gamma_2 g)\rangle} \phi(\gamma_2 g), \qquad (5.110)
\end{aligned}
$$

where the function

$$
\phi(g) = \sum_{\gamma_1 \in B_{L_{i_*}}(\mathbb{Q})\backslash L_{i_*}(\mathbb{Q})} e^{\langle \lambda_{\perp i_*} + \rho_{\perp i_*} | H(\gamma_1 g)\rangle} \qquad (5.111)
$$

is an Eisenstein series on the Levi L_{i_*}, induced from the Borel subgroup $B_{L_{i_*}}$. □

Remark 5.29 Proposition 5.28 can be straightforwardly generalised to give a relation between Eisenstein series $E(\lambda, g)$, induced from the Borel subgroup B, and Eisenstein series $E(\lambda, P, \phi, g)$, induced from an arbitrary parabolic subgroup P (not necessarily maximal).

We note that for special λ the statement of Proposition 5.28 becomes even stronger:

Proposition 5.30 (Maximal parabolic Eisenstein series as special cases) *Let $\lambda = 2s\Lambda_{i_*} - \rho$ be a weight associated with a maximal parabolic P_{i_*} such that the character $e^{\langle \lambda+\rho | H_{P_{i_*}}(\cdot)\rangle}$ is invariant on the left under $P_{i_*}(\mathbb{Q})$. Then the minimal Eisenstein series $E(\lambda, g)$ defined in* (4.133) *is identical to the maximal parabolic Eisenstein series $E(\lambda, P_{i_*}, g)$ defined in* (5.97).

Proof For the specific $\lambda = 2s\Lambda_{i_*} - \rho$ one finds from (5.108) that

$$
(\lambda + \rho)_{\perp i_*} = 0, \qquad (5.112)
$$

since $\lambda + \rho = 2s\Lambda_{i_*}$ is completely in the direction of the fundamental weight Λ_{i_*}. But this implies that the function ϕ appearing in (5.111) becomes trivial according to Proposition 4.26. Using this, relation (5.110) becomes

$$
E(\lambda, g) = \sum_{\gamma \in P_{i_*}(\mathbb{Q})\backslash G(\mathbb{Q})} e^{\langle \lambda_{\parallel i_*} + \rho_{\parallel i_*} | H(\gamma g)\rangle} = E(\lambda, P_{i_*}, g)\,. \qquad (5.113)
$$

□

Remark 5.31 Proposition 5.30 allows us to never explicitly discuss (maximal) parabolic Eisenstein series of the type (5.97) since they can be viewed as special cases of minimal parabolic Eisenstein series $E(\lambda, g)$. This is the approach we will take in this book and derive theorems for $E(\lambda, g)$ that then can be specialised to the cases $\lambda = 2s\Lambda_{i_*} - \rho$ as needed. We also note that the convergence criterion for a maximal parabolic Eisenstein series is $\langle s\Lambda_{i_*} - \rho, \Lambda_{i_*}\rangle > 0$, as can be deduced by carefully studying the consequences of the Godement criterion (4.134).

To illustrate the general analysis of this section, we shall conclude with an example dealing with the case of maximal parabolic subgroups $P(\mathbb{A})$. This is the opposite extreme compared to the Borel subgroup, which we recall is a minimal parabolic.

Example 5.32: Eisenstein series on E_n induced from Heisenberg parabolics

Let $G(\mathbb{A})$ be the adelisation of either split E_6, E_7 or E_8 with Lie algebra $\mathfrak{g}$. We shall analyse the above construction for a very special type of maximal parabolic subgroup of G, known as the *Heisenberg parabolic*, henceforth denoted by P_{Heis}. This parabolic subgroup is associated with the highest root θ of $\mathfrak{g}$. Similar arguments can be made for the ADE-series of simple Lie algebras [395, 393] but then not necessarily resulting in P_{Heis} being maximal.

Associated with θ the Lie algebra exhibits a canonical five-grading

$$\mathfrak{g} = \mathfrak{g}_{-2} \oplus \mathfrak{g}_{-1} \oplus \mathfrak{g}_0 \oplus \mathfrak{g}_1 \oplus \mathfrak{g}_2, \tag{5.114}$$

where the subscript indicates the eigenvalue under the Cartan generator H_θ associated with θ, and $\mathfrak{g}_{\pm 2}$ are one-dimensional subspaces spanned by the corresponding Chevalley generators $E_{\pm\theta}$. The triple $(H_\theta, E_\theta, E_{-\theta})$ generates an $\mathfrak{sl}(2, \mathbb{R})$ subalgebra:

$$[H_\theta, E_\theta] = 2E_\theta, \qquad [H_\theta, F_\theta] = -2F_\theta, \qquad [E_\theta, F_\theta] = H_\theta. \tag{5.115}$$

The zeroth subspace $\mathfrak{g}_0$ is of the form $\mathfrak{m}_{\text{Heis}} \oplus \mathbb{R}H_\theta$, where $\mathfrak{m}_{\text{Heis}} \subset \mathfrak{g}$ is a reductive Lie algebra corresponding to the commutant of the $\mathfrak{sl}(2, \mathbb{R})$-algebra (5.115) inside $\mathfrak{g}$. The nilpotent subspace $\mathfrak{g}_1 \oplus \mathfrak{g}_2$ is a Heisenberg algebra of dimension $\dim_{\mathbb{R}} \mathfrak{g}_1 + 1 \equiv 2d + 1$, with commutator

$$[\mathfrak{g}_1, \mathfrak{g}_1] \subseteq \mathfrak{g}_2. \tag{5.116}$$

We set

$$\mathfrak{p} = \mathfrak{g}_0 \oplus \mathfrak{g}_1 \oplus \mathfrak{g}_2, \tag{5.117}$$

which is the Lie algebra of a maximal parabolic subgroup $P_{\text{Heis}} \subset G$, the *Heisenberg parabolic*. Its Levi and Langlands decompositions are

$$P_{\text{Heis}} = L_{\text{Heis}} U_{\text{Heis}} = M_{\text{Heis}} A_{\text{Heis}} U_{\text{Heis}}, \tag{5.118}$$

where the Levi subgroup $L_{\text{Heis}} = M_{\text{Heis}} A_{\text{Heis}}$ is the exponentiation of $\mathfrak{g}_0$, further decomposing into $\mathfrak{a}_{\text{Heis}} = \mathbb{R}H_\theta$ and $\mathfrak{m}_{\text{Heis}}$ above, and the unipotent radical U_{Heis} is the *Heisenberg group* whose Lie algebra is $\mathfrak{g}_1 \oplus \mathfrak{g}_2$.

For types D and E, we can characterise the Heisenberg parabolics P_{Heis} in terms of maximal parabolics P_{i_*} for specific choices of simple roots α_{i_*} of G in the terminology of Section 3.1.3. The root system of the Levi L_{i_*} is obtained from that of G by removing the node corresponding to α_{i_*}. In the present context we have $i_* = 2, 1, 8$ for E_6, E_7, E_8, respectively. According to Proposition 5.30, we can construct a Heisenberg parabolic Eisenstein series $E(\lambda, P_{i_*}, g)$ by restricting a minimal parabolic Eisenstein series $E(\lambda, g) = E(\lambda, B, g)$ to the special weight $\lambda = 2s\Lambda_{i_*} - \rho$:

$$E(\lambda, P_{i_*}, g) = \sum_{\gamma \in B(\mathbb{Q})\backslash G(\mathbb{Q})} e^{\langle 2s\Lambda_{i_*} | H(\gamma g)\rangle}. \tag{5.119}$$

This Eisenstein series is attached to the *degenerate principal series* $\text{Ind}_{B(\mathbb{A})}^{G(\mathbb{A})} \exp\langle 2s\Lambda_{i_*}|H\rangle$. By Proposition 5.30 this function space is isomorphic to the induced representation $\text{Ind}_{P_{\text{Heis}}(\mathbb{A})}^{G(\mathbb{A})} \exp\langle 2s\Lambda_{i_*}|H_{P_{i_*}}\rangle$, which at the archimedean place has functional dimension

$$\text{GKdim}\,\text{Ind}_{P_{\text{Heis}}(\mathbb{R})}^{G(\mathbb{R})} e^{\langle 2s\Lambda_{i_*}|H_{P_{i_*}}\rangle} = \dim P_{\text{Heis}}(\mathbb{R})\backslash G(\mathbb{R}) = \dim \mathfrak{g}_1 \oplus \mathfrak{g}_2 = 2d + 1, \tag{5.120}$$

and depends on a single complex parameter s. In contrast, the generic principal series $\text{Ind}_{B(\mathbb{A})}^{G(\mathbb{A})} \exp\langle \sum_{i=1}^r s_i\Lambda_i|H\rangle$ depends on $r = \text{rank}\,\mathfrak{g}$ parameters $(s_1, \ldots, s_r) \in \mathbb{C}^r$. Formally one can view $\text{Ind}_{P_{\text{Heis}}(\mathbb{A})}^{G(\mathbb{A})} \exp\langle 2s\Lambda_{i_*}|H_{P_{i_*}}\rangle$ as the limit of $\text{Ind}_{B(\mathbb{A})}^{G(\mathbb{A})} \exp\langle \sum_{i=1}^r s_i\Lambda_i|H\rangle$ when projecting onto the complement of a complex co-dimension one locus in $\mathbb{C}^r$.

For any standard section $\mathfrak{f}_s \in \text{Ind}_{B(\mathbb{A})}^{G(\mathbb{A})} \exp\langle 2s\Lambda_{i_*}|H\rangle$ the Eisenstein series attached to the degenerate principal series is

$$E(\mathfrak{f}_s, P_{i_*}, g) = \sum_{\gamma \in B(\mathbb{Q})\backslash G(\mathbb{Q})} \mathfrak{f}_s(\gamma g). \tag{5.121}$$

This has interesting properties because its residues at the poles in the complex s-plane give rise to automorphic forms attached to special types of (unipotent) representations of G which have very small functional dimensions (typically of dimension less than $2d + 1$). The smallest such representation is known as the *minimal representation* of G and has functional dimension $d + 1$. For a more general discussion of small representations, see Section 6.4.3. Automorphic forms attached to minimal representations were analysed from this point of view in [281], and have also played an important rôle in physical applications [334, 393, 518, 336, 525, 520, 315]. See also Chapter 14 and Section 15.4.

5.8 Induced Representations and Spherical Vectors*

In this section, we elaborate on some aspects of induced representations and present a different construction of Eisenstein series based on global spherical vectors.

5.8.1 Analytic Induction

We first recall briefly the notion of analytic induction of a representation from a parabolic subgroup P of a group G, which was already discussed in Section 5.3. Generalising the set-up there, let

$$\sigma : P \to GL(V) \tag{5.122}$$

be a (smooth) representation of the parabolic subgroup P on a (finite-dimensional) vector space V. In the case of the principal series, the vector space V is one-dimensional, P is a Borel subgroup and the representation σ is given by a character χ.

Definition 5.33 (Induced representation) The *induced representation* of σ from P to G is defined to be the space of smooth functions

$$\mathrm{Ind}_P^G \sigma = \{f : G \to V \mid f(pg) = \sigma(p)f(g)\} . \tag{5.123}$$

The group G acts on this function space by right-translation.

Functions in the induced representation space are by construction fully determined by their values on the coset $P\backslash G$ (and the representation σ). The coset typically has a complicated cell decomposition but often one can learn many important aspects by focussing on the biggest cell. As was already mentioned in (5.96), the functional dimension of the induced representation is given by the dimension of the coset $P\backslash G$; see for instance [75, 382].

Example 5.34: Principal series of $SL(2, \mathbb{R})$

Consider the case $G = SL(2, \mathbb{R})$ and $P = \bar{B} = \left\{ \begin{pmatrix} A & 0 \\ Z & A^{-1} \end{pmatrix} \middle| A \in \mathbb{R}^{\times}, Z \in \mathbb{R} \right\}$, the standard lower parabolic subgroup. In the *line model* of $\bar{B}\backslash G$ one decomposes a matrix $g \in SL(2, \mathbb{R})$ as

$$g = \bar{b}n : \quad \begin{pmatrix} a & b \\ c & d \end{pmatrix} = \begin{pmatrix} A & \\ Z & A^{-1} \end{pmatrix} \begin{pmatrix} 1 & x \\ & 1 \end{pmatrix} = \begin{pmatrix} A & Ax \\ Z & Zx + A^{-1} \end{pmatrix}, \tag{5.124}$$

which is always possible as long as $a \neq 0$. The representative of g is then

$$n = \begin{pmatrix} 1 & x \\ & 1 \end{pmatrix} \quad \text{with} \quad x = \frac{b}{a} \in \mathbb{R}. \tag{5.125}$$

If $a = 0$, this *LU-type decomposition* breaks down. One can check that there is only a single class $\bar{B}\backslash G$ with $a = 0$ and that this can be represented by

$$\begin{pmatrix} 0 & -1 \\ 1 & 0 \end{pmatrix}. \tag{5.126}$$

This point should be thought of as the 'point at infinity' that compactifies the real line $x \in \mathbb{R}$ to a compact space isomorphic to a circle. (The Iwasawa decomposition implies that $\bar{B}\backslash G$ is a quotient of the compact space K and this way of looking at the coset is called the *circle model* or *ball model*.) The set of matrices with $a \neq 0$ corresponds to the big cell, whereas the 'point at infinity' is the single smaller cell in this example.

Suppose that $\bar{B}$ is represented on the one-dimensional vector space $V = \mathbb{C}$ by

$$\sigma\left(\begin{pmatrix} A & \\ Z & A^{-1} \end{pmatrix}\right) = |A|^{-2s}. \tag{5.127}$$

This is a character on $\bar{B}$ and so we are dealing with a principal series representation as discussed in Section 5.3. A given representative (5.125) in the big cell transforms as follows under right multiplication by $SL(2,\mathbb{R})$:

$$ng = \bar{b}'n' \tag{5.128}$$

with

$$n' = \begin{pmatrix} 1 & x' \\ & 1 \end{pmatrix} \quad \text{with} \quad x' = \frac{dx+b}{cx+a}, \quad \bar{b}' = \begin{pmatrix} 1 & \\ \star & 1 \end{pmatrix}\begin{pmatrix} A & \\ & A \end{pmatrix} \tag{5.129}$$

with $A = cx + a$. The $\star$ entry is easy to compute but of no further importance to us. If we think of $\mathrm{Ind}_{\bar{B}}^{G}\,\sigma$ as the space of functions $f(x)$ on the real line since x is a good coordinate on $\bar{B}\backslash G$ almost everywhere, the functions in $\mathrm{Ind}_{\bar{B}}^{G}\,\sigma$ satisfy

$$f(xg) = |cx+a|^{-2s} f\left(\frac{dx+b}{cx+a}\right) \tag{5.130}$$

under $g = \begin{pmatrix} a & b \\ c & d \end{pmatrix} \in SL(2,\mathbb{R})$.

The action above can be linearised to obtain a realisation of the Lie algebra $\mathfrak{sl}(2,\mathbb{R})$ on functions on the real line corresponding to the principal series. The result is

$$h = -2x\partial_x - 2s, \quad e = \partial_x, \quad f = -x^2\partial_x - 2sx, \tag{5.131}$$

where the standard generators (4.2) have been used. The above action can be formally extended by including the point at infinity corresponding to the small cell.

One can also investigate induced representations for p-adic groups.

Example 5.35: Principal series of $SL(2,\mathbb{Q}_p)$

We use the same type of decomposition of parabolic subgroup and line model as in the example above, so that the induced function space becomes

$$\mathrm{Ind}_{\bar{B}(\mathbb{Q}_p)}^{SL(2,\mathbb{Q}_p)}\,\sigma \cong \left\{ f\colon \mathbb{Q}_p \to \mathbb{C} \;\middle|\; f(xg) = |cx+a|_p^{-2s} f\left(\frac{x+b}{cx+a}\right) \right\} \tag{5.132}$$

almost everywhere.

The next example deals with the *conformal realisation* of the orthogonal group.

Example 5.36: Conformal realisation of $SO(d,d)$

Our discussion will be based on the Lie algebra $\mathfrak{so}(d,d)$ that can be presented by generators $M^{IJ} = -M^{JI}$ for $I, J = 1, \dots, 2d$ that satisfy

$$\begin{aligned}\left[M^{IJ}, M^{KL}\right] &= 2\eta^{K[J} M^{I]L} - 2\eta^{L[J} M^{I]K} \\ &= \eta^{JK} M^{IL} - \eta^{IK} M^{JL} - \eta^{JL} M^{IK} + \eta^{IL} M^{JK} . \end{aligned} \tag{5.133}$$

The $SO(d,d)$ metric $\eta = \text{diag}(+-+-\dots+-)$ is fully split.

We consider the Lie algebra $\mathfrak{so}(d,d)$ (for $d \geq 3$) in the graded decomposition with respect to $\mathfrak{so}(d-1,d-1)$:

$$\mathfrak{so}(d,d) = \underbrace{(2(d-1))}_{K^i}{}^{(-1)} \oplus (\underbrace{\mathfrak{so}(d-1,d-1)}_{M^{ij}} \oplus \underbrace{\mathfrak{gl}(1)}_{D})^{(0)} \oplus \underbrace{(2(d-1))}_{P^i}{}^{(+1)}. \tag{5.134}$$

Indices can be raised and lowered freely with the $\mathfrak{so}(d-1,d-1)$ metric η that is obtained by truncating the $\mathfrak{so}(d,d)$ metric. The superscripts denote the eigenvalues under the $\mathfrak{gl}(1)$ generator D. The commutation relations in this decomposition are

$$\begin{aligned}\left[M^{ij}, M^{kl}\right] &= \eta^{jk} M^{il} + \eta^{ik} M^{jl} - \eta^{jl} M^{ik} + \eta^{il} M^{jk}, \\ \left[M^{ij}, P_k\right] &= 2\delta_k^{[j} P^{i]}, \\ \left[M^{ij}, K_k\right] &= 2\delta_k^{[j} K^{i]}, \\ \left[M^{ij}, D\right] &= 0, \\ \left[D, P^i\right] &= P^i, \\ \left[D, K^i\right] &= -K^i, \\ \left[P^i, K^j\right] &= -2M^{ij} - 2\eta^{ij} D. \end{aligned} \tag{5.135}$$

This is the standard *conformal algebra* for split signature. The indices lie in the range $1, \dots, 2d-2$.

We choose the parabolic subgroup P to be generated by all generators with D-eigenvalue ≤ 0, i.e., P has generators K^i, M^{ij} and D. For this example we take the representation σ of P to be again a character, namely such that $\sigma(e^{\log(a)D}) = |a|^{4s}$. The big cell in the coset space $\bar{P}\backslash G$ is then represented by the unipotent elements

$$n = \exp(x^i P_i) \tag{5.136}$$

and has dimension $2d-2$, corresponding to the Gelfand–Kirillov dimension of the induced representation. The Lie algebra $\mathfrak{so}(d,d)$ is then represented on functions of the $2d-2$ variables x^i by the differential operators

$$\begin{aligned} M_{ij} &= x_i\partial_j - x_j\partial_i, \\ P_i &= \partial_i, \\ D &= -x^i\partial_i + 4s, \\ K_i &= 2x_ix^k\partial_k - x^2\partial_i - 8sx_i. \end{aligned} \tag{5.137}$$

We recognise the familiar form of generators of the conformal group on scalar fields defined in $(2d-2)$-dimensional space-time [177], albeit here in a Wick-rotated form. The generators P_i act as standard translation operators, the generators M^{ij} as geometrical rotation generators, the generator D as a dilatation operator and K_i as a special conformal transformation. The value s that appears in the inducing character is related to the conformal weight of the field. It is also possible to extend this representation to non-scalar fields by including a non-trivial representation of the M^{ij} generator in the induction process; this leads to bosonic and fermionic higher spin analogues of the scalar field representation; see for example [206] for a discussion.

In the representation (5.135), the quadratic Casimir operator evaluates to

$$\begin{aligned} -M^{IJ}M_{IJ} &= -M^{ij}M_{ij} - P^iK_i - K^iP_i + 2D^2 \\ &= 8s(2s-2d+2). \end{aligned} \tag{5.138}$$

In language that will be introduced in Section 6.4, the representation (5.135) is associated with a nilpotent orbit of Bala–Carter type $2A_1$. For $\mathfrak{so}(d,d)$ and $d>4$ there are two such orbits, one of dimension $4(d-1)$ and one of dimension $8d-20$. The degenerate principal series above is attached to the smaller one of dimension $4(d-1)$ in those cases.

The degenerate series representation (5.135) admits a subquotient for the value $s=\frac{d-2}{2}$ that corresponds to the minimal unitary representation of $SO(d,d)$ with Gelfand–Kirillov dimension $2d-3$. In conformal field theory, this representation is known as the *singleton* and plays a distinguished rôle in the context of the *AdS/CFT correspondence*; see for example [55] for a discussion.

5.8.2 Spherical Vector Constructions

The importance of spherical elements in a representation has already been emphasised repeatedly in this chapter. Using explicit realisations of induced representations of the type just discussed, one can try to find spherical elements very concretely. This knowledge can also be used in the construction of Eisenstein series and Fourier expansions, as will become more clear in the subsequent chapters. Here, we content ourselves with providing a few very explicit examples and also exhibiting yet a different method of constructing Eisenstein series that extends an approach taken in [393].

Example 5.37: Spherical vector in $SL(2,\mathbb{R})$ and $SL(2,\mathbb{Q}_p)$ principal series

The compact subgroup $SO(2) \subset SL(2,\mathbb{R})$ is generated by the element

$$e - f = (1 + x^2)\partial_x + 2sx, \tag{5.139}$$

where we have used the explicit differential operators (5.131) in the line model. We are therefore asked to look for a function $f_\infty^\circ(x)$ that is annihilated by this differential operator:

$$(1 + x^2)\partial_x f_\infty^\circ(x) + 2sx f_\infty^\circ(x) = 0. \tag{5.140}$$

This ordinary first-order differential equation is easy to solve, with the result

$$f_\infty^\circ(x) = c(1 + x^2)^{-s}, \tag{5.141}$$

where $c \in \mathbb{R}$ is an integration constant. It is useful to normalise by putting $c = 1$. The definition can also be extended to include the point at infinity $x = \infty$ by letting $f_\infty^\circ(\infty) = 1$.

For $SL(2,\mathbb{Q}_p)$ one has to use the action (5.132) and study sphericality at the group level. Invariance under $SL(2,\mathbb{Z}_p)$ is given by the p-adic spherical vector

$$f_p^\circ(x) = \begin{cases} 1 & \text{if } x \in \mathbb{Z}_p \\ |x|_p^{-2s} & \text{if } x \notin \mathbb{Z}_p \end{cases}. \tag{5.142}$$

This corresponds to the inducing character.

A global spherical vector can then be defined by

$$f^\circ(x) = \left[\prod_{p<\infty} f_p^\circ(x)\right] f_\infty^\circ(x). \tag{5.143}$$

It is also interesting to consider the spherical vector in the Fourier transformed version in the line model. According to Section 2.6, the global Fourier transformation is defined generally by

$$\tilde{f}(u) = \int_{\mathbb{A}} e^{-2\pi i u x} f(x)dx. \tag{5.144}$$

The Fourier transforms of the individual pieces occurring in the global spherical vector (5.143) are

$$\tilde{f}_\infty^\circ(u) = \frac{2\pi^s}{\Gamma(s)}|u|^{s-1/2}K_{s-1/2}(2\pi|u|), \tag{5.145a}$$

$$\tilde{f}_p^\circ(u) = (1 - p^{-2s})\gamma_p(u)\frac{1 - p^{-2s+1}|u|^{2s-1}}{1 - p^{-2s+1}}, \tag{5.145b}$$

where we have used the results from Example 2.20 for $p < \infty$ and the standard real Fourier transform of the Bessel function.

We also determine the spherical vector in the conformal realisation of the orthogonal group that was introduced in Example 5.36.

Example 5.38: Spherical vector for $SO(d,d)$ in conformal realisation

In the big cell, functions in the induced representation (5.135) depend on the $2d-2$ variables x^i that are in the vector representation of $SO(d-1,d-1)$. It is useful to split the $2d-2$ variables as $x^a, x^{\bar{a}}$, where $a = 1,\ldots,d-1$ and $\bar{a} = d,\ldots,2d-2$ and both span a Euclidean space. Sphericality then implies that the spherical vector f_∞° only depends on the Euclidean norms $r = ||x^a||$ and $\bar{r} = ||x^{\bar{a}}||$. This makes it sufficient to consider $f_\infty^\circ(r,\bar{r})$ that satisfies

$$x_a(P^a - K^a)f_\infty^\circ(r,\bar{r}) = 0, \quad x_{\bar{a}}(P^{\bar{a}} + K^{\bar{a}})f_\infty^\circ(r,\bar{r}) = 0. \tag{5.146}$$

The spherical vector is

$$f_\infty^\circ(r,\bar{r}) = \left(1 + r^4 + \bar{r}^4 - 2r^2\bar{r}^2 + 2r^2 + 2\bar{r}^2\right)^s = \left((1 + \bar{r}^2 - r^2)^2 + 4r^2\right)^s. \tag{5.147}$$

It does not depend on d and the argument of the power is manifestly positive. We have again chosen a convenient normalisation. This spherical vector corresponds to the inducing character.

We will now outline an approach to Eisenstein series (or automorphic forms more generally) that was advocated in [393, 394, 528]. The setting is global and requires a group $G(\mathbb{A})$ with a discrete subgroup $G(\mathbb{Q})$ as well as an induced representation $\mathrm{Ind}_P^G\,\sigma$. The approach for constructing an automorphic form is to form the following pairing:

$$\varphi(g) = \langle f_{\mathbb{Q}}, \pi(g) f^\circ \rangle, \tag{5.148}$$

where the objects involved in this expression are the following:

- $\langle\cdot,\cdot\rangle$ is an inner product on the function space $\mathrm{Ind}_P^G\,\sigma$ of an induced representation. In a given model of the coset $P\backslash G$, this is an integral over the product of two functions.
- $f_{\mathbb{Q}}$ represents a distribution on the function space $\mathrm{Ind}_P^G\,\sigma$ that is invariant under the discrete subgroup $G(\mathbb{Q})$: $f_{\mathbb{Q}}(\gamma x) = f_{\mathbb{Q}}(x)$ for all $\gamma \in G(\mathbb{Q})$.
- f° is a spherical vector in the representation space. It is acted upon by the group $G(\mathbb{A})$ by right-translation $\pi(g)$; see (5.1).

The function $\varphi(g)$ thus defined is manifestly right K-invariant and left $G(\mathbb{Q})$-invariant.

In more mundane terms, if x are the coordinates on $P\backslash G$ and $d\mu(x)$ an appropriate invariant measure on the induced function space, then

$$\varphi(g) = \int\limits_{P\backslash G} f_{\mathbb{Q}}(x) f^\circ(xg) d\mu(x). \tag{5.149}$$

It is important here that one uses the full coset space and not only the big cell.

Example 5.39: The $SL(2,\mathbb{A})$ Eisenstein series from spherical vectors

Let us carry this out for $SL(2,\mathbb{A})$ using the ingredients already developed in Example 5.37 above. Additionally, we require the $SL(2,\mathbb{Q})$-invariant distribution. Formally, this is given by [393]

$$f_{\mathbb{Q}}(x) = \sum_{q\in\bar{\mathbb{Q}}} \delta(x-q) \tag{5.150}$$

in terms of the standard δ-distribution. The summation here is over the compactified $\bar{\mathbb{Q}}$ that also includes the point at infinity corresponding to the small cell. The measure in the line model is simply $d\mu(x) = dx$. Plugging this into the definition (5.149) leads to

$$\varphi(g) = \sum_{q\in\bar{\mathbb{Q}}} f^\circ(qg). \tag{5.151}$$

Restricting to $g \in SL(2,\mathbb{R})$ and using (5.143) leads to

$$\varphi(g) = \sum_{q\in\bar{\mathbb{Q}}} \left[\prod_{p<\infty} f_p^\circ(q)\right] \frac{y^s}{((x+q)^2+y^2)^s}. \tag{5.152}$$

Here, we have used the coordinate $z = x+iy$ from Section 4.3 on $SL(2,\mathbb{R})/SO(2)$. We can next split the sum into $q = \frac{m}{n} \in \mathbb{Q}$ (with m and n coprime and $n > 0$) and $q = \infty$ related to the small cell. Using that

$$\prod_{p<\infty} f_p^\circ\left(\frac{m}{n}\right) = n^{-2s}, \tag{5.153}$$

we then obtain

$$\begin{aligned}\varphi(g) &= \sum_{\substack{(m,n)=1\\ n>0}} n^{-2s} \frac{y^s}{((x+m/n)^2+y^2)^s} + y^s \\ &= \sum_{\substack{(m,n)=1\\ n>0}} \frac{y^s}{((nx+m)^2+(ny)^2)^s} + y^s \\ &= \frac{1}{2}\sum_{(m,n)=1} \frac{y^s}{|nz+m|^{2s}} = E(\chi_s, g).\end{aligned} \tag{5.154}$$

This agrees precisely with the standard definition (1.12) of the $SL(2,\mathbb{R})$ Eisenstein series, including all constant terms. The extra term y^s is exactly the contribution from the point at infinity in the line model (the small Bruhat cell) since $f_\infty^\circ(\infty g) = y^s f_\infty^\circ(\infty) = y^s$ by using the transformation of the point at infinity and the value of the spherical vector there.

6

Whittaker Functions and Fourier Coefficients

In this chapter, we analyse the general structure of the Fourier expansion of automorphic forms, with particular emphasis on Eisenstein series and the associated theory of Whittaker functions and coefficients. We will discuss both local and global aspects. As advanced topics we introduce the useful notion of wavefront set [470, 486, 484, 485] and discuss the method of Piatetski-Shapiro and Shalika [516, 570]. General references are [374, 116, 290] and nice discussions are also in [315, 476, 280, 415, 377].

6.1 Preliminary Example: $SL(2,\mathbb{R})$ Whittaker Functions

In Section 1.3, we discussed the Fourier expansion of the non-holomorphic Eisenstein series $E(s,z)$ where $z = x + iy$ is on the upper half-plane $\mathbb{H}$. Invariance under $SL(2,\mathbb{Z})$ implies the periodicity of the series in the real x-direction:

$$E(s, x + 1 + iy) = E(s, x + iy); \tag{6.1}$$

hence we have a Fourier expansion of the form

$$E(s, x + iy) = \sum_{m\in\mathbb{Z}} a_m(y)e^{2\pi imx}, \tag{6.2}$$

where the y-dependent Fourier coefficients $a_m(y)$ can be extracted from the explicit expansion stated in (1.17) and will be derived in detail in Chapter 7 using adelic methods. Let us now reinterpret $E(s,z)$ as a function on $SL(2,\mathbb{R}) = N(\mathbb{R})A(\mathbb{R})K(\mathbb{R})$ according to the prescription in Section 4.3.2. To this end we define

$$\begin{aligned}\varphi_E(g) &= \varphi_E(nak)\\ &= \varphi_E\left(\begin{pmatrix}1 & x\\ & 1\end{pmatrix}\begin{pmatrix}y^{1/2} & \\ & y^{-1/2}\end{pmatrix}\begin{pmatrix}\cos\theta & \sin\theta\\ -\sin\theta & \cos\theta\end{pmatrix}\right)\\ &= E(s, x+iy),\end{aligned} \tag{6.3}$$

where $n \in N(\mathbb{R})$, $a \in A(\mathbb{R})$, $k \in K(\mathbb{R}) = SO(2,\mathbb{R})$. From this point of view, the periodicity (6.1) of $E(s,z)$ in the variable x is equivalent to the invariance of $\varphi_E(g)$ under discrete left-translations: $\varphi_E(ng) = \varphi_E(g)$, $n \in N(\mathbb{Z})$. This follows from the simple calculation for $n = \left(\begin{smallmatrix}1 & 1\\ & 1\end{smallmatrix}\right)$:

$$\begin{aligned}\varphi_E\left(\begin{pmatrix}1 & 1\\ & 1\end{pmatrix}g\right) &= \varphi_E\left(\begin{pmatrix}1 & x+1\\ & 1\end{pmatrix}\begin{pmatrix}y^{1/2} & \\ & y^{-1/2}\end{pmatrix}\begin{pmatrix}\cos\theta & \sin\theta\\ -\sin\theta & \cos\theta\end{pmatrix}\right)\\ &= E(s, x+1+iy),\end{aligned} \tag{6.4}$$

which equals $\varphi_E(g) = E(s, x+iy)$ by left $N(\mathbb{Z})$-invariance.

More generally, we can consider an automorphic form φ on $SL(2,\mathbb{Z})\backslash SL(2,\mathbb{R})$, satisfying

$$\varphi(\gamma g k) = \sigma(k)\varphi(g), \qquad \gamma \in SL(2,\mathbb{Z}),\ k \in K(\mathbb{R}) = SO(2,\mathbb{R}), \tag{6.5}$$

where σ can be a non-trivial finite-dimensional representation of $K(\mathbb{R})$. When σ is non-trivial, the function φ depends on all three coordinates (x, y, θ). When σ is trivial and hence φ independent of k, the function is spherical, i.e., θ-independent.

The automorphy of φ includes invariance under $N(\mathbb{Z})$ and therefore $\varphi(g) = \varphi(x, y, \theta)$ will have a Fourier expansion of the same form as the one for $E(s,z)$, although the precise coefficients will of course be different depending on the choice of φ. To pave the way for higher-rank groups, we now wish to recast this expansion in a form that can be easily generalised.

To this end, let $\psi\colon N(\mathbb{Z})\backslash N(\mathbb{R}) \to U(1)$ be a *unitary multiplicative character* on $N(\mathbb{R})$ which is trivial on $N(\mathbb{Z})$. The space of such characters is $\mathrm{Hom}\,(N(\mathbb{Z})\backslash N(\mathbb{R}), U(1)) \cong \mathbb{Z}$ and we can parametrise the choice of character by a single integer m via

$$\psi\left(\begin{pmatrix}1 & x\\ & 1\end{pmatrix}\right) = e^{2\pi i m x}, \qquad m \in \mathbb{Z},\ x \in \mathbb{R}. \tag{6.6}$$

This is therefore nothing but a set of Fourier modes. If ψ is *non-trivial*, i.e., $m \neq 0$, we say that ψ is *generic*. For higher-rank groups, if a character is non-trivial, it does not necessarily mean that it is also generic. In Definition 6.10, we will extend our concept of this notion to the case of higher-rank groups by introducing a more refined notion of generic versus non-generic (or degenerate) characters.

Then, due to the periodicity of the automorphic form, $\varphi(ng) = \varphi(g)$, $n \in N(\mathbb{Z})$, we can write $\varphi(g)$ as a Fourier expansion along $N(\mathbb{R})$:

$$\varphi(g) = \sum_{\psi \in \mathrm{Hom}(N(\mathbb{Z})\backslash N(\mathbb{R}), U(1))} W_\psi(g), \tag{6.7}$$

where the sum runs over all possible characters ψ and hence over $m \in \mathbb{Z}$. We have also defined the *Fourier–Whittaker coefficient*

$$W_\psi(g) = \int_{N(\mathbb{Z})\backslash N(\mathbb{R})} \varphi(ng)\overline{\psi(n)}dn, \tag{6.8}$$

with dn the *Haar measure* on $N(\mathbb{Z})\backslash N(\mathbb{R})$. The Haar measure is normalised such that $\int_{N(\mathbb{Z})\backslash N(\mathbb{R})} dn = 1$.

Remark 6.1 For brevity we shall often refer to $W_\psi(g)$ simply as a *Whittaker coefficient* or sometimes *Whittaker function*, in place of the more cumbersome *Fourier–Whittaker coefficient.*

The expansion (6.7) is a reformulation of (6.2), as we will now illustrate. By the Iwasawa decomposition $g = nak$ it follows that $W_\psi(g)$ is determined by its restriction to $A(\mathbb{R})$:

$$\begin{aligned} W_\psi(nak) &= \int_{N(\mathbb{Z})\backslash N(\mathbb{R})} \varphi(n'nak)\overline{\psi(n')}dn' \\ &= \sigma(k) \int_{N(\mathbb{Z})\backslash N(\mathbb{R})} \varphi(\tilde{n}a)\overline{\psi(\tilde{n}n^{-1})}d\tilde{n} \\ &= \psi(n)\sigma(k)W_\psi(a), \end{aligned} \tag{6.9}$$

where we used the multiplicativity of ψ as well as the invariance of the Haar measure under translations by $N(\mathbb{R})$. In particular, this allows us to rewrite the expansion in a way that is more akin to the classical form (6.2):

$$\varphi(g) = \sum_\psi W_\psi(ak)\psi(n). \tag{6.10}$$

Note that contrary to standard *harmonic analysis* the function $W_\psi(g)$ is not a numerical coefficient, but also contains explicitly the Fourier variable(s) that one is expanding in. This is made explicit by the factor $\psi(n)$ appearing in (6.9) and (6.10).

By an explicit Iwasawa parametrisation of g in terms of the variables (x, y, θ) as in (6.3) and the character ψ in terms of an integer (6.6), the integral (6.8) takes the more familiar form

$$W_\psi(ak) = W_m(y, \theta) = \int_0^1 \varphi(x, y, \theta)e^{-2\pi imx}dx. \tag{6.11}$$

(For trivial $\sigma(k)$ the integral is independent of θ and equal to $a_m(y)$ of (6.2).)

The general $SL(2,\mathbb{R})$ expansion (6.10) contains two types of terms, corresponding to $m = 0$ ($\psi = 1$) and $m \neq 0$ ($\psi \neq 1$), which are useful to distinguish:

Definition 6.2 (Constant term and Whittaker coefficients for $SL(2,\mathbb{R})$) The sum (6.10) can be split into

$$\varphi(g) = W_1(ak) + \sum_{\psi \neq 1} W_\psi(g), \tag{6.12}$$

where the first term is independent of n and is called the *constant term*. It is defined by

$$W_1(ak) = \int_{N(\mathbb{Z})\backslash N(\mathbb{R})} \varphi(nak)dn, \tag{6.13}$$

and we will sometimes also denote it by $C(ak) \equiv W_1(ak)$. The functions $W_\psi(g)$ for non-trivial characters ψ ($m \neq 0$) are the proper Whittaker coefficients.

Remark 6.3 The functions $W_\psi(g)$ were termed Whittaker functions by Jacquet [374] because they reduce to the classical Whittaker function $W_{k,m}(y)$ for the group $GL(2,\mathbb{R})$. For $SL(2,\mathbb{R})$ they are given basically by modified Bessel functions that arise for the special case $k = 0$; see also Appendix B.2. For higher-rank groups $G(\mathbb{R})$, Whittaker functions define more complicated special functions. For example, for $GL(n,\mathbb{R})$ it was shown in [582, 370] how these generalised Whittaker functions can be obtained as nested integrals over standard Whittaker functions. We will encounter an instance of this for $SL(3,\mathbb{R})$ in Sections 6.2 and 9.8 (see, e.g., (9.108)). The general asymptotics of these functions near a cusp will be discussed in Section 14.2.4, where also their relation to string theory effects is mentioned.

Example 6.4: Fourier and q-expansion of holomorphic Eisenstein series

Consider now the example when $\varphi = \varphi_f$ with $f(z) = E_{2w}(z)$ being a weight $2w$ holomorphic Eisenstein series:

$$E_{2w}(z) = \frac{1}{2} \sum_{(c,d)=1} \frac{1}{(cz+d)^{2w}}, \qquad z \in \mathbb{H},\ w > 1,\ w \in \mathbb{Z}. \tag{6.14}$$

This function is spherical (θ-independent) and has a well-known Fourier expansion (sometimes called q-*expansion*),

$$E_{2w}(z) = 1 + \sum_{m=1}^{\infty} a_m q^m, \qquad q = e^{2\pi i z}, \tag{6.15}$$

where the coefficients are given by

$$a_m = \frac{2}{\zeta(1-2w)}\sigma_{2w-1}(m), \tag{6.16}$$

with $\sigma_{2w-1}(m)$ the sum over positive divisors as in (1.19):

$$\sigma_s(m) = \sum_{d|m} d^s. \tag{6.17}$$

In this case the constant term and Whittaker coefficients are given by

$$\begin{aligned} C(a) &\equiv W_1(a) = 1, \\ W_\psi(z) &\equiv W_\psi(na) = W_m(z) = \frac{2}{\zeta(1-2w)}\sigma_{2w-1}(m)q^m, \qquad m > 0. \end{aligned} \tag{6.18}$$

The coefficients can alternatively be expressed in terms of *Bernoulli numbers*; see for example [24]. Even Bernoulli numbers are defined in (15.19). Notice that the holomorphicity of $E_{2w}(z)$ requires that $W_m(z)$ vanish unless $m > 0$. As mentioned in Section 5.6, this is due to the holomorphic Eisenstein series' E_{2w} being associated with the discrete series representation of $SL(2,\mathbb{R})$. (There are other common normalisations of holomorphic Eisenstein series where the constant term is not given by 1 but by $2\zeta(2w)$.)

For completeness, we also recall the constant terms and Whittaker coefficients for the non-holomorphic Eisenstein series $E(s,z)$ on $SL(2,\mathbb{R})$ from the introduction.

Example 6.5: Fourier expansion of non-holomorphic Eisenstein series

In the case when $\varphi = \varphi_E$, with $E(s,z)$ the non-holomorphic Eisenstein series on $\mathbb{H}$, the constant term $W_1(a)$ and Whittaker coefficient $W_\psi(na)$ will be derived in Chapter 7 with the result

$$\begin{aligned} W_1(a) &= W_1(y) = y^s + \frac{\xi(2s-1)}{\xi(2s)}y^{1-s}, \\ W_\psi(na) &= W_m(x,y) = \frac{2y^{1/2}}{\xi(2s)}|m|^{s-1/2}\sigma_{1-2s}(m)K_{s-1/2}(2\pi|m|y)e^{2\pi imx}, \end{aligned} \tag{6.19}$$

with $m \neq 0$. In contrast to the holomorphic case, the 'constant term' here is not really constant; it is a function on the Cartan torus that is parametrised by the imaginary part y of $z = x + iy$. As we will see below, this is in fact a general feature, namely that the constant term of a spherical automorphic function is only constant with respect to the coordinates along the unipotent radical $N(\mathbb{R})$ of the Borel subgroup $B(\mathbb{R}) \subset G(\mathbb{R})$.

6.2 Fourier Expansions and Unitary Characters

We now turn to the general analysis of Fourier coefficients of automorphic forms on semi-simple Lie groups G, and we also switch to the adelic framework. For this we first require the notion of a unitary character ψ on a unipotent subgroup $U \subset G$ that generalises the Fourier mode $e^{2\pi imx}$ in (6.6). This is discussed in detail in Section 6.2.1. We will then discuss the notion of Fourier expansion for different types of unipotent groups U in the sequel.

6.2.1 Unitary Characters

Definition 6.6 (Global unitary character) Let $U(\mathbb{A})$ be a unipotent subgroup of the adelic group $G(\mathbb{A})$. A *unitary character* on $U(\mathbb{A})$ is a group homomorphism

$$\psi\colon U(\mathbb{A}) \to U(1), \tag{6.20}$$

and we also require it to be trivial on the discrete subgroup $U(\mathbb{Q}) = U(\mathbb{A}) \cap G(\mathbb{Q})$ since we will study in the context of automorphic forms on $G(\mathbb{A})$ that are invariant under the discrete subgroup $G(\mathbb{Q})$. The unitary characters on $U(\mathbb{A})$ that are trivial on $U(\mathbb{Q})$ form the integral points of the *character variety*.

Remark 6.7 Unipotent groups are required if one wants to have non-trivial unitary characters. On a semi-simple group $G(\mathbb{A})$ there are no non-trivial unitary characters.

Definition 6.6 generalises (6.6). As ψ is a group homomorphism to the abelian group $U(1)$, it is trivial on the *commutator subgroup*

$$[U, U] = \left\{u_1 u_2 u_1^{-1} u_2^{-1} \;\middle|\; u_1, u_2 \in U\right\}. \tag{6.21}$$

In other words,

$$\psi([U, U]) = 1, \tag{6.22}$$

such that ψ is sensitive only to the *abelianisation* $[U, U]\backslash U$. We note that $[U, U]$ equals the second member of the derived series of U defined in Section 3.1.1. We will discuss the relevance of the derived series for Fourier expansions in more detail below in Section 6.2.3.

It is convenient to have a more explicit parametrisation of possible unitary characters ψ. To this end we restrict to the case where U is the unipotent of a standard parabolic subgroup $P = LU$ as defined in Section 3.1.3. As always we are working with a fixed choice of split Cartan torus $A \subset G$. Such unipotent groups U can be generated from the product of *one-parameter subgroups*

$$U_\alpha = \{x_\alpha(u_\alpha) = \exp(u_\alpha E_\alpha) \mid u_\alpha \in \mathbb{A}\}, \tag{6.23}$$

with α ranging over the subset $\Delta(\mathfrak{u})$ of positive roots of $\mathfrak{g}$ corresponding to the Lie algebra $\mathfrak{u}$ of U:

$$U = \prod_{\alpha \in \Delta(\mathfrak{u})} U_\alpha. \tag{6.24}$$

The restriction of ψ to any of the one-parameter subgroups U_α then yields a unitary character

$$\psi_\alpha \colon U_\alpha(\mathbb{Q})\backslash U_\alpha(\mathbb{A}) \to U(1). \tag{6.25}$$

Because any one-parameter subgroup U_α is abelian and satisfies the isomorphism

$$U_\alpha(\mathbb{Q})\backslash U_\alpha(\mathbb{A}) \cong \mathbb{Q}\backslash\mathbb{A}\,, \tag{6.26}$$

the unitary character ψ_α can be parametrised by a rational number $m_\alpha \in \mathbb{Q}$ as discussed in Section 2.5 (see also [164, Thm. 5.4.3]), and can be thought of as the global function

$$\psi_\alpha\left(x_\alpha(u_\alpha)\right) = e^{2\pi i m_\alpha u_\alpha}, \tag{6.27}$$

and we will sometimes refer to the m_α as *mode numbers* or *instanton charges* as this is their interpretation in a string theory context; see Chapter 13.

The triviality (6.22) of ψ can then be restated as

$$\psi\left(\prod_{\alpha \in \Delta([\mathfrak{u},\mathfrak{u}])} U_\alpha\right) = 1 \tag{6.28}$$

and the non-trivial unitary characters are therefore sensitive only to the one-parameter subgroups U_α such that α is a 'root' of $\mathfrak{u}$ but not of $[\mathfrak{u},\mathfrak{u}]$. This means that the parametrisation of different unitary characters ψ on U only requires knowledge of the mode numbers m_α for the positive roots α that belong to $\Delta(\mathfrak{u})$ but not to $\Delta([\mathfrak{u},\mathfrak{u}])$. We define these roots as

$$\Delta^{(1)}(\mathfrak{u}) := \Delta(\mathfrak{u}) \setminus \Delta([\mathfrak{u},\mathfrak{u}])\,. \tag{6.29}$$

Remark 6.8 The notation $\Delta^{(1)}(\mathfrak{u})$ indicates that these are the 'roots' of the abelianisation $[U,U]\backslash U$ of the degree-one piece $U = U^{(1)}$ in the *derived series* of U, defined in Section 3.1.1. See Section 6.2.3 for a more detailed discussion of the relevance of the derived series of U for Fourier expansions.

The above considerations lead to:

Proposition 6.9 (Parametrisation of unitary characters) *Let $U(\mathbb{A})$ be a unipotent subgroup of $G(\mathbb{A})$. Unitary characters $\psi\colon U(\mathbb{Q})\backslash U(\mathbb{A}) \to U(1)$ can be parametrised uniquely by a set of mode numbers $\left\{m_\alpha \in \mathbb{Q} \,\middle|\, \alpha \in \Delta^{(1)}(\mathfrak{u})\right\}$. The unitary character is then given by*

$$\psi\left(\prod_{\alpha\in\Delta^{(1)}(\mathfrak{u})} x_\alpha(u_\alpha)\right) = \exp\left(2\pi i \sum_{\alpha\in\Delta^{(1)}(\mathfrak{u})} m_\alpha u_\alpha\right). \tag{6.30}$$

It factorises into local places as in (2.75).

Proof The triviality of ψ on the commutator subgroup $[U, U]$ shows that it suffices to define ψ on the abelianisation that is constructed from the one-parameter subgroups U_α with $\alpha \in \Delta^{(1)}(\mathfrak{u})$, for which the characters were determined in (6.27) above. The group homomorphism property of ψ then yields the proposition. □

The following notions will be important in the sequel.

Definition 6.10 (Generic and degenerate characters) For a unipotent subgroup $U(\mathbb{A}) \subset G(\mathbb{A})$ let $\psi\colon U(\mathbb{Q})\backslash U(\mathbb{A}) \to U(1)$ be a *global* character (6.30).

(*i*) ψ is called *generic* if $m_\alpha \neq 0$ for all $\alpha \in \Delta^{(1)}(\mathfrak{u})$, i.e., if the character is non-trivial on each one-parameter subgroup $U_\alpha(\mathbb{A})$ for $\alpha \in \Delta^{(1)}(\mathfrak{u})$.

(*ii*) If $m_\alpha = 0$ for all $\alpha \in \Delta^{(1)}(\mathfrak{u})$, the character ψ is called *trivial*.

(*iii*) Furthermore, if $m_\alpha \neq 0$ for at least one, but not all, $\alpha \in \Delta^{(1)}(\mathfrak{u})$, the character ψ is called *non-generic* or *degenerate*.

Example 6.11: Unitary characters on the maximal unipotent of $SL(n, \mathbb{A})$

Consider the case $G = SL(n, \mathbb{A})$ and $U(\mathbb{A}) = N(\mathbb{A})$ to be the (maximal) unipotent subgroup of the Borel subgroup $B(\mathbb{A})$, implying $\mathfrak{n} = \mathfrak{u}$. The set $\Delta(\mathfrak{n})$ is given by all positive roots Δ_+ of $\mathfrak{sl}(n)$ and the set $\Delta^{(1)}(\mathfrak{n})$ equals the $(n-1)$ simple roots $\Pi \subset \Delta_+$. We can write elements of $n \in N$ as $(n \times n)$ matrices of the form

$$n = \begin{pmatrix} 1 & u_1 & * & * & \cdots \\ & 1 & u_2 & * & \cdots \\ & & \cdots & & \\ & & & 1 & u_{n-1} \\ & & & & 1 \end{pmatrix}. \tag{6.31}$$

The starred entries are of no relevance for the discussion of unitary characters as they are associated with the commutator subgroup $[N, N]$. A character ψ on N is determined by $n-1$ rational numbers m_i $(i = 1, \ldots, n-1)$ such that

$$\psi(n) = \exp\left(2\pi i \sum_{i=1}^{n-1} m_i u_i\right). \tag{6.32}$$

The character ψ is generic when all $m_i \neq 0$. It is degenerate when some m_i vanish and then it does not depend on the corresponding one-parameter subgroups.

We recall from Section 2.5 that a global unitary character ψ_α on $\mathbb{Q}\backslash\mathbb{A}$ as in (6.27) factorises as

$$\psi_\alpha = \prod_{p\leq\infty} \psi_{\alpha,p}, \tag{6.33}$$

where for $p < \infty$

$$\psi_{\alpha,p} \colon U(\mathbb{Z}_p)\backslash U(\mathbb{Q}_p) \to U(1), \qquad \psi_{\alpha,p}(x_\alpha(u)) = e^{-2\pi i [m_\alpha u]} \tag{6.34}$$

in terms of the fractional part (2.28) of a p-adic number, and for $p = \infty$

$$\psi_{\alpha,\infty} \colon U(\mathbb{Z})\backslash U(\mathbb{R}) \to U(1), \qquad \psi_{\alpha,p}(x_\alpha(u)) = e^{2\pi i m_\alpha u}. \tag{6.35}$$

This factorisation extends to characters ψ on unipotent groups U:

$$\psi = \prod_{p\leq\infty} \psi_p. \tag{6.36}$$

Definition 6.10 extends to all local characters ψ_p. Moreover, we have the following notion:

Definition 6.12 (Unramified unitary character) A generic *local* character ψ_p for $p < \infty$ is called *unramified* if for all $\alpha \in \Delta^{(0)}(\mathfrak{u})$ one has

$$\psi_{\alpha,p}\left(e^{uE_\alpha}\right) = e^{-2\pi i [u]}, \qquad u \in \mathbb{Q}_p. \tag{6.37}$$

Equivalently, this means that all instanton charges $|m_\alpha|_p = 1$ in (6.30). We call a global character unramified if $m_\alpha = 1$ for all α.

6.2.2 General Fourier Coefficients versus Whittaker Coefficients

Now that we have the Fourier modes in terms of characters ψ on unipotent subgroups U, it is possible to define Fourier coefficients of automorphic forms.

Definition 6.13 (Fourier coefficient) Let φ be an automorphic form on $G(\mathbb{A})$, i.e., an element of the space $\mathcal{A}(G(\mathbb{Q})\backslash G(\mathbb{A}))$, and $U(\mathbb{A})$ be a unipotent subgroup of $G(\mathbb{A})$. The *Fourier coefficient* of φ with respect to a unitary character ψ on U is given by

$$F_\psi(\varphi, g) = \int\limits_{U(\mathbb{Q})\backslash U(\mathbb{A})} \varphi(ug)\overline{\psi(u)}du, \tag{6.38}$$

where du denotes the invariant Haar measure on U.

Remark 6.14 The Fourier coefficient F_ψ can be viewed either as a function $F_\psi(g)$ on $G(\mathbb{A})$ for fixed φ, or as a functional $F_\psi(\varphi)$ on $\mathcal{A}(G(\mathbb{Q})\backslash G(\mathbb{A}))$. When it is clear from the context which fixed φ is meant, we may write simply $F_\psi(g)$ for conciseness.

A short calculation similar to (6.9) shows that Fourier coefficients satisfy

$$F_\psi(\varphi, ug) = \psi(u)F_\psi(\varphi, g) \qquad \text{for all } u \in U. \tag{6.39}$$

We make the additional definitions for the case $U(\mathbb{A}) = N(\mathbb{A})$.

Definition 6.15 (Whittaker coefficient) Let φ be an automorphic form on $G(\mathbb{A})$, $N(\mathbb{A})$ be the maximal unipotent subgroup of a fixed Borel $B(\mathbb{A})$ and ψ be a unitary character on $N(\mathbb{A})$.

(i) The integral

$$W_\psi(\varphi, g) = \int_{N(\mathbb{Q})\backslash N(\mathbb{A})} \varphi(ug)\overline{\psi(u)}du \tag{6.40}$$

is called the *Whittaker coefficient* of φ with respect to ψ.

(ii) If φ is $K_\mathbb{A}$ invariant, the Whittaker coefficient $W_\psi(\varphi, g)$ is right-invariant under $K_\mathbb{A}$ and the Whittaker coefficient is then called *spherical*. We denote it by $W^\circ_\psi(\varphi, g)$. The spherical Whittaker coefficient is completely determined by its values on the Cartan torus $A(\mathbb{A})$: writing $g = nak$ in Iwasawa decomposed form, one has

$$W^\circ_\psi(\varphi, nak) = \psi(n)W^\circ_\psi(\varphi, a). \tag{6.41}$$

This is the case for Eisenstein series as defined in Definition 4.24.

Remark 6.16 Even though Definition 6.15 is a special case of Definition 6.13, it is useful to distinguish this case notationally. Throughout this work, we will denote Whittaker coefficients (i.e., Fourier coefficients along the maximal unipotent N) by W_ψ and reserve the notation F_ψ for the case when the unipotent U is different from N. Whittaker coefficients will be studied in detail in Chapter 9 for Eisenstein series.

We note that if U is the unipotent of some standard parabolic subgroup $P = LU$ and φ is $K_\mathbb{A}$-invariant, the general Fourier coefficient $F_\psi(\varphi, g)$ is determined by its values on the Levi subgroup L and one could define a spherical Fourier coefficient F°_ψ, but we will not make use of this notion.

Definition 6.17 (Constant term)

(*i*) The Fourier coefficient of an automorphic form φ with respect to the trivial character $\psi = \mathbb{1}$ on U is called the *constant term along* U:

$$C_U(\varphi, g) = \int\limits_{U(\mathbb{Q})\backslash U(\mathbb{A})} \varphi(ug)du. \tag{6.42}$$

It is independent of $u \in U$: $C_U(\varphi, ug) = C_U(\varphi, g)$.

(*ii*) For the case $U = N$, we will call this simply the *constant term* and denote it by

$$C(\varphi, g) \equiv C_N(\varphi, g) = \int\limits_{N(\mathbb{Q})\backslash N(\mathbb{A})} \varphi(ng)dn. \tag{6.43}$$

If φ is spherical, the constant term is a function only of the Cartan torus $A(\mathbb{A})$ by Iwasawa decomposition $C(\varphi, nak) = C(\varphi, a)$.

6.2.3 Abelian versus Non-abelian Fourier Expansions

In the $SL(2)$ example of Section 6.1, the Whittaker coefficients W_ψ were used in (6.7) to give a *complete* Fourier expansion of an automorphic form φ by summing over all possible unitary characters ψ.

It is a natural question as to how this carries over to higher-rank groups $G(\mathbb{A})$. In view of Proposition 6.9, we can already anticipate that the Fourier expansion with unitary characters ψ on a unipotent group U will in general be incomplete since the characters ψ only depend on the abelianisation $[U, U]\backslash U$; see (6.22). For $SL(2)$ the (maximal) unipotent group N is abelian and we did not have to consider this subtlety. The general statement is:

Proposition 6.18 (Partial Fourier sum) *Let $U(\mathbb{A})$ be a unipotent subgroup of $G(\mathbb{A})$ and φ be an automorphic form on $G(\mathbb{A})$. Then the* partial sum of Fourier coefficients *over all unitary characters ψ on U yields*

$$\sum_{\psi} F_\psi(\varphi, g) = \int\limits_{[U,U](\mathbb{Q})\backslash[U,U](\mathbb{A})} \varphi(ug)du. \tag{6.44}$$

In other words, the sum of the Fourier coefficients reconstitutes only the average of the automorphic form over the commutator subgroup $[U, U]$. If U is abelian, the Fourier expansion is complete.

Proof See [476]. □

In order to obtain a complete Fourier expansion when the unipotent U is non-abelian, one has to consider the *derived series* of U (see also Section 3.1.1):

$$U^{(i+1)} = [U^{(i)}, U^{(i)}], \qquad U^{(0)} = U. \tag{6.45}$$

Since U is unipotent, the derived series trivialises after finitely many steps: $U^{(i_0)} = \{\mathbb{1}\}$ for some $i_0 \geq 1$ and we assume i_0 to be the smallest integer for which $U^{(i_0)} = \{\mathbb{1}\}$. If U is abelian, one has $i_0 = 1$. The successive quotients $U^{(i+1)}\backslash U^{(i)}$ are the *abelianisations* of the unipotent groups $U^{(i)}$ for any integer $i \geq 0$. A unitary character $\psi^{(i)}$ on $U^{(i)}$ is trivial on $U^{(i+1)}$. One can define Fourier coefficients for any of the $U^{(i)}$ by the same formula as in Definition 6.13:

$$F_{\psi^{(i)}}(\varphi, g) = \int\limits_{U^{(i)}(\mathbb{Q})\backslash U^{(i)}(\mathbb{A})} \varphi(ug)\overline{\psi^{(i)}(u)}du. \tag{6.46}$$

As an immediate analogue of Proposition 6.18 one has that

$$\sum_{\psi^{(i)}} F_{\psi^{(i)}}(\varphi, g) = \int\limits_{U^{(i+1)}(\mathbb{Q})\backslash U^{(i+1)}(\mathbb{A})} \varphi(ug)du. \tag{6.47}$$

We observe that the right-hand side is nothing but the constant term of φ along $U^{(i+1)}$, corresponding to $\psi^{(i+1)} = 1$. It is therefore natural that the complete *non-abelian Fourier expansion* of φ along U is given by

$$\varphi(g) = C_U(\varphi, g) + \sum_{\psi^{(0)}\neq 1} F_{\psi^{(0)}}(\varphi, g) + \sum_{\psi^{(1)}\neq 1} F_{\psi^{(1)}}(\varphi, g) + \cdots + \sum_{\psi^{(i_0-1)}\neq 1} F_{\psi^{(i_0-1)}}(\varphi, g). \tag{6.48}$$

The trivial character $\psi^{(i)} = 1$ is always excluded because the sum of the preceding terms reconstitutes the constant term along $U^{(i)}$ by (6.47). Note that unitary characters $\psi^{(0)}$ are characters on $U^{(0)} = U$ and therefore equal the unitary characters we have been discussing in Definition 6.13. We will sometimes refer to the Fourier coefficients in (6.48) associated with $U^{(i)}$ and $i \geq 1$ as *non-abelian Fourier coefficients* and to those associated with $U^{(0)} = U$ as the *abelian Fourier coefficient*.

The same structure of the expansion and terminology arises for the case when the unipotent U is given by the maximal unipotent N. Then we have

$$\varphi(g) = \underbrace{C(g)}_{\text{constant term}} + \underbrace{\sum_{\psi^{(0)}\neq 1} W_{\psi^{(0)}}(g)}_{\text{abelian term}} + \underbrace{\sum_{\psi^{(1)}\neq 1} W_{\psi^{(1)}}(g)}_{\text{non-abelian term}} + \cdots, \tag{6.49}$$

where we have suppressed the fixed automorphic function φ on the right-hand side.

Remark 6.19 Our main interest in this work lies with the abelian Whittaker coefficients $W_{\psi^{(0)}}$, and we will discuss them in more detail in the following sections and in particular in Chapter 9. Non-abelian Fourier expansions have been carried out in detail for $SL(3, \mathbb{R})$ in [116, 605, 497, 525], and this will be reviewed in Section 9.8. Non-abelian Fourier expansions for the non-split real group $SU(2,1)$ can be found in Chapter 18 and some further comments on the non-abelian coefficients will be offered in Chapter 16.

6.2.4 Restriction of Fourier Coefficients

We will in this section show that the Fourier coefficients with respect to real groups $U(\mathbb{R})$ can be recovered by restricting the arguments of the adelic counterparts to $g = (g_\infty; \mathbb{1}, \mathbb{1}, \ldots)$.

Let U be a unipotent subgroup of G and ψ be a character on $U(\mathbb{Q})\backslash U(\mathbb{A})$ and be parametrised as in Proposition 6.9, but with integer m_α. Let also $u \in U(\mathbb{A})$ and $k \in U(\hat{\mathbb{Z}}) = \prod_{p<\infty} U(\mathbb{Z}_p)$ parametrised by

$$k = \prod_{\alpha \in \Delta^{(1)}(\mathfrak{u})} x_\alpha(k_\alpha), \qquad k_\alpha \in \hat{\mathbb{Z}}, \tag{6.50}$$

where $\Delta^{(1)}(\mathfrak{u}) = \Delta(\mathfrak{u})\backslash\Delta([\mathfrak{u},\mathfrak{u}])$. Then we have that $\psi(uk) = \psi(u)\psi(k)$ and

$$\psi(k) = \psi_\infty(1) \prod_{p<\infty} \exp\Big(-2\pi i \sum_{\alpha \in \Delta^{(1)}(\mathfrak{u})} [m_\alpha k_{\alpha,p}]_p\Big) = 1, \tag{6.51}$$

since when m_α is an integer, $m_\alpha k_{\alpha,p} \in \mathbb{Z}_p$, implying that $[m_\alpha k_{\alpha,p}]_p = 0$. The character ψ is thus a function on $U(\mathbb{Q})\backslash U(\mathbb{A})/U(\hat{\mathbb{Z}})$ which we can see as the adelisation of the character ψ_∞ on $U(\mathbb{Z})\backslash U(\mathbb{R})$ using the map in Remark 3.13.

Proposition 6.20 (Restriction of Fourier coefficients)
Let $f\colon G(\mathbb{Z})\backslash G(\mathbb{R}) \to \mathbb{C}$ with adelisation $\varphi = f \circ \phi^{-1}\colon G(\mathbb{Q})\backslash G(\mathbb{A})/K_f$, where ϕ is given in Theorem 3.10, and let $\psi = \prod_{p\le\infty} \psi_p$ be a unitary multiplicative character on $U(\mathbb{Q})\backslash U(\mathbb{A})$ parametrised by $m_\alpha \in \mathbb{Z}$ for $\alpha \in \Delta^{(1)}(\mathfrak{u})$ as described in Proposition 6.9. Then, for $g_\infty \in G(\mathbb{R})$,

$$\int_{U(\mathbb{Q})\backslash U(\mathbb{A})} \varphi(u(g_\infty;\mathbb{1}))\overline{\psi(u)}\, du = \int_{U(\mathbb{Z})\backslash U(\mathbb{R})} f(u_\infty g_\infty)\overline{\psi_\infty(u_\infty)}\, du_\infty, \tag{6.52}$$

which is the Fourier coefficient on $U(\mathbb{R})$.

Proof We have that

$$\begin{aligned}
&\int_{U(\mathbb{Q})\backslash U(\mathbb{A})} \varphi(u(g_\infty;\mathbb{1}))\overline{\psi(u)}\, du \\
&\quad= \int_{U(\mathbb{Q})\backslash U(\mathbb{A})/U(\hat{\mathbb{Z}})} du \int_{U(\hat{\mathbb{Z}})} dk\ \varphi(uk(g_\infty;\mathbb{1})\overline{\psi(uk)} \\
&\quad= \int_{U(\mathbb{Q})\backslash U(\mathbb{A})/U(\hat{\mathbb{Z}})} du\ \varphi(u(g_\infty;\mathbb{1})\overline{\psi(u)} \int_{U(\hat{\mathbb{Z}})} dk,
\end{aligned} \tag{6.53}$$

where we have used the fact that φ and ψ are invariant under right-translations by $k \in U(\hat{\mathbb{Z}})$ as seen in (6.51) for the latter.

With the homeomorphism $\Phi\colon U(\mathbb{Z})\backslash U(\mathbb{R}) \to U(\mathbb{Q})\backslash U(\mathbb{A})/U(\hat{\mathbb{Z}})$ defined by $U(\mathbb{Z})x_\infty \mapsto U(\mathbb{Q})(x_\infty;\mathbb{1})U(\hat{\mathbb{Z}})$ from Remark 3.13, we have that, for $x_\infty \in U(\mathbb{R})$,

$$\begin{aligned} &G(\mathbb{Q})\Phi(U(\mathbb{Z})u_\infty)(g_\infty;\mathbb{1})K_f \\ &\quad = G(\mathbb{Q})\big(U(\mathbb{Q})(u_\infty;\mathbb{1})U(\hat{\mathbb{Z}})\big)(g_\infty;\mathbb{1})K_f = G(\mathbb{Q})(u_\infty g_\infty;\mathbb{1})K_f. \end{aligned} \tag{6.54}$$

Recalling that φ is a function on $G(\mathbb{Q})\backslash G(\mathbb{A})/K_f$, and using the homeomorphism from Remark 3.13, the above integral then becomes

$$\int\limits_{U(\mathbb{Z})\backslash U(\mathbb{R})} \varphi((u_\infty g_\infty;\mathbb{1}))\overline{\psi((u_\infty;\mathbb{1}))}\, du_\infty = \int\limits_{U(\mathbb{Z})\backslash U(\mathbb{R})} f(u_\infty g_\infty)\overline{\psi_\infty(u_\infty)}\, du_\infty\,. \tag{6.55}$$

□

6.3 Induced Representations and Whittaker Models

We now specialise to the case when the automorphic form $\varphi \in \mathcal{A}(G(\mathbb{Q})\backslash G(\mathbb{A}))$ is an Eisenstein series

$$E(\mathfrak{f}_\lambda, g) = \sum_{\gamma\in B(\mathbb{Q})\backslash G(\mathbb{Q})} \mathfrak{f}_\lambda(\gamma g), \qquad g \in G(\mathbb{A}), \tag{6.56}$$

constructed from a standard section $\mathfrak{f}_\lambda$ of the (in general, non-unitary) principal series $\mathrm{Ind}_{B(\mathbb{A})}^{G(\mathbb{A})}\chi$; see Section 5.3. Here $\chi = e^{\langle\lambda+\rho|H\rangle}$ is the inducing character on the Borel subgroup $B(\mathbb{A}) = N(\mathbb{A})A(\mathbb{A})$, as defined in Section 4.5.1. For the constant term of $E(\mathfrak{f}_\lambda, g)$ one can derive an explicit formula; this is done in great detail for $SL(2,\mathbb{A})$ in Chapter 7. The formula for arbitrary split groups $G(\mathbb{A})$, due to Langlands, will be derived in Chapter 8. Here we are interested in the representation-theoretic properties of the non-constant (abelian) Whittaker coefficients of $E(\mathfrak{f}_\lambda, g)$.

6.3.1 Global Considerations

For a character ψ on $N(\mathbb{Q})\backslash N(\mathbb{A})$ the abelian coefficients of $E(\mathfrak{f}_\lambda, g)$ are given by the Whittaker coefficient W_ψ of the type (6.40). Plugging $E(\mathfrak{f}_\lambda, g)$ from (6.56) into (6.40) and exchanging the order of summation and integration we obtain the formula

$$W_\psi(\mathfrak{f}_\lambda, g) = \sum_{\gamma\in B(\mathbb{Q})\backslash G(\mathbb{Q})} \int_{N(\mathbb{Q})\backslash N(\mathbb{A})} \mathfrak{f}_\lambda(\gamma n g)\overline{\psi(n)}dn. \tag{6.57}$$

Representation-theoretically, $W_\psi(\mathfrak{f}_\lambda, g)$ belongs to the *induced representation*

$$\mathrm{Ind}_{N(\mathbb{A})}^{G(\mathbb{A})}\psi = \left\{W_\psi : G(\mathbb{A}) \to \mathbb{C} \,\middle|\, W_\psi(ng) = \psi(n)W_\psi(g),\ n \in N(\mathbb{A})\right\}. \tag{6.58}$$

Equation (6.57) thus gives an embedding

$$I(\lambda) = \mathrm{Ind}_{B(\mathbb{A})}^{G(\mathbb{A})}\chi \hookrightarrow \mathrm{Ind}_{N(\mathbb{A})}^{G(\mathbb{A})}\psi. \tag{6.59}$$

Definition 6.21 (Whittaker model) The space

$$Wh_\psi(\lambda) = \{W_\psi(\mathfrak{f}_\lambda, \cdot) \mid \mathfrak{f}_\lambda \in I(\lambda)\} \subset \mathrm{Ind}_{N(\mathbb{A})}^{G(\mathbb{A})}\psi \tag{6.60}$$

is called a *Whittaker model* of $I(\lambda)$; its elements $W_\psi(\mathfrak{f}_\lambda, g)$ are functions on $G(\mathbb{A})$ and are called *Whittaker functions* or *Whittaker coefficients* (see Remark 6.14). The associated map

$$\mathfrak{f}_\lambda \mapsto W_\psi(\mathfrak{f}_\lambda) \tag{6.61}$$

is an *intertwiner* between the principal series $I(\lambda)$ and its Whittaker model $Wh_\psi(\lambda)$.

Remark 6.22 An important result about Whittaker models for some groups is their uniqueness: for each fixed section $\mathfrak{f}_\lambda \in I(\lambda)$ and generic character ψ there exists, up to scale, a *unique* Whittaker function $W_\psi(\mathfrak{f}_\lambda)$ (see, e.g., [116, 140]). This property has been used to show the so-called *multiplicity-one theorem* for $GL(n)$, which states that any irreducible admissible representation of $G(\mathbb{A})$ appears at most once in the decomposition in the space of cusp forms with fixed central character. The multiplicity-one theorem for $GL(n)$ was originally shown locally for archimedean and non-archimedean fields in [375, 572], and we note that it fails for $SL(n)$ when $n > 2$ [62] even though uniqueness of the Whittaker model continues to hold.

In Chapter 9 we will show that, *for generic* ψ, the Whittaker coefficient can be written as a single integral rather than as a sum. The argument relies on the Bruhat decomposition of $G(\mathbb{Q})$, which allows one to trade the sum over $\gamma \in B(\mathbb{Q})\backslash G(\mathbb{Q})$ for a sum over the Weyl group $\mathcal{W}(\mathfrak{g})$. The end result is that the Whittaker coefficient for generic ψ may be written as

$$W_\psi(\mathfrak{f}_\lambda, g) = \int\limits_{N(\mathbb{A})} \mathfrak{f}_\lambda(w_{\mathrm{long}} n g)\overline{\psi(n)}dn. \tag{6.62}$$

The sum over γ has reduced to a single contribution represented by w_{long}, the longest element in the Weyl group $\mathcal{W}(\mathfrak{g})$ (for the details see Chapter 9). Note also that the integral is now over a non-compact domain.

Remark 6.23 The expression (6.62) is sometimes known as a *Jacquet–Whittaker integral* [374]. We will often refer to it simply as the (global) Whittaker function.

6.3.2 Local Considerations

Recall from Section 5.2 that by Flath's tensor product theorem the principal series decomposes into a product over all places [211],

$$\mathrm{Ind}_{B(\mathbb{A})}^{G(\mathbb{A})}\chi = \bigotimes_{p\leq\infty}\mathrm{Ind}_{B(\mathbb{Q}_p)}^{G(\mathbb{Q}_p)}\chi_p, \tag{6.63}$$

and, for generic ψ, we have a similar decomposition for $\mathrm{Ind}_{N(\mathbb{A})}^{G(\mathbb{A})}\psi$:

$$\mathrm{Ind}_{N(\mathbb{A})}^{G(\mathbb{A})}\psi = \bigotimes_{p\leq\infty}\mathrm{Ind}_{N(\mathbb{Q}_p)}^{G(\mathbb{Q}_p)}\psi_p. \tag{6.64}$$

To each standard section $\mathfrak{f}_{\lambda,p} \in \mathrm{Ind}_{B(\mathbb{Q}_p)}^{G(\mathbb{Q}_p)}\chi_p$ we then have a local (p-adic) Whittaker function

$$W_{\psi_p}(\mathfrak{f}_{\lambda,p}, g) = \int_{N(\mathbb{Q}_p)} \mathfrak{f}_{\lambda,p}(w_{\mathrm{long}}ng)\overline{\psi_p(n)}dn, \quad g \in G(\mathbb{Q}_p) \tag{6.65}$$

and the global Whittaker model $Wh_\psi(\chi)$ splits accordingly

$$Wh_\psi(\chi) = \bigotimes_{p\leq\infty} Wh_{\psi_p}(\chi_p). \tag{6.66}$$

In Chapter 9 we will derive an explicit formula (*Casselman–Shalika formula*) for the p-adic Whittaker function $W_{\psi_p}(\mathfrak{f}_{\lambda,p}, g)$, $p < \infty$, in the special case when $W_{\psi_p}(\mathfrak{f}_{\lambda,p}, g)$ is *spherical* and ψ unramified, notions that were defined in Definitions 6.15 and 6.12, respectively.

For generic characters ψ, the global Whittaker function can then be recovered as an Euler product over all places:

$$W_\psi(\mathfrak{f}_\lambda, g) = \prod_{p\leq\infty} W_{\psi_p}(\mathfrak{f}_{\lambda,p}, g_p), \qquad g \in G(\mathbb{A}),\ g_p \in G(\mathbb{Q}_p). \tag{6.67}$$

It is sometimes useful to separate the finite places $p < \infty$ from the infinite place $p = \infty$ and make the following definition:

Definition 6.24 (Finite Whittaker function) Consider the Whittaker function W_ψ^{fin} obtained by taking the product over all the *finite places*:

$$W_\psi^{\mathrm{fin}}(\mathfrak{f}_\lambda^{\mathrm{fin}}, g_f) = \prod_{p<\infty} W_{\psi_p}(\mathfrak{f}_{\lambda,p}, g_p), \qquad g_f = (1; g_2, g_3, \dots) \in G(\mathbb{A}_f). \tag{6.68}$$

We call this the *finite Whittaker function.*

Remark 6.25 The finite Whittaker function plays an important rôle in string theory, where it contributes to the *instanton measure*, as we illustrate in Example 6.28 below. The physical interpretation of the divisor sum as an instanton measure will be elaborated upon in Chapter 13; see in particular Section 13.6.3.

6.3.3 Spherical Whittaker functions

Here we will introduce a special class of Whittaker functions which are spherical in an appropriate sense. Assume that $\mathrm{Ind}_{B(\mathbb{A})}^{G(\mathbb{A})}\chi$ is *unramified*, i.e., for almost all places p the local component $\mathrm{Ind}_{B(\mathbb{Q}_p)}^{G(\mathbb{Q}_p)}\chi_p$ is *spherical*. This implies that there exists a *unique* (up to normalisation) section $\mathsf{f}^\circ_{\lambda,p} \in \mathrm{Ind}_{B(\mathbb{Q}_p)}^{G(\mathbb{Q}_p)}\chi_p$ that satisfies

$$\mathsf{f}^\circ_{\lambda,p}(bk) = \chi_p(b), \qquad \mathsf{f}^\circ_{\lambda,p}(k) = \mathsf{f}^\circ_{\lambda,p}(1) = 1, \tag{6.69}$$

where $b \in B(\mathbb{Q}_p)$ and $k \in G(\mathbb{Z}_p)$.

Definition 6.26 (Spherical vector) We call $\mathsf{f}^\circ_{\lambda,p} \in \mathrm{Ind}_{B(\mathbb{Q}_p)}^{G(\mathbb{Q}_p)}\chi_p$, defined by (6.69), a *spherical vector*.

Definition 6.27 (Spherical Whittaker function) To each spherical vector $\mathsf{f}^\circ_{\lambda,p}$ and generic character ψ_p we can associate a *spherical Whittaker function* $W^\circ_{\psi_p} \in Wh_{\psi_p}(\chi_p)$, defined by

$$W^\circ_{\psi_p}(\lambda, g) = \int_{N(\mathbb{Q}_p)} \mathsf{f}^\circ_{\lambda,p}(w_{\text{long}}ng)\overline{\psi_p(n)}dn, \qquad g \in G(\mathbb{Q}_p). \tag{6.70}$$

As before, the spherical Whittaker function satisfies the relation

$$W^\circ_{\psi_p}(\lambda, nak) = \psi_p(n)W^\circ_{\psi_p}(\lambda, a), \tag{6.71}$$

where $n \in N(\mathbb{Q}_p)$, $a \in A(\mathbb{Q}_p)$, $k \in G(\mathbb{Z}_p)$. This again implies that $W^\circ_{\psi_p}(\lambda, g)$ is completely determined by its restriction to the Cartan torus $A(\mathbb{Q}_p)$, where it equals

$$W^\circ_{\psi_p}(\lambda, a) = \int_{N(\mathbb{Q}_p)} \mathsf{f}^\circ_{\lambda,p}(w_{\text{long}}na)\overline{\psi_p(n)}dn. \tag{6.72}$$

Example 6.28: Spherical Whittaker function for $SL(2,\mathbb{A})$

We now illustrate the discussion for the Eisenstein series $E(s, g)$ on $SL(2,\mathbb{A})$. The results below are all derived in Section 7.3. Recall from Example 5.23 that the Eisenstein series is obtained by choosing the standard section f_λ to be the spherical vector $\mathsf{f}^\circ_\lambda = \mathsf{f}^\circ_s$, such that

$$E(\mathsf{f}^\circ_s, g) = \sum_{\gamma \in B(\mathbb{Q})\backslash SL(2,\mathbb{Q})} \mathsf{f}^\circ_s(\gamma g) = \sum_{\gamma \in B(\mathbb{Q})\backslash SL(2,\mathbb{Q})} \chi_s(\gamma na), \tag{6.73}$$

where $\chi_s = e^{\langle 2s\Lambda|H\rangle}$, $\Lambda = \alpha/2$ with α the simple root of $\mathfrak{sl}(2,\mathbb{R})$. The local spherical Whittaker function (6.72) is

$$W^\circ_{\psi_p}(s, a) = \int_{N(\mathbb{Q}_p)} \chi_s(w_{\text{long}}na)\overline{\psi(n)}dn, \qquad a \in A(\mathbb{Q}_p). \tag{6.74}$$

As will be shown in detail in Section 7.3, the integral equals

$$W^{\circ}_{\psi_\infty}(s,y) = \frac{2\pi^s}{\Gamma(s)} y^{1/2}|m|^{s-1/2}K_{s-1/2}(2\pi|m|y), \tag{6.75}$$

at the archimedean place $p = \infty$ (see (7.73)). Here, $m \in \mathbb{Z}^\times$, and the Cartan torus $A(\mathbb{R})$ is given by

$$\begin{pmatrix} y^{1/2} & \\ & y^{-1/2} \end{pmatrix}, \qquad y \in \mathbb{R}_{>0}. \tag{6.76}$$

At the non-archimedean places, the integral becomes (see (7.77))

$$W^{\circ}_{\psi_p}(s,v) = |v|_p^{-2s+2}\gamma_p(mv^2)(1-p^{-2s})\frac{1-p^{-2s+1}|mv^2|_p^{2s-1}}{1-p^{-2s+1}}, \tag{6.77}$$

with $p < \infty$ and $m \in \mathbb{Q}^\times$, and we have parametrised the torus $A(\mathbb{Q}_p)$ by

$$\begin{pmatrix} v & \\ & v^{-1} \end{pmatrix}, \qquad v \in \mathbb{Q}_p^\times \tag{6.78}$$

for all $p < \infty$.

The associated finite Whittaker function (6.68), evaluated at the identity $v = 1$, is only non-vanishing for $m \in \mathbb{Z}^\times$ because of the γ_p factors, as is seen in Section 2.4. For $m \in \mathbb{Z}^\times$,

$$W^{\circ,\text{fin}}_{\psi}(s,1) = \prod_{p<\infty} W^{\circ}_{\psi_p}(s,1) = \left(\prod_{p<\infty}(1-p^{-2s})\right)\left(\prod_{p<\infty}\frac{1-p^{-(2s-1)}\,|m|_p^{2s-1}}{1-p^{-(2s-1)}}\right). \tag{6.79}$$

The first factor is simply the Euler product (1.22) of the (inverse of the) Riemann zeta function $\zeta(2s)^{-1}$. We will now show that the second factor is actually the divisor sum $\sigma_t(m)$ defined in (1.19), denoting $t = 1-2s$ for brevity.

Assume first that $m = p^a$ for some prime p and positive integer a. Then

$$\sigma_t(m) = \sum_{d|m} d^t = 1 + p^t + p^{2t} + \cdots + p^{at} = \frac{1-p^{(a+1)t}}{1-p^t}. \tag{6.80}$$

For $m = p^a q^b$ we get

$$\begin{aligned}\sigma_t(m) &= 1 + p^t + q^t + p^{2t} + q^{2t} + p^t q^t + \cdots + p^{at}q^{bt}\\ &= (1+p^t+\cdots+p^t)(1+q^t+\cdots+q^{bt}) = \sigma_t(p^a)\sigma_t(q^b).\end{aligned} \tag{6.81}$$

Similarly, for the general case with m having the prime factorisation $m = p_1^{a_1}\cdots p_r^{a_r}$,

$$\sigma_t(m) = \sigma_t(p_1^{a_1})\cdots\sigma_t(p_r^{a_r}) = \prod_{i=1}^{r}\frac{1-p_i^t p_i^{a_i t}}{1-p_i^t} = \prod_{p<\infty}\frac{1-p^t\,|m|_p^{-t}}{1-p^t}, \tag{6.82}$$

since $|m|_p = p_j^{-a_j}$ for $p = p_j$ (some j) and otherwise $|m|_p = 1$. In other words, the finite spherical Whittaker function for $SL(2,\mathbb{A})$ (and the divisor sum σ_t) are *multiplicative*. Thus, for non-zero integer m,

$$W_{\psi}^{\circ,\text{fin}}(s,1) = \frac{1}{\zeta(2s)}\sigma_{1-2s}(m)\,. \tag{6.83}$$

Comparing this with the discussion in Section 1.3 we conclude that the finite Whittaker function $W_{\psi}^{\circ,\text{fin}}$, defined in (6.68), is closely related to the *instanton measure* in string theory (see Section 13.6.3). More precisely, when evaluating the finite Whittaker function at the identify in $SL(2,\mathbb{A}_f)$ we obtain the divisor sum which is characteristic for so-called D(-1)-instanton effects in string theory (see [308]). This in fact also holds for more general groups $G(\mathbb{A})$ and gives a strong physics motivation for the detailed analysis of the Casselman–Shalika formula presented in Section 9.3.

6.4 Wavefront Set and Small Representations

When considering the Fourier expansion along a unipotent radical $U(\mathbb{A})$ that is part of a standard parabolic subgroup $P(\mathbb{A}) = L(\mathbb{A})U(\mathbb{A})$, one can group the Fourier integrals (6.38) into orbits of the Levi factor $L(\mathbb{Q})$; see for example [476, 315]. There is a close connection to the theory of nilpotent orbits of the adjoint action of $G(\mathbb{C})$ on its Lie algebra $\mathfrak{g}(\mathbb{C})$ and the notion of wavefront sets through the work of Mœglin–Waldspurger [486, 484], Matumoto [470], Ginzburg–Rallis–Soudry [283, 278, 279], Gomez–Gourevitch–Sahi [296, 298, 299], Jiang–Liu–Savin [377] and many others.

Remark 6.29 The discussion of the present section only applies to Fourier expansions along unipotent radicals U of non-minimal parabolic subgroups; for expansions along $N(\mathbb{A})$ contained in the (minimal parabolic) Borel subgroup $B(\mathbb{A})$ the orbits under the abelian Levi factor are too trivial.

6.4.1 Character Variety Orbits

Let ψ denote a unitary character on $U(\mathbb{A})$ that is trivial on $U(\mathbb{Q})$ and consider the Fourier integral $F_\psi(g) \equiv F_\psi(\varphi, g)$ of an automorphic form φ as defined in Definition 6.13. We consider φ fixed for the following discussion and will suppress it in the notation $F_\psi(g)$. Under the action of an element $\gamma \in L(\mathbb{Q})$, that is, an element γ of the intersection of the discrete subgroup with the Levi factor, the Fourier coefficient changes as follows:

$$\begin{aligned} F_\psi(\gamma g) &= \int\limits_{U(\mathbb{Q})\backslash U(\mathbb{A})} \varphi(u\gamma g)\overline{\psi(u)}du = \int\limits_{U(\mathbb{Q})\backslash U(\mathbb{A})} \varphi(\gamma^{-1}u\gamma g)\overline{\psi(u)}du \\ &= \int\limits_{U(\mathbb{Q})\backslash U(\mathbb{A})} \varphi(ug)\overline{\psi(\gamma u\gamma^{-1})}du = F_{\psi^\gamma}(g), \end{aligned} \tag{6.84}$$

where we have used the fact that φ is invariant under discrete transformations as well as the fact that the change of coordinates $u \to \gamma^{-1}u\gamma$ is unimodular since γ is in the discrete subgroup. In the last step, we have defined the transformed character

$$\psi^{\gamma}(u) := \psi(\gamma u \gamma^{-1}) \tag{6.85}$$

and identified its Fourier coefficient. The transformed character ψ^{γ} is well-defined since the Levi component $L(\mathbb{Q})$ acts on $U(\mathbb{Q}_p)$ by conjugation. In view of the terminology introduced in Definition 6.6, the orbits thus produced are called *character variety orbits*. We also introduce the following notion:

Definition 6.30 (Stabiliser of a unitary character) Let ψ be a unitary character on the unipotent subgroup $U(\mathbb{A})$ of a standard parabolic subgroup $P(\mathbb{A}) = L(\mathbb{A})U(\mathbb{A})$. The set

$$C_{\psi} = \{\gamma \in L(\mathbb{Q}) \mid \psi^{\gamma} = \psi\} \tag{6.86}$$

is called the *stabiliser* of the character ψ in the Levi subgroup $L(\mathbb{Q})$. We will sometimes use the same terminology when referring to the action of $L(\mathbb{R})$ or $L(\mathbb{C})$ on the corresponding character variety.

The calculation (6.84) shows that the Fourier coefficient F_{ψ} is invariant (automorphic) under the stabiliser subgroup C_{ψ}.

The adjoint action of $L(\mathbb{Q})$ on $U(\mathbb{Q})$ can be described more explicitly by realising the original character ψ in terms of a weight vector, analogous to Proposition 6.9. The Lie algebra $\mathfrak{u}$ consists of nilpotent elements $X \in \mathfrak{u} \subset \mathfrak{g}$, and we can write an element $u \in U$ as $u = e^X$. A unitary character ψ on U is then given by an element ω of the dual space $\mathfrak{u}^*$ via

$$\psi(e^X) = \exp\left(2\pi i \omega(X)\right), \tag{6.87}$$

and the triviality (6.22) of ψ on the commutator subgroup $[U, U]$ enforces that

$$\omega\left([\mathfrak{u}, \mathfrak{u}]\right) = 0, \tag{6.88}$$

so that ω is not an arbitrary element of $\mathfrak{u}^*$ but one associated with the Lie algebra of the abelianisation $[U, U]\backslash U$. Clearly, the abelianisation $[U, U]\backslash U$ is preserved by the adjoint action of $L(\mathbb{Q})$ on $U(\mathbb{A})$, and $L(\mathbb{Q})$ therefore acts dually on the space of allowed ω. By virtue of (6.84), the Fourier coefficients for all characters in one orbit are related and it suffices to calculate the Fourier coefficient of one representative of an orbit. In practice, it is more convenient to take the dual of ω and study the adjoint nilpotent orbits of the action of $L(\mathbb{Q})$ on $\mathfrak{u}(\mathbb{Q})$, where one can also restrict to the abelian quotient $[\mathfrak{u}, \mathfrak{u}]\backslash\mathfrak{u}$.

Remark 6.31 Let Σ be the subset of the simple roots Π that defines a standard parabolic subgroup $P = LU$; see Section 3.1.3. The nilpotent Lie algebra $\mathfrak{u}$ of U has a (finite) graded decomposition

$$\mathfrak{u} = \bigoplus_{j\in\mathbb{Z}} \mathfrak{u}_j, \quad \text{with} \quad \mathfrak{u}_j = \left\langle E_\alpha \,\middle|\, \alpha = \sum_{\beta\in\Pi} n_\beta \beta \in \Delta_+ \quad \text{and} \quad \sum_{\alpha\in\Pi\setminus\Sigma} n_\alpha = j \right\rangle. \tag{6.89}$$

Each space $\mathfrak{u}_j$ is preserved by the adjoint action of L and the space of characters ψ on U is dual to $\mathfrak{u}_1$. The character variety orbits can therefore be viewed dually in $\mathfrak{u}_1$. The space $\mathfrak{u}_1$ is isomorphic (as a vector space) to $[\mathfrak{u}, \mathfrak{u}]\backslash\mathfrak{u}$.

Example 6.32: Mirabolic subgroups of $GL(n, \mathbb{R})$

Consider $G(\mathbb{R}) = SL(n, \mathbb{R})$, which can be represented by $(n \times n)$ matrices. A maximal parabolic subgroup can be chosen with Levi factor $L(\mathbb{R}) = GL(n-1, \mathbb{R})$ through the matrices

$$L(\mathbb{R}) = \left\{ \begin{pmatrix} * & 0 & 0 & \cdots & 0 & 0 \\ 0 & * & * & \cdots & * & * \\ 0 & * & * & \cdots & * & * \\ \vdots & \vdots & & & & \vdots \\ 0 & * & * & \cdots & * & * \end{pmatrix} \right\}$$
$$= \left\{ \begin{pmatrix} r & 0 \\ 0 & m \end{pmatrix} \,\middle|\, r \in \mathbb{R},\ m \in GL(n-1, \mathbb{R}) \text{ with } \det(m) = r^{-1} \right\} \tag{6.90}$$

and associated $(n-1)$-dimensional unipotent radical

$$U(\mathbb{R}) = \left\{ \begin{pmatrix} 1 & * & * & \cdots & * & * \\ 0 & 1 & 0 & \cdots & 0 & 0 \\ 0 & 0 & 1 & \cdots & 0 & 0 \\ \vdots & \vdots & & & & \vdots \\ 0 & 0 & 0 & \cdots & 0 & 1 \end{pmatrix} \right\} = \left\{ \begin{pmatrix} 1 & u^T \\ 0 & \mathbb{1}_{n-1} \end{pmatrix} \,\middle|\, u \in \mathbb{R}^{n-1} \right\}. \tag{6.91}$$

The unipotent radical is abelian in the present case and acted upon by $L = GL(n-1, \mathbb{R})$. Characters ψ can be thought of as being given by $(n-1)$-column vectors ω that contract into $X \in \mathrm{Lie}(U) = \mathfrak{u}$ and define the character via (6.87). These parabolic subgroups are sometimes referred to as *mirabolic subgroups*.

For the local transformation of the Fourier coefficients (6.84) at the archimedean place one needs to restrict to orbits under $L(\mathbb{Z})$ that force $r = 1$ and $m \in SL(n-1, \mathbb{Z})$ in (6.90) (or $r = -1$ and $\det(m) = -1$, but this does not influence the discussion below). The action of the Levi subgroup $L(\mathbb{Z})$ on $U(\mathbb{R})$ is by

$$\begin{pmatrix} 1 & \\ & m \end{pmatrix} \begin{pmatrix} 1 & u^T \\ & 1 \end{pmatrix} \begin{pmatrix} 1 & \\ & m^{-1} \end{pmatrix} = \begin{pmatrix} 1 & u^T m^{-1} \\ & 1 \end{pmatrix}. \tag{6.92}$$

The group $L(\mathbb{Z})$ then acts on the character variety $\mathfrak{u}^*$ modelled by a vector $\omega \in \mathbb{R}^{n-1}$ by

$$\omega \mapsto m^{-1}\omega, \tag{6.93}$$

that is, simply by left multiplication of the column vector. The character variety $\{\omega \in \mathbb{R}^{n-1}\}$ decomposes into infinitely many orbits with representatives

$$\begin{pmatrix} \sigma \\ 0 \\ \vdots \\ 0 \end{pmatrix} \quad \text{for} \quad \sigma \in \mathbb{R}_{\geq 0} \tag{6.94}$$

under this action. If the character is trivial on integral points (as is the case for unitary characters trivial on $U(\mathbb{Z})$) the representatives are labelled by $\sigma \in \mathbb{Z}_{\geq 0}$:

$$\mathbb{Z}^{n-1} = \bigcup_{\sigma \in \mathbb{Z}_{\geq 0}} \left\{ SL(n-1, \mathbb{Z}) \cdot \begin{pmatrix} \sigma \\ 0 \\ \vdots \\ 0 \end{pmatrix} \right\}. \tag{6.95}$$

An arbitrary vector $\omega \in \mathbb{Z}^{n-1}$ belongs to the orbit with $\sigma = \gcd(\omega)$.

Remark 6.33 Classifying the orbits of the action of L on U over $\mathbb{Z}$ or $\mathbb{Q}$ is in general a difficult task; see [378, 60, 546, 425] for some results. A slightly coarser description can be obtained by complexifying the Levi subgroup to $L(\mathbb{C})$ and studying the complex orbits. All such complex orbits have been determined in the literature [452, 162, 476], using the methods of Dynkin [196], Kostant [421, 423], Bala–Carter [36, 37], Vinberg [604] and Kac [256]. The *Bala–Carter classification* in particular allows for a labelling of the complex nilpotent orbits by reductive subalgebras of $\mathfrak{g}$ and their distinguished nilpotent elements.

6.4.2 Wavefront Sets and Vanishing Theorems for Fourier Coefficients

There are many different choices of parabolic subgroup $P(\mathbb{A}) = L(\mathbb{A})U(\mathbb{A})$ and associated Fourier expansions along their unipotents U. All the different character variety orbits of the action of $L(\mathbb{Q})$ on unitary characters on $U(\mathbb{A})$ are associated with nilpotent elements $\omega \in \mathfrak{u}^* \subset \mathfrak{g}^*$. A given character variety orbit therefore lies in a *co-adjoint nilpotent orbit* of the action of $G(\mathbb{Q})$ on elements of $\mathfrak{g}^*$ that are dual to nilpotent elements. Properties of automorphic representations of $G(\mathbb{A})$ are only associated with structures arising from G, implying that the character variety orbits (under the action of L) are less fundamental than the *nilpotent orbits* they embed into. We will not fully develop the theory of nilpotent orbits here but refer the reader to the books [146, 128, 581] for a detailed exposition. Below we will mention only some aspects that are of relevance to our discussion.

The approach using nilpotent orbits is useful because it sometimes allows us to determine that certain Fourier coefficients must vanish identically without actually calculating them. At the heart of this is the notion of the (complexified) *wavefront set* of a local irreducible representation π_p (archimedean or non-archimedean; see Section 5.2).

Definition 6.34 (Wavefront set) Let π_p be an irreducible representation of $G(\mathbb{Q}_p)$ at a local place p. The *wavefront set* of π_p is given by

$$\mathrm{WF}(\pi_p) = \bigcup_{i \in I} O_i, \tag{6.96}$$

where the O_i are a finite collection of complex nilpotent orbits characterised by admitting a non-trivial local orbit Fourier integral [484, 486] of the type defined below in Definition 6.37.

Using the Zariski topology on the set of nilpotent elements of $\mathfrak{g}(\mathbb{C})$, the wavefront set can be viewed as the closure of the collection of orbits that we will refer to as the maximal elements of the wavefront set [383, 75]. At the archimedean place, this is defined as the *annihilator ideal* associated with a local representation π_∞ [470].

Remark 6.35 We will also use the notion of a *global wavefront set* of an adelic representation $\pi = \otimes_{p \leq \infty} \pi_p$ of $G(\mathbb{A})$. It is a priori not guaranteed that the local wavefront set $\mathrm{WF}(\pi_p)$ does not vary as p varies and therefore one has to treat this notion with care. For Eisenstein series induced by characters of the form (4.122) this does not happen. Global wavefront sets have been discussed for example in [377], where it was also shown in many cases that the maximal orbits in wavefront sets have to be so-called *special* orbits. This property was known in the p-adic case for local wavefront sets due to [484]. In general, one expects the maximal orbits of wavefront sets to be *admissible* [295].

A nilpotent orbit for a Lie algebra $\mathfrak{g}$ is the orbit of a nilpotent element $X \in \mathfrak{g}$ under the action of the adjoint group G with Lie algebra $\mathfrak{g}$; see for example [146, 581] for an introduction. The theorems of [383, 76] show that one can associate (the closure of) a unique nilpotent orbit in $\mathfrak{g}$ to any irreducible automorphic representation π of G, meaning that the wavefront set of an irreducible automorphic representation is given by the closure of a unique maximal orbit (with respect to the partial closure ordering). One can also consider the action of the adjoint group G on the dual Lie algebra $\mathfrak{g}^*$ and study co-adjoint nilpotent orbits. Using the non-degenerate Killing form, we can identify *adjoint* and co-adjoint nilpotent orbits.

By the correspondence (6.87) one can view characters ψ on some unipotent U as elements of $\mathfrak{g}^*$ and the character variety orbits lie therefore in co-adjoint nilpotent orbits.

The link to the $L(\mathbb{C})$-orbits of Fourier coefficients F_{ψ_U} of an automorphic function φ is provided by the theorems of Mœglin–Waldspurger and Matumoto [486, 470, 476] that assert that a Fourier coefficient can only be non-zero if its associated character variety orbit in $\mathfrak{u}^*$ (under the action of $L(\mathbb{C})$) intersects a co-adjoint nilpotent orbit in $\mathfrak{g}^* \supset \mathfrak{u}^*$ (under the action of $G(\mathbb{C})$) that belongs to the wavefront set associated with the automorphic representation to which φ belongs.

Example 6.36: Minimal representation of E_6

Suppose φ belongs to the minimal representation of the exceptional Lie group $E_6(\mathbb{R})$ of dimension 78. Then its associated wavefront set is the closure of the minimal orbit (of dimension 22). The minimal representation of E_6 can be realised as a special point in the degenerate principal series representation that is associated with a maximal parabolic subgroup $P = LU$ with Levi factor $L = SO(5,5)\times GL(1)$. The unipotent U in this case is a Heisenberg group of dimension 21. The (dualised) character variety $\mathfrak{u}_1 = [\mathfrak{u},\mathfrak{u}]\backslash\mathfrak{u}$ has dimension 20 and is acted upon by $L(\mathbb{R})$. After complexification one finds that $\mathfrak{u}_1$ breaks up into five different character variety orbits under $GL(6,\mathbb{C})$ [476]. Of these only the trivial and the smallest non-trivial one intersect the closure of the minimal co-adjoint nilpotent orbit. One concludes that the Fourier coefficients in the remaining three character variety orbits must vanish in the minimal representation.

We also note that the Gelfand–Kirillov dimension of the degenerate principal series in this case is $21 = 20 + 1$, corresponding to the dimension of the Heisenberg group; see Example 5.32. At the special point where the minimal representation is realised, the 20-dimensional space can be polarised into 10 ‘coordinates’ and 10 ‘momenta’ and the Heisenberg algebra is realised on functions of the 10 coordinate variables on which the momenta act as derivative operators. This action of the Heisenberg group extends to all of $E_6(\mathbb{R})$ and can also be given an oscillator realisation [334].

This example is based on [281, 520, 476, 315], where more information can be found. The minimal representation discussed here is an example of a small representation that we will discuss in more detail in Sections 6.4.3 and 10.3.2 and in Chapter 14.

The connection between nilpotent orbits and Fourier coefficients is made more concrete in the work of Ginzburg [278]. We follow [340] in the following discussion. To the nilpotent orbit $\mathcal{O} \equiv \mathcal{O}_X$ of a nilpotent element $X \in \mathfrak{g}$ one can associate a *Jacobson–Morozov triple* $H, X, Y \in \mathfrak{g}$ that satisfies the standard $\mathfrak{sl}(2)$ Lie algebra relations. The orbit is uniquely characterised by the (unique) Weyl chamber image of H under the action of the Weyl group. This leads to a labelling of nilpotent orbits by *weighted Dynkin diagrams*, where the weights are non-negative integers. (These integers are less than or equal to 2 but this does not matter for our discussion.) Any integrally weighted Dynkin diagram gives rise to a graded decomposition

$$\mathfrak{g} = \bigoplus_{i\in\mathbb{Z}} \mathfrak{g}_i\,, \tag{6.97}$$

where $\mathfrak{g}_i$ is the space of elements in $\mathfrak{g}$ with eigenvalue i under the adjoint action of the H that lies in the Weyl chamber. All $\mathfrak{g}_i$ are of finite dimension and there are only finitely many non-trivial $\mathfrak{g}_i$ since $\mathfrak{g}$ is finite-dimensional. We define

$$\mathfrak{l}_O = \mathfrak{g}_0, \quad \mathfrak{u}_O = \bigoplus_{i\geq 1} \mathfrak{g}_i \quad \text{and} \quad \mathfrak{v}_O = \bigoplus_{i\geq 2} \mathfrak{g}_i. \tag{6.98}$$

Let L_O, U_O and V_O be corresponding subgroups of G. A nilpotent orbit O has a unique stabiliser $C_O \subset L_O$ that is a reductive group.

Definition 6.37 (Orbit Fourier coefficient) Let O be a non-trivial nilpotent orbit and let $\psi_V\colon V_O(\mathbb{Q})\backslash V_O(\mathbb{A}) \to U(1)$ be a unitary character on V_O. We require ψ_V to have the same stabiliser type under the action of L_O as the stabiliser C_O of the orbit O. Then the *orbit Fourier coefficient* of an automorphic form φ belonging to some automorphic representation π is defined as

$$F_O(\varphi, \psi_V, g) = \int\limits_{V_O(\mathbb{Q})\backslash V_O(\mathbb{A})} \varphi(vg)\overline{\psi_V(v)}dv. \tag{6.99}$$

For the trivial orbit $O = \{0\}$ we define the orbit Fourier coefficient to be the constant term along the maximal unipotent $N(\mathbb{A})$ as in Definition 6.17.

The orbit Fourier coefficients vanish when the orbit does not belong to the wavefront set, and allow a rewriting of the Fourier expansion of an automorphic function in terms of a sum over nilpotent orbits. This is similar to the expansion of the *Howe–Harish-Chandra expansion* of the *character distribution* of an automorphic representation [360, 343]:

$$\mu(\pi) = \sum_{O\in\mathrm{WF}(\pi)} c_O \mu_O. \tag{6.100}$$

For local automorphic representations π_p, the numbers c_O for maximal orbits O and $p < \infty$ are computed by the *Mœglin–Waldspurger theorem* [543, 486].

Remark 6.38 The connection between the *associated variety* of automorphic representations and degenerate Whittaker coefficients was also studied in detail recently in work of Gourevitch and Sahi together with Gomez [298, 299, 296]. Their main emphasis was on local results for the real and p-adic place but some results in the global case can be found in [296]. An important refinement contained in their work is the introduction of so-called Whittaker pairs, which go beyond Jacobson–Morozov triples for nilpotent orbits. In this more general language it is often easier to determine the vanishing (or not) of certain generalised degenerate Whittaker coefficients and from this obtain information about the wavefront set of a representation.

It is often possible to relate the orbit Fourier coefficients to (degenerate) Whittaker coefficients, and this was done for example in [278, 246, 340]. Turning the argument around, one might suspect that the wavefront set can be effectively computed by studying the degenerate Whittaker coefficients with charges defining the parabolic subgroups defining a nilpotent orbit in the Bala–Carter classification. This is borne out for minimal representations [246, 476] and also well supported for some other small representations relevant for string theory [83, 340, 80]; see also the discussion in Sections 14.2.4 and 14.3.3. We will discuss small representations more generally in the following section.

6.4.3 Small Representations

Any automorphic representation has a fixed Gelfand–Kirillov dimension; for generic elements of the principal series this functional dimension is given by (5.34). Of particular interest to string theory are those with a small Gelfand–Kirillov dimension, as we will discuss in more detail in Chapter 15. For the purposes of this book we will define small representations as those with small wavefront sets, such that only very few orbits appear in (6.96).

The smallest possible representation is the *trivial representation.* It is associated with constant functions, or equivalently, with trivial inducing character determined by $\lambda = -\rho$ in (5.31). The Gelfand–Kirillov dimension vanishes and the wavefront set consists only of the trivial nilpotent orbit.

The next-larger representations are more interesting. We recall that every split real Lie group has a unique *minimal nilpotent orbit* $\mathcal{O}_{\text{min}}$ with Bala–Carter label A_1 [146]. It is the smallest non-trivial nilpotent orbit.

Definition 6.39 (Minimal representation) An automorphic representation π of a split real group G is called *minimal* if its wavefront set satisfies $\text{WF}(\pi) = \overline{\mathcal{O}}_{\text{min}}$.

This global definition makes sense since it is known that an automorphic representation that is minimal at some place p has to be minimal at all places [281, 246]. At non-archimedean places, the minimality of a representation π_p can also be expressed in terms of the character expansion (6.100) as the property that only the trivial and the minimal orbit contribute and that $c_{\mathcal{O}_{\text{min}}} = 1$. The annihilator ideal of the minimal representation at the archimedan place is known as the *Joseph ideal.*

As a unitary representation of the real Lie group $G(\mathbb{R})$, minimal automorphic representations are equivalent to the *minimal unitary representation* of smallest non-trivial functional dimension among all $G(\mathbb{R})$ representations [381, 382]. In the physics literature these minimal unitary representations are often constructed using oscillator representations or using the so-called *quasi-conformal action* [362, 381, 333, 334, 337, 338].

Minimal representations of a group G are closely related to minimal nilpotent G-orbits. Specifically, via *Kirillov's 'orbit method'* one can obtain π_{min} through the geometric quantisation of the minimal nilpotent orbit $\mathcal{O}_{\text{min}}$ [114]. More generally, the orbit method exploits the natural symplectic structure on any nilpotent orbit to construct representations of Lie groups [404].

For Lie groups of Cartan type ADE different from $SL(2)$ and $SL(3)$ there is also a well-defined *next-to-minimal nilpotent orbit* $\mathcal{O}_{\text{ntm}}$ with Bala–Carter label $2A_1$. For orthogonal type D there can be several orbits with this Bala–Carter label and we choose the smaller one as $\mathcal{O}_{\text{ntm}}$.

Definition 6.40 (Next-to-minimal representation) An automorphic representation π of a split real Lie group G of ADE type is called *next-to-minimal* if its wavefront set satisfies $\text{WF}(\pi) = \overline{\mathcal{O}}_{\text{ntm}}$.

Remark 6.41 Contrary to the case of minimal representations discussed above it is not known whether a (global) automorphic representation that is next-to-minimal at one place has to be so at all places. The particular examples we are considering will be.

Automorphic forms attached to small representations π are interesting from both a mathematical and a physical perspective. It was shown in the seminal paper by Ginzburg–Rallis–Soudry [281] that automorphic forms in the minimal representation π_{min} have very few non-vanishing Fourier coefficients, a fact that has far-reaching consequences. In particular, it allows us to describe the complete Fourier expansion very explicitly, a task which is generally very difficult for Lie groups beyond $SL(2)$. One of the main applications of the theory of small representations has been in the context of the so-called *theta correspondence*, which is a method of lifting automorphic representations from one group G to another G'; see Chapter 12. Ginzburg has also developed a method which uses small automorphic representations for constructing new automorphic L-functions (see [279] for a survey).

From the automorphic point of view, small representations can arise as special points in degenerate principal series representations where the Gelfand–Kirillov dimension reduces. According to (5.96), the functional dimension of

Table 6.1 The dimensions of the minimal and next-to-minimal representations of ADE type split Lie groups. For $SL(3)$ a next-to-minimal representation with wavefront set equal to the closure of the $2A_1$-orbit does not exist as there is no such orbit.

Lie group	GKdim(π_{min})	GKdim(π_{ntm})
$SL(n)$	$n-1$	$2n-4$
$SO(n,n)$	$2n-3$	$2n-2$
$E_{6(6)}$	11	16
$E_{7(7)}$	17	26
$E_{8(8)}$	29	46

the automorphic representation induced from a parabolic subgroup $P(\mathbb{A}) \subset G(\mathbb{A})$ is given by

$$\text{GKdim}\, I_P(\chi) = \dim G - \dim P = \dim U, \tag{6.101}$$

where χ corresponds to a generic character on the parabolic subgroup $P(\mathbb{A}) = L(\mathbb{A})U(\mathbb{A})$. It turns out that for very special choices of the inducing character χ there may exist unitarizable submodules of $I_P(\chi)$ with smaller functional dimension. These points are often associated with representations contributing to the *residual spectrum*; see for example [474] and the discussions in Sections 5.5 and 10.2.3.

As already explained in Example 5.32, minimal automorphic representations can be constructed using Eisenstein series associated with maximal parabolic subgroups, in particular when choosing the so-called Heisenberg parabolic. At the level of Lie algebras, the Heisenberg parabolic is associated with a five-grading of the Lie algebra $\mathfrak{g}$:

$$\mathfrak{g} = \mathfrak{g}_{-2} \oplus \mathfrak{g}_{-1} \oplus \mathfrak{g}_0 \oplus \mathfrak{g}_1 \oplus \mathfrak{g}_2, \tag{6.102}$$

such that $\mathfrak{g}_2$ is the one-dimensional root space associated with the highest root of $\mathfrak{g}$. Then $\mathfrak{g}_1$ is even-dimensional, say, $\dim \mathfrak{g}_1 = 2d$. The Gelfand–Kirillov dimension of the minimal representation then equals $d+1$ [334]. As an automorphic representation this can be realised for example via a maximal Eisenstein series [281] associated with the Heisenberg parabolic. However, due to the functional relations (4.136), discussed in more detail in Chapter 8.8, many other ways of arriving at an automorphic realisation of the minimal representation exist. This will also be one of the main themes of Chapter 10.

For the next-to-minimal representation a similar uniform realisation is not known, to the best of our knowledge. For any concrete example, however, a realisation can be obtained readily and these play a prominent rôle in string theory, as we will explain in Chapter 14. In general, the 'smallness' of small representations is visible both in their constant terms and in their Fourier

Table 6.2 Summary of some automorphic realisations of minimal and next-to-minimal representations for split ADE groups. We give the value s_{i_*} for realising the representation, where the index i_* denotes the maximal parabolic used in the induction. The labelling follows the 'Bourbaki convention'. The dimensions of the representations are given in Table 6.1.

Lie group	π_{min}	π_{ntm}
$SL(n)$	generic s_1 or generic s_{n-1}	generic s_2 or generic s_{n-2}
$SO(n,n)$	$s_1 = \frac{n-2}{2}$ or $s_n = 1$ or $s_{n-1} = 1$	generic s_1
$E_{6(6)}$	$s_1 = \frac{3}{2}$ or $s_6 = \frac{3}{2}$	$s_1 = \frac{5}{2}$ or $s_5 = 1$ or $s_6 = \frac{7}{2}$
$E_{7(7)}$	$s_1 = \frac{3}{2}$ or $s_7 = 2$	$s_1 = \frac{5}{2}$ or $s_6 = \frac{3}{2}$ or $s_7 = 4$
$E_{8(8)}$	$s_1 = \frac{3}{2}$ or $s_8 = \frac{5}{2}$ or $s_8 = 12$	$s_1 = \frac{5}{2}$ or $s_7 = 2$ or $s_8 = \frac{9}{2}$

coefficients. We list the Gelfand–Kirillov dimensions of all split real Lie groups of ADE type in Table 6.1. For the orthogonal and exceptional groups the minimal unitary representation is unique, whereas for $SL(n)$ a one-parameter family of minimal representations exists.

For further reference, we have listed ways of realising small representations automorphically in Table 6.2. The automorphic realisation depends on choosing a maximal parabolic P_{i_*} and a weight $\lambda = 2s_{i_*}\Lambda_{i_*} - \rho$ for the inducing character as in Proposition 5.28. Table 6.2 then gives the values of s_{i_*} for which the small representations are realised. Due to the functional relations to be discussed in more detail in Section 8.8, there can be multiple realisations for different maximal parabolics. We do not give all possible realisations but only a selection.

Remark 6.42 The value of s_{i_*} depends also on the convention for the inducing character χ or the weight λ. In the seminal work by Ginzburg–Rallis–Soudry on minimal representations [281], the inducing weight was given in terms of multiples of $\rho_{P_{i_*}}$ of (5.85) rather than multiples of the fundamental weight Λ_{i_*} that we use here. This leads to different values of s_{i_*}. For instance, for E_8 one has $\rho_{P_8} = \frac{29}{2}\Lambda_{P_8}$. The relation between the weights in [281] and ours is $(2s_{\text{GRS}} + 1)\,\rho_P = 2s_8\Lambda_8$. For the choice of $s_8 = 12$ for the minimal representation of E_8 one then recovers their value $s_{\text{GRS}} = \frac{19}{58}$.

6.5 Method of Piatetski-Shapiro and Shalika*

The grouping of Fourier coefficients into orbits under a Levi subgroup L discussed in the previous section is a powerful tool for analysing automorphic forms. This is at the heart of the method of Piatetski-Shapiro and Shalika, which expresses an automorphic form on $GL(n,\mathbb{R})$ completely in terms of its Whittaker coefficients (with respect to N) [516, 570]. We briefly explain how this connection between Fourier coefficients along U and Whittaker coefficients along N comes about in the case of $GL(n,\mathbb{R})$, following [241, 476]. Generalisations to some other groups have been discussed by Miller and Sahi [476].

Let $P = LU$ be a parabolic subgroup of G. According to Proposition 6.18 we have for a spherical automorphic form $\varphi(g) = \varphi(gk)$ in an automorphic representation π that

$$\sum_{\psi} F_{\psi}^{\circ}(g) = \int_{U^{(1)}(\mathbb{Q})\backslash U^{(1)}(\mathbb{A})} \varphi(ug)du, \tag{6.103}$$

where $U^{(1)} = [U,U]$ is the derived group of U and

$$F_{\psi}(g) = \int_{U(\mathbb{Q})\backslash U(\mathbb{A})} \varphi(ug)\overline{\psi(u)}du \tag{6.104}$$

is the Fourier coefficient of φ along U for the character ψ on U.

Now we want to group the sum over the characters ψ into complex orbits thanks to (6.84). A given character ψ can have a stabiliser $C_{\psi}(\mathbb{A}) \subset L(\mathbb{A})$ under the action of $L(\mathbb{A})$, and $C_{\psi}(\mathbb{Q}) = C(\mathbb{Q}) \cap L_{\psi_U}(\mathbb{A})$ is a discrete subgroup of it. Writing the set of complex character orbits as $\mathrm{WF}(\pi)$, one can write (6.103) as [476]

$$\sum_{\psi} F_{\psi}^{\circ}(g) = \sum_{\mathcal{O}\in\mathrm{WF}(\pi)} \sum_{\psi\in\mathcal{O}} \sum_{\gamma\in C_{\psi}(\mathbb{Q})\backslash L(\mathbb{Q})} F_{\psi}(\gamma g), \tag{6.105}$$

where the sum over $\psi \in \mathcal{O}$ denotes *single representatives* of the different integral orbits contained in the complex orbit $\mathcal{O} \in \mathrm{WF}(\pi)$. The method of Piatetski-Shapiro and Shalika then uses the fact that $F_{\psi}(\gamma g)$ is a function on the reductive stabiliser $C_{\psi}(\mathbb{A})$ and automorphic under $C_{\psi}(\mathbb{Q})$ and so can be expanded in the same manner, yielding an iterative procedure for determining the Fourier expansion.

In the case of $GL(n,\mathbb{R})$ this can be done very successfully in terms of iterations of parabolic subgroups of the type discussed in Example 6.32. As was explained there, the unipotent subgroup is abelian and therefore the Fourier expansion (6.103) recovers the whole automorphic function φ. Moreover, there is a unique non-trivial complex orbit of $GL(n-1,\mathbb{C})$ acting on the $(n-1)$-dimensional $U(\mathbb{C})$. The trivial orbit corresponds to trivial $\psi = 1$ and corresponds to the constant term in the expansion along U; see Definition 6.17. In order to avoid having to include this term at every iteration step, we now assume until the end of this section that φ is a cusp form. Then the sum over complex orbits in $\mathrm{WF}(\pi)$ has only a single element.

The integral orbits contained in the single complex orbit can also be identified easily in this case. They are represented by (non-zero) integers m_{α_1} and the representatives can be chosen to be such that

$$\psi(u) = e^{2\pi i m_{\alpha_1} u_{\alpha_1}}, \tag{6.106}$$

where α_1 is the first simple root. (Allowing m_{α_1} to be integral instead of integral and positive actually overcounts the integral orbits by a factor of two but this has no impact on the final result.) In terms of matrices as in Example 6.32 this can be written as

$$u = \exp\begin{pmatrix} 0 & u_{\alpha_1} & u_{\alpha_1+\alpha_2} & \cdots & u_{\alpha_1+\ldots+\alpha_{n-1}} \\ 0 & 0 & 0 & \cdots & 0 \\ 0 & 0 & 0 & \cdots & 0 \\ \vdots & \vdots & & & \vdots \\ 0 & 0 & 0 & \cdots & 0 \end{pmatrix},$$

$$\omega = (m_{\alpha_1}, 0, 0, \ldots, 0)^T \in \mathfrak{u}^*. \tag{6.107}$$

The stabiliser of such a character in $L(\mathbb{Q}) = GL(n-1, \mathbb{Q})$ is given by $GL(n-2, \mathbb{Q}) \subset GL(n-1, \mathbb{Q})$. At the present stage we have therefore from (6.105)

$$\varphi(g) = \sum_{m_{\alpha_1}\in\mathbb{Z}} \sum_{\gamma\in GL(n-2,\mathbb{Q})\backslash GL(n-1,\mathbb{Q})} F_\psi(\gamma g). \tag{6.108}$$

The Fourier coefficient $F_\psi(\gamma g)$ appearing in (6.108) is therefore an automorphic form on $GL(n-1, \mathbb{Q})$ automorphic under $GL(n-2, \mathbb{Q})$. The iteration now consists in repeating the same process for this smaller subgroup. What this will produce is a sum over $m_{\alpha_2} \in \mathbb{Z}$ and an automorphic form on $GL(n-2, \mathbb{A})$ and so on. At the end of the iteration we obtain

$$\varphi(g) = \sum_{m_{\alpha_1},\ldots m_{\alpha_{n-1}}\in\mathbb{Z}} \sum_{\gamma\in N(n-1,\mathbb{Q})\backslash GL(n-1,\mathbb{Q})} W_\psi(\gamma g), \tag{6.109}$$

where

$$W_\psi(g) = \int_{N(\mathbb{Q})\backslash N(\mathbb{A})} \varphi(ng)\overline{\psi(n)}dn \tag{6.110}$$

is a standard Whittaker coefficient on N for the character with instanton charges m_α for the simple roots $\alpha \in \Pi$ and as defined in (6.40). (The integration domain is enlarged from U to N by combining some of the intermediate sums over cosets.) Reassembling the sum over all these characters we therefore can also write

$$\varphi(g) = \sum_{\gamma\in N(n-1,\mathbb{Z})\backslash GL(n-1,\mathbb{Z})} \int_{N^{(1)}(\mathbb{Z})\backslash N^{(1)}(\mathbb{R})} \varphi(n\gamma g)dn, \tag{6.111}$$

where $N^{(1)}$ is the derived group of N (see (6.45)) and we have projected back down to $\mathbb{R}$. The power of the formula (6.109) is that it allows us to reconstruct the whole (cuspidal) automorphic form from its standard Whittaker coefficients by taking suitable translates of them. This is important since, according to (6.49), an automorphic function also contains terms beyond the standard Whittaker coefficients in its expansion. The result of Piatetski-Shapiro and Shalika tells us how to compute these non-abelian terms as translates of abelian terms. We will see a similar structure later when we study the case of $SL(3)$ in detail in Section 9.8.

7

Fourier Coefficients of Eisenstein Series on $SL(2, \mathbb{A})$

In this chapter, we apply the formalism developed in Chapter 6 to the classical theory of non-holomorphic Eisenstein series $E(s, z)$ on the double coset $SL(2, \mathbb{Z})\backslash\mathbb{H}$, with $z \in \mathbb{H} = SL(2, \mathbb{R})/SO(2, \mathbb{R})$, which was already presented as a canonical example in the introduction. Following the analysis in the previous chapters, we will consider the adelic treatment of this Eisenstein series. The purpose of this chapter is to give an example for the full Fourier expansion of an Eisenstein series, where the definitions of the previous chapters come to life. The more general case of arbitrary $G(\mathbb{A})$ will then be discussed in the following chapters. We note that in Appendix D we present another derivation of the Fourier expansion of $E(s, z)$ based on Kloosterman sums.

7.1 Statement of Theorem

Before we state the theorem, let us introduce some of the necessary terminology. Recall from Chapter 3 that the adelic group $SL(2, \mathbb{A})$ has the maximal compact subgroup $K_{\mathbb{A}}$, defined by

$$K_{\mathbb{A}} = K(\mathbb{R}) \times \prod_{p<\infty} K(\mathbb{Q}_p) = SO(2, \mathbb{R}) \times \prod_{p<\infty} SL(2, \mathbb{Z}_p)\,; \tag{7.1}$$

see for example [443]. By strong approximation as stated in Theorem 3.10, we have

$$SL(2, \mathbb{Z})\backslash SL(2, \mathbb{R}) \cong SL(2, \mathbb{Q})\backslash SL(2, \mathbb{A})/K_f\,, \tag{7.2}$$

where K_f denotes the factors in $K_{\mathbb{A}}$ for finite primes. This decomposition ensures that any automorphic form φ on the space $SL(2, \mathbb{Z})\backslash SL(2, \mathbb{R})/SO(2, \mathbb{R})$ corresponds to an automorphic form on the space $SL(2, \mathbb{Q})\backslash SL(2, \mathbb{A})/K_{\mathbb{A}}$.

For the adelic group $SL(2, \mathbb{A})$ we have the Iwasawa decomposition

$$SL(2, \mathbb{A}) = N(\mathbb{A})A(\mathbb{A})K_{\mathbb{A}}\,. \tag{7.3}$$

Given a generic group element $g = nak$ of $SL(2, \mathbb{A})$, we define the character χ on $SL(2, \mathbb{A})$, in analogy with the general Definition (4.122), such that

$$\chi(nak) = |a^{\lambda+\rho}|, \tag{7.4}$$

where λ is some weight vector of $\mathfrak{sl}(2)$ and ρ is the Weyl vector. In the case of $SL(2, \mathbb{A})$ the space of (complex) weights is one-dimensional and spanned by the fundamental weight Λ_1 dual to the unique simple root α_1 of $\mathfrak{sl}(2)$. The Weyl vector ρ is also identical to Λ_1. Therefore, we can parametrise the weight appearing in (7.4) with a single parameter $s \in \mathbb{C}$ as

$$\lambda = 2s\Lambda_1 - \rho = (2s-1)\Lambda_1 \quad \Rightarrow \quad \lambda + \rho = 2s\Lambda_1 . \tag{7.5}$$

Furthermore, we make use of the function $H(g)$ of (4.116), which denotes the Lie algebra element associated with the abelian part a in the Iwasawa decomposition of $g = nak$, such that $a = \exp(H(g))$. With this function the character can now be written as

$$\chi_s(g) \equiv |a^{\lambda+\rho}| = e^{\langle \lambda+\rho | H(g) \rangle} = e^{\langle \lambda+\rho | H(a) \rangle} = e^{2s\langle \Lambda_1 | H(a) \rangle}, \tag{7.6}$$

where we have introduced the notation χ_s for the character parametrised by $s \in \mathbb{C}$. In the case of $SL(2, \mathbb{A})$ we can write all these objects explicitly (in the fundamental representation) as (2×2) matrices, as follows:

$$g = nak = \begin{pmatrix} 1 & u \\ & 1 \end{pmatrix} \begin{pmatrix} v & \\ & v^{-1} \end{pmatrix} k, \tag{7.7}$$

with $k \in K_{\mathbb{A}}$ of (7.1). Here, u and v are adelic numbers. We recall the definition of the upper half-plane $\mathbb{H}$ from Section 4.1:

$$\mathbb{H} = \{z = x + iy \in \mathbb{C} \mid \operatorname{Im}(z) > 0\} = SL(2, \mathbb{R})/SO(2, \mathbb{R}). \tag{7.8}$$

This requires that at the archimedean place $p = \infty$ we have to use the following parametrisation:

$$g_\infty = n_\infty a_\infty k_\infty = \begin{pmatrix} 1 & x \\ & 1 \end{pmatrix} \begin{pmatrix} y^{1/2} & \\ & y^{-1/2} \end{pmatrix} k_\infty , \tag{7.9}$$

with $y > 0$ and $k_\infty \in SO(2)$. Evaluated on the group element (7.7), the $SL(2)$ character (7.6) becomes

$$\chi_s(g) = e^{2s\langle \Lambda_1 | H(a) \rangle} = |v|^{2s} \tag{7.10}$$

since $H(a) = \log|v| \cdot H_1$, where H_1 is the Cartan generator of $SL(2)$ and the norm is the adelic one. For the archimedean place this implies with (7.9) that $\chi_s(g_\infty) = y^s$, where we have embedded g_∞ into $G(\mathbb{A})$ as $g = (g_\infty, 1, 1, \ldots)$.

The adelic Eisenstein series $E(\chi, g)$ (see (4.133)) is then defined by summing the character over a coset according to ($g \in SL(2, \mathbb{A})$):

$$E(\chi_s, g) = \sum_{\gamma \in B(\mathbb{Q})\backslash SL(2,\mathbb{Q})} e^{\langle \lambda+\rho | H(\gamma g)\rangle}, \tag{7.11}$$

where the Borel subgroup $B(\mathbb{Q}) = N(\mathbb{Q})A(\mathbb{Q})$. Recall that this is the definition of the $SL(2, \mathbb{A})$ Eisenstein series attached to the induced representation, which was given in Equation (5.43) of Example 5.23. The sum converges absolutely for $\mathrm{Re}(s) > 1$.

We are now ready to state the theorem.

Theorem 7.1 (Fourier expansion of $SL(2, \mathbb{A})$ Eisenstein series) *The Fourier expansion of the $SL(2, \mathbb{A})$ Eisenstein series $E(\chi_s, g)$ with respect to the unipotent radical N of $SL(2, \mathbb{A})$ is given for $g = g_\infty \in SL(2, \mathbb{R}) \subset SL(2, \mathbb{A})$ by*

$$\begin{aligned} E(\chi_s, g) &= \sum_{\psi} W_\psi(s, g) \\ &= y^s + \frac{\xi(2s-1)}{\xi(2s)} y^{1-s} \\ &\quad + \frac{2}{\xi(2s)} y^{1/2} \sum_{m \neq 0} |m|^{s-1/2} \sigma_{1-2s}(m) K_{s-1/2}(2\pi|m|y) e^{2\pi i m x}, \end{aligned} \tag{7.12}$$

where the terms on the right-hand side of the first line in the last equality constitute the constant term and the second line provides the non-constant terms. Here, we have used the parametrisation (7.9) for $g_\infty \in SL(2, \mathbb{R})$.

Furthermore, the Eisenstein series satisfies the functional relation

$$E(\chi_s, g) = \frac{\xi(2s-1)}{\xi(2s)} E(\chi_{1-s}, g). \tag{7.13}$$

Proof The proof of this theorem constitutes the rest of the present chapter.

To prove the theorem we now wish to analyse the Fourier expansion of $E(\chi_s, g)$ along the unipotent radical N. This was already outlined in Section 6.1. According to the general discussion of the previous chapter, we have the following expansion:

$$E(\chi_s, g) = \sum_{\psi \in \mathrm{Hom}(N(\mathbb{Q})\backslash N(\mathbb{A}), U(1))} W^\circ_\psi(s, g). \tag{7.14}$$

We recall from Section 6.2 that the superscript indicates that the Whittaker coefficient W°_ψ is spherical, i.e., K-independent: $W^\circ_\psi(nak) = W^\circ_\psi(na)$. We shall distinguish the 'constant' Whittaker coefficient $W^\circ_1(s, g)$ corresponding to the special case of a trivial character $\psi \equiv 1$,

$$W_1^\circ(s,g) = \int\limits_{N(\mathbb{Q})\backslash N(\mathbb{A})} E(\chi_s, ng)dn, \tag{7.15}$$

and the remaining 'non-constant' coefficients given for $\psi \neq 1$ by

$$W_\psi^\circ(s,g) = \int\limits_{N(\mathbb{Q})\backslash N(\mathbb{A})} E(\chi_s, ng)\overline{\psi(n)}dn. \tag{7.16}$$

The expressions $W_1(s,g)$ and $W_\psi(s,g)$ are sometimes simply referred to as the 'constant term' and the 'Fourier coefficients', respectively.

Plugging in the definition of the Eisenstein series, and interchanging the sum and integration, we can rewrite the coefficients in the following form:

$$W_1^\circ(s,g) = \sum_{\gamma\in B(\mathbb{Q})\backslash SL(2,\mathbb{Q})} \int\limits_{N(\mathbb{Q})\backslash N(\mathbb{A})} \chi_s(\gamma ng)dn, \tag{7.17a}$$

$$W_\psi^\circ(s,g) = \sum_{\gamma\in B(\mathbb{Q})\backslash SL(2,\mathbb{Q})} \int\limits_{N(\mathbb{Q})\backslash N(\mathbb{A})} \chi_s(\gamma ng)\overline{\psi(n)}dn, \tag{7.17b}$$

where we recall that the sums converge absolutely for $\mathrm{Re}(s) > 1$.

We now proceed with the analysis of the constant and non-constant terms, starting with the constant term.

7.2 Constant Term

In order to evaluate the constant term, we rewrite the integral (7.17a) as

$$\begin{aligned} W_1^\circ(s,g) &= \sum_{\gamma\in B(\mathbb{Q})\backslash SL(2,\mathbb{Q})} \int\limits_{N(\mathbb{Q})\backslash N(\mathbb{A})} \chi_s(\gamma ng)dn \\ &= \sum_{\gamma\in B(\mathbb{Q})\backslash G(\mathbb{Q})/B(\mathbb{Q})} \sum_{\delta\in\gamma^{-1}B(\mathbb{Q})\gamma\cap B(\mathbb{Q})\backslash B(\mathbb{Q})} \int\limits_{N(\mathbb{Q})\backslash N(\mathbb{A})} \chi_s(\gamma\delta ng)dn. \end{aligned} \tag{7.18}$$

Here, we have formed coarser classes by going to the double coset $B(\mathbb{Q})\backslash G(\mathbb{Q})/B(\mathbb{Q})$ and then summing over representatives δ that link these double cosets to the original single cosets. We must make sure not to overcount the coset representatives δ, and this is achieved by the restriction on the δ sum.

We can next unfold the δ sum in (7.18) to the integration domain by enlarging it, which yields

$$W_1^\circ(s,g) = \sum_{\gamma\in B(\mathbb{Q})\backslash SL(2,\mathbb{Q})/B(\mathbb{Q})} \int_{\gamma^{-1}B(\mathbb{Q})\gamma\cap N(\mathbb{Q})\backslash N(\mathbb{A})} \chi_s(\gamma ng)dn. \tag{7.19}$$

The measure on this larger space is induced from the embedding $N(\mathbb{Q}) \to N(\mathbb{A})$.

To simplify the expression further we shall need the following result:

Proposition 7.2 (Bruhat decomposition for $SL(2,\mathbb{Q})$) *One has the disjoint union*

$$SL(2,\mathbb{Q}) = \bigcup_{w\in\mathcal{W}} B(\mathbb{Q})wB(\mathbb{Q}). \tag{7.20}$$

Proof Recall that for $SL(2)$ one has $\mathcal{W} = \{\mathbb{1}, w_{\text{long}}\}$. To establish (7.20) we begin by noting that for the first coset representative $w = \mathbb{1}$ we have the double coset $BwB = B$, and for the second coset representative corresponding to the non-trivial $w = w_{\text{long}}$ we get

$$\begin{aligned}
B\begin{pmatrix} & 1\\ -1 & \end{pmatrix}B &= \left\{\begin{pmatrix} a & b\\ & a^{-1}\end{pmatrix}\begin{pmatrix} & 1\\ -1 & \end{pmatrix}\begin{pmatrix} \tilde{a} & \tilde{b}\\ & \tilde{a}^{-1}\end{pmatrix} : b,\tilde{b}\in\mathbb{Q},\ a,\tilde{a}\in\mathbb{Q}^\times\right\}\\
&= \left\{\begin{pmatrix} -b\tilde{a} & a\tilde{a}^{-1} - b\tilde{b}\\ -a^{-1}\tilde{a} & -a^{-1}\tilde{b}\end{pmatrix} : b,\tilde{b}\in\mathbb{Q},\ a,\tilde{a}\in\mathbb{Q}^\times\right\}\\
&= \left\{\begin{pmatrix} a & \frac{ad-1}{c}\\ c & d\end{pmatrix} : a,d\in\mathbb{Q},\ c\in\mathbb{Q}^\times\right\},
\end{aligned} \tag{7.21}$$

and hence the Bruhat decomposition (7.20) corresponds to the division of $SL(2,\mathbb{Q})$ into those matrices with lower-left entry equal to zero and those where it is non-zero. □

Remark 7.3 The Bruhat decomposition exists for any group $G(\mathbb{Q})$ of the type considered in this book and will be central in the following chapters.

Denoting the double coset representatives by Weyl words w, we can therefore write the constant term as

$$W_1^\circ(s,g) = \sum_{w\in\mathcal{W}} C_w = \sum_{w\in\mathcal{W}} \int_{w^{-1}B(\mathbb{Q})w\cap N(\mathbb{Q})\backslash N(\mathbb{A})} \chi_s(wng)dn, \tag{7.22}$$

where we have defined individual contributions C_w to the constant term that are labelled by elements of the Weyl group.

Below we will simplify many expressions even further by using the embedding of the Weyl group $\mathcal{W}$ into $K(\mathbb{Q})$; the sum over cosets has only two contributions, arising from the trivial and non-trivial coset representatives, which can be chosen as

$$w = \begin{pmatrix} 1 & \\ & 1 \end{pmatrix} \quad \text{or} \quad \begin{pmatrix} & 1 \\ -1 & \end{pmatrix}. \tag{7.23}$$

These correspond precisely to the two Weyl elements of the Weyl group $\mathcal{W}$ of the Lie algebra $\mathfrak{sl}(2)$. We note that this embedding gives $w_{\text{long}}^2 = -\mathbb{1}$ and only $w_{\text{long}}^4 = \mathbb{1}$. This is a general feature of embedding the Weyl group in a representation of the group G [386, Chap. 3]. The additional quotient by $-\mathbb{1}$ to arrive at the usual Coxeter relations need not be imposed separately here as the $-\mathbb{1} \in K(\mathbb{R})$ factors out of all calculations.

7.2.1 Trivial Weyl Word

In the case when the Weyl word is the trivial Weyl reflection, i.e., $w = \mathbb{1}$, the integral reduces to

$$\begin{aligned} C_{\mathbb{1}} &= \int\limits_{B(\mathbb{Q})\cap N(\mathbb{Q})\backslash N(\mathbb{A})} \chi_s(ng)dn = \int\limits_{N(\mathbb{Q})\backslash N(\mathbb{A})} \chi_s(ng)dn \\ &= |v|^{2s} \int\limits_{N(\mathbb{Q})\backslash N(\mathbb{A})} dn = |v|^{2s}, \end{aligned} \tag{7.24}$$

where we have used the fact that the Haar measure on the compact space $N(\mathbb{Q})\backslash N(\mathbb{A})$ is normalised to 1, and applied the definition (7.10) of the character χ_s for the Iwasawa decomposed group element g.

7.2.2 Non-trivial Weyl Word

When $w = w_{\text{long}}$ is the non-trivial Weyl reflection in (7.23), it is clear that we have a trivial intersection,

$$w^{-1}B(\mathbb{Q})w \cap N(\mathbb{Q}) = \{\mathbb{1}\}, \tag{7.25}$$

and hence the integral for the non-trivial Weyl word simplifies to

$$C_w = \int\limits_{w^{-1}B(\mathbb{Q})w\cap N(\mathbb{Q})\backslash N(\mathbb{A})} \chi_s(wng)dn = \int\limits_{N(\mathbb{A})} \chi_s(wng)dn. \tag{7.26}$$

To evaluate this non-compact integral we first note that we can restrict the argument to $\chi_s(wng) = \chi_s(wna)$ since we integrate over $N(\mathbb{A})$ and χ_s is trivial on $K_{\mathbb{A}}$. Therefore, we have to evaluate

$$\int\limits_{N(\mathbb{A})} \chi_s(wna)dn. \tag{7.27}$$

Now we choose a parametrisation of $N(\mathbb{A})$ by

$$N(\mathbb{A}) = \left\{ \begin{pmatrix} 1 & u \\ & 1 \end{pmatrix} \mid u \in \mathbb{A} \right\}, \tag{7.28}$$

and of a as in (7.7) to write the integral explicitly as

$$\int_{N(\mathbb{A})} \chi_s(wna)dn = \int_{\mathbb{A}} \chi_s \Bigg(\underbrace{\begin{pmatrix} & 1 \\ -1 & \end{pmatrix}}_{w} \underbrace{\begin{pmatrix} 1 & u \\ & 1 \end{pmatrix}}_{n} \underbrace{\begin{pmatrix} v & \\ & v^{-1} \end{pmatrix}}_{a} \Bigg) du. \tag{7.29}$$

We now want to separate out how the integral depends on a. This is done by writing

$$wna = waa^{-1}na = (waw^{-1})w(a^{-1}na). \tag{7.30}$$

The a-dependence comes from both parentheses in this relation. The factor in the first parenthesis is

$$waw^{-1} = \begin{pmatrix} & 1 \\ -1 & \end{pmatrix} \begin{pmatrix} v & \\ & v^{-1} \end{pmatrix} \begin{pmatrix} & -1 \\ 1 & \end{pmatrix} = \begin{pmatrix} v^{-1} & \\ & v \end{pmatrix} \tag{7.31}$$

and lies in $A(\mathbb{A})$. It can therefore be extracted from the character χ_s using $\chi_s(waw^{-1}) = |v|^{-2s}$ by the definition (7.10) of χ_s and using the multiplicative properties of χ_s.

The factor in the second parenthesis in (7.30) is a conjugation of $N(\mathbb{A})$ by a diagonal element a and can be undone by a change of integration variable. Explicitly, we have

$$a^{-1}na = \begin{pmatrix} v^{-1} & \\ & v \end{pmatrix} \begin{pmatrix} 1 & u \\ & 1 \end{pmatrix} \begin{pmatrix} v & \\ & v^{-1} \end{pmatrix} = \begin{pmatrix} 1 & v^{-2}u \\ & 1 \end{pmatrix}. \tag{7.32}$$

Making the change of variables $u \to v^{-2}u$ that maps the (Haar) measure $du \to |v|^2 du$, we can combine the contributions from the two parentheses in (7.30) to obtain the a-dependence:

$$\int_{\mathbb{A}} \chi_s(wna)dn = \underbrace{|v|^{-2s}}_{\chi_s(waw^{-1})} \underbrace{|v|^2}_{\text{change of } du} \int_{\mathbb{A}} \chi_s \Bigg(\underbrace{\begin{pmatrix} & 1 \\ -1 & \end{pmatrix}}_{w} \underbrace{\begin{pmatrix} 1 & u \\ & 1 \end{pmatrix}}_{n} \Bigg) du. \tag{7.33}$$

In order to evaluate the remaining integral, we rewrite the character according to

$$\chi_s(wn) = \chi_s(wnw^{-1}w) = \chi(wnw^{-1}), \tag{7.34}$$

where we have used that we have embedded the Weyl group in $K_{\mathbb{A}}$ and the fact that χ_s is right-invariant under $K_{\mathbb{A}}$. Inserting the explicit parametrisations for w and n we find

$$wnw^{-1} = \begin{pmatrix} 1 & \\ -u & 1 \end{pmatrix}. \tag{7.35}$$

We see that the Weyl transformation w maps the upper triangular element into a lower triangular element as expected since the non-trivial w maps the (unique) positive root of $SL(2,\mathbb{A})$ to the unique negative root. To evaluate the character χ_s we will need to perform an Iwasawa decomposition of its argument; see Example 3.8. By Langlands' theory (see [443]) the integral (7.33) enjoys complete factorisation into a product

$$\int_{\mathbb{A}} \chi_s\left(\begin{pmatrix} & 1 \\ -1 & \end{pmatrix}\begin{pmatrix} 1 & u \\ & 1\end{pmatrix}\right)du = \prod_{p\leq\infty}\int_{\mathbb{Q}_p}\chi_{s,p}\left(\begin{pmatrix} 1 & 0 \\ -u & 1\end{pmatrix}\right)du, \tag{7.36}$$

such that one can analyse the integrals for each prime p separately.

Archimedean place $p = \infty$. We first prove the following result for the archimedean integral corresponding to the real prime at infinity $\mathbb{Q}_\infty = \mathbb{R}$.

Lemma 7.4 ($SL(2,\mathbb{R})$ **Gindikin–Karpelevich formula)** *The above integral (7.36) evaluates at the archimedean place to*

$$\int_{\mathbb{R}} \chi_{s,\infty}\left(\begin{pmatrix} 1 & 0 \\ -u & 1\end{pmatrix}\right)du = \sqrt{\pi}\frac{\Gamma(s-1/2)}{\Gamma(s)}. \tag{7.37}$$

Proof At the archimedean place, we denote the coordinates on the space $SL(2,\mathbb{R})/SO(2,\mathbb{R})$ by x and $y^{1/2}$ rather than u and v as shown in (7.9). The integral then becomes

$$\int_{-\infty}^{\infty} \chi_{s,\infty}\left(\begin{pmatrix} 1 & 0 \\ -x & 1\end{pmatrix}\right)dx. \tag{7.38}$$

In order to evaluate this we must bring the argument of the character into Iwasawa form, i.e., we must find $n \in N$ and $a \in A$ such that

$$\begin{pmatrix} 1 & 0 \\ -x & 1\end{pmatrix} = nak, \tag{7.39}$$

for some $k \in K_\infty = SO(2,\mathbb{R})$. This was done in Example 3.8 with the result (see (3.59)) that

$$\begin{pmatrix} 1 & 0 \\ -x & 1\end{pmatrix} = \begin{pmatrix} 1 & \frac{-x}{1+x^2} \\ & 1\end{pmatrix}\begin{pmatrix} \sqrt{1+x^2}^{-1} & \\ & \sqrt{1+x^2}\end{pmatrix}k, \tag{7.40}$$

and the character therefore evaluates to

$$\chi_{s,\infty}\left(\begin{pmatrix} 1 & 0 \\ -x & 1\end{pmatrix}\right) = \chi_{s,\infty}\left(\begin{pmatrix} \sqrt{1+x^2}^{-1} & \\ & \sqrt{1+x^2}\end{pmatrix}\right) = \left(\sqrt{1+x^2}\right)^{-2s}. \tag{7.41}$$

We then find for the integral

$$\int_{-\infty}^{\infty} \chi_{s,\infty}\left(\begin{pmatrix} 1 & 0 \\ -x & 1 \end{pmatrix}\right)dx = \int_{-\infty}^{\infty} \left(\sqrt{1+x^2}\right)^{-2s} dx = \sqrt{\pi}\frac{\Gamma(s-1/2)}{\Gamma(s)}. \tag{7.42}$$

□

Non-archimedean places $p < \infty$. We shall now prove the corresponding result for finite primes.

Lemma 7.5 ($SL(2,\mathbb{Q}_p)$ **Gindikin–Karpelevich formula)**

$$\int_{\mathbb{Q}_p} \chi_{s,p}\left(\begin{pmatrix} 1 & 0 \\ -u & 1 \end{pmatrix}\right)du = 1 + \frac{p-1}{p}\frac{p^{-2s+1}}{1-p^{-2s+1}} = \frac{1-p^{-2s}}{1-p^{-2s+1}}. \tag{7.43}$$

Proof We now consider the terms in (7.36) for which p is a finite prime:

$$\int_{\mathbb{Q}_p} \chi_{s,p}\left(\begin{pmatrix} 1 & 0 \\ -u & 1 \end{pmatrix}\right)du. \tag{7.44}$$

The Iwasawa decomposition of this element was also discussed in Example 3.8. When $u \in \mathbb{Z}_p$ we have

$$\begin{pmatrix} 1 & 0 \\ -u & 1 \end{pmatrix} \in SL(2,\mathbb{Z}_p). \tag{7.45}$$

The compact part of $SL(2,\mathbb{Q}_p)$ is $K_p = SL(2,\mathbb{Z}_p)$ and hence, since χ_s is trivial on K, we have

$$\chi_{s,p}\left(\begin{pmatrix} 1 & 0 \\ -u & 1 \end{pmatrix}\right) = 1 \quad \text{for } u \in \mathbb{Z}_p. \tag{7.46}$$

We may thus split the integral into

$$\int_{\mathbb{Z}_p} du + \int_{\mathbb{Q}_p\setminus\mathbb{Z}_p} \chi_{s,p}\left(\begin{pmatrix} 1 & 0 \\ -u & 1 \end{pmatrix}\right)du, \tag{7.47}$$

where the first term is unity by choice of normalisation (2.18) for the measure du. When $u \in \mathbb{Q}_p$ but not in $\mathbb{Z}_p$, i.e., $|u|_p > 1$, we write the matrix in an Iwasawa decomposition:

$$\begin{pmatrix} 1 & 0 \\ -u & 1 \end{pmatrix} = \begin{pmatrix} 1 & * \\ & 1 \end{pmatrix}\begin{pmatrix} u^{-1} & \\ & u \end{pmatrix}k. \tag{7.48}$$

(The ambiguity in the Iwasawa decomposition mentioned in Example 3.8 can be seen to not affect the result below.) The remaining integral becomes

$$\int_{\mathbb{Q}_p\setminus\mathbb{Z}_p} \chi_{s,p}\left(\begin{pmatrix} 1 & * \\ & 1 \end{pmatrix}\begin{pmatrix} u^{-1} & \\ & u \end{pmatrix}k\right)du = \int_{\mathbb{Q}_p\setminus\mathbb{Z}_p} |u|_p^{-2s}du. \tag{7.49}$$

We recognise this as an integral of the type (2.25) that we already evaluated, so the result is

$$\int_{\mathbb{Q}_p\setminus\mathbb{Z}_p} |u|_p^{-2s} du = \frac{p-1}{p}\frac{p^{-2s+1}}{1-p^{-2s+1}}. \tag{7.50}$$

Combining this with the first term in (7.47) proves the claim. □

Remark 7.6 Integrals of this type will be evaluated more generally in (8.36), leading to a more general *Gindikin–Karpelevich formula.*

7.2.3 The Global Form of the Full Constant Term

We are now ready to assemble all the pieces and write down the complete result for the constant term $W_1^\circ(s,g)$. The only remaining step is to compute the product over all finite primes in (7.36). Recalling the Euler product representation of the Riemann zeta function (1.22), we find

$$\prod_{p<\infty}\int_{\mathbb{Q}_p}\chi_{s,p}\left(\begin{pmatrix}1&0\\-u&1\end{pmatrix}\right)du = \prod_{p<\infty}\frac{1-p^{-2s}}{1-p^{-2s+1}} = \frac{\zeta(2s-1)}{\zeta(2s)}. \tag{7.51}$$

Combining this with the result from the archimedean integral (7.42), including the overall prefactor from (7.33), as well as the contribution from the trivial Weyl word in (7.24), we finally find

$$W_1^\circ(s,g) = |v|^{2s} + \sqrt{\pi}\frac{\Gamma(s-1/2)}{\Gamma(s)}\frac{\zeta(2s-1)}{\zeta(2s)}|v|^{-2s+2}. \tag{7.52}$$

Here, we have left $v\in\mathbb{A}$. Restricting to $g\in SL(2,\mathbb{R})$, we can write this in terms of $v=y^{1/2}$ instead, with the result that the first term scales like y^s while the second term scales as y^{1-s}. The relation between the exponents is that induced from the non-trivial Weyl reflection:

$$s\to 1-s\,. \tag{7.53}$$

Referring back to our particular parametrisation (7.4) and (7.5) of the character χ_s, we recall that for w being trivial we have $\lambda+\rho=2s\Lambda_1$, while for w being the non-trivial Weyl word we obtain $w\lambda+\rho=2(1-s)\Lambda_1$. Hence the parameter s is seen to be related by the above transformation. Recall from Section 2.7.1 that the completed zeta function is given by

$$\xi(s)=\pi^{-s/2}\Gamma(s/2)\zeta(s) \tag{7.54}$$

and satisfies the functional relation

$$\xi(s)=\xi(1-s). \tag{7.55}$$

Using this we can write the constant term in the following compact way:

$$W_1^\circ(s,g) = |v|^{2s} + \frac{\xi(2s-1)}{\xi(2s)}|v|^{-2s+2}. \tag{7.56}$$

For $v = y^{1/2}$ this agrees with the constant term in (1.17) and the statement of Theorem 7.1. We note that constant terms therefore satisfy the functional relation

$$W_1^\circ(s,g) = \frac{\xi(2-2s)}{\xi(1-2s)}W_1^\circ(1-s,g) = \frac{\xi(2s-1)}{\xi(2s)}W_1^\circ(1-s,g), \tag{7.57}$$

where the functional relation for the completed Riemann zeta function (2.88) has been used.

7.3 The Non-constant Fourier Coefficients

The Whittaker coefficient W_ψ° of (7.17b) we want to compute is given by

$$W_\psi^\circ(s,g) = \sum_{w\in\mathcal{W}} \int\limits_{w^{-1}B(\mathbb{Q})w\cap N(\mathbb{Q})\backslash N(\mathbb{A})} \chi_s(wng)\overline{\psi(n)}dn = \sum_{w\in\mathcal{W}} F_{w,\psi}, \tag{7.58}$$

where we have used the Bruhat trick (7.18) and have defined the individual contributions $F_{w,\psi}$ for each of the Weyl words. Using the Iwasawa decomposition of $g = n'ak$ and performing a change of variables, this expression can be re-written as

$$W_\psi^\circ(s,g) = \overline{\psi(n'^{-1})} \sum_{w\in\mathcal{W}} \int\limits_{w^{-1}B(\mathbb{Q})w\cap N(\mathbb{Q})\backslash N(\mathbb{A})} \chi_s(wna)\overline{\psi(n)}dn\,. \tag{7.59}$$

As for the constant term case, the sum over the Weyl group has two contributions, one each for when the Weyl word is trivial, $\mathbb{1}$, and non-trivial, w:

$$W_\psi^\circ = F_{\mathbb{1},\psi} + F_{w,\psi}\,. \tag{7.60}$$

The two contributions will be treated separately below.

Given our standard parametrisation of n by the variable u, we define the character ψ (against which we integrate) as in (2.75) as a direct product,

$$\psi = \prod_{p\le\infty} \psi_p, \tag{7.61}$$

with, for $m \in \mathbb{Q}$,

$$\psi_p(u) = \begin{cases} e^{2\pi i m u} & \text{for } p = \infty\,, \\ e^{-2\pi i [mu]} & \text{for } p < \infty\,. \end{cases} \tag{7.62}$$

The function $[\,\cdot\,]$ returns the fractional part of a p-adic number as defined in (2.28). An important point here is that we are interested in characters ψ of the continuous group $N(\mathbb{Q})$ embedded diagonally in $N(\mathbb{A})$. Therefore the coefficient m is a rational number and identical in all ψ_p.

In (7.62), we have not indicated the conductor m as a subscript on the character ψ_p, in contrast to Section 2.3, in order to keep the notation light. It is always understood that the conductor is m. Note that for the prefactor in (7.59) we have $\overline{\psi(n'^{-1})} = \psi(n')$.

7.3.1 Trivial Weyl Word

In the trivial case when w is equal to the identity matrix, the integral in (7.59) takes the form

$$\int\limits_{N(\mathbb{Q})\backslash N(\mathbb{A})} \chi_s(na)\overline{\psi(n)}dn\,. \tag{7.63}$$

As before, we use the definition $\chi_s(na) = |v|^{2s}$, which is independent of n.

The complete expression for the 'trivial' term of the Fourier coefficient is then given by

$$F_{\mathbb{1},\psi}(s,g) = \psi(n')|v|^{2s} \int\limits_{N(\mathbb{Q})\backslash N(\mathbb{A})} \overline{\psi(n)}dn\,. \tag{7.64}$$

We now proceed to write this expression as a product over all primes, including the place at infinity, as

$$F_{\mathbb{1},\psi}(s,g) = \prod_{p\leq\infty} \psi_p(n')|v|_p^{2s} \int\limits_{N(\mathbb{Z}_p)\backslash N(\mathbb{Q}_p)} \overline{\psi_p(u)}du\,. \tag{7.65}$$

This has to be evaluated separately for each prime $p \leq \infty$. Starting with the archimedean $p = \infty$, the domain of integration is $\mathbb{Z}\backslash\mathbb{R} \cong [0,1]$. This leads to the integral

$$F_{\mathbb{1},\psi,\infty}(s,g) = \psi_\infty(n')|v|_\infty^{2s} \int_0^1 e^{-2\pi i m u}du = 0, \tag{7.66}$$

since this is the integral of a periodic function (with mean value zero) over a full period. Therefore, the full Fourier coefficient vanishes for the trivial Weyl word:

$$F_{\mathbb{1},\psi}(s,g) = 0. \tag{7.67}$$

This is the reflection of a general phenomenon that will be discussed in Section 9.1 below.

7.3.2 Non-trivial Weyl Word

In the case when the Weyl word is non-trivial, with representative

$$w = \begin{pmatrix} 0 & 1 \\ -1 & 0 \end{pmatrix}, \tag{7.68}$$

the corresponding term in the Fourier coefficient reads

$$F_{w,\psi}(s,g) = \psi(n') \int_{N(\mathbb{A})} \chi_s(wna)\overline{\psi(n)}dn\,. \tag{7.69}$$

We now perform the same transformation (7.30) to remove the a-dependence from χ_s. Under a change of variables $n \to ana^{-1}$, the integration measure transforms as $dn \to |v|^2 dn$, and we obtain

$$\psi(n')|v|^2 \int_{N(\mathbb{A})} \chi_s(wan)\overline{\psi(ana^{-1})}dn\,. \tag{7.70}$$

Inserting $w^{-1}w$ in the argument before and after n, we find for the character $\chi_s(wan) = \chi_s(waw^{-1})\chi_s(wnw^{-1}w) = |v|^{-2s}\chi_s(wnw^{-1})$, where we have again used the fact that χ_s is right-invariant under a Weyl group transformation. The full expression then takes the form

$$\psi(n')|v|^{-2s+2} \int_{N(\mathbb{A})} \chi_s(wnw^{-1})\overline{\psi(ana^{-1})}dn\,. \tag{7.71}$$

Now we write the expression in the standard way as a product over all places,

$$F_{w,\psi}(s,g) = \prod_{p\leq\infty} \psi_p(n')|v|_p^{-2s+2} \int_{N(\mathbb{Q}_p)} \chi_{s,p}(wnw^{-1})\overline{\psi_p(ana^{-1})}dn, \tag{7.72}$$

and evaluate the archimedean and non-archimedean places separately.

The archimedean place $p = \infty$: The Iwasawa decomposition of wnw^{-1} is as in (7.41) and again leads to $\chi_{s,\infty}(wnw^{-1}) = \sqrt{1+x^2}^{\,-2s}$. Furthermore, the character evaluates to $\psi_\infty(ana^{-1}) = \exp(2\pi imyx)$, such that overall we obtain

$$\begin{aligned} F_{w,\psi,\infty} &= \psi_\infty(n')|y|_\infty^{-s+1} \int_{-\infty}^{\infty} \left(1+x^2\right)^{-s} e^{-2\pi imyx}dx \\ &= \frac{2\pi^s}{\Gamma(s)} y^{1/2}|m|_\infty^{s-1/2} K_{s-1/2}(2\pi|m|_\infty y)\psi_\infty(n'), \end{aligned} \tag{7.73}$$

where we have used the integral representation of the modified Bessel function given in (2.54) and $y > 0$.

The non-archimedean places $p<\infty$: We have to analyse the integral

$$F_{w,\psi,p} = \psi_p(n')|v|_p^{-2s+2} \int\limits_{N(\mathbb{Q}_p)} \chi_{s,p}(wnw^{-1})\overline{\psi_p(ana^{-1})}dn\,. \tag{7.74}$$

We will set $a = n' = 1$ along the finite primes since we are interested in the Eisenstein series as a function on $SL(2,\mathbb{R})$ only. From the Iwasawa decomposition of wnw^{-1} following (7.44) we know that $\chi_{s,p}(wnw^{-1})$ is given by

$$\chi_{s,p}(wnw^{-1}) = \max(1,|u|_p)^{-2s}, \tag{7.75}$$

where u parametrises $N(\mathbb{Q}_p)$ as in (7.28) and we have to integrate this against the appropriate character:

$$\int\limits_{\mathbb{Q}_p} \max(1,|u|_p)^{-2s} e^{2\pi i[mu]}du\,. \tag{7.76}$$

Using Example 2.20, this integral evaluates to

$$\int\limits_{\mathbb{Q}_p} \max(1,|u|_p)^{-2s} e^{2\pi i[mu]}du = \gamma_p(m)(1-p^{-2s})\frac{1-p^{-2s+1}|m|_p^{2s-1}}{1-p^{-2s+1}}\,. \tag{7.77}$$

Taking the product over all finite places yields

$$\prod_{p<\infty} F_{w,\psi,p} = \left(\prod_{p<\infty}(1-p^{-2s})\right)\left(\prod_{p<\infty}\gamma_p(m)\frac{1-p^{-(2s-1)}|m|_p^{2s-1}}{1-p^{-(2s-1)}}\right). \tag{7.78}$$

The first factor is equal to $\zeta(2s)^{-1}$ by virtue of (1.22). We can restrict to $m\in\mathbb{Z}$ due to the occurrence of the p-adic Gaussian $\gamma_p(m)$ for all $p<\infty$, as seen in Section 2.4. Writing m then in terms of its unique prime factorisation $m=\prod q_i^{k_i}$ with q_i primes, we can rewrite the second factor (see Example 6.28). Consider first the case when $m=q^k$ for a single prime q. Then (2.16) implies that the second factor can be written as a sum over (positive) divisors of q^k:

$$\left(\prod_{p<\infty}\frac{1-p^{-(2s-1)}|q^k|_p^{2s-1}}{1-p^{-(2s-1)}}\right) = \frac{1-q^{-(2s-1)}q^{-k(2s-1)}}{1-q^{-(2s-1)}} = \frac{1-q^{-(k+1)(2s-1)}}{1-q^{-(2s-1)}}$$
$$= \sum_{d|q^k} d^{-(2s-1)}. \tag{7.79}$$

By multiplicativity of the expressions, we therefore obtain, for a general integral m,

$$\left(\prod_{p<\infty} \gamma_p(m) \frac{1-p^{-(2s-1)}|m|_p^{2s-1}}{1-p^{-(2s-1)}}\right) = \sum_{d|m} d^{-2s+1} =: \sigma_{1-2s}(m), \tag{7.80}$$

where we have used the general *divisor sum* $\sigma_{1-2s}(m)$ over positive divisor of an integer (see (1.19)).

Putting everything together we therefore obtain a non-vanishing coefficient only for integral $m \neq 0$, with value

$$W_\psi^\circ(s,g) = F_{w,\psi}(s,g) = \frac{2}{\xi(2s)} y^{1/2}|m|^{s-1/2}\sigma_{1-2s}(m)K_{s-1/2}(2\pi|m|y)\psi_\infty(n'), \tag{7.81}$$

where we have used the definition $\xi(s) = \pi^{-s/2}\Gamma(s/2)\zeta(s)$ for the completed Riemann zeta function (1.18) and where the norm on $m \in \mathbb{Z}$ is the archimedean norm, which we suppress here and in the two equations below for ease of notation.

Finally, we address the functional relation (7.13) for the non-zero Fourier coefficients. The modified Bessel function has the property $K_{s-1/2}(w) = K_{1/2-s}(w)$ for all $w > 0$. For the divisor sum one finds similarly that

$$\sigma_{1-2s}(m) = \sum_{d|m} d^{1-2s} = |m|^{1-2s} \sum_{d|m} \left(\frac{|m|}{d}\right)^{2s-1} = |m|^{1-2s}\sigma_{2s-1}(m). \tag{7.82}$$

Putting this together, we obtain

$$W_\psi^\circ(s,g) = \frac{\xi(2s-1)}{\xi(2s)} W_\psi^\circ(1-s,g), \tag{7.83}$$

where again the functional relation (2.88) of the Riemann zeta function was used. This concludes the proof of Theorem 7.1. □

8

Langlands Constant Term Formula

In this chapter, we shall provide a detailed proof of the Langlands constant term formula for Eisenstein series on an arbitrary split semi-simple and simply-connected group G. This generalises the results of the previous chapter for $G = SL(2)$. We will also discuss the general functional relation satisfied by Eisenstein series, and explain how to define and evaluate constant terms with respect to non-maximal unipotent radicals.

8.1 Statement of Theorem

We start from Definition 4.24 of the minimal parabolic Eisenstein series:

$$E(\chi, g) = \sum_{\gamma \in B(\mathbb{Q})\backslash G(\mathbb{Q})} \chi(\gamma g) \,. \tag{8.1}$$

When writing (8.1), we can allow $g \in G(\mathbb{A})$. A function on $G(\mathbb{R})$ as in (1.13) is re-obtained by setting $g = (g_\infty, 1, 1, \ldots)$, i.e., by setting the components along $G(\mathbb{Q}_p)$ equal to the identity for $p \neq \infty$. The Eisenstein series can also be thought of as a sum over the cosets $B(\mathbb{Z})\backslash G(\mathbb{Z})$ since the cosets $B(\mathbb{Q})\backslash G(\mathbb{Q})$ are in bijection with those of $B(\mathbb{Z})\backslash G(\mathbb{Z})$ (see Example 3.16 for a proof of this for $SL(2)$). However, treating the function (8.1) adelically provides us with additional tools for the analysis.

As in Section 4.5 we parametrise the character χ by

$$\chi(nak) = a^{\lambda+\rho}, \tag{8.2}$$

in terms of a weight λ of the Lie algebra; ρ is the Weyl vector defined in (3.6). Compared to Section 4.5 we mostly do not write out the global norm on the torus elements in order to keep the notation uncluttered.

Our interest in the present chapter is to evaluate the so-called constant term in the minimal parabolic subalgebra B (standard Borel). We shall prove:

Theorem 8.1 (Langlands constant term formula) *The constant term of $E(\lambda, g)$ with respect to the unipotent radical $N \subset B$ is given by*

$$\int_{N(\mathbb{Q})\backslash N(\mathbb{A})} E(\lambda, ng)dn = \sum_{w\in\mathcal{W}} a^{w\lambda+\rho} \prod_{\alpha>0\,|\,w\alpha<0} \frac{\xi(\langle\lambda|\alpha\rangle)}{\xi(1+\langle\lambda|\alpha\rangle)}, \tag{8.3}$$

where dn is the Haar measure which is normalised such that $N(\mathbb{Q})\backslash N(\mathbb{A})$ has unit volume. $\mathcal{W}$ is the Weyl group acting on the roots α as in Section 3.1.

Proof The proof of this theorem constitutes the greater part of the present chapter and is contained in Sections 8.2 to 8.7.

Clearly, the constant term (8.3) depends only on a: for $g = n'ak$ in Iwasawa form, right K-invariance and a change of integration variation reduce the dependence to a. In what follows we shall therefore define

$$C(\chi, a) = \int_{N(\mathbb{Q})\backslash N(\mathbb{A})} E(\chi, ng)dn = \int_{N(\mathbb{Q})\backslash N(\mathbb{A})} E(\chi, na)dn, \tag{8.4}$$

and we view the integral as a function on the Cartan torus. We recall at this point from (4.134) that the series (8.1) converges absolutely in the *Godement range* when λ satisfies $\mathrm{Re}\langle\lambda|H_\alpha\rangle > 1$ for all simple roots α. We assume for Sections 8.2 to 8.7 that we are in this range and discuss analytic continuation afterwards.

8.2 Bruhat Decomposition

The first step in evaluating (8.3) is to use the *Bruhat decomposition* [72]:

$$G(\mathbb{Q}) = \bigcup_{w\in\mathcal{W}} B(\mathbb{Q})wB(\mathbb{Q}), \tag{8.5}$$

which describes the group $G(\mathbb{Q})$ as a disjoint union of double cosets by the Borel subgroup $B(\mathbb{Q}) \subset G(\mathbb{Q})$. The group $\mathcal{W}$ is the Weyl group of $G(\mathbb{R})$. Clearly, we could restrict the group $B(\mathbb{Q})$ on the right to the subgroup generated by those positive step operators that are mapped to negative step operators by the action of w. The ones that stay positive are already contained in the Borel subgroup on the left. One can think of the Bruhat decomposition as the extension of the tessellation of the Cartan subalgebra into Weyl chambers to the full group.

Using the same trick as in Section 7.2 we can rewrite the constant term as

$$\begin{aligned} C(\chi, a) &= \sum_{\gamma \in B(\mathbb{Q})\backslash G(\mathbb{Q})} \int_{N(\mathbb{Q})\backslash N(\mathbb{A})} \chi(\gamma n a) dn \\ &= \sum_{w \in \mathcal{W}} \int_{w^{-1}B(\mathbb{Q})w \cap N(\mathbb{Q})\backslash N(\mathbb{A})} \chi(wna) dn\,. \end{aligned} \tag{8.6}$$

Continuing from (8.6) we look at the individual terms

$$C_w = \int_{w^{-1}B(\mathbb{Q})w \cap N(\mathbb{Q})\backslash N(\mathbb{A})} \chi(wna) dn \tag{8.7}$$

and note that the integration domain can be simplified to

$$C_w = \int_{N_w(\mathbb{A})} \chi(wna) dn, \tag{8.8}$$

where $N_w(\mathbb{A})$ is generated from a product over the positive roots that are mapped to negative roots by the given Weyl word w:

$$N_w(\mathbb{A}) = \prod_{\alpha > 0 \,|\, w\alpha < 0} N_\alpha(\mathbb{A}). \tag{8.9}$$

$N_\alpha(\mathbb{A})$ is the subgroup of $G(\mathbb{A})$ generated by the step operator E_α and the dimension of $N_w(\mathbb{A})$ is given by the length ℓ of the reduced Weyl word w. This simplification of (8.7) uses two facts:

- (i) (Upper) Borel elements that get mapped to lower Borel elements by w have trivial intersection with $N(\mathbb{Q})$ and therefore the quotient becomes trivial and leaves an integral over all of $\mathbb{A}$ in that direction.
- (ii) If an (upper) Borel element is mapped to another upper Borel element by the action of w, one is left with the integral over the corresponding quotient. However, since the part wn of the argument is then still a Borel element, the character is insensitive to it and one is left with the volume of corresponding Borel directions which is normalised to unity.

Therefore, in (8.8) we have carried out many trivial integrals and are only left with the non-trivial integrals where wn is really a lower Borel element. These integrals are non-compact.

8.3 Parametrising the Integral

We will eventually evaluate integral (8.8) using a recursive method, and we start by parametrising it conveniently. First, we need to know something about

$N_w(\mathbb{A})$ defined in (8.9). We fix a reduced expression $w = w_{i_1} w_{i_2} \cdots w_{i_\ell}$ for the Weyl word w of length $\ell = \ell(w)$ as defined in Section 3.1.1. The subscripts refer to the nodes of the Dynkin diagram of $G(\mathbb{Q})$ and w_i are the fundamental reflections that generate the Weyl group. Then one can explicitly enumerate all positive roots that are mapped to negative roots by the action of w as follows. Define

$$\gamma_k = w_{i_\ell} w_{i_{\ell-1}} \cdots w_{i_{k+1}} \alpha_{i_k}, \tag{8.10}$$

where α_{i_k} is the i_kth simple root. That this gives a valid description of the positive roots generating N_w can be checked easily by induction. Therefore we have

$$\{\alpha > 0 \mid w\alpha < 0\} = \{\gamma_i \mid i = 1, \dots, \ell\}. \tag{8.11}$$

We also note that there is a simple expression for the sum of all these roots:

$$\gamma_1 + \cdots + \gamma_\ell = \rho - w^{-1}\rho, \tag{8.12}$$

which can again be checked by induction. We note in particular $\gamma_\ell = \alpha_{i_\ell}$.

In the next step, we use the Chevalley basis notation of Section 3.1.4 to write elements $n \in N_w(\mathbb{A})$ as

$$n = x_{\gamma_1}(u_1) \cdots x_{\gamma_\ell}(u_\ell), \tag{8.13}$$

with the *Chevalley generator* $x_\alpha(u)$ being defined by

$$x_\alpha(u) = e^{uE_\alpha}, \tag{8.14}$$

where E_α is the generator of the α root space normalised to unity (for both short and long roots) and $u \in \mathbb{A}$ is the parameter of the group element. With this parametrisation, we can rewrite our individual term C_w of (8.8) as

$$C_w = \int_{\mathbb{A}^\ell} \chi\left(w x_{\gamma_1}(u_1) \cdots x_{\gamma_\ell}(u_\ell) a\right) du_1 \cdots du_\ell. \tag{8.15}$$

8.4 Obtaining the a-dependence of the Integral

It is now possible to extract the dependence on a from the integral (8.15). This is done by conjugating the abelian element a through to the left in the argument of χ. The result is

$$C_w = \int_{\mathbb{A}^\ell} \chi\left(w a x_{\gamma_1}(a^{-\gamma_1} u_1) \cdots x_{\gamma_\ell}(a^{-\gamma_\ell} u_\ell)\right) du_1 \cdots du_\ell, \tag{8.16}$$

where we have used the fact that Cartan elements act diagonally on the E_{γ_i} root spaces. In the next step we can move w past a in the argument of χ and employ the multiplicative property (4.112) of the character to obtain

$$\begin{aligned} C_w &= \chi(waw^{-1}) \int_{\mathbb{A}^\ell} \chi\left(w x_{\gamma_1}(a^{-\gamma_1} u_1) \cdots x_{\gamma_\ell}(a^{-\gamma_\ell} u_\ell)\right) du_1 \cdots du_\ell \\ &= \chi(waw^{-1}) a^{\gamma_1 + \ldots + \gamma_\ell} \int_{\mathbb{A}^\ell} \chi\left(w x_{\gamma_1}(u_1) \cdots x_{\gamma_\ell}(u_\ell)\right) du_1 \cdots du_\ell , \end{aligned} \tag{8.17}$$

where we have also rescaled the u-variables and moved the Jacobi factor outside the integral. In the form (8.17), one can read off the full a-dependence of the constant term. Using (8.2) and (8.12) we can rewrite the a-dependence of the constant term, remembering that χ is $\mathcal{W}$-invariant from the right, as

$$\begin{aligned} C_w &= a^{w^{-1}(\lambda+\rho)} a^{\rho - w^{-1}\rho} I_w \\ &= a^{w^{-1}\lambda+\rho} I_w , \end{aligned} \tag{8.18}$$

where the remaining, a-independent integral is

$$I_w = \int_{\mathbb{A}^\ell} \chi\left(x_{w\gamma_1}(u_1) \cdots x_{w\gamma_\ell}(u_\ell)\right) du_1 \cdots du_\ell , \tag{8.19}$$

and we have applied w to all the Chevalley elements and again used K-invariance of χ on the right.

8.5 Solving the Remaining Integral by Induction

We will now solve (8.19) by induction. First, we note that $w\gamma_i$ is a negative root for all i by virtue of the definition of γ_i. Therefore, the factors $x_{w\gamma_i}(u_i)$ appearing in (8.19) are elements of the lower triangular Borel subgroup of $G(\mathbb{A})$. To evaluate the character χ in (8.19) according to (8.2), we need to perform an Iwasawa decomposition and isolate the $A(\mathbb{A})$ part of the argument of the character χ. We start by Iwasawa decomposing the last Chevalley factor in the argument of the character according to

$$x_{w\gamma_\ell}(u_\ell) = n(u_\ell) a(u_\ell) k(u_\ell) . \tag{8.20}$$

As discussed around (3.16), the (negative) step operator $E_{w\gamma_\ell}$ that enters in $x_{w\gamma_\ell}(u_\ell)$ is part of an $SL(2,\mathbb{A})$ subgroup of $G(\mathbb{A})$ and the Iwasawa decomposition (8.20) takes place in that subgroup. We choose to label the $SL(2,\mathbb{A})$ subgroup by its *positive* root $-w\gamma_\ell$, so that the corresponding Cartan generator is proportional to $H_{-w\gamma_\ell}$. The problem of Iwasawa decomposing the

$SL(2,\mathbb{A})$ associated with $-w\gamma_\ell$ is different for $p=\infty$ and $p<\infty$, and these will be treated separately in Sections 8.6.1 and 8.6.2 below.

Inserting the Iwasawa decomposed (8.20) into (8.19), we can again drop the compact element on the right. Then we can conjugate $n(u_\ell)$ through to the left. For this we have to pass through the negative step operators $x_{w\gamma_i}(u_i)$ for $i=1,\ldots,\ell-1$. This produces an element that can be arranged as a product of nilpotent elements on the left times negative step operators $x_{w\gamma_i}(u_i')$ for $i=1,\ldots,\ell-1$ with different u_i'. The nilpotent elements disappear in the character and the transformation of the space of parameters u_1 to $u_{\ell-1}$ is unimodular. We tacitly perform the corresponding change of variables $u_i\to u_i'$. We therefore obtain that $n(u_\ell)$ can be completely absorbed and we are left with

$$I_w=\int_{\mathbb{A}^\ell}\chi\left(x_{w\gamma_1}(u_1)\cdots x_{w\gamma_{\ell-1}}(u_{\ell-1})a(u_\ell)\right)du_1\cdots du_\ell. \tag{8.21}$$

In the next step, we conjugate $a(u_\ell)$ to the left. This again rescales the u variables with the result that

$$\begin{aligned}I_w&=\int_{\mathbb{A}^\ell}\chi\big(a(u_\ell)x_{w\gamma_1}(a(u_\ell)^{-w\gamma_1}u_1)\cdots x_{w\gamma_{\ell-1}}(a(u_\ell)^{-w\gamma_{\ell-1}}u_{\ell-1})\big)\,du_1\cdots du_\ell\\&=\int_{\mathbb{A}}\chi(a(u_\ell))a(u_\ell)^{w(\gamma_1+\ldots+\gamma_{\ell-1})}du_\ell\cdot I_{w'},\end{aligned} \tag{8.22}$$

where we have undone the scaling of the variables at the cost of introducing a Jacobi factor and introduced w' through

$$w=w'w_{i_\ell}, \tag{8.23}$$

i.e., it is obtained from the Weyl word w by removing the right-most fundamental reflection. Relation (8.22) is the recursion formula we are after. All that remains now is to evaluate one integral over $\mathbb{A}$.

8.6 The Gindikin–Karpelevich Formula

Using the expression (8.2) for the character χ, the desired integral is

$$I_\ell=\int_{\mathbb{A}}a(u_\ell)^{\lambda+\rho+w(\gamma_1+\ldots+\gamma_{\ell-1})}du_\ell=\prod_{p\le\infty}\int_{\mathbb{Q}_p}a(u_\ell)^{\lambda+\rho+w(\gamma_1+\ldots+\gamma_{\ell-1})}du_\ell, \tag{8.24}$$

and can be evaluated for each finite and infinite prime $\le\infty$ as follows. For this one needs the explicit Iwasawa decomposition expressions for $a(u_\ell)$ that will be derived below for $p = \infty$ and $p < \infty$. We also introduce the notation

$$a(u_\ell)^{\lambda+\rho+w(\gamma_1+\ldots+\gamma_{\ell-1})} = |\phi_\ell|^{z_\ell+1} \tag{8.25}$$

with (8.2) and

$$z_\ell = \langle \lambda + w'\rho | H_{-w\gamma_\ell} \rangle - 1 = -\langle \lambda | w\gamma_\ell \rangle, \tag{8.26}$$

where we have used $a(u_\ell) = e^{\log(\phi_\ell) H_{-w\gamma_\ell}}$ to introduce the Cartan generator $H_{-w\gamma_\ell}$ of the $SL(2)$ subgroup associated with the $-w\gamma_\ell$ positive root space as in (3.16). We have also used

$$w(\gamma_1 + \cdots + \gamma_{\ell-1}) = w'\rho - \rho, \tag{8.27}$$

which follows from (8.12) together with $w = w' w_{i_\ell}$, and

$$\langle w'\rho | H_{-w\gamma_\ell} \rangle = -\langle w'\rho | w\gamma_\ell \rangle = -\langle \rho | w_\ell \gamma_\ell \rangle = \langle \rho | \alpha_{i_\ell} \rangle = 1\,, \tag{8.28}$$

since $\gamma_\ell = \alpha_{i_\ell}$, and we have normalised the symmetric bilinear form such that ρ has unit inner product with all simple roots. The precise value of ϕ_ℓ depends on whether one is at $Q_\infty = \mathbb{R} \subset \mathbb{A}$ or at $Q_p \subset \mathbb{A}$ for $p < \infty$.

The calculations for the inductive step take place in an $SL(2, \mathbb{A})$ subgroup and are therefore very similar to the ones performed in Chapter 7.

8.6.1 Integral over $\mathbb{R}$: $p = \infty$

At the archimedean place we have the result:

Lemma 8.2 (Archimedean Gindikin–Karpelevich formula)

$$\int_{\mathbb{R}} a(u_\ell)^{\lambda+\rho+w(\gamma_1+\ldots+\gamma_{\ell-1})} du_\ell = \sqrt{\pi} \frac{\Gamma(z_\ell/2)}{\Gamma\left((z_\ell+1)/2\right)}. \tag{8.29}$$

Proof If $u = u_\ell \in \mathbb{R}$, the Iwasawa decomposition (8.20) is (see Example 3.8)

$$\begin{pmatrix} 1 & 0 \\ u & 1 \end{pmatrix} = \begin{pmatrix} 1 & \frac{u}{1+u^2} \\ 0 & 1 \end{pmatrix} \begin{pmatrix} 1/\sqrt{1+u^2} & 0 \\ 0 & \sqrt{1+u^2} \end{pmatrix} k, \tag{8.30}$$

with

$$k = \frac{1}{\sqrt{1+u^2}} \begin{pmatrix} 1 & -u \\ u & 1 \end{pmatrix} \in SO(2, \mathbb{R}). \tag{8.31}$$

The diagonal matrix is $a(u_\ell)$. Substituting this into the integral (8.24) for $u = u_\ell \in \mathbb{R}$, one obtains

$$\int_{\mathbb{R}} \left(1 + u^2\right)^{-(z_\ell+1)/2} du = \sqrt{\pi} \frac{\Gamma(z_\ell/2)}{\Gamma\left((z_\ell+1)/2\right)}, \tag{8.32}$$

where (8.25) was used. □

8.6.2 Integral over $\mathbb{Q}_p$ for Finite p

We now prove the corresponding result for finite primes.

Lemma 8.3 (Non-archimedean Gindikin–Karpelevich formula) *In* (8.24), *the non-archimedean integral evaluates to*

$$\int_{\mathbb{Q}_p} a(u_\ell)^{\lambda+\rho+w(\gamma_1+\ldots+\gamma_{\ell-1})} du_\ell = \frac{1-p^{-z_\ell-1}}{1-p^{-z_\ell}}, \tag{8.33}$$

where $z_\ell > 0$.

Proof If $u = u_\ell \in \mathbb{Z}_p$, the matrix

$$\begin{pmatrix} 1 & 0 \\ u & 1 \end{pmatrix} \tag{8.34}$$

is in $SL(2,\mathbb{Z}_p)$, which is the compact part of $SL(2,\mathbb{Q}_p)$. Therefore $a(u_\ell) = 1$ in this case and the integral is trivial.

If $u \in \mathbb{Q}_p \backslash \mathbb{Z}_p$, the (non-unique) Iwasawa decomposition yields (see Example 3.8)

$$\begin{pmatrix} 1 & 0 \\ u & 1 \end{pmatrix} = \begin{pmatrix} 1 & * \\ 0 & 1 \end{pmatrix} \begin{pmatrix} u^{-1} & 0 \\ 0 & u \end{pmatrix} k\,. \tag{8.35}$$

Even though the Iwasawa decomposition is not unique, the norm of $a(u_\ell)$ is defined uniquely. We therefore obtain

$$\begin{aligned} \int_{\mathbb{Q}_p} |\phi_\ell|_p^{z_\ell+1} du &= \int_{\mathbb{Z}_p} dx + \int_{\mathbb{Q}_p\backslash\mathbb{Z}_p} |u|_p^{-z_\ell-1} du \\ &= 1 + \frac{p-1}{p}\frac{p^{-z_\ell}}{1-p^{-z_\ell}} = \frac{1-p^{-z_\ell-1}}{1-p^{-z_\ell}}, \end{aligned} \tag{8.36}$$

where we have used the integral (2.25) and the normalisation of the measure (2.18). □

8.6.3 The Global Formula

Putting the archimedean and non-archimedean contributions together, we therefore obtain the *global Gindikin–Karpelevich formula*:

$$I_\ell = \sqrt{\pi}\frac{\Gamma(z_\ell/2)}{\Gamma((z_\ell+1)/2)} \prod_{p<\infty} \frac{1-p^{-z_\ell-1}}{1-p^{-z_\ell}} = \frac{\xi(z_\ell)}{\xi(z_\ell+1)}, \tag{8.37}$$

with z_ℓ as in (8.26). We have used the Euler product formula (1.22) for the Riemann zeta function and its completion

$$\xi(s) = \pi^{-s/2}\Gamma(s/2)\zeta(s) \tag{8.38}$$

that satisfies the functional relation (see (2.88))

$$\xi(s) = \xi(1-s)\,. \tag{8.39}$$

8.7 Assembling the Constant Term

We can now write the final formula for the constant term (8.3) by assembling (8.18) and the result (8.37) inserted into the recursion relation (8.22). The answer is

$$\int_{N(\mathbb{Q})\backslash N(\mathbb{A})} E(\chi, ng)dn = \sum_{w\in\mathcal{W}} a^{w^{-1}\lambda+\rho} \prod_{\alpha>0\,|\,w\alpha<0} \frac{\xi(-\langle\lambda|w\alpha\rangle)}{\xi(1-\langle\lambda|w\alpha\rangle)}\,. \tag{8.40}$$

By relabelling the sum by $w \to w^{-1}$ we obtain the standard Langlands formula for the constant term in the minimal parabolic:

$$\int_{N(\mathbb{Q})\backslash N(\mathbb{A})} E(\chi, ng)dn = \sum_{w\in\mathcal{W}} a^{w\lambda+\rho} \prod_{\alpha>0\,|\,w\alpha<0} \frac{\xi(\langle\lambda|\alpha\rangle)}{\xi(1+\langle\lambda|\alpha\rangle)}\,. \tag{8.41}$$

We note that the inner product here is normalised such that $\langle\rho|\alpha_i\rangle = 1$ for all simple roots α_i. Often one denotes the *intertwining coefficient* or short *intertwiner* by

$$M(w,\lambda) = \prod_{\alpha>0\,|\,w\alpha<0} \frac{\xi(\langle\lambda|\alpha\rangle)}{\xi(1+\langle\lambda|\alpha\rangle)}\,. \tag{8.42}$$

This concludes the proof of Theorem 8.1. □

Remark 8.4 The derivation of the constant term formula above made heavy use of the adeles $\mathbb{A}$. This was most noticeable when evaluating the integral (8.24) that yielded the completed Riemann zeta functions in their Euler product form. Still, one may wonder whether this level of abstraction was really necessary. For $SL(2,\mathbb{R})$ one can obtain the constant terms alternatively by *Poisson resummation* techniques (see Appendix A for a summary) without ever making reference to p-adic numbers. What this requires, however, is an explicit understanding of the sum over the cosets $B(\mathbb{Z})\backslash G(\mathbb{Z})$ in the definition of the Eisenstein series and their relation to sums over integer lattices. In the general case, the description of these cosets is not easy to characterise and the lattice sum descriptions typically involve representation-theoretic constraints on the sum; this is discussed in more detail in Section 15.3. None of these details are required for obtaining the constant term formula when using the p-adic description, and this is where the power of the method lies.

8.8 Functional Relations for Eisenstein Series

The definition of the Eisenstein series (8.1) is initially restricted to the domain of (absolute) convergence of the defining sum. As mentioned before in Section 4.5.2, this requires that the weight λ entering in the definition of the character χ through the relation (8.2) must have sufficiently large real parts. More precisely, it must lie in the *Godement range*

$$\mathrm{Re}\langle\lambda|\alpha_i\rangle > 1 \qquad \text{for all simple roots } \alpha_i \tag{8.43}$$

given in (4.134). This criterion of absolute convergence can now be understood better from the intertwiners $M(w, \lambda)$ occurring in the constant term formula (8.41): the integrals giving rise to the completed Riemann zeta function $\xi(s)$ only converge absolutely for $\mathrm{Re}(s) > 1$; this has to be satisfied generically for all positive roots $\alpha > 0$ and thus $\mathrm{Re}\langle\lambda|\alpha\rangle > 1$ for all positive roots. Since all positive roots can be written as positive linear combinations of simple roots it is sufficient to require this criterion only for the simple roots α_i. For non-minimal Eisenstein series of the type discussed in Section 5.7 one has to require $\mathrm{Re}\langle\lambda|\alpha\rangle > 1$ for only those positive roots α associated with the unipotent of the non-minimal standard parabolic subgroup.

For weights λ outside the Godement range (8.43), the Eisenstein series $E(\lambda, g)$ can be defined by analytic continuation in the complexified weight λ to almost all values of λ. Note that a complex weight λ corresponds to a character χ taking values not in $U(1)$ but in $\mathbb{C}^\times$. We are mainly interested in real weights. As shown in [442, 447], this continuation is possible everywhere except for certain hyperplanes that are related to the integral weight lattice.

An important property of the Eisenstein series is that they obey functional relations. More precisely one has:

Theorem 8.5 (Functional relation for Eisenstein series) *For each $w \in \mathcal{W}$ the Eisenstein series $E(\lambda, g)$ satisfies the functional relation* [447]

$$E(\lambda, g) = M(w, \lambda)E(w\lambda, g)\,. \tag{8.44}$$

In other words, the Eisenstein series along the Weyl orbit of a character are all proportional to each other.

Proof To prove this, note first that the coefficients $M(w, \lambda)$ given by (8.42) satisfy the following property:

Lemma 8.6 (Multiplicative property of intertwiners $M(w, \lambda)$) *For all w_1, $w_2 \in \mathcal{W}$ one has*

$$M(w_1 w_2, \lambda) = M(w_1, w_2\lambda)M(w_2, \lambda)\,. \tag{8.45}$$

Proof (of Lemma 8.2) Consider the case when $w_1 = w'$ is an arbitrary Weyl word and $w_2 = w_{i_\ell}$ a fundamental reflection. Proving the relation (8.45) for this case suffices to show it for arbitrary products $w_1 w_2$ since one can always separate the letters one by one from the right and then recombine them. The product $w = w_1 w_2 = w' w_{i_\ell}$ is of the type discussed in Section 8.3 if this is a reduced expression of length $\ell = \ell(w)$. We have

$$\{\alpha > 0 \mid w\alpha < 0\} = \{\gamma_k \mid k = 1, \dots, \ell\} = \left[w_{i_\ell} \{\alpha > 0 \mid w'\alpha < 0\} \right] \cup \{\alpha_{i_\ell}\} \tag{8.46}$$

by using the explicit form of γ_k in (8.10), where w_{i_ℓ} always appears on the left in any γ_k for $k < \ell$. Therefore

$$\begin{aligned} M(w' w_{i_\ell}, \lambda) &= \left[\prod_{\substack{\alpha<0 \\ w'\alpha<0}} \frac{\xi(\langle \lambda | w_{i_\ell} \alpha \rangle)}{\xi(1 + \langle \lambda | w_{i_\ell} \alpha \rangle)} \right] \frac{\xi(\langle \lambda | \alpha_{i_\ell} \rangle)}{\xi(1 + \langle \lambda | \alpha_{i_\ell} \rangle)} \\ &= \left[\prod_{\substack{\alpha<0 \\ w'\alpha<0}} \frac{\xi(\langle w_{i_\ell} \lambda | \alpha \rangle)}{\xi(1 + \langle w_{i_\ell} \lambda | \alpha \rangle)} \right] \prod_{\substack{\alpha>0 \\ w_{i_\ell}\alpha<0}} \frac{\xi(\langle \lambda | \alpha \rangle)}{\xi(1 + \langle \lambda | \alpha \rangle)} \\ &= M(w', w_{i_\ell} \lambda) M(w_{i_\ell}, \lambda), \end{aligned} \tag{8.47}$$

where we have used the Weyl invariance of the bilinear form $\langle \lambda | w_{i_\ell} \alpha \rangle = \langle w_{i_\ell} \lambda | \alpha \rangle$ and the fact that the only positive root that is mapped to a negative root by the fundamental Weyl reflection w_{i_ℓ} is the simple root α_{i_ℓ} itself. This is the desired relation.

If the expression $w = w' w_{i_\ell}$ is not reduced, there are common factors between the two intertwiners that cancel. □

The functional relation (8.44) for the constant term of the Eisenstein series, now follows from this result applied to the constant term formula (8.41). The fact that this also extends to the non-constant terms was shown by Langlands [447]. This completes the proof of Theorem 8.5. □

Remark 8.7 The functional relation (8.44) shows the limitations of the analytic continuation: for weights λ and Weyl words $w \in \mathcal{W}$ for which $M(w, \lambda)$ is not a non-zero finite number, the relation appears ill-defined. This can only happen for λ on certain hyperplanes as indicated above. Another apparent limitation of the functional relation is that, if one chooses a Weyl word w that stabilises the weight λ, one would require $M(w, \lambda) = 1$. This is not guaranteed to be true. For generic λ, (8.44) provides a valid relation. The remaining cases are those when $E(\lambda, g)$ actually develops poles or zeroes (as a function of λ) and one has to consider appropriate normalising factors in the functional relation to make it well-defined. This topic will be discussed in more detail in Chapter 10.

Remark 8.8 The quantity $M(w,\lambda)$ is called an *intertwiner* because it intertwines between different isomorphic principal series representations. The principal series representation $I(\lambda)$ was defined in Section 5.3 and it turns out that the representations $I(\lambda)$ and $I(w\lambda)$ are isomorphic for $w \in \mathcal{W}$. The spherical vectors in these representations are given by the Eisenstein series $E(\lambda,g)$ and $E(w\lambda,g)$. The intertwining isomorphism of the representations reduces to multiplication by $M(w,\lambda)$ on the spherical vectors and this is expressed in the functional relation (8.44).

8.9 Expansion in Maximal Parabolics*

In the previous sections, we have explained how to expand Eisenstein series along a minimal parabolic subgroup, with unipotent radical N. It is also possible to make an expansion along different parabolic subgroups with smaller unipotent radical. The analogue of the constant term (8.3) then retains a dependence on some of the coordinates on $N(\mathbb{A})$ since only a subset of $N(\mathbb{A})$ is integrated out. In this section, we state and prove a theorem giving the formula for the constant terms of $E(\lambda,g)$ in an expansion along a maximal parabolic subgroup.

The maximal parabolic subgroup, which we denote by $P_{j_\circ}$ similar to Section 3.1.3, is defined with respect to a particular choice of simple root $\alpha_{j_\circ}$ (i.e., node $j_\circ$ in the Dynkin diagram) of G. The maximal parabolic subgroup has a Levi and Langlands decomposition $P_{j_\circ} = L_{j_\circ}U_{j_\circ} = M_{j_\circ}GL(1)U_{j_\circ}$.

According to Definition 6.17, the constant term along $U_{j_\circ}$ is given by

$$C_{U_{j_\circ}}(\lambda,g) = \int\limits_{U_{j_\circ}(\mathbb{Q})\backslash U_{j_\circ}(\mathbb{A})} E(\lambda,ug)du\,. \tag{8.48}$$

The subscript $j_\circ$ indicates the restriction to the maximal parabolic subgroup. As in the case of the minimal parabolic expansion, a similar constant term formula can be derived for this case. Upon deleting the $j_\circ$th node from the Dynkin diagram of G, we will denote by $M_{j_\circ}$ the group associated with the Dynkin diagram which is left. We also note that the Levi factor $L_{j_\circ}$ of $P_{j_\circ}$ can then be written as $L_{j_\circ} = GL(1) \times M_{j_\circ}$, where the one-parameter group $GL(1)$ is parametrised by a single variable in $\mathbb{A}^\times$.

Let us state the formula for the constant term in this maximal parabolic:

Theorem 8.9 (Constant terms in maximal parabolics) *The constant term of $E(\lambda, g)$ with respect to the unipotent radical $U_{j_\circ} \subset P_{j_\circ}$ is given by* [488]

$$\int\limits_{U_{j_\circ}(\mathbb{Q})\backslash U_{j_\circ}(\mathbb{A})} E(\lambda, ug)du = \sum_{w\in\mathcal{W}_{j_\circ}\backslash\mathcal{W}} e^{\langle (w\lambda+\rho)_{\parallel j_\circ}|H(g)\rangle} M(w,\lambda)E^{M_{j_\circ}}(\chi_w, g)\ . \tag{8.49}$$

Below we provide a concise explanation of the notation used and give a sketch of the proof of the formula.

In Equation (8.49), the Weyl group of $M_{j_\circ}$ is denoted by $\mathcal{W}_{j_\circ}$ and the sum on the right-hand side is then restricted to a sum over a coset of the Weyl group, in contrast to the minimal parabolic case. Furthermore the projection operators $(\cdot)_{\parallel j_\circ}$ and $(\cdot)_{\perp j_\circ}$ were defined in (5.108) as follows when acting on a weight $\lambda \in \mathfrak{h}^*$:

$$\lambda_{\parallel j_\circ} := \frac{\langle \Lambda_{j_\circ}|\lambda\rangle}{\langle \Lambda_{j_\circ}|\Lambda_{j_\circ}\rangle}\Lambda_{j_\circ}, \tag{8.50a}$$

$$\lambda_{\perp j_\circ} := \lambda - (\lambda)_{\parallel j_\circ}. \tag{8.50b}$$

These correspond, respectively, to the components of λ parallel and orthogonal to the fundamental weight $\Lambda_{j_\circ}$. The orthogonal component is given by a linear combination of simple roots of $M_{j_\circ}$. The character χ_w follows the definition (8.2), but with λ now replaced by the weight $(w\lambda)_{\perp j_\circ}$. The $M_{j_\circ}(\mathbb{Q})$-invariant Eisenstein series on the right-hand-side of the equation is independent of the $GL(1)$ factor in the decomposition of the Levi subgroup $L_{j_\circ}$, as this dependence is projected out using the $(\cdot)_{\perp j_\circ}$ operator and appears solely through the exponential prefactor. Note that for simplicity of notation we have put g in the argument of the Eisenstein series on the right, even though g lies effectively in $M_{j_\circ}$.

We also note that the formula (8.49) is well-defined and independent of the choice of coset representative due to the functional relation (8.44).

Proof Sources for the analysis presented here are [488, 281]. Consider two parabolic subgroups $P_1(\mathbb{A}) = L_1(\mathbb{A})U_1(\mathbb{A})$ and $P_2(\mathbb{A}) = L_2(\mathbb{A})U_2(\mathbb{A})$ of $G(\mathbb{A})$. The first one we take to be the one defining an Eisenstein series through

$$E(\chi, g) = \sum_{\gamma\in P_1(\mathbb{Q})\backslash G(\mathbb{Q})} \chi(\gamma g), \tag{8.51}$$

where $\chi\colon P_1(\mathbb{Q})\backslash P_1(\mathbb{A}) \to \mathbb{C}^\times$ is a character on the parabolic $P_1(\mathbb{A})$. The parabolic subgroup $P_2(\mathbb{A}) = L_2(\mathbb{A})U_2(\mathbb{A})$ is used to define the constant term along $U_2(\mathbb{A})$ via

$$C_{U_2}(\chi,g) = \int\limits_{U_2(\mathbb{Q})\backslash U_2(\mathbb{A})} E(\chi,ug)du. \tag{8.52}$$

By the definition of the integral, the result is determined by its dependence on $g = l \in L_2(\mathbb{A})$ and we will restrict to the Levi factor now. Most of the steps in the evaluation of (8.52) will be very similar to those in Section 8.2.

Using the Bruhat decomposition, we can rewrite the integral (8.52) as

$$C_{U_2}(\chi,l) = \sum_{w\in\mathcal{W}_1\backslash\mathcal{W}/\mathcal{W}_2} C_{w,U_2}(\chi,l)\,. \tag{8.53}$$

Here, $\mathcal{W}_1$ and $\mathcal{W}_2$ are the Weyl groups of L_1 and L_2, respectively. The individual term of the double coset of the Weyl group $\mathcal{W}$ is

$$C_{w,U_2}(\chi,l) = \sum_{\delta\in w^{-1}P_1(\mathbb{Q})w\cap P_2(\mathbb{Q})\backslash P_2(\mathbb{Q})} \int\limits_{U_2(\mathbb{Q})\backslash U_2(\mathbb{A})} \chi(w\delta ul)du. \tag{8.54}$$

Now the sum over $\delta \in P_2(\mathbb{Q})$ can be split into the Levi and unipotent part according to $\delta = \gamma_l\gamma_u$ and then one can unfold the sum over γ_u onto the integration domain as in (8.6). The result is

$$C_{w,U_2}(\chi,l) = \sum_{\gamma_l\in w^{-1}P_1(\mathbb{Q})w\cap L_2(\mathbb{Q})\backslash L_2(\mathbb{Q})} \int\limits_{w^{-1}P_1(\mathbb{Q})w\cap U_2(\mathbb{Q})\backslash U_2(\mathbb{A})} \chi(w\gamma_l ul)du. \tag{8.55}$$

We now specialise to $P_2(\mathbb{A})$ being maximal parabolic. Then the Levi factor takes the form (see (3.43))

$$L_2(\mathbb{A}) = GL(1,\mathbb{A}) \times M_2(\mathbb{A}) \tag{8.56}$$

with M_2 semi-simple, and we parametrise the $L_2(\mathbb{A})$ element as $l = rm$. We next separate out the dependence on the $GL(1,\mathbb{A})$ element r by moving it to the left within χ. This leads to

$$\begin{aligned} C_{w,U_2}(\chi,rm) \\ = r^{w^{-1}\lambda+\rho} \sum_{\gamma_m\in w^{-1}P_1(\mathbb{Q})w\cap M_2(\mathbb{Q})\backslash M_2(\mathbb{Q})} \int\limits_{w^{-1}P_1(\mathbb{Q})w\cap U_2(\mathbb{Q})\backslash U_2(\mathbb{A})} \chi(wu\gamma_m m)du, \end{aligned} \tag{8.57}$$

by combining the contribution from $\chi(wrw^{-1})$ and the change of the measure du. Note also that we have changed the summation over the $L_2(\mathbb{Q})$ cosets to one over $M_2(\mathbb{Q})$ cosets since the two agree. We have also interchanged γ_m and u as the corresponding change of variables is unimodular ($\gamma_m \in M_2(\mathbb{Q})$ is discrete).

Let us analyse the properties of the integral

$$I = \int_{U_2^w(\mathbb{Q})\backslash U_2(\mathbb{A})} \chi(wu\gamma_m m)du, \tag{8.58}$$

which is a function from $M_2(\mathbb{A}) \to \mathbb{C}$. We also define for simplicity

$$M_2^w(\mathbb{Q}) := w^{-1}P_1(\mathbb{Q})w \cap M_2(\mathbb{Q}) \quad \text{and} \quad U_2^w(\mathbb{Q}) := w^{-1}P_1(\mathbb{Q})w \cap U_2(\mathbb{Q}), \tag{8.59}$$

and we note that $M_2^w(\mathbb{Q})$ is a parabolic subgroup of $M_2(\mathbb{Q})$. The integral (8.58) is invariant under $\epsilon \in M_2^w(\mathbb{Q})$ according to

$$\int_{U_2^w(\mathbb{Q})\backslash U_2(\mathbb{A})} \chi(wu\epsilon\gamma_m m)du = \int_{U_2^w(\mathbb{Q})\backslash U_2(\mathbb{A})} \chi(w\epsilon u\gamma_m m)du = I, \tag{8.60}$$

since $\epsilon = w^{-1}p_1 w$ for some $p_1 \in P_1(\mathbb{Q})$ and the character $\chi\colon P_1(\mathbb{Q})\backslash P_1(\mathbb{A}) \to \mathbb{C}^\times$ is invariant under $P_1(\mathbb{Q})$. But this means that

$$\chi_2^w(\gamma_m m) = \int_{U_2^w(\mathbb{Q})\backslash U_2(\mathbb{A})} \chi(wu\gamma_m m)du \tag{8.61}$$

is a character $\chi_2^w\colon M_2^w(\mathbb{Q})\backslash M_2(\mathbb{A}) \to \mathbb{C}^\times$ if it is non-zero. (The integral I serves as an intertwiner from characters on $P_1(\mathbb{A})$ to characters on $M_2^w(\mathbb{A})$.)

Inserting this back into the individual constant term (8.57), we obtain

$$C_{w,U_2}(\chi, rm) = r^{w^{-1}\lambda+\rho} \sum_{\gamma_m \in M_w^2(\mathbb{Q})\backslash M_2(\mathbb{Q})} \chi_2^w(\gamma_m m). \tag{8.62}$$

The sum over γ_m now produces an Eisenstein series on $M_2(\mathbb{A})$ so that

$$C_{U_2}(\chi, rm) = \sum_{w\in\mathcal{W}_1\backslash\mathcal{W}/\mathcal{W}_2} r^{w^{-1}\lambda+\rho} E^{M_2}(\chi_2^w, m), \tag{8.63}$$

where we have indicated that the Eisenstein series is on $M_2(\mathbb{A})$. This expression can be simplified a bit more by identifying the character χ_2^w in terms of a weight of $\mathfrak{m}_2 = \mathrm{Lie}(M_2)$. To this end we evaluate (8.61) at a semi-simple element of $M_2(\mathbb{A})$, i.e., $m = a$. This leads to

$$\begin{aligned}\chi_2^w(a) &= \int_{U_2^w(\mathbb{Q})\backslash U_2(\mathbb{A})} \chi(wua)du = \chi(waw^{-1})\delta_{\bar{U}_2^w}(a) \int_{U_2^w(\mathbb{Q})\backslash U_2(\mathbb{A})} \chi(wu)du \\ &= M(w^{-1},\lambda)a^{(w^{-1}\lambda+\rho)_{M_2}},\end{aligned} \tag{8.64}$$

where the last symbol denotes the orthogonal projection onto the space of M_2 weights. The exponent comes about as follows. The character $\chi(waw^{-1})$ evaluates to $a^{w^{-1}(\lambda+\rho)}$ and the modulus character on $\bar{U}_2^w$ is determined by the sum over all roots of U_2 that are not mapped to roots of P_1, i.e., the total exponent of a is

$$w^{-1}(\lambda + \rho) + \sum_{\alpha \in \Delta(\mathfrak{u}_2) | w\alpha \notin \Delta(\mathfrak{p}_1)} \alpha = (w^{-1}\lambda + \rho)_{M_2}. \tag{8.65}$$

We have furthermore made use of our knowledge of the Gindikin–Karpelevich type integral; see the evaluation of (8.19). In summary we arrive at

$$C_{U_2}(\chi, rm) = \sum_{w \in \mathcal{W}_1 \backslash \mathcal{W} / \mathcal{W}_2} r^{w^{-1}\lambda+\rho} M(w^{-1}, \lambda) E^{M_2}((w^{-1}\lambda)_{M_2}, m), \tag{8.66}$$

in agreement with (8.49) if one replaces w by w^{-1} which maps the double coset to $\mathcal{W}_2 \backslash \mathcal{W} / \mathcal{W}_1$. □

9

Whittaker Coefficients of Eisenstein Series

In this chapter, we derive Theorem 9.1, which states the formula of Casselman–Shalika [130] (see also [569]) for the *local* abelian Fourier coefficients in the minimal parabolic (Borel) subgroup $B(\mathbb{A}) \subset G(\mathbb{A})$ for the Eisenstein series $E(\chi, g)$. This formula is used to evaluate Fourier integrals with a generic character ψ. By the discussion in Chapter 6 the global form of these Fourier coefficients is captured by the spherical Whittaker coefficient

$$W_\psi^\circ(\chi, g) = \int_{N(\mathbb{Q})\backslash N(\mathbb{A})} E(\chi, ng)\overline{\psi(n)}dn \tag{9.1}$$

for a (quasi-)character $\chi\colon B(\mathbb{A}) \to \mathbb{C}^\times$ and a general unitary character $\psi\colon N(\mathbb{Q})\backslash N(\mathbb{A}) \to U(1)$. For $SL(2, \mathbb{A})$ we have already evaluated this integral in Section 7.3. As in that section, a useful strategy is to factorise the integral and perform it at all places separately. It will turn out that only for the finite primes $p < \infty$ and so-called generic and *unramified characters* ψ (to be defined below) a nice and compact formula exists. In Sections 9.4 and 9.5, we will explain how to also evaluate (9.1) for arbitrary generic or even degenerate characters ψ.

9.1 Reduction of the Integral and the Longest Weyl Word

To begin with, we bring the integral (9.1) into a form that is more amenable to evaluation. As discussed in Section 6.3.3, the spherical Whittaker coefficient satisfies

$$W_\psi^\circ(\chi, ngk) = \psi(n)W_\psi^\circ(\chi, g) \tag{9.2}$$

and is therefore determined by its values on the abelian group $A(\mathbb{A})$ due to the Iwasawa decomposition (3.22) and (3.52). Hence, we will only consider it for elements $g = a \in A(\mathbb{A})$ in the sequel. For the discussion in this section we assume the character ψ to be generic (see Definition 6.10), i.e., it does not vanish on any simple root generator.

We start evaluating (9.1) by applying the Bruhat decomposition as for the constant term, to obtain

$$\begin{aligned} W^{\circ}_{\psi}(\chi, a) &= \sum_{\gamma\in B(\mathbb{Q})\backslash G(\mathbb{Q})} \int_{N(\mathbb{Q})\backslash N(\mathbb{A})} \chi(\gamma n a)\overline{\psi(n)}dn \\ &= \sum_{w\in\mathcal{W}} \int_{w^{-1}B(\mathbb{Q})w\cap N(\mathbb{Q})\backslash N(\mathbb{A})} \chi(wna)\overline{\psi(n)}dn\,. \end{aligned} \tag{9.3}$$

From the last line we define $W^{\circ}_{\psi}(\chi, a) = \sum_{w\in\mathcal{W}} F_{w,\psi}$, where

$$F_{w,\psi} = \int_{w^{-1}B(\mathbb{Q})w\cap N(\mathbb{Q})\backslash N(\mathbb{A})} \chi(wna)\overline{\psi(n)}dn \tag{9.4}$$

is the contribution from the Weyl word w.

Let us start with an analysis of the integration range of the Fourier integral (9.4), given by the coset $w^{-1}B(\mathbb{Q})w\cap N(\mathbb{Q})\backslash N(\mathbb{A})$, and the corresponding contribution to $W^{\circ}_{\psi}(\chi, a)$. It is clear that the intersection in the denominator of this coset consists of those elements of the (upper) Borel subgroup that are mapped to upper Borel elements under the action of the Weyl element w. For the whole denominator we can therefore write

$$w^{-1}B(\mathbb{Q})w\cap N(\mathbb{Q}) = \prod_{\substack{\alpha>0\\ w\alpha>0}} N_{\alpha}(\mathbb{Q})\,. \tag{9.5}$$

With this, the integration range conveniently splits up in the following way:

$$w^{-1}B(\mathbb{Q})w\cap N(\mathbb{Q})\backslash N(\mathbb{A}) \simeq \left(\prod_{\substack{\beta>0\\ w\beta>0}} N_{\beta}(\mathbb{Q})\backslash N_{\beta}(\mathbb{A})\right)\cdot\left(\prod_{\substack{\gamma>0\\ w\gamma<0}} N_{\gamma}(\mathbb{A})\right). \tag{9.6}$$

Let us introduce some notation. We denote the union in the first parenthesis as

$$N^{w}_{\{\beta\}} := \left(\prod_{\substack{\beta>0\\ w\beta>0}} N_{\beta}(\mathbb{Q})\backslash N_{\beta}(\mathbb{A})\right) \tag{9.7}$$

and the union in the second parenthesis as

$$N^{w}_{\{\gamma\}} := \left(\prod_{\substack{\gamma>0\\ w\gamma<0}} N_{\gamma}(\mathbb{A})\right). \tag{9.8}$$

Here, the sets of roots $\{\beta\}$ and $\{\gamma\}$ contain precisely those roots which satisfy the conditions imposed on the products in (9.7) and (9.8), respectively. It is important to note that there is a qualitative difference in the two sets: $N^w_{\{\gamma\}}$ is non-compact while $N^w_{\{\beta\}}$ is *compact*. With this splitting of the integration range the contribution $F_{w,\psi}$ then takes the following form:

$$\begin{aligned} F_{w,\psi} &= \int\limits_{N^w_{\{\beta\}}N^w_{\{\gamma\}}} \chi(wna)\overline{\psi(n)}dn \\ &= \int\limits_{N^w_{\{\beta\}}} \int\limits_{N^w_{\{\gamma\}}} \chi(wn_\beta n_\gamma a)\overline{\psi(n_\beta n_\gamma)}dn_\beta dn_\gamma \,. \end{aligned} \tag{9.9}$$

Inserting $w^{-1}w$ between n_β and n_γ and splitting the unitary character up into two factors, we obtain

$$F_{w,\psi} = \int\limits_{N^w_{\{\beta\}}} \int\limits_{N^w_{\{\gamma\}}} \chi(wn_\beta w^{-1}wn_\gamma a)\overline{\psi(n_\beta)}\,\overline{\psi(n_\gamma)}dn_\beta dn_\gamma \,. \tag{9.10}$$

We recall that the character χ is left-invariant under any subgroup that is given by the exponential of positive root generators. In particular this applies to elements n of the unipotent subgroup of the (upper) Borel subgroup. Hence, by definition of $N^w_{\{\beta\}}$, the character χ is insensitive to the factor $wn_\beta w^{-1}$ in the argument and we can split off the integral over n_β, leaving us with

$$F_{w,\psi} = \int\limits_{N^w_{\{\beta\}}} \overline{\psi(n_\beta)}dn_\beta \int\limits_{N^w_{\{\gamma\}}} \chi(wn_\gamma a)\overline{\psi(n_\gamma)}dn_\gamma \,. \tag{9.11}$$

Given the form of (9.11), we see that, in the integral over n_β, effectively a periodic function is integrated over a full period in the compact space $N^w_{\{\beta\}}$. Provided that the character ψ is non-trivial along at least one simple root contained in $\{\beta\}$, this means that the whole integral will vanish. Since this is always true for generic ψ, we arrive at the conclusion

$$F_{w,\psi} = 0 \quad \text{unless } w = w_{\text{long}}. \tag{9.12}$$

This is true since all Weyl transformations except for the longest Weyl word w_{long} leave at least one simple root positive. Therefore we obtain the following expression for the Whittaker function for a generic ψ:

$$W^\circ_\psi(\chi, a) = \int\limits_{N(\mathbb{A})} \chi(w_{\text{long}}na)\overline{\psi(n)}dn \,. \tag{9.13}$$

We will also refer to this expression as the *Jacquet integral*; see [374].

9.2 Unramified Local Whittaker Functions

The integral (9.13) should now be evaluated for all places separately, that is,

$$\int_{N(\mathbb{A})} \chi(w_{\text{long}}na)\overline{\psi(n)}dn = \int_{N(\mathbb{R})} \chi_\infty(w_{\text{long}}na)\overline{\psi_\infty(n)}dn \\ \times \prod_{p<\infty} \int_{N(\mathbb{Q}_p)} \chi_p(w_{\text{long}}na)\overline{\psi_p(n)}dn \\ = W^\circ_{\psi_\infty} \times \prod_{p<\infty} W^\circ_{\psi_p}. \tag{9.14}$$

This factorisation was already discussed in Section 6.3.2. It holds for generic unitary character ψ.

However, while we will be able to derive a nice closed formula for the local places $p < \infty$, a general expression for the real place is not known, to the best of our knowledge. In the case of $SL(2, \mathbb{A})$, the resulting expression was given by the Bessel function (7.73) and for $SL(3, \mathbb{A})$ it is known that the triple integral over the three unipotent generators gives convoluted integrals of Bessel functions [605, 120]; see also Section 9.8. For general groups with 'more non-abelian' unipotent subgroups, a proliferation of this nested structure of special functions is to be expected. On the other hand, the Whittaker functions for finite primes $p < \infty$ contain the essential number-theoretic information that is reflected as instanton measures in string theory applications [308, 493] (see Section 13.6.3). Therefore we will from now on consider only the group $G(\mathbb{Q}_p)$ for $p < \infty$ and calculate the local Whittaker functions. To ease notation we shall suppress all subscripts involving primes, and in the remainder of this section write ψ for ψ_p and χ for χ_p.

9.2.1 Unramified Characters ψ

The formula of Casselman and Shalika for the local Whittaker functions is most conveniently stated when one restricts the character ψ to be *unramified* (see Definition 6.12 and [130, p. 219]). Recall from Definition 6.12 that this means that the character $\psi\colon N(\mathbb{Q}_p) \to U(1)$ has what a physicist might call 'unit instanton charges'. Suppose the element n is expanded in terms of positive step operators in a Chevalley basis as

$$n = \Big(\prod_{\alpha\in\Delta_+\setminus\Pi} x_\alpha(u_\alpha)\Big)\Big(\prod_{\alpha\in\Pi} x_\alpha(u_\alpha)\Big) \in N(\mathbb{Q}_p), \tag{9.15}$$

where we have ordered the individual factors in a convenient way. Note that for evaluating $\psi(n)$ the order does not matter since ψ is a homomorphism between abelian(ised) groups. An *unramified character* ψ is then one that satisfies

$$\psi(n) = \exp\left(2\pi i\left[\sum_{\alpha\in\Pi} m_\alpha u_\alpha\right]\right), \tag{9.16}$$

with $m_\alpha = 1$ for all simple roots $\alpha \in \Pi$. An unramified character is automatically generic.

The local Whittaker function for an unramified vector will be denoted simply by

$$W^\circ(\chi, a) = \int\limits_{N(\mathbb{Q}_p)} \chi(w_{\text{long}}na)\overline{\psi(n)}dn, \tag{9.17}$$

where the reference to ψ has been suppressed for notational convenience and we do not display the fact that we are using a fixed $p < \infty$. Standard manipulations on (9.17) similar to the ones in (8.17) lead to

$$\begin{aligned} W^\circ(\chi, a) &= \chi(w_{\text{long}}aw_{\text{long}}^{-1}) \int\limits_{N(\mathbb{Q}_p)} \chi(w_{\text{long}}a^{-1}na)\overline{\psi(n)}dn \\ &= \chi(w_{\text{long}}aw_{\text{long}}^{-1})\delta(a) \int\limits_{N(\mathbb{Q}_p)} \chi(w_{\text{long}}n)\overline{\psi(ana^{-1})}dn \\ &= |a^{w_{\text{long}}\lambda+\rho}| \int\limits_{N(\mathbb{Q}_p)} \chi(w_{\text{long}}n)\overline{\psi^a(n)}dn, \end{aligned} \tag{9.18}$$

where we have defined $\psi^a(n) := \psi(ana^{-1})$.

9.2.2 Vanishing Properties of Whittaker Coefficients

One advantage of restricting to unramified characters is that it is very simple to determine the support of the local Whittaker function $W^\circ(\chi, a)$ (see also [130, Lemma 5.1]). Consider an element $n \in N(\mathbb{Z}_p) \subset K_p$; then, by right K_p-invariance and the transformation properties (9.2),

$$W^\circ(\chi, a) = W^\circ(\chi, an) = W^\circ(\chi, ana^{-1}a) = \psi(ana^{-1})W^\circ(\chi, a). \tag{9.19}$$

Therefore, $W^\circ(\chi, a)$ can only be non-vanishing if $\psi(ana^{-1}) = 1$, which requires that $a^\alpha \in \mathbb{Z}_p$ for all positive roots α since ψ is unramified.

The vanishing properties of the Whittaker coefficient $W^\circ(\chi, a)$ are also very important in view of the non-uniqueness of the p-adic Iwasawa decomposition that was discussed in Section 3.2. Consider two Iwasawa decompositions $g = n_1a_1k_1 = n_2a_2k_2$ of the same group element $g \in G(\mathbb{Q}_p)$. The ambiguity can be parametrised by $n_2 = n_1 a_1 n a_1^{-1}$ with $n \in N(\mathbb{Z}_p)$, and we assume for simplicity that $a_1 = a_2 = a$. The Whittaker integral (9.1) being well-defined requires that

$$W^\circ(\chi, n_2 a) = W^\circ(\chi, n_1 a) \quad \Leftrightarrow \quad \psi(n_1) W^\circ(\chi, a) = \psi(n_2) W^\circ(\chi, a)\,. \tag{9.20}$$

So either the Whittaker coefficient $W^\circ(\chi, a)$ has to vanish or $\psi(n_1) = \psi(n_2)$ for the two different Iwasawa decompositions. Plugging in the relation between n_1 and n_2 leads to

$$\psi(ana^{-1}) = 1 \quad \text{or} \quad W^\circ(\chi, a) = 0\,. \tag{9.21}$$

Since $n \in N(\mathbb{Z}_p)$ this is completely in agreement with the vanishing condition (9.19). The ambiguity in the $A(\mathbb{Q}_p)$ component of the Iwasawa decomposition is also of no consequence as the Whittaker coefficient only depends on the norm of $A(\mathbb{Q}_p)$ components, as we can see from (9.18), and they are unambiguous.

9.3 The Casselman–Shalika Formula

We are now ready to state the main theorem of this chapter:

Theorem 9.1 (Casselman–Shalika formula) *The local unramified Whittaker function $W^\circ(\chi, a)$, defined by the integral (9.14) for each $p < \infty$, is given by*

$$\int_{N(\mathbb{Q}_p)} \chi(w_{\text{long}} n a) \overline{\psi(n)} dn = \frac{\epsilon(\lambda)}{\zeta(\lambda)} \sum_{w \in \mathcal{W}} (\det(w)) |a^{w\lambda+\rho}| \prod_{\substack{\alpha>0 \\ w\alpha<0}} p^{\langle\lambda|\alpha\rangle}. \tag{9.22}$$

Proof The proof of this theorem will constitute the remainder of this section.

When translated to our notation, the Casselman–Shalika formula found for $W^\circ(\chi, a)$ in [130, Thm. 5.4] takes the form

$$\begin{aligned} W^\circ(\chi, a) &= \frac{\epsilon(\lambda)}{\zeta(\lambda)} \sum_{w \in \mathcal{W}} (\det(w)) \Big(\prod_{\substack{\alpha>0 \\ w\alpha<0}} p^{\langle\lambda|\alpha\rangle} \Big) |a^{w\lambda+\rho}| \\ &= \frac{1}{\zeta(\lambda)} \sum_{w \in \mathcal{W}} \epsilon(w\lambda) |a^{w\lambda+\rho}|, \end{aligned} \tag{9.23}$$

with

$$\zeta(\lambda) = \prod_{\alpha>0} \frac{1}{1 - p^{-(\langle\lambda|\alpha\rangle+1)}}, \quad \epsilon(\lambda) = \prod_{\alpha>0} \frac{1}{1 - p^{\langle\lambda|\alpha\rangle}}. \tag{9.24}$$

The latter identity of (9.23) follows from

$$\epsilon(w_i\lambda) = \prod_{\alpha>0} \frac{1}{1-p^{\langle\lambda|w_i\alpha\rangle}} = \frac{1}{1-p^{-\langle\lambda|\alpha_i\rangle}} \prod_{\substack{\alpha>0\\ \alpha\neq\alpha_i}} \frac{1}{1-p^{\langle\lambda|\alpha\rangle}} = -p^{\langle\lambda|\alpha_i\rangle}\epsilon(\lambda), \tag{9.25}$$

where w_i is a fundamental reflection switching the sign of α_i and permuting the remaining positive roots. Recall that $\det w = (-1)^{\ell(w)}$, where $\ell(w)$ is the length of w as introduced in Section 3.1.1. Formula (9.23) is only valid for unramified ψ and we have used χ and λ interchangeably.

We also record the following from [130, Thm. 5.3]:

Corollary 9.2 (Unramified Whittaker function at the identity) *Evaluated at the identity $a = \mathbb{1}$, the Whittaker function* (9.23) *becomes*

$$W^\circ(\chi, \mathbb{1}) = \frac{1}{\zeta(\lambda)}. \tag{9.26}$$

Proof The proof will follow immediately from a rewriting of the Casselman–Shalika formula that we derive in (9.81) in Section 9.6. □

Remark 9.3 The formula (9.23) above is correct for split groups. Casselman and Shalika [130] also proved a version for quasi-split $SU(2,1)$ over a local field, which we will present in Section 18.3.

Our strategy for proving Theorem 9.1 will be a mixture of the works of Jacquet [374] and Casselman–Shalika [130]. The argument consists of the following steps:

1. Derive a functional equation for the Whittaker function under Weyl transformations on χ.
2. Use this to show that a suitable multiple of the Whittaker function is Weyl invariant and write it as a sum over Weyl images.
3. Determine one term in this sum and derive all other terms from it. This will yield formula (9.23).

Finally, we will also show in Section 9.4 how the formula (9.23) can be used to derive the Whittaker functions for all generic characters ψ.

However, as a preparatory 'step 0', we first recall and slightly extend some results from Chapter 7 where the Fourier coefficients for Eisenstein series on $SL(2,\mathbb{A})$ were discussed. After Equation (7.74) we derived the Whittaker function at finite places, evaluated at the identity $a = 1 \in A(\mathbb{Q}_p)$, for general ψ. Repeating the same steps but (i) keeping a arbitrary and (ii) choosing an unramified character ($m = 1$) leads to

$[G(\mathbb{Q}_p) = SL(2, \mathbb{Q}_p)]$

$$
\begin{aligned}
F_{w_{\text{long}},\psi,p} &= \int_{N(\mathbb{Q}_p)} \chi(w_{\text{long}} n a)\overline{\psi(n)} dn \\
&= \chi(w_{\text{long}} a w_{\text{long}}^{-1})\delta(a) \int_{N(\mathbb{Q}_p)} \chi(w_{\text{long}} n)\overline{\psi(ana^{-1})} dn \\
&= \gamma_p(v^2)(1 - p^{-2s})\frac{|v|^{-2s+2} - p^{-2s+1}|v|^{2s}}{1 - p^{-2s+1}},
\end{aligned}
\tag{9.27}
$$

with $\chi(a) = |a|^{2s}$ and $a = \text{diag}(v, v^{-1}) = v^{H_\alpha}$ in terms of the unique positive root α of $\mathfrak{sl}(2, \mathbb{R})$. This formula, after dividing by $(1-p^{-2s})$, exhibits invariance under the Weyl reflection $s \leftrightarrow 1 - s$. We will see how this feature generalises to arbitrary G and why it is basically a consequence of this $SL(2, \mathbb{Q}_p)$ calculation. Equation (9.27) also manifestly exhibits the vanishing property of Section 9.2.2 since the p-adic Gaussian vanishes unless $|v^2| = |a^\alpha| \leq 1$.

Before embarking on the proof proper, we also record the following [130]:

Proposition 9.4 (Holomorphy of local Whittaker function) *The local Whittaker function $W^\circ(\chi, a)$ depends holomorphically on the quasi-character χ.*

Proof Inspection of formula (9.23) immediately reveals holomorphy when χ is in the Godement domain (4.134) of absolute convergence. This extends to all χ by virtue of the functional relation derived below. □

The holomorphy of the Whittaker function in the case of $SL(2, \mathbb{Q}_p)$ (as a function of s) can also be seen from the explicit expression (9.27) above. For $s \to 1/2$, the expression stays finite. We will comment in much more detail on the behaviour of Eisenstein series as a function of χ in Chapter 10.

9.3.1 Functional Relation for the Local Whittaker Function

We follow Jacquet's thesis [374]. First one defines a function associated to the Whittaker function by

$$
F(\lambda, g) = W_\psi^\circ(\lambda, w_{\text{long}}^{-1} g) \tag{9.28}
$$

for $g \in G(\mathbb{Q}_p)$. This leads to the integral expression

$$
F(\lambda, g) = \int_{N_-(\mathbb{Q}_p)} \chi(n_- g)\overline{\psi_-(n_-)} dn_- \tag{9.29}
$$

for the associated function. Here, objects with a minus subscript refer to the unipotent opposite to the standard unipotent $N(\mathbb{Q}_p)$. In other words, $N_-(\mathbb{Q}_p)$ designates the subgroup of $G(\mathbb{Q}_p)$ generated by the exponentials of the *negative* roots, whereas the usual $N(\mathbb{Q}_p)$ is associated with the positive roots. The reason that the opposite group arises here is because w_{long} maps all positive roots to negative ones (possibly combined with an outer automorphism). We will derive a functional relation for F under Weyl transformations, which by (9.28) will imply a functional relation for the Whittaker function.

The method for deriving the functional relation will be to reduce to the functional relation for $SL(2,\mathbb{Q}_p)$ that is manifest in (9.27) and then use the fact that $G(\mathbb{Q}_p)$ is made up of $SL(2,\mathbb{Q}_p)$ subgroups.

Let α_i be a simple positive root of $G(\mathbb{Q}_p)$. Then define, for $g \in \mathbb{Q}_p$,

$$F_i(\lambda, g) = \int\limits_{N_{i,-}(\mathbb{Q}_p)} \chi(n_{i,-}g)\overline{\psi_{i,-}(n_{i,-})}dn_{i,-}, \tag{9.30}$$

where the integral is now only over the one-dimensional subgroup generated by $x_{-\alpha_i}(u)$, and similarly the character $\psi_{i,-}$ is a character on the (lower) unipotent of the $SL(2,\mathbb{Q}_p)$ associated with α_i and can be obtained from ψ_- by restriction to the subgroup $N_{i,-}$. The function F_i is useful because for any α_i we can write

$$N_-(\mathbb{Q}_p) = N_{i,-}(\mathbb{Q}_p)\hat{N}_-(\mathbb{Q}_p), \tag{9.31}$$

where $\hat{N}_-(\mathbb{Q}_p)$ are the lower unipotent elements that are not of the form $x_{-\alpha_i}(u)$ for some $u \in \mathbb{Q}_p$. Associated with the factorisation above is a unique decomposition $n_- = n_{i,-}\hat{n}_-$, and then the integral (9.29) leads to

$$F(\lambda, g) = \int\limits_{\hat{N}_-(\mathbb{Q}_p)} F_i(\hat{n}_-g)\overline{\psi_-(\hat{n}_-)}d\hat{n}_- \tag{9.32}$$

by carrying out the integral over $dn_{i,-}$.

The $SL(2,\mathbb{Q}_p)$ projected function (9.30) has the following invariances:

$$F_i(\lambda, \hat{n}gk) = F_i(\lambda, g) \quad \text{for } \hat{n} \in \hat{N}(\mathbb{Q}_p) \text{ and } k \in K_p = G(\mathbb{Z}_p), \tag{9.33}$$

where $\hat{N}(\mathbb{Q}_p)$ is the unipotent subgroup opposite to $\hat{N}_-(\mathbb{Q}_p)$. It is generated by all positive roots but α_i. The set of these roots is invariant under the Weyl reflection w_i. Let P_i be the next-to-minimal parabolic subgroup defined by the (non-unique) decomposition

$$G(\mathbb{Q}_p) = P_i(\mathbb{Q}_p)K_p = \hat{N}(\mathbb{Q}_p)L_i(\mathbb{Q}_p)K_p = \hat{N}(\mathbb{Q}_p)\hat{A}(\mathbb{Q}_p)SL(2,\mathbb{Q}_p)_{\alpha_i}K_p, \tag{9.34}$$

with $\hat{A}(\mathbb{Q}_p)$ the part of the split torus $A(\mathbb{Q}_p)$ that is not contained in the torus of the embedded $SL(2,\mathbb{Q}_p)_{\alpha_i}$. Using this decomposition and the invariances of

F_i, one finds that the function $F_i(g)$ is determined by its values on elements of the form $g = \hat{a}g_i$ with $\hat{a} \in \hat{A}(\mathbb{Q}_p)$ and $g_i \in SL(2, \mathbb{Q}_p)_{\alpha_i}$. On such values one has that

$$F_i(\lambda, \hat{a}g_i) = |\hat{a}^{\lambda+\rho-\alpha_i}| \int_{N_{i,-}(\mathbb{Q}_p)} \chi(n_{i,-}g_i)\overline{\psi^{\hat{a}}_{i,-}(n_{i,-})}dn_{i,-} \tag{9.35}$$

with $\psi^{\hat{a}}_{-}(n_{i,-}) = \psi_{-}(\hat{a}n_{i,-}\hat{a}^{-1})$. The integral is basically the integral we have done in (9.27), the only change being that χ is now defined on all of $G(\mathbb{Q}_p)$ in which $SL(2, \mathbb{Q}_p)_{\alpha_i}$ is embedded. The result is determined by diagonal a_i and reads for $\hat{a} = 1$

$$\begin{aligned} &\int_{N_{i,-}(\mathbb{Q}_p)} \chi(n_{i,-}a_i)\overline{\psi_{i,-}(n_{i,-})}dn_{i,-} \\ &\quad = \gamma_p(a_i^{-\alpha_i})(1 - p^{-(\langle\lambda|\alpha_i\rangle+1)})\frac{1 - p^{-\langle\lambda|\alpha_i\rangle}|a_i^{-\alpha_i}|^{\langle\lambda|\alpha_i\rangle}}{1 - p^{-\langle\lambda|\alpha_i\rangle}}|a_i^{\lambda+\rho-\alpha_i}|. \end{aligned} \tag{9.36}$$

Under the Weyl reflection w_i one has $w_i\lambda = \lambda - \langle\lambda|\alpha_i\rangle\alpha_i$ and the function F_i therefore satisfies

$$F_i(w_i\lambda, g) = F_i(\lambda, g)\frac{1 - p^{-(1+\langle w_i\lambda|\alpha_i\rangle)}}{1 - p^{-(1+\langle\lambda|\alpha_i\rangle)}}, \tag{9.37}$$

where one also must keep track of the non-trivial $\hat{a}$ given by the prefactor in (9.35). The relation (9.32) then immediately gives the same transformation under w_i for $F(\lambda, g)$ and therefore for the unramified Whittaker function:

$$W^\circ(w_i\lambda, a) = \frac{\zeta_p(w_i, \lambda)}{\zeta_p(w_i, -\lambda)}W^\circ(\lambda, a), \tag{9.38}$$

where we defined the local zeta factor

$$\zeta_p(w, \lambda) = \prod_{\substack{\alpha>0 \\ w\alpha<0}} \frac{1}{1 - p^{-(1+\langle\lambda|\alpha\rangle)}}. \tag{9.39}$$

By an argument completely analogous to that in the proof of Lemma 8.6, a general Weyl transformation $w \in \mathcal{W}$ therefore leads to the functional relation

$$W^\circ(w\lambda, a) = \frac{\zeta_p(w, \lambda)}{\zeta_p(w, -\lambda)}W^\circ(\lambda, a). \tag{9.40}$$

This is, not surprisingly, the same factor that appeared in (the functional relation for) the constant term; see (8.36).

9.3.2 Weyl Invariant Combination

As for the constant term (and the full Eisenstein series), one can obtain a Weyl invariant form by multiplying through by the denominator of zeta factors associated with the longest Weyl word. Denoting

$$\zeta(\lambda) \equiv \zeta_p(w_{\text{long}}, \lambda) = \prod_{\alpha>0} \frac{1}{1 - p^{-(1+\langle\lambda|\alpha\rangle)}}, \tag{9.41}$$

one has that the function

$$\zeta(\lambda)W^\circ(\lambda, a) \tag{9.42}$$

is Weyl invariant. This is checked simply by combining (9.40) with the transformation of $\zeta_p(\lambda)$, which can be derived straightforwardly.

Because of the Weyl invariance of (9.42), we write it as a sum over Weyl images as

$$\zeta(\lambda)W^\circ(\lambda, a) = \sum_{w\in\mathcal{W}} c(w\lambda)|a^{w\lambda+\rho}|, \tag{9.43}$$

since the invariant function has to be a polynomial in $a^{\lambda+\rho}$ (and its images). The fact that the local Whittaker function is a single Weyl orbit follows from the considerations in [130].

9.3.3 Determining a Special Coefficient

Next we determine $c(w\lambda)$ for $w = w_{\text{long}}$, which is the coefficient of $|a^{w_{\text{long}}\lambda+\rho}|$ in (9.43). Referring back to (9.18) we see that the coefficient of $|a^{w_{\text{long}}\lambda+\rho}|$ in $W^\circ(\lambda, a)$ is obtained as the a-independent part of the integral

$$\int_{N(\mathbb{Q}_p)} \chi(w_{\text{long}}n)\overline{\psi^a(n)}dn\,. \tag{9.44}$$

The integral is a polynomial in a and we can obtain its a-independent part formally by sending a to zero. (This is only formal because, of course, $0 \notin A(\mathbb{Q}_p)$.) Therefore, the a-independent part of this integral can be obtained by removing the character ψ^a from the integral and then one is left with the same integral as in the constant term (8.19) for $w = w_{\text{long}}$. The result then is the same as the local factor for $\mathbb{Q}_p$ in the constant term formula (8.41), namely

$$\begin{aligned} c(w_{\text{long}}\lambda) &= \zeta(\lambda)\prod_{\alpha>0}\frac{1-p^{-(\langle\lambda|\alpha\rangle+1)}}{1-p^{-\langle\lambda|\alpha\rangle}} = \prod_{\alpha>0}\frac{1}{1-p^{-\langle\lambda|\alpha\rangle}} = \prod_{\alpha>0}\frac{1}{1-p^{\langle w_{\text{long}}\lambda|\alpha\rangle}} \\ &= \epsilon(w_{\text{long}}\lambda). \end{aligned} \tag{9.45}$$

This means that the general coefficient is given by

$$c(\lambda) = \epsilon(\lambda) = \prod_{\alpha>0} \frac{1}{1 - p^{\langle\lambda|\alpha\rangle}} \tag{9.46}$$

and the general formula for the Whittaker function for an unramified character is

$$W^{\circ}(\lambda, a) = \frac{1}{\zeta(\lambda)} \sum_{w\in\mathcal{W}} \epsilon(w\lambda)|a^{w\lambda+\rho}|, \tag{9.47}$$

thus demonstrating (9.23). This concludes the proof of Theorem 9.1. □

9.4 Whittaker Functions for Generic Characters ψ

Theorem 9.1 is only valid for unramified characters ψ, but we will now show that it can also be used indirectly for generic characters. Recall from Definition 6.12 that, for an unramified character, $m_\alpha = 1$ for all simple roots $\alpha \in \Pi$ and that a generic character has $m_\alpha \neq 0$ for all $\alpha \in \Pi$.

Let us take a closer look at the so-called 'twisted' character $\psi^a(n) = \psi(ana^{-1})$ introduced above, where ψ without superscript a is the unramified character. We note that periodicity of ψ^a is of course different from that of ψ, but this will not influence our reasoning.

From (9.15) and (9.16) we have that

$$\psi(n) = \exp\left(-2\pi i \left[\sum_{\alpha\in\Pi} u_\alpha\right]\right), \qquad n = \Big(\prod_{\alpha\in\Delta_+\setminus\Pi} x_\alpha(u_\alpha)\Big)\Big(\prod_{\alpha\in\Pi} x_\alpha(u_\alpha)\Big), \tag{9.48}$$

where $x_\alpha(u_\alpha) = \exp(u_\alpha E_\alpha)$.

Let a be parametrised as

$$a = \prod_{\beta\in\Pi} h_\beta(v_\beta)\,. \tag{9.49}$$

With insertions of aa^{-1}, the expression for ana^{-1} splits into factors of $ax_\alpha(u_\alpha)a^{-1}$. From (3.49) we then have that

$$\begin{aligned} ana^{-1} &= \Big(\prod_{\alpha\in\Delta_+\setminus\Pi} x_\alpha(u'_\alpha)\Big)\Big(\prod_{\alpha\in\Pi} x_\alpha(u'_\alpha)\Big), \\ u'_\alpha &= \Big(\prod_{\beta\in\Pi} (v_\beta)^{\alpha(H_\beta)}\Big) u_\alpha = a^\alpha u_\alpha, \end{aligned} \tag{9.50}$$

and, finally, by listing the simple roots as $\alpha_i \in \Pi$ for $i = 1, \ldots, r$ and denoting the associated elements u_{α_i} and v_{α_j} as u_i and v_j, respectively,

$$\psi^a(n) = \exp\left(-2\pi i \left[\sum_{i=1}^{r} u_i'\right]\right) = \exp\left(-2\pi i \left[\sum_{i=1}^{r}\left(\prod_{j=1}^{r}(v_j)^{A_{ji}}\right)u_i\right]\right), \tag{9.51}$$

where we have introduced the Cartan matrix A_{ij} defined in (3.19).

We now note that this is really a generic character with

$$m_i = m_{\alpha_i} = \prod_{j=1}^{r}(v_j)^{A_{ji}} \tag{9.52}$$

and that any generic character can be expressed in this way with the inverse relation

$$v_j = \prod_{i=1}^{r}(m_i)^{A_{ij}^{-1}}, \tag{9.53}$$

where A_{ij}^{-1} is the inverse Cartan matrix.

Now that we can express a generic character in terms of the unramified character, we would like to find the Whittaker function for ψ^a using (9.47) indirectly. More specifically, we ultimatelly want to find $W^\circ_{\psi^a}(\chi, a')$ with $a' = 1$ along the finite primes where $g = na'k$ and $a' \in A(\mathbb{Q}_p) \subset G(\mathbb{Q}_p)$. For each p this gives a contributing factor to the instanton measure, as discussed in Example 6.28.

This will bring us one step closer to finding the Fourier coefficients of the Eisenstein series with general instanton charges m_α in (9.13), and not only the restricted case of an unramified character.

Using similar steps as in (9.18), but in reverse order, we obtain

$$W^\circ_{\psi^a}(\chi, \mathbb{1}) = \left(\chi(w_{\text{long}} a w_{\text{long}}^{-1})\delta(a)\right)^{-1} W^\circ(\chi, a) = |a^{-(w_{\text{long}}\lambda+\rho)}| W^\circ(\chi, a). \tag{9.54}$$

Therefore, the local instanton measure for a generic character ψ^a with instanton charges m_α can be expressed through the local instanton measure evaluated for an unramified character at non-trivial $a = \prod_{\alpha\in\Pi} v_\alpha^{H_\alpha} \in A(\mathbb{Q}_p)$.

A more general formula than (9.54) can be shown for $a' \in A(\mathbb{Q}_p)$ to be

$$W^\circ_{\psi^a}(\chi, a') = |a^{-(w_{\text{long}}\lambda+\rho)}| W^\circ(\chi, aa'), \tag{9.55}$$

but we will not use it elsewhere.

Example 9.5: $SL(2,\mathbb{Q}_p)$ Whittaker function for generic character

We illustrate formula (9.54) by recovering the result (7.77) for $SL(2,\mathbb{Q}_p)$. In this case, there is only one simple root α and $A_{\alpha\alpha} = 2$. The unramified Whittaker function is as given in (9.27). If we want to get the Whittaker function for a character ψ^a with instanton charge m, then (9.53) tells us that we have $v = m^{1/2}$ and from (9.54) we find that

$$\begin{aligned} W^\circ_{\psi^a}(\chi,\mathbb{1}) &= |v|^{2s-2}\gamma_p(v^2)(1-p^{-2s})\frac{|v|^{-2s+2} - p^{-2s+1}|v|^{2s}}{1-p^{-2s+1}} \\ &= \gamma_p(m)(1-p^{-2s})\frac{1-p^{-2s+1}|m|^{2s-1}}{1-p^{-2s+1}}, \end{aligned} \tag{9.56}$$

in agreement with (7.77).

9.5 Degenerate Whittaker Coefficients

While the Casselman–Shalika formula (9.47) provides an elegant expression for unramified *local* characters and, via (9.54), also for Fourier coefficients of generic characters, it is also desirable to understand Fourier coefficients for *non*-generic characters ψ. These are also sometimes referred to as *degenerate Whittaker functions* in the literature [631, 486, 298] and have the property that the character ψ only depends on a subset of the simple roots of $G(\mathbb{R})$ rather than all simple roots, see Definition 6.10.

In this section, we will prove the following theorem that holds for *global* degenerate characters ψ [214, 349]:

Theorem 9.6 (Degenerate Whittaker functions) *Let $\psi\colon N(\mathbb{Q})\backslash N(\mathbb{A}) \to U(1)$ be a degenerate character with* $\mathrm{supp}(\psi) = \Pi' \neq \Pi$ *with associated subgroup $G'(\mathbb{A}) \subset G(\mathbb{A})$. Let $w_c w'_{\mathrm{long}}$ be the representatives of the coset $\mathcal{W}/\mathcal{W}'$ defined below in* (9.63). *Then the degenerate Whittaker coefficient on $G(\mathbb{A})$ is given by*

$$W^\circ_\psi(\chi,a) = \sum_{w_c w'_{\mathrm{long}} \in \mathcal{W}/\mathcal{W}'} a^{(w_c w'_{\mathrm{long}})^{-1}\lambda+\rho} M(w_c^{-1},\lambda) W'^\circ_{\psi^a}(w_c^{-1}\lambda,\mathbb{1}), \tag{9.57}$$

where W'°_ψ denotes a Whittaker function on the $G'(\mathbb{A})$ subgroup of $G(\mathbb{A})$. The weight $w_c^{-1}\lambda$ is given as a weight of $G'(\mathbb{A})$ by orthogonal projection.

By $W'^\circ_{\psi^a}(w_c^{-1}\lambda,\mathbb{1})$ we mean the Jacquet integral (9.13) evaluated for ψ^a. Before embarking on the proof, we explain the notation used here. For a global character

$$\psi\left(\prod_{\alpha\in\Pi} x_\alpha(u_\alpha)\right) = \exp\left(2\pi i \sum_{\alpha\in\Pi} m_\alpha u_\alpha\right) \tag{9.58}$$

we call

$$\mathrm{supp}(\psi) = \{\alpha \in \Pi \mid m_\alpha \neq 0\} \subset \Pi, \tag{9.59}$$

determined by the non-vanishing m_α, the *support of the character* ψ. With this notion, the Definition 6.10 becomes

$$\begin{aligned} \mathrm{supp}(\psi) = \Pi \quad &\Longleftrightarrow \quad \psi \text{ generic},\\ \mathrm{supp}(\psi) \neq \Pi \quad &\Longleftrightarrow \quad \psi \text{ non-generic or degenerate}. \end{aligned}$$

We note that a degenerate character $\psi\colon N(\mathbb{Q})\backslash N(\mathbb{A}) \to U(1)$ canonically defines a semi-simple proper subgroup $G' \subset G$. This subgroup G' is the one with simple root system $\Pi' = \mathrm{supp}(\psi)$; its Dynkin diagram is the subdiagram of the Dynkin diagram of G obtained by restricting to the nodes corresponding to $\mathrm{supp}(\psi)$. The subgroup G' has a Weyl group $\mathcal{W}'$ with longest Weyl word w'_{long}.

Proof Using the Bruhat decomposition, the spherical Whittaker coefficient $W^\circ_\psi(\chi, a)$ can be written as a sum over the Weyl group $\mathcal{W}$ of G as in (9.3):

$$W^\circ_\psi(\chi, a) = \int\limits_{N(\mathbb{Q})\backslash N(\mathbb{A})} E(\chi, na)\overline{\psi(n)}dn = \sum_{w\in\mathcal{W}} F_{w,\psi}(\chi, a), \tag{9.60}$$

with

$$\begin{aligned} F_{w,\psi}(\chi, a) &= \int\limits_{w^{-1}B(\mathbb{Q})w\cap N(\mathbb{Q})\backslash N(\mathbb{A})} \chi(wna)\overline{\psi(n)}dn \\ &= \int\limits_{N^w_{\{\beta\}}} \overline{\psi(n_\beta)}dn_\beta \int\limits_{N^w_{\{\gamma\}}} \chi(wn_\gamma a)\overline{\psi(n_\gamma)}dn_\gamma\,. \end{aligned} \tag{9.61}$$

The various $F_{w,\psi}$ can be analysed as in Section 9.1 and we have used the w-dependent split of positive roots of N into two sets of $\{\beta\}$ and $\{\gamma\}$ as in (9.7) and (9.8). Importantly, for degenerate ψ the integral over the compact domain $N^w_{\{\beta\}}$ can be non-vanishing for Weyl words w different from w_{long}: if ψ is trivial on all the n_β in (9.61), the corresponding integral yields unity rather than zero. This means that the Weyl word w must map all elements in $\mathrm{supp}(\psi)$ to negative roots in order to avoid the vanishing of $F_{w,\psi}$ and the sum over $w \in \mathcal{W}$ in (9.60) can be restricted to the subset

$$C_\psi = \{w \in \mathcal{W} \mid w\alpha < 0 \quad \text{for all } \alpha \in \mathrm{supp}(\psi)\}\,. \tag{9.62}$$

(If ψ is generic, one recovers $C_\psi = \{w_{\text{long}}\}$, in agreement with the discussion of Section 9.1.)

We will now parametrise the set C_ψ explicitly. Denote by $\mathcal{W}'$ the Weyl subgroup generated by the fundamental reflections associated with $\Pi' = \text{supp}(\psi)$ only. It is the Weyl group of G' and has its own longest Weyl word that we denote by w'_{long}. The longest Weyl word w'_{long} has the desired property that it maps all elements in $\text{supp}(\psi)$ to negative roots and it is the only Weyl word in $\mathcal{W}'$ with this property. In fact, any element in C_ψ can be represented in a form that involves the longest word w'_{long} of $\mathcal{W}'$:

$$w \in C_\psi \quad \Longleftrightarrow \quad w = w_c w'_{\text{long}}. \tag{9.63}$$

Here, $w_c \in \mathcal{W}$ must satisfy

$$w_c \alpha > 0 \quad \text{for all } \alpha \in \text{supp}(\psi) \tag{9.64}$$

in order for $w = w_c w'_{\text{long}}$ to belong to C_ψ.

The words $w_c \in \mathcal{W}$ can be constructed as carefully chosen representatives of the coset $\mathcal{W}/\mathcal{W}'$. Consider the weight

$$\Lambda_\psi = \sum_{i:\, \alpha_i \notin \text{supp}(\psi)} \Lambda_i, \tag{9.65}$$

i.e., the sum of fundamental weights of G that are not associated with the support of the degenerate character ψ. The weight Λ_ψ is stabilised by $\mathcal{W}'$ and its $\mathcal{W}$-orbit is in bijection with the coset $\mathcal{W}/\mathcal{W}'$. A standard result for Weyl groups is that if $w(\alpha_i) < 0$ for some simple root, then $\ell(ww_i) < \ell(w)$ [386, Lemma 3.11]. Therefore, if $w(\alpha_i) < 0$ and $\alpha_i \in \text{supp}(\psi)$ we have

$$w(\Lambda_\psi) = (ww_i)(\Lambda_\psi), \tag{9.66}$$

since Λ_ψ is stabilised by the fundamental reflections from $\text{supp}(\psi)$ (these generate $\mathcal{W}'$). This means that if $w(\alpha_i) < 0$ there is a shorter Weyl word ww_i leading to the same point in the $\mathcal{W}$-orbit of Λ_ψ as the word w does. By induction, the shortest word leading to a given point in the Weyl orbit of Λ_ψ must be those $w_c \in \mathcal{W}$ that satisfy $w_c\alpha > 0$ for all $\alpha \in \text{supp}(\psi)$. Hence, the words w_c appearing in (9.64) are the shortest words leading to the points of the $\mathcal{W}$-orbit of Λ_ψ. Such shortest words are not necessarily unique; for a given $\mathcal{W}$-orbit point any shortest word w_c will do. An explicit construction of the w_c can be achieved by the same orbit method as in Section 10.3 below; see also [212, 214].

With the parametrisation $w = w_c w'_{\text{long}}$ of the elements of C_ψ we thus arrive at the following expression for the degenerate Whittaker integral (9.60):

$$W^\circ_\psi(\chi, a) = \sum_{w_c w'_{\text{long}} \in \mathcal{W}/\mathcal{W}'} F_{w_c w'_{\text{long}}, \psi}(\chi, a), \tag{9.67}$$

where it is understood that $w_c w'_{\text{long}}$ is the specific coset representative described above.

The quantities $F_{w_c w'_{\text{long}},\psi}(\chi, a)$ can be evaluated by reducing them to Whittaker functions of the subgroup $G'(\mathbb{A}) \subset G(\mathbb{A})$ associated with $\text{supp}(\psi)$ as follows. First, we separate out the a-dependence as usual by conjugating it to the left and using the multiplicativity of χ:

$$\begin{aligned}
F_{w_c w'_{\text{long}},\psi}(\chi, a) &= \int\limits_{(w_c w'_{\text{long}})^{-1} B(\mathbb{Q}) w_c w'_{\text{long}} \cap N(\mathbb{Q}) \backslash N(\mathbb{A})} \chi(w_c w'_{\text{long}} n a) \overline{\psi(n)} dn \\
&= a^{(w_c w'_{\text{long}})^{-1}\lambda+\rho} \int\limits_{(w_c w'_{\text{long}})^{-1} B(\mathbb{Q}) w_c w'_{\text{long}} \cap N(\mathbb{Q}) \backslash N(\mathbb{A})} \chi(w_c w'_{\text{long}} n) \overline{\psi^a(n)} dn,
\end{aligned} \tag{9.68}$$

with $\psi^a(n) = \psi(ana^{-1})$ as before. We can also rewrite the integration into the two sets $N^w_{\{\beta\}}$ and $N^w_{\{\gamma\}}$ (for $w = w_c w'_{\text{long}}$) as in (9.61), and we know that by construction the integral over $N^w_{\{\beta\}}$ gives unity.

The remaining integral over $N^w_{\{\gamma\}}$ is then over all positive roots γ that are mapped to negative roots by the action of $w = w_c w'_{\text{long}}$, and we drop the γ subscript for ease of notation. The particular form of w implies that we can parametrise the unipotent element as $n = n_c n'$, where $n' \in N'(\mathbb{A})$ is the (full) unipotent radical of the standard minimal Borel subgroup $B'(\mathbb{A})$ of $G'(\mathbb{A})$ that is determined by ψ; and n_c are the remaining elements whose total space we call $N_c(\mathbb{A})$. We note also that $w'_{\text{long}} n_c (w'_{\text{long}})^{-1}$ is generated exactly by the positive roots that are mapped to negative roots by w_c alone. The degenerate character ψ only depends on n', i.e., $\psi^a(n_c n') = \psi^a(n')$.

Putting these observations together one obtains

$$F_{w_c w'_{\text{long}},\psi}(\chi, a) = a^{(w_c w'_{\text{long}})^{-1}\lambda+\rho} \int\limits_{N_c(\mathbb{A})} \int\limits_{N'(\mathbb{A})} \chi(w_c w'_{\text{long}} n_c n') \overline{\psi^a(n')} dn_c dn'. \tag{9.69}$$

As the next step one can rewrite the argument of the character χ as

$$\begin{aligned}
\chi(w_c w'_{\text{long}} n_c n') &= \chi\left(w_c w'_{\text{long}} n_c (w_c w'_{\text{long}})^{-1} w_c w'_{\text{long}} n'\right) \\
&= \chi\left(w_c w'_{\text{long}} n_c (w_c w'_{\text{long}})^{-1} w_c \tilde{n}\tilde{a}\right),
\end{aligned} \tag{9.70}$$

where we have performed an Iwasawa decomposition (in $G'(\mathbb{A})$) of $w'_{\text{long}} n' = \tilde{n}\tilde{a}\tilde{k}$ and used left-invariance of χ under $K'(\mathbb{A}) \subset K(\mathbb{A})$ in the last step. In the next step we want to perform another Iwasawa decomposition (now in $G(\mathbb{A})$) of

$$w_c \tilde{n}\tilde{a} = \hat{n}\hat{a}\hat{k}\,. \tag{9.71}$$

The important observation now is that $\tilde{n} \in N'(\mathbb{A})$ and w_c satisfies (9.64), which implies that $w_c N'(\mathbb{A}) w_c^{-1} \subset N(\mathbb{A})$. Therefore, the Iwasawa decomposition (9.71) has

$$\hat{n} = w_c \tilde{n} w_c^{-1}, \quad \hat{a} = w_c \tilde{a} w_c^{-1}, \quad \hat{k} = w_c\,. \tag{9.72}$$

Inserting this back into the integral (9.69) one can bring the element $\hat{n} \in N(\mathbb{A})$ to the left. This will induce a unimodular change of the integration variables dn_c as in Section 8.5. Conjugating the element $\hat{a} \in A(\mathbb{A})$ to the left will induce a non-trivial change of measure ($w = w_c w'_{\text{long}}$):

$$\begin{aligned}
\int\limits_{N_c(\mathbb{A})} \chi(w n_c w^{-1} \hat{n}\hat{a}) dn_c &= \int\limits_{N_c(\mathbb{A})} \chi(\hat{n}\hat{a} w n_c w^{-1}) \hat{a}^{w_c\rho - \rho} dn_c \\
&= \int\limits_{N_c(\mathbb{A})} \chi(w n_c w^{-1}) \tilde{a}^{w_c^{-1}\lambda - \rho} dn_c \\
&= \tilde{a}^{w_c^{-1}\lambda - \rho} \int\limits_{N_c(\mathbb{A})} \chi(w n_c w^{-1}) dn_c \\
&= \chi'(\tilde{a}) \int\limits_{N_c(\mathbb{A})} \chi(w n_c w^{-1}) dn_c.
\end{aligned} \tag{9.73}$$

We have evaluated the character χ on $\hat{n}\hat{a}$ in the second step according to $\chi(\hat{n}\hat{a}) = \hat{a}^{\lambda+\rho} = \tilde{a}^{w_c^{-1}\lambda + w_c^{-1}\rho}$ due to (9.72). In the last step, we have used that $\tilde{a}$ does not depend on n_c and can therefore be taken out of the integral, and have defined the character

$$\chi'(\tilde{a}) = \tilde{a}^{w_c^{-1}\lambda + \rho} = \chi'(w'_{\text{long}} n'), \tag{9.74}$$

on the group $G'(\mathbb{A})$. In the last step we have used the definition of $\tilde{a}$.

Putting everything together in (9.69) one obtains the factorised expression

$$\begin{aligned}
F_{w_c w'_{\text{long}}, \psi}(\chi, a) = a^{(w_c w'_{\text{long}})^{-1}\lambda + \rho} &\int\limits_{N_c(\mathbb{A})} \chi(w_c w'_{\text{long}} n_c) dn_c \\
\times &\int\limits_{N'(\mathbb{A})} \chi'(w'_{\text{long}} n') \overline{\psi^a(n')} dn'.
\end{aligned} \tag{9.75}$$

The two separate integrals are both of types we have encountered before: the $N_c(\mathbb{A})$ integral is precisely the Gindikin–Karpelevich expression (8.19) for the Weyl word $w_c \in \mathcal{W}$ and so gives a factor $M(w_c^{-1}, \lambda)$ defined in (8.42), and the second integral is the generic Whittaker function (9.13) for the subgroup $G'(\mathbb{A}) \subset G(\mathbb{A})$ with *generic* unitary character ψ^a, in the representation given by the weight $w_c^{-1}\lambda$, projected orthogonally to $G'(\mathbb{A})$ and evaluated at the identity $\mathbb{1} \in A'(\mathbb{A})$. This completes the proof of Theorem 9.6. □

As a consequence of the theorem, Whittaker functions of non-generic characters ψ can be evaluated as sums over Whittaker functions of subgroups on which the character is generic. We stress again that the choice of coset representative of $\mathcal{W}/\mathcal{W}'$ is important here. If the full Whittaker function on the subgroup is known, the above formula provides the explicit expression for any character ψ. Thanks to the Casselman–Shalika formula, this means that the *local* Whittaker function ($p < \infty$) can be calculated for *any* character, generic or not. The archimedean part is typically more intricate.

Remark 9.7 Theorem 9.6 of course also remains true in the case of generic ψ since then $\mathcal{W}' = \mathcal{W}$ and the sum on the right-hand side just contains the single element w_{long}, leading to the standard Jacquet–Whittaker integral. The power of the theorem arises in cases where one deals with Eisenstein series that do not have any generic Whittaker coefficients and one can then use (9.57) to determine the degenerate ones. This will be explored in more detail in Section 10.4 below.

9.6 The Casselman–Shalika Formula and Langlands Duality*

In this section we provide an alternative view on the Casselman–Shalika formula (9.23), which hints at a deeper representation-theoretic structure that will be further elaborated upon in Chapter 11. For the present analysis it is useful to separate out the modulus character contribution $a^\rho = \delta^{1/2}(a)$ in formula (9.23) and write

$$a^{w\lambda+\rho} = a^\rho a^{w\lambda} = \delta^{1/2}(a) a^{w\lambda}. \tag{9.76}$$

Let ψ be an unramified character on N. The Casselman–Shalika formula (9.23) for the p-adic spherical Whittaker function on $\mathbb{Q}_p$ evaluated at $a \in A(\mathbb{Q}_p)$ is

$$W^\circ(\lambda, a) = \frac{1}{\zeta(\lambda)} \delta^{1/2}(a) \sum_{w \in \mathcal{W}} w \left(\frac{|a^\lambda|}{\prod_{\alpha>0}(1 - p^{\langle\lambda|\alpha\rangle})} \right), \tag{9.77}$$

with

$$\frac{1}{\zeta(\lambda)} = \prod_{\alpha>0} \left(1 - p^{-1} p^{-\langle\lambda|\alpha\rangle}\right). \tag{9.78}$$

The sum over the Weyl group in (9.77) resembles closely the Weyl character formula (3.29) for highest weight modules. In order to make this resemblance exact, we compare with the rewritten character formula in (3.30), which we reproduce here for convenience:

$$\text{ch}_\Lambda(b) = \sum_{w \in \mathcal{W}} w \left(\frac{b^\Lambda}{\prod_{\beta>0}(1 - b^{-\beta})} \right), \tag{9.79}$$

where β runs over the positive roots of the group whose representation is being constructed and b is an element of its Cartan torus.

An important first observation now is that because of the way λ appears in the numerator and in the denominator of (9.77) the comparison can only work if the character we are trying to match onto is one of the *Langlands dual group* ${}^LG$, or *L-group* for short, which is a complex algebraic group canonically associated to G [444]; see also Sections 11.6 and 16.1 for more details. The L-group is obtained by interchanging roots and co-roots [412]; see also [285] for a realisation in physics. The root systems of G and ${}^LG$ are in bijection and the two groups have isomorphic Weyl groups.

Denoting the roots of the Langlands dual group by $\alpha^\vee$ instead of β, we are therefore looking for an element b of the dual torus ${}^LA$ such that $|b^{-\alpha^\vee}| = p^{\langle\alpha|\lambda\rangle}$. This condition fixes uniquely an element $b = a_\lambda \in {}^LA$, where we emphasise that the particular element depends on λ. To ensure that the numerator matches the character of an irreducible highest weight module V_Λ of ${}^LG$ we also need to evaluate (9.77) at a very specific point $a \equiv a_\Lambda$ of $A(\mathbb{Q}_p)$. This element a_Λ is fixed by the requirement that the following (duality) relation hold:

$$a_\lambda^\Lambda = a_\Lambda^\lambda, \tag{9.80}$$

where the left-hand side derives from evaluating the character ch_Λ at the place a_λ and the right-hand side is what one obtains by evaluating the spherical Whittaker function at the special point $a_\Lambda \in A$.

We observe that Λ parametrises points in the space of co-roots $\mathfrak{h}$ of G. By contrast, λ is an element of the space of roots (or weights) $\mathfrak{h}^*$ of G from the start, so that one has to consider a_λ as an element of the *dual torus* ${}^LA$ of the *Langlands dual group* ${}^LG$. Putting everything together, we can write the spherical Whittaker function evaluated at a_Λ in terms of the character of the highest weight representation V_Λ of ${}^LG$ as

$$W^\circ(\lambda, a_\Lambda) = \begin{cases} \frac{1}{\zeta(\lambda)} \delta^{1/2}(a_\Lambda)\text{ch}_\Lambda(a_\lambda) & \text{if } \Lambda \text{ a dominant int. weight of } {}^LG, \\ 0 & \text{otherwise.} \end{cases} \tag{9.81}$$

The vanishing for weights Λ of ${}^LG$ that are not dominant integrals is a consequence of the vanishing properties of Whittaker functions, discussed in Section 9.2.2. We also note that for $a_\Lambda = \mathbb{1}$ one has the trivial representation $\Lambda = 0$, which immediately gives Corollary 9.2.

Remark 9.8 The formula (9.81) for $GL(n)$ was first proven by Shintani in 1976 [577], and it was subsequently generalised by Casselman–Shalika in 1980 to (9.1), which holds for any G. Remarkably, the general formula was in fact already conjectured by Langlands in 1967 in a letter to Godement [445], a fact that was apparently unknown to Casselman and Shalika at the time of their proof [129].

Remark 9.9 Using formula (9.54) we can also reinterpret (9.81) in terms of a Whittaker function for the twisted character

$$\psi_\Lambda(n) := \psi(a_\Lambda n a_\Lambda^{-1}) \tag{9.82}$$

as

$$W^\circ_{\psi_\Lambda}(\lambda, \mathbb{1}) = a_\Lambda^{-w_{\text{long}}\lambda-\rho} W^\circ(\lambda, a_\Lambda) = \frac{1}{\zeta(\lambda)} a_\lambda^{-w_{\text{long}}\Lambda} \mathrm{ch}_\Lambda(a_\lambda), \tag{9.83}$$

where we have used (9.80).

To summarise the main result of this section: local spherical Whittaker functions for a principal series representation parametrised by a weight λ of G and evaluated at special points a_Λ associated with dominant weights Λ of the Langlands dual group ${}^LG$ can be evaluated in terms of the character ch_Λ of the irreducible highest weight V_Λ of ${}^LG$ evaluated at a point a_λ determined by the parameter of the principal series. We will come back to this point of view on the Casselman–Shalika formula in Section 11.7, where we elaborate on the connection with Hecke algebras.

The parameter $a_\lambda \in {}^LA$ is called the *Satake–Langlands parameter* of the principal series representation of $G(\mathbb{Q}_p)$ determined by the weight λ, and we will come back to it in a slightly different guise in Section 11.6. We also note that the element $a_\Lambda \in A(\mathbb{Q}_p)$ actually corresponds to an equivalence class $A(\mathbb{Q}_p)/A(\mathbb{Z}_p)$ due to sphericality (right $K(\mathbb{Q}_p)$ invariance) of the Whittaker function.

For the case of $GL(n, \mathbb{Q}_p)$ one has ${}^LG = GL(n, \mathbb{C})$. If one considers the case when Λ is the highest weight of the fundamental n-dimensional representation $\mathbb{C}^n$, then the character ch_Λ is given by the degree n *Schur polynomial* S_n [231]:

$$W^\circ(\lambda, a_\Lambda) = \frac{1}{\zeta(\lambda)} \delta^{1/2}(a_\Lambda) S_n(\alpha_1, \ldots, \alpha_n), \qquad G = GL(n, \mathbb{Q}_p). \tag{9.84}$$

Here, λ is thought of as the diagonal matrix $\lambda = \mathrm{diag}(\alpha_1, \ldots, \alpha_n)$. In Example 9.10 we illustrate the Casselman–Shalika formula for $SL(2, \mathbb{Q}_p)$ in terms of characters for the Langlands dual group $SL(2, \mathbb{C})$.

Example 9.10: $SL(2,\mathbb{Q}_p)$ spherical Whittaker function and $SL(2,\mathbb{C})$ characters

For the case $SL(2,\mathbb{Q}_p)$ the spherical Whittaker function for unramified ψ was given explicitly in (9.27) for $\lambda = (2s-1)\rho$ and general $a = v^{H_\alpha}$ as

$$W^\circ(\lambda, a) = \gamma_p(v^2)(1-p^{-2s})\frac{|v|^{-2s+2} - p^{-2s+1}|v|^{2s}}{1-p^{-2s+1}}. \tag{9.85}$$

In order to verify the expression (9.81) we need to evaluate it at the special values a_Λ where $\Lambda = NH_\alpha/2$ is a dominant integral weight of ${}^LSL(2,\mathbb{Q}_p) = SL(2,\mathbb{C})$ for $N \in \mathbb{Z}_{\geq 0}$. This means $v^2 = p^N$ and the Whittaker function evaluates to

$$W^\circ(\lambda, a_\Lambda) = (1-p^{-2s})\frac{p^{\frac{N}{2}(2s-2)} - p^{-2s+1-Ns}}{1-p^{-2s+1}}. \tag{9.86}$$

The Whittaker function vanishes if N is not in $\mathbb{Z}_{\geq 0}$ because of the factor $\gamma_p(v^2)$.

Let us now determine the right-hand side of (9.81). For $\Lambda = \frac{N}{2}H_\alpha$, the character of the $(N+1)$-dimensional highest weight representation of ${}^LSL(2,\mathbb{Q}_p) \cong PSL(2,\mathbb{C})$ is

$$\mathrm{ch}_\Lambda = e^{NH_\alpha/2} + e^{(N-2)H_\alpha/2} + \cdots + e^{-NH_\alpha/2} = \frac{e^{-NH_\alpha/2} - e^{(N+2)H_\alpha/2}}{1-e^{H_\alpha}}. \tag{9.87}$$

This has to be evaluated at $a_\lambda = p^\lambda = p^{(2s-1)\Lambda_\alpha}$, which leads to

$$\mathrm{ch}_\Lambda(a_\lambda) = \frac{p^{N(2s-1)/2} - p^{-(N+2)(2s-1)/2}}{1-p^{-2s+1}}, \tag{9.88}$$

where we recall that the p-adic characters are evaluated with the p-adic norm such that for instance $e^{H_\alpha}(a_\lambda) = |p^{2s-1}| = p^{-2s+1}$. For $v^2 = p^N$, the modulus character evaluates to $\delta^{1/2}(a_\Lambda) = |p^{N/2}| = p^{-N/2}$ and one also has $\frac{1}{\zeta(\lambda)} = 1 - p^{-2s}$ from (9.78). Putting everything together in (9.81) leads to

$$\begin{aligned} W^\circ(\lambda, a_\Lambda) &= (1-p^{2s})p^{-N/2}\frac{p^{N(2s-1)/2} - p^{-(N+2)(2s-1)/2}}{1-p^{-2s+1}} \\ &= (1-p^{-2s})\frac{p^{N(2s-2)/2} - p^{-2s+1-sN}}{1-p^{-2s+1}}, \end{aligned} \tag{9.89}$$

which equals (9.86).

9.7 Quantum Whittaker Functions*

In this section we will briefly discuss an interesting generalisation of Whittaker functions developed by Gerasimov–Lebedev–Oblezin [271, 272, 273, 276] which are called *q-deformed* (or *quantum*) *Whittaker functions*, and denoted Ψ_q. As we will see, the quantum Whittaker functions interpolate between the

p-adic and the archimedean Whittaker functions. Although one can define the Ψ_qs for all the classical Lie groups, we shall focus on $GL(\ell+1)$.

The quantum Whittaker function for $GL(\ell+1)$ can be written in the following explicit form [269]:

$$\Psi_q(p) = \sum_{p_{k,i}\in\mathcal{P}^{(\ell+1)}} \prod_{k=1}^{\ell+1} q^{l_k(\sum_{i=1}^k p_{k,i}-\sum_{i=1}^{k-1} p_{k-1,i})} \times \frac{\prod_{k=2}^{\ell}\prod_{i=2}^{k-1}(p_{k,i+1}-p_{k,i})_q!}{\prod_{k=1}^{\ell}\prod_{i=1}^{\ell}(p_{k,i}-p_{k+1,i})_q!\,(p_{k+1,i+1}-p_{k,i})_q!}. \tag{9.90}$$

Let us briefly explain the notation. The sum runs over a collection of integers $p_{i,j}\in\mathbb{Z}$, $i=1,\ldots,\ell+1$, $j=1,\ldots,i$, belonging to a set of *Gelfand–Tsetlin patterns* $\mathcal{P}^{(\ell+1)}$ associated to an irreducible representation of $GL(\ell+1,\mathbb{C})$ [405, 108]. The q-deformed factorial function is defined as $(n)_q! = (1-q)(1-q^2)\cdots(1-q^n)$.

In the limit $q\to 1$ the quantum Toda Hamiltonians reduce to the ordinary Toda system, whose solution is given by Givental's integral representation of the archimedean Whittaker function:

$$\Psi_{\mathbb{R}}(x_1,\ldots,x_{\ell+1}) = \int_{\mathbb{R}^{\ell(\ell+1)/2}} \prod_{k=1}^{\ell}\prod_{i=1}^{k} dx_{k,i}e^{\mathcal{F}(x)}, \tag{9.91}$$

where

$$\mathcal{F}(x) = \sum_{k=1}^{\ell+1} l_k\Big(\sum_{i=1}^{k} x_{k,i} - \sum_{i=1}^{k-1} x_{k-1,i}\Big) - \sum_{k=1}^{\ell}\sum_{i=1}^{k}\left(e^{x_{k,i}-x_{k+1,i}} + e^{x_{k+1,i+1}-x_{k,i}}\right). \tag{9.92}$$

The new variables are obtained by setting $x_j \equiv q^{p_{\ell+1,j}+j-1}$ while taking the limit $q\to 1$.

In the other limit $q\to 0$, the quantum Whittaker function can be reduced to

$$\Psi_q(p)\big|_{q\to 0} = \sum_{p_{k,i}\in\mathcal{P}^{(\ell+1)}} \prod_{k=1}^{\ell+1} z_k^{(\sum_{i=1}^k p_{k,i}-\sum_{i=1}^{k-1} p_{k-1,i})}, \tag{9.93}$$

where we have defined $z_i = q^{l_i}$, $i=1,\ldots,\ell+1$. This has precisely the form of the characters of irreducible representations of $GL(\ell+1)$ in the Gelfand–Tsetlin bases.

To get a better handle on the somewhat unwieldy expressions above we now consider a simple example.

Example 9.11: The $GL(2)$ quantum Whittaker function

For $\ell = 1$ we have simply $GL(2)$, for which the lattice is $(p_{2,1}, p_{2,2}) \in \mathbb{Z}^2$. Then the general quantum Whittaker function becomes

$$\Psi_q(p_{2,1}, p_{2,2}) = \sum_{p_{2,1} \leq p_{1,1} \leq p_{2,2}} \frac{q^{l_1 p_{1,1}} q^{l_2(p_{2,1}+p_{2,2}-p_{1,1})}}{(p_{1,1}-p_{2,1})_q!(p_{2,2}-p_{1,1})_q!}. \tag{9.94}$$

In the $q \to 1$ limit the integral representation (9.91) reduces to a Bessel integral, and up to normalisation one obtains the real spherical vector $\tilde{f}_\infty^\circ$ in (5.145). Similarly, for $q \to 0$ the expression (9.93) can be readily identified with a standard character of $GL(2,\mathbb{C})$, and after appropriate identification of variables one obtains the p-adic spherical vector $\tilde{f}_p^\circ$ in (5.145).

Remark 9.12 The development of quantum Whittaker functions followed Givental's influential work on generalised mirror symmetry for flag varieties [284]. By now, the theory of quantum Whittaker functions has found a variety of interesting applications in mathematics and theoretical physics. In particular, they play a rôle in quantum integrable systems [270, 275], topological string theory and mirror symmetry [274, 297].

Remark 9.13 As mentioned above, Ψ_q interpolates between the archimedean and non-archimedean Whittaker functions. These functions can be obtained as eigenmodes of the so-called q-deformed Toda Hamiltonian associated with $GL(\ell+1)$, and reduce in the limit $q \to 1$ to Givental's integral representation of the archimedean Whittaker function [284], while the limit $q \to 0$ recovers the Casselman–Shalika formula for the p-adic Whittaker function. Since archimedean and non-archimedean Whittaker functions appear as local factors in the Fourier coefficient of an Eisenstein series, the following question naturally presents itself: *Is there some kind of 'quantum automorphic form' for which Ψ_q appears naturally as a Fourier coefficient, or some generalisation thereof?*

9.8 Whittaker Coefficients on $SL(3,\mathbb{A})$*

We illustrate the general considerations above through the explicit example of $SL(3,\mathbb{A})$. The Eisenstein series on $SL(3,\mathbb{R})$, $GL(3,\mathbb{R})$ and this group have been studied in great detail in the literature [605, 120] by various techniques.

The split real group $SL(3,\mathbb{R})$ has rank two and we denote the two simple roots by α_1 and α_2. The corresponding Cartan generators will be called $H_1 \equiv H_{\alpha_1}$ and $H_2 \equiv H_{\alpha_2}$. A general element $a \in A(\mathbb{A})$ will be written as

$$a = v_1^{H_1} v_2^{H_2}. \tag{9.95}$$

The Eisenstein series is determined by the weight

$$\lambda = (2s_1 - 1)\Lambda_1 + (2s_2 - 1)\Lambda_2 \tag{9.96}$$

in terms of the fundamental weights dual to the simple roots.

The Weyl group consists of six elements:

$$\mathcal{W} = \{\mathbb{1}, w_1, w_2, w_1w_2, w_2w_1, w_1w_2w_1\}. \tag{9.97}$$

We will first compute the constant term using the Langlands constant term formula of Chapter 8. Then, using the results of Sections 9.1–9.3, we find the local part of a Whittaker function with an unramified character, which, with the help of Section 9.4, can then be used to compute the local part of any Whittaker function with a generic character. The remaining, degenerate Whittaker functions are then found following the arguments of Section 9.5. Lastly, we will comment on the non-abelian Whittaker coefficients in the sense of Section 6.2.3.

9.8.1 Constant Terms

We first evaluate the Langlands constant term formula (8.41). This yields a sum of six terms:

$$\begin{aligned}\int\limits_{N(\mathbb{Q})\backslash N(\mathbb{A})} E(\chi, ng)dn &= v_1^{2s_1}v_2^{2s_2} + \frac{\xi(2s_1-1)}{\xi(2s_1)}v_1^{2-2s_1}v_2^{2s_1+2s_2-1}\\
&+ \frac{\xi(2s_2-1)}{\xi(2s_2)}v_1^{2s_1+2s_2-1}v_2^{2-2s_2} + \frac{\xi(2s_1-1)\xi(2s_1+2s_2-2)}{\xi(2s_1)\xi(2s_1+2s_2-1)}v_1^{2s_2}v_2^{3-2s_1-2s_2}\\
&+ \frac{\xi(2s_2-1)\xi(2s_1+2s_2-2)}{\xi(2s_2)\xi(2s_1+2s_2-1)}v_1^{3-2s_1-2s_2}v_2^{2s_1}\\
&+ \frac{\xi(2s_1-1)\xi(2s_2-1)\xi(2s_1+2s_2-2)}{\xi(2s_1)\xi(2s_2)\xi(2s_1+2s_2-1)}v_1^{2-2s_2}v_2^{2-2s_1}.\end{aligned} \tag{9.98}$$

Here, v_1 and v_2 are real positive parameters when we restrict the torus element a to the real group $SL(3,\mathbb{R})$.

9.8.2 Generic Whittaker Coefficients

We start by determining the *local* Whittaker function for an unramified character ψ by using the Casselman–Shalika formula in the form (9.23). The quantities $1/\zeta(\lambda)$ and $\epsilon(\lambda)$ of (9.24) evaluate to

$$\frac{1}{\zeta(\lambda)} = (1-p^{-2s_1})(1-p^{-2s_2})(1-p^{1-2s_1-2s_2}), \tag{9.99a}$$

$$\epsilon(\lambda) = \frac{1}{(1-p^{2s_1-1})(1-p^{2s_2-1})(1-p^{2s_1+2s_2-2})} \tag{9.99b}$$

and the full unramified p-adic Whittaker coefficient is then

$$\begin{aligned} W^\circ(\chi, a) &= \int\limits_{N(\mathbb{Q}_p)} \chi(w_{\text{long}} n a)\overline{\psi(n)} dn \\ &= \frac{\epsilon(\lambda)}{\zeta(\lambda)} \Big(|v_1|^{2s_1} |v_2|^{2s_2} - p^{2s_1-1} |v_1|^{2-2s_1} |v_2|^{2s_1+2s_2-1} \\ &\qquad - p^{2s_2-1} |v_1|^{2s_1+2s_2-1} |v_2|^{2-2s_2} + p^{4s_1+2s_2-3} |v_1|^{2s_2} |v_2|^{3-2s_1-2s_2} \\ &\qquad + p^{2s_1+4s_2-3} |v_1|^{3-2s_1-2s_2} |v_2|^{2s_1} - p^{4s_1+4s_2-4} |v_1|^{2-2s_2} |v_2|^{2-2s_1} \Big), \end{aligned} \tag{9.100}$$

where $v_1, v_2 \in \mathbb{Q}_p$.

From (9.100) we can deduce the Whittaker coefficient for a generic character with non-zero instanton charges m_1 and m_2, i.e., one that satisfies

$$\psi^a\left(x_{\alpha_1}(u_1) x_{\alpha_2}(u_2)\right) = \exp\left(2\pi i [m_1 u_1 + m_2 u_2]\right), \tag{9.101}$$

by exploiting (9.54). For this we require $v_1 = m_1^{2/3} m_2^{1/3}$ and $v_2 = m_1^{1/3} m_2^{2/3}$ in the expression above as well as the prefactor $|a^{-(w_{\text{long}}\lambda+\rho)}| = |v_1|^{2s_2-2} |v_2|^{2s_1-2}$. The result is

$$\begin{aligned} W^\circ_{\psi^a}(\chi, \mathbb{1}) = \frac{\epsilon(\lambda)}{\zeta(\lambda)} \Big(& |m_1|^{2s_1+2s_2-2} |m_2|^{2s_1+2s_2-2} - p^{2s_1-1} |m_1|^{2s_2-1} |m_2|^{2s_1+2s_2-2} \\ & - p^{2s_2-1} |m_1|^{2s_1+2s_2-2} |m_2|^{2s_1-1} + p^{4s_1+2s_2-3} |m_1|^{2s_2-1} \\ & + p^{2s_1+4s_2-3} |m_2|^{2s_1-1} - p^{4s_1+4s_2-4} \Big). \end{aligned} \tag{9.102}$$

As explained in Section 9.6 this can also be expressed in terms of a Schur polynomial in (m_1, m_2), which here encodes the character of a highest weight representation of $\mathfrak{sl}(3,\mathbb{C})$. Taking the product over all $p < \infty$ produces double divisor sums [120].

In this case, we can also work out the archimedean Whittaker function. The Whittaker function at $p = \infty$ can be explicitly written as a convoluted integral of two modified Bessel functions, as we will now show.

Starting from (9.14) and using the same standard manipulations as in (9.18) we have that

$$\begin{aligned} W^\circ_{\psi_\infty}(\chi_\infty, a) &= \int_{N(\mathbb{R})} \chi_\infty(w_{\text{long}} n a) \overline{\psi_\infty(n)} dn \\ &= |a^{w_{\text{long}}\lambda+\rho}| \int_{N(\mathbb{R})} \chi_\infty(w_{\text{long}} n) \overline{\psi_\infty(a n a^{-1})} dn, \end{aligned} \tag{9.103}$$

where

$$
\begin{aligned}
\chi_\infty(v_1^{H_1} v_2^{H_2}) &= |v_1|^{2s_1} |v_2|^{2s_2} \\
\left|a^{w_{\text{long}}\lambda+\rho}\right| &= |v_1|^{2-2s_2} |v_2|^{2-2s_1}
\end{aligned} \tag{9.104}
$$

$$
n = \begin{pmatrix} 1 & u_1 & z \\ 0 & 1 & u_2 \\ 0 & 0 & 1 \end{pmatrix}, \qquad w_{\text{long}} = \begin{pmatrix} 0 & 0 & 1 \\ 0 & 1 & 0 \\ 1 & 0 & 0 \end{pmatrix}.
$$

The generic character ψ_∞ is given by two integers m_1 and m_2 through

$$
\psi_\infty(x_{\alpha_1}(u_1)x_{\alpha_2}(u_2)) = \exp\left(2\pi i(m_1u_1 + m_2u_2)\right). \tag{9.105}
$$

Evaluating the integrand we obtain

$$
\begin{aligned}
&\int_{N(\mathbb{R})} \chi_\infty(w_{\text{long}}n)\overline{\psi(ana^{-1})}dn \\
&\quad = \int_{\mathbb{R}^3} (1 + (1+u_1^2)u_2^2 - 2u_1u_2z + z^2)^{-s_1}(1 + u_1^2 + z^2)^{-s_2} \\
&\qquad \times \exp\left(-2\pi i \frac{v_1^3 m_1 u_1 + v_2^3 m_2 u_2}{v_1 v_2}\right) du_1 du_2 dz\,.
\end{aligned} \tag{9.106}
$$

Using the variable substitution $u_2 \to (u_2 + u_1 z)/(1+u_1^2)$ and integrating over u_2 we get

$$
\begin{aligned}
&\frac{2\pi^{s_1}}{\Gamma(s_1)} \left|\frac{m_2 v_2^2}{v_1}\right|^{s_1-1/2} \int_{\mathbb{R}^2} \frac{(1+u_1^2+z^2)^{\frac{1}{4}-\frac{1}{2}s_1-s_2}}{\sqrt{1+u_1^2}} K_{s_1-1/2}\left(2\pi\left|\frac{m_2v_2^2}{v_1}\right|\frac{\sqrt{1+u_1^2+z^2}}{1+u_1^2}\right) \\
&\qquad \times \exp\left(-2\pi i\left(\frac{m_1 v_1^2}{v_2}u_1 + \frac{m_2 v_2^2}{v_1}\frac{u_1 z}{1+u_1^2}\right)\right) du_1 dz\,.
\end{aligned} \tag{9.107}
$$

With standard manipulations (see for example [605, Lemma 7]), this integral can be expressed as a convoluted integral of two Bessel functions, giving $W^\circ_{\psi_\infty}$ as

$$
\begin{aligned}
W^\circ_{\psi_\infty}(\chi_\infty, a) &= \frac{4\pi^{2s_3+1/2}|v_1v_2|}{\Gamma(s_1)\Gamma(s_2)\Gamma(s_3)} |m_1m_2|^{s_3-1/2} \left|\frac{v_1}{v_2}\right|^{s_1-s_2} \\
&\times \int_0^\infty K_{s_3-1/2}\left(2\pi\left|\frac{m_1v_1^2}{v_2}\right|\sqrt{1+1/x}\right) K_{s_3-1/2}\left(2\pi\left|\frac{m_2v_2^2}{v_1}\right|\sqrt{1+x}\right) x^{\frac{s_2-s_1}{2}}\frac{dx}{x},
\end{aligned} \tag{9.108}
$$

where we have introduced $s_3 = s_1 + s_2 - 1/2$ for compactness.

9.8.3 Degenerate Whittaker Coefficients

We now evaluate the Whittaker coefficients for non-generic characters, i.e., those where either m_1 or m_2 vanishes. Note that it is not trivially possible to obtain this result from the generic one above by setting some parameters to zero.

We will employ Theorem 9.6 and perform this for the example $m_2 = 0$. Then the support of the character is only on the first simple root, so that $w'_{\text{long}} = w_1$ and the subgroup $G'(\mathbb{A})$ is the one associated with the first simple root only. The possible Weyl words that contribute to (9.57) are

$$w = w_c w'_{\text{long}} \in \left\{\mathbb{1}w'_{\text{long}}, w_2 w'_{\text{long}}, w_1 w_2 w'_{\text{long}}\right\} = \{w_1, w_2 w_1, w_1 w_2 w_1\}. \tag{9.109}$$

As a first step, we calculate the projected weights $w_c^{-1}\lambda$ and $M(w_c^{-1},\lambda)$ factors that appear in (9.57) for the three choices:

$$w_c = \mathbb{1}: \qquad \lambda' = (w_c^{-1}\lambda)_{G'} = (2s_1 - 1)\Lambda'_1, \quad M(w_c^{-1},\lambda) = 1, \tag{9.110a}$$

$$w_c = w_2: \qquad \lambda' = (2s_1 + 2s_2 - 2)\Lambda'_1, \qquad M(w_c^{-1},\lambda) = \frac{\xi(2s_2 - 1)}{\xi(2s_2)}, \tag{9.110b}$$

$$w_c = w_1 w_2: \quad \lambda' = (2s_2 - 1)\Lambda'_1,$$

$$M(w_c^{-1},\lambda) = \frac{\xi(2s_1 - 1)\xi(2s_1 + 2s_2 - 2)}{\xi(2s_1)\xi(2s_1 + 2s_2 - 1)}, \tag{9.110c}$$

where $\Lambda'_1 = \alpha_1/2$ is the fundamental weight for $G'(\mathbb{A})$. This will need to be combined with

$$\psi^a(x_{\alpha_1}(u_1)) = \psi(a x_{\alpha_1}(u_1) a^{-1}) = \exp\left(2\pi i a^{\alpha_1} u_1 m_1\right) = \exp\left(2\pi i v_1^2 v_2^{-1} m_1 u_1\right) \tag{9.111}$$

and the $SL(2,\mathbb{A})$ Whittaker function for $\lambda' = (2s' - 1)\Lambda'_1$ given by (see (7.81))

$$W'^{\circ}_{\psi^a}(\lambda', \mathbb{1}) = \frac{2(2\pi)^{1/2-s'}}{\xi(2s')}\sigma_{2s'-1}(m_1)\mathcal{K}_{1/2-s'}\left(2\pi|m_1|v_1^2 v_2^{-1}\right), \tag{9.112}$$

where we have introduced the shorthand $\mathcal{K}_t(x) = x^{-t}K_{-t}(x)$ in order to facilitate comparison with [525]. Recall also the compact notation $s_3 = s_1 + s_2 - 1/2$. The resulting expression for the $(m_1, 0)$ degenerate Whittaker function is then

$$\begin{aligned}
W^{\circ}_{\psi}(\chi, a) = {} & \frac{2(2\pi)^{1/2-s_1}}{\xi(2s_1)} v_1^{2-2s_1} v_2^{2s_1+2s_2-1} \sigma_{2s_1-1}(m_1)\mathcal{K}_{1/2-s_1}\left(2\pi|m_1|v_1^2 v_2^{-1}\right) \\
& + \frac{2(2\pi)^{1/2-s_3}}{\xi(2s_3)} \frac{\xi(2s_2 - 1)}{\xi(2s_2)} v_1^{3-2s_1-2s_2} v_2^{2s_1} \\
& \times \sigma_{2s_3-1}(m_1)\mathcal{K}_{1/2-s_3}\left(2\pi|m_1|v_1^2 v_2^{-1}\right) \\
& + \frac{2(2\pi)^{1/2-s_2}}{\xi(2s_2)} \frac{\xi(2s_1 - 1)\xi(2s_3 - 1)}{\xi(2s_1)\xi(2s_3)} v_1^{2-2s_2} v_2^{2-2s_1} \\
& \times \sigma_{2s_2-1}(m_1)\mathcal{K}_{1/2-s_2}\left(2\pi|m_1|v_1^2 v_2^{-1}\right).
\end{aligned} \tag{9.113}$$

This also matches the expressions in [525] if one adapts the conventions. More precisely, one uses $\nu_1 = \nu^{-1/6}\tau_2^{1/2}$, $\nu_2 = \nu^{-1/3}$ and exchanges s_1 and s_2 to find the $\Psi_{0,q}$ coefficient in [525, Eq. (3.45)]. A similar calculation can be carried out for the degenerate Whittaker coefficient associated with instanton charges $(0, m_2)$; it simply amounts to interchanging the subscripts 1 and 2 everywhere thanks to the Dynkin diagram automorphism of $\mathfrak{sl}(3, \mathbb{R})$.

9.8.4 Non-abelian Fourier Coefficients

So far in Section 9.8 we have only studied Fourier coefficients on N, but since the characters on N are trivial on the centre $Z = N^{(2)} = [N, N]$ they do not capture the complete Fourier expansion of $E(\chi, g)$ as discussed in Section 6.2.3. To obtain a complete expansion we also need Fourier coefficients on Z with non-trivial characters $\psi_Z\colon Z(\mathbb{Q})\backslash Z(\mathbb{A}) \to U(1)$ parametrised by $k \in \mathbb{Q}^\times$:

$$\psi_Z(n_{(2)}) = e^{2\pi i k z}, \qquad n_{(2)} = \begin{pmatrix} 1 & 0 & z \\ 0 & 1 & 0 \\ 0 & 0 & 1 \end{pmatrix} \in Z(\mathbb{A}) . \tag{9.114}$$

To avoid ambiguities, we will denote the character on Z by ψ_Z and the characters on N by ψ_N.

Recalling (6.2.3), the Fourier coefficients on Z are defined by

$$W^\circ_{\psi_Z}(\chi, g) = \int\limits_{Z(\mathbb{Q})\backslash Z(\mathbb{A})} E(\chi, n_{(2)} g)\overline{\psi_Z(n_{(2)})}\, dn_{(2)} . \tag{9.115}$$

For the remaining parts of Section 9.8, we will drop the superscript for the spherical property and write the mode number explicitly as $W^{(k)}_{\psi_Z}$ for clarity.

We will now show that these Fourier coefficients on Z are determined by the Whittaker coefficients (i.e., Fourier coefficients on N), but before we can make an exact statement we need to make a few definitions.

Let $k, m_2 \in \mathbb{Q}$ with $k = a_1/b_1$ and $m_2 = a_2/b_2$ in shortened form, where $a_i \in \mathbb{Z}$ and $b_i \in \mathbb{N}$. Define

$$d = d(k, m_2) := \frac{\gcd(a_1 b_2, a_2 b_1)}{b_1 b_2}, \tag{9.116}$$

which is then strictly positive since $k \neq 0$. Furthermore, let $k' := k/d = a_1 b_2 / \gcd(a_1 b_2, a_2 b_1) \in \mathbb{Z}$ and $m_2' := m_2/d = a_2 b_1 / \gcd(a_1 b_2, a_2 b_1) \in \mathbb{Z}$. Then, there exist integers α and β such that

$$\alpha m_2' - \beta k' = \gcd(k', m_2') = 1 . \tag{9.117}$$

The ambiguity in the definition of α and β is discussed in the proof of the following proposition.

Proposition 9.14 (Non-abelian $SL(3)$ Whittaker function) *Let $k \in \mathbb{Q}^\times$ with α, β, k' and m_2' defined as above. Then*

$$W_{\psi_Z}^{(k)}(\chi, g) = \sum_{m_1, m_2 \in \mathbb{Q}} W_{\psi_N}^{(m_1,d)}(\chi, lg), \qquad l = \begin{pmatrix} \alpha & \beta & 0 \\ k' & m_2' & 0 \\ 0 & 0 & 1 \end{pmatrix} \in SL(3,\mathbb{Z}), \tag{9.118}$$

where $g = (g_\infty, g_2, g_3, \ldots)$ is an arbitrary element of $G(\mathbb{A})$.

We will consider the restriction $g = (g_\infty, \mathbb{1}, \mathbb{1}, \ldots)$ yielding integer charges in Proposition 9.15. By $W_{\psi_N}^{(m_1,d)}$ we mean the Whittaker coefficients on N given by ψ_N with instanton charges m_1 and d for the simple roots, which were calculated in (9.102), (9.108) and (9.113).

Proof To show (9.118), first let l be defined as in that equation. We can expand $W_{\psi_Z}^{(k)}(\chi, g)$ further as

$$\begin{aligned} W_{\psi_Z}^{(k)}(\chi, g) &= \int_{\mathbb{Q}\backslash\mathbb{A}} E(\chi, \begin{pmatrix} 1 & 0 & z \\ 0 & 1 & 0 \\ 0 & 0 & 1 \end{pmatrix} g) e^{-2\pi i k z} \, dz \\ &= \sum_{m_2 \in \mathbb{Q}} \int_{(\mathbb{Q}\backslash\mathbb{A})^2} E(\chi, \begin{pmatrix} 1 & 0 & z \\ 0 & 1 & x_2 \\ 0 & 0 & 1 \end{pmatrix} g) e^{-2\pi i (kz + m_2 x_2)} \, dz dx_2 \, . \end{aligned} \tag{9.119}$$

Using the automorphic invariance of $E(\chi, g)$ we can make the following conjugation with l:

$$\begin{aligned} W_{\psi_Z}^{(k)}(\chi, g) &= \sum_{m_2} \int_{(\mathbb{Q}\backslash\mathbb{A})^2} E(\chi, l \begin{pmatrix} 1 & 0 & z \\ 0 & 1 & x_2 \\ 0 & 0 & 1 \end{pmatrix} l^{-1} l g) e^{-2\pi i (kz + m_2 x_2)} dz dx_2 \\ &= \sum_{m_2} \int_{(\mathbb{Q}\backslash\mathbb{A})^2} E(\chi, \begin{pmatrix} 1 & 0 & -d(x_2 - \alpha(kz + m_2 x_2)/d)/k \\ 0 & 1 & (kz + m_2 x_2)/d \\ 0 & 0 & 1 \end{pmatrix} l g) e^{-2\pi i (kz + m_2 x_2)} dz dx_2 \\ &= \sum_{m_2} \int_{(\mathbb{Q}\backslash\mathbb{A})^2} E(\chi, \begin{pmatrix} 1 & 0 & x_2 \\ 0 & 1 & x_3 \\ 0 & 0 & 1 \end{pmatrix} l g) e^{-2\pi i d x_3} dx_2 dx_3 \, , \end{aligned} \tag{9.120}$$

where we have made the substitution $(kz + m_2 x_2)/d \to x_3$ and then $-d(x_2 - \alpha x_3)/k \to x_2$, leaving the integration domain the same. According to (2.17) and (2.67) the measure is also unchanged. We note that the ambiguity in α simply results in an extra shift in the periodic variable x_2.

We once more expand

$$\begin{aligned}
W_{\psi_Z}^{(k)}(\chi,g) &= \sum_{m_1,m_2\in\mathbb{Q}} \int_{(\mathbb{Q}\backslash\mathbb{A})^3} E(\chi,\left(\begin{smallmatrix}1&x_1&0\\0&1&0\\0&0&1\end{smallmatrix}\right)\left(\begin{smallmatrix}1&0&x_2\\0&1&x_3\\0&0&1\end{smallmatrix}\right)lg)e^{-2\pi i(m_1x_1+dx_3)}\,d^3x \\
&= \sum_{m_1,m_2} \int_{(\mathbb{Q}\backslash\mathbb{A})^3} E(\chi,\left(\begin{smallmatrix}1&x_1&x_2\\0&1&x_3\\0&0&1\end{smallmatrix}\right)lg)e^{-2\pi i(m_1x_1+dx_3)}\,d^3x \\
&= \sum_{m_1,m_2} W_{\psi_N}^{(m_1,d)}(\chi,lg), \qquad (9.121)
\end{aligned}$$

where, in the second step, we have made the substitution $x_2 + x_1x_3 \to x_2$. □

Note that when inserting $g = (g_\infty, \mathbb{1}, \mathbb{1}, \ldots) \in G(\mathbb{A})$ into (9.118), reducing the adelic Eisenstein series on the left-hand side to the real Eisenstein series, the arguments on the right-hand side become non-trivial at the finite places. To be able to use the above expressions for W_{ψ_N}, which require trivial arguments at the finite places, we need to factor out these effects.

Proposition 9.15 (Handling non-trivial arguments of W_{ψ_N}) *Let $\tau = u_1 + iv_1^2/v_2 \in \mathbb{H}$ and*

$$\gamma = \left(\begin{smallmatrix}\alpha&\beta\\k'&m_2'\end{smallmatrix}\right) \in SL(2,\mathbb{Z}), \qquad \gamma(\tau) = \frac{\alpha\tau+\beta}{k'\tau+m_2'}, \qquad a'_{\operatorname{Im}\gamma(\tau)} = \begin{pmatrix}v_1'&0&0\\0&v_2'/v_1'&0\\0&0&1/v_2'\end{pmatrix}, \qquad (9.122)$$

with $v_1' = \sqrt{v_2' \operatorname{Im}\gamma(\tau)}$ and $v_2' = v_2$, and $g = (g_\infty, \mathbb{1}, \ldots) \in G(\mathbb{A})$. Then $W_{\psi_Z}^{(k)}$ is non-vanishing only for $k \in \mathbb{Z}$ for which

$$W_{\psi_Z}^{(k)}(\chi,(g_\infty,\mathbb{1},\ldots)) = \sum_{m_1,m_2\in\mathbb{Z}} W_{\psi_N}^{(m_1,d)}(\chi,(a'_{\operatorname{Im}\gamma(\tau)},\mathbb{1},\ldots)) \times e^{-2\pi i(m_1\operatorname{Re}\gamma(\tau)+m_2u_2+kz)}. \qquad (9.123)$$

Note that the sums over rationals have collapsed to sums over integers and that

$$l = \left(\begin{smallmatrix}\gamma&0\\0&1\end{smallmatrix}\right). \qquad (9.124)$$

This proves the results of [605] and [538] reviewed in [525, 510] with only a few manipulations using the compact framework of adelic automorphic forms.

Proof In (9.118) the argument for $W_{\psi_N}^{(m_1,d)}$ is $lg = (lg_\infty; l, l, \ldots)$ and, since $W_{\psi_N}(\chi, n'a'k') = \psi_N(n')W_{\psi_N}(\chi, a')$, we factorise lg at the archimedean and non-archimedean places into their respective Iwasawa decompositions. We have that $l \in SL(3,\mathbb{Z})$, which makes the p-adic Iwasawa decomposition trivial with $l \in K_p$. This was the reason for choosing l in this particular form.

We then use the following relation, similar to (9.19), to obtain conditions for m_1 and m_2. For $\hat{n} = (\mathbb{1}; \hat{n}_2, \hat{n}_3, \ldots)$ with $\hat{n}_p \in N(\mathbb{Z}_p) \subset K_p$ we have that $\hat{n} \in K_{\mathbb{A}}$ and

$$W_{\psi_N}(\chi, a) = W_{\psi_N}(\chi, a\hat{n}) = W_{\psi_N}(\chi, a\hat{n}a^{-1}a) = \psi_N(a\hat{n}a^{-1})W_{\psi_N}(\chi, a), \tag{9.125}$$

which requires that $\psi_N(a\hat{n}a^{-1}) = 1$ for $W_{\psi_N}(\chi, a)$ to be non-vanishing.

Specifically, for $W_{\psi_N}^{(m_1,d)}(\chi, a)$ with $a = (a_\infty; \mathbb{1}, \mathbb{1}, \ldots)$ we require that

$$1 = \psi_N(a\hat{n}a^{-1}) = \psi_{N,\infty}(\mathbb{1}) \prod_{p<\infty} \psi_{N,p}(\hat{n}_p) = \exp\Big(-2\pi i \sum_{p<\infty} [m_1 u_1 + d u_2]_p \Big) \tag{9.126}$$

for all $u_1, u_2 \in \mathbb{Z}_p$, where

$$\hat{n}_p = \begin{pmatrix} 1 & u_1 & z \\ 0 & 1 & u_2 \\ 0 & 0 & 1 \end{pmatrix}. \tag{9.127}$$

This implies that $\sum_{p<\infty}[m_1]_p \in \mathbb{Z}$ and $\sum_{p<\infty}[d]_p \in \mathbb{Z}$, which, according to Proposition 2.13, gives that $m_1, d \in \mathbb{Z}$. That d is integer means that, for all primes p,

$$1 \geq |d|_p = \frac{\max(|a_1 b_2|_p, |a_2 b_1|_p)}{|b_1 b_2|_p} = \max(|k|_p, |m_2|_p) \tag{9.128}$$

according to (2.11), and hence that k and m_2 are also integers.

For the archimedean place we have the Iwasawa decomposition

$$l g_\infty = l n_\infty a_\infty k_\infty = l \begin{pmatrix} 1 & u_1 & z \\ 0 & 1 & u_2 \\ 0 & 0 & 1 \end{pmatrix} \begin{pmatrix} v_1 & 0 & 0 \\ 0 & v_2/v_1 & 0 \\ 0 & 0 & 1/v_2 \end{pmatrix} k_\infty = n'_\infty a'_\infty k'_\infty, \qquad \text{with}$$

$$u'_1 = -\frac{d^2(m_2 + k u_1) v_2^2}{k^3 v_1^4 + k(m_2 + k u_1)^2 + v_2^2} + \frac{d\alpha}{k}, \qquad u'_2 = \frac{m_2 u_2 + k z}{d},$$

$$v'_1 = \frac{v_1 v_2 d}{\sqrt{k^2 v_1^4 + (m_2 + k u_1)^2 v_2^2}}, \qquad v'_2 = v_2. \tag{9.129}$$

We define $\tau = u_1 + i v_1^2/v_2 \in \mathbb{H}$, which, under the l-translation on g_∞ above, transforms as $\tau \to \tau'$ with

$$\tau' = u'_1 + i\frac{(v'_1)^2}{v'_2} = \gamma(\tau), \qquad \gamma = \begin{pmatrix} \alpha & \beta \\ k' & m'_2 \end{pmatrix} \in SL(2, \mathbb{Z}). \tag{9.130}$$

Putting everything together we obtain for $k \in \mathbb{Z}^\times$ and $g = (g_\infty; \mathbb{1}, \mathbb{1}, \ldots)$

$$\begin{aligned}
W_{\psi_Z}^{(k)}&(\chi, (g_\infty; \mathbb{1}, \ldots)) \\
&= \sum_{m_1, m_2 \in \mathbb{Z}} W_{\psi_N}^{(m_1,d)}(\chi, a') \psi_N^{(m_1,d)}(n') \\
&= \sum_{m_1, m_2 \in \mathbb{Z}} W_{\psi_N}^{(m_1,d)}(\chi, (a'_\infty; \mathbb{1}, \ldots)) \Big(\prod_{p<\infty} \psi_{N,p}^{(m_1,d)}(n'_p) \Big) \psi_{N,\infty}^{(m_1,d)}(n'_\infty) \\
&= \sum_{m_1, m_2 \in \mathbb{Z}} W_{\psi_N}^{(m_1,d)}(\chi, (a'_{\operatorname{Im}\gamma(\tau)}; \mathbb{1}, \ldots)) e^{-2\pi i (m_1 \operatorname{Re}\gamma(\tau) + m_2 u_2 + kz)},
\end{aligned} \tag{9.131}$$

where $a'_{\operatorname{Im}\gamma(\tau)}$ is defined in (9.122). □

The remaining Whittaker coefficients on N with trivial arguments at the non-archimedean places were computed in Sections 9.8.2 and 9.8.3.

Remark 9.16 The physical intepretation of the $SL(2, \mathbb{Z})$ action described by γ is described by S-duality of type IIB string theory compactified on a Calabi–Yau threefold [525, 542, 511]. In this setting the parameters z and u_2 are scalar fields sourced by D5- and NS5-branes with charges m_2 and k (more generally denoted by p and q). The branes form bound states that are often refered to as (p, q) 5-branes. The two scalar fields transform as an $SL(2)$-doublet under S-duality mirrored by their transformation under $g \to lg$ and the charges p and q, which appear as m_2 and k in (9.123), break the classical $SL(2, \mathbb{R})$ symmetry of the supergravity theory to the discrete $SL(2, \mathbb{Z})$ symmetry of the quantum corrected effective action described by Eisenstein series. In short, this tells us that, if we can compute the effects from a $(p, 0)$ 5-brane, the results for any (p, q) 5-brane follow from S-duality which is mirrored in the sum over matrices γ in (9.123).

10

Analysing Eisenstein Series and Small Representations

After having developed the formal theory of Eisenstein series and their Fourier expansion in the previous chapters, we would like to discuss Eisenstein series from a more practical point of view in this chapter. In concrete examples this typically means obtaining as much information as possible for a particular Eisenstein series $E(\chi, g)$, that is, a particular given χ. Many of the general theorems either simplify for such a χ or have to be evaluated with much care as $E(\chi, g)$ might be divergent for the chosen χ. This chapter deals with developing methods for addressing these issues. In particular, we exhibit methods for efficiently evaluating the constant term formula (8.41) and formula (9.57) for the Whittaker coefficients of a given Eisenstein series. We will also discuss the pole structure of Eisenstein series (as a function of χ) in examples, their residues as well as different normalisations. In this chapter, the emphasis is on illustrating different methods through many examples; for proofs of general statements we will typically refer to the appropriate literature. The simplifications that arise are associated with a reduction of the Gelfand–Kirillov dimension of the associated automorphic representation and thus with small representations in the sense of Section 6.4.3. This will be a recurring theme in this chapter.

Many of the properties of Eisenstein series are controlled by the completed Riemann zeta function whose properties we briefly recall.

Proposition 10.1 (Properties of completed Riemann zeta function) *As a function of $s \in \mathbb{C}$, the completed Riemann zeta function $\xi(s) = \pi^{-s/2}\Gamma(s/2)\zeta(s)$ converges absolutely for* $\mathrm{Re}(s) > 1$ *and has simple poles at $s = 0$ and $s = 1$ with residues -1 and $+1$, respectively. It is non-zero everywhere else on the real line and it satisfies the functional relation $\xi(s) = \xi(1 - s)$.*

Proof The first statements follow directly from the definition and the properties of gamma and zeta functions. The functional relation was shown originally by Riemann using analytic continuation [541]. □

Note that the proposition does not say anything about zeroes of the Riemann zeta function off the real axis. The famous *Riemann conjecture* states that all non-trivial zeroes of $\zeta(s)$ lie on the critical line $s = 1/2 + it$ with $t \in \mathbb{R}$.

10.1 The $SL(2,\mathbb{R})$ Eisenstein Series as a Function of s

We begin with the $SL(2,\mathbb{R})$ Eisenstein series $E(s,z)$ that was analysed in great detail in Chapter 7, with its complete Fourier expansion given in Theorem 7.1. We repeat the result here for convenience:

$$\begin{aligned} E(s,z) =& y^s + \frac{\xi(2s-1)}{\xi(2s)} y^{1-s} \\ &+ \frac{2}{\xi(2s)} y^{1/2} \sum_{m\neq 0} |m|^{s-1/2} \sigma_{1-2s}(m) K_{s-1/2}(2\pi|m|y) e^{2\pi i m x}, \end{aligned} \tag{10.1}$$

where $z = x + iy$ is an element of the upper half-plane $\mathbb{H} = SL(2,\mathbb{R})/SO(2)$. The original definition of $E(s,z)$ only converged for $\mathrm{Re}(s) > 1$ but by virtue of the functional relation (see Theorem 7.1)

$$E(s,z) = \frac{\xi(2s-1)}{\xi(2s)} E(1-s,z), \tag{10.2}$$

or through analytic continuation of the Fourier expansion (10.1), one can define $E(s,z)$ for almost all complex s. We restrict our discussion to real s for simplicity.

10.1.1 Limiting Values in Original Normalisation

From the explicit form (10.1) one sees that special things might happen for the values $s = 0$, $s = 1/2$ and $s = 1$. All of them are outside the original domain of convergence. Let us note that the region $0 \leq \mathrm{Re}(s) \leq 1$ is often called the *critical strip*.

- $s = 0$: This is the limit where the inducing character $\chi_s(z) = y^s$ becomes trivial. Taking the limit in the expression (10.1) for the Fourier expansion one also sees that all terms go to zero except for the first. This is due to the factors $\frac{1}{\xi(2s)}$ that vanish linearly for $s \to 0$ while everything else stays bounded. The proper limiting behaviour is therefore

$$E(s,z) = 1 + O(s). \tag{10.3}$$

The constant value 1 could have been expected from the triviality of the inducing character but the definition in terms of a *Poincaré sum* is ill-defined. Only after analytic continuation of the sum does one obtain the constant $E(0, z) = 1$.

Representation-theoretically, the function $E(s, z)$ in the limit $s \to 0$ belongs to the trivial representation of $SL(2,\mathbb{R})$.

- $s = 1/2$: Inspection of the Fourier expansion (10.1) shows that the non-zero Fourier modes disappear in this limit due to the $\frac{1}{\xi(2s)}$ prefactor. For the constant terms one has to take the limit of the quotient of completed Riemann zeta functions, which is found to be -1, and the two contributions to the constant term cancel, leading to

$$E(s, z) = 0 + O(s - 1/2)\,. \tag{10.4}$$

The first-order term is a member of the principal series. It is on the critical line and is therefore almost unitary. The value $s = 1/2$ also has special properties from the point of view of the eigenvalue equation under the Laplace operator; see (B.19b) in Appendix B for more details.

- $s = 1$: This is the most interesting case. The Fourier expansion (10.1) shows that the second constant term diverges in the limit $s \to 1$ while all other terms remain finite. The residue at the simple pole can be calculated easily from the completed Riemann zeta functions:

$$E(s, z) = \frac{3}{\pi(s-1)} + O\left((s-1)^0\right). \tag{10.5}$$

The residue is a constant function and is therefore also of the same type as the limit $s \to 0$ discussed above. This is not surprising since the functional relation (10.2) relates the values $s = 0$ and $s = 1$ and one sees that the prefactor introduces the additional pole. Representation-theoretically, the residue of the series $E(s, z)$ at the simple pole $s = 1$ belongs to the trivial representation of $SL(2,\mathbb{R})$.

The term at order $(s-1)^0$ can also be evaluated from the Fourier expansion using the fact that the modified Bessel function $K_{1/2}$ has an exact asymptotic expansion in terms of a simple exponential. One finds

$$\begin{aligned} E(s, z) = \frac{3}{\pi(s-1)} - \frac{6}{\pi}\Bigg(& -\frac{\pi}{6}y + \log(4\pi\sqrt{y}) - 12\log A \\ & - \sum_{m>0} \sigma_{-1}(m) e^{2\pi i m(x+iy)} \\ & - \sum_{m>0} \sigma_{-1}(m) e^{2\pi i m(x-iy)}\Bigg) \\ + O(s-1). \end{aligned} \tag{10.6}$$

Here, $A \approx 1.2824\ldots$ is the *Glaisher–Kinkelin constant* which satisfies $\log A = \frac{1}{12} - \zeta'(1)$. The expression can be rewritten by using the *Dedekind eta function*

$$\eta(z) = q^{1/24} \prod_{n=1}^{\infty} (1 - q^n), \tag{10.7}$$

where $q = e^{2\pi i z} = e^{2\pi i(x+iy)}$ on the right-hand side. From the product formula for $\eta(z)$ one concludes

$$\begin{aligned}
\log \eta(z) &= \frac{1}{24} \log q + \sum_{n>0} \log(1 - q^n) \\
&= \frac{\pi i}{12}(x + iy) - \sum_{n>0} \sum_{k>0} k^{-1} q^{kn} \\
&= \frac{\pi i}{12}(x + iy) - \sum_{m>0} \sum_{d|m} d^{-1} q^m \\
&= \frac{\pi i}{12}(x + iy) - \sum_{m>0} \sigma_{-1}(m) q^m.
\end{aligned} \tag{10.8}$$

The s-independent term in (10.6) can therefore be written as

$$\begin{aligned}
&-\frac{6}{\pi}\bigg(-\frac{\pi}{6} y + \log(4\pi\sqrt{y}) - 12 \log A \\
&\qquad - \sum_{m>0} \sigma_{-1}(m) q^m - \sum_{m>0} \sigma_{-1}(m) \bar{q}^m \bigg) \\
&= -\frac{6}{\pi}\Big(-12 \log A + \log(4\pi) + \log\left(\sqrt{y}|\eta(z)|^2\right) \Big),
\end{aligned} \tag{10.9}$$

leading to

$$\begin{aligned}
E(s, z) = {}& \frac{3}{\pi(s-1)} + \frac{6}{\pi}\Big(12 \log A - \log(4\pi) - \log\left(\sqrt{y}|\eta(z)|^2\right)\Big) \\
&+ O(s-1).
\end{aligned} \tag{10.10}$$

This formula is known as the *(first) Kronecker limit formula*. Even though neither $\eta(z)$ nor $|\eta(z)|^2$ are $SL(2,\mathbb{Z})$-invariant, the particular combination appearing in this expression is invariant.

Remark 10.2 The particular combination of constants in the Kronecker limit formula (10.10) depends on the way the Eisenstein series is normalised. The formula is more commonly stated for the $SL(2,\mathbb{Z})$-invariant lattice sum (see (1.2)), for which one finds

$$\sum_{\substack{(c,d)\in\mathbb{Z}^2\\(c,d)\neq(0,0)}} \frac{y^s}{|cz+d|^{2s}} = 2\zeta(2s)E(s,z) \tag{10.11}$$

$$= \frac{\pi}{s-1} + 2\pi\Big(\gamma_{\rm E} - \log(2) - \log\left(\sqrt{y}|\eta(z)|^2\right)\Big) + O(s-1) \tag{10.12}$$

if one uses the following relation between the Glaisher–Kinkelin constant A and the *Euler–Mascheroni constant* $\gamma_{\rm E}$: $12\log A - \log(4\pi) = \gamma_{\rm E} - \log 2 - \frac{\zeta'(2)}{\zeta(2)}$.

10.1.2 Weyl Symmetric Normalisation

The functional relation (10.2) suggests defining a *completed Eisenstein series* in analogy with the completed Riemann zeta function by the definition

$$E^\star(s,z) = \xi(2s)E(s,z). \tag{10.13}$$

This then has the simple property that

$$E^\star(s,z) = E^\star(1-s,z), \tag{10.14}$$

and we call this the *Weyl symmetric normalisation* as it yields a function invariant under Weyl transformations acting on the character. Indeed, the non-trivial Weyl reflection w of $SL(2,\mathbb{R})$ acts on the weight $\lambda_s = (2s-1)\rho$ by

$$w\lambda_s = -(2s-1)\rho = (2(1-s)-1)\rho = \lambda_{1-s} \tag{10.15}$$

and so exchanges s and $1-s$. This was of course already used and apparent in the constant terms in (10.1).

Since the normalising factor has poles and zeroes of its own, the discussion of the behaviour of $E^\star(s,z)$ as a function of s is slightly changed from the one above. More precisely, the completed function $E^\star(s,z)$ has simple poles at $s=0$ and $s=1$, whereas it has a non-trivial limit for $s=1/2$. Representation-theoretically, $E^\star(1/2,z)$ belongs to the principal series.

10.2 Properties of Eisenstein Series

The behaviour of the $SL(2,\mathbb{R})$ Eisenstein series at the special values of s above was completely controlled by the constant terms. This is a general feature due to the holomorphy of the Fourier coefficients; see Proposition 9.4. As we have full control of the constant terms thanks to the Langlands constant term formula (Theorem 8.1), we can in principle completely determine the behaviour of an Eisenstein series $E(\lambda,g)$ on a group $G(\mathbb{R})$ as a function of λ. As the number of constant terms is generically equal to the order of the Weyl group $\mathcal{W}$ of G, this can be quite tedious due to the large number of terms that have to be considered. In Section 10.3, we will present a method that makes the problem more tractable for the case of non-generic λ when the Eisenstein series $E(\lambda, g)$ is not attached to the full principal series but to a degenerate principal series. The prime example of this is when it becomes a maximal parabolic Eisenstein series as defined in Section 5.7. Before focussing on these cases in Section 10.3, we offer a few general and cautionary remarks.

10.2.1 Validity of Functional Relation

As Langlands showed in his seminal work [447], the functional Equation (8.44), repeated here for convenience:

$$E(\lambda, g) = M(w, \lambda) E(w\lambda, g), \tag{10.16}$$

is valid for almost all $\lambda \in \mathfrak{h}^*(\mathbb{C})$. The exceptions are affine hyperplanes in the complex vector space $\mathfrak{h}^*(\mathbb{C})$. These affine hyperplanes are associated with poles and zeroes of the intertwining factor $M(w, \lambda)$. Since all poles and zeroes are of finite order, one can make sense of the functional relation even on these planes by also treating the Eisenstein series as meromorphic functions with finite order poles and singularities.

Example 10.3: Functional relation at a simple zero and pole

For $SL(2, \mathbb{R})$ and general $\lambda = (2s-1)\rho$, the intertwining factor for the non-trivial Weyl element $w = w_{\text{long}}$ is

$$M(w_{\text{long}}, \lambda) = \frac{\xi(2s-1)}{\xi(2s)} \tag{10.17}$$

and has a simple zero at $s = 0$ and a simple pole at $s = 1$. The functional relation (10.16) remains valid even at these places if one considers

$$E(s, z) = 1 + s\hat{E}_0(z) + O\left(s^2\right),$$
$$E(s, z) = \frac{3}{\pi(s-1)} + \hat{E}_1(z) + O(s-1) \tag{10.18}$$

around $s = 0$ and $s = 1$, respectively. The expansion of the intertwining factor around these values is

$$\frac{\xi(2s-1)}{\xi(2s)} = -\frac{\pi s}{3} + O(s^2) = \frac{3}{\pi(s-1)} + O\left((s-1)^0\right), \tag{10.19}$$

such that

$$E(s,z) = 1 + s\hat{E}_0(z) + O(s^2) = \left(-\frac{\pi s}{3} + O(s^2)\right)\left(-\frac{3}{\pi s} + \hat{E}_1 + O(s)\right)$$
$$= 1 - \frac{\pi s}{3}\hat{E}_1 + O(s^2) \qquad (10.20)$$

and so the functional relation relates $\hat{E}_0(z)$ and $\hat{E}_1(z)$ (as well as all higher-order terms).

Also of interest are fixed planes of the action of the Weyl group action. In these cases, the functional relation (10.16) constrains the Eisenstein series on the fixed plane.

Example 10.4: Functional relation for $SL(2,\mathbb{R})$ Eisenstein series with $\lambda = 0$

For $SL(2,\mathbb{R})$ and $E(s,z)$ the fixed plane is $s = 1/2$, corresponding to $\lambda = \lambda_{1/2} = 0$. The intertwining factor at this place takes the value $M(w,0) = -1$ such that the functional relation implies

$$E(1/2,z) = -E(1/2,z) \quad \Longrightarrow \quad E(1/2,z) = 0, \qquad (10.21)$$

consistent with the analysis in Section 10.1.1.

We consider also a few examples of functional relations for higher-rank groups.

Example 10.5: Functional relation for $SL(3,\mathbb{R})$ Eisenstein series

The most general Eisenstein series on $G = SL(3,\mathbb{R})$ is given by a weight

$$\lambda_{s_1,s_2} = 2s_1\Lambda_1 + 2s_2\Lambda_2 - \rho \qquad (10.22)$$

that is parametrised by two complex parameters s_1 and s_2. The Λ_i are as always the fundamental weights. We denote the corresponding character by $\chi_{s_1,s_2}(a) = a^{\lambda_{s_1,s_2}+\rho}$ and the Eisenstein series by

$$E(s_1,s_2,g) = \sum_{B(\mathbb{Z})\backslash SL(3,\mathbb{Z})} \chi_{s_1,s_2}(\gamma g). \qquad (10.23)$$

The sum is absolutely convergent for $\mathrm{Re}(s_1) > 1$ and $\mathrm{Re}(s_2) > 1$ [120]. The Weyl group of $SL(3,\mathbb{R})$ is isomorphic to the symmetric group on three letters and hence consists of six elements, and the constant terms were already given in (9.98). Denoting the fundamental reflections by w_1 and w_2 one finds that

$$w_1\lambda_{s_1,s_2} = (1-2s_1)\Lambda_1 + 2(s_1+s_2-1)\Lambda_2 = \lambda_{1-s_1,s_1+s_2-1/2}, \qquad (10.24a)$$
$$w_2\lambda_{s_1,s_2} = \lambda_{s_1+s_2-1/2,1-s_2}\,. \qquad (10.24b)$$

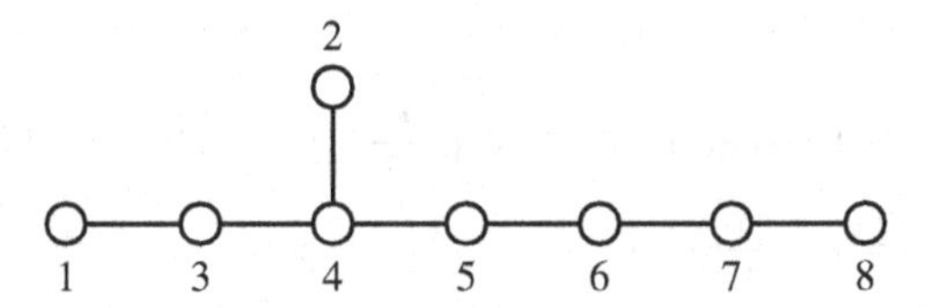

Figure 10.1 The Dynkin diagram of E_8 with labelling of nodes in the 'Bourbaki convention'.

The other Weyl images can be obtained similarly. One functional relation is therefore

$$\begin{aligned} E(s_1, s_2, g) &= M(w_1, \lambda_{s_1,s_2}) E(1 - s_1, s_1 + s_2 - 1/2, g) \\ &= \frac{\xi(\langle \alpha_1 | \lambda_{s_1,s_2} \rangle)}{\xi(\langle \alpha_1 | \lambda_{s_1,s_2} \rangle + 1)} E(1 - s_1, s_1 + s_2 - 1/2, g) \\ &= \frac{\xi(2s_1 - 1)}{\xi(2s_1)} E(1 - s_1, s_1 + s_2 - 1/2, g). \end{aligned} \tag{10.25}$$

That this is a valid relation can be checked on the constant terms from (9.98). We can consider the limit $s_1 \to 1/2$ to conclude

$$E(1/2, s_2, g) = -E(1/2, s_2, g) \quad \Longrightarrow \quad E(1/2, s_2, g) = 0\,. \tag{10.26}$$

This is exactly as in the $SL(2, \mathbb{R})$ case in Example 10.4 above. Again $s_1 = 1/2$ corresponds to a fixed plane of a fundamental reflection and in this case one always obtains a vanishing Eisenstein series.

More involved examples are obtained for exceptional groups. These will play an important rôle in Chapter 14 in the context of string theory.

Example 10.6: Functional relation for $E_8(\mathbb{R})$ parabolic Eisenstein series

Consider a (maximal parabolic) Eisenstein series on E_8 with Dynkin diagram given in Figure 10.1. For the weight

$$\lambda_s = 2s\Lambda_8 - \rho \tag{10.27}$$

the associated character $\chi_s(a) = a^{\lambda_s + \rho} = a^{2s\Lambda_8}$ is invariant under the maximal parabolic subgroup with semi-simple part E_7. (Here, Λ_8 denotes as always the fundamental weight associated with node 8.) We therefore have a family of maximal parabolic Eisenstein series

$$E(s, g) \equiv E(\lambda_s, P, g) = \sum_{\gamma \in P(\mathbb{Z}) \backslash G(\mathbb{Z})} \chi_s(\gamma g). \tag{10.28}$$

There are many functionally related Eisenstein series. The Weyl group of E_8 has order $|\mathcal{W}(E_8)| = 696\,729\,600$, but for the particular choice of parameter λ_s in (10.27) not all Weyl images of λ_s give different Eisenstein series. Instead it suffices to consider elements of the coset $\mathcal{W}(E_8)/\mathcal{W}(E_7)$ and representatives that do not end on an element of $\mathcal{W}(E_7)$. (This will be the main theme of Section 10.3.) We consider as an example the Weyl word

$$w = w_1 w_3 w_4 w_5 w_6 w_7 w_8 \,. \tag{10.29}$$

Then

$$w(2s\Lambda_8 - \rho) = (8-2s)\Lambda_1 + (2s-5)\Lambda_2 - \rho \,. \tag{10.30}$$

The intertwining factor is

$$M(w, 2s\Lambda_8 - \rho) = \frac{\xi(2s-7)}{\xi(2s)} \tag{10.31}$$

and therefore

$$E(2s\Lambda_8 - \rho, g) = \frac{\xi(2s-7)}{\xi(2s)} E((8-2s)\Lambda_1 + (2s-5)\Lambda_2 - \rho, g). \tag{10.32}$$

At $s = \frac{5}{2}$ this specialises to

$$E(5\Lambda_8 - \rho, g) = \frac{\xi(3)}{\xi(5)} E(3\Lambda_1 - \rho, g) \tag{10.33}$$

and therefore relates a specific maximal parabolic Eisenstein series 'on node 8' to another specific maximal parabolic Eisenstein series, this time 'on node 1'. The former appears in the discussion of minimal theta series for E_8 [281] while the latter version appears commonly in string theory (see Chapters 13 and 14) and both are related to the minimal unitary representation, as we will discuss more in Section 10.3.2 and Chapter 14 below.

10.2.2 Weyl Symmetric Normalisation

For a general (minimal) Eisenstein series $E(\lambda, g)$ on a split real simple group G one can define a completed version according to

$$E^\star(\lambda, g) = \underbrace{\left[\prod_{\alpha>0} \xi(\langle\lambda|\alpha\rangle + 1)\right]}_{N_\lambda} E(\lambda, g). \tag{10.34}$$

The *normalising factor* N_λ is the denominator in $M(w_{\text{long}}, \lambda)$. This function is completely invariant under the action of the Weyl group $\mathcal{W}$: for any $w \in \mathcal{W}$ one has

$$E^\star(w\lambda, g) = E^\star(\lambda, g). \tag{10.35}$$

To see this it is sufficient to consider the action of a fundamental reflection $w_i \in \mathcal{W}$:

$$\begin{aligned} E^\star(w_i\lambda, g) &= N_{w_i\lambda} E(w_i\lambda, g) = N_{w_i\lambda} M(w_i, \lambda)^{-1} E(\lambda, g) = N_\lambda E(\lambda, g) \\ &= E^\star(\lambda, g). \end{aligned} \tag{10.36}$$

The prefactor works as follows:

$$\begin{aligned} N_{w_i\lambda} M(w_i, \lambda)^{-1} &= \frac{\xi(\langle\lambda|\alpha_i\rangle + 1)}{\xi(\langle\lambda|\alpha_i\rangle)} \prod_{\alpha>0} \xi(\langle\lambda|w_i\alpha\rangle + 1) \\ &= \frac{\xi(\langle\lambda|\alpha_i\rangle + 1)}{\xi(\langle\lambda|\alpha_i\rangle)} \xi(-\langle\lambda|\alpha_i\rangle + 1) \prod_{\substack{\alpha>0 \\ \alpha\neq\alpha_i}} \xi(\langle\lambda|\alpha\rangle + 1) \\ &= \prod_{\alpha>0} \xi(\langle\lambda|\alpha\rangle + 1) = N_\lambda, \end{aligned} \tag{10.37}$$

where we have used the fact that w_i permutes the set of positive roots $\Delta_+ \setminus \{\alpha_i\}$ as well as the functional relation of the completed Riemann zeta function.

The normalising factor for minimal Eisenstein series has as many factors as the order of $\mathcal{W}$. When the character defined by a weight λ has a stabiliser that is larger than the Borel subgroup B, it is sufficient to use a normalising factor with fewer factors to obtain a suitably symmetric combination. Consider the case of a non-minimal parabolic Eisenstein series $E(\lambda, g)$ where the stabiliser is given by a parabolic subgroup $P(\mathbb{Z})$; see Section 5.7. Then the semi-simple part M of the Levi subgroup L of $P = LU$ has a Weyl group $\mathcal{W}(M)$. The normalising factor $N_{P,\lambda}$ in this case can be chosen to be the denominator (after cancelling all factors) of $M(w, \lambda)$ where w is defined by $w_{\text{long}}(G) = w w_{\text{long}}(M)$ through the longest words in $\mathcal{W}(G)$ and $\mathcal{W}(M)$. An alternative definition of w is as the longest Weyl word in the $\mathcal{W}(G)$-orbit of $\lambda + \rho$. The *normalised Eisenstein series* is

$$E^\star(\lambda, P, g) = N_{P,\lambda} E(\lambda, P, g). \tag{10.38}$$

This completed function then is either invariant under a reflection group isomorphic to the Weyl group generated by the simple reflections in $\mathcal{W}(G)$ that do not belong to $\mathcal{W}(M)$, or it maps to a similar one that is obtained by intertwining additionally by an (outer) Dynkin diagram automorphism. Furthermore, the Weyl normalised series has a different pole structure compared to the Eisenstein series $E(\lambda, P, g)$ with standard normalisation as the normalising factor has zeroes and poles.

In the case of maximal parabolic Eisenstein series with weight

$$\lambda = 2s\Lambda_{i_*} - \rho, \tag{10.39}$$

this implies a reflection symmetry $s \leftrightarrow \frac{\langle\rho|\Lambda_{i_*}\rangle}{\langle\Lambda_{i_*}|\Lambda_{i_*}\rangle} - s$. We illustrate this by two examples.

Example 10.7: Weyl normalisation of $SL(3,\mathbb{R})$ parabolic Eisenstein series

The first example contains a non-trivial diagram automorphism. Consider the group $G = SL(3,\mathbb{R})$ and the weight

$$\lambda_s = 2s\Lambda_1 - \rho. \tag{10.40}$$

The associated character $\chi_s(a) = a^{\lambda_s+\rho} = a^{2s\Lambda_1}$ is invariant under a maximal parabolic subgroup P with Levi factor $L = GL(1,\mathbb{R}) \times M$ with $M = SL(2,\mathbb{R})$. Denote the associated maximal parabolic Eisenstein series by

$$E(s,g) \equiv E(\lambda_s, P, g) = \sum_{\gamma\in P(\mathbb{Z})\backslash G(\mathbb{Z})} \chi_s(\gamma g). \tag{10.41}$$

The Weyl word w that enters in the definition of the normalising factor N_{P,λ_s} is given by the relation

$$\underbrace{w_2 w_1 w_2}_{w_{\text{long}}(G)} = \underbrace{w_2 w_1}_{w} \underbrace{w_2}_{w_{\text{long}}(M)} \tag{10.42}$$

such that

$$M(w,\lambda_s) = \frac{\xi(2s-2)\xi(2s-1)}{\xi(2s-1)\xi(2s)} = \frac{\xi(2s-2)}{\xi(2s)} \quad \Longrightarrow N_{P,\lambda_s} = \xi(2s). \tag{10.43}$$

The normalised series then has a reflection symmetry $s \leftrightarrow \frac{3}{2} - s$ but it maps to the maximal parabolic Eisenstein series associated with the parabolic subgroup P' obtained by the diagram automorphism. In other words, the characters that are being related are

$$2s\Lambda_1 - 1 \quad \leftrightarrow 2\left(\tfrac{3}{2} - s\right)\Lambda_2 - \rho. \tag{10.44}$$

The second example does not have any non-trivial automorphisms.

Example 10.8: Weyl normalisation of $E_8(\mathbb{R})$ parabolic Eisenstein series

We consider the E_8 Eisenstein series from Example 10.6 with weight

$$\lambda_s = 2s\Lambda_8 - \rho. \tag{10.45}$$

The parabolic subgroup leaving the associated character invariant has semi-simple part E_7. The normalising factor in this case is associated with a Weyl word w of length $\ell(w) = 57$ that we do not spell out. The normalising factor turns out to be

$$N_{P,\lambda_s} = \xi(2s)\xi(2s-5)\xi(2s-9)\xi(4s-28) \tag{10.46}$$

and the thus normalised Eisenstein series $E^\star(s,g) = N_{P,\lambda_s} E(s,g)$ is invariant under the reflection w_8 which leads to the reflection law

$$E^\star(s,g) = E^\star\left(\tfrac{29}{2} - s, g\right). \tag{10.47}$$

This example is also discussed in [281] and we will say more about this below in Example 10.20. Here, we note that the normalising factor (10.46) has introduced a pole at $s = \frac{5}{2}$ (and also for other values). This means that the special Eisenstein series from Example 10.6 now appears as a *residue of an Eisenstein series.*

10.2.3 Square-Integrability of Eisenstein Series

Langlands provided a criterion for Eisenstein series to be *square-integrable* [447, Sec. 5]. We state the criterion for Eisenstein series $E(\lambda, g)$ such that they are finite at λ, meaning that they do not have a zero or pole as a meromorphic function of λ for the λ chosen.

Remark 10.9 If $E(\lambda, g)$ has a pole or zero at a given λ one has to consider a one-parameter family in the neighbourhood of λ and study a suitably normalised version such that the zeroth-order term becomes finite [474]. In the case of $SL(2,\mathbb{R})$ and $E(s,z)$ this means multiplying by $(s-1)$ if one wants to study the square-integrability of $E(s,z)$ at $s = 1$. Similarly, one would have to multiply by s^{-1} for the $s = 0$ case.

Under the assumption of a finite $E(\lambda, g)$, the constant term formula of Theorem 8.1 implies that

$$\int\limits_{N(\mathbb{Q})\backslash N(\mathbb{A})} E(\lambda, ng)dn = \sum_{w\in\mathcal{W}} M(w,\lambda) a^{w\lambda+\rho} \tag{10.48}$$

is a well-defined and non-vanishing function of a.

Proposition 10.10 (Square-integrability of Eisenstein series) *A finite Eisenstein series $E(\lambda, g)$ in the sense just described is square-integrable if and only if* [447]

$$\mathrm{Re}\langle w\lambda|\Lambda_i\rangle < 0 \qquad \textit{for all } i = 1,\dots,\mathrm{rank}(G) \tag{10.49}$$

for all $w \in \mathcal{W}$ such that $M(w,\lambda) \neq 0$.

The intuition behind this proposition is that the condition ensures that all terms fall off fast enough as one approaches any cusp of $G(\mathbb{Z})\backslash G(\mathbb{R})$. A proof can be found in [447] and we content ourselves here with some examples.

Example 10.11: Non-square-integrability of $SL(2,\mathbb{R})$ Eisenstein series

Consider square-integrability of Eisenstein series on $SL(2,\mathbb{R})$. For $\lambda_s = (2s-1)\rho$ one has to check the condition (10.49) for the Weyl words $w = \mathbb{1}$ and $w = w_{\text{long}}$. Plugging in the explicit expressions leads to

$$\mathrm{Re}\langle\lambda_s|\Lambda_1\rangle = \mathrm{Re}\,\frac{1}{2}(2s-1) < 0 \quad\text{and}\quad \mathrm{Re}\langle w_{\text{long}}\lambda_s|\Lambda_1\rangle = -\,\mathrm{Re}\,\frac{1}{2}(2s-1) < 0. \tag{10.50}$$

Clearly, these two conditions cannot be satisfied simultaneously and therefore we recover the well-known result that non-holomorphic Eisenstein series on $SL(2,\mathbb{R})$ are never square-integrable. There is, however, a limiting case $\mathrm{Re}\,s = 1/2$, where the conditions are almost satisfied. This corresponds to the Eisenstein series on the *critical line* $s = 1/2 + it$ (for $t \in \mathbb{R}$) that are δ-function normalisable. See for instance [263, 598] for a discussion of these properties of Eisenstein series on $SL(2,\mathbb{R})$, and Section 5.1 for a discussion in connection with the Selberg trace formula.

More interesting is the case when there are non-trivial square-integrable functions within a degenerate principal series.

Example 10.12: Square-integrability of E_8 maximal parabolic Eisenstein series

Consider the maximal parabolic E_8 Eisenstein series $E(s,g)$ with weight

$$\lambda_s = 2s\Lambda_8 - \rho \tag{10.51}$$

that was introduced in Example 10.6. Computing the constant term one finds 240 non-vanishing $M(w,\lambda_s)$. Checking the criterion (10.49) one finds that it is satisfied for the values

$$s = \frac{5}{2}, \quad s = \frac{9}{2}, \quad s = 7, \tag{10.52}$$

and hence these are normalisable Eisenstein series for E_8 that belong to the (residual) discrete spectrum of the Laplacian on $G(\mathbb{R})/K(\mathbb{R})$ for $G = E_8$.

The value $s = \frac{5}{2}$ was also discussed in Example 10.6 and it was shown there that $E(s,g)$ for this value is functionally related to another known normalisable maximal parabolic Eisenstein series [315] that appears in string theory for the R^4 curvature correction and which will be discussed in great detail in Chapter 14.

The value $s = \frac{9}{2}$ can be analysed using the functional relation

$$E(2s\Lambda_8 - \rho) = \frac{\xi(2s-10)\xi(2s-13)}{\xi(2s)\xi(2s-5)} E(2(7-s)\Lambda_1 + 2(s-9/2)\Lambda_2 - \rho), \tag{10.53}$$

which shows that for $s = \frac{9}{2}$ the adjoint E_8 series is connected to the maximal parabolic series on node 1 with $s = \frac{5}{2}$. This is the case that appears in string theory for the $\nabla^4 R^4$ correction and it is known that the function is associated with the next-to-minimal series [315].

The value $s = 7$ is interesting because it does not represent any simplification in the wavefront set (compared to generic s) and so is just part of the residual discrete spectrum with orbit type A_2. These cases were also analysed in [474].

10.3 Evaluating Constant Term Formulas

The Langlands constant term formula (see Theorem 8.1)

$$\int_{N(\mathbb{Z})\backslash N(\mathbb{R})} E(\lambda, ng)dn = \sum_{w\in\mathcal{W}} M(w,\lambda)a^{w\lambda+\rho} \tag{10.54}$$

is nice and compact but evaluating it will a priori produce as many terms as there are different elements in the Weyl group. Since the order of the Weyl group becomes large very quickly as the rank of $G(\mathbb{R})$ grows, this can render the resulting expressions rather unwieldy. However, by dint of choice of the parameter λ of the (degenerate) principal series the sum over Weyl elements may simplify as then some of the coefficients $M(w, \lambda)$ appearing in (10.54) vanish. For convenience we also recall the definition of the *intertwiner*

$$M(w,\lambda) = \prod_{\substack{\alpha>0\\ w\alpha<0}} \frac{\xi(\langle\lambda|\alpha\rangle)}{\xi(\langle\lambda|\alpha\rangle+1)} \tag{10.55}$$

and its multiplicative property

$$M(w_1w_2,\lambda) = M(w_1, w_2\lambda)M(w_2,\lambda) \quad \text{for any } w_1, w_2 \in \mathcal{W}. \tag{10.56}$$

A convenient method for evaluating the Langlands constant term formula can then be developed by exploiting the multiplicative relation (10.56). We first adumbrate this method, which we will refer to as the *orbit method*. Then we discuss a number of examples and finally mention further simplifications that arise for constant terms in non-maximal unipotent subgroups $U \subset N$. The corresponding constant term formula was given in Section 8.9.

10.3.1 The Orbit Method

The factor $M(w, \lambda)$ is, by its definition in (10.55), given by the product of factors of the form

$$c(k) = \frac{\xi(k)}{\xi(k+1)}, \tag{10.57}$$

where $k = \langle\lambda|\alpha\rangle$ and α runs over all positive roots that satisfy $w\alpha < 0$. The function $c(k)$ is sometimes referred to as the *Harish-Chandra c-function*. It has a simple pole at $k = 1$ and a simple zero at $k = -1$; otherwise it takes finite non-zero values for real k and satisfies $c(k)c(-k) = 1$ as well as $c(0) = -1$. For vanishing $M(w, \lambda)$, we are therefore particularly interested in roots α which satisfy $w\alpha < 0$ and $\langle\lambda|\alpha\rangle = -1$.

To characterise these α further, let us define the *stabiliser of the weight* λ

$$\operatorname{stab}(\lambda) = \{\alpha \in \Pi \mid \langle\lambda + \rho|\alpha\rangle = 0\}, \tag{10.58}$$

so that it is the subset of the simple roots Π for which $\lambda + \rho$ has vanishing Dynkin labels.

Remark 10.13 If $\operatorname{stab}(\lambda) \neq \{\}$, the corresponding Eisenstein series $E(\lambda, g)$ belongs to a *degenerate* principal series. Let $P \subset G$ be the parabolic subgroup corresponding to $\operatorname{stab}(\lambda) \subset \Pi$ as defined in Section 3.1.3. Then

$$E(\lambda, g) = E(\lambda, P, g) = \sum_{\gamma \in P(\mathbb{Z})\backslash G(\mathbb{Z})} e^{\langle\lambda+\rho_P|H_P(\gamma g)\rangle}, \tag{10.59}$$

as explained in Section 5.7.2.

Example 10.14: Stabiliser of a maximal parabolic λ

As an example and referring back to Section 5.7 we note that maximal parabolic Eisenstein series have very large stabilisers. According to Proposition 5.30, maximal parabolic Eisenstein series can be thought of as having $\lambda = 2s\Lambda_{i_*} - \rho$, corresponding to $\operatorname{stab}(\lambda) = \Pi \setminus \{\alpha_{i_*}\}$ for the value i_* that determines the maximal parabolic subgroup under which $\chi(a) = a^{\lambda+\rho}$ is left-invariant.

If w_i is the fundamental Weyl reflection in the simple root α_i defined in (3.15), it clearly maps $w_i(\alpha_i) = -\alpha_i$, and this is the only positive root that is mapped to a negative root by the fundamental reflection w_i [386]. If furthermore $\alpha_i \in \operatorname{stab}(\lambda)$, then

$$\alpha_i \in \operatorname{stab}(\lambda) \Leftrightarrow \langle\lambda|\alpha_i\rangle = -1 \quad \Rightarrow \quad M(w_i, \lambda) = c(-1) = 0. \tag{10.60}$$

By the multiplicative property (10.56) one can then deduce that all Weyl words w that end (on the right) on a fundamental reflection w_i with α_i in $\operatorname{stab}(\lambda)$ obey $M(w, \lambda) = 0$; see also [315]. Another way of putting this is that only those w can have non-vanishing $M(w, \lambda)$ that lie in

$$C(\lambda) = \{w \in \mathcal{W} \mid w\alpha > 0 \quad \text{for all } \alpha \in \operatorname{stab}(\lambda)\}. \tag{10.61}$$

Depending on $\operatorname{stab}(\lambda)$ this set can be much smaller than $\mathcal{W}$. In fact, its order is given by $|C(\lambda)| = |\mathcal{W}|/|\mathcal{W}(\operatorname{stab}(\lambda))|$, where $\mathcal{W}(\operatorname{stab}(\lambda))$ is the subgroup of $\mathcal{W}$ that is generated by taking only words in the fundamental reflections associated with $\text{stab}(\lambda) \subset \Pi$. Moreover, one then has the simplified constant term formula

$$\int_{N(\mathbb{Q})\backslash N(\mathbb{A})} E(\chi, ng)dn = \sum_{w\in C(\lambda)} a^{w\lambda+\rho} M(w, \lambda). \tag{10.62}$$

The elements in $C(\lambda)$ can be constructed using the Weyl orbit of a dominant weight Λ that is defined as follows:

Definition 10.15 (Dominant weight associated to λ) Let $\lambda \in \mathfrak{h}^*$ be a weight with stabiliser $\text{stab}(\lambda)$ as defined in (10.58), and let $r = \dim \mathfrak{h}^*$ denote the rank of the underlying group. Let $I \subset \{1, \ldots, r\}$ be such that a simple root α_i of $\mathfrak{g}$ belongs to $\text{stab}(\lambda)$ if and only if $i \in I$. Let $\bar{I}$ be the complement of I in $\{1, \ldots, r\}$. Then the dominant weight Λ associated to λ is defined as a sum over fundamental weights as

$$\Lambda = \sum_{i\in\bar{I}} \Lambda_i. \tag{10.63}$$

In other words, one considers the weight $\lambda+\rho$ and replaces all non-zero Dynkin labels by 1 to obtain Λ.

Clearly, $\mathcal{W}(\text{stab}(\lambda))$ stabilises Λ thus defined and the number of distinct points in the orbit $\mathcal{W} \cdot \Lambda$ equals $|\mathcal{W}|/|\mathcal{W}(\text{stab}(\lambda))|$. Therefore the points in the Weyl orbit are in bijection with the set $C(\lambda)$.

In order to establish the bijection, we use the fact that for each element μ in the Weyl orbit of Λ there is a shortest element $w \in \mathcal{W}$ that satisfies $w\Lambda = \mu$. These elements w are exactly the Weyl words that make up the set $C(\lambda)$. They can also be seen as specific representatives of the coset $\mathcal{W}/\mathcal{W}(\text{stab}(\lambda))$ whose size was already argued above to determine the number of summands in (10.62). The shortest element leading to an element μ is not necessarily unique but all choices of the same shortest length yield the same factor $M(w, \lambda)$.

A standard algorithm for constructing the Weyl orbit $\mathcal{W} \cdot \Lambda$ of a dominant weight Λ is as follows:

1. Define the initial set of orbit points as $\mathcal{O} = \{\Lambda\}$. This is the 'highest' element (with respect to the *height function* $\text{ht}(\mu) = \langle\rho|\mu\rangle$ on $\mathfrak{h}^*$) in the orbit and others will be constructed by using lowering Weyl reflections.
2. For a given $\mu \in \mathcal{O}$ compute the *Dynkin labels* $p_i = \langle\mu|\alpha_i\rangle$ with respect to all simple roots α_i.
3. If $p_i > 0$ for some $i = 1, \ldots, \text{rank}(G)$, then construct $\mu' = w_i\mu$ where w_i is the fundamental Weyl reflection in the simple root α_i. If μ' is not already in the orbit $\mathcal{O}$, add it.
4. For any weight μ in $\mathcal{O}$ for which steps 2 and 3 have not been carried out go to step 2.

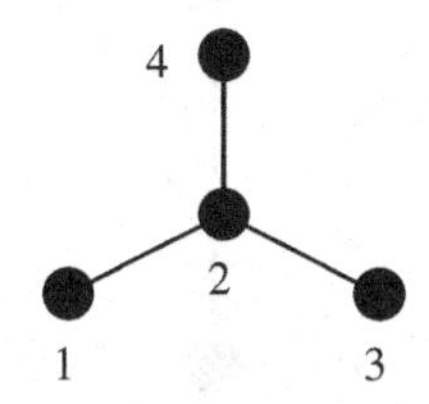

Figure 10.2 The Dynkin diagram of $SO(4,4)$ with labelling of simple roots.

Remark 10.16 Since for the initial dominant weight Λ the Dynkin labels p_i are zero for all simple roots $\alpha_i \in \text{stab}(\lambda)$, any Weyl word thus constructed will end on a letter w_i (fundamental Weyl reflection) that does not belong to $\mathcal{W}(\text{stab}(\lambda))$.

Remark 10.17 In practice, it is very useful to think of the Weyl orbit of Λ in terms of a graph where nodes correspond to weights μ that lie in the orbit O and links are labelled by the fundamental reflections that relate two such weights. This graph can be constructed algorithmically starting from Λ, which corresponds to the identity element $\mathbb{1} \in \mathcal{W}$ by the above algorithm, and one also keeps track of the corresponding Weyl words in this way. Elements μ that are farther from the dominant weight in this graph correspond to longer Weyl words.

With the Weyl orbit $\mathcal{W} \cdot \Lambda$ one has constructed all Weyl words that belong to $C(\lambda)$ and can therefore evaluate the constant term formula (10.62). We consider an example to illustrate the method.

Example 10.18: Orbit method for $SO(4,4;\mathbb{R})$ parabolic Eisenstein series

We consider the group $SO(4,4;\mathbb{R})$ with Dynkin diagram of type D_4 shown in Figure 10.2. The Weyl group $\mathcal{W}$ is of order 192 in this case. Taking

$$\lambda = 2s\Lambda_1 - \rho \tag{10.64}$$

yields a maximal parabolic Eisenstein series and

$$\text{stab}(\lambda) = \{\alpha_2, \alpha_3, \alpha_4\}, \tag{10.65}$$

such that $\mathcal{W}(\text{stab}(\lambda))$ is of type $\mathcal{W}(A_3)$ and order 24. The dominant weight Λ of (10.63) equals the sum of all fundamental weights Λ_i such that $\alpha_i \notin \text{stab}(\lambda)$ and thus $\Lambda = \Lambda_1$. The Weyl orbit of $\Lambda = \Lambda_1$ consists of only eight points. Figure 10.3 shows the graph of this orbit.

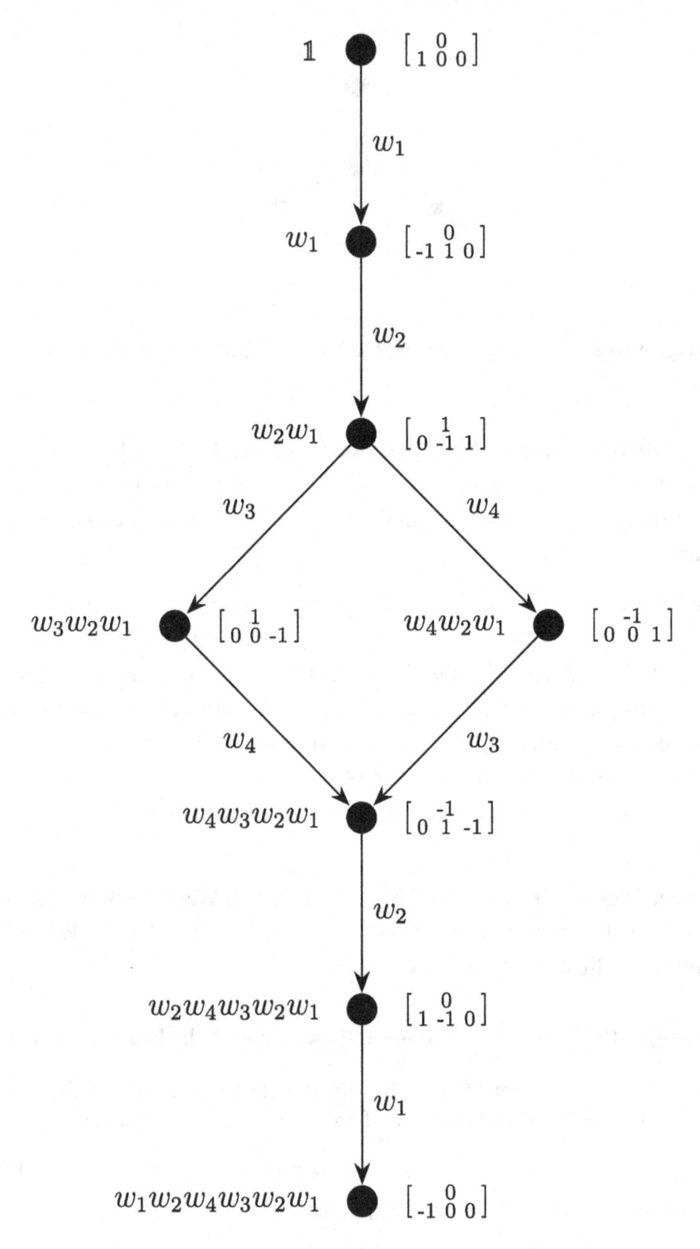

Figure 10.3 The Weyl orbit of the fundamental weight $\Lambda_1 = \left[\begin{smallmatrix} & 0 & \\ 1 & 0 & 0 \end{smallmatrix}\right]$ under the D_4 Weyl group. For each image point in the orbit, we have listed the Dynkin labels and a choice of shortest Weyl word that leads to the given point. The shortest Weyl words are those that make up the set $C(\lambda)$ for $\lambda = 2s\Lambda_1 - \rho$ that contribute to the constant terms in (10.62).

The Weyl orbit can be calculated by starting from the highest weight Λ_1 and applying fundamental Weyl reflections in those simple roots whose Dynkin labels are positive. This is the implementation of the algorithm above. For the example shown in Figure 10.3 this allows only w_1 acting on Λ_1.

Considering now the weight $\lambda = 2s\Lambda_1 - \rho$ that defines the $SO(4,4)$ Eisenstein series $E(\lambda, g)$, we see that the eight Weyl elements potentially contributing to the constant term formula (10.62) are

$$C(\lambda) = \{\mathbb{1}, w_1, w_2w_1, w_3w_2w_1, w_4w_2w_1, w_4w_3w_2w_1, \\ w_2w_4w_3w_2w_1, w_1w_2w_4w_3w_2w_1\}. \tag{10.66}$$

The corresponding factors $M(w, \lambda)$ are

w	$M(w,\lambda)$
$\mathbb{1}$	1
w_1	$c(2s-1)$
w_2w_1	$c(2s-1)c(2s-2)$
$w_3w_2w_1$	$c(2s-1)c(2s-2)c(2s-3)$
$w_4w_2w_1$	$c(2s-1)c(2s-2)c(2s-3)$
$w_4w_3w_2w_1$	$c(2s-1)c(2s-2)^2c(2s-3)^2$
$w_2w_4w_3w_2w_1$	$c(2s-1)c(2s-2)^2c(2s-3)^2c(2s-4)$
$w_1w_2w_4w_3w_2w_1$	$c(2s-1)c(2s-2)^2c(2s-3)^2c(2s-4)c(2s-5)$

(10.67)

The table clearly reflects the multiplicative property (10.56) of the factors $M(w, \lambda)$: moving one step down the Weyl orbit adds a single factor $c(k)$ to $M(w, \lambda)$.

Depending on the value of s some of the factors $M(w, \lambda)$ can vanish, leading to a further reduction in the number of constant terms in (10.62). As already argued based on the multiplicative property (10.56) we should start at the top of the Weyl orbit. Let us look at a few examples, keeping in mind that we are looking for values of s where there are more factors $c(-1)$ than $c(+1)$ in the product.

The simplest case is of course $s = 0$. Then only $w = \mathbb{1}$ has a non-vanishing $M(w, \lambda) = 1$ and this is the whole constant term. This is not surprising since $s = 0$ corresponds to $\lambda = -\rho$, yielding the trivial constant automorphic function $E(-\rho, g) \equiv 1$.

The next-simplest case is $s = 1/2$. For this choice the two Weyl words $w = \mathbb{1}$ and $w = w_1$ contribute to the constant term (10.62). Working out their contributions one finds that they cancel (using $c(0) = -1$) and the constant term vanishes. (The same thing happens for the $SL(2,\mathbb{R})$ series; see Section 10.1.1.)

For the value $s = 1$ the factor $c(2s - 3)$ leads to a vanishing contribution but the factor $c(2s - 1)$ has a pole so one needs to take the limit carefully. In all there are five non-vanishing contributions to the constant term since the last three orbit points (out of the total eight) contain the factor $c(2s - 3)^2$. Summing up the non-vanishing contributions leads to

$$v_1^2 + \frac{6v_2}{\pi}\left(\gamma_E - \log(4\pi) - \log\left(v_1v_2^{-2}v_3v_4\right)\right) + v_3^2 + v_4^2, \tag{10.68}$$

where we have parametrised $a = v_1^{h_1} v_2^{h_2} v_3^{h_3} v_4^{h_4}$. The logarithms arise when taking the limit $s \to 1$ and reflect the confluence of the eigenvalues of two polynomial eigenfunctions of the Laplace operator.

Further simplifications occur for $s = \frac{3}{2}$ and $s = 2$, which we leave to the reader to evaluate.

If one had started with a fixed value of s for which simplifications occur, it would have been sufficient to construct the Weyl orbit up to the points where the $M(w, \lambda) = 0$.

10.3.2 Special λ-Values and $E(\lambda, g)$

As we have seen in the $SO(4, 4)$ Example 10.18 just now and in Section 10.1.1, there can be special points $\lambda \in \mathfrak{h}^*$ where the constant terms (and the whole Eisenstein series) simplify. Parametrising the weight λ in terms of (complex) parameters s_i, these special points correspond to specific values for the s_i. These simplifications were already observed for the exceptional group G_2 by Langlands in his original work [447]; see also [396].

In order to detect such simplifications, it is not efficient to calculate the whole set $C(\lambda)$ and then the coefficients $M(w, \lambda)$ as in (10.67). Due to the partially ordered structure of the Weyl orbit and the multiplicative property (10.56), it suffices to also calculate the factor $M(w, \lambda)$ at the same time as one constructs w, using the Weyl orbit method of Section 10.3.1. One need not construct further any path of the graph (of increasing word length) where one of the intermediate words satisfies $M(w, \lambda) = 0$. This simplifies the calculation of the constant term formula considerably [212].

Remark 10.19 Simplifications in the constant term have corresponding simplifications in the Whittaker coefficients, as we will see in Section 10.4. They are typically associated with subrepresentations in the (degenerate) principal series called *small representations* that we already encountered in Section 6.4.3 (see Table 6.2). At these places the functional dimension of the automorphic representation reduces. This is also discussed in more detail in Section 14.3.3 in the context of string theory.

Example 10.20: Orbit method for minimal E_8 Eisenstein series

Consider again the maximal parabolic E_8 Eisenstein series of Example 10.6 with weight $\lambda_s = 2s\Lambda_8 - \rho$. For calculating the constant term, we require the Weyl orbit of the dominant weight $\Lambda_8 = [0, 0, 0, 0, 0, 0, 0, 1]$ in Dynkin label notation. In total, the Weyl orbit of Λ_8 has 240 elements. The beginning of the Weyl orbit, computed with the orbit method, is depicted in Figure 10.4.

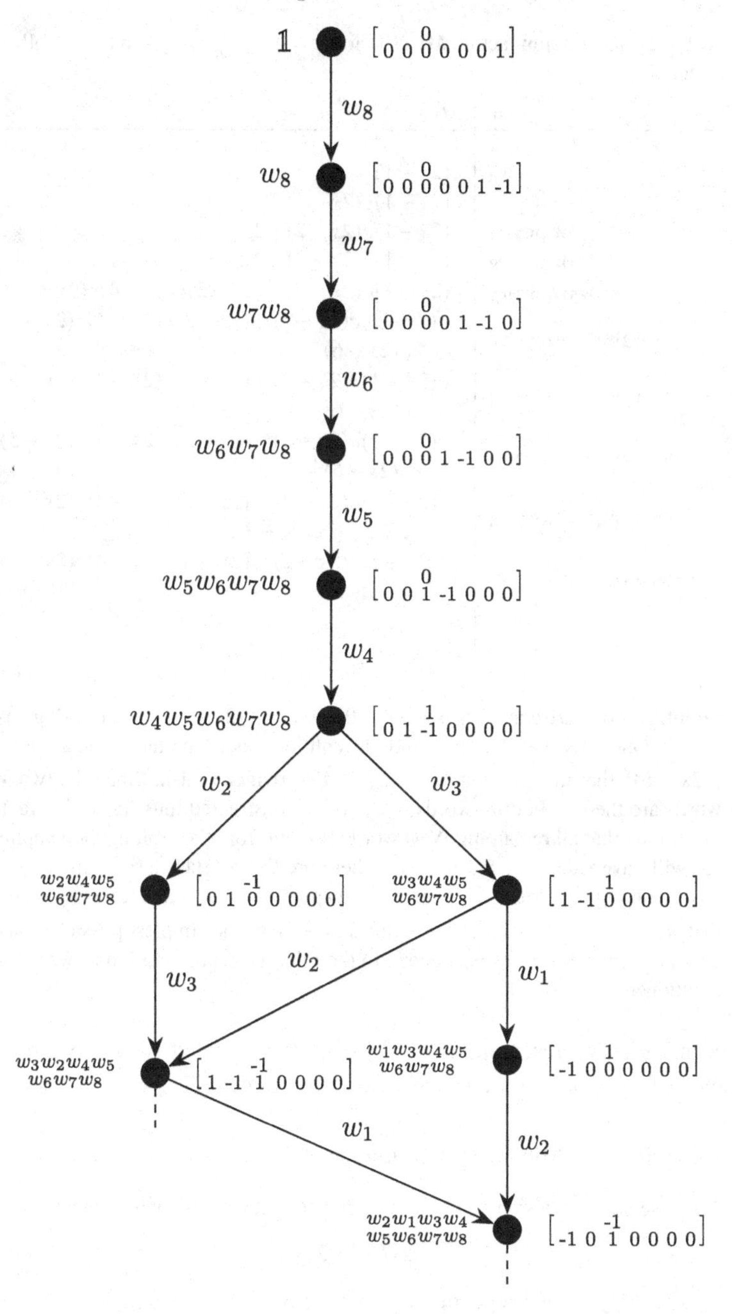

Figure 10.4 Part of the Weyl orbit of the fundamental weight $\Lambda_1 = \left[{}^{\quad\;\; 0}_{0\,0\,0\,0\,0\,0\,1}\right]$ under the E_8 Weyl group. For each image point in the orbit, we have listed the Dynkin labels and a choice of shortest Weyl word that leads to the given point. The shortest Weyl words are those that make up the set $C(\lambda)$ that contribute to the constant terms in Example 10.20.

The corresponding factors $M(w,\lambda)$ for $\lambda = 2s\Lambda - \rho$ are given by the following table:

w	$M(w,\lambda)$
$\mathbb{1}$	1
w_8	$c(2s-1)$
w_7w_8	$c(2s-1)c(2s-2)$
$w_6w_7w_8$	$c(2s-1)c(2s-2)c(2s-3)$
$w_5w_6w_7w_8$	$c(2s-1)c(2s-2)c(2s-3)c(2s-4)$
$w_4w_5w_6w_7w_8$	$c(2s-1)c(2s-2)^2c(2s-3)c(2s-4)c(2s-5)$
$w_2w_4w_5w_6w_7w_8$	$c(2s-1)c(2s-2)^2c(2s-3)c(2s-4)c(2s-5)$ $\times c(2s-6)$
$w_3w_4w_5w_6w_7w_8$	$c(2s-1)c(2s-2)^2c(2s-3)c(2s-4)c(2s-5)$ $\times c(2s-6)$
$w_3w_2w_4w_5w_6w_7w_8$	$c(2s-1)c(2s-2)^2c(2s-3)c(2s-4)c(2s-5)$ $\times c(2s-6)^2$
$w_1w_3w_4w_5w_6w_7w_8$	$c(2s-1)c(2s-2)^2c(2s-3)c(2s-4)c(2s-5)$ $\times c(2s-6)c(2s-7)$
$w_2w_1w_3w_4w_5w_6w_7w_8$	$c(2s-1)c(2s-2)^2c(2s-3)c(2s-4)c(2s-5)$ $\times c(2s-6)^2c(2s-7)$
$\vdots$	$\vdots$

(10.69)

Simplifications arise as always for $s = 0$ and $s = 1/2$. Another interesting case is $s = \frac{5}{2}$. One sees that the ninth and eleventh entries in the table contain a factor $c(2s-6)^2$ that makes the corresponding intertwiner vanish. Since the two Weyl words are the two bottom words in the orbit constructed thus far in Figure 10.4, one knows that all remaining Weyl words coming from the orbit method applied to Λ_8 will have vanishing $M(w,\lambda)$ and therefore the constant term consists of nine terms. Taking the limit $s \to \frac{5}{2}$ carefully again gives logarithmic terms as in (10.68) that we do not display here. The value $s = \frac{5}{2}$ gives the simplest possible constant and the Eisenstein series is attached to the minimal representation as was already mentioned.

The minimal representation of $SL(3,\mathbb{R})$ is associated with a whole one-parameter family of Eisenstein series; see also Table 6.2.

Example 10.21: Minimal representation of $SL(3,\mathbb{R})$

The Eisenstein series on $SL(3,\mathbb{A})$ introduced in Section 9.8 with weight

$$\lambda = 2s_1\Lambda_1 + 2s_2\Lambda_2 - \rho \tag{10.70}$$

simplifies for special values of the parameters s_i. Putting

$$s_1 = 0 \quad \text{or} \quad s_2 = 0 \quad \text{or} \quad s_3 = s_1 + s_2 - \frac{1}{2} = 0 \tag{10.71}$$

or

$$s_1 = 1 \quad \text{or} \quad s_2 = 1 \quad \text{or} \quad s_3 = s_1 + s_2 - \frac{1}{2} = 1 \tag{10.72}$$

makes the generic Eisenstein series into one on a maximal parabolic subgroup as defined in Sections 3.1.3 and 5.7; the case $s_1 = 0$ corresponds to inducing from the maximal parabolic subgroup $P_1(\mathbb{A})$. In this case the Fourier expansion simplifies considerably, as can be seen be inspecting the expression of Section 9.8. This is already manifest from (9.99a) which appears in the expression of any generic Whittaker coefficient. Precisely for the choices above, $1/\xi(\lambda)$ vanishes identically, implying that all generic Whittaker coefficients vanish.

For the degenerate Whittaker coefficients one also obtains shorter expressions: out of the three Weyl elements displayed in (9.109) two have a vanishing expression, and one is left with a single modified Bessel function with associated divisor sum. The non-abelian Fourier coefficient also simplifies, as is shown in [525].

10.3.3 Constant Terms in Maximal Parabolic Subgroups

Given a unipotent $U \subset N$ and an Eisenstein series $E(\lambda, g)$ on a group G, one can define the constant term along U (see Equation (6.17) and Section 8.9) by

$$C_U = \int\limits_{U(\mathbb{Z})\backslash U(\mathbb{R})} E(\lambda, ug) du. \tag{10.73}$$

When $U = U_{j_\circ}$ is the unipotent of a maximal parabolic subgroup $P_{j_\circ} = L_{j_\circ} U_{j_\circ}$ associated with node $j_\circ$, a general formula for this constant term was given in Theorem 8.9:

$$\int\limits_{U_{j_\circ}(\mathbb{Z})\backslash U_{j_\circ}(\mathbb{R})} E(\lambda, ug) du = \sum_{w \in \mathcal{W}_{j_\circ}\backslash \mathcal{W}} e^{\langle (w\lambda+\rho)_{\parallel j_\circ} | H(g) \rangle} M(w, \lambda) E^{M_{j_\circ}}((w\lambda)_{\perp j_\circ}, m). \tag{10.74}$$

The Eisenstein series on the right-hand side is one on the semi-simple part $M_{j_\circ}$ of the Levi subgroup $L_{j_\circ} = GL(1, \mathbb{R}) \times M_{j_\circ}$ and the exponential prefactor is a function only on the $GL(1, \mathbb{R})$ factor. We note that it is the same numerical coefficient $M(w, \lambda)$ as in (10.54) that controls this constant term. As we have explained in Section 10.3.1, one can restrict the Weyl words to the set $C(\lambda)$ that is in bijection with the Weyl orbit of $\lambda + \rho$ (or an equivalent dominant weight Λ defined in (10.63)). This bijection implies that it suffices to consider Weyl words w in the left coset $\mathcal{W}/\mathcal{W}(\text{stab}(\lambda)) \cong C(\lambda)$. The additional quotient in formula (10.74) then allows the restriction to the double coset [281, 212]

$$w \in \mathcal{W}_{j_\circ}\backslash \mathcal{W}/\mathcal{W}(\text{stab}(\lambda)). \tag{10.75}$$

This double coset typically has very few representatives, allowing for a swift evaluation of formula (10.74).

Remark 10.22 The double coset (10.75) also depends on λ and there can therefore be simplifications similar to those discussed in Section 10.3.2.

Example 10.23: Constant term of $SO(4,4;\mathbb{R})$ Eisenstein series

For the $SO(4,4;\mathbb{R})$ Eisenstein series considered in Example 10.18 with weight $\lambda = 2s\Lambda_1 - \rho$ we compute the constant term along the unipotent U_3 of the maximal parabolic subgroup P_3. The Weyl group of the semi-simple Levi part is again of type $\mathcal{W}(A_3)$ and generated by the fundamental reflections w_1, w_2 and w_4. Inspecting the list (10.66) of elements of $C(\lambda)$ shows that the double coset (10.75) in this case has only two classes for which we choose the representatives

$$\mathbb{1} \quad \text{and} \quad w_3 w_2 w_1 . \tag{10.76}$$

Projecting $w\lambda + \rho$ we obtain for the trivial representative

$$\begin{aligned}(2s\Lambda_1)_{\parallel 3} &= \frac{\langle 2s\Lambda_1|\Lambda_3\rangle}{\langle\Lambda_3|\Lambda_3\rangle}\Lambda_3 = s\Lambda_3 ,\\ (2s\Lambda_1 - \rho)_{\perp 3} &= (2s-1)\Lambda_1^{M_3} - \Lambda_2^{M_3} - \Lambda_3^{M_3}\end{aligned} \tag{10.77}$$

(where $\Lambda_i^{M_3}$ are the three fundamental weights of $M_3(\mathbb{R}) = SL(4,\mathbb{R})$) and a similar decomposition for the other representative. Denoting the coordinate on $GL(1,\mathbb{R})$ by

$$r = e^{\langle\Lambda_3|H(g)\rangle}, \tag{10.78}$$

we get the following constant term:

$$\begin{aligned}\int\limits_{U_3(\mathbb{Z})\backslash U_3(\mathbb{R})} E(\lambda, ug)du &= r^s E([2s-1,-1,-1],m)\\ &\quad + r^{3-s}\frac{\xi(2s-3)}{\xi(2s)}E([-1,-1,2(s-1)-1],m),\end{aligned} \tag{10.79}$$

where we have indicated the weight on the semi-simple subgroup M_3 by its Dynkin labels and have evaluated the intertwiner $M(w_3w_2w_1,\lambda)$ using (10.67).

10.4 Evaluating Spherical Whittaker Coefficients

We now turn to the question of efficiently evaluating degenerate Whittaker coefficients that are given by Theorem 9.6, whose result we briefly recall. The final formula there was

$$W_\psi^\circ(\lambda, a) = \sum_{w_c w'_{\text{long}} \in \mathcal{W}/\mathcal{W}'} a^{(w_c w'_{\text{long}})^{-1}\lambda+\rho} M(w_c^{-1},\lambda) W'^\circ_{\psi^a}(w_c^{-1}\lambda, \mathbb{1}) . \tag{10.80}$$

We briefly recall the notation used in this formula. The symbol ψ denotes a degenerate character on the maximal unipotent $N \subset B \subset G$. It has support

$\mathrm{supp}(\psi) \subset \Pi$ given by (9.59) and this subset of simple roots defines a semi-simple subgroup $G' \subset G$ with Weyl group $\mathcal{W}' = \mathcal{W}(\mathrm{supp}(\psi))$. The longest element $\mathcal{W}'$ is called w'_{long} and $w_c \in \mathcal{W}$ satisfies $w_c\alpha > 0$ for all $\alpha \in \mathrm{supp}(\psi)$. The representative w_c of the coset $\mathcal{W}/\mathcal{W}'$ can be constructed using the orbit method of Section 10.3.1. For any $a \in A(\mathbb{A})$, the twisted character $\psi^a(n) = \psi(ana^{-1})$ restricted to the unipotent $N' \subset G'$ is generic and $W'^{\circ}_{\psi^a}(w_c^{-1}\lambda, \mathbb{1})$ is the (generic) spherical Whittaker function on the G' of the prinicipal series representation given by the restriction of the weight $w_c^{-1}\lambda$ to G' and evaluated at the identity. An example of this formula was worked out for $SL(3, \mathbb{A})$ in Section 9.8.

One sees from formula (10.80) that it is again an intertwining coefficient $M(w_c^{-1}, \lambda)$ that controls possible simplifications in the degenerate Whittaker coefficients. We know from the discussion of the orbit method in Section 10.3.1 that only those w_c^{-1} give a non-trivial $M(w_c^{-1}, \lambda)$ that are the minimal representatives of the coset $\mathcal{W}/\mathcal{W}(\mathrm{stab}(\lambda))$ (constructed by the orbit method). Due to the inverse w_c^{-1} we are therefore again faced with a double coset

$$\mathcal{W}(\mathrm{supp}(\psi))\backslash\mathcal{W}/\mathcal{W}(\mathrm{stab}(\lambda)). \tag{10.81}$$

This has very few representatives one has to consider when evaluating (10.80) for a non-generic λ and ψ.

For which Eisenstein series $E(\lambda, g)$ does formula (10.80) actually offer the prospect of helping find complete information about the Fourier expansion? As mentioned already in Remark 9.7 this will happen when λ is such that $E(\lambda, g)$ is not in the generic principal series but in a degenerate one or even at one of the special λ values discussed in Section 10.3.2.

10.4.1 Degenerate Principal Series and Degenerate Whittaker Coefficients

If λ is such that $\mathrm{stab}(\lambda) \neq \{\}$, the Eisenstein series $E(\lambda, g)$ is associated to a degenerate principal series; see Section 5.7. Prime examples are maximal parabolic Eisenstein series when $\mathrm{stab}(\lambda) = \Pi \setminus \{\alpha_{j_*}\}$ where α_{j_*} is a single simple root that defines a maximal parabolic subgroup $P_{j_*} \subset G$. More generally, we define a parabolic subgroup

$$P_\lambda = L_\lambda U_\lambda, \tag{10.82}$$

such that the semi-simple part $M_\lambda \subset L_\lambda$ has the simple root system given by $\mathrm{stab}(\lambda)$. As mentioned in Section 5.7.2, the Eisenstein series $E(\lambda, g)$ belongs to the degenerate principal series with Gelfand–Kirillov dimension

$$\mathrm{GKdim}(I(\lambda)) = \dim(P_\lambda\backslash G) \tag{10.83}$$

or even a *subrepresentation* or *subquotient* of this in case λ sits at a special value.

For automorphic functions in a degenerate principal series one has that typically not all Whittaker coefficients are non-zero and often the generic ones are absent. In order to determine which Whittaker coefficients are non-zero we recall the notion of a wavefront set introduced in Section 6.4. The wavefront set is the set of nilpotent orbits of $G(\mathbb{R})$ such that there are non-trivial Fourier coefficients (or Whittaker coefficients) associated with it. Characters ψ are associated with nilpotent elements of $\mathfrak{g} = \mathfrak{g}(\mathbb{R})$ and one has to consider their $G(\mathbb{R})$-orbits in an automorphic representation.

Nilpotent orbits of $\mathfrak{g}$ under $G(\mathbb{R})$ come with a certain (even) dimension and they must be able to 'fit into' the automorphic representation of $E(\lambda, g)$ for a non-trivial (non-vanishing) Fourier coefficient to exist. There is a symplectic structure on a nilpotent orbit [146] and only a Lagrangian subspace corresponds to the character ψ of a Whittaker coefficient. Let $X \in \mathfrak{g}$ be a nilpotent element and $\mathcal{O}_X$ its corresponding nilpotent orbit under $G(\mathbb{R})$. The constraint just explained means that there can be non-trivial Fourier coefficients only if

$$\frac{1}{2} \dim_{\mathbb{C}} \mathcal{O}_X \leq \dim (P_\lambda \backslash G) . \tag{10.84}$$

Remark 10.24 If λ is generic such that $E(\lambda, g)$ is in the full principal series, then P_λ equals the standard Borel subgroup $B \subset G$ and the Gelfand–Kirillov dimension equals $1/2\,(\dim \mathfrak{g} - \dim \mathfrak{h})$. At the same time, the largest nilpotent orbit (called the principal or regular orbit) has dimension $\dim \mathfrak{g} - \dim \mathfrak{h}$, confirming the fact that such a generic Eisenstein series will have generic Fourier coefficients in general.

The condition (10.84) puts strong constraints on the orbits one has to consider for a degenerate principal series representation. Moreover, if a nilpotent orbit $\mathcal{O}_X$ has a representative X that lies completely in $[N, N]\backslash N$ such that there is a character $\psi_X : N \to U(1)$ associated with it, one can test whether $\mathcal{O}_X$ belongs to the wavefront set by computing the (degenerate) Whittaker coefficient for ψ_X with formula (10.80).

Example 10.25: Minimal orbit and A_1-type Whittaker coefficients

Any simple group $G(\mathbb{R})$ has a unique *minimal non-trivial nilpotent orbit* $\mathcal{O}_{\text{min}}$ that is given by the orbit of a generator E_θ from the root space of the highest root θ. If $G(\mathbb{R})$ is simply-laced, one can alternatively choose as a nilpotent representative any simple step operator $X = E_{\alpha_i}$ where α_i is a simple root. The corresponding character ψ_X is maximally degenerate and has $\text{supp}(\psi_X) = \{\alpha_i\}$ and one can compute the associated degenerate Whittaker coefficients using (10.80) in terms of Whittaker coefficients on $SL(2, \mathbb{R})$ (which are completely known). The minimal orbit is called type A_1 in *Bala–Carter terminology* [36] and this relates to the subgroup G' that appears in the formula for degenerate Whittaker coefficients. Note that the minimal nilpotent orbit is closely associated to the Heisenberg parabolic that appeared in Example 5.32.

As another example we consider the consequences of the condition (10.84) on a degenerate principal series of the group E_8.

Example 10.26: Wavefront sets of the adjoint E_8 series

Consider the maximal parabolic Eisenstein series $E(\lambda, g)$ of $E_8(\mathbb{R})$ given by the weight

$$\lambda = 2s\Lambda_8 - \rho \tag{10.85}$$

as in Example 10.6. The degenerate principal series that $E(\lambda, g)$ belongs to is of Gelfand–Kirillov dimension (for generic s)

$$\text{GKdim}\, I(\lambda) = \dim(E_8) - \dim(P_\lambda) = 248 - (133 + 1 + 56 + 1) = 57. \tag{10.86}$$

According to (10.84), the largest nilpotent orbit that can contribute to the wavefront set of $E(\lambda, g)$ is therefore of dimension 114. Here is a list of nilpotent orbits of (split) $E_8(\mathbb{R})$ of small dimension [186, 146]:

$\dim \mathcal{O}$	Bala–Carter label	weighted diagram over $\mathbb{C}$
0	0	$\begin{bmatrix} & & 0 & & & & \\ 0 & 0 & 0 & 0 & 0 & 0 & 0 \end{bmatrix}$
58	A_1	$\begin{bmatrix} & & 0 & & & & \\ 0 & 0 & 0 & 0 & 0 & 0 & 1 \end{bmatrix}$
92	$2A_1$	$\begin{bmatrix} & & 0 & & & & \\ 1 & 0 & 0 & 0 & 0 & 0 & 0 \end{bmatrix}$
112	$3A_1$	$\begin{bmatrix} & & 0 & & & & \\ 0 & 0 & 0 & 0 & 0 & 1 & 0 \end{bmatrix}$
114	A_2 and $(4A_1)''$	$\begin{bmatrix} & & 0 & & & & \\ 0 & 0 & 0 & 0 & 0 & 0 & 2 \end{bmatrix}$
$\vdots$	$\vdots$	$\vdots$

The last entry corresponds to a single complex orbit of type A_2 that splits into two real orbits [186]. All these orbits have representatives X in $[N, N]\backslash N$ and the associated Whittaker coefficients can be calculated using (10.80). More complicated Whittaker coefficients are absent in this degenerate principal series. For special s values, not all the orbits in the above table appear in the wavefront set.

10.4.2 Whittaker Coefficients of Maximal Parabolic Eisenstein Series

For maximal parabolic Eisenstein series one can also make statements about the vanishing of some Whittaker coefficients. Consider an Eisenstein series on $G(\mathbb{A})$ induced from a character $\chi\colon P_{i_*}(\mathbb{A}) \to \mathbb{C}^\times$, i.e., one that is in the degenerate principal series and is parametrised by a single complex parameter $s \in \mathbb{C}$ through the weight $\lambda = 2s\Lambda_{i_*} - \rho$; see Proposition 5.28. The Whittaker integral for an arbitrary character $\psi\colon N(\mathbb{Q})\backslash N(\mathbb{A}) \to U(1)$ leads to (see (9.11))

$$W_\psi(\chi, a) = \sum_{w \in \mathcal{W}(P_{i_*})\backslash\mathcal{W}} \int_{N^w_{\{\beta\}}} \overline{\psi(n_\beta)} dn_\beta \cdot \int_{N^w_{\{\gamma\}}} \chi(wn_\gamma a)\overline{\psi(n_\gamma)} dn_\gamma, \quad (10.87)$$

where the important point is now that the set of contributing Weyl words is restricted to the quotient $\mathcal{W}(P_{i_*})\backslash\mathcal{W}$ from the outset. Again, the integral over $N^w_{\{\beta\}}$ can make the whole expression vanish and imposes constraints on w and ψ. Now the set of positive roots β that appear in that integral is

$$\left\{\beta > 0 \,|\, w\beta \in \Delta_{P_{i_*}}\right\}, \quad (10.88)$$

where $\Delta_{P_{i_*}}$ denotes the subset of all roots in Δ that belong to the maximal parabolic $P_{i_*}(\mathbb{A})$; it involves all positive roots and some negative roots. The set of β now always involves a simple root for any w; therefore for a generic ψ the integral over n_β will vanish and we conclude that the generic Whittaker coefficient for a maximal parabolic Eisenstein series vanishes. Note that the non-generic terms in (10.87) do not enjoy Euler factorisation.

Another way to see this is by noting that the factor $1/\zeta(\lambda)$ that appears in (9.24) generally contains factors $(1 - p^{-(\langle\lambda|\alpha_i\rangle+1)}) = (1 - p^{-2s_i})$ for all simple roots α_i. For any degenerate principal series Eisenstein series, one of these factors vanishes identically, and there is no pole at the same s_i values in the factor $\epsilon(\lambda)$ (see (9.24)). This is guaranteed by the holomorphy of the local Whittaker function discussed in Proposition 9.4.

We will come back to this in the discussion in Section 10.4.4.

10.4.3 Examples of Degenerate Whittaker Coefficients

We now present some explicit expressions for degenerate Whittaker coefficients calculated with the help of (10.80). The examples are taken mainly from [214]. The following notation will be used:

$$B_{s,m}(v) = \frac{1}{\xi(2s)}\tilde{B}_{s,m}(v) = \frac{2}{\xi(2s)}|v|^{s-1/2}|m|^{s-1/2}\sigma_{1-2s}(m)K_{s-1/2}(2\pi|m|v), \quad (10.89)$$

for Whittaker coefficients on an $SL(2,\mathbb{R})$ subgroup. For an $SL(3,\mathbb{R})$ subgroup we write similarly

$$B_{s_1,s_2,m_1,m_2}(v_1, v_2) = \frac{1}{\xi(2s_1)\xi(2s_2)\xi(2s_1+2s_2-1)}\tilde{B}_{s_1,s_2,m_1,m_2}(v_1, v_2). \quad (10.90)$$

The explicit expression for $\tilde{B}_{s_1,s_2,m_1,m_2}(v_1, v_2)$ in terms of an integral over two Bessel functions can be found in [120, 525]; see also the $SL(3, \mathbb{A})$ example in Section 9.8. For our purposes, we only need to know that it is finite and non-zero for all values of s_1 and s_2. The same is true for $\tilde{B}_{s,m}(v)$.

Example 10.27: Degenerate Whittaker coefficients for $E_n(\mathbb{R})$ Eisenstein series

We consider maximal parabolic Eisenstein series of the finite-dimensional exceptional groups $E_n(\mathbb{R})$ with weight vector $\lambda = 2 \cdot \frac{3}{2}\Lambda_1 - \rho$ for $n = 6, 7, 8$ in Bourbaki labelling. This is associated with the minimal representation. In these examples all non-vanishing (abelian) Whittaker coefficients turn out to be given by a finite sum of n Whittaker coefficients on the $SL(2, \mathbb{R})$ subgroup associated with each node of the Dynkin diagram. The full expression for the Whittaker coefficients will be given by [214]

$$\sum_{\psi \neq 0} W_{\psi}^{\circ}(\lambda, na) = \sum_{\alpha \in \Pi} \sum_{\psi_\alpha} c_\alpha(a) W'^{\circ}_{\psi_\alpha^a}(\lambda'_\alpha, \mathbb{1}) \psi_\alpha(n), \tag{10.91}$$

where W_{ψ}° on the left-hand side is given by (10.80). In the language of Section 6.2.3 this is only the abelian part of the Fourier expansion along the maximal unipotent N. For the maximally degenerate character ψ_α associated with the simple roots α, m_α is the only non-zero charge and $c_\alpha(a)$ is a function of the variables parametrising the Cartan torus. Furthermore, $W'^{\circ}_{\psi_\alpha^a}$ is a generic $SL(2, \mathbb{R})$ Whittaker coefficient associated with the simple root α and λ'_α is the projection of λ onto this subgroup.

We can then provide lists of the degenerate Whittaker coefficients for each case where the results are taken from [214]:

- E_6:

ψ_α	$c_\alpha(a) W'^{\circ}_{\psi_\alpha^a}(\chi'_\alpha, \mathbb{1})$
$(m, 0, 0, 0, 0, 0)$	$v_3^2 v_1^{-1} B_{3/2,m}\left(v_1^2 v_3^{-1}\right)$
$(0, m, 0, 0, 0, 0)$	$v_2^2 \tilde{B}_{0,m}(v_2^2 v_4^{-1})/\xi(3)$
$(0, 0, m, 0, 0, 0)$	$\xi(2) v_4 B_{1,m}\left(v_3^2 v_1^{-1} v_4^{-1}\right)/\xi(3)$
$(0, 0, 0, m, 0, 0)$	$v_4 \tilde{B}_{1/2,m}(v_4^2 v_2^{-1} v_3^{-1} v_5^{-1})/\xi(3)$
$(0, 0, 0, 0, m, 0)$	$v_5^2 v_6^{-1} \tilde{B}_{0,m}(v_5^2 v_4^{-1} v_6^{-1})/\xi(3)$
$(0, 0, 0, 0, 0, m)$	$\xi(2) v_6^3 B_{-1/2,m}\left(v_6^2 v_5^{-1}\right)/\xi(3)$

- E_7:

ψ_α	$c_\alpha(a)W'^{\circ}_{\psi^a_\alpha}(\chi'_\alpha, \mathbb{1})$
$(m,0,0,0,0,0,0)$	$v_3^2v_1^{-1}B_{\frac{3}{2},m}\left(v_1^2v_3^{-1}\right)$
$(0,m,0,0,0,0,0)$	$v_2^2\tilde{B}_{0,m}(v_2^2v_4^{-1})/\xi(3)$
$(0,0,m,0,0,0,0)$	$\xi(2)v_4B_{1,m}\left(v_3^2v_1^{-1}v_4^{-1}\right)/\xi(3)$
$(0,0,0,m,0,0,0)$	$v_4\tilde{B}_{1/2,m}(v_4^2v_2^{-1}v_3^{-1}v_5^{-1})/\xi(3)$
$(0,0,0,0,m,0,0)$	$v_5^2v_6^{-1}\tilde{B}_{0,m}(v_5^2v_4^{-1}v_6^{-1})/\xi(3)$
$(0,0,0,0,0,m,0)$	$\xi(2)v_6^3v_7^{-2}B_{-1/2,m}\left(v_6^2v_5^{-1}v_7^{-1}\right)/\xi(3)$
$(0,0,0,0,0,0,m)$	$v_7^4B_{-1,m}\left(v_7^2v_6^{-1}\right)$

- E_8:

ψ_α	$c_\alpha(a)W'^{\circ}_{\psi^a_\alpha}(\chi'_\alpha, \mathbb{1})$
$(m,0,0,0,0,0,0,0)$	$v_3^2v_1^{-1}B_{3/2,m}\left(v_1^2v_3^{-1}\right)$
$(0,m,0,0,0,0,0,0)$	$v_2^2\tilde{B}_{0,m}(v_2^2v_4^{-1})/\xi(3)$
$(0,0,m,0,0,0,0,0)$	$\xi(2)v_4B_{1,m}\left(v_3^2v_1^{-1}v_4^{-1}\right)/\xi(3)$
$(0,0,0,m,0,0,0,0)$	$v_4\tilde{B}_{1/2,m}(v_4^2v_2^{-1}v_3^{-1}v_5^{-1})/\xi(3)$
$(0,0,0,0,m,0,0,0)$	$v_5^2v_6^{-1}\tilde{B}_{0,m}(v_5^2v_4^{-1}v_6^{-1})/\xi(3)$
$(0,0,0,0,0,m,0,0)$	$\xi(2)v_6^3v_7^{-2}B_{-1/2,m}\left(v_6^2v_5^{-1}v_7^{-1}\right)/\xi(3)$
$(0,0,0,0,0,0,m,0)$	$v_7^4v_8^{-3}B_{-1,m}\left(v_7^2v_6^{-1}v_8^{-1}\right)$
$(0,0,0,0,0,0,0,m)$	$\xi(4)v_8^5B_{-3/2,m}\left(v_8^2v_7^{-1}\right)/\xi(3)$

Clearly, there is a simple pattern involving the weight of the Bessel function and the number of Weyl reflections needed to relate a given degenerate character ψ to the one in the first line where the unipotent of the Whittaker expansion is aligned with the unipotent of the maximal parabolic defining the Eisenstein series. In cases where the function $B_{s,m}$ would be ill-defined due to a pole in the ξ prefactor, one has that consistently only the differently normalised and finite function $\tilde{B}_{s,m}$ appears.

The following provides an example of a degenerate Whittaker coefficient of type A_2.

Example 10.28: A_2-type Whittaker coefficients for an $E_8(\mathbb{R})$ Eisenstein series

For the E_8 series of Example 10.26 with $\lambda = 2s\Lambda_8 - \rho$ we compute the Whittaker coefficient associated with the degenerate character ψ on N with 'charges'

$$\psi \quad \leftrightarrow \quad \left[\begin{smallmatrix} & & 0 & & & & \\ 0 & 0 & 0 & 0 & 0 & m & n \end{smallmatrix}\right]. \tag{10.92}$$

This choice of character ψ is associated with the 114-dimensional nilpotent orbit of type A_2 from the table in Example 10.26. For simplicity we put the torus element $a = \mathbb{1}$. Then formula (10.80) gives

$$W_\psi(\lambda, \mathbb{1}) = \frac{\xi(2s-11)\xi(2s-14)\xi(2s-18)\xi(4s-29)}{\xi(2s)\xi(2s-5)\xi(2s-9)\xi(4s-28)} B_{6-s,\frac{19}{2}-s,m,n}\left(1,1\right). \tag{10.93}$$

(The contributing Weyl word has length 30 and we do not spell it out here.) As was argued in Example 10.20, the value $s = \frac{5}{2}$ corresponds to a simpler Eisenstein series where the constant term simplifies. From the above formula we can see this also in the Whittaker coefficient of type A_2. The prefactor tends to zero for $s \to \frac{5}{2}$ while the $SL(3,\mathbb{R})$ Whittaker coefficient stays finite and hence the degenerate coefficient (10.93) disappears.

In the case $s = \frac{5}{2}$ one similarly checks that all Whittaker coefficients but the ones of type A_1 vanish, consistent with the fact that the corresponding Eisenstein series belongs to an automorphic realisation of the minimal representation.

10.4.4 Relation Between General Fourier Coefficients and Whittaker Coefficients

In this section we investigate, based on the methods of Miller–Sahi [476] and Ginzburg [278], how to compute Fourier coefficients F_{ψ_U} on unipotent subgroups U (from Definition 6.13) in terms of Whittaker coefficients W_{ψ_N}, the latter of which are known using the methods of Chapter 9. Details can be found in [340] and similar ideas were discussed in Section 6.5. As we are discussing characters and Fourier coefficients on different subgroups we will adopt the subscript notation used in Section 9.8.4 to avoid ambiguities.

Since both Whittaker coefficients on N and the constant term simplify for automorphic forms in small representations, as seen in the examples above, it is natural to also expect simplifications for more general Fourier coefficients. Using the *wavefront set* of Section 6.4.2 and the arguments in the previous sections one can tell for which representations a Fourier coefficient is non-vanishing, but by rewriting F_{ψ_U} in terms of W_{ψ_N} it is also possible to see how a non-vanishing F_{ψ_U} simplifies for smaller representations.

We have already seen an example of such a computation in Proposition 9.14 where we showed for $G = SL(3)$ how the Fourier coefficients on $Z = [N, N]$ can be expressed as a sum of G-translated Whittaker coefficients on N. When restricting to Eisenstein series in the minimal representation this expression simplifies as follows.

Example 10.29: Non-abelian coefficient in $SL(3)$ minimal representation

For the example $G = SL(3)$ with $\lambda = (2s_1-1)\Lambda_1+(2s_2-1)\Lambda_2$ the generic principal series has Gelfand–Kirillov dimension $\text{GKdim}(I(\lambda)) = \dim G - \dim B = 3$, but for $(s_1, s_2) = (s, 0)$ or $(s_1, s_2) = (0, s)$ it reduces to $\text{GKdim}(I(\lambda_{\text{min}})) = 2$ for which the Eisenstein series belongs to the minimal automorphic representation. The orbits of $SL(3, \mathbb{C})$ are [146]

$\dim \mathcal{O}$	Bala–Carter label	weighted diagram
0	0	0 0 •—•
4	A_1	2 2 •—•
6	A_2	1 1 •—•

This means that for π_{min} we only have Whittaker coefficients of type A_1, that is, maximally degenerate Whittaker coefficients charged under a single simple root.

In Proposition 9.14 it was shown that

$$W_{\psi_Z}^{(k)}(\chi, g) = \sum_{m_1, m_2 \in \mathbb{Q}} W_{\psi_N}^{(m_1, d)}(\chi, lg), \tag{10.94}$$

where $d = d(k, m_2)$ as defined in (9.116) and l depends on k and m_2 as described in the proposition. Recall from Section 9.8 that $W_{\psi_N}^{(m_1,d)}$ for $m_1 \neq 0$ contains a convoluted integral of two Bessel functions, while $W_{\psi_N}^{(0,d)}$ is simpler, being proportional to a single Bessel function.

When restricting χ to χ_{min} (parametrised by s), we get that the sum over charges in (10.94) collapses to $m_1 = 0$ since $d \neq 0$, simplifying the expression for $W_{\psi_Z}^{(k)}$, which then only contains single Bessel functions. Note though that the sum over m_2

(that is, over l-translates) still remains, that is,

$$W_{\psi_Z}^{(k)}(\chi_{\min}, g) = \sum_{m_2 \in \mathbb{Q}} W_{\psi_N}^{(0,d)}(\chi_{\min}, lg) \,. \tag{10.95}$$

When inserting the argument $g = (g_\infty, \mathbb{1}, \mathbb{1}, \dots)$ the sum over rational charges becomes a sum over integers, similar to what happens in Proposition 9.15, since $l \in SL(3, \mathbb{Z})$ by design.

Let us now consider another example for $G = SL(3)$, but with Fourier coefficients on a maximal parabolic subgroup. Here, the expression simplifies even further in the minimal representation, resulting in a single translated maximally degenerate Whittaker coefficient on N.

Example 10.30: Unipotent Fourier coefficient in $SL(3)$ minimal representation

Continuing with $G = SL(3)$, we will now see that a Fourier coefficient F_{ψ_U} on the maximal parabolic subgroup corresponding to the first simple root

$$P = P_1 = LU = \begin{pmatrix} * & * & * \\ 0 & * & * \\ 0 & * & * \end{pmatrix}, \qquad L = \begin{pmatrix} * & 0 & 0 \\ 0 & * & * \\ 0 & * & * \end{pmatrix}, \qquad U = \begin{pmatrix} 1 & * & * \\ 0 & 1 & 0 \\ 0 & 0 & 1 \end{pmatrix} \tag{10.96}$$

can be expressed as a single L-translated, maximally degenerate Whittaker coefficient in the minimal representation. Let $m_1, m_2 \in \mathbb{Q}$ be such that $m_1 m_2 \neq 0$ and define the unitary character ψ_U by

$$u = \begin{pmatrix} 1 & u_1 & u_2 \\ 0 & 1 & 0 \\ 0 & 0 & 1 \end{pmatrix} \in U \,, \qquad \psi_U(u) = e^{2\pi i (m_1 u_1 + m_2 u_2)} \,. \tag{10.97}$$

Then, $d = d(m_1, m_2)$ as defined in (9.116) is strictly positive and $m_i' := m_i / d \in \mathbb{Z}$ with $\gcd(m_1', m_2') = 1$, which tells us that there exist integers α and β such that

$$l = \begin{pmatrix} 1 & 0 & 0 \\ 0 & \alpha & \beta \\ 0 & -m_2' & m_1' \end{pmatrix} \in L(\mathbb{Z}) \,. \tag{10.98}$$

Now we conjugate the Fourier coefficient with l as follows (see (6.84)):

$$\begin{aligned} F_{\psi_U}^{(m_1,m_2)}(\chi, g) &:= \int_{(\mathbb{Q}\backslash\mathbb{A})^2} E(\chi, \begin{pmatrix} 1 & u_1 & u_2 \\ 0 & 1 & 0 \\ 0 & 0 & 1 \end{pmatrix} g) e^{-2\pi i (m_1 u_1 + m_2 u_2)} \, du^2 \\ &= \int_{(\mathbb{Q}\backslash\mathbb{A})^2} E(\chi, \begin{pmatrix} 1 & (m_1 u_1 + m_2 u_2)/d & -b u_1 + a u_2 \\ 0 & 1 & 0 \\ 0 & 0 & 1 \end{pmatrix} lg) e^{-2\pi i (m_1 u_2 + m_2 u_2)} \, du^2 \\ &= \int_{(\mathbb{Q}\backslash\mathbb{A})^2} E(\chi, \begin{pmatrix} 1 & x_1 & x_2 \\ 0 & 1 & 0 \\ 0 & 0 & 1 \end{pmatrix} lg) e^{-2\pi i d x_1} \, dx^2 = F_{\psi_U}^{(d,0)}(\chi, lg) \,, \end{aligned} \tag{10.99}$$

where we have made the substitutions $(m_1 u_1 + m_2 u_2)/d \to x_1$ and $-b u_1 + a u_2 \to x_2$. Note that there are other matrices $l \in L(\mathbb{Q})$ (explicitly given by m_1 and m_2) that would accomplish similar results, but if $l \in L(\mathbb{Z})$ the p-adic Iwasawa decomposition simplifies when inserting $g = (g_\infty, \mathbb{1}, \ldots, \mathbb{1})$ as in Section 9.8.4.

Expanding further we obtain

$$F_{\psi_U}^{(m_1,m_2)}(\chi, g) = \sum_{m_3 \in \mathbb{Q}} \int_{(\mathbb{Q}\backslash\mathbb{A})^3} E(\chi, \begin{pmatrix} 1 & x_1 & x_2 \\ 0 & 1 & x_3 \\ 0 & 0 & 1 \end{pmatrix} lg) e^{-2\pi i(dx_1+m_3x_3)} \, d^3x$$
$$= \sum_{m_3 \in \mathbb{Q}} W_{\psi_N}^{(d,m_3)}(\chi, lg), \tag{10.100}$$

with $d > 0$ and where $W_{\psi_N}^{(d,m_3)}$ is a Whittaker coefficient on N with charges d and m_3 for the two simple roots.

Now restricting to $\chi_{\min}$, only the maximally degenerate Whittaker coefficients are non-vanishing. This collapses the sum above to $m_3 = 0$ giving

$$F_{\psi_U}^{(m_1,m_2)}(\chi_{\min}, g) = W_{\psi_N}^{(d,0)}(\chi_{\min}, lg) \,. \tag{10.101}$$

We note that for $m_2 = 0$ and positive m_1 we have that $d = m_1$ and $l = \mathbb{1}$ giving

$$F_{\psi_U}^{(m_1,0)}(\chi_{\min}, g) = W_{\psi_N}^{(m_1,0)}(\chi_{\min}, g) \,. \tag{10.102}$$

The same statement can be made for negative m_1 as well and can be derived by directly making a further expansion as in (10.100) without a conjugation with l. We conclude that, in the minimal representation, a maximal parabolic Fourier coefficient charged only on the simple root α_1 simplifies to the maximally degenerate Whittaker coefficient charged on the same root.

In [340] it was similarly shown for $SL(3)$ and $SL(4)$ that all non-trivial Fourier coefficients on any maximal parabolic subgroup automorphic forms in the minimal representation simplify to a single translated maximally degenerate Whittaker coefficient on N.

This was accomplished by relating Fourier coefficients to the orbit Fourier coefficients defined in Definition 6.37 and which vanish when the orbit does not belong to the wavefront set of the considered automorphic representation. Then, the orbit coefficients were expanded as sums of translated Whittaker coefficients which were found to be maximally degenerate for the minimal orbit coefficients and Whittaker coefficients charged under two strongly orthogonal roots for the next-to-minimal orbit coefficients. In the minimal representation, the maximal parabolic Fourier coefficients picked up only one of these maximally degenerate Whittaker coefficients in the minimal orbit coefficient. In the same paper the next-to-minimal representation for $SL(4)$ is discussed as well.

Also, it was shown in [476] that, for E_6 and E_7, Fourier coefficients on certain maximal parabolic subgroups of automorphic forms in $\pi_{\min}$ are determined by maximally degenerate Whittaker coefficients. From their proof, one may also deduce that, concretely, such a Fourier coefficient is exactly a translate of a maximally degenerate Whittaker coefficient, similar to the results of [340, 2] for $SL(n)$. We also refer to Section 6.5 where similar ideas were discussed.

From this it was conjectured in [340] that Fourier coefficients on maximal parabolic subgroups for other simply-laced, simple Lie groups simplify in a similar way and that each may be given in terms of a single, translated, maximally degenerate Whittaker coefficient on N. In the remaining parts of this section we will explore some applications and verifications of this statement.

We have seen in Sections 9.2 and 9.4 that generic Whittaker coefficients factorise:

$$W_{\psi_N}(\chi, g) = \prod_{p \leq \infty} W_{\psi_{N,p}}(\chi_p, g), \tag{10.103}$$

with $W_{\psi_{N,p}} \in \mathrm{Ind}_{N(\mathbb{Q}_p)}^{G(\mathbb{Q}_p)} \psi_{N,p}$, but that degenerate Whittaker coefficients, in general, do not and are expressed as sums of factorising terms as seen in (10.80). As such, we cannot expect that all Fourier coefficients on any parabolic subgroup should factorise; that is, we *cannot* a priori expect that

$$\mathrm{Ind}_{U(\mathbb{A})}^{G(\mathbb{A})} \psi_U = \bigotimes_{p \leq \infty} \mathrm{Ind}_{U(\mathbb{Q}_p)}^{G(\mathbb{Q}_p)} \psi_{U,p} \qquad \text{(wrong in general!)}. \tag{10.104}$$

However, in the minimal representation all but the maximally degenerate Whittaker coefficients on N vanish and the remaining simplify, becoming factorisable as seen for E_6, E_7 and E_8 in the tables of Example 10.27 in [214, App. A]. This means that if a Fourier coefficient on a maximal parabolic subgroup can be expressed as a single translated maximally degenerate Whittaker coefficient (as in (10.101) above), then it does indeed factorise.

This is interesting since, although not much is known in general about $\mathrm{Ind}_{U(\mathbb{Q}_p)}^{G(\mathbb{Q}_p)} \psi_{U,p}$ for non-minimal parabolic subgroups, the image under the embedding

$$\pi_{\min,p} \subset \mathrm{Ind}_{P(\mathbb{Q}_p)}^{G(\mathbb{Q}_p)} \chi_{\min,p} \hookrightarrow \mathrm{Ind}_{U(\mathbb{Q}_p)}^{G(\mathbb{Q}_p)} \psi_{U,p}, \tag{10.105}$$

with $\chi_{\min,p}$ spherical, has multiplicity one [246]; the corresponding spherical vectors $f^\circ_{\psi_{U,p}}$ have been computed in several cases using representation theory [195, 393, 394, 546].

Assuming the factorisation of maximal parabolic Fourier coefficients discussed above, it is possible to rederive and extend these results by considering the spherical vectors induced from $\pi_{\min}$ as coming from local factors of global Fourier coefficients.

Indeed, in [340] it was shown that the products of known spherical vectors $f^\circ_{\psi_{U,p}}$ in maximal parabolics for E_6, E_7 and E_8 give exactly the expected translated maximally degenerate Whittaker coefficients in $\pi_{\min}$, giving strong support for the claim that (10.101) can be generalised to all simply-laced simple Lie groups. Let us consider the case $G = E_7$ below.

Example 10.31: E_7 spherical vectors

Let $G = E_7$ and $P_7 = LU$ be the maximal parabolic subgroup obtained by removing the simple root α_7, using the Bourbaki labelling in Figure 10.1. This was one of the parabolic subgroups studied in [476]. Then U is abelian and can also be obtained from the three-grading

$$\mathfrak{e}_7 = \mathfrak{g}_{-1} \oplus \mathfrak{g}_0 \oplus \mathfrak{g}_1 = \mathbf{27} \oplus (\mathfrak{e}_6 \oplus \mathbf{1}) \oplus \mathbf{27}, \tag{10.106}$$

with $\mathfrak{u} = \mathfrak{g}_1$.

The unique spherical vectors in $\mathrm{Ind}_{U(\mathbb{Q}_p)}^{G(\mathbb{Q}_p)} \psi_{U,p}$ in the minimal representation have been computed at the non-archimedean places by [546] and are here shown evaluated at the identity in $G(\mathbb{Q}_p)$

$$f^\circ_{\psi_{U,p}} = \frac{1 - p^3\,|m|_p^{-3}}{1 - p^3}, \tag{10.107}$$

where $m \in \mathbb{Q}^\times$ is the charge of $\psi_{U,p}$ conjugated to the simple root α_7 in U.

At the archimedean place we have instead, from [195], that

$$f^\circ_{\psi_{U,\infty}} = m^{-3/2} K_{3/2}(m), \tag{10.108}$$

when evaluated at the identity in $G(\mathbb{R})$.

We will now rederive these results by instead viewing the spherical vectors as coming from local factors of a global Fourier coefficient F_{ψ_U} of a spherical Eisenstein series in $\pi_{\min}$. Such an Eisenstein series may be realised from a parabolically induced representation $I_{P_1}(\lambda_{\min})$, with the maximal parabolic subgroup obtained by removing α_1 and $\lambda_{\min} = 2s\Lambda_1 - \rho$ with $s = 3/2$. We consider the Fourier coefficient F_{ψ_U} with a character non-trivial only on the simple root α_7, associated with the abelian unipotent in (10.106). Similar to the examples above, it simplifies to the single maximally degenerate Whittaker coefficient $W^{(\alpha_7)}_{\psi_N}$ charged only on the same root which, in turn, factorises.

From Example 10.27 we get that

$$\begin{aligned} W^{(\alpha_7)}_{\psi_N}(\lambda_{\min}, \mathbb{1}) &= \frac{2}{\xi(4)}\,|m|^{-3/2}\,\sigma_3(m) K_{3/2}(m) \\ &= \frac{2}{\xi(4)} \left(\prod_{p<\infty} \frac{1 - p^3\,|m|_p^{-3}}{1 - p^3} \right) \left(|m|^{-3/2}\, K_{3/2}(m) \right), \end{aligned} \tag{10.109}$$

where we recognise the first parenthesis as a product of the non-archimedean spherical vectors in (10.107) and the second as the archimedean spherical vector in (10.108).

In [340], the spherical vectors are rederived in a similar way for E_6, E_7 and E_8 in both the abelian and Heisenberg realisations of the minimal representation, with complete agreement.

11

Hecke Theory and Automorphic L-functions

In this chapter, we outline the theory of Hecke operators and Hecke algebras. In a nutshell, Hecke operators act on the space of automorphic forms on a group G, forming a commutative ring called the Hecke algebra. The representation theory of this algebra carries a wealth of information about automorphic forms and automorphic representations that connect with many of the structures discussed in the preceding chapters. We begin by outlining the Hecke theory in the case of automorphic forms on real arithmetic quotients $G(\mathbb{Z})\backslash G(\mathbb{R})$, providing detailed examples for the case of $SL(2,\mathbb{R})$. After this treatment of the classical Hecke theory, we consider the counterpart in the adelic context. The key object here is the local spherical Hecke algebra $\mathcal{H}_p^\circ$ which acts on the space of adelic automorphic forms $\mathcal{A}(G(\mathbb{Q})\backslash G(\mathbb{A}))$. We study the representation theory of the spherical Hecke algebra and show how this relates to automorphic representations via the Satake isomorphism. Our treatment is mainly done in the context of $SL(2,\mathbb{A})$ and $GL(2,\mathbb{A})$, but many results carry over to arbitrary reductive groups. In particular, in Section 11.6 we give some details on the generalisation to $GL(n,\mathbb{A})$ and we make contact with the Langlands program. The starting point is the rewriting of the Casselman–Shalika formula that we encountered in Section 9.6. Finally, we end this chapter with a brief discussion of automorphic L-functions, which form a cornerstone of the Langlands program. This also connects directly to a more detailed discussion of the so-called Langlands–Shahidi method.

11.1 Classical Hecke Operators and the Hecke Ring

Besides the ring of invariant differential operators there is another set of operators that act on the space $\mathcal{A}(G(\mathbb{Z})\backslash G(\mathbb{R}))$ of automorphic functions on the group $G(\mathbb{R})$ invariant under the discrete group $G(\mathbb{Z})$. These additional operators are called *Hecke operators* and we sketch their general definition following [290]. Their power is worked out for $SL(2,\mathbb{R})$ in Section 11.2 in the classical setting. Hecke operators and algebras can also be introduced in the adelic setting and this will be the topic of Sections 11.2.4 and beyond.

Let $g \in G(\mathbb{R})$ be a fixed element *commensurable* with $G(\mathbb{Z})$, i.e., the intersection $g^{-1}G(\mathbb{Z})g \cap G(\mathbb{Z})$ has finite index in both $G(\mathbb{Z})$ and $g^{-1}G(\mathbb{Z})g$. We rewrite its double coset with respect to the discrete group $G(\mathbb{Z})$ as

$$G(\mathbb{Z})gG(\mathbb{Z}) = \bigcup_{i=1}^{d} G(\mathbb{Z})g\delta_i. \tag{11.1}$$

On the right-hand side we have written the double coset as a finite disjoint union of single cosets with representatives $\delta_i \in G(\mathbb{Z})$ for $i = 1, \ldots, d$. The finiteness of this decomposition follows from the commensurability of g.

Definition 11.1 (Hecke operator) The classical *Hecke operator* T_g associated with a $G(\mathbb{Z})$-commensurable $g \in G(\mathbb{R})$ acting on an automorphic function φ is defined by

$$(T_g\varphi)(h) = \sum_{i=1}^{d} \varphi(g\delta_i h) \qquad \text{with } h \in G(\mathbb{R}), \tag{11.2}$$

where the δ_i are representatives of the double coset decomposition (11.1).

This operator is well-defined as a finite sum. One can check easily that T_g maps $G(\mathbb{Z})$-invariant functions to $G(\mathbb{Z})$-invariant functions.

Remark 11.2 It is often useful to take a slightly larger group than the original $G(\mathbb{R})$ if it acts on the same space. For $SL(2,\mathbb{R})$ acting on spherical automorphic functions that are defined on the upper half-plane $SL(2,\mathbb{R})/SO(2,\mathbb{R})$, one can also consider the action of $GL(2,\mathbb{R})$ on $\mathbb{H}$ and define Hecke operators for elements $g \in GL(2,\mathbb{R})$ with respect to $SL(2,\mathbb{Z})$. This viewpoint will be useful in Section 11.2 below.

The Hecke ring is formed by also allowing integer multiples mT_g of Hecke operators for $m \in \mathbb{Z}$ and defining the product of two Hecke operators T_{g_1} and T_{g_2} by representing the combined double coset $G(\mathbb{Z})g_1G(\mathbb{Z}) \cdot G(\mathbb{Z})g_2G(\mathbb{Z})$ as the union of double cosets $G(\mathbb{Z})hG(\mathbb{Z})$, possibly with multiplicity. It turns out that a finite union suffices and the product of T_{g_1} and T_{g_2} is then the sum over the T_h with integer coefficients. This operation turns the set of Hecke operators into a *Hecke ring*.

Remark 11.3 The normalisation of the Hecke operators in (11.2) is not uniquely fixed. The one used there yields the Hecke ring over $\mathbb{Z}$. It can be useful to change the normalisation and then obtain a Hecke algebra over the field $\mathbb{Q}$.

The Hecke ring is usually defined together with a given choice of *semi-group* $\mathcal{S}$ of commensurable elements g. A semi-group is a set with an associative product, but not all elements in $\mathcal{S}$ need to be invertible. The example to have in mind here is the set of matrices with determinant equal to some positive integer. The semi-group needs to be chosen such that $G(\mathbb{Z})$ is a (proper) subgroup of $\mathcal{S}$. Importantly, the Hecke ring (for the cases of interest here) turns out to be *commutative*. For the precise statement see [290, Thm. 3.10.10].

Furthermore, the Hecke operators also commute with the ring of differential operators. This means that we can seek common automorphic eigenfunctions of the ring of differential operators and the ring of Hecke operators. The action of the operators then puts additional constraints on the Fourier coefficients that appear in the analysis of the automorphic function, and in fact captures much of the number-theoretic structure of these coefficients. An example of this is worked out in the following section for the case of $SL(2,\mathbb{R})$.

11.2 Hecke Operators for $SL(2,\mathbb{R})$

In this section, we illustrate some basic features of the Hecke algebra as sketched in the previous section and the way it interacts with the Fourier expansion in the case of $SL(2,\mathbb{R})$. The presentation here is based on [290, 24].

11.2.1 Definition of Hecke Operators

Let $f\colon \mathbb{H} \to \mathbb{R}$ be a *Maaß wave form*, i.e., an $SL(2,\mathbb{Z})$ left-invariant function on the upper half-plane $\mathbb{H} = SL(2,\mathbb{R})/SO(2)$ that is also an eigenfunction of the $SL(2,\mathbb{R})$-invariant Laplace operator Δ, defined in (4.68). For example f could be the non-holomorphic Eisenstein series $E(s,z)$ as considered in (4.76). As explained in Section 4.3.2, Maaß wave forms can also be considered as spherical automorphic forms on $SL(2,\mathbb{R})$.

According to the general discussion of Hecke operators in Section 11.1, we have to choose a semi-group $\mathcal{S}$ of $SL(2,\mathbb{Z})$ commensurable elements. This we do by letting $\mathcal{S}$ be the group of integer (2×2) matrices with positive integer determinant n. For fixed $n>0$ let

$$g = \begin{pmatrix} m_1 m_2 & 0 \\ 0 & m_2 \end{pmatrix} \tag{11.3}$$

be a parametrisation of such matrices up to left- and right-multiplication by $SL(2,\mathbb{Z})$. We will define a Hecke operator T_n not to a single such element but to the union of all diagonal g with determinant equal to n. This T_n can be thought of as the sum of *all* the individual T_g defined according to the formula (11.2).

According to (11.1), we require the double coset decomposition into right cosets [290, Eq. (3.12.2)]

$$\bigcup_{m_1^2 m_2 = n} SL(2,\mathbb{Z}) \begin{pmatrix} m_1 m_2 & 0 \\ 0 & m_2 \end{pmatrix} SL(2,\mathbb{Z}) = \bigcup_{\substack{ad=n \\ 0 \le b < d}} SL(2,\mathbb{Z}) \begin{pmatrix} a & b \\ 0 & d \end{pmatrix}, \qquad (11.4)$$

in order to define T_n. Then to each $n > 0$ we can associate a Hecke operator T_n acting on a Maaß wave form $f(z)$:

$$\begin{aligned}(T_n f)(z) &:= \frac{1}{\sqrt{n}} \sum_{\substack{a \ge 1;\, ad=n \\ 0 \le b < d}} f\left(\begin{pmatrix} a & b \\ 0 & d \end{pmatrix} \cdot z \right) \\ &= \frac{1}{\sqrt{n}} \sum_{d|n} \sum_{b=0}^{d-1} f\left(\frac{nz + bd}{d^2} \right), \end{aligned} \qquad (11.5)$$

which maps f to a new function $T_n f$ on the upper half-plane. Here, we have slightly changed the normalisation of the operator compared to the general discussion as anticipated in Remark 11.3. Note that the transformation of the argument is not in $SL(2,\mathbb{Z})$ but has determinant n. Defining the set

$$M_2(n) = \left\{ \begin{pmatrix} a & b \\ 0 & d \end{pmatrix} \middle|\, a, b, d \in \mathbb{Z} \text{ with } ad = n \right\} \qquad (11.6)$$

of upper triangular integer (2×2) matrices with determinant n, we can rewrite the Hecke operator also as

$$(T_n f)(z) = \frac{1}{\sqrt{n}} \sum_{\gamma_n \in B(\mathbb{Z}) \backslash M_2(n)} f(\gamma_n \cdot z) = \frac{1}{\sqrt{n}} \sum_{\substack{ad=n \\ 0 \le b < d}} f\left(\frac{az + b}{d} \right). \qquad (11.7)$$

The resulting function $T_n f$ is also a Maaß wave form since it is (i) invariant and (ii) an eigenfunction of the Laplacian, as we will now show.

(i) Invariance requires evaluating

$$(T_n f)(\gamma \cdot z) = \frac{1}{\sqrt{n}} \sum_{\gamma_n \in B(\mathbb{Z}) \backslash M_2(n)} f(\gamma_n \gamma \cdot z) \qquad (11.8)$$

for $\gamma \in SL(2,\mathbb{Z})$. Using

$$\gamma_n \gamma = \tilde{\gamma} \tilde{\gamma}_n \qquad (11.9)$$

for some other $\tilde{\gamma} \in SL(2,\mathbb{Z})$ and $\tilde{\gamma}_n \in M_2(n)$ according to (11.4) together with invariance of φ under $\tilde{\gamma}$, one arrives at

$$(T_n f)(\gamma \cdot z) = \frac{1}{\sqrt{n}} \sum_{\tilde{\gamma}_n \in B(\mathbb{Z}) \backslash M_2(n)} f(\tilde{\gamma}_n \cdot z) = (T_n f)(z), \qquad (11.10)$$

and the function $T_n f$ is $SL(2,\mathbb{Z})$-invariant for any positive n. (See also chapters 6.8 and 6.9 of [24].)

(ii) Consider the action of the Laplacian (4.68) on $T_n\varphi$. It is straightforward to check that

$$[\Delta(T_n f)](z) = [T_n(\Delta f)](z). \tag{11.11}$$

Therefore, *the Laplacian commutes with all the Hecke operators* and if f is an eigenfunction of Δ, so is $T_n f$ and with the same eigenvalue.

Finally, we study the Fourier expansion of $T_n f$. Suppose that f has a Fourier expansion of the form (see (6.2))

$$f(z) = f(x+iy) = \sum_{m\in\mathbb{Z}} a_m(y)e^{2\pi imx}. \tag{11.12}$$

Then one finds for $T_n f$

$$\begin{aligned}
(T_n f)(z) &= \frac{1}{\sqrt{n}}\sum_{d|n}\sum_{b=0}^{d-1} f\left(\frac{n}{d^2}x + \frac{b}{d} + i\frac{n}{d^2}y\right)\\
&= \frac{1}{\sqrt{n}}\sum_{d|n}\sum_{m\in\mathbb{Z}} a_m\left(\frac{n}{d^2}y\right)e^{2\pi imnx/d^2}\sum_{b=0}^{d-1} e^{2\pi imb/d}\\
&= \frac{1}{\sqrt{n}}\sum_{m\in\mathbb{Z}}\sum_{d|n,d|m} d\, a_m\left(\frac{n}{d^2}y\right)e^{2\pi imnx/d^2}\\
&= \frac{1}{\sqrt{n}}\sum_{m\in\mathbb{Z}}\sum_{d|(n,m)} \frac{n}{d}a_{mn/d^2}\left(\frac{d^2}{n}y\right)e^{2\pi imx},
\end{aligned} \tag{11.13}$$

where we have changed the divisor sum variable from d to $\frac{n}{d}$ in the last step and have relabelled the sum over m in between. The Fourier expansion of $T_n f$ is therefore

$$(T_n f)(z) = \sum_{m\in\mathbb{Z}} \tilde{a}_m(y)e^{2\pi imx} \quad \text{with} \quad \tilde{a}_m(y) = \frac{1}{\sqrt{n}}\sum_{d|(n,m)}\frac{n}{d}a_{mn/d^2}\left(\frac{d^2}{n}y\right), \tag{11.14}$$

where $d|(n,m)$ means that $d|n$ and $d|m$, i.e., divides the greatest common divisor of n and m which is denoted by $(m,n) = \gcd(m,n)$ as usual.

11.2.2 Algebra of Hecke Operators

Importantly, the Hecke operators T_n satisfy a simple algebra on the space of Maaß wave forms: *they all commute*. Moreover, they satisfy the *Hecke algebra*

$$T_m T_n = T_n T_m = \sum_{d|(m,n)} T_{mn/d^2} \,. \tag{11.15}$$

Commutativity is manifest in this expression. To prove (11.15), one can first consider the case $(m,n) = 1$ and use the explicit definition. In the next step one can consider the case when both m and n are powers of the same prime. A proof can be found in [24, Sec. 6.10], where a different normalisation is used.

11.2.3 Common Eigenfunctions of T_n and Δ

Suppose $f\colon \mathbb{H} \to \mathbb{R}$ is an eigenfunction of all Hecke operators

$$T_n f = c_n f \tag{11.16}$$

for some eigenvalues c_n, and at the same time a cuspidal Maaß wave form with Fourier expansion

$$f(z) = \sum_{m\neq 0} a_m(y) e^{2\pi i m x} \,. \tag{11.17}$$

Applying T_n to f gives, with (11.14),

$$c_n f(z) = \frac{1}{\sqrt{n}} \sum_{m\neq 0} \sum_{d|(n,m)} \frac{n}{d} a_{mn/d^2}\left(\frac{d^2}{n} y\right) e^{2\pi i m x} \,. \tag{11.18}$$

Comparing the individual Fourier modes on both sides leads to

$$c_n a_m(y) = \frac{1}{\sqrt{n}} \sum_{d|(n,m)} \frac{n}{d} a_{nm/d^2}\left(\frac{d^2}{n} y\right) . \tag{11.19}$$

Setting $n = 1$ gives $c_1 a_m(y) = a_m(y)$ for all $m \neq 0$, implying $c_1 = 1$ unless f vanishes. Setting $m = 1$ implies

$$c_n a_1(y) = \sqrt{n} a_n \left(\frac{y}{n}\right) . \tag{11.20}$$

If f is not constant, one has $a_1(y) \neq 0$, otherwise all the Fourier coefficients would vanish.

From solving the Laplace condition on $\Delta f = s(s-1)f$ (see Appendix B.2) one knows that the dependence of the Fourier coefficient $a_m(y)$ on y is through the modified Bessel function as

$$a_m(y) = a_m y^{1/2} K_{s-1/2}(2\pi|m|y) \tag{11.21}$$

for some purely numerical coefficient a_m that we will now relate to the Hecke eigenvalues c_n. Rescaling f such that $a_1 = 1$ (*Hecke normalisation*) the relation (11.20) above implies that the Hecke eigenvalues equal the Fourier coefficients:

$$c_n = a_n. \tag{11.22}$$

Obtaining this simple relation was the reason for the choice of normalisation of the Hecke operator T_n. Note that the c_n are only defined for positive n but a_n for any n. The reality of f relates a_n to a_{-n}.

By virtue of the Hecke algebra (11.15) we have

$$T_m T_n f = c_m c_n f = \sum_{d|(m,n)} c_{mn/d^2} f, \tag{11.23}$$

so that the Fourier coefficients of a normalised simultaneous eigenfunction satisfy

$$a_m a_n = \sum_{d|(m,n)} a_{mn/d^2}. \tag{11.24}$$

In particular, they must be *multiplicative*, i.e., for coprime m and n one has $a_m a_n = a_{mn}$. This number-theoretic property of the Fourier coefficients follows from the action of the Hecke operators and would not have been apparent from $SL(2,\mathbb{Z})$ invariance alone. Note that the constant term is not captured by these considerations.

The algebra (11.24) allows the determination of *all* Fourier coefficients in terms of the ones for prime numbers a_p. We note for later reference that powers of primes can be calculated recursively using the relation

$$a_{p^{k+1}} = a_{p^k} a_p - a_{p^{k-1}} \tag{11.25}$$

for $k > 1$, where again Hecke normalisation $a_1 = 1$ enters. The Hecke operators T_p determine the full structure of the Hecke algebra and hence are the only relevant ones for the development of the theory. We will see soon that they fit naturally into an adelic framework.

Example 11.4: Fourier series of $SL(2,\mathbb{Z})$ Eisenstein series and Hecke algebra

We now use the Hecke algebra to rederive the Fourier expansion (1.17) of the non-holomorphic Eisenstein series $E(s,z)$. Since the Eisenstein series $E(s,z)$ is defined as a sum over an $SL(2,\mathbb{Z})$-orbit it is easy to evaluate the Hecke operators by multiplying the acting matrices:

$$
\begin{aligned}
(T_n E)(s,z) &= \frac{1}{2}\frac{1}{\sqrt{n}} \sum_{d|n} \sum_{b=0}^{d-1} \sum_{\gcd(p,q)=1} \left[\operatorname{Im}\left(\begin{pmatrix} n/d & b \\ 0 & d \end{pmatrix} \begin{pmatrix} * & * \\ p & q \end{pmatrix} \cdot z \right) \right]^s \\
&= \frac{1}{\sqrt{n}} \sum_{d|n} \sum_{b=0}^{d-1} \left(\frac{n}{d^2} \right)^s E(s,z) = \sum_{d|n} \left(\frac{n}{d^2} \right)^{s-1/2} E(s,z) \\
&= \underbrace{n^{s-1/2} \sigma_{1-2s}(n)}_{c_n} E(s,z) \,.
\end{aligned} \tag{11.26}
$$

We have used that the coset sum $B(\mathbb{Z})\backslash SL(2,\mathbb{Z})$ can be parametrised by two coprime integers p and q and the unspecified top row corresponds to an arbitrary representative of the coset. In particular, the Eisenstein series is an eigenfunction of all Hecke operators and the relation (11.22) between the Fourier coefficients and the Hecke eigenvalues immediately implies the form (1.17) for the non-zero Fourier coefficients up to a normalisation factor. The constant term is not fixed by these considerations. However, this method of deriving the Fourier modes did not require any Poisson resummation nor adelic technology.

Let us verify the relation (11.24) for the explicit example of the Eisenstein series (1.17) to check whether it is a simultaneous eigenfunction. There one has for $n > 0$

$$
a_n = c_n = \sum_{d|n} d^{1-2s} n^{s-1/2} \,, \tag{11.27}
$$

where Hecke normalisation was used. Let m and n be coprime, then

$$
a_m a_n = \sum_{d|m} \sum_{\tilde{d}|n} d^{1-2s} m^{s-1/2} \tilde{d}^{1-2s} n^{s-1/2} = \sum_{d|mn} d^{1-2s} (mn)^{s-1/2} = a_{mn} \,. \tag{11.28}
$$

The more general relation (11.24) can also be verified and the Eisenstein series is an eigenfunction of the Hecke operators (with eigenvalues given by the Fourier coefficients).

Remark 11.5 (Hecke operators for holomorphic modular forms) For holomorphic modular forms $f : \mathbb{H} \to \mathbb{C}$ of weight k one can also define Hecke operators; see for example [24, Chap. 6]. In this case, they act by

$$
(T_n f)(z) = n^{k-1} \sum_{d|n} d^{-k} \sum_{b=0}^{d-1} f\left(\frac{bz + bd}{d^2} \right) \qquad (f \text{ holomorphic of weight } k) \tag{11.29}
$$

and map holomorphic modular forms to homomorphic modular forms. Note that the normalisation convention here is slightly different from the non-holomorphic case. The multiplicative law (11.15) in this case reads

$$
T_m T_n = T_n T_m = \sum_{d|(m,n)} d^{k-1} T_{mn/d^2} \qquad (\text{weight } k \text{ Hecke algebra}). \tag{11.30}
$$

One can again define Hecke normalised common eigenfunctions. If f is a common eigenfunction with the Fourier expansion $f(z) = \sum_{m\geq 0} a_m q^n$ (with $q = e^{2\pi i z}$ as always), then one has again

$$T_n f = a_n f \tag{11.31}$$

when $a_1 = 1$, i.e., the modular form is Hecke normalised. In this case the Fourier coefficients satisfy

$$a_m a_n = \sum_{d|(m,n)} d^{k-1} a_{mn/d^2} \qquad (f \text{ holomorphic of weight } k) \tag{11.32}$$

because of (11.29). If f is a cusp form and its Fourier coefficients satisfy the above relation (11.32), one can show that it is automatically a common eigenfunction [24, Thm. 6.15]. For non-cuspidal forms this is not guaranteed. We record the following consequence of (11.32) for later use:

$$a_{p^\ell} a_p = a_{p^{\ell+1}} + p^{k-1} a_{p^{\ell-1}} \quad \text{(Fourier coefficients of weight } k \text{ modular form)} \tag{11.33}$$

for $\ell \geq 0$. This is to be contrasted with (11.25) for non-holomorphic forms.

11.2.4 Hecke Operators and Dirichlet Series

Given the powerful applications of Hecke operators demonstrated in the previous sections, it is natural to wonder about the action of Hecke operators in the adelic setting of automorphic forms on $SL(2,\mathbb{Q})\backslash SL(2,\mathbb{A})$. It turns out that this gives rise to an even richer structure, and provides a link to the theory of automorphic representations. In this section, we take the first steps toward such a theory by studying the Hecke operators T_p for p a prime, based on [116, 290].

Hecke's original motivation to study Hecke operators was to find a way to encode the properties of a holomorphic modular form in terms of its associated *Dirichlet series* [351, 352]. Given a weight k modular form $f(z) = \sum_{n\geq 0} a_n(f) q^n$ (with $q = e^{2\pi i z}$) one may form the series

$$L(s,f) = \sum_{n\geq 1} \frac{a_n(f)}{n^s} = \prod_{p<\infty} \sum_{\ell\geq 0} \frac{a_{p^\ell}(f)}{p^{\ell s}}, \tag{11.34}$$

which is called the Dirichlet series attached to f. The rewriting in the second step is the application of prime factorisation under the assumption of absolute convergence of the *Hecke L-function of the modular form*. Rewriting Dirichlet series $\sum_{n\geq 1} a_n/n^s$ as an *Euler product* as above is always possible when the coefficients a_n are multiplicative. In the special case when the Fourier coefficients $a_n(f)$ are *completely multiplicative*, i.e., satisfy $a_m a_n = a_{mn}$ for any $m, n \in \mathbb{Z}$, then the L-function leads to the following Euler product via geometric series:

$$L(s,f) = \prod_{p<\infty} \frac{1}{1 - a_p(f)p^{-s}}, \qquad a_m(f) \text{ completely multiplicative.} \tag{11.35}$$

This is called a *degree-one Euler product* since the denominator contains at most the power p^{-s}. The prime example of a degree-one Euler product is the Riemann zeta function $\zeta(s) = \prod_{p<\infty}(1 - p^{-s})^{-1}$ already seen in (1.22), corresponding to $a_m = 1$ for all $m \geq 1$, which, however, is not associated with a holomorphic modular form on $SL(2,\mathbb{R})$ but rather with $GL(1,\mathbb{A})$ as was explained in Section 2.7.

Hecke showed that whenever the Fourier coefficients $a_m(f)$ are multiplicative according to (11.33) then the Dirichlet series can be written as an Euler product

$$L(s,f) = \prod_{p<\infty} L_p(s,f) = \prod_{p<\infty} \frac{1}{1 - a_p(f)p^{-s} + p^{k-1-2s}}. \tag{11.36}$$

The derivation of this formula is as follows. Let $L_p(s,f)$ be a factor in the Euler product as above. Then (11.33) implies

$$\begin{aligned} L_p(s,f) = \sum_{\ell \geq 0} \frac{a_{p^\ell}}{p^{\ell s}} &= \frac{1}{a_p}\left[p^s \sum_{\ell \geq 0} \frac{a_{p^{\ell+1}}}{p^{(\ell+1)s}} + p^{k-1-s} \sum_{\ell \geq 0} \frac{a_{p^{\ell-1}}}{p^{(\ell-1)s}}\right] \\ &= \frac{1}{a_p}\left[p^s\left(L_p(s,f) - 1\right) + p^{k-1-s} L_p(s,f)\right], \end{aligned} \tag{11.37}$$

which yields (11.36) after solving for $L_p(s,f)$. The series $L(s,f)$ is also called the *Hecke L-function* of f and $L_p(s,f)$ the *local L-factor*. The L-function $L(s,f)$ in (11.36) is of degree two, due to the factor p^{-2s}.

Remark 11.6 In the case of non-holomorphic automorphic forms f one can go through the same derivation of an L-function. Using the normalisation of Hecke operators defined in (11.5) and the Fourier coefficients a_m defined in (11.21) one obtains an L-function for a common eigenfunction f in Hecke normalisation of the form

$$L(s,f) = \prod_{p<\infty} \frac{1}{1 - a_p p^{-s} + p^{-2s}}. \tag{11.38}$$

The shifted exponent on the last term in the denominator is due to the normalisation of the Hecke operators. We assume for simplicity that f is even, i.e., $f(-z) = f(z)$. Then one can define a *completed L-function* via

$$L^\star(s,f) = \pi^{-s}\Gamma\left(\frac{2s + 2\nu - 1}{4}\right)\Gamma\left(\frac{2s - 2\nu + 1}{4}\right)L(s,f), \tag{11.39}$$

where ν is the eigenvalue under the Laplacian $\Delta f = \nu(\nu - 1)f$. The completed L-function satisfies the simple functional relation

$$L^\star(s, f) = L^\star(1 - s, f). \tag{11.40}$$

For a proof of this and extensions to odd Maaß forms, see [290, Prop. 3.13.5]. One should think of the normalising factors in (11.39) as arising from the archimedean place $p = \infty$, and the completed L-function as a global one.

The L-function (11.36) attached to a modular form f therefore characterises whether or not the Fourier coefficients exhibit a multiplicative behaviour, something which is certainly not guaranteed. When does this happen? It turns out that the Fourier coefficients of a modular form f are multiplicative if and only if f is a *Hecke eigenform*, i.e., an eigenfunction of the entire ring of Hecke operators T_n [24, Thm. 6.15]. As was emphasised above, the ring of Hecke operators is generated by the T_p for p prime, and we will now focus on these.

Remark 11.7 Weil [613, 614] has resolved the problem of generalising the L-function to automorphic forms for congruence subgroups $\Gamma_0(N)$ of $SL(2,\mathbb{Z})$. In this case one needs to twist the L-function by a Dirichlet character. We will encounter these L-functions again in Chapter 16 when we discuss the modularity theorem and the Langlands correspondence.

11.3 The Spherical Hecke Algebra

Recall from Section 4.3 that for each modular form $f(z)$ of weight k on the upper half-plane $\mathbb{H}$ we have a corresponding automorphic form $\varphi_f(g)$ on $SL(2,\mathbb{Q})\backslash SL(2,\mathbb{A})$. We now want to find out how the action of the Hecke operator T_p lifts to the space of automorphic forms $\mathcal{A}(SL(2,\mathbb{Q})\backslash SL(2,\mathbb{A}))$.

As for the classical case in Section 11.2, the Hecke operators in the adelic context are associated with double cosets of matrices of determinant different from 1 and hence outside of $SL(2,\mathbb{Q}_p)$. For this reason, we consider the group $GL(2,\mathbb{Q}_p)$. The spherical Hecke algebra is a *convolution algebra* on $GL(2,\mathbb{Q}_p)$. It is given by the space of locally constant $\mathbb{C}$-valued functions on $GL(2,\mathbb{Q}_p)$ with the (commutative) product given by convolution:

$$\begin{aligned}(\Phi_1 \star \Phi_2)(g) &= \int_{GL(2,\mathbb{Q}_p)} \Phi_1(gh)\Phi_2(h^{-1})dh \\ &= \int_{GL(2,\mathbb{Q}_p)} \Phi_1(h)\Phi_2(h^{-1}g)dh,\end{aligned} \tag{11.41}$$

where dh denotes the bi-invariant Haar measure on the unimodular group $GL(2,\mathbb{Q}_p)$. Convolution turns the space of such functions into a ring, called the *(local) Hecke algebra*, commonly denoted by $\mathcal{H}(GL(2, \mathbb{Q}_p))$ or simply $\mathcal{H}_p$ for short. Although it is a ring it has no unit. To see the connection with the classical Hecke algebra generated by the T_p's, we now restrict to *bi-invariant functions* with respect to the maximal compact subgroup $K_p = GL(2, \mathbb{Z}_p)$, i.e., we consider functions in $\mathcal{H}_p$ that satisfy

$$\Phi(kgk') = \Phi(g), \qquad k, k' \in K_p,\ g \in GL(2, \mathbb{Q}_p). \tag{11.42}$$

We then obtain the *spherical Hecke algebra* $\mathcal{H}(GL(2, \mathbb{Q}_p))^{K_p}$ of K_p-bi-invariant functions, which we denote by $\mathcal{H}_p^\circ$. It is a central result that $\mathcal{H}_p^\circ$ forms a commutative ring (see, e.g., [116]). If we fix the Haar measure on $GL(2, \mathbb{Q}_p)$ such that K_p has unit volume, then $\mathcal{H}_p^\circ$ also has a unit given by the characteristic function on K_p:

$$\mathrm{char}_{K_p}(g) = \begin{cases} 1 & g \in K_p, \\ 0 & \text{otherwise}. \end{cases} \tag{11.43}$$

To see this we calculate the convolution product of the characteristic function with any $\Phi \in \mathcal{H}_p^\circ$:

$$(\Phi \star \mathrm{char}_{K_p})(g) = \int_{SL(2,\mathbb{Q}_p)} \Phi(gh)\mathrm{char}_{K_p}(h^{-1})dh = \int_{K_p} \Phi(gh)dh = \Phi(g), \tag{11.44}$$

where in the last step we have used that f is bi-invariant under K_p and K_p has unit volume. One says that the spherical Hecke algebra is *idempotented*.

The spherical Hecke algebra $\mathcal{H}_p^\circ$ acts on the space of K_p-spherical functions on $GL(2, \mathbb{Q}_p)$ via right-translation. For any $\Phi \in \mathcal{H}_p^\circ$ and K_p-spherical function φ on $GL(2, \mathbb{Q}_p)$ we define a new function on $GL(2, \mathbb{Q}_p)$ by

$$(\pi(\Phi)\varphi)(g) = \int_{GL(2,\mathbb{Q}_p)} \Phi(h)\varphi(gh)dh. \tag{11.45}$$

One can check easily that this maps the right-regular action of $GL(2, \mathbb{Q}_p)$ on functions on $GL(2, \mathbb{Q}_p)$ to a representation of the spherical Hecke algebra (with convolution product (11.41)) according to

$$\pi(\Phi_1 \star \Phi_2)\varphi = \pi(\Phi_1)(\pi(\Phi_2)\varphi). \tag{11.46}$$

The space of K_p-spherical functions is therefore a representation of the spherical Hecke algebra $\mathcal{H}_p^\circ$.

Remark 11.8 By taking the restricted direct product (with respect to K_p; see Section 2.5) over all the local algebras $\mathcal{H}_p^\circ$, we obtain the *global, or adelic,* *spherical Hecke algebra*

$$\mathcal{H}^\circ = \bigotimes_{p\leq\infty}{}' \mathcal{H}_p^\circ. \tag{11.47}$$

For $p = \infty$, the spherical Hecke algebra $\mathcal{H}_\infty^\circ$ is given by $K(\mathbb{R})$-bi-finite distributions supported on $K(\mathbb{R})$ [140, Lecture 3.1]. This includes the invariant differential operators on $G(\mathbb{R})$ lying in the universal enveloping algebra $U(\mathfrak{g})$. The global Hecke algebra $\mathcal{H}^\circ$ acts on $\mathcal{A}(GL(2,\mathbb{Q})\backslash GL(2,\mathbb{A}))$ by the same formula (11.45). Our main interest in the following lies with the spherical Hecke algebra $\mathcal{H}_p^\circ$ at the finite primes $p < \infty$.

We now investigate the structure of the (local) spherical Hecke algebra $\mathcal{H}_p^\circ$ in more detail. More explicitly, we define the elements $\mathbb{T}_p$ and $\mathbb{R}_p \in \mathcal{H}_p^\circ$ by the K_p-bi-invariant functions

$$\mathbb{T}_p = \mathrm{char}_{K_p\left(\begin{smallmatrix} p & \\ & 1\end{smallmatrix}\right)K_p}, \qquad \mathbb{R}_p = \mathrm{char}_{K_p\left(\begin{smallmatrix} p & \\ & p\end{smallmatrix}\right)K_p}. \tag{11.48}$$

It is an important result that $\mathbb{T}_p$, $\mathbb{R}_p$ and $\mathbb{R}_p^{-1}$ together generate the spherical Hecke algebra $\mathcal{H}_p^\circ$. A proof of this statement can be found for example in [116, Prop. 4.6.5].

On functions $\varphi\colon GL(2,\mathbb{Q}_p) \to \mathbb{C}$ they act according to (11.45). To ease notation we shall simply continue to call them $\mathbb{T}_p$ and $\mathbb{R}_p$ also when acting on spherical functions:

$$(\mathbb{T}_p\varphi)(g) = \int_{K_p\left(\begin{smallmatrix} p & \\ & 1\end{smallmatrix}\right)K_p} \varphi(gh)dh, \tag{11.49}$$

$$(\mathbb{R}_p\varphi)(g) = \int_{K_p\left(\begin{smallmatrix} p & \\ & p\end{smallmatrix}\right)K_p} \varphi(gh)dh. \tag{11.50}$$

Even though written in terms of integrals, they act on functions on $GL(2,\mathbb{Q}_p)$ by finite sums after performing a decomposition of the double cosets into a finite union of left cosets, similar to (11.4). This decomposition for the operator $\mathbb{T}_p$ is [116, Prop. 4.6.4]

$$K_p\begin{pmatrix} p & \\ & 1\end{pmatrix}K_p = \begin{pmatrix} 1 & \\ & p\end{pmatrix}K_p \cup \bigcup_{i=0}^{p-1}\begin{pmatrix} p & i \\ & 1\end{pmatrix}K_p, \tag{11.51}$$

such that for K_p-spherical φ

$$(\mathbb{T}_p\varphi)(g) = \varphi\left(g\left(\begin{smallmatrix} 1 & \\ & p\end{smallmatrix}\right)\right) + \sum_{i=0}^{p-1}\varphi\left(g\left(\begin{smallmatrix} p & i \\ & 1\end{smallmatrix}\right)\right). \tag{11.52}$$

The connection with the classical Hecke operators now follows from the fact that if $f\colon \mathbb{H} \to \mathbb{R}$ is a Maaß wave form with eigenvalue a_p under T_p, then the associated adelic automorphic form $\varphi_f \in \mathcal{A}(SL(2,\mathbb{Q})\backslash SL(2,\mathbb{A}))$ defined in Section 4.3.2 is an eigenform under $\mathbb{T}_p$ with the same eigenvalue, up to a (convention-dependent) factor:

$$T_p f = a_p f \quad \longleftrightarrow \quad (\mathbb{T}_p \varphi_f)(g) = p^{1/2} a_p \varphi_f. \tag{11.53}$$

This will be verified for Eisenstein series in Example 11.10 below but it is valid in general.

Remark 11.9 There is also a classical Hecke operator R_p acting on f which lifts to $\mathbb{R}_p$, but we shall not discuss this further here (see [116] for more details).

11.4 Hecke Algebras and Automorphic Representations

This and the following sections make use of the theory of automorphic representations which was introduced in Chapter 5.

Recall that $GL(2,\mathbb{A})$ acts by right-translation on $\mathcal{A}(GL(2,\mathbb{Q})\backslash GL(2,\mathbb{A}))$, such that at the archimedean places it has the form of a $(\mathfrak{g}_\infty, K_\infty)$-module, while the finite places carry a representation of $GL(2,\mathbb{A}_f)$. The irreducible constituents (π, V) in the decomposition of $\mathcal{A}(GL(2,\mathbb{Q})\backslash GL(2,\mathbb{A}))$ are called automorphic representations. But we have also just seen that $\mathcal{A}(SL(2,\mathbb{Q})\backslash SL(2,\mathbb{A}))$ carries an action of the adelic spherical Hecke algebra $\mathcal{H}^\circ$. A natural question is then: is there a relation between these representations? Not surprisingly, the answer is yes, and we shall now sketch how to see this.

Suppose that $(\pi, V) = \otimes_{p\leq\infty}(\pi_p, V_p)$ is an unramified automorphic representation (see Definition 5.14); this implies that V_p contains a spherical vector f_p° (unique up to multiplication by a complex scalar; see Remark 5.17), satisfying $\mathsf{f}_p^\circ(k) = 1$ for all $k \in K_p$. The spherical vector therefore spans the complex one-dimensional space $V_p^{K_p}$ consisting of K_p-invariant vectors in V_p.

We can for example take π_p to be the local induced representation with module

$$V_p = \mathrm{Ind}_{B(\mathbb{Q}_p)}^{GL(2,\mathbb{Q}_p)} \delta^{1/2}\mu, \tag{11.54}$$

where δ is the modulus character of the Borel subgroup and the quasi-character $\mu : B(\mathbb{Z}_p)\backslash B(\mathbb{Q}_p) \to \mathbb{C}^\times$ is defined by

$$\mu(b) = \mu(na) = \mu(a), \qquad n \in N(\mathbb{Q}_p), \quad a \in A(\mathbb{Q}_p). \tag{11.55}$$

In the notation of (4.128) we have therefore $\chi(g) = \mu(g)\delta^{1/2}(g)$. The explicit separation of the modulus character in (11.54) turns out to be convenient for the forthcoming analysis, and also facilitates comparison with the literature. In the notation of that section we would write $\chi(a) = a^{\lambda+\rho}$, so that $\delta^{1/2}(a) = a^\rho$

and $\mu(a) = a^\lambda$, where ρ is the Weyl vector of the Lie algebra $\mathfrak{sl}(2)$ and λ is a (complex) weight.

The spherical vector $f_p^\circ \in V_p$ is the standard section defined by the extension of $\delta^{1/2}\mu$ to all of $GL(2,\mathbb{Q}_p)$ via the Iwasawa decomposition (see also Section 5.6):

$$f_p^\circ(g) = f_p^\circ(nak) = \delta^{1/2}(a)\mu(a). \tag{11.56}$$

The local spherical Hecke algebra $\mathcal{H}_p^\circ$ acts on V_p via the action (11.45). By construction this action preserves the one-dimensional space $V_p^{K_p}$ of spherical vectors: indeed for any $\Phi \in \mathcal{H}_p^\circ$ we have for all $k \in K_p$

$$\begin{aligned}(\pi(\Phi)f_p^\circ)(gk) &= \int_{GL(2,\mathbb{Q}_p)} f_p^\circ(gkh)\Phi(h)dh = \int_{GL(2,\mathbb{Q}_p)} f_p^\circ(ghk)\Phi(khk^{-1})dh,\\ &= (\pi(\Phi)f_p^\circ)(g),\end{aligned} \tag{11.57}$$

since f_p° is spherical and Φ K_p-bi-invariant. This implies that $V_p^{K_p}$ furnishes a representation of $\mathcal{H}_p^\circ$. Since the spherical vector f_p° spans the one-dimensional space $V_p^{K_p}$, we conclude that the action of $\Phi \in \mathcal{H}_p^\circ$ must give back f_p°, up to a complex scalar:

$$(\pi(\Phi)f_p^\circ)(g) = \lambda_\mu(\Phi)f_p^\circ(g), \tag{11.58}$$

where the eigenvalue $\lambda_\mu(\Phi)$ determines a (quasi-)character of the spherical Hecke algebra

$$\lambda_\mu : \mathcal{H}_p^\circ \longrightarrow \mathbb{C}^\times. \tag{11.59}$$

As we have indicated, this character depends on the choice of μ in (11.54).

To find an explicit description of the characters λ_μ we shall work out the action of the Hecke operator $\mathbb{T}_p$ defined in (11.48). To proceed, we parametrise the Cartan torus $A(\mathbb{Q}_p) \subset GL(2,\mathbb{Q}_p)$ by

$$a = \begin{pmatrix} v_1 & \\ & v_2 \end{pmatrix}, \qquad v_1, v_2 \in \mathbb{Q}_p^\times. \tag{11.60}$$

We can further describe the unramified character μ explicitly by

$$\mu(a) = \mu\left(\begin{pmatrix} v_1 & \\ & v_2 \end{pmatrix}\right), = |v_1|_p^{s_1}|v_2|_p^{s_2}, \tag{11.61}$$

where $s_1, s_2 \in \mathbb{C}$. Note that the parametrisation in terms of s_1 and s_2 differs from the one used in Chapter 7. The reason here is to simplify some of the following expressions. The corresponding value of the modulus character on

$B \subset GL(2, \mathbb{Q}_p)$ is

$$\delta\begin{pmatrix} v_1 & \\ & v_2 \end{pmatrix} = \left|\frac{v_1}{v_2}\right|_p. \tag{11.62}$$

This implies that the representation $\mathrm{Ind}_{B(\mathbb{Q}_p)}^{SL(2,\mathbb{Q}_p)} \delta^{1/2}\mu$ is in fact completely determined by

$$\alpha_p \equiv p^{-s_1}, \qquad \beta_p \equiv p^{-s_2}. \tag{11.63}$$

Now we wish to compute

$$(\mathbb{T}_p \mathrm{f}_p^\circ)(g) = \int_{K_p \begin{pmatrix} p & \\ & 1 \end{pmatrix} K_p} \mathrm{f}_p^\circ(gh)dh. \tag{11.64}$$

Since we know that f_p° is an eigenfunction and is normalised so that $\mathrm{f}_p^\circ(1) = 1$, it suffices to evaluate the action at the identity $1 \in GL(2, \mathbb{Q}_p)$, which then, via (11.58), directly corresponds to the value of the character λ_μ:

$$\lambda_\mu(\mathbb{T}_p) = \int_{K_p \begin{pmatrix} p & \\ & 1 \end{pmatrix} K_p} \mathrm{f}_p^\circ(h)dh. \tag{11.65}$$

To evaluate this we decompose the double coset space as in (11.51). Plugging the decomposition into the integral (11.65) yields

$$\begin{aligned} \lambda_\mu(\mathbb{T}_p) &= \sum_{i=0}^{p-1} \mathrm{f}_p^\circ\left(\begin{pmatrix} p & i \\ & 1 \end{pmatrix}\right) + \mathrm{f}_p^\circ\left(\begin{pmatrix} 1 & \\ & p \end{pmatrix}\right) \\ &= \underbrace{(\delta^{1/2}\mu)\left(\begin{pmatrix} p & \\ & 1 \end{pmatrix}\right) + \cdots + (\delta^{1/2}\mu)\left(\begin{pmatrix} p & \\ & 1 \end{pmatrix}\right)}_{p \text{ terms}} + (\delta^{1/2}\mu)\left(\begin{pmatrix} 1 & \\ & p \end{pmatrix}\right) \\ &= pp^{-1/2}p^{-s_1} + p^{1/2}p^{-s_2} \\ &= p^{1/2}(\alpha_p + \beta_p). \end{aligned} \tag{11.66}$$

By a similar analysis one also shows that

$$\lambda_\mu(\mathbb{R}_p) = \int_{K_p \begin{pmatrix} p & \\ & p \end{pmatrix} K_p} \mathrm{f}_p^\circ(h)dh = \alpha_p\beta_p, \tag{11.67}$$

$$\lambda_\mu(\mathbb{R}_p^{-1}) = \int_{K_p \begin{pmatrix} p^{-1} & \\ & p^{-1} \end{pmatrix} K_p} \mathrm{f}_p^\circ(h)dh = \alpha_p^{-1}\beta_p^{-1}. \tag{11.68}$$

These results imply that the one-dimensional representation λ_μ of the Hecke algebra $\mathcal{H}_p^\circ$ acting on $V_p^{K_p}$ completely determines the unramified character μ, and thereby the automorphic representation $\mathrm{Ind}_{B(\mathbb{Q}_p)}^{GL(2,\mathbb{Q}_p)} \delta^{1/2}\mu$. It is quite remarkable that this infinite-dimensional automorphic representation can be encoded in the finite-dimensional representations of $\mathcal{H}_p^\circ$. In the next section we shall further investigate the consequences of this fact.

Example 11.10: Classical and p-adic Hecke operators for $SL(2)$

In this example, we come back to the mentioned relation (11.53) between the Hecke eigenvalue of a non-holomorphic Eisenstein series $E(s,z)$ on $\mathbb{H}$ under the classical T_p calculated in (11.26) and the action of $\mathbb{T}_p \in \mathcal{H}_p^\circ$ on the associated adelic Eisenstein series $E(\chi_s, g) \in \mathcal{A}(SL(2,\mathbb{Q})\backslash SL(2,\mathbb{A}))$ defined in (5.43).

To begin with, we need to relate the parameters s_1 and s_2 of the $GL(2,\mathbb{Q}_p)$ principal series (11.61) to the parameter s occuring in $E(s,z)$ and $E(\chi_s, g)$ via (7.10). Elements of the Cartan torus in $SL(2,\mathbb{Q}_p)$ are of the form $a = \operatorname{diag}(v, v^{-1})$, so that (11.61) yields

$$\mu\left(\begin{pmatrix} v & \\ & v^{-1} \end{pmatrix}\right) = |v|_p^{s_1 - s_2}. \tag{11.69}$$

This has to be contrasted with

$$(\delta^{-1/2}\chi_s)\left(\begin{pmatrix} v & \\ & v^{-1} \end{pmatrix}\right) = |v|_p^{2s-1} \tag{11.70}$$

that follows from (7.10) (recall the general $\chi = \delta^{1/2}\mu$). For symmetry reasons one therefore deduces

$$s_1 = -s_2 = s - \frac{1}{2}. \tag{11.71}$$

Plugging this into α_p and β_p in (11.63) one therefore finds from (11.66) that

$$\lambda_\mu(\mathbb{T}_p) = p^{1/2}\left(\alpha_p + \beta_p\right) = p^{1/2}\left(p^{-s+1/2} + p^{s-1/2}\right). \tag{11.72}$$

This is the eigenvalue of the adelic Eisenstein series under $\mathbb{T}_p$. From (11.26) one finds that for the classical Hecke operator T_p acting on the classical $E(s,z)$ the eigenvalue is

$$(T_p E)(s,z) = p^{s-1/2}\left(1 + p^{1-2s}\right)E(s,z) = \left(p^{s-1/2} + p^{-s+1/2}\right)E(s,z). \tag{11.73}$$

This confirms the claimed relation (11.53) that

$$\mathbb{T}_p \sim p^{1/2} T_p, \tag{11.74}$$

where we reiterate that the prefactor is convention dependent.

11.5 The Satake Isomorphism

We recall from Section 3.1.1 that the Weyl group $\mathcal{W} = \mathcal{W}(\mathfrak{g})$ acts on the Cartan torus A, and consequently it also acts on the characters μ via

$$w\mu(a) = \mu(w^{-1}aw), \qquad w \in \mathcal{W}. \tag{11.75}$$

Under this action the unramified principal series remains invariant:

$$\mathrm{Ind}_{B(\mathbb{Q}_p)}^{GL(2,\mathbb{Q}_p)}\delta^{1/2}w(\mu) \cong \mathrm{Ind}_{B(\mathbb{Q}_p)}^{GL(2,\mathbb{Q}_p)}\delta^{1/2}\mu. \tag{11.76}$$

This is what the functional relation (8.44) for Eisenstein series expresses; see also Remark 8.8 discussing the associated *intertwining operators*.

In terms of the parametrisation of μ by the complex numbers (α_p, β_p) the Weyl group $\mathcal{W} = \mathbb{Z}/2\mathbb{Z}$ simply acts by $(\alpha_p, \beta_p) \mapsto (\beta_p, \alpha_p)$. Now notice that the characters $\lambda_\mu(\mathbb{T}_p) = p^{1/2}(\alpha_p + \beta_p)$, $\lambda_\mu(\mathbb{R}_p) = \alpha_p\beta_p$ and $\lambda_\mu(\mathbb{R}_p^{-1}) = \alpha_p^{-1}\beta_p^{-1}$ are Weyl invariant. Hence, at the level of the representations of the spherical Hecke algebra we have

$$\lambda_{w\mu} = \lambda_\mu, \qquad \forall\, w \in \mathcal{W}. \tag{11.77}$$

As a consequence, the image of the homomorphism $\mathcal{H}_p^\circ \to \mathbb{C}$ lies in the polynomial $\mathbb{C}$-ring of Weyl invariants

$$\mathbb{C}[\alpha_p^{\pm 1}, \beta_p^{\pm 1}]^{\mathcal{W}} \cong \mathbb{C}[\alpha_p + \beta_p, \alpha_p\beta_p, \alpha_p^{-1}\beta_p^{-1}]. \tag{11.78}$$

It is an important result of Satake [544] that this homomorphism in fact yields an isomorphism between the spherical Hecke algebra and the ring of Weyl invariant polynomials in (α_p, β_p):

$$\mathcal{H}_p^\circ \cong \mathbb{C}[\alpha_p + \beta_p, \alpha_p\beta_p, \alpha_p^{-1}\beta_p^{-1}]. \tag{11.79}$$

See [325] for a nice survey of the *Satake isomorphism* and its applications.

The key step in Satake's analysis was to introduce the *Satake transform*

$$\mathcal{S} : \mathcal{H}_p^\circ(GL(2, \mathbb{Q}_p)) \longrightarrow \mathcal{H}_p^\circ(A(\mathbb{Q}_p)) \tag{11.80}$$

from the spherical Hecke algebra of $GL(2, \mathbb{Q}_p)$ to the spherical Hecke algebra of the Cartan torus $A(\mathbb{Q}_p)$. The Satake transform is defined by

$$(\mathcal{S}\Phi)(a) = \delta^{-1/2}(a)\int_{N(\mathbb{Q}_p)} \Phi(na)dn, \qquad \Phi \in \mathcal{H}_p^\circ(GL(2, \mathbb{Q}_p)), \tag{11.81}$$

where $N(\mathbb{Q}_p)$ is the unipotent radical of the Borel subgroup $B(\mathbb{Q}_p) \subset GL(2, \mathbb{Q}_p)$. Satake then proved that the image of $\mathcal{S}$ lies in $\mathcal{H}_p^\circ(A(\mathbb{Q}_p))^{\mathcal{W}}$, the Weyl invariant elements of the spherical Hecke algebra of $A(\mathbb{Q}_p)$. To see the connection with our previous analysis, we consider again the formula for the eigenvalues λ_μ:

$$\lambda_\mu(\Phi) = \int_{GL(2,\mathbb{Q}_p)} \mathrm{f}_p^\circ(h)\Phi(h)dh, \tag{11.82}$$

which is (11.58) evaluated at the identity $g = 1$. We shall now manipulate this expression in order to elucidate the rôle played by the Satake transform. To the best of our knowledge, this calculation was first outlined by Langlands in [445], but here we follow the more detailed exposition by Garrett [254]. We begin by splitting the integral according to the Iwasawa decomposition $GL(2,\mathbb{Q}_p) = B(\mathbb{Q}_p)K_p$:

$$\int_{GL(2,\mathbb{Q}_p)} \mathfrak{f}_p^\circ(h)\Phi(h)dh = \int_{B(\mathbb{Q}_p)}\int_{K_p} \mathfrak{f}_p^\circ(b^{-1}k)\Phi(b^{-1}k)dbdk, \tag{11.83}$$

where dk and db are *right*-invariant Haar measures on K_p and $B(\mathbb{Q}_p)$, respectively. Next, we make the change of variables $b \to b^{-1}$, which brings out a factor of δ^{-1} from the measure:

$$\int_{B(\mathbb{Q}_p)}\int_{K_p} \mathfrak{f}_p^\circ(bk)\Phi(bk)\delta(b)^{-1}dbdk. \tag{11.84}$$

Using right K_p-invariance of $\mathfrak{f}_p^\circ$ and Φ as well as $\int_{K_p} dk = 1$ this further simplifies to

$$\int_{B(\mathbb{Q}_p)} \mathfrak{f}_p^\circ(b)\Phi(b)\delta(b)^{-1}db = \int_{B(\mathbb{Q}_p)} \Phi(b)(\delta^{-1/2}\mu)(b)db, \tag{11.85}$$

where we have used (11.56). To proceed we split the integral according to $B(\mathbb{Q}_p) = N(\mathbb{Q}_p)A(\mathbb{Q}_p)$ and use the fact that $\delta^{-1/2}\mu$ is trivial on $N(\mathbb{Q}_p)$ acting on the left:

$$\int\limits_{A(\mathbb{Q}_p)}\int\limits_{N(\mathbb{Q}_p)} \Phi(na)(\delta^{-1/2}\mu)(na)dnda = \int\limits_{A(\mathbb{Q}_p)}\int\limits_{N(\mathbb{Q}_p)} \Phi(na)(\delta^{-1/2}\mu)(a)dnda. \tag{11.86}$$

After reshuffling the integrand we finally arrive at the result

$$\begin{aligned}\lambda_\mu(\Phi) &= \int_{A(\mathbb{Q}_p)} \mu(a)\left[\delta^{-1/2}(a)\int_{N(\mathbb{Q}_p)} \Phi(na)dn\right]da \\ &= \int_{A(\mathbb{Q}_p)} \mu(a)(\mathcal{S}\Phi)(a)da.\end{aligned} \tag{11.87}$$

This clearly shows that the Satake transform lies at the heart of the relation between the unramified automorphic representation $\mathrm{Ind}_{B(\mathbb{Q}_p)}^{GL(2,\mathbb{Q}_p)}\delta^{1/2}\mu$ and the one-dimensional representation λ_μ of the spherical Hecke algebra $\mathcal{H}_p^\circ$, the essence of which is the Satake isomorphism (11.79).

11.6 The L-group and Generalisation to $GL(n)$

It is illuminating to assemble the parameters (α_p, β_p) in a matrix

$$A_{\pi_p} = \begin{pmatrix} \alpha_p & \\ & \beta_p \end{pmatrix}. \tag{11.88}$$

This matrix belongs to $GL(2,\mathbb{C})$ and since conjugation $A_{\pi_p} \mapsto wA_{\pi_p}w^{-1}$ by an element $w \in \mathcal{W}$ will not alter the result (11.79) we find that the representation π_p determines a (semi-simple) *conjugacy class* $[A_{\pi_p}] \subset GL(2,\mathbb{C})$. This conjugacy class is called the *Satake parameter* of the local representation π_p.

The conclusion of the discussion in this and the previous sections is that *unramified automorphic representations* π_p *of* $GL(2,\mathbb{Q}_p)$ *are in bijection with semi-simple conjugacy classes* $[A_{\pi_p}] \subset GL(2,\mathbb{C})$. The appearance of $GL(2,\mathbb{C})$ in the context of local representations of $GL(2,\mathbb{Q}_p)$ may seem surprising, but is in fact a simple instance of a more general phenomenon envisioned by Langlands [446]. Langlands suggested that to each reductive algebraic group G over a number field $\mathbb{F}$ there exists an associated complex group ${}^LG(\mathbb{C})$, called the *L-group*, or *Langlands dual group*. We have already encountered the group ${}^LG$ briefly in Section 9.6 in our discussion of the Casselman–Shalika formula but we will now put this group into a more general context.

A precise definition of ${}^LG$ can be found in [71]; here we only recall the salient features. For simple groups G the root system of ${}^LG$ is obtained from that of G by interchanging the short and long roots. In other words, the co-weight lattice $\Lambda^\vee$ of the Lie algebra $\mathfrak{g} = \operatorname{Lie} G$ is identified with the weight lattice ${}^L\Lambda$ of the dual Lie algebra ${}^L\mathfrak{g} = \operatorname{Lie} {}^LG$. This is captured by the isomorphism

$$\operatorname{Hom}({}^LA, U(1)) \cong \operatorname{Hom}(U(1), A), \tag{11.89}$$

between the lattice of characters on ${}^LA$ and the lattice of co-characters on A. For example, in the case of $G = GL(n,\mathbb{Q}_p)$ the L-group is $GL(n,\mathbb{C})$, and for $G = SL(n,\mathbb{Q}_p)$ we have ${}^LG = PGL(n,\mathbb{C})$. The duality is even more drastic in the case when $G = Sp(n)$, where we have ${}^LG = SO(2n+1)$. See also [115, 129] for details.

Remark 11.11 The group ${}^LG$ we have introduced here is sometimes called the *connected L-group* in order to distinguish it from the L-group in the more general context of field extensions. If one considers a finite field extension $\mathbb{E}$ of $\mathbb{F} = \mathbb{Q}_p$ then the L-group ${}^LG$ is defined with the inclusion of the (finite) Galois group $\operatorname{Gal}(\mathbb{E}/\mathbb{F})$ of the field extension. This more general viewpoint is relevant for the global Langlands conjectures and will be discussed in Section 16.1.

The Satake parameter A_{π_p} associated with the automorphic representation π_p should thus be viewed as an element of the Cartan torus ${}^LA(\mathbb{C}) \subset GL(2,\mathbb{C})$ dual to the original Cartan torus $A(\mathbb{Q}_p)$. In fact, this holds more generally for any (split) reductive algebraic group G. From this perspective, one gets the following reformulation of the Satake isomorphism (adapted from [129]):

Theorem 11.12 (Reformulated Satake isomorphism) *There is a natural bijection between the Weyl invariant homomorphism $\mathcal{H}_p^\circ(G) \to \mathbb{C}$ and semi simple ${}^LG(\mathbb{C})$-conjugacy classes in the dual torus ${}^LA(\mathbb{C})$.*

Remark 11.13 The Satake parameter $A_{\pi_p} \in {}^LA(\mathbb{C})$ already appeared in Section 9.6 where it was denoted by a_λ, where λ parametrises an element of the principal series representation of $G(\mathbb{Q}_p)$ which is here denoted abstractly by π_p.

Let us briefly discuss some details regarding the generalisation of our analysis to $G = GL(n, \mathbb{Q}_p)$. We take π_p to be the unramified principal series with module

$$V_p = \mathrm{Ind}_{B(\mathbb{Q}_p)}^{GL(n,\mathbb{Q}_p)} \delta^{1/2}\mu, \tag{11.90}$$

where the inducing character is a straightforward generalisation of (11.61):

$$\mu(a) = \mu\begin{pmatrix} v_1 & & \\ & \ddots & \\ & & v_n \end{pmatrix} = \prod_{i=1}^{n} |v_i|_p^{s_i}. \tag{11.91}$$

As before, this representation is determined by the n complex numbers

$$\alpha_i := p^{-s_i}, \qquad i = 1, \ldots, n. \tag{11.92}$$

(Note that the α_i here are for fixed prime p, which we do not indicate explicitly unlike in (11.88).) Associated with the representation π_p we then have the Satake parameter

$$A_{\pi_p} = \begin{pmatrix} \alpha_1 & & \\ & \ddots & \\ & & \alpha_n \end{pmatrix} \in {}^LA(\mathbb{C}) \subset GL(n, \mathbb{C}) = {}^LGL(n, \mathbb{Q}_p), \tag{11.93}$$

on which the Weyl group $\mathcal{W}$ acts by permuting the α_i's. The generators of the spherical Hecke algebra $\mathcal{H}_p^\circ(G)$ act on elements $\varphi \in V_p$ by (11.45), i.e.,

$$\begin{aligned}(\Phi_i\varphi)(g) &= \int_{GL(n,\mathbb{Q}_p)} \varphi(gh)\mathrm{char}_{K_p\tau_iK_p}(h)dh \\ &= \int_{K_p\tau_iK_p} \varphi(gh)dh,\end{aligned} \tag{11.94}$$

where we have defined [115]

$$\tau_i = \begin{pmatrix} p\mathbb{1}_i & \\ & \mathbb{1}_{n-i} \end{pmatrix}, \tag{11.95}$$

with $\mathbb{1}_r$ the $r \times r$ identity matrix. We use the convention that for $i = n$ the double coset is $K_p(p\mathbb{1}_n)K_p$. Thus, in the special case of $n = 2$ the definition

(11.94) reduces to the generators in Section 11.2.4, i.e., $\Phi_1 = \mathbb{T}_p$ and $\Phi_2 = \mathbb{R}_p$. Tamagawa has shown [593] that the operators $\Phi_1, \ldots, \Phi_n$ together with Φ_n^{-1} (which is the only invertible Φ_i) generate the spherical Hecke algebra $\mathcal{H}_p^\circ$ of $GL(n, \mathbb{Q}_p)$.

As before, the one-dimensional space $V_p^{K_p} = \mathbb{C} \cdot \mathfrak{f}_p^\circ$ of K_p-invariant vectors in V_p furnishes a representation of the spherical Hecke algebra, such that for any $\Phi \in \mathcal{H}_p^\circ$ and any $v^\circ \in V_p^{K_p}$ one has

$$\pi(\Phi)v^\circ = \lambda_\mu(\Phi)v^\circ, \tag{11.96}$$

where $\lambda_\mu : \mathcal{H}_p^\circ \to \mathbb{C}^\times$ is a (quasi-)character. To evaluate the eigenvalue $\lambda_\mu(\Phi)$ on all the generators Φ_i we must decompose the double cosets in (11.94). The result can be written as follows using the finite *Cartan decomposition* (see for instance [115] for a nice proof):

$$K_p \tau_i K_p = \bigcup_j \beta_{i,j} K_p, \tag{11.97}$$

where the matrices $\beta_{,j}$ are all integral and upper-triangular with diagonal entries of the form p^η, where $\eta \in \{1, \ldots, n\}$ and j ranges over some finite set. These generalise the matrices on the first line of (11.66) and, similarly to that calculation, we must evaluate the spherical vector $\mathfrak{f}_p^\circ$ on all $\beta_{i,j}$. Bump shows that this takes the form [115]

$$\mathfrak{f}_p^\circ(\beta_{i,j}) = (\delta^{1/2}\mu)(\beta_{i,j}) = p^{-\frac{i(n+1)}{2}} \prod_{\ell=1}^{i} p^{\eta_\ell} \alpha_{\eta_\ell}, \tag{11.98}$$

where $\eta_\ell \in \{1, \ldots, n\}$ are determined by j and ordered such that $\eta_1 < \eta_2 < \cdots < \eta_i$. The α_{η_ℓ} are the complex parameters (11.92) that determine the representation π_p. There are furthermore a total of

$$p^{i(n-i-1/2) - \sum_{\ell=1}^{i} \eta_\ell} \tag{11.99}$$

$\beta_{i,j}$'s for each $i \in \{1, \ldots, n\}$. Combining everything, we find that the eigenvalue of the Hecke operator Φ_i is given by

$$\begin{aligned}
\lambda_\mu(\Phi_i) &= \int_{K_p \tau_i K_p} \mathfrak{f}_p^\circ(h) dh = \sum_{\beta_i \in \Lambda_i} \mathfrak{f}_p^\circ(\beta_i) \\
&= \sum_{\eta_1 < \cdots < \eta_i} p^{i(n-i-1/2) - \sum_{j=1}^{i} \eta_j} p^{-\frac{i(n+1)}{2}} \prod_{\ell=1}^{i} p^{\eta_\ell} \alpha_{\eta_\ell} \\
&= p^{i(n-i)/2} \sum_{\eta_1 < \cdots < \eta_i} \alpha_{\eta_1} \cdots \alpha_{\eta_n} \\
&= p^{i(n-i)/2} e_i(\alpha_1, \ldots, \alpha_n),
\end{aligned} \tag{11.100}$$

where $e_i(\alpha_1, \dots, \alpha_n)$ is the ith elementary symmetric polynomial in n variables. In fact, the Satake isomorphism can be written in terms of these elementary symmetric polynomials

$$\mathcal{H}_p^\circ(GL(n, \mathbb{Q}_p)) \cong \mathbb{C}[e_1(\alpha_1, \dots, \alpha_n), \dots, e_n(\alpha_1, \dots, \alpha_n), e_n(\alpha_1, \dots, \alpha_n)^{-1}], \tag{11.101}$$

corresponding to the values on the generators $\Phi_1, \dots, \Phi_n$ and Φ_n^{-1} of the spherical Hecke algebra of $GL(n, \mathbb{Q}_p)$. Indeed, for $n = 2$ we have

$$e_1(\alpha_1, \alpha_2) = \alpha_1 + \alpha_2, \qquad e_2(\alpha_1, \alpha_2) = \alpha_1\alpha_2, \tag{11.102}$$

thus recovering (11.79).

Let us end this section with a comment on how these results fit into the general theory of automorphic forms. Recall from Definition 4.13 that an automorphic form φ on the adelic quotient $G(\mathbb{Q})\backslash G(\mathbb{A})$ is required to be $\mathcal{Z}(\mathfrak{g})$-finite, i.e., that φ is an eigenfunction of the entire ring of invariant differential operators in the centre of $U(\mathfrak{g})$. This can be viewed as a statement about the behaviour of φ under the action of differential operators in the *real* group $G_\infty = G(\mathbb{R})$. For the case of automorphic forms attached to unramified automorphic representations $\pi = \pi_\infty \otimes \bigotimes_{p<\infty} \pi_p$, the spherical Hecke algebra provides the non-archimedean analogue of this: for each finite place p, φ is an eigenfunction of the ring of Hecke operators generated by $\Phi_i \in \mathcal{H}_p^\circ$. These statements combine together in the global Hecke algebra as mentioned in Remark 11.8.

11.7 The Casselman–Shalika Formula Revisited

There is a close relation between the discussion above and the Casselman–Shalika formula for the p-adic spherical Whittaker function W_ψ°. Spherical Whittaker functions were the central objects in Chapter 9 and here we recall and extend some of the notions in a more general context. According to Definition 6.21, for an unramified character $\psi\colon N(\mathbb{Z}_p)\backslash N(\mathbb{Q}_p) \to U(1)$ we have an embedding

$$Wh_\psi\colon \mathrm{Ind}_{B(\mathbb{Q}_p)}^{GL(n,\mathbb{Q}_p)} \delta^{1/2}\mu \longrightarrow \mathrm{Ind}_{N(\mathbb{Q}_p)}^{GL(n,\mathbb{Q}_p)} \psi \tag{11.103}$$

of the unramified principal series into the space of functions $W_\psi\colon GL(n, \mathbb{Q}_p) \to \mathbb{C}$ satisfying

$$W_\psi(\delta^{1/2}\mu, ng) = \psi(n) W_\psi(\delta^{1/2}\mu, g), \qquad \forall n \in N(\mathbb{Q}_p), \tag{11.104}$$

where, as in Chapter 9, the first argument indicates the dependence on the inducing character μ in the unramified principal series that was written there in terms of $\chi = \delta^{1/2}\mu$. The image of the space $V_p^{K_p}$ of K_p-fixed vectors in V_p is a one-dimensional space of *spherical Whittaker functions*. In particular, for the generator $\mathsf{f}_p^\circ \in V_p^{K_p}$ we obtain a canonical spherical Whittaker function via the explicit *Jacquet integral* (see Chapter 9 for details):

$$W_\psi^\circ(\delta^{1/2}\mu, g) = \int_{N(\mathbb{Q}_p)} \mathsf{f}_p^\circ(w_0 n g)\overline{\psi(n)}dn, \tag{11.105}$$

where we have used $\mathsf{f}_p^\circ = \delta^{1/2}\mu$ and that ψ is unramified. This satisfies

$$W_\psi^\circ(\delta^{1/2}\mu, nak) = \psi(n)W_\psi^\circ(\delta^{1/2}\mu, a), \tag{11.106}$$

and so is completely determined by its restriction to the Cartan torus $A(\mathbb{Q}_p)$. For $GL(n, \mathbb{Q}_p)$ the vanishing properties of W_ψ° analysed in Section 9.2.2 can be simplified as follows. Parametrising a according to

$$a = \varpi^J := \begin{pmatrix} p^{j_1} & & \\ & \ddots & \\ & & p^{j_n} \end{pmatrix} \in A(\mathbb{Q}_p)/A(\mathbb{Z}_p), \tag{11.107}$$

with $J = (j_1, \dots, j_n) \in \mathbb{Z}^n$, one finds that (see, e.g., [140])

$$W_\psi^\circ(\delta^{1/2}\mu, a) = 0, \qquad \text{unless} \quad j_1 \geq j_2 \geq \cdots \geq j_n\,. \tag{11.108}$$

The map (11.103) commutes with the Hecke action and therefore the spherical Whittaker function is an eigenfunction of all the Hecke operators with the same eigenvalue (11.100) as before:

$$\Phi_i W_\psi^\circ(\delta^{1/2}\mu, a) = \lambda_\mu(\Phi_i) W_\psi^\circ(\delta^{1/2}\mu, a). \tag{11.109}$$

This fact can be used to derive a recursive formula for the value $W_\psi^\circ(\delta^{1/2}\mu, a)$, as we will now show. This will give the connection with the Casselman–Shalika formula that we are after.

The main difference with the calculation (11.100) is of course that a priori we do not know the explicit value of W_ψ° on $A(\mathbb{Q}_p)$, in contrast to the case of the original spherical vector f_p° where we had the formula (11.56) at hand. The key is that we should parametrise the decomposition of the cosets $K_p\tau_i K_p$ in such a way that we can make use of the defining relation (11.106). Such a parametrisation was given by Shintani [577]; here we follow the treatment by Cogdell [140], which reads

$$K_p \tau_i K_p = \bigcup_{\epsilon \in I_i} \bigcup_{n \in N_\epsilon} n\varpi^\epsilon K_p\,, \tag{11.110}$$

where the set I_i is defined as

$$I_i = \Big\{\epsilon = (\epsilon_1, \dots, \epsilon_n) \in \mathbb{Z}^n \mid \epsilon_j \in \{0, 1\}, \sum_{j=1}^n \epsilon_j = i\Big\}, \tag{11.111}$$

and

$$N_\epsilon = N(\mathbb{Z}_p)/(N(\mathbb{Z}_p) \cap \varpi^\epsilon K_p \varpi^{-\epsilon}). \tag{11.112}$$

Using this result we can compute the left-hand side of (11.109) explicitly:

$$\begin{aligned}
\int_{K_p \tau_i K_p} W_\psi^\circ(\delta^{1/2}\mu, \varpi^J h) dh &= \sum_{\epsilon \in I_i} \sum_{n \in N_\epsilon} W_\psi^\circ(\delta^{1/2}\mu, \varpi^J n \varpi^\epsilon) \qquad (11.113) \\
&= \sum_{\epsilon \in I_i} \sum_{n \in N_\epsilon} W_\psi^\circ(\delta^{1/2}\mu, \varpi^J n \varpi^{-J} \varpi^J \varpi^\epsilon) \\
&= \sum_{\epsilon \in I_i} \sum_{n \in N_\epsilon} \psi(\varpi^J n \varpi^{-J}) W_\psi^\circ(\delta^{1/2}\mu, \varpi^J \varpi^\epsilon),
\end{aligned}$$

where we have used that $\varpi^J n \varpi^{-J} \in N(\mathbb{Q}_p)$ combined with (11.106). In fact, because of the constraint (11.108), which requires $j_1 \geq \cdots \geq j_n$, we have that $\varpi^J n \varpi^{-J} \in N(\mathbb{Z}_p)$ and consequently $\psi(\varpi^J n \varpi^{-J}) = 1$. The summand is therefore independent of n and the sum yields only a factor corresponding to the size of the coset space (11.112). Cogdell shows that [140]

$$|N_\epsilon| = p^{i(n-i)/2} \delta^{-1/2}(\varpi^\epsilon), \tag{11.114}$$

so we obtain, for all i [140, Prop. 7.3],

$$\lambda_\mu(\Phi_i) W_\psi^\circ(\delta^{1/2}\mu, \varpi^J) = \sum_{\epsilon \in I_i} p^{i(n-i)/2} \delta^{-1/2}(\varpi^\epsilon) W_\psi^\circ(\delta^{1/2}\mu, \varpi^{J+\epsilon}). \tag{11.115}$$

This is a recursive formula for the spherical Whittaker function $W_\psi^\circ(\delta^{1/2}\mu, \varpi^J)$! We recall that all the Hecke eigenvalues $\lambda_\mu(\Phi_i)$ are known from (11.100).

Example 11.14: Unramified Whittaker functions for $GL(2, \mathbb{Q}_p)$

Let us determine some unramified spherical Whittaker functions for $GL(2, \mathbb{Q}_p)$ using the recursion relation (11.115), starting from $J = (0, 0)$. The recursion relation then reads, for the two values $i = 1, 2$,

$$p^{1/2}(\alpha_p + \beta_p) W^{(0,0)} = p^{1/2} \delta^{-1/2} \left(\varpi^{(0,1)}\right) W^{(0,1)} + p^{1/2} \delta^{-1/2} \left(\varpi^{(1,0)}\right) W^{(1,0)}, \tag{11.116}$$

$$\alpha_p \beta_p W^{(0,0)} = \delta^{-1/2} \left(\varpi^{(1,1)}\right) W^{(1,1)}, \tag{11.117}$$

where (11.66) and (11.67) were used and we have introduced the shorthand notations

$$W^{(j_1, j_2)} \equiv W_\psi^\circ(\delta^{1/2}\mu, \varpi^J) \quad \text{and} \quad \varpi^{(j_1, j_2)} \equiv \varpi^J. \tag{11.118}$$

Since $W^{(0,1)} = 0$ according to (11.108) we can solve for $W^{(1,0)}$ and $W^{(1,1)}$ in terms of $W^{(0,0)}$ to obtain

$$W^{(1,0)} = W^{(0,0)}\delta^{1/2}\left(\varpi^{(1,0)}\right)(\alpha_p + \beta_p),$$
$$W^{(1,1)} = W^{(0,0)}\delta^{1/2}\left(\varpi^{(1,1)}\right)\alpha_p\beta_p. \tag{11.119}$$

We note that

$$\alpha_p + \beta_p = \mathrm{Tr}\begin{pmatrix}\alpha_p & \\ & \beta_p\end{pmatrix} = \mathrm{Tr}_{(1,0)}(A_{\pi_p})$$
$$\text{and} \qquad \alpha_p\beta_p = \mathrm{Tr}\left(\alpha_p\beta_p\right) = \mathrm{Tr}_{(1,1)}(A_{\pi_p}) \tag{11.120}$$

are the characters of the Satake parameter A_{π_p} in the two- and one-dimensional representations of $GL(2,\mathbb{C}) = {}^LGL(2,\mathbb{Q}_p)$, respectively, which are labelled here by their Young tableaux indexed by $J = (1,0)$ and $J = (1,1)$. The translation from non-increasing tuples $(j_1, \ldots, j_n)$ to a *Young tableau* is such that the ith row has j_i boxes. Therefore, we have

$$J = (1,0) \longleftrightarrow \square \qquad \text{and} \qquad J = (1,1) \longleftrightarrow \begin{array}{|c|}\hline \\ \hline \\ \hline\end{array},$$

such that $(1,0)$ corresponds to the fundamental two-dimensional representation and $(1,1)$ to the one-dimensional $GL(2,\mathbb{C})$ representation of weight 1 (tensor density). The relation between spherical Whittaker functions and characters is no coincidence, as we explain in the text.

The key to solving the recursion relation (11.115) for $GL(n,\mathbb{Q}_p)$ is to note that the set of integers $J = (j_1, \ldots, j_n)$, subject to the condition $j_1 \geq j_2 \geq \cdots \geq j_n$, is well known to parametrise the highest weights of irreducible representations V_J of $GL(n,\mathbb{C})$, which, we recall, is the L-group ${}^LG$ of $GL(n,\mathbb{Q}_p)$. But the analogy goes even further than that. Let $\chi_J = \mathrm{Tr}_{V_J}$ be the character of the representation V_J. This is a *class function*, meaning that it is invariant under conjugation:

$$\mathrm{Tr}_{V_J}(g) = \mathrm{Tr}_{V_J}(hgh^{-1}), \qquad g, h \in GL(n,\mathbb{C}), \tag{11.121}$$

and so only depends on the conjugacy class of V_J. If we take V_J to be the fundamental n-dimensional representation of $GL(n,\mathbb{C})$ then we already have a conjugacy class at hand, namely the Satake parameter $A_{\pi_p} \in {}^LA(\mathbb{C})$ (11.93) of π_p. From this perspective J is a dominant weight in the weight lattice $\Lambda^\vee$ of ${}^L\mathfrak{g}$, which is the co-weight lattice of $\mathfrak{g}$. One can then solve the recursion (11.115) in terms of the characters χ_J, with the result [140]

$$W_\psi^\circ(\delta^{1/2}\mu, \varpi^J) = \begin{cases}\text{const} \times \delta^{1/2}(\varpi^J)\chi_J(A_{\pi_p}) & \text{if } J \in \Lambda^\vee \text{ is dominant,}\\ 0 & \text{otherwise.}\end{cases} \tag{11.122}$$

We note that the recursion relation only determines the spherical Whittaker function up to a constant. At first sight this looks very different from the Casselman–Shalika formula (9.23) we derived in Chapter 9. To see that they indeed coincide we shall rewrite the formula given there in a way similar to what was done in Section 9.6. Setting $a = \varpi^J$ in (9.23) and doing some reshuffling we arrive at

$$\begin{aligned} \frac{1}{\zeta(\lambda)} \sum_{w\in\mathcal{W}} \epsilon(w\lambda)|a^{w\lambda+\rho}| &= \frac{1}{\zeta(\delta^{1/2}\mu)} a^{\rho} \sum_{w\in\mathcal{W}} w\left[\frac{a^{\lambda}}{\prod_{\alpha>0}(1-p^{\langle\lambda|\alpha\rangle})}\right] \\ &= \frac{\delta^{1/2}(\varpi^J)}{\zeta(\delta^{1/2}\mu)} \sum_{w\in\mathcal{W}} w\left[\frac{\mu(\varpi^J)}{\prod_{\alpha>0}(1-\mu(\varpi^{-\alpha}))}\right], \end{aligned} \tag{11.123}$$

where we have rewritten the arguments as follows:

$$a^{\rho} = e^{\langle\rho|H(\varpi^J)\rangle} = \delta^{1/2}(\varpi^J), \tag{11.124a}$$

$$a^{\lambda} = e^{\langle\lambda|H(\varpi^J)\rangle} = p^{-\langle\lambda|J\rangle} = \mu(\varpi^J), \tag{11.124b}$$

$$p^{\langle\lambda|\alpha\rangle} = \mu(\varpi^{-\alpha}). \tag{11.124c}$$

To interpret the new form (11.123) of the Casselman–Shalika formula we recall that the weight lattice of the L-group ${}^LG(\mathbb{C})$ is $\Lambda^\vee$, the co-weight lattice of G. We now identify this with the *character lattice* of ${}^LA$ according to

$$\Lambda^\vee \cong \mathrm{Hom}({}^LA, U(1)) \cong \mathbb{Z}^n. \tag{11.125}$$

Under this identification a weight $J = (j_1, \ldots, j_n) \in \Lambda^\vee$ can be interpreted as a character $J : {}^LA(\mathbb{C}) \to U(1)$. We can in particular evaluate this character on the Satake parameter $A_{\pi_p} \in {}^LA(\mathbb{C})$ with the result

$$A_{\pi_p}^J = J(A_{\pi_p}) = J\begin{pmatrix} \alpha_1 & & \\ & \ddots & \\ & & \alpha_n \end{pmatrix} = \prod_{i=1}^n \alpha_i^{j_i}, \tag{11.126}$$

which further implies the equality

$$A_{\pi_p}^J = \mu(\varpi^J). \tag{11.127}$$

Remark 11.15 The standard notation being used here might be a source for confusion: in general we denote the value of the character J on $a \in {}^LA$ by a^J or $J(a)$ as in (11.126); however, this should *not* be confused with the *matrix* ϖ^J, which is defined in (11.107). We trust that this will not cause any trouble since it should be clear from the context which definition is referred to.

Next we compare (11.123) with the Weyl character formula for a representation V_J of a Lie group G with highest weight J. According to (3.30), the character χ_J evaluated at $z \in A$ is explicitly given by

$$\chi_J(z) = \sum_{w\in\mathcal{W}} w\left[\frac{z^J}{\prod_{\alpha>0}(1-z^{-\alpha})}\right], \qquad z \in A. \tag{11.128}$$

We can therefore rewrite (11.123) as

$$\begin{aligned}\frac{1}{\zeta(\delta^{1/2}\mu)}\delta^{1/2}(\varpi^J)\sum_{w\in\mathcal{W}} w\left[\frac{\mu(\varpi^J)}{\prod_{\alpha>0}(1-\mu(\varpi^{-\alpha}))}\right] \\ = \frac{1}{\zeta(\delta^{1/2}\mu)}\delta^{1/2}(\varpi^J)\chi_J(A_{\pi_p}).\end{aligned} \tag{11.129}$$

Comparing this with (11.122) we indeed find perfect agreement, provided that we fix the overall constant there to be $\zeta(\mu)^{-1}$. We conclude that the Casselman–Shalika formula for the spherical Whittaker function $W_\psi^\circ \in (\mathrm{Ind}_{N(\mathbb{Q}_p)}^{G(\mathbb{Q}_p)}\psi)^{K_p}$ can be written in terms of the Weyl character formula for an irreducible representation V_J of the Langlands dual group ${}^LG(\mathbb{C})$:

$$W_\psi^\circ(\delta^{1/2}\mu, \varpi^J) = \frac{1}{\zeta(\delta^{1/2}\mu)}\delta^{1/2}(\varpi^J)\chi_J(A_{\pi_p}). \tag{11.130}$$

For $GL(n,\mathbb{C})$ the characters of V_J are well known to be given by symmetric polynomials that can be expressed in the basis of Schur polynomials. Instances of this for $GL(2)$ can be found in Example 11.14.

11.8 Automorphic L-functions

Equipped with the adelic Hecke technology of the previous sections we shall now revisit the discussion of Dirichlet series of Section 11.2.4 in the more general context of $GL(n,\mathbb{A})$.

Suppose first that f is a Maaß form on the upper half-plane $\mathbb{H}$ which is an eigenfunction of the classical Hecke operator T_p with eigenvalue a_p. For instance, f could be a non-holomorphic Eisenstein series. This lifts to an automorphic form $\varphi_f \in \mathcal{A}(SL(2,\mathbb{Q})\backslash SL(2,\mathbb{A}))$ which is an eigenfunction of $\mathbb{T}_p$ with eigenvalue $\lambda_\mu(\mathbb{T}_p) = p^{1/2}(\alpha_p+\beta_p)$ as we found in (11.66). According to (11.53) the relation between the eigenvalues is thus

$$a_p = \alpha_p + \beta_p. \tag{11.131}$$

This implies that we can rewrite the local factor in the Dirichlet series (11.38) as follows:

$$(1 - a_p p^{-s} + p^{-2s})^{-1} = \left[(1 - \alpha_p p^{-s})(1 - \beta_p p^{-s})\right]^{-1} = \det\left(\mathbb{1} - A_{\pi_p} p^{-s}\right)^{-1}, \tag{11.132}$$

where A_{π_p} is the semi-simple Satake parameter (11.88) in the fundamental matrix representation.

The relation (11.131) has a natural generalisation to higher-rank groups. Suppose $\varphi \in \mathcal{A}(G(\mathbb{Q})\backslash G(\mathbb{A}))^{K_{\mathbb{A}}}$ (i.e., φ is a spherical automorphic form) is attached to an unramified automorphic representation π. Suppose also that φ is an eigenfunction of the spherical Hecke algebras $\mathcal{H}_p^\circ = \mathcal{H}(\mathbb{Q}_p)^{K_p}$. This implies that for $\Phi \in \mathcal{H}_p^\circ$ we have $\pi(\Phi)\varphi = \lambda_\pi(\Phi)\varphi$ according to (11.58). In this situation there exists a unique Satake class $[A_{\pi_p}] \subset {}^LG(\mathbb{C})$ such that

$$\lambda_\pi(\Phi) = p^\sharp \mathrm{Tr}_\pi(A_{\pi_p}), \tag{11.133}$$

where the prefactor is some power of the prime p. In particular, for $G = GL(n)$ we see from (11.100) that

$$\lambda_\pi(\Phi_1) = \lambda_\mu(\Phi_1) = p^{(n-1)/2}(\alpha_1 + \cdots + \alpha_n) = p^{(n-1)/2}\mathrm{Tr}_\pi(A_{\pi_p}), \tag{11.134}$$

where the semi-simple conjugacy class A_{π_p} is given in (11.93).

We can now generalise the construction of the Dirichlet series to $GL(n, \mathbb{A})$. To this end let $\pi = \bigotimes_{p\leq\infty} \pi_p$ be the unramified principal series $\mathrm{Ind}_{B(\mathbb{A})}^{GL(n,\mathbb{A})} \delta^{1/2}\mu$ and A_{π_p} be the corresponding Satake parameter associated with each local factor π_p. To this data we attach the following *local L-factor*:

$$L_p(\pi_p, s) = \det\left(\mathbb{1} - A_{\pi_p} p^{-s}\right)^{-1}, \tag{11.135}$$

and we define the *standard L-function* as

$$L(\pi, s) = \prod_{p<\infty} L_p(\pi_p, s). \tag{11.136}$$

Langlands has proven [447] that this can be completed by adding a certain factor for the prime at infinity:

$$L^\star(\pi, s) = L_\infty(\pi_\infty, s) \prod_{p<\infty} L_p(\pi_p, s), \tag{11.137}$$

which has an analytic continuation to a meromorphic function in the entire complex s-plane, and satisfying a functional equation. This is a vast generalisation of the completed Riemann zeta function $\xi(s) = \xi_\infty(s) \prod_{p<\infty}(1-p^{-s})^{-1}$, where the prime at infinity corresponds to the gamma factor $\xi_\infty(s) = \pi^{-s/2}\Gamma(s/2)$. For Maaß wave forms on $\mathbb{H}$ the factors at infinity were given in (11.39).

But Langlands suggested generalising this even further. Suppose G is a reductive algebraic group over $\mathbb{Q}_p$ and π_p is an unramified automorphic representation of $G(\mathbb{Q}_p)$. Let A_{π_p} be the associated Satake parameter, giving a semi-simple conjugacy class $[A_{\pi_p}] \subset {}^LG(\mathbb{C})$. Further, let

$$\rho : {}^LG(\mathbb{C}) \longrightarrow GL(n, \mathbb{C}) = \mathrm{Aut}(\mathbb{C}^n) \tag{11.138}$$

be an n-dimensional representation of the L-group. Note that the representation does *not* depend on the prime p. In the case of $G = GL(n, \mathbb{Q}_p)$ and ρ the fundamental representation, $\rho(A_{\pi_p})$ will just be the diagonal matrix (11.93), but in general this need not be the case.

Moreover, in general one has that, for an unramified global representation π of $G(\mathbb{A})$, only for *all but finitely many* p the local representations π_p are spherical, i.e., contain vectors $\mathfrak{f}_p^\circ$ fixed under $K_p = G(\mathbb{Z}_p)$. To take care of this complication we let S be a finite set of places such that if $p \notin S$, π_p is spherical. The set S always includes the archimedean place $p = \infty$. For this data we now construct the *(partial) Langlands L-function*

$$L_S(\pi, s, \rho) = \prod_{p \notin S} \frac{1}{\det\left(\mathbb{1} - \rho(A_{\pi_p})p^{-s}\right)}. \tag{11.139}$$

In this situation the analytic continuation is more involved but Langlands has conjectured that $L_S(\pi, s, \rho)$ can be completed at the unramified places S to obtain a meromorphic function $L^\star(\pi, s, \rho)$ of s, called the *global Langlands L-function*.

Example 11.16: L-function for $G = GL(2, \mathbb{Q}_p)$

To give a simple example of what such an L-function would look like, let us consider $G = GL(2, \mathbb{Q}_p)$ but now take ρ to be the kth symmetric power $Sym^k(\mathbb{C}^2)$ of the fundamental representation $\mathbb{C}^2$ of $GL(2, \mathbb{C})$ (see, e.g., [265] for a nice discussion of this and other examples). The resulting L-function reads

$$L(\pi, s, Sym^k) = \prod_{p<\infty} \frac{1}{\det(\mathbb{1} - \rho(A_{\pi_p})p^{-s})} = \prod_{p<\infty} \prod_{j=0}^{k} \frac{1}{1 - \alpha_p^j \beta_p^{k-j} p^{-s}}. \tag{11.140}$$

Using the formalism outlined above, Langlands thus provided a systematic procedure for attaching L-functions to automorphic forms, a task that had previously only been understood in special cases. The relation between automorphic forms on G, the L-group ${}^LG$ and automorphic L-functions provides the cornerstone of the Langlands program, which is a set of far-reaching conjectures put forward by Langlands, of which only a tiny fraction have been proven. In Chapter 16, we discuss some of the ideas in the Langlands program, and how they relate and extend the theory we have presented in this work.

11.9 The Langlands–Shahidi Method*

The functional properties of L-functions such as (11.140) are important since these can be used to give estimates on Hecke eigenvalues (or Fourier coefficients) of cusp forms. This application to number theory is reviewed for example in [566, 569, 265]; we will content ourselves here with explaining the basic construction and its relation to Eisenstein series and Whittaker coefficients.

The starting point for the investigation of the functional relation of many L-functions is knowledge of the functional Equation (8.44) for Eisenstein series on G induced from a representation of the Levi subgroup L of some parabolic subgroup $P = LU$ of G. From this functional equation and knowledge how the L-function of interest arises in the Fourier expansion one can then deduce properties of the L-function. This method was suggested by Langlands in [447, 443] and then developed in detail by Shahidi [563, 562, 564, 566, 569].

To motivate the procedure, we look once more at the Fourier expansion of the $SL(2,\mathbb{R})$ Eisenstein series (see (1.17)):

$$\begin{aligned}E(s,z) &= y^s + \frac{\xi(2s-1)}{\xi(2s)}y^{1-s}\\ &\quad + \frac{2y^{1/2}}{\xi(2s)}\sum_{n\neq 0}|n|^{s-1/2}\sigma_{1-2s}(n)K_{s-1/2}(2\pi|n|y)e^{2\pi inx}.\end{aligned} \tag{11.141}$$

The Eisenstein series satisfies the functional equation (see (7.13))

$$E(1-s,z) = \frac{\xi(2s)}{\xi(2s-1)}E(s,z), \tag{11.142}$$

and the L-function whose properties one is interested in is the completed Riemann zeta function $\xi(k)$. As we have seen in Chapters 7 and 8, this functional equation can be read off from the constant terms of the Eisenstein series and does not require knowledge of the completed Riemann zeta function beyond its definition in terms of an Euler product.

Additional properties of $\xi(s)$ can be inferred from the first Fourier coefficient ($n = 1$). The functional relation (11.142) for this Fourier coefficient reads

$$\frac{1}{\xi(2(1-s))}K_{1/2-s}(2\pi y) = \frac{\xi(2s)}{\xi(2s-1)}\frac{1}{\xi(2s)}K_{s-1/2}(2\pi y). \tag{11.143}$$

Using the property $K_t(x) = K_{-t}(x)$ of the modified Bessel function, one deduces that

$$\xi(2s-1) = \xi(2-2s) \quad \Leftrightarrow \quad \xi(k) = \xi(1-k). \tag{11.144}$$

Thus, the functional equation of the completed Riemann zeta function $\xi(s)$ is a consequence of the functional equation of Eisenstein series. One can also deduce the non-vanishing of $\zeta(s)$ on the line $\mathrm{Re}(s) = 1$ from the holomorphy (in s) of $E(s,z)$ on the line $\mathrm{Re}(s) = 0$ and further properties of $\zeta(s)$ from the study of $E(s,z)$ [265]. (The higher Fourier coefficients $n > 1$ provide no additional information.)

The more general realisation of this method relies on Eisenstein series on G induced from a cuspidal automorphic representation π_L of the Levi factor L of a maximal parabolic subgroup $P = LU \subset G$. We assume that the representation π_L is spherical at almost all places p.

As before, we have that at the spherical finite places p one can characterise the representation by means of its Satake parameter $A_{\pi_p} \in {}^LA$; see Section 11.6. Let S be a set of places that includes all the non-spherical places and the archimedean one. In the everywhere-unramified case one would have $S = \{\infty\}$. Since ${}^LL$ is a complex linear group, it admits standard finite-dimensional complex representations $\rho_L \colon {}^LL \to GL(n,\mathbb{C})$ where n is the dimension of the representation. For any such pair (π_L, ρ_L), the *partial Langlands L-function* is given by

$$L_S(s,\pi_L,\rho_L) = \prod_{p\notin S} L_p(s,\pi_L,\rho_L) = \prod_{p\notin S} \frac{1}{\det(\mathbb{1} - \rho_L(A_{\pi_p})p^{-s})}, \tag{11.145}$$

where the determinant is taken in the representation associated with ρ_L. Formally, this is the same as the definition (11.139) above, but this time we have emphasised that this is for the Levi part L of a parabolic subgroup P of G. The *global Langlands L-function* requires the definition of factors for the places S that is less uniform and not known in full generality. Important progress for the global L-functions for $GL(n)$ and $SO(2n+1)$ can be found in [344, 346, 355, 379].

The virtue of these L-functions is that they arise in the Fourier expansion of Eisenstein series induced from a *cuspidal* representation π_L of L. For an automorphic form $\phi \in \pi_L$ we let

$$E(s,\phi,g) = \sum_{\gamma\in P(\mathbb{Q})\backslash G(\mathbb{Q})} \phi(\gamma g)\delta_P(\gamma g)^s \tag{11.146}$$

be the *Eisenstein series on G induced from* $\phi \in \pi_L$. $\delta_P(g)$ is here (the trivial extension to G of) the modulus character on $P \subset G$ defined by

$$d(lul^{-1}) = \delta_P(l)du \tag{11.147}$$

and $\delta_P(ulk) = \delta_P(l)$. It can be given explicitly by $\delta_P(l) = l^{2\rho_P}$, where ρ_P is half the sum of the positive roots contained in U. In the discussion in Section 5.7, we had taken $\phi = 1$ in the non-cuspidal trivial representation.

The Eisenstein series $E(s, \phi, g)$ on G has a Fourier expansion with respect to the unipotent U that is simpler because ϕ is taken from a *cuspidal* representation of the Levi factor L. This arises because in the Bruhat decomposition of G most classes have a vanishing contribution as ϕ is cuspidal. This is a collapse mechanism not unsimilar to the one discussed for constant terms in Section 10.3 and for Whittaker coefficients in Section 9.5.

Langlands showed [443] that the constant term of $E(s, \phi, g)$ along $\tilde{P}$ (the opposite of P) is controlled by partial L-functions (11.145) and Shahidi extended this to non-trivial Fourier coefficients [563, 562, 564, 566, 569]. Shahidi's work relies also on the Casselman–Shalika formula for (generic) Whittaker coefficients of an Eisenstein series $E(\lambda, g)$ at unramified places (11.130).

We first explain why Langlands L-functions arise from the Casselman–Shalika formula (11.130). If (11.130) is evaluated for the special case of $\varpi^J = \mathbb{1}$, corresponding to the trivial representation, one obtains from Corollary 9.2

$$W^\circ(\delta^{1/2}\mu, \mathbb{1}) = \frac{1}{\zeta(\delta^{1/2}\mu)} = \prod_{\alpha^\vee>0}(1 - p^{-1-\langle\lambda|\alpha\rangle}) = \prod_{\alpha^\vee>0}(1 - p^{-1}\mu(\varpi^{\alpha^\vee})). \tag{11.148}$$

Now each $\mu(\varpi^{\alpha^\vee})$ corresponds to the adjoint action of the Satake parameter A_{π_p} on the root space of $\alpha^\vee$, which is nothing but the representation of the split torus ${}^LA$ on the Lie algebra ${}^L\mathfrak{n}$. Denoting this action by $\rho\colon {}^LA \to \mathrm{End}({}^L\mathfrak{n})$, we have that

$$\prod_{\alpha^\vee>0}(1 - p^{-1}\mu(\varpi^{\alpha^\vee})) = \det(\mathbb{1} - \rho(A_{\pi_p})p^{-1}), \tag{11.149}$$

since the representation ρ of ${}^LA$ decomposes into the direct sum of one-dimensional representations labelled by the positive roots, and the determinant therefore factorises. Hence, for a fixed place p the Whittaker–Jacquet integral can also be written as

$$W^\circ(\lambda, \mathbb{1}) = \int_{N(\mathbb{Q}_p)} \chi(w_{\text{long}}n)\overline{\psi(n)}dn = \frac{1}{L_p(1, \lambda, \rho)}, \tag{11.150}$$

i.e., yielding an L-function $L_S(1, \lambda, \rho)$ of the type (11.145) when multiplied over all places $p \notin S$. Here, we have labelled the representation π_L of the Levi ${}^LA$ of the minimal parabolic (Borel) ${}^LB$ by its quasi-character λ.

In the more general case (11.146) of the Eisenstein series $E(s, \phi, g)$ induced from a cuspidal representation π_L of the Levi subgroup of a maximal parabolic $P = UL \subset G$, one has to consider the adjoint action ρ of ${}^LL$ on the Lie algebra ${}^L\mathfrak{u}$ of the unipotent ${}^LU$. Under this action, ${}^L\mathfrak{u}$ decomposes into a finite number of irreducible representations according to

$$\rho = \bigoplus_{j=1}^{m} \rho_j, \tag{11.151}$$

where m is the maximum coefficient (among all roots of G) of the simple root defining the maximal parabolic subgroup $P \subset G$. Shahidi showed [563, 569] that it is possible to choose $\phi \in \pi_L$ such that the generic Fourier coefficient of $E(s, \phi, g)$ at $g = \mathbb{1}$ is given globally by

$$\int\limits_{U(\mathbb{Q})\backslash U(\mathbb{A})} E(s, \phi, u)\overline{\psi(u)}du = \prod_{p\in S} W_p^{\circ}(\mathbb{1}) \cdot \prod_{j=1}^{m} \frac{1}{L_S(1 + ajs, \pi_L, \rho_j)}, \tag{11.152}$$

where a is a fixed number that depends on the choice of parabolic subgroup and we have separated out the contribution from the 'bad' primes $p \in S$. The shifts by s in the argument of the L-function (compared to (11.150)) are due to the factor $\delta_P(\gamma g)^s$ in the definition of the Eisenstein series.

In the constant term, the same L-functions appear; see the intertwining factors $M(w, \lambda)$ in (8.42). Langlands showed [443] that the intertwiner appearing in the constant term for a place $p \notin S$ is

$$\prod_{j=1}^{m} \frac{L_p(ajs, \pi_L, \rho_j)}{L_p(1 + ajs, \pi_L, \rho_j)}. \tag{11.153}$$

Due to the cuspidality of π_L this is the only non-trivial coefficient appearing in the constant term and it plays the rôle of the coefficient in front of y^{1-s} in (11.141) above. It also appears in the functional equation satisfied by $E(s, \phi, g)$ and allows one to deduce a functional equation for the partial L-function L_S obtained from all the places $p \in S$:

$$\prod_{j=1}^{m} L_S(ajs, \pi_L, r_j) = \prod_{j=1}^{m} L_S(1 - ajs, \pi_L, \tilde{r}_j) \cdot \prod_{p\in S} C(s, \tilde{\pi}_v), \tag{11.154}$$

which is called the *crude functional relation* [563, 562]. The factors $C(s, \tilde{\pi}_v)$ appearing in this functional relation are called *local coefficients* and they can be determined from the study of the intertwining operator for $p \in S$ [563, 562].

The tildes in the above formula refer to the parabolic subgroup $\tilde{P}$ *opposite* to P (that is obtained by applying the longest Weyl element w_{long}). From this identification of the product of m partial L-functions as a Fourier coefficient of an Eisenstein series one can also deduce that the product extends to a meromorphic function (in s) and does not vanish on the imaginary axis [569]. Moreover, it is possible to perform induction on m to deduce the same statements for each of the individual factors. This produces a host of non-trivial results for generalised L-functions in various representations r_i that arise from all maximal parabolic subgroups [565]. The results described here for split groups can also be extended to so-called *quasi-split groups* [569, 436].

Remark 11.17 The Langlands–Shahidi method just outlined makes it possible to study the analytic properties of L-functions [262]. Additional information about the analytic properties at potential singularities comes from knowledge of the unitary dual of G in connection with residues of Eisenstein series [398, 401]. This knowledge can then be used for *Langlands functoriality* in some examples. As will be discussed more in Section 16.1, Langlands functoriality deals with the question of transferring automorphic forms from a group G to another group G'.

Concretely, one starts from an L-function that is tentatively associated with the group G' and takes the *Rankin–Selberg product* with automorphic L-functions on subgroups $G \subset G'$. If certain technical conditions are fulfilled, one can conclude that there must be a (cuspidal) automorphic representation of G' whose L-function is the one under study, thereby lifting the representations from G to G'. This is the content of various *converse theorems*, most notably those of Cogdell and Piatetski-Shapiro for the general linear group [144, 143], which state that an L-function satisfying certain technical conditions must be a global L-function arising from an automorphic form on the general linear group. Such converse theorems can be seen as the extensions of Hecke's results for L-functions that were discussed at the end of Section 11.2.4.

The assumptions for the L-functions of the converse theorems can be established using the Langlands–Shahidi method and this has produced many non-trivial examples of functoriality. We refer the reader to [142, 399, 400, 401] for some examples and to [567, 265, 569] for recent overviews.

12

Theta Correspondences

A theta correspondence (or theta lift) is a method for lifting (or transferring) an automorphic form φ on a group H to an automorphic form φ' on another group H', where $H \times H'$ is a subgroup of a bigger group G. In the prototypical example, H is an orthogonal group, H' is a symplectic group and G is a larger symplectic group. The correspondence is realised through an integral formula with kernel given by a classical theta function on G, restricted to the product $H \times H'$, thereby explaining the appearance of the word 'theta' in the name of the correspondence. Theta correspondences play an important rôle in establishing concrete examples of the Langlands functoriality principle (see Section 16.1), and also appear ubiquitously in string theory through one-loop scattering amplitudes (see Section 13.4).

In this chapter, we shall review the theory of classical theta series and some basic features of the theta correspondence, and make various remarks on connections with other parts of this book, both physical and mathematical.

12.1 Classical Theta Series

Perhaps the simplest example of a theta series is the *Jacobi theta function*, which, for $z \in \mathbb{C}$ and $\tau \in \mathbb{H}$, is defined as

$$\theta(z|\tau) = \sum_{n\in\mathbb{Z}} e^{i\pi(n^2\tau+2nz)}. \tag{12.1}$$

Remark 12.1 In string theory one also encounters the *generalised Jacobi theta functions* defined by *characteristics* $\alpha, \beta \in \mathbb{R}$:

$$\begin{aligned}\theta\left[\begin{smallmatrix}\alpha\\ \beta\end{smallmatrix}\right](z|\tau) &= \sum_{n\in\mathbb{Z}} e^{i\pi\left[(n+\alpha)^2\tau+2(n+\alpha)(z+\beta)\right]} \\ &= e^{i\pi\left[\alpha^2\tau+2\alpha(z+\beta)\right]}\theta(z+\beta+\alpha\tau|\tau).\end{aligned} \tag{12.2}$$

They will enter in the superstring partition function in (13.102). Sometimes, the Jacobi theta function is defined as the restriction $\theta\left[\begin{smallmatrix}\alpha\\ \beta\end{smallmatrix}\right](0|\tau)$, a so-called *theta nullwert* or *theta constant*, but we will in this section mainly consider the function $\theta(z|\tau)$.

For $a, b \in \mathbb{Z}$, we see that, by changing the summation index in (12.2),

$$\theta\left[\begin{smallmatrix}a\\ b\end{smallmatrix}\right](z|\tau) = \theta\left[\begin{smallmatrix}0\\ 0\end{smallmatrix}\right](z|\tau) = \theta(z|\tau), \tag{12.3}$$

and thus the Jacobi theta function satisfies the functional identity

$$\theta(z + b + a\tau|\tau) = e^{-i\pi(a^2\tau+2az)}\theta(z|\tau), \tag{12.4}$$

which, in particular, includes $\theta(z + 1|\tau) = \theta(z|\tau)$.

We will now consider the modular properties of $\theta(z|\tau)$ with respect to τ. Recall that the modular group $SL(2,\mathbb{Z})$ acting on $\tau \in \mathbb{H}$ is generated by the transformations $\tau \to \tau + 1$ and $\tau \to -1/\tau$.

Using the fact that $n^2 - n = n(n-1) \equiv 0 \pmod 2$ and the definition (12.1) we see that

$$\theta(z|\tau + 1) = \sum_{n\in\mathbb{Z}} e^{i\pi(n^2\tau+n+2nz)} = \theta\left(z + \tfrac{1}{2}|\tau\right) = \theta\left[\begin{smallmatrix}0\\ 1/2\end{smallmatrix}\right](z|\tau). \tag{12.5}$$

For the transformation $\tau \to -1/\tau$ we rewrite the Jacobi theta function using *Poisson resummation* (see also Lemma 2.30 and Appendix A):

$$\theta(z|\tau) = \sum_{n\in\mathbb{Z}} f(n) = \sum_{n\in\mathbb{Z}} \hat{f}(n), \tag{12.6}$$

where $f(n) = e^{i\pi(n^2\tau+2nz)}$ and

$$\hat{f}(n) = \int_{-\infty}^{\infty} f(x)e^{-2\pi inx}\, dx = \frac{1}{\sqrt{-i\tau}}e^{-i\pi(n-z)^2/\tau}. \tag{12.7}$$

This gives us that

$$\theta(z|\tau) = \frac{e^{-i\pi z^2/\tau}}{\sqrt{-i\tau}} \sum_{n\in\mathbb{Z}} e^{i\pi[n^2(-1/\tau)+2kz/\tau]} = \frac{e^{-i\pi z^2/\tau}}{\sqrt{-i\tau}}\theta\left(\frac{z}{\tau}\Big|-\frac{1}{\tau}\right). \tag{12.8}$$

The two Equations (12.5) and (12.8) are called the *Jacobi identities* and we note that they also define specialisations

$$\begin{aligned}
\theta_1(\tau) &= \theta\left[\begin{smallmatrix}1/2\\ 1/2\end{smallmatrix}\right](0|\tau), & \theta_3(\tau) &= \theta\left[\begin{smallmatrix}0\\ 0\end{smallmatrix}\right](0|\tau),\\
\theta_2(\tau) &= \theta\left[\begin{smallmatrix}1/2\\ 0\end{smallmatrix}\right](0|\tau), & \theta_4(\tau) &= \theta\left[\begin{smallmatrix}0\\ 1/2\end{smallmatrix}\right](0|\tau),
\end{aligned} \tag{12.9}$$

that we will encounter again in superstring theory in (13.103).

Following [249, 630, 149, 550], we will now study theta series attached to a lattice. Let V be a $2d$-dimensional space over $\mathbb{Q}$ with an inner product $\langle x, y\rangle$. We say that $L \subseteq V$ is an *integral lattice* of V if $\langle x|x\rangle \in \mathbb{Z}$ for all $x \in L$, of full rank if $\mathbb{Q} \otimes L \cong V$ and *even* if $\langle x|x\rangle \in 2\mathbb{Z}$ for all $x \in L$. We will henceforth assume that our lattice satisfies all these properties. We also define the dual lattice as

$$L^* = \{x \in V \mid \langle x|L\rangle \subseteq \mathbb{Z}\}. \tag{12.10}$$

For an element $\lambda \in V = \mathbb{Q} \otimes L$ we define the *theta series attached to the lattice* as [249]

$$\theta_L[\lambda](z|\tau) = \sum_{l \in L} e^{i\pi\left(\langle l+\lambda|l+\lambda\rangle\tau + 2\langle l+\lambda|z\rangle\right)}, \tag{12.11}$$

where $z \in \mathbb{C} \otimes L$ and $\tau \in \mathbb{H}$. Let also $\theta_L(z|\tau) = \theta_L[0](z|\tau)$. ($\lambda$ is sometimes referred to as *glue* and generalises the characteristics in (12.2).)

Similar to (12.2) in Remark 12.1 we have that

$$\theta_L[\lambda](z|\tau) = e^{i\pi\left(\langle\lambda|\lambda\rangle\tau + 2\langle\lambda|z\rangle\right)}\theta_L(z + \lambda\tau|\tau)\,. \tag{12.12}$$

It can also be shown that, for $a \in L$ and $b \in L^*$,

$$\theta_L(z + b + a\tau|\tau) = e^{-i\pi\left(\langle a,a\rangle\tau + 2\langle a|z\rangle\right)}\theta_L(z|\tau)\,. \tag{12.13}$$

Let us now study the modular properties with respect to τ. Since L is even we have that $\langle l|l\rangle \in 2\mathbb{Z}$ in the definition of θ_L and thus

$$\theta_L(z|\tau + 1) = \theta_L(z|\tau)\,. \tag{12.14}$$

Lastly, we have that

$$\theta_L\left(\frac{z}{\tau}\Big|-\frac{1}{\tau}\right) = \frac{(-i\tau)^d}{\sqrt{|L|}}e^{i\pi\langle z|z\rangle/\tau}\theta_{L^*}(z|\tau), \tag{12.15}$$

where $\mathrm{rank}(L) = 2d$, we identify $\mathbb{C} \otimes L \cong \mathbb{C} \otimes L^*$ and $|L|$ is the determinant of the Gram matrix B defined in a basis x_i by $(B)_{ij} = \langle x_i|x_j\rangle$, corresponding to the volume of a single cell of the lattice L. The last transformation formula (12.15) follows again from Poisson resummation.

Let us now restrict to a special class of lattices. An integral lattice satisfies [149]

$$L \subseteq L^* \subseteq \frac{1}{|L|}L, \tag{12.16}$$

and we say that a lattice with $|L| = 1$, or, equivalently $L = L^*$, is called *unimodular*. One can show that even unimodular Euclidean lattices exist only for lattices whose rank is divisible by eight, that is $4 \mid d$. We shall come back to the Lorentzian case in Section 13.4.3.

For an even unimodular lattice, (12.15) gives that the theta nullwert $\theta_L(\tau) = \theta_L(0|\tau)$ satisfies

$$\theta_L\left(-\frac{1}{\tau}\right) = (-i\tau)^d\theta_L(\tau) = \tau^d\theta_L(\tau), \tag{12.17}$$

which, together with (12.14), means that $\theta_L(\tau)$ is a modular form of weight d.

Remark 12.2 There is, up to isomorphism, a unique even unimodular Euclidean lattice of rank 8, namely the root lattice of E_8. For rank 16, there are two such lattices: $E_8 \oplus E_8$ and D_{16}^+, where the latter is constructed from the D_{16} root lattice. For rank 24 we have the 24 *Niemeier lattices* with 23 of them having minimal norm 2 and one with minimal norm 4, namely the *Leech lattice* Λ_{24} [149]. The Niemeier lattices play a key rôle in *umbral moonshine*; see Remark 15.20.

We may rewrite $\theta_L(0|\tau)$ as follows:

$$\theta_L(z|\tau) = \sum_{l \in L} e^{i\pi\langle l|l\rangle\tau} = \sum_{n \in \mathbb{N}} r_L(n) q^{\frac{n}{2}}, \tag{12.18}$$

where $q = \exp(2\pi i\tau)$ and $r_L(n)$ counts the number of elements l of L with (square) length $\langle l|l\rangle = n$ and is called the *representation number*. Recall that for even lattices $\langle l|l\rangle \in 2\mathbb{Z}$. Knowledge of the q-expansion of a theta series θ_L, for example via methods from the theory of modular forms, would then tell us about *sphere packing* within the lattice L (see e.g. [149]).

12.2 Representation Theory of Classical Theta Functions

Before we plunge into the general theory of theta correspondences over adelic groups, we shall first take a short glimpse at the representation-theoretic underpinnings of the classical theta series. Consider the simplest Jacobi theta series discussed above, namely the nullwert

$$\theta(\tau) \equiv \theta(0|\tau) = \sum_{n \in \mathbb{Z}} e^{i\pi n^2\tau}. \tag{12.19}$$

This satisfies

$$\theta\left(\frac{a\tau + b}{c\tau + d}\right) = \left(\frac{c}{d}\right) \sqrt{c\tau + d}\,\theta(\tau), \tag{12.20}$$

for all matrices $\left(\begin{smallmatrix} a & b \\ c & d \end{smallmatrix}\right) \in \Gamma_1(4) \subset SL(2, \mathbb{Z})$, where $\left(\frac{c}{d}\right)$ is the *Jacobi symbol* defined in terms of the prime factorisation $d = p_1^{k_1} \cdots p_r^{k_r}$ of d by

$$\left(\frac{c}{d}\right) = \left(\frac{c}{p_1}\right)^{k_1} \cdots \left(\frac{c}{p_r}\right)^{k_r}, \tag{12.21}$$

where the *Legendre symbols* for prime lower argument are

$$\left(\frac{c}{p}\right) = a^{(p-1)/2} \mod p \tag{12.22}$$

that characterise quadratic residues mod p and take values in $\{-1, 0, 1\}$. By (12.20), the Jacobi theta series $\theta(\tau)$ is a modular form of weight $1/2$, belonging to the space $\mathcal{M}_{1/2}(\Gamma_1(4))$ (see Section 4.2.2).

Recall from Section 4.3 that we can lift a weight w holomorphic modular form f to a function φ_f on $SL(2,\mathbb{R})$ using the formula (4.47):

$$\varphi_f(g) = \varphi_f(x, y, \theta) = e^{iw\theta} y^{w/2} f(x + iy), \tag{12.23}$$

where the element $g \in SL(2,\mathbb{R})$ is parametrised as follows using the Iwasawa decomposition

$$g = nak = \begin{pmatrix} 1 & x \\ & 1 \end{pmatrix} \begin{pmatrix} y^{1/2} & \\ & y^{-1/2} \end{pmatrix} \begin{pmatrix} \cos\theta & \sin\theta \\ -\sin\theta & \cos\theta \end{pmatrix}. \tag{12.24}$$

We will now construct a similar lift of $\theta(\tau)$. However, due to the Jacobi symbol in (12.20) it is actually more natural to lift θ to a function on the double cover $\widetilde{SL(2,\mathbb{R})}$ (*metaplectic cover*), i.e.,

$$1 \longrightarrow \mathbb{Z}_2 \longrightarrow \widetilde{SL(2,\mathbb{R})} \longrightarrow SL(2,\mathbb{R}) \longrightarrow 1. \tag{12.25}$$

Viewed as a set we have simply $\widetilde{SL(2,\mathbb{R})} = SL(2,\mathbb{R}) \times \{\pm 1\}$. The group law involves a *2-cocycle* Ω of order two [264]:

$$(g_1, \epsilon_1) \cdot (g_2, \epsilon_2) = (g_1 g_2, \Omega(g_1, g_2)\epsilon_1\epsilon_2). \tag{12.26}$$

Instead of representations of the metaplectic group $\widetilde{SL(2,\mathbb{R})}$, one can consider *projective representations* of $SL(2,\mathbb{R})$ that are defined up to sign.

We shall first define an action of $SL(2,\mathbb{R})$ on functions in $L^2(\mathbb{R})$. Let $\psi\colon \mathbb{Z}\backslash\mathbb{R} \to U(1)$ be defined by $\psi(x) = e^{2\pi i x}$ with $x \in \mathbb{R}$. We now define a representation

$$\rho\colon SL(2,\mathbb{R}) \longrightarrow L^2(\mathbb{R}) \tag{12.27}$$

by the following action on $f \in L^2(\mathbb{R})$:

$$\begin{aligned} \rho\left(\left(\begin{smallmatrix} 1 & x \\ & 1 \end{smallmatrix}\right)\right) f(t) &= \psi(xt^2)\, f(t), \\ \rho\left(\left(\begin{smallmatrix} \sqrt{y} & \\ & 1/\sqrt{y} \end{smallmatrix}\right)\right) f(t) &= y^{1/4} f(\sqrt{y}t), \\ \rho\left(\left(\begin{smallmatrix} & -1 \\ 1 & \end{smallmatrix}\right)\right) f(t) &= \lambda\, \widehat{f}(t), \end{aligned} \tag{12.28}$$

where λ is an 8th root of unity and $\widehat{f}$ is the usual Fourier transform

$$\widehat{f}(t) = \int_{\mathbb{R}} f(r) e^{-2\pi i r t} dr. \tag{12.29}$$

This defines a projective representation of $SL(2,\mathbb{R})$ in the sense that

$$\rho(g_1)\rho(g_2) = c(g_1, g_2)\rho(g_1 g_2), \tag{12.30}$$

where $c(g_1, g_2)$ is a 2-cocycle (the *Schur multiplier*) compatible with the group 2-cocycle such that this is an honest representation of the metaplectic cover

$$\rho\colon \widetilde{SL(2,\mathbb{R})} \longrightarrow L^2(\mathbb{R}), \tag{12.31}$$

called the *Weil representation*. Using the Weil representation we now construct a function Θ_f on $\widetilde{SL(2,\mathbb{R})}$ as follows:

$$\Theta_f(\tilde{g}) := \sum_{n\in\mathbb{Z}} \rho(\tilde{g})\, f(n). \tag{12.32}$$

Since the cover $\widetilde{\Gamma_1(4)} \to \Gamma_1(4)$ is trivial, the representation splits and one can readily show that Θ_f satisfies

$$\Theta_f(\tilde{\gamma}\tilde{g}) = \Theta_f(\tilde{g}), \tag{12.33}$$

where $\tilde{\gamma} = (\gamma, \epsilon_\gamma)$ with $\gamma \in \Gamma_1(4)$ and $\epsilon_\gamma \in \{\pm 1\}$. Hence, we have constructed a function

$$\Theta_f\colon \Gamma_1(4)\backslash\widetilde{SL(2,\mathbb{R})} \longrightarrow \mathbb{C}, \tag{12.34}$$

to wit, an automorphic form.

Now let us see how this construction is related to the Jacobi theta function $\theta(\tau)$. Choose the function $f \in L^2(\mathbb{R})$ as $f(r) = e^{-\pi r^2}$. This implies

$$\rho\left(\begin{pmatrix} 1 & x \\ & 1 \end{pmatrix}\begin{pmatrix} \sqrt{y} & \\ & 1/\sqrt{y} \end{pmatrix}, 1\right) f(n) = y^{1/4} e^{\pi i n^2 x} e^{-\pi n^2 y} = y^{1/4} e^{\pi i n^2 \tau}, \tag{12.35}$$

with $\tau = x + iy$. In this case the function Θ_f becomes

$$\Theta_f\left(\begin{pmatrix} 1 & x \\ & 1 \end{pmatrix}\begin{pmatrix} \sqrt{y} & \\ & 1/\sqrt{y} \end{pmatrix}\right) = y^{1/4} \sum_{n\in\mathbb{Z}} e^{\pi i n^2 \tau} = y^{1/4}\theta(\tau).$$

Comparing this with (12.23) we conclude that Θ_f precisely yields the automorphic lift φ_θ of the Jacobi theta function. Below we will see how to generalise this to the adelic setting and for arbitrary Lie groups.

Remark 12.3 A similar lift can be performed for the lattice theta functions $\theta_L(\tau)$ defined in (12.17) for even unimodular lattices L. The corresponding lifted function is simply

$$\Theta_L(\tau) = \tau_2^{d/2}\theta_L(\tau) = \tau_2^{d/2} \sum_{l\in L} e^{i\pi\langle l|l\rangle \tau}. \tag{12.36}$$

It is invariant under $SL(2,\mathbb{Z})$. In Section 13.4.3, we shall encounter versions of this where the inner product $\langle l|l\rangle$ is a function of the moduli space of even unimodular lattices.

12.3 Theta Correspondence

We phrase the results in this section slightly more generally, for G an algebraic group defined over an arbitrary number field $\mathbb{F}$.

Definition 12.4 (Dual pair) A *dual pair* (H, H') is a pair of subgroups of G, such that $H \times H' \subset G$ and where H (resp. H') is the *centraliser* of H' (resp. H) inside G [361, 363]. The dual pair is called *reductive* if, under the action of H (and of H'), the defining representation of G splits into the direct sum of two complementary invariant subspaces.

Let π be an automorphic representation of $G(\mathbb{A}_\mathbb{F})$ of small Gelfand–Kirillov dimension. Let $\theta \in \pi$ be an automorphic form on $G(F)\backslash G(\mathbb{A}_\mathbb{F})$. Restricting the representation π to the product $H \times H' \subset G$ then yields an $H(\mathbb{F}) \times H'(\mathbb{F})$-invariant function $\theta(h, h')$ on $H(\mathbb{A}_\mathbb{F}) \times H'(\mathbb{A}_\mathbb{F})$ known as the *theta kernel.* Starting from the theta kernel we can construct two different theta lifts:

Definition 12.5 (Theta lift) For φ an automorphic form on $H(\mathbb{F})\backslash H(\mathbb{A}_\mathbb{F})$ we define the *theta lift* from H to H' by the integral

$$\Theta_\varphi(h') = \int_{H(\mathbb{F})\backslash H(\mathbb{A}_\mathbb{F})} \varphi(h)\theta(h, h')dh, \tag{12.37}$$

where dh is the invariant measure on $H(\mathbb{F})\backslash H(\mathbb{A}_\mathbb{F})$. Similarly, for φ' an automorphic form on $H'(\mathbb{F})\backslash H'(\mathbb{A}_\mathbb{F})$ we can define the theta lift in the other direction, i.e., from H' to H, by the integral

$$\Theta_{\varphi'}(h) = \int_{H'(\mathbb{F})\backslash H'(\mathbb{A}_\mathbb{F})} \varphi'(h')\theta(h, h')dh'. \tag{12.38}$$

Clearly, the function Θ_φ (resp. $\Theta_{\varphi'}$) is defined on $H'(\mathbb{A}_\mathbb{F})$ (resp. $H(\mathbb{A}_\mathbb{F})$) and is by construction invariant under $H'(\mathbb{F})$ (resp. $H(\mathbb{F})$). The prototypical questions that one wishes to study in this context are the following (similar questions apply to $\Theta_{\varphi'}$):

- Does Θ_φ converge?
- Is Θ_φ non-zero?
- Is Θ_φ cuspidal?
- If $\Theta_\varphi \neq 0$, what kind of object is it?

If Θ_φ can be identified with an automorphic form on $H'(\mathbb{A}_\mathbb{F})$ then the theta lift can be viewed as a map,

$$\Theta\colon \mathcal{A}(H(\mathbb{F})\backslash H(\mathbb{A}_\mathbb{F})) \longrightarrow \mathcal{A}(H'(\mathbb{F})\backslash H'(\mathbb{A}_\mathbb{F})), \tag{12.39}$$

and therefore may in principle provide a concrete realisation of Langlands functoriality (see [326, 345] for some examples).

Currently these problems cannot be addressed in general, but below we shall look at some examples where much is known.

12.4 Theta Series and the Weil Representation

We shall now consider global theta series attached to the *Weil representation*, generalising the construction in Section 12.2, following [264]. Let V be an orthogonal vector space, i.e., a finite-dimensional vector space over $\mathbb{F}$ with non-degenerate inner product $\langle\cdot|\cdot\rangle$. Let V' be a symplectic vector space. Take $H = O(V)$, the orthogonal group of V, and $H' = Sp(V')$, the symplectic group over $\mathbb{F}$. Then $W = V \otimes V'$ is naturally a symplectic vector space, and $(O(V), Sp(V'))$ forms a dual reductive pair inside $G = Sp(W)$. The (metaplectic cover of) the adelic group $Sp(W, \mathbb{A}_\mathbb{F})$ acts on the space of Schwartz–Bruhat functions $\mathcal{S}(V(\mathbb{A}_\mathbb{F}))$ via the Weil representation ρ:

$$\rho\colon Sp(W, \mathbb{A}_\mathbb{F}) \longrightarrow \mathcal{S}(V(\mathbb{A}_\mathbb{F})). \tag{12.40}$$

Generalising the discussion in Section 12.2, this representation can be realised as follows:

$$\begin{aligned}
\rho\left(\begin{pmatrix} A & 0 \\ 0 & (A^{-1})^T \end{pmatrix}\right) f(X) &= \sqrt{|\det(A)|}\, f(A^T X), \\
\rho\left(\begin{pmatrix} \mathbb{1} & B \\ 0 & \mathbb{1} \end{pmatrix}\right) f(X) &= \psi\left(\frac{X^T B X}{2}\right) f(X), \\
\rho\left(\begin{pmatrix} 0 & \mathbb{1} \\ -\mathbb{1} & 0 \end{pmatrix}\right) f(X) &= \gamma \widehat{f(X)},
\end{aligned} \tag{12.41}$$

where $\psi : \mathbb{F}\backslash\mathbb{A}_\mathbb{F} \to U(1)$ is a non-trivial character, and γ is a root of unity. This action commutes with the natural action of $O(V)$ on V.

Remark 12.6 Strictly speaking this only defines a projective representation of $Sp(W, \mathbb{A}_\mathbb{F})$, but we can lift this to an honest representation of the metaplectic cover $\widetilde{Sp(W, \mathbb{A}_\mathbb{F})}$ (see Section 12.2) [264]. We shall often abuse notation and nevertheless refer to the projective representation ρ on $Sp(W, \mathbb{A}_\mathbb{F})$ as the Weil representation.

Now, for any $\phi \in \mathcal{S}(V(\mathbb{A}_\mathbb{F}))$, define a distribution

$$\begin{aligned}
\theta\colon \mathcal{S}(V(\mathbb{A}_\mathbb{F})) &\longrightarrow \mathbb{C} \\
\phi &\longmapsto \sum_{x \in V(\mathbb{F})} \phi(x).
\end{aligned} \tag{12.42}$$

From this distribution we define a function, the *theta kernel*, by restriction from $Sp(W)$ to $Sp(V') \times O(V)$ as discussed above:

$$\theta_\phi\colon Sp(V', \mathbb{F})\backslash Sp(V', \mathbb{A}_\mathbb{F}) \times O(V, \mathbb{F})\backslash O(V, \mathbb{A}_\mathbb{F}) \longrightarrow \mathbb{C} \tag{12.43}$$

by the formula

$$\theta_\phi(g,h) = \sum_{x\in V(\mathbb{F})} \rho(g)\cdot\phi(h^{-1}x), \qquad g\in Sp(V',\mathbb{A}_\mathbb{F}),\ h\in O(V,\mathbb{A}_\mathbb{F}). \tag{12.44}$$

Example 12.7: From global theta series to classical Jacobi theta series

Let us consider the simplest example of theta series on $\widetilde{SL(2,\mathbb{A})} \cong \widetilde{Sp(2,\mathbb{A})}$, implying $V(\mathbb{F})\cong\mathbb{F}$, to see how they are related to the classical Jacobi theta series. To this end we fix $h=1$ and parametrise the element $g\in SL(2,\mathbb{A})$ by $g=(g_\mathbb{R};\mathbb{1})$, where $g_\mathbb{R} = \begin{pmatrix}1 & x\\ & 1\end{pmatrix}\begin{pmatrix}\sqrt{y} & \\ & 1/\sqrt{y}\end{pmatrix}$. We further let $\mathbb{F}=\mathbb{Q}$ and suppose that $\phi\in\mathcal{S}(\mathbb{A})$ is Eulerian:

$$\phi = \phi_\infty \times \prod_{p<\infty}\phi_p. \tag{12.45}$$

For all finite primes, let $\phi_p = \phi_p^\circ$, the unique spherical vector given by the p-adic Gaussian γ_p defined in (2.51). At the archimedean place we furthermore fix the real Gaussian

$$\phi_\infty(x) = e^{-\pi x^2}. \tag{12.46}$$

The global unramified unitary character $\psi\colon \mathbb{Q}\backslash\mathbb{A}\to U(1)$ also factorises into $\psi = \psi_\infty\times\prod_{p<\infty}\psi_p$, where ψ_p is trivial on $\mathbb{Z}_p$ and

$$\psi_\infty(x) = e^{2\pi i x}, \qquad x\in\mathbb{R}. \tag{12.47}$$

Using the Weil representation one then finds that the global theta function (12.44) reduces as follows:

$$\begin{aligned}\theta_\phi(g) &= \theta\Big(\begin{pmatrix}1 & x\\ & 1\end{pmatrix}\begin{pmatrix}\sqrt{y} & \\ & 1/\sqrt{y}\end{pmatrix},\mathbb{1}\Big)\\ &= y^{1/4}\sum_{n\in\mathbb{Q}}\Big[\prod_{p<\infty}\gamma_p(n)\Big]e^{\pi i n^2 x}e^{-\pi n^2 y}\\ &= y^{1/4}\sum_{n\in\mathbb{Z}} e^{\pi i n^2 x}e^{-\pi n^2 y}\\ &= y^{1/4}\theta(\tau),\end{aligned} \tag{12.48}$$

where the restriction to a summation over integers is imposed by γ_p.

12.5 The Siegel–Weil Formula

Let $\theta_\phi(g,h)$ be a global theta series defined as in (12.44). In this section we shall study the theta lift

$$\Theta_\varphi(g,\phi) = \int_{O(V)(\mathbb{F})\backslash O(V)(\mathbb{A}_\mathbb{F})} \varphi(h)\theta_\phi(g,h)dh, \tag{12.49}$$

where $\varphi \in \mathcal{A}(O(V,\mathbb{F})\backslash O(V,\mathbb{A}_\mathbb{F}))$. The Siegel–Weil formula is a special case of this theta lift where one takes φ to be the identity function [427, 428]. To state it we need some additional data on Eisenstein series on symplectic groups.

From now on we fix $Sp(V',\mathbb{A}_\mathbb{F})$ to be $Sp(2n,\mathbb{A}_F)$ as well as $m = \dim V$. Let $P = LU$ be the Siegel maximal parabolic subgroup of $Sp(2n)$ with Levi factor

$$L = \left\{ l(a) = \begin{pmatrix} a & \\ & (a^T)^{-1} \end{pmatrix} \,\middle|\, a \in GL(n) \right\} \tag{12.50}$$

and unipotent radical

$$U = \left\{ u(b) = \begin{pmatrix} \mathbb{1}_n & b \\ & \mathbb{1}_n \end{pmatrix} \,\middle|\, b = b^T \right\}. \tag{12.51}$$

We have the (non-unique) decomposition $Sp(2n,\mathbb{A}_\mathbb{F}) = U(\mathbb{A}_\mathbb{F})L(\mathbb{A}_\mathbb{F})K_{\mathbb{A}_F}$ and we write

$$g = u\, l(a)\, k \in Sp(2n,\mathbb{A}_\mathbb{F}), \tag{12.52}$$

with $a \in GL(n,\mathbb{A}_\mathbb{F})$. We further introduce the quantity

$$|a(g)| = |\det(a)|. \tag{12.53}$$

Now define a standard section of the adelic principal series according to

$$\Phi(g,s,\phi) = \rho(g)\cdot\phi(0)|a(g)|^{s-s_0(m,n)}, \tag{12.54}$$

where we introduced the parameter

$$s_0(m,n) = \frac{m}{2} - \frac{n+1}{2}. \tag{12.55}$$

We now have all the data required to define the *Siegel–Eisenstein series*,

$$E(g,s,\phi) = \sum_{\gamma\in P(\mathbb{F})\backslash Sp(2n,\mathbb{F})} \Phi(\gamma g,s,\phi). \tag{12.56}$$

Provided that the function ϕ is $K_{\mathbb{A}_\mathbb{F}}$-finite this converges absolutely for $\mathrm{Re}(s) > (n+1)/2$ and admits a meromorphic continuation and functional equation [427].

Finally we can state the *Siegel–Weil formula*:

Theorem 12.8 (Siegel–Weil formula) *Let $\Theta_\varphi(g,\phi)$ be the theta lift defined in (12.49) and let $E(g,s,\phi)$ be the Siegel–Eisenstein series defined above. Then we have* [612, 427, 428]

$$\Theta_1(g,\phi) = \kappa E(g,s_0(m,n),\phi), \tag{12.57}$$

where

$$\kappa = \begin{cases} 1 & m > n+1, \\ 2 & m \le n+1. \end{cases} \tag{12.58}$$

The Siegel–Weil formula thus states that the theta lift of the trivial function $\varphi = 1$ on $O(V,\mathbb{F})\backslash O(V,\mathbb{A}_\mathbb{F})$ is a Siegel–Eisenstein series on $Sp(2n,\mathbb{F})\backslash Sp(2n,\mathbb{A}_\mathbb{F})$.

Example 12.9: Siegel–Weil formula for the dual pair $SL(2)\times O(V)$

Consider now the special case of the Siegel–Weil formula when $n = 1$, corresponding to the dual reductive pair $SL(2)\times O(V)$. In this case the Siegel–Weil formula provides a transfer of automorphic forms on the orthogonal group $O(V)$ to $SL(2)$. The Siegel–Eisenstein series reduces to the standard principal Eisenstein series on $SL(2,\mathbb{A}_\mathbb{F})$ discussed at length in Section 5.6.1:

$$E(\mathfrak{f}_\lambda, g) = \sum_{\gamma\in B(\mathbb{F})\backslash SL(2,\mathbb{F})} \mathfrak{f}_\lambda(\gamma g), \tag{12.59}$$

where $\mathfrak{f}_\lambda = \Pi_v \mathfrak{f}_{\lambda,v}$ is a standard section of the adelic principal series which is non-spherical at the archimedean place:

$$\mathfrak{f}_{\lambda,\infty}\left(g_\infty\begin{pmatrix}\cos\theta & \sin\theta\\ -\sin\theta & \cos\theta\end{pmatrix}\right) = e^{iw\theta}\mathfrak{f}_{\lambda,\infty}(g_\infty). \tag{12.60}$$

The Siegel–Weil formula now reads explicitly

$$\sum_{\gamma\in B(\mathbb{F})\backslash SL(2,\mathbb{F})} \mathfrak{f}_\lambda(\gamma g) = \int_{O(V)(\mathbb{F})\backslash O(V)(\mathbb{A}_\mathbb{F})} \theta_\phi(g,h)dh. \tag{12.61}$$

Remark 12.10 The Siegel–Weil formula was proven by Weil [612] assuming certain convergence properties of the theta integral. In a series of works, Kudla and Rallis [427, 428, 429] developed the theory of the *regularised Siegel–Weil formula*, which holds when these convergence properties are not satisfied. Recently, the results of Kudla–Rallis were extended by Gan–Qiu–Takeda [245], who proved the regularised Siegel–Weil formula in full generality.

12.6 Exceptional Theta Correspondences

There exist versions of theta correspondences which do not correspond to orthogonal-symplectic dual pairs. In particular, one can consider the case when the ambient group G is an exceptional Lie group [282, 466]. At the level of complex Lie groups we have the following examples of dual pairs:

$$\begin{aligned} PGL(3)\times G_2 &\subset E_6,\\ PGSp(6)\times G_2 &\subset E_7,\\ F_4\times G_2 &\subset E_8. \end{aligned} \tag{12.62}$$

These examples fit into the ‘tower of theta correspondences’ constructed in [282]. The tower consists of a sequence of dual pairs (H, H'), where the first member is always $H = G_2$ and the second varies. In this setting the Weil representation is replaced by the minimal automorphic representation π_{min} of G. This representation is unique, factorisable $\pi_{\text{min}} = \otimes_v \pi_{\text{min},v}$ and has functional dimension (see also Table 6.1)

$$\begin{aligned} E_6 &: \text{GKdim}(\pi_{\text{min}}) = 11, \\ E_7 &: \text{GKdim}(\pi_{\text{min}}) = 17, \\ E_8 &: \text{GKdim}(\pi_{\text{min}}) = 29. \end{aligned} \tag{12.63}$$

Let $\theta_G(g)$, $g \in G(\mathbb{A}_\mathbb{F})$, be an automorphic form in the space of π_{min}. This may be viewed as the exceptional analogue of the Siegel theta series of $Sp(2n)$ considered in the previous section. By restriction to the dual pair $H \times H' \subset G$, this yields a function $\theta_G(h, h')$ on $H(\mathbb{F})\backslash H(\mathbb{A}_\mathbb{F}) \times H'(\mathbb{F})\backslash H'(\mathbb{A}_\mathbb{F})$, i.e., the analogue of the theta kernel. Let π be an automorphic representation of $H'(\mathbb{A}_\mathbb{F})$. For any $\varphi \in \pi$ we then have the exceptional theta lift

$$\Theta_\varphi(h) = \int_{H(\mathbb{F})\backslash H(\mathbb{A}_\mathbb{F})} \varphi(h)\theta_G(h, h')dh. \tag{12.64}$$

For $H' = G_2$, this yields a tower of maps

$$\Theta_H^{G_2} : \mathcal{A}(H(\mathbb{F})\backslash H(\mathbb{A}_\mathbb{F})) \longrightarrow \mathcal{A}(G_2(\mathbb{F})\backslash G_2(\mathbb{A}_\mathbb{F})), \tag{12.65}$$

and we denote the space of such theta lifts by $\Theta_G^{G_2}(\pi)$ as f varies in π and θ_G varies in π_{min}. The space $\Theta_G^{G_2}(\pi)$ is thus a subspace of the space of automorphic forms on $G_2(\mathbb{A}_\mathbb{F})$.

Consider now the case when G is the quaternionic real form of either E_7 or E_8. These real forms are characterised by having real rank four, and the restricted root system is of type F_4 (in the literature they are sometimes denoted by $E_{7(-5)}$ and $E_{8(-24)}$). The maximal compact subgroup is of the form $SU(2) \times M$, where M is the compact form of D_6 for $G = E_7$ and the compact form of E_7 for $G = E_8$. The first member H in the dual pair corresponds to the automorphism group of the exceptional Jordan algebra of 3×3 Hermitian matrices (see [364] for details) with coefficients in the quaternionic division algebra $\mathbb{H}$ (for E_7), or the octonionic division algebra $\mathbb{O}$ (for E_8). For both E_7 and E_8 the second member H' in the dual pair is the split real form of G_2.

Gan [242] has constructed an automorphic realisation of the minimal representation π_{min} of $G(\mathbb{A}_\mathbb{F})$, where $G(\mathbb{F}_v)$ is split for all finite places v, while the archimedean factor $G(\mathbb{F}_\infty)$ is the quaternionic real form of E_7 or E_8. (Quaternionic real forms will also be discussed in Section 18.7.) This was then used to study the theta lift from automorphic forms on $H(\mathbb{F})\backslash H(\mathbb{A}_\mathbb{F})$ to $G_2(\mathbb{F})\backslash G_2(\mathbb{A}_\mathbb{F})$. Let $P = LU$ be the Heisenberg parabolic subgroup of G with modulus character δ_P. The Levi factor L has derived group of type

D_6 for $G = E_7$ and type E_7 for $G = E_8$. For $s \in \mathbb{C}$, let $\mathfrak{f}_s = \prod_v \mathfrak{f}_{s,v}$ be a standard section of the induced representation $\mathrm{Ind}_{P(\mathbb{A}_\mathbb{F})}^{G(\mathbb{A}_\mathbb{F})} \delta_P^s$ and construct the corresponding Eisenstein series

$$E(g, \mathfrak{f}_s) = \sum_{\gamma \in P(\mathbb{F}) \backslash G(\mathbb{F})} \mathfrak{f}_s(\gamma g). \tag{12.66}$$

Gan then proved the following:

Theorem 12.11 (Automorphic realisation of $\pi_{\min}$ for quaternionic real forms) *Let $G(\mathbb{A}_\mathbb{F})$ be the adelic group associated with the quaternionic real form of E_7 or E_8 as above. For any standard section $\mathfrak{f}_s \in \mathrm{Ind}_{P(\mathbb{A}_\mathbb{F})}^{G(\mathbb{A}_\mathbb{F})} \delta_P^s$ the Eisenstein series $E(g, \mathfrak{f}_s)$ has at most a simple pole at $s = s_0$, with*

$$s_0 = \begin{cases} 3/17 & \text{for} \quad G = E_7, \\ 5/29 & \text{for} \quad G = E_8. \end{cases} \tag{12.67}$$

At this value of s the global minimal representations $\pi_{\min}$ of the quaternionic real forms of the exceptional groups arise as a square-integrable subspace [242, 243]

$$\pi_{\min} \subset \mathrm{Ind}_{P(\mathbb{A}_\mathbb{F})}^{G(\mathbb{A}_\mathbb{F})} \delta_P^s \Big|_{s=s_0}, \tag{12.68}$$

and its automorphic realisation is spanned by the residues of the Eisenstein series $E(g, \mathfrak{f}_s)$ at $s = s_0$.

Remark 12.12 The above theorem may be viewed as the quaternionic version of an earlier result by Ginzburg–Rallis–Soudry [281], who constructed an automorphic realisation of the minimal representation of $E_8(\mathbb{A}_\mathbb{F})$ when the archimedean component is the split real form of E_8.

Remark 12.13 In contrast to the split case considered in [281], the minimal representation $\pi_{\min} = \otimes_v \pi_{\min,v}$ of the $G(\mathbb{A}_\mathbb{F})$ constructed by Gan may be non-spherical for some finite places v.

We shall now proceed to discuss theta correspondences in the context of quaternionic exceptional groups. Let us denote by $\theta_G(g)$ the (residue of the) Eisenstein series $E(g, \mathfrak{f}_s)$ at $s = s_0$. By restriction of $g \in G$ to the dual pair $H \times G_2$, through $g = hh'$ we obtain the theta kernel

$$\theta_G(h, h') = E(hh', \mathfrak{f}_{s_0}). \tag{12.69}$$

For π an automorphic representation of $H(\mathbb{A}_\mathbb{F})$ and $\varphi \in \pi$ we then write the theta lift as

$$\Theta_\varphi(h', \mathfrak{f}_{s_0}) = \int_{H(\mathbb{F}) \backslash H(\mathbb{A}_\mathbb{F})} \varphi(h) E(hh', \mathfrak{f}_{s_0}) dh. \tag{12.70}$$

Gan proved that Θ_φ is non-zero and cuspidal under certain conditions on π. In particular:

- Θ_φ is cuspidal if and only if

$$\int_{C(\mathbb{F})\backslash C(\mathbb{A}_\mathbb{F})} \varphi(c)dc = 0, \tag{12.71}$$

  where C is the stabiliser of a particular character ψ of the unipotent radical of a maximal parabolic of G (see [242] for the precise definition).
- Θ_φ is zero if and only if

$$\int_{Z(\mathbb{F})\backslash Z(\mathbb{A}_\mathbb{F})} \varphi(zh)dz = 0, \tag{12.72}$$

  where $Z = [U, U]$ is the centre of the Heisenberg unipotent radical U of P.

For the case of $G = E_8$, Gan has also established a Siegel–Weil formula for the dual pair $H \times G_2 \subset E_8$ [243]. In this case, H is a real form of F_4. The Siegel–Weil formula arises from the theta lift of the trivial function $\varphi = 1$ on $F_4(\mathbb{A}_\mathbb{F})$ to $G_2(\mathbb{A}_\mathbb{F})$:

$$\Theta_1(h', \mathfrak{f}_{s_0}) = \int_{F_4(\mathbb{F})\backslash F_4(\mathbb{A}_\mathbb{F})} E(hh', \mathfrak{f}_{s_0})dh. \tag{12.73}$$

In this situation the theta lift does not give rise to cusp forms, but, analogously to the classical case of orthogonal-symplectic pairs, it yields certain Eisenstein series. To state it we first need some information on Eisenstein series on G_2. Let $P_2 = L_2U_2$ be the maximal parabolic subgroup of G_2 with unipotent radical U_2 a five-dimensional Heisenberg group, and Levi subgroup $L_2 = SL(2)\times GL(1)$. Denote the modulus character of P_2 by δ_{P_2} and for $s \in \mathbb{C}$ consider the global degenerate principal series

$$I(s) = \mathrm{Ind}_{P_2(\mathbb{A}_\mathbb{F})}^{G_2(\mathbb{A}_\mathbb{F})} \delta_{P_2}^s = \bigotimes_\nu I_\nu(s). \tag{12.74}$$

For any standard section $\mathfrak{f}_s \in I(s)$ we have the Eisenstein series

$$E(g, \mathfrak{f}_s) = \sum_{\gamma\in P_2(\mathbb{F})\backslash G_2(\mathbb{F})} \mathfrak{f}_s(\gamma g). \tag{12.75}$$

We will not be needing the Eisenstein series for any such section, but rather a special case thereof, obtained by restriction from the minimal representation π_{min} of E_8. Since π_{min} is a submodule of the degenerate principal series $\mathrm{Ind}_{P(\mathbb{A}_\mathbb{F})}^{E_8(\mathbb{A}_\mathbb{F})} \delta_P^{5/29}$ one can restrict it to a subspace of $\mathrm{Ind}_{P_2(\mathbb{A}_\mathbb{F})}^{G_2(\mathbb{A}_\mathbb{F})} \delta_{P_2}^{5/2}$. More precisely, one has a restriction map [243]:

$$\begin{aligned} \mathrm{Res}\colon \pi_{\mathrm{min}} &\longrightarrow \mathrm{Ind}_{P_2(\mathbb{A}_\mathbb{F})}^{G_2(\mathbb{A}_\mathbb{F})} \delta_{P_2}^{5/2} \\ f &\longmapsto \mathrm{Res}(f). \end{aligned} \tag{12.76}$$

A standard section obtained by restriction in this way is called a *Siegel–Weil section* [427]. These are the relevant ones for the Siegel–Weil formula. To understand the structure of these sections we need the following result [364]:

Proposition 12.14 (Restriction of the minimal representation of E_8) *Let $\pi_{\min} = \pi_{\min,\infty} \otimes \hat{\bigotimes}_\nu \pi_{\min,\nu}$ be the minimal representation of $E_8(\mathbb{A}_\mathbb{F})$, with archimedean component being the quaternionic real form of E_8. Then the restriction of $\pi_{\min}$ to its F_4-invariant subspace is a quaternionic discrete series representation π of the split real form $G_2(\mathbb{R})$, in the sense of Gross–Wallach* [327, 328].

Given the above result one has the following consequence:

Corollary 12.15 (Siegel–Weil section) *A standard section $\mathsf{f} = \mathsf{f}_\infty \times \hat{\prod}_\nu \mathsf{f}_\nu \in Ind_{P_2(\mathbb{A}_\mathbb{F})}^{G_2(\mathbb{A}_\mathbb{F})} \delta_{P_2}^{5/2}$ is a Siegel–Weil section if and only if $\mathsf{f}_\infty \in \pi$.*

A priori the theta lift $\Theta_1(h)$ in (12.73) can have a cuspidal component (in the discrete spectrum) and an Eisenstein component (in the continuous spectrum):

$$\Theta_1(h, \mathsf{f}) = (\Theta_1(h, \mathsf{f}))_{\text{Eisenstein}} + (\Theta_1(h, \mathsf{f}))_{\text{cusp}}. \tag{12.77}$$

The main result of [243] is then:

Theorem 12.16 (Exceptional Siegel–Weil formula for G_2) *For a section $\mathsf{f} \in \pi_{\min}$ of $E_8(\mathbb{A}_F)$, one has* [243]

$$\begin{aligned}(\Theta_1(h, \mathsf{f}))_{\text{cusp}} &= 0,\\ (\Theta_1(h, \mathsf{f}))_{\text{Eisenstein}} &= E(h, \text{Res}(\mathsf{f})).\end{aligned} \tag{12.78}$$

That is, the theorem implies the following explicit form of the Siegel–Weil formula:

$$\int_{F_4(\mathbb{F})\backslash F_4(\mathbb{A}_\mathbb{F})} E(hh', \mathsf{f}) dh = \sum_{\gamma \in P_2(\mathbb{F})\backslash G_2(\mathbb{F})} \text{Res}(\mathsf{f})(\gamma h'). \tag{12.79}$$

Remark 12.17 The above result is expected to have applications in string theory. In fact, the Eisenstein series $E(h, \text{Res}(\mathsf{f}))$ is a natural candidate for the type of automorphic form on G_2 that is relevant for Conjecture 15.13 below that comes out of string theory.

PART TWO

APPLICATIONS IN STRING THEORY

13

Elements of String Theory

This chapter marks the beginning of PART TWO, which discusses how the mathematical structures of the preceding PART ONE occur in string theory. The present chapter aims to give a rough understanding of some of the basic string theory concepts and terminology while Chapter 14 shows in much more detail how automorphic forms and representations arise in the study of certain string theory scattering amplitudes. Scattering amplitudes are but one aspect of the relation between string theory and automorphic forms, and in Chapter 15 we shall then discuss many other connections, including generalisations of the mathematics of PART ONE that string theory predicts.

After a brief outline of the elementary ingredients of string theory in Section 13.1, including the various types of string theories and their connections, we then introduce one of the main objects of interest for us, namely the scattering amplitude for four so-called gravitons in Section 13.2. We provide a discussion of its general structure and what properties are expected of it. String theory, when viewed as a perturbative series around a non-interacting theory, comes with a so-called loop expansion. In a starred section, we calculate the four-graviton amplitude for zero loops (a.k.a. tree-level), mainly in order to illustrate some of the techniques that are needed for determining amplitudes. Similarly, we discuss one-loop amplitudes and show how they are related to theta lifts that were discussed in Chapter 12. Beyond string perturbation theory there are non-perturbative effects in string theory that are called D-branes and instantons, which we discuss in Sections 13.5 and 13.6.

Many of the concepts in the later half of this chapter are fairly advanced and the corresponding sections are starred. We sometimes use the associated terminology in subsequent chapters but will also provide a 'dictionary' to the mathematical concepts of PART ONE. For example, the non-perturbative effects can often be understood as non-trivial Fourier coefficients as we shall explain in Section 14.2.4.

String theory is far too vast a subject to do it justice in just a single chapter, but our sometimes heuristic description of its basic features will hopefully provide a sufficient glimpse to spark the motivation for further studies. For more information about string theory we recommend the books [320, 634, 535, 63] and, for a brief introduction, the lecture notes [167, 601].

13.1 String Theory Concepts

Many aspects of string theory are phrased using the language of quantum field theory. As we shall use these terms, we first fix some of this very basic terminology.

Quantum field theory deals with the processes of interacting elementary point particles. The spectrum of elementary point particles is classified by irreducible unitary representations of the space-time *Poincaré group* and possibly internal symmetry groups according to the *Wigner classification*. A particle is described in terms of a field, that is, a section or a connection of some bundle over space-time. Very heuristically, higher values of the field describe where the particle is localised. These fields have to satisfy partial differential equations that typically follow from the variation of a functional called the *action*.

Elementary point particles can be of one of two types: they are either *bosons* and associated with the forces that exist between different types of matter or they are *fermions* or *spinors* and are the constituents of matter. An example of a (fermionic) matter particle is the electron while an example of a (bosonic) force particle could be the photon that is associated with light and the electromagnetic force. Another example of a bosonic force particle is the *graviton* that is the assumed carrier of the gravitational force. Unlike photons, gravitons have not been directly detected but they are part of most proposals for the elusive and much sought-after theory of *quantum gravity*. String theory is such a proposal that transcends quantum field theory but still admits gravitons, and the investigation of their properties is one place where automorphic forms arise.

A fundamental difference between bosons and fermions is their statistics. Fermions obey odd statistics while bosons are even. In mathematical terms this means that the space of many (non-interacting) fermions is formed as the *anti-symmetric* tensor power of the space of a single fermion. By contrast, the space of many (non-interacting) bosons is the *symmetric* tensor power of the space of a single boson. This implies that there cannot be two fermions in exactly the same state (*Pauli exclusion principle*) while there can be infinitely many bosons in the same state (*Bose–Einstein condensation*), and these properties are well established experimentally. Mathematically, fermions are modelled using anti-commuting Grassmann variables.

In many theories with bosons and fermions, one tries to implement *supersymmetry*, which is a symmetry that exchanges bosons and fermions and thus its symmetry generators are odd. The underlying mathematical structure is that of a $\mathbb{Z}_2$-graded superalgebra or supergroup, for example the super-Poincaré group that extends the normal Poincaré group by odd generators implementing the supersymmetry transformations. The number of independent supersymmetry generators is often called the number of *supercharges* and has to be a multiple $\mathcal{N}$ of the smallest real dimension of the corresponding Lorentz group. Supersymmetric theories are of theoretical interest as they often allow for much stronger statements about their physical properties. Supersymmetry has not been confirmed experimentally. It features prominently in many formulations of string theory.

Interactions in quantum field theory are often treated using *Feynman path integral techniques* where one sums (integrates) over all possible field configurations with a weight provided by evaluating the action on the configuration. Though formal and mathematically not well-defined, this procedure provides many statements that can be proven by different means, as well as providing an intuitive depiction of interactions in terms of Feynman diagrams.

Finally, in the context of quantum field theory, D-dimensional space-time is often denoted by $\mathbb{R}^{1,D-1}$, where the notation indicates that one direction has different properties from the remaining ones. As a Lorentzian metric space, $\mathbb{R}^{1,D-1}$ is the vector space $\mathbb{R}^D$ equipped with a (pseudo-)metric of signature $(-, +, +, \dots, +)$ and the minus direction is called the time direction. The space is commonly referred to as *(flat) Minkowski space* and the Lorentzian or Minkowski metric written as $\eta_{\mu\nu}$. Sometimes we shall also encounter *(flat) Euclidean space* $\mathbb{R}^D$, where the inner product is understood to be the standard positive definite inner product $\delta_{\mu\nu}$. In the expressions for the metric, μ and ν label the components of the bilinear form thought of as a matrix. The stability group of the Minkowski metric on $\mathbb{R}^{1,D-1}$ is the *Lorentz group* $SO(1, D-1)$.

13.1.1 From Point Particles to Strings

Instead of point particles, the fundamental objects in string theory are one-dimensional strings which can be either open or closed, and whereas a point particle traces out a trajectory over time, called a world-line, a string sweeps out a *world-sheet*, as pictured in Figure 13.1.

The world-sheet is a smooth, two-dimensional Lorentzian manifold Σ embedded in a Lorentzian manifold M of dimension D called *space-time* or the *target manifold*. To simplify computations, one often starts with Riemannian manifolds and then makes analytical continuations in the Euclidean time to

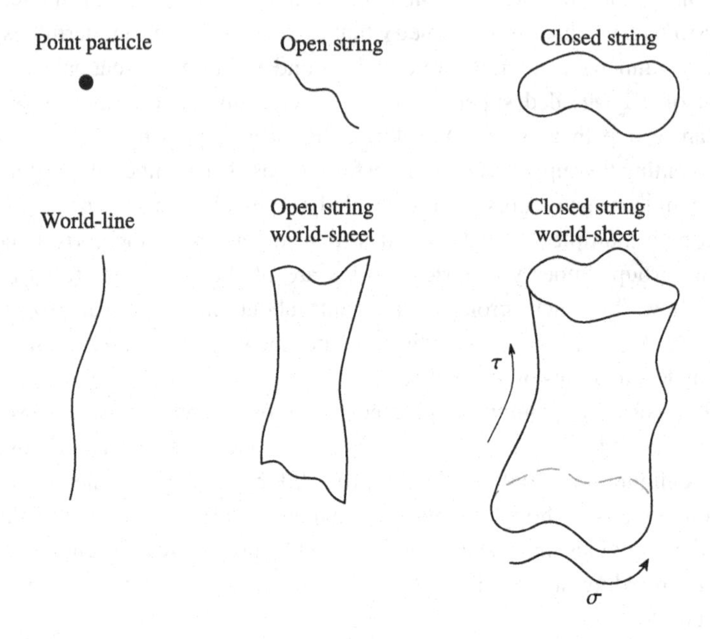

Figure 13.1 Point particle, open string and closed string as seen at an instant in time (top), and their world-line and world-sheets swept over time (bottom).

obtain physical results (where possible). The process of going from ordinary time to Euclidean time $t \to -it$ is called a *Wick rotation*.

String theory is then essentially a description of the dynamics of the embedding map

$$X : \Sigma \to M \tag{13.1}$$

or the coordinate maps X^μ where $\mu = 0, \ldots, D-1$. A Lorentzian world-sheet Σ is usually parametrised by σ and τ as illustrated in Figure 13.1, where the former is either periodic for closed strings or bounded for open strings and the latter is time-like. For a Riemannian world-sheet with Euclidean time σ^0 we will denote the coordinates collectively as σ^m, $m = 0, 1$, which are often combined into a complex variable z, allowing us to interpret Σ as a Riemann surface.

The classical solutions to the theory, that is, the classical shapes of the world-sheet, are found by extremising the action (13.4) discussed in Section 13.1.3. The fluctuations around a classical solution are later quantised and, according to the basic idea of string theory, the different modes should give rise to what we see as different elementary particles. We note that there is currently no experimental evidence for string theory but its mathematical richness – as exemplified by the connections in this book – has led to continuing research.

13.1.2 The Different Types of String Theories

There are several different string theories and depending on which one we study both Σ and M are endowed with additional structure (like a metric and a spinor bundle) and they have different particle spectra. The string theories that contain closed strings all have the *graviton* in their spectrum, which is one reason why string theory is called a quantum theory of gravity. Another important particle in the spectrum is the *dilaton*, which transforms as a scalar under the Lorentz group and will be discussed more in Section 13.1.3.

Unlike a canonical quantisation of Einstein's classical theory of gravity, the physical observables in string theory are finite for high energies, that is, string theory is *UV-finite*. 'UV' here stands for ultraviolet, which refers to very high frequency and has very energetic states of matter. If a theory is not UV-finite it is a sign that, at some large energy scale (or, equivalently, some small length scale), new physics emerges and requires a more detailed theory.

The bosonic string theories contain only bosons in their particle spectra and have several consistency issues due to the existence of a particle called the *tachyon*, which travels faster than the speed of light and thus violates causality. The tachyon is removed from the spectrum by introducing *supersymmetry*, which relates bosons with fermionic non-trivially extending the Poincaré group to the *super-Poincaré group*. The fact that we need to turn to a $\mathbb{Z}_2$-graded algebra is motivated by the *Coleman–Mandula* and *Haag–Łopuszański–Sohnius theorems* about possible symmetries of scattering processes, which put restrictions on how space-time symmetries can be combined with internal symmetries.

There are three ways of incorporating supersymmetry into string theory: either with space-time supersymmetry by using the so-called *Green–Schwarz formalism* or the *pure spinor formalism* where one lets M be a supermanifold introducing extra Grassmanian coordinates and embedding maps, or with world-sheet supersymmetry in the *Ramond–Neveu–Schwarz (RNS) formalism* with extra world-sheet spinors ψ^μ for each bosonic embedding coordinate X^μ. They are believed to be equivalent, but have different strengths highlighting different aspects of the theory. It is sometimes convenient to make the world-sheet supersymmetry manifest in the RNS formalism by letting Σ be a supermanifold. The fields X^μ and ψ^μ would then be coefficients in a series expansion of a superfield similar to what is discussed in Section 14.4. The RNS formalism is used in the scattering amplitude computation in Section 13.3.

Table 13.1 String theories with critical dimension D, the number of (left-handed, right-handed) world-sheet supersymmetry generators and whether they contain open strings. When introducing D-branes (discussed below in Section 13.5), the type II theories may also have open strings whose endpoints are attached to these. The fourth and fifth entries are heterotic string theories, which are hybrids of bosonic and type I strings and differ by their gauge groups: $SO(32)$ or $E_8 \times E_8$.

String theory	D	World-sheet supersymmetry	Open strings
Type IIA	10	$(1, 1)$	Only attached to D-branes
Type IIB	10	$(2, 0)$	Only attached to D-branes
Type I	10	$(1, 0)$	Yes
Het $SO(32)$	10	$(1, 0)$	No
Het $E_8 \times E_8$	10	$(1, 0)$	No
bosonic	26	0	Yes/no

A supersymmetric string theory is strictly called a *superstring theory*, which we almost exclusively consider in this book. We follow common usage and for simplicity call it string theory. The space of allowed world-sheets Σ is then the space of all closed, orientable super-Riemann surfaces if we specialise to so-called type IIA or type IIB superstrings; otherwise one might have to include boundaries and non-orientable surfaces as well.

For completeness, we have listed the different types of string theories in Table 13.1. They differ, for example, in which target spaces they can live on, how much supersymmetry they have and whether they contain open strings or not. Note that interacting string theories with open strings must also contain closed strings since the two endpoints of an open string may join together, and that if we include D-branes, which will be discussed in Section 13.5, we may also have open strings with endpoints attached to these branes.

The amount of world-sheet supersymmetry is given by the number of world-sheet supersymmetry generators on the string world-sheet, that is, the number of irreducible real spin representations. The generators are often separated into left and right chirality where applicable, that is determined by their eigenvalues under the unique (up to normalisation) highest-rank Clifford algebra element, also called the orientation operator.

When the target space is flat, we will see in Section 13.1.5 that string theory can only be consistent for a certain critical dimension of space-time. For bosonic string theories, this dimension is $D = 26$, while for superstring theory it is $D = 10$. An important feature of the RNS formulation is that, even though one started with supersymmetry on the world-sheet Σ, one obtains a supersymmetric spectrum in the space-time $\mathbb{R}^{1,9}$.

To study physics in lower dimensions, one considers target spaces that are compactified in certain directions. In order to preserve much of the supersymmetry of the superstring theories one often compactifies on d-dimensional tori T^d giving $M = \mathbb{R}^{1,9-d} \times T^d$ or on compact $K3$ surfaces with $M = \mathbb{R}^{1,3} \times X$ and $X = K3 \times T^2$, or lets X be certain Calabi–Yau threefolds. These compactifications preserve all, half and a quarter of the supersymmetry charges, respectively. In the remainder of this chapter we will mainly focus on toroidal compactifications, but other compactifications are discussed, for example, in Chapter 15.

The superstring theories in Table 13.1 are intricately related by certain *string dualities*, meaning that physical results, seen as functions on the parameters defining the quantum theory, map to those of another string theory when restricting to certain subsets of the different parameter spaces.

An important example is the *T-duality* relating type IIA and IIB on $M = \mathbb{R}^{1,8} \times S^1$ which states that a type IIA theory on a circle with radius R is equivalent to a type IIB theory on a circle with radius α'/R, where the parameter α' describes a typical area scale of the embedded world-sheet and is called the *Regge slope* or the inverse *string tension*. It is related to the characteristic *string length* ℓ_s by $\alpha' = \ell_s^2$.

Additionally, there is *S-duality* which relates a strongly interacting theory with a weakly interacting theory, an important example being type I with string coupling g_s which is S-dual to $SO(32)$ heterotic string theory with string coupling $1/g_s$. More importantly for us in the following is that type IIB string theory is S-dual to itself, thus relating weak coupling expansions to strong coupling expansions. The union of T-duality and S-duality is called *U-duality*. In Section 13.1.6, we will discuss *U-dualities* for toroidal compactifications, and later we shall see that they play the rôle of the discrete symmetry group of automorphic forms.

The web of dualities binding together the various superstring theories pictured in Figure 13.2 is seen as an indication that there is another fundamental theory behind these relations and that taking different limits of this fundamental theory leads to the different superstring theories [620]. This theory is called *M-theory* and although its full formulation is not known, the fundamental objects are, instead of strings, higher-dimensional membranes on an eleven-dimensional space-time. Its low-energy effective action (a concept that will be discussed more in Section 13.2) is the unique eleven-dimensional maximal *supergravity*. Note that this is the largest space-time dimension for which there exists a supergravity theory [218].

M-theory compactified on a circle S^1 gives type IIA string theory with a string coupling g_s proportional to the radius of the circle (in string units), and if we instead compactified on $S^1/\mathbb{Z}_2$ we would similarly get $E_8 \times E_8$ heterotic string theory. Conversely, the strong coupling limit of type IIA (and one of the possible strong coupling limits of the $E_8 \times E_8$ heterotic theory) is dual to

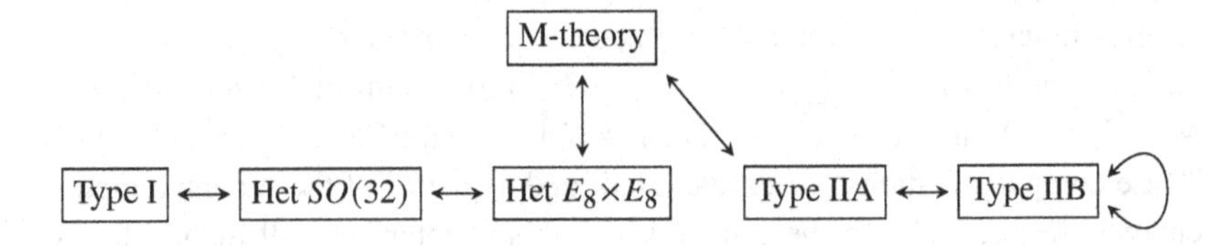

Figure 13.2 Web of dualities forming M-theory. The two type II theories are related by T-duality and the same also for the two heterotic theories. The remaining dualities are S-dualities relating a strongly interacting theory to a weakly interacting theory.

M-theory, where we note that $g_s \to \infty$ means that the radius of the circle goes to infinity and thus decompactifies.

13.1.3 Action and Symmetries

Since many principles and features from bosonic string theory carry over to superstring theory, we will, in this section, study the action and symmetries of the bosonic theory.

We fix a (Riemannian) space-time manifold M with a metric

$$G = G_{\mu\nu}(x)dx^\mu \otimes dx^\nu\,, \tag{13.2}$$

where x^μ are coordinates on M and repeated indices are summed over according to the *Einstein summation convention*. The gravitons in the string spectrum discussed above would then give rise to small perturbations around this background space-time. The metric G on M induces a metric γ on Σ through the pull-back of the embedding map X of (13.1):

$$\gamma = X^*(G) = \gamma_{mn} d\sigma^m \otimes d\sigma^n, \qquad \gamma_{mn} = G_{\mu\nu}(X)\frac{\partial X^\mu}{\partial\sigma^m}\frac{\partial X^\nu}{\partial\sigma^n}\,. \tag{13.3}$$

Analogous to a point particle freely propagating from a point A to a point B in a curved space-time, where a classical solution for the geodesic world-line is found by extremising the proper length of the trajectory, classical string solutions are found by extremising the world-sheet area

$$\text{Area}(X(\Sigma)) = \int_\Sigma d^2\sigma\sqrt{\det\gamma}\,. \tag{13.4}$$

This expression is called the *Nambu–Goto action*, but since the square root makes it difficult to quantise we introduce instead an (auxiliary) intrinsic metric on Σ which is initially independent of the embedding map X:

$$g = g_{mn}(\sigma)d\sigma^m \otimes d\sigma^n, \tag{13.5}$$

where by $g_{mn}(\sigma)$ we mean a dependence on both σ^0 and σ^1. We then take the *Polyakov action* as our bosonic string theory action:

$$S_P[g, G, X] = \frac{1}{4\pi\alpha'} \int_\Sigma d^2\sigma \sqrt{g}\, g^{mn} \partial_m X^\mu \partial_n X^\nu G_{\mu\nu}(X), \qquad (13.6)$$

where $\partial_m = \frac{\partial}{\partial \sigma^m}$, $\sqrt{g} = \sqrt{\det g}$ and the parameter $\alpha' = \ell_s^2$ involving the string length scale ℓ_s is used to give the correct dimension for S_P. Extremising the Polyakov action S_P with respect to the world-sheet metric gives the *equation of motion* for g,

$$\partial_m X^\mu \partial_n X^\nu G_{\mu\nu} - \frac{1}{2} g_{mn} g^{pq} \partial_p X^\mu \partial_q X^\nu G_{\mu\nu} = 0, \qquad (13.7)$$

which tells us that $g \propto \gamma$, and this makes $S_P[\gamma, G, X] \propto \text{Area}(X(\Sigma))$. The Polyakov action is, however, much easier to quantise, which we will later do in the *path integral formalism*. We thus choose to not replace g with the solution of the above equation, but will then have to impose (13.7), called the *Virasoro constraint*, at a later stage.

Let us also consider the equations of motion for the embedding maps X^μ of the string world-sheet Σ. For simplicity, we do this for the case when the space-time metric G is constant with respect to X, for which the Polyakov action (13.6) becomes extremal when

$$\Delta_\Sigma X^\mu = 0, \qquad (13.8)$$

where Δ_Σ is the Laplace–Beltrami operator on Σ. This means that the embedding is given by harmonic maps.

Besides (13.6) there are several more terms that can be added to the total action which, for example, describe different background fields such as the *dilaton* $\phi(X)$, with

$$S_{\text{dilaton}} = \frac{1}{4\pi} \int_\Sigma d^2\sigma \sqrt{g}\, \phi(X) R^{(2)}(g), \qquad (13.9)$$

where $R^{(2)}$ is the Ricci curvature scalar of the world-sheet.

We can also pick a two-form $B = \frac{1}{2} B_{\mu\nu} dx^\mu \wedge dx^\nu$ on M, often referred to as the *B-field*. The pull-back of B induces a two-form on Σ:

$$X^*(B) = \frac{1}{2} B_{mn} d\sigma^m \wedge d\sigma^n. \qquad (13.10)$$

This induces a term in the Polyakov action, generalising (13.6). The full action can then be compactly written

$$S[g, G, B, \phi, X] = S_P[g, G, X] + \frac{i}{2\pi\alpha'} \int_\Sigma X^*(B) + \frac{1}{4\pi} \int_\Sigma d^2\sigma \sqrt{g}\, \phi(X) R^{(2)}, \qquad (13.11)$$

which is a *non-linear sigma model* and is sometimes called the full Polyakov action.

Let us also consider some important symmetries of the Polyakov action (13.6). It is manifestly diffeomorphism invariant on both Σ and M, but it is also invariant under world-sheet *Weyl transformations*

$$g_{mn} \to e^{2\omega(\sigma)} g_{mn} \,. \tag{13.12}$$

If we also (locally) choose coordinates on the world-sheet where the metric is *conformally flat* $g_{mn} = e^{2\omega(\sigma)}\delta_{mn}$ the Weyl invariance allows us to work with a flat metric. The fact that these coordinates cannot always be used globally is important when we later need to study equivalence classes of world-sheet metrics under diffeomorphisms and Weyl transformations in Section 13.1.4. Because of these properties, perturbative string theory is an example of a *conformal field theory*.

Even after fixing a conformally flat metric, we have some residual Diff$\ltimes$Weyl symmetry which is exactly the *conformal symmetry*. For a conformal field theory in two dimensions, the generators of conformal transformations acting on the quantum fields are (two copies of) the *Virasoro generators* L_m satisfying

$$[L_m, L_n] = (m-n)L_{m+n} + \frac{c}{12}(m^3 - m)\delta_{m+n,0}, \tag{13.13}$$

where c is the *central charge*.

Finally, by Stokes' theorem the full action (13.11) is invariant under the *gauge transformation*

$$B \to B + d\Lambda, \tag{13.14}$$

where Λ is a one-form.

13.1.4 Interactions and the Path Integral Formalism

Among the important quantities that can be computed using path integrals are *scattering amplitudes*, which describe probabilities of different interactions. Strings interact by splitting and joining as shown in Figure 13.3, the strength (or, loosely speaking, probability) of which is given by a parameter g_s called the *string coupling constant*. As will be discussed later, g_s is closely related to constant mode ϕ_0 of the dilaton background field (usually taken to be the asymptotic value at infinity). Note also that if we let the typical string length be very small, i.e., let $\alpha' \to 0$, the strings look effectively like point particles and the world-sheet diagrams look a lot like *Feynman diagrams* commonly used in the *quantum field theories* of point particles. The limit $\alpha' \to 0$ is therefore sometimes called the point particle limit or field theory limit.

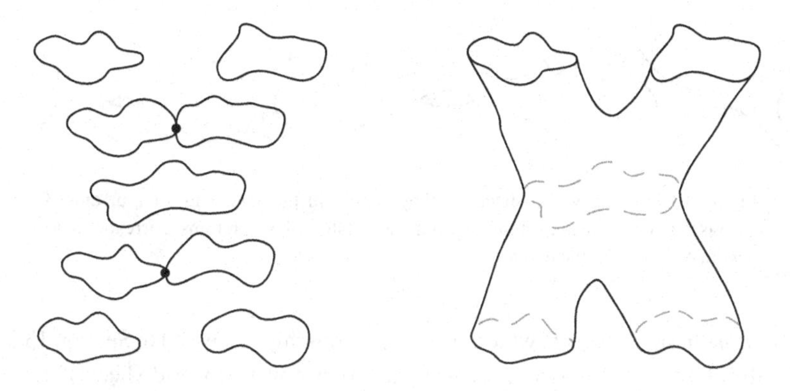

Figure 13.3 Two strings joining and splitting seen as snapshots in time (left) and as a world-sheet (right).

The bosonic spectrum of string excitations in flat ten-dimensional Minkowski space $M = \mathbb{R}^{1,9}$ is given by states of mass (squared)

$$m^2 = \frac{2}{\alpha'} N, \tag{13.15}$$

where $N \in \mathbb{Z}_{\geq 0}$ is the excitation number of the string states. The lightest string states (which include the graviton) are massless, and there is an infinite sequence of massive states with quantised masses with the separation between the masses set by α'. At every mass level one has a finite number of degrees of freedom. In the limit $\alpha' \to 0$ the massive states ($N > 0$) in (13.15) become infinitely heavy compared to the massless states ($N = 0$).

This also implies that when one considers only scattering processes with small energies compared to the energy scale defined by α' one will not be able to detect much of the effect of the massive states, since Einstein's relation $E = mc^2$ tells us that energy and mass are the same when using the standard convention of setting $c = 1$. In other words, when taking the limit of scattering amplitudes where dimensionless quantities like the energies of the scattering excitations measured in units of α' are small, we are in the regime of a quantum field theory description of the massless string theory states, called the *effective action of string theory*. The $\alpha' \to 0$ limit is also called the *low-energy limit of string theory*. We will be more precise on this point when discussing explicit scattering amplitudes in Section 13.2, where we will expand in the dimensionless quantities defined in (13.34).

When computing scattering amplitudes, the external states (that is, the four legs shown to the right in Figure 13.3) are stretched out towards spatial infinity and can be mapped to four punctures on a sphere, as seen in Figure 13.6 in Section 13.3 using Weyl invariance. These punctures are represented by *vertex*

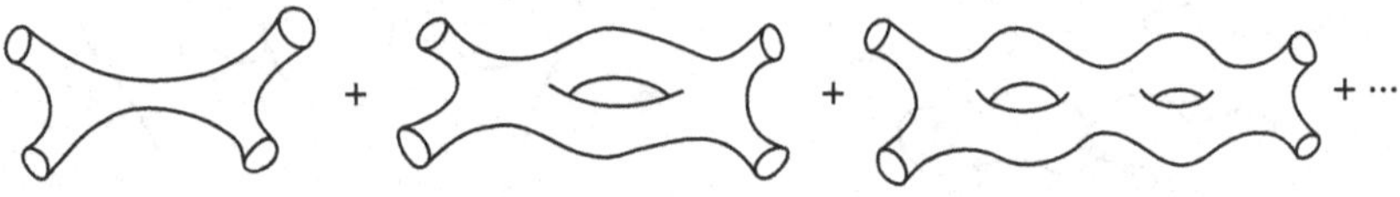

Figure 13.4 String world-sheets as they appear in the scattering of four closed strings. Ignoring the asymptotic boundary states, the diagrams correspond to genera $h = 0, 1, 2$, respectively.

operators in string theory, which are more thoroughly discussed in Section 13.3. For the closed bosonic string, we need then only consider world-sheets that are closed (orientable) Riemann surfaces with punctures, but without boundaries.

The path integral does not only take the classical solutions, or the stationary points of the action S, into account. It considers all the possible paths (or, in this case, world-sheets) and weights them by e^{-S}. For the closed bosonic string we should then sum over all Riemann surfaces (Σ, g) and possible embeddings $X : \Sigma \to M$ modulo diffeomorphisms and Weyl transformations to avoid overcounting physically equivalent configurations. We can separate this sum into terms with different world-sheet topologies, which are here given by the genus h, a few examples being shown in Figure 13.4.

If we separate the constant mode ϕ_0 from the dilaton action (13.9), its contribution is purely topological:

$$\frac{1}{4\pi} \int_\Sigma d^2\sigma \sqrt{g}\, \phi_0 R^{(2)} = \phi_0 \chi(\Sigma), \tag{13.16}$$

where $\chi(\Sigma) = 2 - 2h$ is the Euler characteristic of the world-sheet Riemann surface without punctures. Defining the string coupling constant as $g_s = e^{\phi_0}$ we see that, from the factor e^{-S} in the path integral, each topology is weighted by $g_s^{-\chi(\Sigma)}$. The powers of g_s associated to the punctures are usually incorporated into the string theory vertex operators.

A *perturbative scattering amplitude* $\mathcal{A}$ for n states of the bosonic string can then be computed as

$$\mathcal{A} = \sum_{\text{genus } h} g_s^{-\chi(\Sigma)} \int_{\text{Maps}(\Sigma, M) \times \text{Met}(\Sigma)} \frac{DX\, Dg}{\text{Vol}(\text{Diff} \ltimes \text{Weyl})} V_1 \cdots V_n\, e^{-S[g,G,X]}, \tag{13.17}$$

where $\text{Met}(\Sigma)$ is the space of all metrics on Σ, $\text{Vol}(\text{Diff} \ltimes \text{Weyl})$ is the (infinite) volume of the group of diffeomorphisms and Weyl transformations on Σ, and the V_i are vertex operators for each asymptotic state. The quotient space $\mathcal{M}_h = \text{Met}(\Sigma)/(\text{Diff} \ltimes \text{Weyl})$ is called the *moduli space of Riemann surfaces* with genus h, and its simply-connected covering space is the complex finite-dimensional *Teichmüller space* $\mathcal{T}_h = \text{Met}(\Sigma)/(\text{Diff}_0 \ltimes \text{Weyl})$.

In physics terms, the process of picking a slice of the integration domain that cuts through each equivalence class of (Diff ⋉ Weyl) only once is called *gauge-fixing* and the integration measure on this slice is then determined by the *Faddeev–Popov procedure* which computes the Jacobian by introducing extra virtual particles called *ghost particles*.

In addition, the placements of the punctures should be integrated over to recover a Poincaré invariant result, and it is customary to include these in the moduli space of surfaces. Thus, we need to integrate over $\mathcal{M}_{h,n}$ which is the moduli space of Riemann surfaces with genus h and n punctures.

The expansion over genera in (13.17) giving a power series in g_s is called *string perturbation theory*, and we will see in Section 13.6 that the amplitudes will also get corrected by *non-perturbative terms* of the form $\exp(-1/g_s)$ that are not visible in the above expansion about $g_s = 0$. Note that in string theory it is conventional to use the term *loop* when referring to the genus of a world-sheet and that the probability of a certain string scattering process is given by the modulus square of the scattering amplitude.

Remark 13.1 For the discussion above, we implicitly assumed that the target space-time M was Wick-rotated to a Euclidean space. This is necessary for having formally Gaussian integrals in e^{-S} where S is expressed as a positive definite bilinear form on the tangent space of embedding maps. The path integral formalism then gives the amplitudes in Euclidean space. The amplitudes for Minkowski space-time are obtained from the Euclidean amplitudes by analytic continuation in the momentum vectors and polarisation tensors that appear in the vertex operators for the external particles.

13.1.5 Critical Dimensions and Einstein's Equations

As we have already discussed, the action (13.6) is invariant under Weyl transformations, but to ensure that this is also a symmetry of the quantum string theory (which is required for a consistent theory without negative norm states) we need to make sure that the integration measure $DXDg$ in (13.17) is also Weyl invariant. For flat space-time, this implies that the transformation of DX depends on the dimension D since there are D coordinates X^μ and the transformation of Dg depends on the *central charge* c of the two-dimensional conformal field theory, which for the bosonic theory is $c = 26$. Together they transform as

$$DXDg \to e^{(D-26)(\dots)} DXDg, \tag{13.18}$$

which shows that the critical dimension is $D = 26$. A similar argument follows for superstring theory, where we also need to take the fermion measure into account giving $D = 10$ as the critical dimension.

The Weyl invariance of the Polyakov action (13.6) extends to the full Polyakov action (13.11), and we also have a metric G, a B-field and a dilaton ϕ on space-time M, if appropriate Weyl-transformation laws are imposed. The critical dimension stays the same. However, string theory also enforces equations of motion on these space-time fields. The Weyl invariance of the Polyakov action extends in the classical string to the full conformal group in two dimensions. At the quantum level the classical gauge invariance of the B-field (13.14) is replaced by a discrete symmetry. Namely, for any integer-valued closed two-form $\Lambda_{(2)} \in H^2(M, \mathbb{Z})$ the transformation

$$B \to B + 4\pi^2 \Lambda_{(2)} \tag{13.19}$$

is not a symmetry of the action, but transforms it in such a way that the path integral picks up a phase:

$$\mathcal{A} \to \exp\left[2\pi i \int_\Sigma \Lambda_{(2)}\right] \mathcal{A}, \tag{13.20}$$

which is trivial since $\Lambda_{(2)}$ has integer periods. This implies that the B-field can be viewed as an element in a torus:

$$B \in H^2(M, \mathbb{R})/H^2(M, \mathbb{Z}). \tag{13.21}$$

Conformal invariance at the quantum level is obstructed by the fact that the measure in the path integral is metric dependent. This dependence is measured by the so-called *beta functional* $\beta_{\mu\nu}$, which is a functional on the space of fields that measures the scale dependence of the theory. In order to preserve conformal invariance one must require that, as a functional of the metric G, the beta functional must vanish:

$$\beta_{\mu\nu}(G) = 0. \tag{13.22}$$

To leading order in α' one finds for the full Polyakov action

$$\beta_{\mu\nu}(G) = R_{\mu\nu} + 2\nabla_\mu \nabla_\nu \phi - \frac{1}{4} H_{\mu\kappa\sigma} H_\nu{}^{\kappa\sigma} + \alpha' \frac{1}{2} R_{\mu\kappa\lambda\rho} R_\nu{}^{\kappa\lambda\rho} + \cdots. \tag{13.23}$$

Here $H = dB$ is the 3-form field strength of the B-field. In particular, if we restrict to constant dilaton and B-field, quantum conformal invariance of string theory enforces that to leading order in α' the space-time metric should be Ricci flat:

$$R_{\mu\nu} = 0. \tag{13.24}$$

These are precisely Einstein's equations in vacuum, revealing that string theory includes general relativity in its α'-expansion. One can also derive the beta functional for the B-field and to leading order in α' its vanishing requires

$$dB = 0. \tag{13.25}$$

There is a similar equation for the dilaton field, and the lowest-order equations in α' following from the conformal invariance analysis are just those of Einstein gravity coupled to specific point particles.

13.1.6 String Moduli and U-duality

A scattering amplitude depends on the data of the asymptotic states, such as their momenta, the string coupling g_s and potentially other so-called *moduli fields*, as well as the dimensionful parameter α'. Except for α', these can be thought of as aspects of the target space M in the form of additional (scalar) fields living on them. Much as in the case of the dilaton with $g_s = e^{\phi_0}$, only their vacuum expectation values matter for the present discussion and we will denote these by

$$g \in \mathcal{M} \qquad \text{(moduli expectation values).} \tag{13.26}$$

The so-called *moduli space* of string theory $\mathcal{M}$ should not be confused with the moduli space $\mathcal{M}_h$ of Riemann surfaces with genus h in (13.17). As seen in Section 13.1.4, the string coupling constant g_s is related to one of the moduli fields called the *dilaton*, but in general there are many more moduli fields.

The structure of the moduli space is of central importance in understanding the possible forms of string scattering amplitudes. Much is known for flat target spaces of the type $M = \mathbb{R}^{1,9}$ (*flat Minkowski space*) or $M = \mathbb{R}^{1,9-d} \times T^d$ (*toroidal compactification*). As discussed above, one retains, in both cases, *maximal supersymmetry* strongly constraining the moduli space. The classical low-energy moduli space is a symmetric space of the form

$$\mathcal{M}_{\text{class.}} = G(\mathbb{R})/K(\mathbb{R}) = E_{d+1}(\mathbb{R})/K(E_{d+1}(\mathbb{R})), \tag{13.27}$$

where E_{d+1} is the Cremmer–Julia sequence of hidden symmetry groups [150, 151, 385, 152], which are listed in Table 13.2; their Dynkin diagrams are shown in Figure 13.5, and $K(E_{d+1}(\mathbb{R}))$ are their maximal compact subgroups.

In particular, for ten dimensions the classical moduli space is $G(\mathbb{R})/K(\mathbb{R}) = SL(2,\mathbb{R})/SO(2,\mathbb{R})$, which is isomorphic to the Poincaré upper half-plane $\mathbb{H} = \{z = x + iy \in \mathbb{C} \mid y = \operatorname{Im} z > 0\}$. We parametrise $\mathbb{H}$ by a complex scalar field called the *axio-dilaton* that is denoted by $z = \chi + ie^{-\phi}$ with χ being the (Ramond–Ramond) axion and ϕ the dilaton.

Here, E_{d+1} denote the algebraically simply-connected groups with $E_{d+1}(\mathbb{R})$ being the split real forms. Up to $d \leq 7$, these groups are finite-dimensional reductive groups and we restrict to this range first. We will come back to $d \geq 8$ in Chapter 19. For other internal manifolds one can get a large variety of different Lie groups.

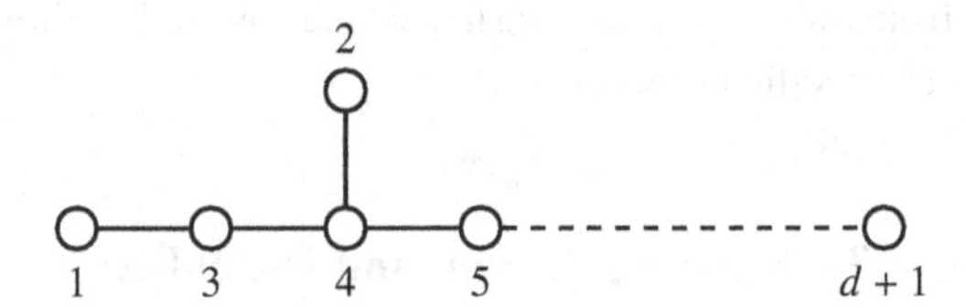

Figure 13.5 The Dynkin diagram of the Cremmer–Julia symmetry group E_{d+1} with labelling of nodes in the 'Bourbaki convention'.

Table 13.2 Cremmer–Julia symmetry groups $G(\mathbb{R})$ with maximal compact subgroup K and corresponding U-duality groups $G(\mathbb{Z})$ for compactifications of type IIB string theory on a d-dimensional torus T^d to $D = 10 - d$ dimensions.

d	$G(\mathbb{R})$	K	$G(\mathbb{Z})$	D
0	$SL(2,\mathbb{R})$	$SO(2)$	$SL(2,\mathbb{Z})$	10
1	$SL(2,\mathbb{R}) \times \mathbb{R}^+$	$SO(2)$	$SL(2,\mathbb{Z}) \times \mathbb{Z}_2$	9
2	$SL(3,\mathbb{R}) \times SL(2,\mathbb{R})$	$SO(3) \times SO(2)$	$SL(3,\mathbb{Z}) \times SL(2,\mathbb{Z})$	8
3	$SL(5,\mathbb{R})$	$SO(5)$	$SL(5,\mathbb{Z})$	7
4	$Spin(5,5;\mathbb{R})$	$(Spin(5) \times Spin(5))/\mathbb{Z}_2$	$Spin(5,5;\mathbb{Z})$	6
5	$E_6(\mathbb{R})$	$USp(8)/\mathbb{Z}_2$	$E_6(\mathbb{Z})$	5
6	$E_7(\mathbb{R})$	$SU(8)/\mathbb{Z}_2$	$E_7(\mathbb{Z})$	4
7	$E_8(\mathbb{R})$	$Spin(16)/\mathbb{Z}_2$	$E_8(\mathbb{Z})$	3

The classical low-energy effective theory of type II string theory in $D = 10-d$ dimensions is described by D-dimensional *maximal ungauged supergravity*. This is the supersymmetric extension of Einstein's theory of gravity in D dimensions. The extension uses 32 supercharges and this is commonly believed to be the maximal number. Maximal supergravity has a symmetry given by the non-compact real Lie group $G(\mathbb{R})$ [365, 502]. However, as mentioned in Section 1.2.2, when passing to the quantum theory, the classical symmetries are generically broken because the (generalised) electro-magnetic charges of physical states become quantised according to the *Dirac–Schwinger–Zwanziger quantisation condition*, and take values in some integral lattice Γ [181, 596, 597, 499]. Although the classical symmetry group $G(\mathbb{R})$ is broken, there is a discrete subgroup of $G(\mathbb{R})$ that survives and remains a symmetry of the full quantum theory. This quantum symmetry is defined as the subgroup of $G(\mathbb{R})$ that preserves the lattice Γ [365]:

$$\{g \in G(\mathbb{R}) \mid g\Gamma = \Gamma\}. \tag{13.28}$$

This quantum symmetry group is generally referred to as a *U-duality group*, which unifies the previously known existing dualities called S- and T-duality and agrees with the Chevalley subgroup $G(\mathbb{Z})$ [580]. The U-duality groups $G(\mathbb{Z})$ for toroidal compactifications of type IIB string theory on a torus T^d are also listed in Table 13.2.

Remark 13.2 In $D = 4$ we can provide a more explicit description of the U-duality group [365]. When compactifying type II string theory on T^6 the low-energy theory contains 28 abelian gauge fields A^I, $I = 1, \ldots, 28$, with associated electric-magnetic charges (q^I, p_I) taking values in a lattice $\Gamma = \Gamma_e \oplus \Gamma_m \cong \mathbb{Z}^{28} \oplus \mathbb{Z}^{28}$. The Dirac–Schwinger–Zwanziger quantisation condition for two charge vectors (q^I, p_I) and (q'^I, p'_I) then reads explicitly

$$q^I p'_I - q'^I p_I \in \mathbb{Z}. \tag{13.29}$$

This condition is invariant with respect to the action of $Sp(56;\mathbb{Z})$, under which (q^I, p_I) transforms as a vector. The definition (13.28) of the U-duality group can now be equivalently presented as the intersection in the fundamental representation of $E_7(\mathbb{R})$ given by

$$E_7(\mathbb{Z}) = E_7(\mathbb{R}) \cap Sp(56;\mathbb{Z}). \tag{13.30}$$

In higher dimensions one can then define the U-duality groups iteratively through the intersection of $E_7(\mathbb{Z})$ with the classical symmetry groups:

$$E_{d+1}(\mathbb{Z}) = E_{d+1}(\mathbb{R}) \cap E_7(\mathbb{Z}), \qquad 0 \le d \le 5. \tag{13.31}$$

The $E_8(\mathbb{Z})$ U-duality in $D = 3$ is defined by adjoining to $E_7(\mathbb{Z})$ certain discrete translational symmetries appearing in the compactification from $D = 4$ [482]. Mathematically, these discrete translations generate the unipotent radical $U(\mathbb{Z})$ of the maximal parabolic of E_8 with Levi $L = E_7 \times GL(1)$, and the generators are precisely the Chevalley generators that we discussed in Section 3.1.4 [482]. Thus, the physical definition of $E_8(\mathbb{Z})$ coincides with the one given there.

Points in moduli space related by U-duality transformations give rise to equivalent string theories. This implies that the correct moduli space of quantum string theory is not the classical symmetric space (13.27) but

$$\mathcal{M} \equiv \mathcal{M}_{\text{quantum}} = G(\mathbb{Z})\backslash G(\mathbb{R})/K(\mathbb{R}), \tag{13.32}$$

and all observables, like string scattering amplitudes, that are functions of the expectation values of the moduli are functions on this space.

Remark 13.3 Although the emphasis in this book lies on the U-duality groups listed in Table 13.2, there is a variety of other examples that also play a rôle in string theory. In particular, a large class of moduli spaces occurring in string compactifications is of the form

$$SO(m,n;\mathbb{Z})\backslash SO(m,n;\mathbb{R})/(SO(m) \times SO(n)). \tag{13.33}$$

The simplest instance is the case when $m = n = d$, in which case $SO(d, d; \mathbb{Z}) \subset E_{d+1}(\mathbb{Z})$ is the so-called T-duality group arising for compactifications on T^d. This case will be discussed more closely in Section 13.4.3. Another interesting example occurs for compactifications of heterotic string theory on T^d, in which case the T-duality group is $SO(d, d + 16)$. In particular, in $D = 4$ this yields $O(6, 22; \mathbb{Z})$, which combines with the S-duality group $SL(2, \mathbb{Z})$ into the semi-simple U-duality group $G(\mathbb{Z}) = SO(6, 22; \mathbb{Z}) \times SL(2, \mathbb{Z})$. When compactifying further to $D = 3$ the $SL(2, \mathbb{Z})$ symmetry is adjoined to the T-duality group, leading to the enhanced U-duality group $SO(8, 24; \mathbb{Z})$. These two latter examples also arise for type II compactifications on $K3 \times T^2$ or $K3 \times T^3$, respectively. See for example [30, 31] for more details on these moduli spaces.

On general grounds physical observables are $G(\mathbb{Z})$-invariant functions on $G(\mathbb{R})/K(\mathbb{R})$. In addition, physical constraints, such as supersymmetry, typically force these observables to satisfy differential equations and have a prescribed asymptotic behaviour at infinity, thereby satisfying the conditions (1)–(4) of Definition 4.13, characterising automorphic forms. We will explain the origin of these additional constraints in more detail in Section 14.4, but first we study a particularly relevant example.

13.2 Four-Graviton Scattering Amplitudes

We now turn to a more detailed discussion of string scattering amplitudes, with particular emphasis on the amplitude for the scattering process of four massless graviton states in type IIB string theory in $D = 10$ space-time dimension. String theory predicts two main types of corrections to the standard gravitational interactions in general relativity that are computed without the inclusion of string theory degrees of freedom. We discuss these two types in general first and then focus on the four-graviton amplitude.

13.2.1 Expansions of Scattering Amplitudes

As was mentioned at the beginning of Section 13.1.6, a scattering amplitude depends on the moduli of string theory that include the dimensionless string coupling g_s as well as the dimensionful string scale $\alpha' = \ell_s^2$. The exact form as a function of all these arguments is not known in any example. There one typically performs a series expansion of the amplitude with respect to some of its arguments. There are two common expansions of the amplitude.

The first expansion is *string perturbation theory* and was already discussed in Section 13.1.4. In string perturbation theory, one treats the string coupling constant g_s as small and computes the contributions to the full amplitude from Riemann surfaces of increasing genus; see Figure 13.4 for an illustration. This involves an integral over the moduli space of all Riemann surfaces of a given genus and with a number of punctures corresponding to the number of asymptotic scattering states. These integrals have been computed up to two loops (see see for instance [174, 176, 173]) and become increasingly hard for increasing genus [294]. One complication arises from the *Schottky problem* of not having a convenient description of the moduli space $\mathcal{M}_h$ of Riemann surfaces in terms of Siegel upper half spaces for $h \geq 4$. Another serious complication is that one should actually integrate over the moduli space of *super*-Riemann surfaces since one is dealing with superstring theory and it is known that this integral cannot be reduced in a simple manner to an integral over ordinary Riemann surfaces for $h \geq 5$ [187]. Finally, the amplitude is not expected to be a convergent series in g_s (or even analytic), meaning that there are *non-perturbative effects* arising from instanton configurations [571, 308, 527]. These are roughly of the form e^{-1/g_s} and do not admit a Taylor series expansion around weak coupling $g_s = 0$ and therefore cannot be captured by string perturbation theory. The string coupling g_s is one of the coordinates on the moduli space $\mathcal{M}$ and the limit $g_s \to 0$ corresponds to approaching a cusp on $\mathcal{M}$.

The second expansion of the amplitude is the *low-energy expansion*, in which one considers the energies of all the scattering states to be small in units of α', leading to an expansion in derivatives of the fields. Dimensionless expansion parameters are formed by multiplying momenta squared with α', which is why the expansion is also called the *α'-expansion*. It has already appeared in (13.23) for the bosonic string, where one can see that higher orders in α' are accompanied by higher powers of curvature and thus are higher derivatives of the metric whose fluctuations are the graviton. It is the low-energy expansion that makes contact to automorphic forms on $\mathcal{M}$, as we will motivate now and study in detail in Chapter 14.

13.2.2 Low-energy Expansion of the Four-graviton Amplitude

Gravitons are fluctuations of the metric tensor $g_{\mu\nu}$ on the target space M, and we consider the standard case of fluctuations around flat Minkowski space. Classically, these fluctuations correspond to *gravitational waves* with momenta (or wave-vectors) $k \in \mathbb{R}^{1,9}$ and polarisations $\epsilon \in S^2(\mathbb{R}^{1,9})$, which are symmetric second-rank tensors subject to some constraints whose detail we do not require for the present discussion. In an interacting quantum theory picture that is necessary for computing scattering amplitudes, the gravitons are asymptotic states at spatial infinity and in string theory they correspond to specific vibration patterns of the string.

Here, we consider the case of four gravitons, such that we have four momenta k_i and four polarisation tensors ϵ_i ($i = 1, \ldots, 4$). Since gravitons are massless, the momenta satisfy $k_i^2 = 0$, where the norm-squared is computed using the Lorentzian metric on $\mathbb{R}^{1,9}$. Out of the four momenta k_i one forms the dimensionless Lorentz invariant *Mandelstam variables*

$$s = -\frac{\alpha'}{4}(k_1 + k_2)^2, \quad t = -\frac{\alpha'}{4}(k_1 + k_3)^2 \quad \text{and} \quad u = -\frac{\alpha'}{4}(k_1 + k_4)^2. \tag{13.34}$$

Identical constructions also apply when compactifying the theory on a torus T^d to $D = 10 - d$ space-time dimensions. Momentum conservation ($k_1 + k_2 + k_3 + k_4 = 0$) and masslessness imply that $s + t + u = 0$. Any symmetric polynomial in s, t, u is then a polynomial in

$$\sigma_2 = s^2 + t^2 + u^2 \quad \text{and} \quad \sigma_3 = s^3 + t^3 + u^3. \tag{13.35}$$

The string scattering amplitude will therefore be a function of the momenta only through σ_2 and σ_3. Similar simplifications arise for the polarisation tensors ϵ_i that enter the final answer only in a particular combination that we will denote by $\mathcal{R}^4$. It can be expressed as the contraction of four copies of the linearised curvature tensor $\mathcal{R}_{\mu\nu\rho\sigma} \propto k_\mu \epsilon_{\nu\rho} k_\sigma$ (with permutations), and two copies of a standard rank 8 tensor t_8, which contracts four powers of an antisymmetric matrix $M_{\mu\nu}$ according to

$$t^{\mu_1 \ldots \mu_8} M_{\mu_1\mu_2} \cdots M_{\mu_7\mu_8} = 4\mathrm{Tr}(M^4) - \left(\mathrm{Tr}(M^2)\right)^2. \tag{13.36}$$

The combination $\mathcal{R}^4$ then is more precisely

$$\mathcal{R}^4 = t_8 t_8 \mathcal{R}\mathcal{R}\mathcal{R}\mathcal{R} = \mathcal{R}_{\mu_1\mu_2\nu_1\nu_2} \cdots \mathcal{R}_{\mu_7\mu_8\nu_7\nu_8} t^{\mu_1\ldots\mu_8} t^{\nu_1\ldots\nu_8}, \tag{13.37}$$

but we will not make use of this expression as it factorises from the structure that we shall focus on.

Our four-graviton amplitude in $D = 10 - d$ dimensions is therefore of the form

$$\mathcal{A}^{(D)}(s, t, u, \epsilon_i; g), \tag{13.38}$$

with $g \in \mathcal{M}$ denoting the moduli in the compactified theory; see (13.32). We recall that the string scale α' was absorbed into the Mandelstam variables s, t, u.

The (analytic part of the) α'-expansion of the four-graviton amplitude (in Einstein frame) takes the form [318]

$$\mathcal{A}^{(D)}(s, t, u, \epsilon_i; g) = \left[\mathcal{E}^{(D)}_{(0,-1)}(g)\frac{1}{\sigma_3} + \sum_{p\geq 0}\sum_{q\geq 0} \mathcal{E}^{(D)}_{(p,q)}(g)\sigma_2^p\sigma_3^q\right]\mathcal{R}^4. \tag{13.39}$$

The interesting objects in this expression are the coefficient functions $\mathcal{E}^{(D)}_{(p,q)}(g)$ that are functions on the quantum moduli space $\mathcal{M}_{\text{quantum}} = G(\mathbb{Z})\backslash G(\mathbb{R})/K(\mathbb{R})$ in (13.32).

The first term in (13.39) plays a special rôle in that it is the only term that is not polynomial in σ_2 and σ_3. It is the lowest-order term in the α'-expansion and it agrees with what one would calculate in a standard theory of gravity with action given by the integral of the Ricci scalar only,

$$S_{\text{class.}} = \int d^D x \sqrt{-G}\, R, \tag{13.40}$$

called the *Einstein–Hilbert action*. The coefficient function $\mathcal{E}^{(D)}_{(0,-1)}(g) = 3$ is constant. By contrast, the infinite series of terms in p and q come with higher powers of α' and they reflect the contribution of *massive* string states to the graviton scattering process [330].

The *low-energy effective theory* is obtained by writing the field theory action whose classical interactions give rise to the same quantum corrected amplitudes obtained from string theory, order by order in α'. It gets corrections from the four-graviton amplitudes of the form [315]

$$\mathcal{E}^{(D)}_{(p,q)}(g) \nabla^{4p+6q} R^4, \tag{13.41}$$

where ∇ denotes a covariant derivative on the target space-time M, and R^4 (not to be confused with the Ricci curvature scalar) is a contraction of two t_8 tensors and four Riemann curvature tensors like the linearised version for the polarisation term $\mathcal{R}^4$ in the amplitude.

In other words, for the first few orders in α' we get the corrections

$$S = S_{\text{class.}} + \int d^D x \sqrt{-G} \Big((\alpha')^3 \mathcal{E}^{(D)}_{(0,0)}(g) R^4 + (\alpha')^5 \mathcal{E}^{(D)}_{(1,0)}(g) \nabla^4 R^4 + (\alpha')^6 \mathcal{E}^{(D)}_{(0,1)}(g) \nabla^6 R^4 + \cdots \Big), \tag{13.42}$$

where $S_{\text{class.}}$ is the classical, zeroth-order low-energy effective action described by (13.40). We see why this expansion is also called the *derivative expansion*. In writing (13.40) and (13.42) we have suppressed an overall dimensionful normalisation of S since we wanted to emphasise the relevant derivative orders.

As the functions $\mathcal{E}^{(D)}_{(p,q)}(g)$ depend on the moduli $g \in \mathcal{M}$, they in particular depend on the string dilaton and thus on the string coupling g_s that controls string perturbation theory in terms of Riemann surfaces as discussed above. However, there is no reason that the dependence on g_s be analytic. Non-analytic terms in g_s are known as *non-perturbative effects* and they appear in $\mathcal{E}^{(D)}_{(p,q)}(g)$

through so-called *instanton contributions*. Their direct determination in terms of a string theory calculation is typically very hard. The action of the U-duality group $G(\mathbb{Z})$ also includes a transformation that mixes perturbative and non-perturbative effects, and therefore using U-duality opens up the opportunity to access non-perturbative effects indirectly.

The fact that the coefficients $\mathcal{E}^{(D)}_{(p,q)}$ are functions on the (quantum) moduli space $\mathcal{M}$ can be rephrased as follows: *they are functions on the symmetric space* $\mathcal{M}_{\text{class.}} = G(\mathbb{R})/K(\mathbb{R})$ *invariant under the left-action of the discrete (U-duality) group* $G(\mathbb{Z})$.

Contributions to $\mathcal{E}^{(D)}_{(p,q)}$ can be calculated in string perturbation theory as discussed in Section 13.1.4. We shall now discuss in more detail the two lowest-order terms in string perturbation theory, corresponding to tree-level and one-loop amplitudes. In addition, we will discuss non-perturbative (in g_s) contributions to string scattering amplitudes due to instantons in Section 13.6. The full function $\mathcal{E}^{(D)}_{(p,q)}$ and its connection to automorphic forms will be the content of Chapter 14.

13.3 The Four-Graviton Tree-Level Amplitude*

A comprehensive discussion of string theory and its scattering amplitudes is beyond the scope of this work. We only give one indicative and, hopefully, illustrative example and refer to the string theory literature [320, 535] for more information.

The example is the four-graviton amplitude at string *tree level*. The closed string tree-level topology is that of a sphere, and the four asymptotic graviton states correspond to four punctures in this sphere as pictured in Figure 13.6. This configuration can be obtained by a homeomorphism of the left-most diagram in Figure 13.4. By definition, the string scattering amplitude is given by an integral over the moduli space of all Riemann spheres and all possible insertion points for four punctures. The discussion below uses the *Ramond–Neveu–Schwarz formalism* that was already mentioned in Section 13.1.2.

In terms of complex geometry, we can describe the sphere $S^2 = \mathbb{CP}^1$ by one complex variable $z \in \mathbb{C}$ everywhere except at the 'north pole'. An asymptotic graviton state at a puncture z_i corresponds to a *vertex operator*. Due to a complication called *ghost picture* [227] that arises for the superstring, we actually require it in two related forms, called ghost picture 0 and ghost picture -1:

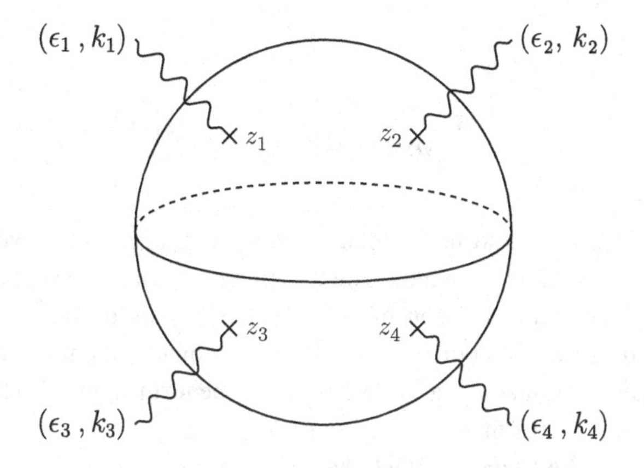

Figure 13.6 Riemann sphere with four punctures as it appears in the four-graviton scattering amplitude. An external massless graviton with polarisation ϵ_i and momentum k_i is located at each of the four punctures $z_i \in \mathbb{CP}^1$.

$$V_0(z_i; k_i, \epsilon_i)$$
$$= \frac{2}{\alpha'} g_s :\epsilon_{i,\mu\nu} \left(\partial X^\mu + \frac{\alpha'}{2} k_{i,\rho}\psi^\rho\psi^\mu\right)\left(\bar{\partial} X^\nu + \frac{\alpha'}{2} k_{i,\sigma}\bar{\psi}^\sigma\bar{\psi}^\nu\right) e^{ik_{i,\mu}X^\mu}:, \tag{13.43a}$$

$$V_{-1}(z_i; k_i, \epsilon_i) = g_s :\epsilon_{i,\mu\nu}\psi^\mu\bar{\psi}^\nu e^{ik_{i,\mu}X^\mu}: . \tag{13.43b}$$

As before, we are using the Einstein summation convention for repeated Lorentz indices $\mu = 0, \ldots, 9$ that label the ten directions of the Minkoswki target space in which the sphere is embedded via the embedding coordinates $X^\mu \equiv X^\mu(z, \bar{z})$. The polarisation tensor $\epsilon_i \equiv \epsilon_{i,\mu\nu}$ is a second-rank symmetric tensor of the Lorentz group $SO(1,9)$ and the external momentum of the graviton $k_i \equiv k_{i,\mu}$ is light-like in the Minkowski metric: $k_i^2 = k_{i,\mu}k_{i,\nu}\eta^{\mu\nu} = 0$ as the scattering gravitons are massless. The colons surrounding these expressions indicate a specific normal ordering procedure necessary for the vertex operators to be well-defined on a Fock space vacuum. The field ψ^μ and its (Dirac) conjugate $\bar{\psi}^\mu$ correspond to the fermionic coordinates that accompany the bosonic X^μ in superstring theory. The derivatives ∂ and $\bar{\partial}$ are with respect to the world-sheet coordinate z and its complex conjugate $\bar{z}$.

Following the discussion in Section 13.1.4, the desired expression for the scattering amplitude is then roughly of the following form for $D = 10$ (in string frame):

$$\mathcal{A}^{(10)}_{\text{tree}}(s,t,u,\epsilon_i;g) = \int_{\text{Met}(\Sigma)} \frac{Dg_{\mathbb{CP}^1}}{\text{Vol}(\text{Diff} \ltimes \text{Weyl})} \left\langle \prod_{i=1}^{4} V(z_i;k_i,\epsilon_i) \right\rangle_{S^2}, \tag{13.44}$$

where the integral is also understood to include an integration over the four punctures z_i, $g_{\mathbb{CP}^1}$ indicates a metric on the sphere $\mathbb{CP}^1$, and the angled brackets denote the *correlation function* of the vertex operators on the given sphere S^2 that we detail below. For reasons of ghost number saturation, two vertex operators have to be taken in ghost picture 0 (of the form V_0 in (13.43)) and two in ghost picture -1 (of the form V_{-1} in (13.43)).

According to the *Riemann–Roch theorem*, the sphere as a Riemann surface has no metric moduli and any metric $g_{\mathbb{CP}^1}$ can be brought into the form of the round sphere

$$ds^2 = \frac{dzd\bar{z}}{(1+|z|^2)^2} \tag{13.45}$$

by diffeomorphisms and Weyl rescalings. There is no modulus in this expression and even this form is left-invariant by the *conformal Killing group* of the sphere $PSL(2,\mathbb{C})$ that acts by

$$\begin{pmatrix} \alpha & \beta \\ \gamma & \delta \end{pmatrix} \in PSL(2,\mathbb{C}): \qquad z \mapsto \frac{\alpha z + \beta}{\gamma z + \delta} \tag{13.46}$$

on the coordinate z. So, even after fixing the Diff $\ltimes$ Weyl gauge freedom to bring the metric to the above form (13.45), one still has the freedom to perform transformations from the conformal Killing group $PSL(2,\mathbb{C})$ that represent the residual gauge freedom. Without fixing this residual freedom, the integral in (13.44), which we recall contains an integration over the placement of punctures z_i, is of the form

$$\int_{\text{Met}(\Sigma)} \frac{Dg_{\mathbb{CP}^1}}{\text{Vol}(\text{Diff} \ltimes \text{Weyl})} = \prod_{j=1}^{4} \int_{\mathbb{CP}^1} d^2 z_j \, \Delta_{\text{FP}}, \tag{13.47}$$

where Δ_{FP} is a *Faddeev–Popov determinant* arising from the gauge-fixing and that is treated by introducing (super-)ghost systems. We will not be more specific about it here (see for example [535]) and only mention its effect on the calculation below. We see that, due to the absence of metric moduli for the sphere, the integral over the moduli space of the four-punctured sphere reduces to an integral over the locations z_i of the four punctures. The complex three-dimensional conformal Killing group can be used to fix three of the puncture positions to one's favourite values; a standard choice is $0, 1, \infty$ and only a single integral over a single puncture position remains.

Let us return to the correlation function appearing in the expression (13.44) above. It is formally given by a path integral over all (super-)embeddings X^μ of the (super-)Riemann sphere into the ten-dimensional target space. Schematically, one has

$$\langle\prod_{i=1}^{4} V(z_i;k_i,\epsilon_i)\rangle_{S^2} = g_s^{-\chi(S^2)}\int DXD\psi D\bar{\psi}\, e^{-S[X,\psi,\bar{\psi}]}\prod_{i=1}^{4} V(z_i;k_i,\epsilon_i), \tag{13.48}$$

where $S[X,\psi,\bar{\psi}]$ denotes the two-dimensional *σ-model* action that is basically the induced volume under the embedding. The *Euler number* $\chi(S^2) = 2$ of the sphere provides the standard topological weighting of different string diagrams that was mentioned above (see Figure 13.4) and here evaluates to g_s^{-2}. The string coupling g_s is the only string theory modulus g of Equation (13.26) of relevance in the present discussion.

The correlation function can be evaluated on the sphere explicitly in terms of the Green's function on the sphere. First, we note that for vertex operators in ghost picture 0 the fermionic integrals (over ψ and $\bar{\psi}$) pick out the contribution

$$\alpha'\epsilon_{i,\mu\nu}k_\rho k_\sigma e^{ik_{i,\tau}X^\tau} \propto \alpha'\mathcal{R}_{\rho\mu\nu\sigma}e^{ik_{i,\tau}X^\tau} \tag{13.49}$$

from $V(z_i;k_i,\epsilon_i)$ in (13.43a) and we recognise the linearised Riemann curvature tensor mentioned in Section 13.2, and the full integral also provides the necessary contractions of the four Riemann tensors. For the vertex operator in picture -1 of (13.43b), the integral yields contributions that are roughly of the form

$$\epsilon_{i,\mu\nu}e^{ik_{i,\tau}X^\tau} \propto \frac{\mathcal{R}_{\rho\mu\nu\sigma}}{k_\rho k_\sigma}e^{ik_{i,\tau}X^\tau}. \tag{13.50}$$

The Green's function on the sphere now evaluates the product of two normal ordered exponentials to

$$:e^{ik_{i,\mu}X^\mu(z_i)}:\,:e^{ik_{j,\nu}X^\nu(z_j)}: \propto |z_i-z_j|^{\alpha' k_i\cdot k_j}\,:e^{i(k_1+k_2)_\mu X^\mu(z_i)}: + \cdots, \tag{13.51}$$

which is called the *operator product expansion* (or *Wick contraction*) and the dots denote subdominant terms in an expansion around $z_i \to z_j$. In this expression we have used the Lorentz inner product $k_i\cdot k_j = k_i^\mu k_j^\nu \eta_{\mu\nu}$ between the two external momenta k_i and k_j. Since the external momenta are massless, we can rewrite this product as $k_i\cdot k_j = \frac{1}{2}(k_i+k_j)^2$ and we already begin to recognise a connection to the Mandelstam variables s, t and u defined in (13.34). One has to perform the above operator product expansion of e^{ikX} for all four factors appearing in the correlation function and this will lead to a permutation invariant expression of the three Mandelstam variables. The factors $|z_i-z_j|^{\alpha' k_i\cdot k_j}$ occurring in the operator product expansion (13.51) are sometimes called *Koba–Nielsen factors*.

We can now put all the pieces together. (i) The integral over the moduli space reduces to an integral over the four punctures; (ii) the conformal Killing group allows us to fix three of four punctures to fixed values that we choose to be 0, 1 and ∞; (iii) the integral over the fermionic variables in the correlation function produces the linearised curvature tensors times exponentials of the form e^{ikX}; (iv) these exponentials get converted into factors of the form $|z_i - z_j|^{\alpha' k_i \cdot k_j}$ in all possible ways and (v) include additional contributions from the ghost sector. The ultimate integral is of the form

$$\begin{aligned}\mathcal{A}^{(10)}_{\text{tree}}(s,t,u,\epsilon_i;g) &= \frac{4(\alpha')^2 g_s^2 \mathcal{R}^4}{\pi (k_1\cdot k_3)^2}\delta(k_{\text{tot}})\int_{\mathbb{C}} d^2z|z|^{\alpha' k_1\cdot k_2 - 2}|1-z|^{\alpha' k_2\cdot k_3 - 2}\\ &= (\alpha')^4 g_s^2 \delta(k_{\text{tot}})\frac{1}{stu}\frac{\Gamma(1-s)\Gamma(1-t)\Gamma(1-u)}{\Gamma(1+s)\Gamma(1+u)\Gamma(1+t)}\mathcal{R}^4 .\end{aligned} \tag{13.52}$$

We have introduced a few factors related to the normalisations of the various measures and vertex operators introduced above. The prefactor g_s^2 is the combination of the tree-level value g_s^{-2} and the g_s^4 from the four vertex operators of the external graviton states. The shift in the α' power is due to the conformal ghost sector that we have not discussed explicitly, and the denominator $(k_1 \cdot k_3)^2 = \frac{4}{(\alpha')^2}t^2$ is due to (13.50). We note that the factor $\delta(k_{\text{tot}}) = \delta(k_1 + k_2 + k_3 + k_4)$ expresses momentum conservation. As stated in Section 13.2, the linearised curvature tensor $\mathcal{R}_{\mu\nu\rho\sigma} \propto k_\mu \epsilon_{\nu\rho} k_\sigma$ is determined by the graviton's momentum and polarisation, and $\mathcal{R}^4$ is a specific contraction of the linearised curvatures of the four gravitons. More precisely, it is given by $t_8 t_8 \mathcal{R}^4$ where the t_8 is defined in (13.36).

The amplitude (13.52) can be expanded for small values of s, t and u which we recall from (13.34) contain the string length $\alpha' = \ell_s^2$. The result is (in string frame)

$$\begin{aligned}\mathcal{A}^{(10)}_{\text{tree}}(s,t,u,\epsilon_i;g) &= g_s^2(\alpha')^4\delta(k_1+k_2+k_3+k_4)\\ &\quad\times\left(\frac{3}{\sigma_3} + 2\zeta(3) + \zeta(5)\sigma_2 + \frac{2\zeta(3)^2}{3}\sigma_3 + \cdots\right)\mathcal{R}^4,\end{aligned} \tag{13.53}$$

where we have used $stu = \frac{1}{3}\sigma_3$ up to momentum conservation implying $s + t + u = 0$ for massless states. The above expression provides the tree-level contributions to the functions $\mathcal{E}^{(10)}_{(0,-1)}$, $\mathcal{E}^{(10)}_{(0,0)}$, $\mathcal{E}^{(10)}_{(1,0)}$ and $\mathcal{E}^{(10)}_{(0,1)}$ appearing in (13.39) in ten dimensions. We note also that $g_s^2(\alpha')^4 = \ell_{10}^8 = \kappa_{10}^2$ in terms of the ten-dimensional Planck scale ℓ_{10} and the ten-dimensional gravitational coupling κ_{10}. The fact that this amplitude is proportional to the square of the coupling is characteristic for a tree-level scattering of four particles (in a theory that has cubic vertices).

We note also that, upon toroidal compactification to $D < 10$ dimensions, integrals such as (13.52) receive additional contributions related to the structure of states on the torus T^d $(d = 10 - D)$ that is used in the compactification. In the context of loop amplitudes, this leads naturally to a relation of string amplitudes to *theta correspondences* of Chapter 12, as we shall see next.

13.4 One-Loop String Amplitudes and Theta Lifts*

In this section, we take a closer look at the structure of the genus-one contribution to a perturbative scattering amplitude $\mathcal{A}$ (13.17). We can write

$$\mathcal{A} = \sum_{h=0}^{\infty} g_s^{2h-2} \mathcal{A}_h, \tag{13.54}$$

with

$$\mathcal{A}_h = \int_{\mathrm{Met}(\Sigma)} \frac{Dg}{\mathrm{Vol}(\mathrm{Diff} \ltimes \mathrm{Weyl})} \langle V_1 \cdots V_n \rangle_h, \tag{13.55}$$

where it is understood that the positions of the vertex operators V_i should also be integrated over, and

$$\langle V_1 \ldots V_n \rangle_h = \int_{\mathrm{Maps}(\Sigma,\mathrm{M})} DX\, V_1 \cdots V_n\, e^{-S_E[g,X]} \tag{13.56}$$

in terms of the *Euclidean* action S_E. We will mainly focus on the one-loop amplitude $\mathcal{A}_1$ in what follows. The amplitude (and the vertex operators) depend on the kinematical data of the states that are being scattered and the moduli of the theory. Every state has a space-time momentum and can have a polarisation tensor as well. This dependence can be seen for instance in the detailed tree-level example discussed in Section 13.3 but is not displayed above for conciseness of notation. Moreover, we will now focus on the simplest case where the vertex operators are absent.

13.4.1 The Bosonic Vacuum to Vacuum Amplitude

We begin by examining the vacuum to vacuum amplitude, which is often referred to as the one-loop partition function. This corresponds to the case $\mathcal{A}_1$ with no insertions of vertex operators, i.e.,

$$\mathcal{A}_1 = \int_{\mathrm{Met}(\Sigma)\times\mathrm{Maps}(\Sigma,\mathrm{M})} \frac{DgDX}{\mathrm{Vol}(\mathrm{Diff} \ltimes \mathrm{Weyl})} e^{-S_E[g,X]}. \tag{13.57}$$

At one-loop the world-sheet can be represented as the torus

$$\Sigma = \mathbb{C}/(\mathbb{Z} + \tau\mathbb{Z}), \tag{13.58}$$

with $\tau_2 = \text{Im}\,\tau > 0$. We shall also refer to Σ as an *elliptic curve* for its algebro-geometric properties; see also Section 16.4. We can take a flat metric on the torus $ds^2 = 2|dz|^2$ for $z \in \Sigma$. There is an action of the mapping class group $SL(2,\mathbb{Z})$ on τ in the upper half-plane $\mathbb{H}$ of (4.6) given by the standard fractional transformation

$$\tau \longmapsto \frac{a\tau + b}{c\tau + d}, \qquad \begin{pmatrix} a & b \\ c & d \end{pmatrix} \in SL(2,\mathbb{Z}). \tag{13.59}$$

The moduli space of inequivalent tori is therefore

$$\mathcal{M}_1 = SL(2,\mathbb{Z})\backslash\mathbb{H}. \tag{13.60}$$

After gauge-fixing and evaluating the resulting Jacobian the integral (13.57) reduces to an integral over (a fundamental domain of) $\mathcal{M}_1$ (see, for instance, [175, 217, 167] for details):

$$\mathcal{A}_1 = \int_{\mathcal{M}_1} \frac{d^2\tau}{\tau_2^2} \left(\frac{\det' \Delta_{z\bar{z}}}{\tau_2} \right)^{-12}, \tag{13.61}$$

where $\det' \Delta_{z\bar{z}}$ denotes the regularised determinant of the Laplacian

$$\Delta_{z\bar{z}} = -4\partial_z\partial_{\bar{z}} \tag{13.62}$$

on the compact surface Σ. The Hermitian Laplacian is positive semi-definite and the prime excludes the zero eigenvalue, which has to be treated separately and gives rise to the contribution of τ_2. The Laplacian comes up since, for Euclidean world-sheets, this is the operator in the kinetic term of a Gaussian integral; see (13.86) below. This is also clearly evident since the string coordinates satisfy the harmonic Equation (13.8).

The (operator) determinant can be evaluated as follows. We write the coordinate z on Σ as $z = u + \tau v$ with $u, v \in [0, 1]$. Thus

$$u = \frac{\tau\bar{z} - \bar{\tau}z}{\tau - \bar{\tau}}, \qquad v = \frac{z - \bar{z}}{\tau - \bar{\tau}}. \tag{13.63}$$

The normalisable eigenfunctions of $\Delta_{z\bar{z}}$ are then of the form

$$\psi_{m,n}(u + \tau v) = e^{2\pi i(mu - nv)} \tag{13.64}$$

for any integers $m, n \in \mathbb{Z}$. Their eigenvalues are

$$\Delta_{z\bar{z}}\psi_{m,n} = \lambda_{m,n}\psi_{m,n}, \tag{13.65}$$

with

$$\lambda_{m,n} = \frac{4\pi^2}{\tau_2^2}|n + m\tau|^2. \tag{13.66}$$

The value $(m, n) = (0, 0)$ corresponds to the zero mode.

Defining then the usual *spectral zeta function* by

$$\zeta_\tau(s) = \sum_{(m,n)\in\mathbb{Z}^2}{}' \lambda_{m,n}^{-s} = \sum_{(m,n)\in\mathbb{Z}^2}{}' e^{-s\log\lambda_{m,n}}, \tag{13.67}$$

one has as always

$$\det{}' \Delta_{z\bar{z}} = \exp\left(-\zeta'_\tau(0)\right) = \prod_{(m,n)\in\mathbb{Z}^2}{}' \lambda_{m,n}, \tag{13.68}$$

where $\zeta'_\tau(0)$ is the s-derivative of $\zeta_\tau(s)$ evaluated at $s = 0$ and the other primes denote the removal of the zero mode.

In the present case we obtain

$$\zeta_\tau(s) = \left(\frac{\tau_2}{4\pi^2}\right)^s \sum_{(m,n)\in\mathbb{Z}^2}{}' \frac{\tau_2^s}{|n+m\tau|^{2s}} = 2\left(\frac{\tau_2}{4\pi}\right)^s \frac{\xi(2-2s)}{\Gamma(s)} E(1-s,\tau), \tag{13.69}$$

where we have recognised the $SL(2)$ Eisenstein series using (1.11) and employed the functional relation (7.13). The spectral zeta function can be expanded around $s = 0$ using the (first) Kronecker limit formula (10.10) to obtain

$$\zeta_\tau(s) = -1 - s\log(\tau_2^2|\eta(\tau)|^4) + O(s^2). \tag{13.70}$$

From this one finds the determinant of the Laplacian on Σ as

$$\det{}' \Delta_{z\bar{z}} = \tau_2^2|\eta(\tau)|^4, \tag{13.71}$$

where $\eta(\tau)$ is the *Dedekind eta function*:

$$\eta(\tau) = e^{i\pi\tau/12}\prod_{n=1}^{\infty}(1-e^{2\pi i\tau}). \tag{13.72}$$

This function is analytic and non-zero everywhere in the upper half-plane $\mathbb{H} = \{\tau \in \mathbb{C} \mid \mathrm{Im}(\tau) > 0\}$. Its 24th power is a weight 12 modular form under $SL(2,\mathbb{Z})$, known as the *discriminant* of the elliptic curve and commonly denoted by

$$\Delta(\tau) = \eta(\tau)^{24}. \tag{13.73}$$

It has already appeared in Section 4.2.1.

Remark 13.4 The operator determinants can also be understood as follows [18, 32]. The holomorphic derivative ∂_z maps holomorphic functions into $(1,0)$-forms on the elliptic curve Σ. The kernel $\ker\partial$ is the space of constant functions, while the cokernel $\operatorname{coker}\partial = H^0(\Sigma)$ is the space of holomorphic $(1,0)$-forms on Σ. The *determinant line bundle* $\mathscr{D}$ is then defined as

$$\mathscr{D} = (\Lambda^{\max}\ker\partial_z)^{-1} \otimes (\Lambda^{\max}\operatorname{coker}\partial_z), \tag{13.74}$$

where $\Lambda^{\max}$ denotes the highest exterior power. The line bundle $\mathscr{D}$ has a canonical holomorphic section identified with the (operator) determinant $\mathrm{Det}\,\partial_z$, defined as

$$\mathrm{Det}\,\partial_z := \exp\left[-\tfrac{1}{2}\zeta_\tau'(0)\right], \tag{13.75}$$

where $\zeta_\tau'(0)$ is the derivative of the spectral zeta function as above. The determinant line bundle is further equipped with a natural smooth metric $||\cdot||_Q$, known as the *Quillen metric*, defined as

$$||\cdot||_Q^2 := ||\cdot||^2 \exp\left[-\zeta_\tau'(0)\right], \tag{13.76}$$

where $||\cdot||$ is the natural Hermitian norm on $\mathscr{D}$. Note that it follows that the regularised determinant of the Laplacian on the elliptic curve Σ may now be reinterpreted as the Quillen norm of the canonical holomorphic section $\mathrm{Det}\,\partial$ of the determinant line bundle:

$$\det{}'\Delta_{z\bar{z}} = ||\mathrm{Det}\,\partial||_Q^2 = \exp\left[-\zeta_\tau'(0)\right], \tag{13.77}$$

leading to the same result (13.71) as above.

We can now write the one-loop partition function (13.61) as

$$\mathcal{A}_1 = \int_{\mathcal{M}_1} \frac{d^2\tau}{\tau_2^2} \frac{1}{\tau_2^{12}|\Delta(\tau)|^2}. \tag{13.78}$$

Let us also comment on the interpretation of this result from the point of view of the world-sheet conformal field theory (CFT) of the bosonic string. On general grounds, for a conformal field theory on a torus parametrised by τ the *partition function* is given by [535]

$$Z[\tau,\bar{\tau}] = \mathrm{Tr}\, q^{L_0-c/24}\bar{q}^{\bar{L}_0-\bar{c}/24}, \tag{13.79}$$

where we have defined $q = e^{2\pi i\tau}$, $\bar{q} = e^{-2\pi i\bar{\tau}}$. This expression can be understood as follows. Let L_0 be the zero mode of the Virasoro algebra (13.13), restricted to the transverse directions; see Remark 13.5 below. The trace is taken over the spectrum of L_0. The Hamiltonian generating time translations along the Euclidean time circle of the world-sheet CFT is given by

$$H = L_0 + \bar{L}_0 - (c+\bar{c})/24, \tag{13.80}$$

while the momentum operator generating rotations of the torus reads

$$P = L_0 - \bar{L}_0. \tag{13.81}$$

In the simplest case of a rectangular torus, $\tau_1 = 0$, the partition function takes the familiar form from statistical mechanics

$$Z[\beta] = \mathrm{Tr}\, e^{-\beta H}, \tag{13.82}$$

where β is the inverse temperature T:

$$\beta = 2\pi/T = 2\pi\tau_2. \tag{13.83}$$

The partition function $Z[\tau]$ is thus the natural generalisation of this to the case of a non-rectangular torus. This partition function can also be represented as a path integral over the maps

$$X : \Sigma \longrightarrow M, \tag{13.84}$$

from the world-sheet torus Σ into space-time M with periodic boundary conditions along the torus:

$$X(z+\tau) = X(z), \qquad X(z+1) = X(z). \tag{13.85}$$

These are just free bosonic fields on Σ with Gaussian action

$$S_E[X] = \frac{1}{2}\int d^2z X \Delta_{z\bar{z}} X \tag{13.86}$$

for a Euclidean target manifold, and we set $\alpha' = 1$ for the remainder of this section. Thus the equivalence between the operator and the path integral representations of the partition function can be formally written

$$\operatorname{Tr} q^{L_0 - c/24} \bar{q}^{\bar{L}_0 - \bar{c}/24} = \int \mathcal{D}X e^{-\frac{1}{2}\int d^2z X \Delta_{z\bar{z}} X}. \tag{13.87}$$

The left-hand side can be computed using standard techniques in conformal field theory by simply evaluating the sum over the harmonic oscillators (see [177]). The path integral expression can be computed by Fourier expanding the fields along the real part z_1 of $z = z_1 + iz_2 \in \Sigma$:

$$X(z_1, z_2) = X_0(z_2) + \sum_{n \neq 0} X_n(z_2) e^{2\pi i n z_1}. \tag{13.88}$$

Integrating out the dependence on z_1 the action becomes

$$S_E[X] = -\int dz_2 \left[\frac{1}{2}\dot{X}_0^2 + \sum_{n\neq 0}\left(\dot{X}_n \dot{X}_{-n} + (2\pi n)^2 X_n X_{-n}\right)\right], \tag{13.89}$$

where

$$\dot{X} \equiv \frac{\partial}{\partial z_2} X. \tag{13.90}$$

Evaluating this in either picture then gives the result

$$Z[\tau, \bar{\tau}] = \frac{1}{\tau_2^{12} |\Delta(\tau)|^2}, \tag{13.91}$$

which is precisely the integrand in (13.78).

We note that $Z[\tau, \overline{\tau}]$ is a modular invariant function, and, togeteher with the modular invariant measure $d^2\tau/\tau_2^2$ in (13.78), this makes the amplitude $\mathcal{A}_1$ modular invariant as well. This is expected since the amplitude should remain unchanged on equivalent tori.

Remark 13.5 It is illuminating to write the partition function as

$$Z[\tau, \bar{\tau}] = |Z[\tau]|^2 = \left| \frac{1}{(\sqrt{\tau_2})^{12}\eta(\tau)^{24}} \right|^2 . \tag{13.92}$$

The expression inside the modulus is the contribution to the path integral from a single chiral boson on the torus. The structure of the partition function thus reflects the fact that we have 24 left-moving and 24 right-moving free bosons on Σ. The factor of $1/\sqrt{\tau_2}$ comes from the integration over zero modes which is common to the left- and right-moving sectors. (This is the case for non-compact target manifolds; for compact targets the argument has to be modified, as we shall see in Section 13.4.3.) Since space-time is 26-dimensional for the bosonic string we actually have 26 free bosons in the theory, but the contributions from two of those (corresponding to the light-cone directions) cancel out agains the integral over the ghost fields in going from (13.58) to (13.61), leaving us with only 24 bosons in the final answer. The reduction to 24 can also be understood as the number of transverse directions relevant for massless particles; this is most evident in the so-called *light-cone formulation* of string theory [320, 535, 63].

To summarise, we have learned that the one-loop vacuum to vacuum amplitude can be written as

$$\mathcal{A}_1 = \int_{\mathcal{M}_1} \frac{d^2\tau}{\tau_2^2} \operatorname{Tr} q^{L_0 - c/24} \bar{q}^{\bar{L}_0 - \bar{c}/24} . \tag{13.93}$$

The general structure of this expression will be preserved when we include fermions below, but the details will differ. We stress that L_0 here only contains the 24 directions transverse to a fixed light-like vector.

13.4.2 Including Fermions

We will now study the vacuum to vacuum amplitude $\mathcal{A}_1$ for the superstring. In the full superstring theory we also have fermionic degrees of freedom. The critical dimension is $D = 10$ and we thus have $10 + 10$ free bosons on the world-sheet. Similarly, as before, the contribution from two of these will cancel out against the ghost fields after gauge-fixing. The bosons therefore contribute a factor

$$Z_b[\tau, \bar{\tau}] = \left|Z_b[\tau]\right|^2 = \left| \frac{1}{(\sqrt{\tau_2})^8 \eta(\tau)^8} \right|^2 \tag{13.94}$$

to the integrand. We can therefore write the vacuum to vacuum amplitude as

$$\mathcal{A}_1 = \int_{\mathcal{M}_1} \frac{d^2\tau}{\tau_2^2} \left|Z_b[\tau]\right|^2 \left|Z_f[\tau]\right|^2, \tag{13.95}$$

where $Z_f[\tau]$ is the contribution from the fermions, which remains to be determined. To construct a supersymmetric theory we use the *Ramond–Neveu–Schwarz formalism*, discussed in Section 13.1.2, introducing additional fermionic fields (ψ_1, ψ_2) on the world-sheet. The path integral for the fermionic fields takes the form

$$\int \mathcal{D}\psi_1\, \mathcal{D}\psi_2 e^{-\frac{1}{2}\int d^2z \psi_1(\partial_{z_1}+i\partial_{z_2})\psi_2}. \tag{13.96}$$

In contrast to the bosonic case, the fermionic path integral allows for twisted periodic boundary conditions on the torus:

$$\begin{aligned} \psi_1(z+1) &= -e^{-2\pi i\theta}\psi_1(z), & \psi_2(z+1) &= -e^{-2\pi i\theta}\psi_2(z), \\ \psi_1(z+\tau) &= -e^{-2\pi i\phi}\psi_1(z), & \psi_2(z+\tau) &= -e^{-2\pi i\phi}\psi_2(z). \end{aligned} \tag{13.97}$$

These are called *spin structures*. There are four possibilities for the twisting parameters (θ, ϕ):

$$(0,0),\ (\tfrac{1}{2},0),\ (0,\tfrac{1}{2}),\ (\tfrac{1}{2},\tfrac{1}{2}). \tag{13.98}$$

The *mapping class group* $SL(2,\mathbb{Z})$ of the torus acts on the spin structures and mixes them. Therefore, to achieve a modular invariant result one must sum over all the spin structures. On general grounds the fermionic partition function will thus be of the form

$$Z_f[\tau] = aZ_{(0,0)}[\tau]^4 + bZ_{(1/2,0)}[\tau]^4 + cZ_{(0,1/2)}[\tau]^4 + dZ_{(1/2,1/2)}[\tau]^4, \tag{13.99}$$

where a, b, c, d are phases that must be fixed in order to ensure modular invariance. It will turn out that these are only relative signs. Each contribution can be written as follows in the operator formalism:

$$\begin{aligned} Z_{(0,0)}[\tau] &= \mathrm{Tr}_{\mathrm{NS}} q^{L_0-c/24}, & Z_{(1/2,0)}[\tau] &= \mathrm{Tr}_{\mathrm{R}} q^{L_0-c/24}, \\ Z_{(0,1/2)}[\tau] &= \mathrm{Tr}_{\mathrm{NS}}(-1)^F q^{L_0-c/24}, & Z_{(1/2,1/2)}[\tau] &= \mathrm{Tr}_{\mathrm{R}}(-1)^F q^{L_0-c/24}. \end{aligned} \tag{13.100}$$

Here, F is the (world-sheet) fermion number operator which has eigenvalue $+1$ on each state with an even number of fermions and -1 on each state with an odd number of fermions. The traces are taken over the states in the Neveu–Schwarz (NS) sector or the Ramond (R) sector, which are characterised by having, respectively, anti-periodic and periodic boundary conditions in the z_1-direction according to (13.97). The partition functions for the various spin structures can be calculated by standard methods (see, e.g., [177]) with the result

$$
\begin{aligned}
Z_{(0,0)}[\tau] &= \frac{\theta_3(\tau)}{\eta(\tau)}, & Z_{(1/2,0)}[\tau] &= \frac{\theta_2(\tau)}{\eta(\tau)}, \\
Z_{(0,1/2)}[\tau] &= \frac{\theta_4(\tau)}{\eta(\tau)}, & Z_{(1/2,1/2)}[\tau] &= \frac{\theta_1(\tau)}{\eta(\tau)} = 0 .
\end{aligned}
\tag{13.101}
$$

The functions appearing on the right-hand side are Jacobi theta functions. To define them, we first write a general Jacobi theta function as follows:

$$
\theta\left[\begin{smallmatrix}\alpha\\ \beta\end{smallmatrix}\right](z|\tau) = \sum_{n\in\mathbb{Z}} e^{i\pi[(n+\alpha)^2\tau + 2(n+\alpha)(z+\beta)]}. \tag{13.102}
$$

The parameters (α, β) are called *characteristics* of the theta function. The theta functions appearing in (13.101) correspond to the specialisations

$$
\begin{aligned}
\theta_1(\tau) &= \theta\left[\begin{smallmatrix}1/2\\ 1/2\end{smallmatrix}\right](0|\tau), & \theta_2(\tau) &= \theta\left[\begin{smallmatrix}1/2\\ 0\end{smallmatrix}\right](0|\tau), \\
\theta_3(\tau) &= \theta\left[\begin{smallmatrix}0\\ 0\end{smallmatrix}\right](0|\tau), & \theta_4(\tau) &= \theta\left[\begin{smallmatrix}0\\ 1/2\end{smallmatrix}\right](0|\tau).
\end{aligned}
\tag{13.103}
$$

Remark 13.6 Properties of the classical Jacobi theta functions occurring above were also discussed in Section 12.1, with explicit formulas for their modular properties. The representation theory of theta functions and the generalisation to higher-rank groups are mentioned in Section 12.2.

To obtain a modular invariant partition function for the fermionic contributions we can choose the relative phases in (13.99) as

$$
a = -b = -c = 1. \tag{13.104}
$$

Since $\theta_1(\tau) = 0$ there is no need to fix d. We then have

$$
Z_f[\tau] = \frac{1}{\eta(\tau)^4}\left(\theta_3(\tau)^4 - \theta_2(\tau)^4 - \theta_4(\tau)^4\right). \tag{13.105}
$$

Combining everything, the full one-loop amplitude (13.95) can now be written in the form

$$
\mathcal{A}_1 = \int_{\mathcal{M}_1} \frac{d^2\tau}{\tau_2^2}\frac{1}{\tau_2{}^4}\frac{1}{|\Delta(\tau)|^2}\left|\theta_3(\tau)^4 - \theta_2(\tau)^4 - \theta_4(\tau)^4\right|^2. \tag{13.106}
$$

Remark 13.7 The integrand can be understood in terms of determinant line bundles in a similar vein as in the bosonic case [18]. The chiral Dirac operator

$$
\mathcal{D} = \partial_{z_1} + i\partial_{z_2} \tag{13.107}
$$

occurring in (13.96) defines the Dirac determinant line bundle $\mathscr{D} \to \Sigma$ via

$$
\mathscr{D} = (\Lambda^{\max}\ker\mathcal{D})^* \otimes (\Lambda^{\max}\operatorname{coker}\mathcal{D}). \tag{13.108}
$$

Although we have not displayed it explicitly, this bundle depends on a choice of spin structure on Σ. The bundle $\mathscr{D}$ has a canonical holomorphic section which is the determinant Det $\mathcal{D}$. One can show that this section is given by

$$\mathrm{Det}\,\mathcal{D} = \theta\left[\begin{smallmatrix}\alpha\\ \beta\end{smallmatrix}\right](0|\tau), \tag{13.109}$$

where the choice of spin structure of $\mathscr{D}$ is encoded in the characteristics (α, β) of the theta function, as already mentioned above.

Remark 13.8 One can show using theta function identities that the precise combination

$$\theta_3(\tau)^4 - \theta_2(\tau)^4 - \theta_4(\tau)^4, \tag{13.110}$$

occurring in (13.106), actually vanishes. As a mathematical identity this is known as Jacobi's *abstruse identity*. From a physical point of view this vanishing is natural, as it implies that the cosmological constant vanishes at one-loop, which is expected in a supersymmetric theory. For a thorough discussion of the physical interpretation of $\mathscr{A}_1$ see, e.g., [63].

13.4.3 Toroidal Compactification

Let us now consider the one-loop amplitude of a closed bosonic string when space-time is compactified on a torus $T^d = \mathbb{R}^d/\Gamma$, where the lattice $\Gamma \cong \mathbb{Z}^d$. That is, the fields on the world-sheet Σ are maps

$$X : \Sigma \longrightarrow \mathbb{R}^d/\Gamma \times \mathbb{R}^{1,25-d}. \tag{13.111}$$

The main new feature is now that some directions in the target manifold are compact and periodic. In the uncompactified case the momenta of the string were continuous quantities, while for strings moving in a torus the momenta p become quantised and take values in the dual lattice:

$$p \in \Gamma^*. \tag{13.112}$$

In addition, we also must take into account the winding number w, which roughly counts the number of times the closed string can wind around a circle in T^d. It takes values in the lattice itself:

$$w \in \Gamma. \tag{13.113}$$

We can view this mathematically as distinguishing the various connected components of the loop space

$$\mathcal{L}T^d = \{x : S^1 \to T^d\}, \tag{13.114}$$

since

$$\pi_0(\mathcal{L}T^d) = \Gamma. \tag{13.115}$$

It is natural to form the vector (w, p) which takes values in the *Narain lattice*

$$\Gamma^{d,d} = \Gamma \oplus \Gamma^*. \tag{13.116}$$

This is an even, self-dual lattice of signature (n, n) with inner product

$$(w, p)^2 = 2w \cdot p, \tag{13.117}$$

where $w \cdot p$ denotes the natural pairing betwen Γ and Γ^*. The lattice $\Gamma^{d,d}$ is unique up to isomorphism

$$\Gamma^{d,d} \cong \Pi^{1,1} \oplus \cdots \oplus \Pi^{1,1}, \tag{13.118}$$

where $\Pi^{1,1}$ is the unique even and self-dual Lorentzian lattice in two dimensions whose bilinear form can be chosen as

$$\begin{pmatrix} 0 & 1 \\ 1 & 0 \end{pmatrix}. \tag{13.119}$$

The automorphisms of the Narain lattice lift to symmetries of the full string theory on T^d. This is the T-duality group

$$SO(d, d; \mathbb{Z}) = \mathrm{Aut}(\Gamma^{d,d}), \tag{13.120}$$

and contains the subgroup

$$SL(d, \mathbb{Z}) \subset SO(d, d; \mathbb{Z}) \tag{13.121}$$

which represents large diffeomorphisms of T^d, i.e., diffeomorphisms not connected to the identity.

Remark 13.9 The elements of $O(d, d; \mathbb{Z})$ with negative determinant interchange the chirality of fermions and therefore the full $O(d, d; \mathbb{Z})$ do not leave the lattice $\Gamma^{d,d}$ invariant.

The inner product (13.117) is of split signature and one can parametrise all such choices by $SO(d, d; \mathbb{R})$. Since $SO(d) \times SO(d) \subset SO(d, d; \mathbb{R})$ leaves the individual subspaces of positive and negative norm invariant, one deals with the Grassmannian $SO(d, d; \mathbb{R})/(SO(d) \times SO(d))$. Points in this Grassmannian can be parametrised by a combination of the metric and the B-field, $G + B$. The quantum theory must also respect the T-duality symmetry, and hence the moduli space is the arithmetic quotient

$$\mathcal{M}_{d,d} = SO(d, d; \mathbb{Z}) \backslash SO(d, d; \mathbb{R})/(SO(d) \times SO(d)). \tag{13.122}$$

This is the moduli space of inequivalent lattices $\Gamma_{d,d}$. In the full superstring theory, the T-duality group is furthermore contained in the U-duality group displayed in Table 13.2:

$$SO(d, d; \mathbb{Z}) \subset E_{d+1}(\mathbb{Z}), \tag{13.123}$$

and hence the moduli space (13.122) is contained in the full quantum moduli space (13.32):

$$\mathcal{M}_{d,d} \subset E_{d+1}(\mathbb{Z})\backslash E_{d+1}(\mathbb{R})/K(E_{d+1}). \tag{13.124}$$

The choice of metric G and B-field induces a splitting of an arbitrary vector in $\Gamma^{d,d}$ of the form

$$p_+ \oplus p_- \in \Gamma^{d,d}. \tag{13.125}$$

If we introduce indices $i, j = 1, \ldots, d$ for the torus directions we can parametrise this as

$$p_{i,\pm} = \frac{1}{\sqrt{2}}\left(k_i \pm G_{ij}w^j + B_{ij}w^j\right) \tag{13.126}$$

for integer co-vectors k^i and vectors w_i. This implies that the $O(d,d)$-invariant bilinear form is

$$(p_+ \oplus p_-)^2 = p_+^2 - p_-^2 = p_{i,+}G^{ij}p_{j,+} - p_{i,-}G^{ij}p_{j,-} = 2k_iw^i\,. \tag{13.127}$$

Moreover, we note for later reference that for $\tau = \tau_1 + i\tau_2 \in \mathbb{H}$ one has

$$\begin{aligned} i\pi\tau p_+^2 - i\pi\bar{\tau}p_-^2 &= i\pi\tau_1(p_+^2 - p_-^2) - \pi\tau_2(p_+^2 + p_-^2) \\ &= -\pi\tau_2\left[(k_i + B_{im}w^m)G^{ij}(k_j + B_{jn}w^n) + w^iG_{ij}w^j\right] \\ &\quad + 2\pi i\tau_1 k^i w_i. \end{aligned} \tag{13.128}$$

Remark 13.10 The expression

$$M^2 = (k_i + B_{im}w^m)G^{ij}(k_j + B_{jn}w^n) + w^iG_{ij}w^j \tag{13.129}$$

represents the squared mass of a 1/2-BPS state with momentum and winding $(k, w) \in \Gamma^{d,d}$. See also Example 13.13.

The main difference in the calculation of the partition function comes from the structure of the zero mode contribution to the path integral. We discuss this in the spirit of Remark 13.5 in the Hamiltonian picture and by focussing on the 24 directions transverse to a fixed light-like vector. For a non-compact target space one has continuous momentum $p \in \mathbb{R}$ and integrals of the form

$$\int_{\mathbb{R}} dp\, e^{-\pi\tau_2 p^2} = \tau_2^{-1/2}\,. \tag{13.130}$$

By contrast, for a compact target space direction, the momentum is quantised and the integral above is replaced by a discrete sum over the allowed momenta. In addition, there is a sum over the topologically distinct winding configurations of the closed string. These two discrete variables can be parametrised by $(p_+, p_-) \in \Gamma_{d,d}$ as we just saw.

In the Hamiltonian picture with Hamiltonian given by (13.80), one then obtains for the zero mode contribution to $\mathrm{Tr}\, q^{L_0-c/24}\bar{q}^{\bar{L}_0-\bar{c}/24}$ instead of the continuous integral (13.130) yielding $\tau_2^{-1/2}$ the sum

$$\begin{aligned}\theta_{\Gamma_{d,d}}(\tau) &= \sum_{(p_+,p_-)\in\Gamma^{d,d}} q^{p_+^2/2}\bar{q}^{p_-^2/2} \\ &= \sum_{k\in\Gamma^*,n\in\Gamma} \exp\left[-\pi\tau_2\left[(k+Bn)G^{-1}(k+Bn)+nGn\right]+2\pi i\tau_1 kn\right],\end{aligned} \tag{13.131}$$

which is a lattice theta function of the type discussed in Section 12.1. Putting this together with the zero modes in the remaining non-compact directions and including also the oscillating contributions, one finds the full partition function for the critical closed bosonic string when the target includes a torus T^d to be

$$Z_{T^d}[\tau;G,B] = \tau_2^{(d-24)/2}|\eta(\tau)|^{-48}\theta_{\Gamma_{d,d}}(\tau), \tag{13.132}$$

where the dependence on G and B appears in the norms p_+^2 and p_-^2 as we saw in (13.128).

It is also instructive to write the partition function as

$$Z_{T^d}[\tau;G,B] = \tau_2^{-12}|\eta(\tau)|^{-48}\Theta_{\Gamma_{d,d}}(\tau) \tag{13.133}$$

in terms of the *Siegel–Narain theta function*

$$\begin{aligned}\Theta_{\Gamma^{d,d}}(\tau) &= \tau_2^{d/2} \sum_{(p_+,p_-)\in\Gamma^{d,d}} q^{p_+^2/2}\bar{q}^{p_-^2/2} \\ &= \tau_2^{d/2}\sum_{k\in\Gamma^*,n\in\Gamma} \exp\left[-\pi\tau_2\left[(k+Bn)G^{-1}(k+Bn)+nGn\right]+2\pi i\tau_1 kn\right],\end{aligned} \tag{13.134}$$

which was already defined in (12.36). The Siegel–Narain theta function is manifestly invariant under T-duality $SO(d,d;\mathbb{Z})$ acting on the metric G and the B-field since this just reorders the lattice points in the sum. To bring out the modular invariance of $\Theta_{\Gamma_{d,d}}$, we perform a Poisson resummation:

$$\begin{aligned}&\tau_2^{d/2}\sum_{k\in\Gamma^*,n\in\Gamma} \exp\left[-\pi\tau_2\left[(k+Bn)G^{-1}(k+Bn)+nGn\right]+2\pi i\tau_1 kn\right] \\ &= \sqrt{\det G}\sum_{m,n\in\Gamma}\exp\left[-\frac{\pi}{\tau_2}(m+\tau n)(G+B)(m+\bar{\tau}n)\right].\end{aligned} \tag{13.135}$$

Performing a modular transformation $\tau \to -1/\tau$ in the second form is easily seen to leave this invariant by just relabelling the summands. The variables $m, n \in \Gamma$ appearing in the Poisson resummed form have the interpretation of the windings of the two circles in the string world-sheet torus Σ around the space-time torus T^d in a Lagrangian description. The full partition function

(13.133) is thus the product of two modular invariant factors and hence modular invariant.

Remark 13.11 Based on the above discussion we can easily deduce the toroidal partition function of the superstring. The Narain theta function arises from the quantised momenta along T^d and is therefore unchanged. The only modification comes from the uncompactified directions and this factor is the same as in (13.106) except for a different power of τ_2. Combining everything, the result is

$$Z_{T^d}[\tau; G, B] = \frac{1}{\tau_2^4} \frac{1}{|\Delta(\tau)|^2} \left|\theta_3(\tau)^4 - \theta_2(\tau)^4 - \theta_4(\tau)^4\right|^2 \Theta_{\Gamma^{d,d}}. \qquad (13.136)$$

Of course, this result also vanishes by the same arguments as in Remark 13.8.

13.4.4 Theta Lifts

We have seen that quite generally the one-loop amplitude $\mathcal{A}_1$ for string theory compactified on a torus T^d is of the following form:

$$\mathcal{A}_1 = \int_{SL(2,\mathbb{Z})\backslash\mathbb{H}} \frac{d^2\tau}{\tau_2^2} Z_{T^d}[\tau; G, B], \qquad (13.137)$$

where $Z_{T^d}[\tau; G, B]$ is the partition function of the world-sheet conformal field theory. Since we are integrating over a fundamental domain of the moduli space $\mathcal{M}_1$ of genus-one Riemann surfaces, the resulting amplitude is independent of τ. By construction it is furthermore invariant under $SO(d, d; \mathbb{Z})$ and hence is an automorphic function:

$$\mathcal{A}_1 : SO(d, d; \mathbb{Z})\backslash SO(d, d; \mathbb{R})/(SO(d) \times SO(d)) \longrightarrow \mathbb{C}. \qquad (13.138)$$

To understand what kind of object this is we rewrite the integrand slightly. Since the metric and the B-field parametrise a point in the Grassmannian, we can replace the dependence on G and B by some group element $g \in SO(d, d; \mathbb{R})$. Because of the invariance under the maximal compact subgroup $SO(d) \times SO(d)$ we can always use the Iwasawa decomposition $g = nak$ to remove the dependence on k from the integrand. The dependence on g is completely captured by the Narain theta function so it is fruitful to split the integrand and write the one-loop amplitude as

$$\Theta_f(g) = \int_{SL(2,\mathbb{Z})\backslash\mathbb{H}} \frac{d^2\tau}{\tau_2^2} f(\tau)\, \Theta_{\Gamma^{d,d}}(\tau; g), \qquad (13.139)$$

where $f(\tau)$ is some (not necessarily holomorphic) modular function, independent of $g \in SO(d, d; \mathbb{R})$. It is now clear that one can view the one-loop integral as a kind of integral transform from modular forms on the upper half-plane

$\mathbb{H}$ to automorphic forms on $SO(d,d;\mathbb{R})$. From this point of view the Narain theta function is the kernel of the transform. We may then recognise this as an example of a *theta lift*.

As discussed in detail in Chapter 12, the general process of theta lifting an automorphic form on a group G to an automorphic form on another group G' requires (G, G') to form a dual reductive pair inside some larger group H. In the present case, the dual reductive pair is

$$SL(2,\mathbb{R}) \times SO(d,d;\mathbb{R}) \subset Sp(2d;\mathbb{R}), \tag{13.140}$$

such that $SL(2,\mathbb{R})$ is the normaliser of $SO(d,d;\mathbb{R})$ inside the symplectic group $Sp(2d;\mathbb{R})$, and vice versa. There also exist higher-genus generalisations.

In general, it is not easy to determine what kind of object one can get as the image of a theta lift. In what follows we shall give some indication of the answer by looking at some explicit examples that arise in string theory.

Example 13.12: One-loop on T^2 and the Dedekind eta function

Consider the case of $d = 2$. This corresponds to compactification of string theory on a two-dimensional torus T^2. In this case the moduli space decomposes:

$$\mathcal{M}_{2,2} = SO(2,2;\mathbb{Z})\backslash SO(2,2)/(SO(2)\times SO(2)) = (SL(2,\mathbb{Z})\backslash\mathbb{H}\times SL(2,\mathbb{Z})\backslash\mathbb{H})/\mathbb{Z}_2. \tag{13.141}$$

Let us parametrise the moduli of the two factors by (T, U) with $\operatorname{Im} T > 0$ and $\operatorname{Im} U > 0$. They correspond, respectively, to the Kähler and complex structure moduli of the space-time torus T^2. We consider the following one-loop integral:

$$\int_{SL(2,\mathbb{Z})\backslash\mathbb{H}} \frac{d^2\tau}{\tau_2^2}\Theta_{2,2}(\tau;T,U). \tag{13.142}$$

This object plays a rôle in several places in string theory. It arises as the one-loop coefficient of the R^4 term in type II string theory on T^2, as well as in the R^2-couplings of type II string theory on $K3 \times T^2$ [184, 503]. The integral can be evaluated by the *Rankin–Selberg method*, which unfolds the fundamental domain and divides the integration into orbits under the $SL(2,\mathbb{Z})$ action on the winding numbers (m, n). See, e.g., [503] for a detailed explanation of this method. One obtains

$$\int_{SL(2,\mathbb{Z})\backslash\mathbb{H}} \frac{d^2\tau}{\tau_2^2}\Theta_{2,2}(\tau;T,U) = -2\pi \log|\sqrt{T_2}\eta(T)^2|^2 - 2\pi\log|\sqrt{U_2}\eta(U)^2|^2. \tag{13.143}$$

This is actually invariant under the larger group $O(2,2;\mathbb{Z})$, where the additional transformation is the central $\mathbb{Z}_2$ that exchanges $T \leftrightarrow U$.

After this warmup example we now consider the same type of integral on an arbitrary d-dimensional torus T^d and its relation to $O(d,d)$ Eisenstein series. As we shall see, this gives rise to interesting specialisations of Eisenstein series on $O(d,d)$.

Example 13.13: One-loop on T^d and constrained Epstein series

The previous example can actually be extended to the lattice $\Gamma^{d,d}$ for any $d > 2$. By similar unfolding techniques as in the $d = 2$ case one can show that the one-loop integral evaluates to [503]

$$\int_{SL(2,\mathbb{Z})\backslash\mathbb{H}} \frac{d^2\tau}{\tau_2^2}\left(\Theta_{d,d}(\tau; G, B) - \tau_2^{d/2}\right) = \frac{\Gamma(d/2-1)}{\pi^{d/2-1}}\mathcal{E}^{SO(d,d)}_{d/2-1}(G, B), \tag{13.144}$$

where $\mathcal{E}^{SO(d,d)}_s(G, B)$ is a constrained *Epstein series* defined by

$$\mathcal{E}^{SO(d,d)}_s(G, B) = \sum_{n\in\Gamma,\, k\in\Gamma^*}{}' \delta(nk)\left[(k + Bn)G^{-1}(k + Bn) + nGn\right]^{-2s}, \tag{13.145}$$

where $\delta(nk)$ enforces the 1/2-BPS constraint $nk = n^i k_i = 0$ and we exclude $n = k = 0$. The insertion of the term $\tau_2^{d/2}$ in (13.144) regulates the integral to ensure convergence in the (infrared) limit $\tau_2 \to \infty$ [184, 318, 19]. We recognise the same structure in the summand as in the resummed theta function (13.135). Physically, this is the formula for the mass (squared) of the 1/2-BPS state corresponding to a fundamental string ground state with charges $(k, n) \in \Gamma^{d,d}$ [523]:

$$M^2(k, n) = p_+^2 + p_-^2 = (k + Bn)G^{-1}(k + Bn) + nGn. \tag{13.146}$$

One can show that the constrained Epstein series (13.145) is in fact proportional to a particular degenerate Langlands Eisenstein series on $SO(d, d; \mathbb{R})$ [318, 19]:

$$\mathcal{E}^{SO(d,d)}_s(G, B) = 2\zeta(2s)E(\lambda_s, g), \qquad g \in SO(d, d; \mathbb{R}). \tag{13.147}$$

The weight appearing for the maximal parabolic Eisenstein series in this formula is $\lambda_s = 2s\Lambda_1 - \rho$ and is associated with the 'vector node' of the D_d Dynkin diagram in Bourbaki labelling. The Eisenstein series $E(\lambda_s, g)$ were defined and discussed in detail in Section 4.5.2, while the relation between Eisenstein series and constrained lattice sums is treated in Section 15.3.

Remark 13.14 The preceding two examples illustrate the theta lift of a trivial modular function $f = 1$ on $\mathbb{H}$ to an automorphic form on $SO(d, d; \mathbb{R})$. Representation-theoretically this is an example of a theta correspondence where the trivial automorphic representation of $SL(2, \mathbb{R})$ is lifted to a particular (minimal) automorphic representation of $SO(d, d; \mathbb{R})$. This point of view was also taken in Chapter 12 where these two examples were interpreted in terms of the so-called *Siegel–Weil formula*.

We have now seen explicit examples of theta lifts where the image is an Eisenstein series. However, in string theory one often encounters one-loop amplitudes like (13.139) with an insertion of a non-trivial modular form $f(\tau)$. Below we will give various examples of this which illustrate that one can get much more exotic objects in the image of a theta lift, some of which lie beyond the scope of this book.

Example 13.15: One-loop on $K3 \times T^2$ and the J-function

In heterotic string theory on $K3 \times T^2$ (with no Wilson lines) there are one-loop threshold corrections to gauge couplings that take the form of a theta lift [347]:

$$\Theta_j(T,U) = \int_{SL(2,\mathbb{Z})\backslash\mathbb{H}} \frac{d^2\tau}{\tau_2^2} \left[j(\tau)\Theta_{2,2}(\tau;U,V) - c(0) \right]. \tag{13.148}$$

This depends only on the moduli (U,V) of the space-time torus T^2, just as in Example 13.12. The result will therefore be an automorphic form on $\mathcal{M}_{2,2}$ (13.141). The function $j(\tau)$ is the modular invariant j-function (see Section 4.2.1) whose Fourier expansion reads

$$j(\tau) = \sum_{n=-1}^{\infty} c(n)q^n = \frac{1}{q} + c(0) + 196884q + 21493760q + \cdots. \tag{13.149}$$

This function is unique up to the choice of constant term $c(0)$. The above integral was evaluated by Harvey and Moore with the result [347]

$$\begin{aligned}\Theta_j(T,U) &= -2\log|j(T)-j(U)|^2 - c(0)\log|(\eta(T)\eta(U))^2|^2 \\ &\quad -c(0)\log(2T_2U_2) - c(0)\log\left(\tfrac{4\pi}{\sqrt{27}}e^{1-\gamma_E}\right),\end{aligned} \tag{13.150}$$

where γ_E is the Euler–Mascheroni constant. This result is not in the form of any kind of Eisenstein series. In fact, the resulting object does not obviously fit into the standard framework of automorphic representations which is the main focus of this book. See Section 15.5.4 for some remarks on this.

Notice that for the special choice $c(0) = 0$ the function $j(\tau)$ has no constant term. This case is usually denoted by $J(\tau)$ and for this the theta lift simplifies to

$$\Theta_J(T,U) = -2\log|J(T) - J(U)|^2. \tag{13.151}$$

This equation lies at the heart of Borcherds' proof of monstrous moonshine; see Section 15.5 for more on this.

Remark 13.16 It is often the case in string theory that the function $f(\tau)$ does not exhibit rapid decay in the limit $\tau \to i\infty$. For instance, the j-function in the previous example has a pole at infinity and so grows exponentially. Physically, this leads to infrared divergences which must be regulated away. In Borcherds' general treatment of such integrals he therefore referred to them as *singular theta lifts*. This is discussed in Section 15.5.2. Recently, such one-loop integrals in string theory were revisited in a series of papers [19, 20, 21, 22, 215]. The key idea is to represent the modular form $f(\tau)$ in terms of a Poincaré series

$$f(\tau) = \sum_{\gamma\in B(\mathbb{Z})\backslash SL(2,\mathbb{Z})} \psi(\gamma\cdot\tau), \tag{13.152}$$

for some suitable function ψ. Thus, one writes the theta lift as

$$\Theta_f(g) = \sum_{\gamma \in B(\mathbb{Z})\backslash SL(2,\mathbb{Z})} \int_{SL(2,\mathbb{Z})\backslash \mathbb{H}} \frac{d^2\tau}{\tau_2^2} \psi(\gamma \cdot \tau)\, \Theta_{\Gamma^{d,d}}(\tau; g), \tag{13.153}$$

and then unfolds the sum in the standard way [184]. This method has the benefit of working quite generally and keeps manifest the $SO(d, d; \mathbb{Z})$ symmetry of the integrand. See [19] for details of this procedure.

13.5 D-branes*

In addition to strings, string theory also contains extended objects, called *branes*, as was first emphasised by Polchinski [532, 533, 534]. These correspond to submanifolds of space-time M on which open strings can end. They are not visible in the perturbative formulation of string theory based on sigma models of the form (13.11) but show up as effects that are non-perturbative in the string coupling g_s. We will restrict the discussion in the following to superstring theory where M is ten-dimensional and mainly analyse the bosonic states of the theory.

As we have seen, in the absence of the B-field the critical points of the Polyakov action (13.6) are just harmonic maps

$$X : \Sigma \longrightarrow M. \tag{13.154}$$

A closed string has to be a periodic function in the spatial variable $\sigma \equiv \sigma^1$ (see Section 13.1), but for an open string one must specify boundary conditions at the end-points, or, equivalently, at the boundary $\partial\Sigma$ of the string world-sheet. Writing the components of the map X as X^μ in some coordinate system for M $(\mu = 0, 1, \ldots, 9)$, we have two basic options for each space-time direction μ and for each of the two end-points separately:

$$\begin{aligned} &\text{Dirichlet :} && X^\mu(\sigma)|_{\partial\Sigma} = X_0^\mu \in \mathbb{R}, \\ &\text{Neumann :} && n^m \partial_m X^\mu(\sigma)|_{\partial\Sigma} = 0, \end{aligned} \tag{13.155}$$

where X_0^μ is a constant while n^m is the normal vector field to the boundary of Σ. The *Neumann boundary condition* implies that the string endpoints move freely in this direction while the *Dirichlet boundary condition* requires that the value at the end-point is fixed to some constant X_0^μ in the μth direction. Therefore, if we impose Neumann boundary conditions in $p+1$ directions of the ten-dimensional space-time M, while imposing Dirichlet boundary conditions in the remaining $9 - p$ directions, the open string end-point is confined to move along a $(p + 1)$-dimensional submanifold $\mathcal{D} \subset M$, as depicted in Figure 13.7. Such a submanifold is by convention called a Dp-brane, where the letter 'D' represents 'Dirichlet'. In the context of string theory, Dirichlet

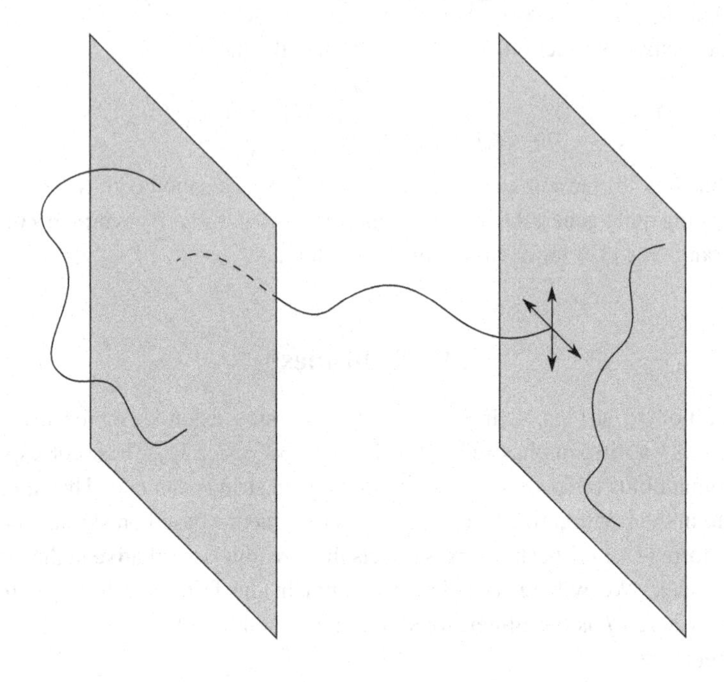

Figure 13.7 Open strings on D-branes.

and Neumann boundary conditions are interchanged when performing a T-duality transformation, thus changing the dimensionality of a Dp-brane by one unit.

Remark 13.17 From the point of view of the variational problem, one could also consider, more generally than (13.155), *Robin boundary conditions* where a linear combination of the value of X^μ at the end-point of the open string and its normal derivative there is set to zero. Unlike Dirichlet and Neumann boundary conditions, Robin boundary conditions are less common in string theory. They appear for example for strings in electric fields [35], moving boundary conditions [34] and for certain bound states [91]. Another variant encountered sometimes arises from the fact that open strings have two end-points and boundary conditions have to be fixed at both ends and can be fixed independently. For example, one could have one end of the open string fixed to a certain point in space-time, hence ending on a D-brane, while the other end is free to move anywhere. These conditions are sometimes referred to as mixed boundary conditions in the context of string theory.

Open string states interact with the closed string states that can be formed by interactions as described in Section 13.1.4, and the closed strings are not confined to the brane in the same way as open strings but can move freely through space-time. The interactions, moreover, make the brane a dynamical object: its position that originated as a Dirichlet boundary condition can fluctuate. One should not envisage the D-brane as perfectly rigid; it will be deformed by strings attached to it and near it that pull or push on it.

Moving strings attached to a Dp-brane give rise to a quantum field theory in $p + 1$ dimensions, called the *world-volume theory* of the Dp-brane. This quantum field theory is the low-energy effective action of the massless string states in the presence of the Dp-brane. These massless states for a single Dp-brane are a $U(1)$ gauge field and $9 - p$ scalar fields. The $U(1)$ gauge field can be visualised as being the particle associated with the one-dimensional world-line within the brane traced out by the end-point of the open string, where the gauge symmetry is inherited from the so-called BRST symmetry of the string spectrum. The scalar fields describe the geometric fluctuations of the brane in its transverse directions.

Using local coordinates ζ^a, $a = 0, 1, \ldots, p$ along the $(p + 1)$-dimensional brane $\mathcal{D}$ one can describe its position in ten-dimensional space-time in terms of the (brane) embedding $X^\mu(\zeta)$. In so-called static gauge for the coordinates of the brane one has

$$X^a(\zeta) = \zeta^a, \quad X^i(\zeta) = X_0^i + 2\pi\alpha'\Phi^i(\zeta) + \cdots . \tag{13.156}$$

Here, $i = p + 1, \ldots, 9$ are the transverse directions and the brane is aligned (statically) along the first $p+1$ space-time directions, while the $9-p$ transverse directions carry the fluctuations mentioned above and denoted Φ^i here. The interaction of the brane with the closed string states is expressed through this embedding while the interaction with the open string states is in terms of the $U(1)$ gauge field $A_a(\zeta)$ that is only defined on the brane. A $U(1)$ gauge field is a connection of a principal $U(1)$ bundle over $\mathcal{D}$.

Remark 13.18 We note that the number of physical *degrees of freedom* does not depend on the type of boundary condition chosen for the open string. With Neumann conditions in every space-time direction, the only massless open string state is a $U(1)$ gauge field that lives on all of space-time. At low energies this is effectively like *Maxwell's theory* of electrodynamics in D space-time dimensions, where $D = 10$ is the dimension of space-time. It is well known that, due to gauge invariance and the Maxwell equations, there are $D-2$ propagating degrees of freedom contained in such a $U(1)$ gauge field. (In $D = 4$ the gauge field corresponds to the standard photon that has two independent polarisations.) With a Dp-brane one only has $p + 1$ Neumann conditions, such that the Maxwell theory on the $(p + 1)$-dimensional brane has $p - 1$ degrees of freedom. In addition to the $U(1)$ gauge field on the brane one has to consider the $D-p-1$ transverse fluctuations, each of which corresponds to a single degree of freedom. In total one ends up again with $D-2$ degrees of freedom as it should be.

We will simply state the form of the low-energy effective action for the D-brane without deriving it; details on the derivation can be found for instance in [535, 534, 63, 634, 188]. The effective world-volume action has two distinct pieces:

$$S_{\mathrm{D}p}[G,\phi,B,A,C_{p+1}] = S_{\mathrm{DBI}}[\gamma,b,\phi,A] + S_{\mathrm{WZ}}[C_{p+1}], \tag{13.157}$$

with the *Dirac–Born–Infeld (DBI) action*

$$S_{\mathrm{DBI}}[G,B,\phi,A] = T_p \int_{\mathcal{D}} d^{p+1}\zeta\, e^{-\phi}\sqrt{\det(\gamma_{ab}+b_{ab}+2\pi\alpha' F_{ab})} \tag{13.158}$$

and the *Wess–Zumino term*

$$S_{\mathrm{WZ}}[C_{p+1}] = -i\int_{\mathcal{D}} C_{p+1}\,. \tag{13.159}$$

The notation used in these equations is as follows. The quantities γ_{ab}, b_{ab} and ϕ are the pull-backs of the target space data $G_{\mu\nu}$, $B_{\mu\nu}$ and ϕ discussed in Section 13.1.3:

$$\gamma_{ab}(\zeta) = \frac{\partial X^\mu(\zeta)}{\partial\zeta^a}\frac{\partial X^\nu(\zeta)}{\partial\zeta^b}G_{\mu\nu}(X(\zeta)), \tag{13.160}$$

$$b_{ab}(\zeta) = \frac{\partial X^\mu(\zeta)}{\partial\zeta^a}\frac{\partial X^\nu(\zeta)}{\partial\zeta^b}B_{\mu\nu}(X(\zeta)), \tag{13.161}$$

$$\phi(\zeta) = \phi(X(\zeta))\,. \tag{13.162}$$

These are the metric, B-field and dilaton. For the type II superstring they arise in the so-called massless Neveu–Schwarz–Neveu–Schwarz sector; see Section 13.4.2 for more details. Their contribution to the DBI action (13.158) describes the interaction of the D-brane with these closed string states. The field strength

$$F_{ab}(\zeta) = \frac{\partial A_b(\zeta)}{\partial\zeta^a} - \frac{\partial A_a(\zeta)}{\partial\zeta^b} \tag{13.163}$$

of the $U(1)$ gauge field is not the pull-back of a space-time field and describes the interaction of the brane with the open string states. The *brane tension* T_p appearing in (13.158) generalises the coefficient $\frac{1}{4\pi\alpha'}$ appearing in the Polyakov action (13.6). One can show for the type II string [535, 63] that it is also expressed in terms of α' as

$$T_p = \frac{2\pi}{(2\pi\alpha')^{(p+1)/2}}\,. \tag{13.164}$$

The Wess–Zumino term (13.159) contains the coupling to the remaining bosonic massless close type II states that arise from the complementary Ramond–Ramond sector (see Section 13.4.2). These are differential forms on space-time and the only one a Dp-brane couples to is a $(p+1)$-form that is denoted C_{p+1} in (13.159). The Wess–Zumino term can be thought of as a generalisation of the coupling of an electric point particle to the electromagnetic field: the D-brane is an extended object and couples similarly to a higher-rank form. This concludes the description of the effective action (13.157) for a Dp-brane coupling to the massless open and closed string states; we now offer some additional remarks on the rôle and importance of branes.

The combination of the two terms in (13.157) is fixed by supersymmetry. Very important is the occurrence of the factor $e^{-\phi}$ in the DBI term, which indicates that D-branes are open string effects since the simplest worldsheet topology of an open string is a disc with Euler number $\chi = 1$ and hence contributes with $e^{-\phi}$; see (13.17). With the discovery of D-branes, a long-standing puzzle concerning the rôle of the Ramond–Ramond fields was resolved: in the sigma models of the type discussed in (13.11) that describe *perturbative* (closed) string theory, only the Neveu–Schwarz–Neveu–Schwarz fields enter and it was not known what couples to the Ramond–Ramond fields. The Wess–Zumino term shows that these are the D-branes and further analysis shows that they are *non-perturbative* objects from the point of view of the string coupling g_s, meaning that they will not be visible in a series expansion around $g_s = 0$. For constant dilaton $\phi = \phi_0$ this follows directly from (13.158) since then the total DBI action becomes proportional to $e^{-\phi_0} = 1/g_s$. Branes are nevertheless needed for the consistency of the full theory [571, 525] and therefore, even when starting with a perturbative picture for type II strings as put forward in Sections 13.1.2–13.1.4, one is inevitably led to the conclusion that, at the non-perturbative level, one also has to include open strings in the picture.

There are two special cases of the DBI action (13.158) that are of interest. First, we can consider the case without B-field, without gauge field and for constant dilaton $\phi = \phi_0$ such that $g_s = e^{\phi_0}$. Then the DBI action reduces to

$$S_{\text{DBI}}[G_{\mu\nu}, \phi_0, 0, 0] = \frac{T_p}{g_s} \int_{\mathcal{D}} d^{p+1}\zeta \sqrt{\det \gamma_{ab}} \tag{13.165}$$

and so is proportional to the volume of the embedded brane $\mathcal{D} \subset M$. This is a generalisation of the Nambu–Goto action (13.4) considered by Dirac (independently of string theory) [182]. We observe that the action of such a simple D-brane is proportional to $1/g_s$, indicating its non-perturbative nature. The second special case brings out the gauge theory aspect of the single D-brane and corresponds to expanding (13.158) around a flat background $G_{\mu\nu} = \eta_{\mu\nu}$

and $B = 0$, again for constant dilaton, but for arbitrary gauge field A_a. Using $\det M = e^{\operatorname{tr}\ln(M)}$ one finds

$$S_{\rm DBI}[\eta_{\mu\nu}, \phi_0, 0, A_a] = \frac{T_p}{g_s} \int d^{p+1}\zeta \left(1 + (2\pi\alpha')^2 F_{ab}F^{ab} + O(||\alpha' F||^4)\right), \tag{13.166}$$

where we have also set the fluctuating scalars to zero for simplicity. The first term corresponds to a vacuum energy contribution and the second term is indeed the action for a $U(1)$ gauge theory (a.k.a. Maxwell theory) in $p + 1$ dimensions. The higher-order terms (in the field strength F) provide the non-linear extension of Maxwell theory considered by Born and Infeld [77] and become relevant only at higher energies due to the scale α'. For a stack of N D-branes the world-volume theory becomes a non-abelian $U(N)$ *Yang–Mills gauge theory*; including the fluctuating scalars and fermionic fields leads to super Yang–Mills theories. The non-abelian generalisation of the Dirac–Born–Infeld action is more complicated and will not be needed for our purposes; see [602, 495] for details on this.

Dp-branes have the property that they are partially supersymmetric; they are what is commonly called 1/2-BPS, where BPS stands for *Bogomol'nyi–Prasad–Sommerfield*. Superstring theory in the Ramond–Neveu–Schwarz formalism is given by a two-dimensional sigma model with supersymmetry on the world-sheet; see for instance Table 13.1. In the critical dimension discussed in Section 13.1.5 there is a well-defined Hilbert space of string states and, moreover, supersymmetry not only on the world-sheet but also in the target space-time M. The algebra of supersymmetry transformation is a Lie superalgebra whose generators can be written in terms of the world-sheet fields, and there are 32 independent fermionic generators, sometimes referred to as supersymmetries, that map bosonic space-time fields to fermionic ones and vice versa. As discussed in Sections 13.1.4 and 13.2, the low-energy limit ($\alpha' \to 0$) of type II string theory leads to gravity, and a proper inclusion of all the space-time supersymmetry transformations extends gravity to *supergravity*. The branes discussed in this section can also be realised as purely solutions to the field equations of supergravity [358, 191, 584, 26]. One can then ask the question how many of the 32 space-time supersymmetries leave the brane solution invariant; such supersymmetry transformations are called unbroken. The fraction of unbroken supersymmetries is what labels the BPS-ness of the brane. For the simplest Dp-branes of the type discussed above one can show [191, 584] from the explicit supergravity solution that 16 supersymmetries are unbroken, hence the Dp-branes are 1/2-BPS. This is also consistent with the observation that imposing *no* Dirichlet conditions corresponds to a D9-brane and the open string is then free to move anywhere in space-time. This is very close to type I string theory viewed as coupled to type II string theory. As was noted already in Section 13.1.2, type I strings have half the supersymmetry of type II strings. We remark for completeness that this picture is only suggestive since a single D9-brane in type II theory is inconsistent and needs to be supplemented by other so-called orientifold planes. An explicit example of a space-time solution will be given below in (13.170) and the above D-brane action (13.157) will be evaluated for this solution. There it will be seen that the imaginary contribution from the Wess–Zumino term (13.159) is related to the phase of the non-zero Fourier modes in the expansion of automorphic forms.

Remark 13.19 Besides the space-time picture, where Dp-branes are viewed as hypersurfaces $\mathcal{D}$ in space-time on which open strings can end, one can also develop a world-sheet perspective. As the usual perturbative string is formulated using conformal field theory one has to introduce appropriate boundary conditions in the conformal field theory. This subject is known as *boundary conformal field theory* and the D-branes are in the closure of the states of the perturbative conformal field theory. As we will not make use of this language in this book, we refer the interested reader to [536, 123, 533, 247, 124] for more information.

13.6 Non-perturbative Corrections from Instantons*

In previous sections, we have analysed the structure of tree-level and one-loop contributions to the genus expansion of string amplitudes. We will now show that in addition there can be non-perturbative contributions arising from so-called D-brane instantons. These effects are crucial for the purposes of this book since in many cases they have direct interpretations as Fourier coefficients of automorphic forms. The connection with automorphic forms will be pursued in more detail in Chapter 14.

13.6.1 Non-perturbative Effects in String Theory

Instantons in a (Wick rotated, Euclidean) field theory are non-trivial solutions to the equations of motion with finite values of the action. In the semi-classical approximation the partition function Z, which is (loosely speaking) a generating function of scattering amplitudes, is approximated as a sum over local extrema to the action:

$$Z \approx \sum_{\text{extrema}} e^{-S[\text{extremum}]} \times (\text{quantum fluctuation effects around extremum})\,. \tag{13.167}$$

The global minimum corresponds to the true vacuum with the usual perturbative corrections while other local minima give instanton corrections with their own perturbative corrections around them.

The existence of similar non-perturbative effects in string theory was first suggested by Shenker [571] based on the structure of the large-order behaviour of string perturbation theory. Because of the growth of the volume of the moduli space of genus g Riemann surfaces $\mathcal{M}_g$, the string amplitudes $\mathcal{A}_g$ (at fixed order in α') grow factorially:

$$\mathcal{A}_g \sim (2g)! \quad \text{as} \quad g \to \infty. \tag{13.168}$$

This implies that the genus expansion of string theory is not convergent but should be treated as an asymptotic series. A Borel resummation argument then suggests that these divergences in the genus expansion can be cured by adding effects which are of the order e^{-1/g_s}. Hence, the genus expansion (13.54) should really be understood in the form

$$\mathcal{A} = \sum_{h=0}^{\infty} g_s^{2h-2} \mathcal{A}_h + O(e^{-1/g_s}). \tag{13.169}$$

These effects are suppressed in the weak coupling limit $g_s \to 0$ but contribute for small but finite values of the coupling. In the strong coupling region $g_s \to \infty$ perturbation theory is not valid and these effects will completely dominate. This behaviour is the hallmark of non-perturbative effects. Shenker's prediction was given a concrete realisation with the discovery of D-branes [532, 533], which are extended objects with a tension of the order g_s^{-1} (see Section 13.5). D-branes therefore become very heavy at weak coupling but behave like fundamental massless states at strong coupling.

We will now consider the simplest example of this, corresponding to string amplitudes in type IIB string theory in the presence of D-instantons, following [532, 306].

13.6.2 Case Study: D-instantons in Type IIB String Theory

In this section, we will provide a detailed analysis of instantons in type IIB string theory. This is a very special example since instanton effects arise already in the uncompactified ten-dimensional theory, in contrast to the general situation where instanton effects arise due to D-branes wrapping some submanifold in a compactified space-time (see Section 13.6.4 for more details on this). As we will see, the reason for this is that type IIB string theory contains in its spectrum a (pseudo)-scalar field χ which is sourced by a D(-1)-brane. By convention a Dp-brane is a $(p+1)$-dimensional object in space-time, and so a D(-1)-brane is a point, both in space and in time. Therefore it behaves like an instanton already in ten dimensions. We now turn to the details of this.

Instanton Action

In ten-dimensional type IIB supergravity, the following configurations are spherical instanton solutions to the equations of motion [277]:

$$g_{\mu\nu} = \delta_{\mu\nu}, \qquad e^{\phi} - e^{\phi_0} = \frac{|q|}{8\,\mathrm{Vol}(S^9)}\frac{1}{r^8}, \qquad \chi - \chi_0 = \mathrm{sgn}(q)(e^{-\phi} - e^{-\phi_0}), \tag{13.170}$$

where ϕ_0 is the value of ϕ at $r = \infty$ with $g_s = e^{\phi_0}$ and similarly for χ_0. $\mathrm{Vol}(S^9)$ is the volume of S^9 with unit radius, and q is the Ramond–Ramond charge of the solution. For negative q the solution is called an *anti-instanton*, while for positive q it is called an *instanton*. This q is the Noether-charge with respect to translations in χ and is obtained by a closed hypersurface integral in the Euclidean space-time, enveloping the origin. The integral is invariant under deformations of the surface as long as we do not cross the origin, and the instanton is in this way localised to a single point in space-time. Due to *Dirac–Schwinger–Zwanziger quantisation* this charge becomes quantised as $q = 2\pi m$, where m is an integer which we colloquially also call the instanton charge. In the conventions of Section 13.5, this instanton is also called D(-1)-instanton or D(-1)-brane.

The value of the action for these solutions is [308]

$$S_{\text{inst}} = 2\pi\,|m|\,g_s^{-1} - 2\pi i m \chi_0, \tag{13.171}$$

which for an instanton and anti-instanton becomes $-2\pi i\,|m|\,z_0$, and $2\pi i\,|m|\,\bar{z}_0$ respectively, where $z_0 = \chi_0 + i g_s^{-1}$. This is the value for $S[\text{extremum}]$ that enters in (13.167). This is nothing but the instanton action discussed in (13.157).

Instanton Effects in String Amplitudes

The contributions to the four-graviton scattering amplitude in string theory shown schematically in Figure 13.4 only include closed string world-sheets, that is, world-sheets without boundaries (but with punctures for external states). However, when adding D-branes into the picture, the sum over topologies should also include open strings whose world-sheets have boundaries on the D-branes. While the closed string world-sheets give perturbative corrections in g_s to the amplitudes, the open string world-sheets give non-perturbative corrections.

Both the perturbative and non-perturbative contributions are present in the ten-dimensional four-graviton scattering amplitude $\mathcal{A}^{(10)}$ of (13.39) for which we computed the tree-level, perturbative term in Section 13.3. To go beyond the perturbative corrections, we write the amplitude as a sum over the number

n of D-instantons at positions $\{y_i\}_{i=1}^n$ in space-time:

$$\mathcal{A}^{(10)} = \sum_{n=0}^{\infty} \mathcal{A}_n , \qquad (13.172)$$

where $\mathcal{A}_0$ contains all the perturbative contributions shown in Figure 13.4 including the tree-level term. The positions y_i are later integrated over to restore translational invariance. This integration also imposes momentum conservation of the four-graviton states for the non-perturbative corrections [308].

Each $\mathcal{A}_n$ is, in turn, a sum over different world-sheets, which we now allow to be disconnected and have boundaries which are attached to the points y_i in space-time. We will call the disjoint parts *world-sheet components*, which are, by themselves, connected but can have holes and handles.

Remark 13.20 We may have a world-sheet with boundaries attached to the same D-instantons that seems disconnected in the view of the world-sheet but that is connected from the space-time point of view – the different components being connected by the D-instantons. As in ordinary field theory, the scattering amplitudes can be shown to only involve diagrams which are connected when viewed in space-time [306].

Focussing on the single D-instanton case ($n = 1$) at position y_1, the contribution $\mathcal{A}_1$ is a sum over disconnected world-sheets with boundaries at the point y_1 together with appropriate symmetry factors for exchanging identical world-sheet components and boundaries. Similarly to closed strings, the world-sheet components are weighted by the Euler characteristic as $g_s^{2(h-1)+b}$, where h is the genus of the surface and b is the number of boundaries. The leading-order world-sheet components, together with their symmetry factors for exchanging identical boundaries, are shown in Figure 13.8. Besides these components we will also have surfaces with punctures corresponding to insertions of external states.

For a given disconnected world-sheet, each component gives a contributing factor to the amplitude $\mathcal{A}_1$ which we will denote by $\langle \bigcirc \rangle$ for the disc and $\langle \circledcirc \rangle$ for the annulus. Summing over the number of possible discs d_1, annuli d_2 etc. together with symmetry factors for exchanging these identical components, we obtain an exponentiation [532, 306]

$$\begin{aligned} \mathcal{A}_1 &= \int d^{10}y_1 \sum_{d_1=0}^{\infty} \frac{1}{d_1!} \left(g_s^{-1} \langle \bigcirc \rangle \right)^{d_1} \sum_{d_2=0}^{\infty} \frac{1}{d_2!} \left(\frac{1}{2!} \langle \circledcirc \rangle \right)^{d_2} \cdots \mathcal{A}_1^{(\text{external})} \\ &= \int d^{10}y_1 \exp\!\left(g_s^{-1} \langle \bigcirc \rangle + \frac{1}{2!} \langle \circledcirc \rangle + \cdots \right) \mathcal{A}_1^{(\text{external})} , \end{aligned} \qquad (13.173)$$

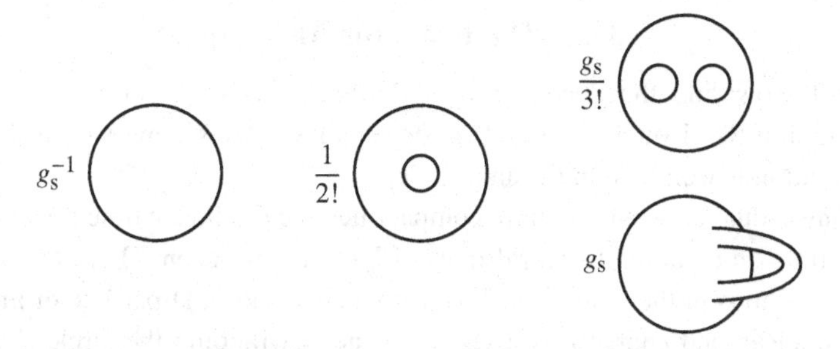

Figure 13.8 Leading-order open world-sheet components with topological weights and symmetry factors for identical boundaries. Each boundary is contracted to the same D-instanton at a point y_1 in space-time. The last, lower diagram is a disc with a single boundary and a handle.

where $\mathcal{A}_1^{(\text{external})}$ contains the contributions from world-sheet components with punctures for external states. With $\langle \bigcirc \rangle$ negative, we see that this gives the expected non-perturbative contribution of the form e^{-1/g_s}.

The leading-order contribution to $\mathcal{A}_1^{(\text{external})}$ for the four-graviton scattering amplitude is obtained from four discs with one puncture on each corresponding to an external state. This gives exactly the $\mathcal{R}^4$ combination of the momenta and polarisations as in (13.52). Also including a non-zero axion, Green and Gutperle [308] found the single, unit charged instanton leading-order correction as

$$\mathcal{A}_1 = Ce^{2\pi i z_0} \int d^{10} y_1 e^{i y_1 \cdot (k_1+k_2+k_3+k_4)} \mathcal{R}^4 + \cdots , \tag{13.174}$$

where $z_0 = \chi_0 + i g_s^{-1}$, k_j are external momenta and C an unspecified constant. This matches our expectations from the field theory arguments above with corrections of the form $e^{-S_{\text{inst}}}$ using $m = 1$ in (13.171). The contribution (13.174) also agrees with the corresponding term in the Fourier expansion of the Eisenstein series $\mathcal{E}_{(0,0)}^{(10)}(z)$ in (14.7) which is the coefficient function to the $\mathcal{R}^4$ interaction. This, together with the $SL(2,\mathbb{Z})$-invariance and the perturbative corrections, in fact motivated the expression for $\mathcal{E}_{(0,0)}^{(10)}(z)$ as an Eisenstein series in [308]. (Note that when computing the effective action one also includes the fluctuations in the dilaton and the axion, and the exponent in the instanton contribution should then become $2\pi i z$ with the full axio-dilaton z, in contrast to the constant background z_0 used in the amplitude calculation above.)

13.6.3 The Instanton Measure

We will now find the degeneracy for D-instantons of a certain charge m by studying the T-dual picture. This degeneracy affects how each instanton contribution is weighted in the amplitude.

Using T-duality, where we have compactified the Euclidean time direction in space-time on a circle, the world-lines of D-particles become D-instantons in ten dimensions in the dual limit. The T-dual action for a D-particle of integer Ramond–Ramond charge n whose world-line is wrapping the circle d times can be shown to be [308]

$$S_{\text{inst}} = 2\pi\,|nd|\,e^{-\phi} - 2\pi i n d \chi\,, \tag{13.175}$$

which is exactly S_{inst} in (13.171) with $m = nd$ and $z = z_0$. Thus, the number of D-instanton states with charge m is related to the number of divisors of m. This implies that each factor $e^{-S_{\text{inst}}}$ associated with an instanton of charge m should be weighted with a combinatorial factor $\mu(m)$ which takes into account the degeneracy. The contribution from a charge m instanton to the amplitude is thus of the general form

$$\mu(m)\exp\left[-\frac{2\pi|m|}{g_s} + 2\pi i m\chi\right]. \tag{13.176}$$

To obtain the full contribution we must sum over all instanton charges. The formula above then represents the overall contribution to the sum for each charge. In addition there can be perturbative corrections to the amplitude in the instanton background. Thus the sum over instanton charges then takes the schematic form

$$\sum_{n\neq 0}\mu(m)\exp\left[-\frac{2\pi|m|}{g_s} + 2\pi i m\chi\right]\Big(1 + O(g_s)\Big). \tag{13.177}$$

To understand the physical interpretation of the instanton measure it is useful to use T-duality, which relates the D-instantons to *D-particles*. More precisely, in the context of type IIB string theory on $S^1 \times \mathbb{R}^{8,1}$, T-duality relates D(−1)-instantons to D0-branes in type IIA string theory on $S^1 \times \mathbb{R}^{8,1}$, whose world-line wraps around the S^1. Equivalently, these D0-branes correspond to *Kaluza–Klein modes* of eleven-dimensional supergravity compactified on a circle. The mass of such a Kaluza–Klein mode of charge n is then

$$M_{\text{KK}}(n) = nM_{\text{D0}}(1) = M_{\text{D0}}(1) + \cdots M_{\text{D0}}(1), \tag{13.178}$$

where $M_{\text{D0}}(1)$ is the mass of a unit charged D0-brane. This implies that we can view the charge n Kaluza–Klein mode as a 'threshold bound state', i.e., a bound state with zero binding energy, of n D0-particles. There is exactly one such threshold bound state for each charge n. These bound states can be described by a supersymmetric $SU(n)$ matrix model that is obtained by compactification to zero dimensions of $SU(n)$ super Yang–Mills theory in ten Euclidean dimensions [621]. This consists of bosonic and fermionic $n \times n$ unitary matrices, A and ψ, respectively, whose action is given by

$$S[A, \psi] = \frac{1}{4} \sum_{a,b=1}^{10} \mathrm{Tr}\left([A_a, A_b]^2\right) + \frac{i}{2} \sum_{a=1}^{10} \mathrm{Tr}\left(\bar{\psi}\Gamma^a [A_a, \psi]\right), \qquad (13.179)$$

where Γ^a are gamma matrices for $Spin(10)$. The partition function of this matrix model is then given by the path integral

$$Z_n = \frac{1}{\mathrm{Vol}(SU(n))} \int_{SU(n)} \mathcal{D}A\, \mathcal{D}\psi\, \exp\left(-S[A, \psi]\right), \qquad (13.180)$$

where the integral is defined using the Haar measure on $SU(n)$. This path integral should precisely reproduce the instanton measure $\mu(n)$:

$$Z_n = \mu(n). \qquad (13.181)$$

To evaluate the partition function Z_n is difficult in general. One approach is to relate it to the *Witten index* of n D-particles, which is defined as

$$\mathcal{I}^{(n)}[\beta] = \mathrm{tr}(-1)^F e^{-\beta H}, \qquad (13.182)$$

where β is the inverse radius (or 'temperature') of the S^1, and F is the fermion number operator, which is 0 when acting on bosons and 1 on fermions. On general grounds the Witten index computes the difference between the number of fermionic and bosonic ground states. The trace is taken over all physical, gauge-invariant states in the spectrum. By the previous arguments there should only exist exactly one threshold bound state for each n, and this predicts that in the low-temperature limit

$$\mathcal{I}^{(n)}[\infty] = 1 \qquad (13.183)$$

for all n. If the spectrum is discrete, the index is independent of β and can be computed in the simpler limit $\beta \to 0$, while in general when the spectrum has a continuous contribution one has

$$\mathcal{I}^{(n)}[\infty] = \mathcal{I}^{(n)}[0] + \delta\mathcal{I}, \qquad (13.184)$$

where

$$\mathcal{I}^{(n)}[0] = \lim_{\beta \to 0} \mathrm{tr}(-1)^F e^{-\beta H} \qquad (13.185)$$

is often referred to as the 'bulk' contribution to the index and contains contributions from the continuous spectrum, while the term $\delta\mathcal{I}$ corrects this in order to ensure the result (13.183). T-duality predicts that the bulk contribution to the index should equal the matrix partition function Z_n,

$$Z_n = \lim_{\beta \to 0} \mathrm{tr}(-1)^F e^{-\beta H}, \qquad (13.186)$$

and thereby also the instanton measure $\mu(n)$. The partition function Z_n was evaluated for $n = 2$ in [626, 561] with the result

$$Z_2 = \frac{5}{4}. \tag{13.187}$$

Subsequently, Green and Gutperle used $SL(2,\mathbb{Z})$-invariance and the structure of the Fourier coefficients of the Eisenstein series $E_s(\tau,\bar{\tau})$ at $s = 3/2$ (see Section 14.2.1 for more details on this) to conjecture the exact formula

$$\mu(n) = \sum_{\substack{d|n \\ d>0}} \frac{1}{d^2}. \tag{13.188}$$

This formula was proven in [493, 424] using different and independent techniques.

Remark 13.21 The mathematical interpretation of the instanton measure in the context of p-adic Whittaker coefficients is discussed in Example 6.28.

13.6.4 General D-instanton Effects

Consider a split of space-time according to $M = \mathbb{R}^{1,3} \times X$ where X is some compact manifold. By convention we say that a Dp-brane is a $(p+1)$-dimensional extended object in M. Suppose X has a $(p+1)$-dimensional compact submanifold C. If we choose to put the Dp-brane completely along C we say that the brane is *wrapping* C. If the volume of X is small then string theory on M is said to be compactified on X and the physics is well approximated by the theory on $\mathbb{R}^{1,3}$. Since the Dp-brane is completely localised along C it looks like a point in Minkowski space $\mathbb{R}^{1,3}$. Therefore, from the point of view of the four-dimensional physics, the Dp-brane wrapped on C behaves like an *instanton*. In general, such instanton effects can contribute to terms in the effective action in $\mathbb{R}^{1,3}$ by factors of the order

$$\exp\left[-\mathrm{vol}(C)/g_s\right]. \tag{13.189}$$

Just like a string wrapped on a torus T^d gives rise to a lattice of charges (see Section 13.4.3), a Dp-brane wrapped on a cycle $C \subset X$ yields a lattice of charges given by the integer homology

$$\Gamma_{Dp} = H_{p+1}(X,\mathbb{Z}). \tag{13.190}$$

Roughly, the path integral over the Dp-brane is then summing over embeddings $C \supset X$ in the homology class $[\gamma] \in H_{p+1}(X,\mathbb{Z})$. The D$p$-brane acts as a source for a $(p+1)$-form C_{p+1} (Ramond–Ramond form), which is a space-time field that enters in the effective action. The full instanton action associated with this Dp-brane is then given by [622, 51]

$$S_{\mathrm{D}p} = \frac{\mathrm{vol}(C)}{g_s} - i \int_C C_{p+1}. \tag{13.191}$$

Remark 13.22 This is nothing but the Dp-brane action (13.157) evaluated on the classical instanton solution to the supergravity equations of motion.

Each instanton contribution is weighted by $e^{-S_{\mathrm{D}p}}$ together with a numerical factor that counts the degeneracies, i.e., the instanton measure μ. On general grounds, this is a function

$$\mu_{\mathrm{D}p}: H_{p+1}(X, \mathbb{Z}) \longrightarrow \mathbb{Q}, \tag{13.192}$$

the precise form of which will depend on the details, such as the type of string theory and the choice of manifold X. This generalises the instanton measure discussed in Section 13.6.3 above. In the present context the instanton measure counts (with signs) the (supersymmetric) embeddings:

$$x : C \longrightarrow X. \tag{13.193}$$

Combining everything, we deduce that a Dp-instanton of charge $\gamma \in H_{p+1}(X, \mathbb{Z})$ yields a correction in the effective action which is weighted by

$$\mu_{\mathrm{D}p}(\gamma) \exp\left[-\frac{\mathrm{vol}(C)}{g_s} + i \int_C C_{p+1} \right], \tag{13.194}$$

generalising Equation (13.189).

Let us now consider some examples where we can be a little more explicit.

Example 13.23: Instanton action on T^6

Consider type II string theory on $M = \mathbb{R}^{1,3} \times X$, where $X = T^6$, the six-dimensional torus. From Section 13.4.3 we know that the Narain lattice is

$$\Gamma^{6,6} \cong \Gamma \oplus \Gamma^* \tag{13.195}$$

and the T-duality group is $SO(6,6;\mathbb{Z})$. In type IIA we have the Ramond–Ramond forms (C_1, C_3, C_5, C_7) which are sourced by D0, D2, D4, D6-branes, respectively. In type IIB the situation is reversed; we have Ramond–Ramond forms (C_0, C_2, C_4, C_6), sourced by D(-1), D1, D3, D5-branes. The associated D-brane charge lattices are then

$$\begin{aligned} \Gamma_{\mathrm{IIA}} &= H_{\mathrm{odd}}(T^6, \mathbb{Z}) \cong \Lambda^{\mathrm{odd}} \Gamma^*, \\ \Gamma_{\mathrm{IIB}} &= H_{\mathrm{even}}(T^6, \mathbb{Z}) \cong \Lambda^{\mathrm{even}} \Gamma^*. \end{aligned} \tag{13.196}$$

When considering the theory on T^6 we also get non-perturbative contributions from so-called *NS5-instantons*, i.e., Euclidean NS5-branes wrapping the entire six-dimensional torus. The associated NS5-instanton charge $k \in \mathbb{Z}$ ensures that the complete charge lattice in type IIA/IIB string theory is given by

$$\Gamma^{6,6} \oplus H_{\text{even/odd}}(T^6, \mathbb{Z}) \oplus \mathbb{Z}. \tag{13.197}$$

The full symmetry that leaves the charge lattice invariant is the U-duality symmetry $E_7(\mathbb{Z})$, and the moduli space is given by

$$E_7(\mathbb{Z})\backslash E_7(\mathbb{R})/(SU(8)/\mathbb{Z}_2). \tag{13.198}$$

Thus, it is the presence of D-branes *and* NS5-branes which is responsible for the enhancement from T-duality to U-duality. The decomposition of the lattice (13.197) corresponds to the following decomposition of the adjoint representation of the classical symmetry algebra $\mathfrak{e}_7(\mathbb{R})$ into representations of $\mathfrak{so}(6,6)$:

$$\mathfrak{e}_7(\mathbb{R}) = \mathbf{1}_{-2} \oplus \mathbf{32}_{-1} \oplus (\mathfrak{so}(6,6) \oplus \mathfrak{gl}(1))_0 \oplus \mathbf{32}_1 \oplus \mathbf{1}_2. \tag{13.199}$$

Here, the grading is induced by the $\mathfrak{gl}(1)$-factor in the Levi $\mathfrak{so}(6,6) \oplus \mathfrak{gl}(1) \subset \mathfrak{e}_7$, obtained by removing the left-most node in the Dynkin diagram of $\mathfrak{e}_7$ given in Figure 13.5. The moduli space restricted to degree 0 is simply the Narain moduli space (13.122) and the lattices in the positive degrees give the D-instanton and NS5-instanton charges, respectively. In string theory, the decomposition (13.199) is referred to as the *string theory perturbation limit* since the $\mathfrak{gl}(1)$ which is taking care of the grading is parametrised by the string coupling g_s. See Section 14.2.2 for a more detailed discussion of the various limits of string theory and the associated parabolic subgroups.

We now turn to a discussion of the instanton actions corresponding to D-branes on T^6. Let us consider the case of the D0-brane in type IIA for illustration. This is a one-dimensional extended object and is therefore sometimes also called *D-particle* if that dimension is taken to be time; see also the discussion in Section 13.6.3. In our case, however, we wish to treat the D0-brane as an instanton in effective theory on $\mathbb{R}^{1,3}$ and so we take it to be wrapped on some cycle $S^1 \subset T^6$. Let γ_i, $i = 1, \ldots, 6$ denote a basis of one-cycles in $H_1(T^6, \mathbb{Z})$. Then an arbitrary one-cycle C can be represented by the homology class

$$\gamma = \sum_{i=1}^{6} n^i \gamma_i \in H_1(T^6, \mathbb{Z}) \cong \mathbb{Z}^{32}, \tag{13.200}$$

where $n^i \in \mathbb{Z}$. Evaluating the Dp-brane action (13.157) on the D0-brane background, the instanton action becomes

$$S_{\text{D0}} = \frac{|m|}{g_s} \int_\gamma \text{vol}(x^*(g)) + im \int_\gamma C_1, \tag{13.201}$$

where $\text{vol}(x^*(g))$ denotes the volume form on C with metric $x^*(g)$, the pull-back of the metric g on T^6 with respect to the embedding $x\colon C \to T^6$. The variable $m \in \mathbb{Z}$ corresponds to the number of bound states of D0-branes. If we introduce the 'Ramond–Ramond' scalars (or 'axions')

$$\mathcal{A}_i = \int_{\gamma_i} C_1, \tag{13.202}$$

we can write the action a little more explicitly, similar to (13.157), as

$$S_{\mathrm{D}0} = \frac{|m|}{g_s}\sqrt{\sum_{i,j} n^i g_{ij} n^j} - im\sum_i n^i \mathcal{A}_i. \tag{13.203}$$

Writing this in terms of the winding number $m^i = mn^i$ also induces an instanton measure $\mu(m^i)$ which is a sum over divisors of the m^i's, similarly to the discussion after Equation (13.175).

One can perform a similar analysis for all the other Dp-instanton effects. See [523] for more details on how this works.

So far we have mainly restricted the discussion to compactifications on tori. In the following example we shall consider what happens when generalising this to Calabi–Yau manifolds. In particular this opens an interesting connection between automorphic forms and enumerative geometry, as we shall explore further in Sections 15.4.4 and 18.6.

Example 13.24: Instanton action on a Calabi–Yau threefold

Let us now consider type IIA string theory compactified on a Calabi–Yau threefold X. This is a complex three-dimensional Kähler manifold with $c_1(X) = 0$ (trivial canonical bundle). The integer homologies decompose according to

$$\begin{aligned} H_{\mathrm{odd}}(X,\mathbb{Z}) &= H_3(X,\mathbb{Z}),\\ H_{\mathrm{even}}(X,\mathbb{Z}) &= H_0(X,\mathbb{Z}) \oplus H_2(X,\mathbb{Z}) \oplus H_4(X,\mathbb{Z}) \oplus H_6(X,\mathbb{Z}). \end{aligned} \tag{13.204}$$

Since the Dp-branes in type IIA are all odd-dimensional, the only way to make an instanton in this case is by wrapping the D2-brane on a three-cycle $C \subset X$ in homology class $[\gamma] \in H_3(X,\mathbb{Z})$. The number of such three-cycles is the dimension of $H_3(X,\mathbb{Z})$ which is $2h_{2,1}+2$, where $h_{2,1}$ is the Hodge number. If we introduce a symplectic basis (A^I, B_J), $I,J = 0,\ldots,h_{2,1}$, of the charge lattice $H_3(X,\mathbb{Z})$ we can write

$$\gamma = \sum_{I=0}^{h_{2,1}} (q_I A^I + p^I B_I) \in H_3(X,\mathbb{Z}), \tag{13.205}$$

where (q_I, p^J) are integers that should be interpreted as electro-magnetic charges of the D2-instanton. A Calabi–Yau threefold has a unique (up to scaling) holomorphic 3-form $\Omega \in H^{3,0}(X,\mathbb{C})$ such that $\Omega \wedge \bar{\Omega}$ is a volume form on X. Using this fact, the instanton action of the D2-brane can be written as [51]

$$S_{\rm D2} = \frac{1}{g_{\rm s}} \frac{\left|\int_\gamma \Omega\right|}{\sqrt{i\int_X \Omega\wedge\bar{\Omega}}} - i\int_\gamma C_3, \tag{13.206}$$

generalising (13.157). In this case the instanton measure

$$\mu\colon H_3(X,\mathbb{Z}) \longrightarrow \mathbb{Q} \tag{13.207}$$

is counting supersymmetric embeddings of C into X. The requirement that the instanton preserves supersymmetry translates to the constraint that C must be a so-called *special Lagrangian* submanifold of X [51]. This implies that $\mu(\gamma)$ is related to the enumerative geometry of X; namely it should equal the so-called *generalised Donaldson–Thomas invariants* (see [420, 511, 10] for discussions on this connection). The D2-instantons will therefore contribute to the effective action with terms of the form

$$\sum_{\gamma\in H_3(X,\mathbb{Z})} \mu(\gamma)\exp\left[-\frac{1}{g_{\rm s}} \frac{\left|\int_\gamma \Omega\right|}{\sqrt{i\int_X \Omega\wedge\bar{\Omega}}} + i\int_\gamma C_3\right] + \cdots, \tag{13.208}$$

where the ellipsis includes multi-instanton contributions as well as perturbative fluctuations in the instanton background. In Sections 15.4.4 and 18.6, we will come back to this example for the special case of Calabi–Yau manifolds which are rigid, that is, when $h_{2,1} = 0$. We will then see how the instanton sum can be directly interpreted as the Fourier coefficient of an automorphic form.

14

Automorphic Scattering Amplitudes

In this chapter, we return in more detail to the low-energy expansion of string theory scattering amplitudes, a topic that was already mentioned in Sections 13.1.2 and 13.2. We shall see more directly the power of U-duality of Section 13.1.6 acting on the coefficient functions $\mathcal{E}^{(D)}_{(p,q)}$ that arise in the low-energy α'-expansion. Together with constraints from supersymmetry we shall argue that the coefficient functions have to be automorphic forms in the sense of PART ONE of the book. Generalisations also arise but their proper discussion is deferred to the next chapter. As automorphic forms, the functions $\mathcal{E}^{(D)}_{(p,q)}$ have Fourier expansions that can be accessed with the techniques of PART ONE, and we explain how this gives valuable information on perturbative and non-perturbative effects in string theory. The automorphic representations that arise are shown to be consistent with string theory expectations.

14.1 U-duality Constraints in the α'-expansion

In Section 13.2, we discussed the α'-expansion of the four-graviton scattering amplitude in type IIB string theory compactified on T^d to $D = 10-d$ space-time dimensions and studied the coefficients $\mathcal{E}^{(D)}_{(p,q)}$ in (13.39) which are functions on the moduli space invariant under U-duality transformations.

This means that the coefficients $\mathcal{E}^{(D)}_{(p,q)}$ satisfy condition (1) of the definition of automorphic forms in Section 1.1 and, since they should also have perturbative expansions in the weak string coupling limit $g_s \to 0$ and other similar limits of the moduli space corresponding to cusps in G/K, they also satisfy the growth condition (3).

The remaining condition (2) from the introduction requires that an automorphic form satisfy appropriate differential equations under the action of G-invariant differential operators. It turns out that supersymmetry in string theory imposes precisely such differential conditions on the coefficient functions. This was analysed in detail first by Green and Sethi in the case of ten-dimensional ($D = 10$) type IIB string theory and $p = q = 0$ [321], and we review aspects of their arguments in Section 14.4. More recent work on supersymmetry constraints includes [318, 47, 83, 84, 85, 609]. They found that $\mathcal{E}^{(10)}_{(0,0)}(g)$ has to satisfy a Laplace equation with an eigenvalue determined by supersymmetry considerations. The solution to this Laplace equation is the one proposed by Green and Gutperle [308] and Green and Vanhove [322, 311] in terms of a non-holomorphic Eisenstein series on $SL(2,\mathbb{R})$ and was shown to be the unique solution by Pioline [517]

In other dimensions D and for small values of p and q, there are strong arguments that the coefficient functions $\mathcal{E}^{(D)}_{(p,q)}(g)$ satisfy the differential equations [318, 521]

$$R^4: \left(\Delta - \frac{3(11-D)(D-8)}{D-2}\right)\mathcal{E}^{(D)}_{(0,0)} = 6\pi\delta_{D,8}, \tag{14.1a}$$

$$\nabla^4 R^4: \left(\Delta - \frac{5(12-D)(D-7)}{D-2}\right)\mathcal{E}^{(D)}_{(1,0)} = 40\zeta(2)\delta_{D,7} + 7\mathcal{E}^{(6)}_{(0,0)}\delta_{D,6}, \tag{14.1b}$$

$$\nabla^6 R^4: \left(\Delta - \frac{6(14-D)(D-6)}{D-2}\right)\mathcal{E}^{(D)}_{(0,1)} = -\left(\mathcal{E}^{(D)}_{(0,0)}\right)^2 + 40\zeta(3)\delta_{D,6} + \frac{55}{3}\mathcal{E}^{(5)}_{(0,0)}\delta_{D,5} + \frac{85}{2\pi}\mathcal{E}^{(4)}_{(1,0)}\delta_{D,4}, \tag{14.1c}$$

where $\Delta \equiv \Delta_{G/K}$ is the *Laplace–Beltrami operator* on G/K that is also discussed more in Appendix B.

We see that the third equation is qualitatively very different from the first two since it has the square of a non-constant function as a source on the right-hand side. This would take us out of the standard domain of automorphic forms, and we will discuss this case in more detail in Section 15.1.

The Kronecker delta contributions in all three equations in (14.1) are related to the existence of ultraviolet and infrared divergences in the underlying supergravity theory and the existence of supersymmetric counterterms. The ultraviolet divergences arise in those dimensions where also the eigenvalue vanishes and signal logarithmic terms in the coefficient function $\mathcal{E}^{(D)}_{(p,q)}$. We refer the reader to [319] for further discussion of this point. The additional Kronecker delta contributions proportional to coefficient functions associated with fewer derivatives are related to form factor divergences, and these are discussed in [521, 80].

Besides these special cases, Equations (14.1a) and (14.1b) correspond to eigenfunction conditions from (2) in Section 1.1. For dimensions lower than ten, there are additional G-invariant differential operators other than $\Delta_{G/K}$. A linearised superspace analysis that generates additional tensorial differential equations was pioneered in [83, 84] where also the possible solutions of these differential constraints were analysed. When there are several solutions, the coefficient functions for R^4 and $\nabla^4 R^4$ are specific linear combinations fixed by perturbative boundary conditions, and they satisfy the tensorial differential equations that contain information about all G-invariant differential operators. In particular, the solutions are automorphic forms.

As seen in Section 1.1 in the case of $G(\mathbb{R}) = SL(2, \mathbb{R})$, Eisenstein series are eigenfunctions of the Laplace–Beltrami operator, and comparing with scattering amplitudes computed in string theory it has been possible to conjecture the exact forms of the coefficients $\mathcal{E}^{(D)}_{(0,0)}$ and $\mathcal{E}^{(D)}_{(1,0)}$ in terms of maximal parabolic Eisenstein series, which were defined in Chapter 4. Parabolic subgroups, denoted by P, were introduced in Section 3.1.3.

More precisely, in five, four and three space-time dimensions, with symmetry groups E_6, E_7 and E_8 according to Table 13.2, if one considers the maximal parabolic subgroups P that have semi-simple Levi parts $SO(5, 5)$, $SO(6, 6)$ and $SO(7, 7)$, respectively, then the solutions

$$R^4: \qquad \mathcal{E}^{(D)}_{(0,0)}(g) = 2\zeta(3)E(\lambda_{s=3/2}, P, g), \tag{14.2a}$$

$$\nabla^4 R^4: \qquad \mathcal{E}^{(D)}_{(1,0)}(g) = \zeta(5)E(\lambda_{s=5/2}, P, g) \tag{14.2b}$$

to Equations (14.1a) and (14.1b) are the conjectured coefficient functions appearing in the four-graviton amplitudes, or equivalently, as corrections to the effective action (13.42).

The parabolic Eisenstein series $E(\lambda, P, g)$ was defined in Section 5.7. The weight λ_s, which specifies the character χ_s on P, is given by

$$\lambda_s = 2s\Lambda_P - \rho, \tag{14.3}$$

where Λ_P denotes the fundamental weight orthogonal to the semi-simple factor M in the Levi subgroup $L = MA_P$ of $P = LU$. By Proposition 5.30, we can think of these maximal parabolic Eisenstein series as special cases of Borel–Eisenstein series, and we shall do so in the following.

Remark 14.1 Looking at the coefficients of (14.2) we recognise the corresponding values from the tree-level amplitudes which are computed in Section 13.3. This means that the above functions are nothing but the (single) U-duality orbit of the tree-level results. This is no longer true for the higher functions $\mathcal{E}^{(D)}_{(p,q)}$ as we will discuss in more detail in Section 15.1.

These conjectures have been subjected to numerous consistency checks [313, 520, 315] and, in particular, capture the known results of scattering amplitudes in the weak coupling limit $g_s \to 0$ which we will discuss in the following section.

Remark 14.2 We would also like to mention that recent investigations of superstring scattering amplitudes at tree-level and one-loop for more than four particles have revealed very interesting different connections to number theory. Instead of single zeta values like $\zeta(3)$ one will typically have so-called (elliptic) *multiple zeta values* governed by *Drinfeld associators* [548, 97, 94, 204]. We note that this structure is at fixed order in string perturbation theory whereas the U-duality invariant functions we are discussing here include all perturbative and non-perturbative effects. Further comments on the connection to multiple zeta values can be found in Section 15.2.

14.2 Physical Interpretation of the Fourier Expansion

We will now study the functions $\mathcal{E}^{(D)}_{(p,q)}$ which were found above as the quantum corrections to the low-energy effective action in type IIB string theory on tori T^d with $d = 10 - D$. Since the $\mathcal{E}^{(D)}_{(p,q)}$ are invariant under the discrete subgroup $G(\mathbb{Z})$, they are periodic functions and we can extract physical information from their Fourier expansions.

14.2.1 Case Study: Type IIB in $D = 10$

For concreteness, let us consider the R^4 and $\nabla^4 R^4$ coefficients $\mathcal{E}^{(10)}_{(0,0)}$ and $\mathcal{E}^{(10)}_{(1,0)}$ in ten dimensions, where $G(\mathbb{R}) = SL(2,\mathbb{R})$ – although the physical interpretations hold for general dimensions and coefficients.

As stated in Section 13.1.6, the classical moduli space of (13.27) in this case is $G(\mathbb{R})/K(\mathbb{R}) = SL(2,\mathbb{R})/SO(2,\mathbb{R})$ and is isomorphic to the Poincaré upper half-plane $\mathbb{H} = \{z = x + iy \in \mathbb{C} \mid y = \operatorname{Im} z > 0\}$ parametrised by the *axio-dilaton* $z = \chi + ie^{-\phi}$ where χ is the axion and ϕ the dilaton. When ϕ is constant we have that $g_s = e^{\phi} = y^{-1}$. More generally, the string coupling constant is the asymptotic value at infinity denoted by e^{ϕ_0}. The U-duality group $G(\mathbb{Z}) = SL(2,\mathbb{Z})$ acts on z by (1.3), including the translation $z \to z + 1$.

The four-graviton amplitude at tree level together with its α'-expansion was given in (13.53). This means that, in order to reproduce the right perturbative behaviour in the weak coupling limit $g_s \to 0$, i.e., $y \to \infty$, the leading-order terms of the coefficient functions $\mathcal{E}^{(10)}_{(0,0)}$ and $\mathcal{E}^{(10)}_{(1,0)}$ should be (in Einstein frame):

$$\begin{aligned} \mathcal{E}^{(10)}_{(0,0)}(x+iy), &\sim 2\zeta(3)y^{3/2}, \\ \mathcal{E}^{(10)}_{(1,0)}(x+iy), &\sim \zeta(5)y^{5/2}, \end{aligned} \qquad \text{as} \quad y \to \infty. \tag{14.4}$$

These weak coupling limits correspond to the tree-level contribution to the scattering amplitude that were computed in Section 13.3.

The eigenvalue Equations (14.1) for the coefficient functions are

$$\Delta\mathcal{E}^{(10)}_{(0,0)}(z) = \frac{3}{4}\mathcal{E}^{(10)}_{(0,0)}(z),$$
$$\Delta\mathcal{E}^{(10)}_{(1,0)}(z) = \frac{15}{4}\mathcal{E}^{(10)}_{(1,0)}(z), \tag{14.5}$$

where $\Delta = y^2(\partial_x^2 + \partial_y^2)$ is the Laplace–Beltrami operator on $\mathbb{H}$ from (1.4). The coefficient functions are thus automorphic forms as defined in Section 1.1.

It was first realised by Green et al. in [308] and [312] that these conditions are solved by

$$\mathcal{E}^{(10)}_{(0,0)}(z) = f_{3/2}(z) = 2\zeta(3)E(s = 3/2, z),$$
$$\mathcal{E}^{(10)}_{(1,0)}(z) = \frac{1}{2}f_{5/2}(z) = \zeta(5)E(s = 5/2, z), \tag{14.6}$$

as seen from (1.17) and (1.5) with $f_s(z)$ the lattice sum defined in (1.2). This is exactly the $SL(2,\mathbb{R})$ variant of (14.2).

Remark 14.3 A priori, the solutions (14.6) are only unique up to the addition of cusp forms but they were subsequently ruled out in [517] for the R^4 coupling by using the known bounds on the spectrum of cusp forms. One can similarly uniquely fix the $\nabla^4 R^4$ coupling [579].

The Fourier expansion (1.17) then has a direct physical interpretation: the first two terms (constant terms) correspond to the perturbative quantum corrections (tree-level and one-loop), while the infinite series of the remaining Fourier modes encode non-perturbative effects. To see this we can expand the Bessel function $K_{s-1/2}(2\pi|n|y)$ in the limit $y \to \infty$ which, for the R^4 coefficient, yields

$$\mathcal{E}^{(10)}_{(0,0)}(z) = \overbrace{\underbrace{2\zeta(3)y^{3/2}}_{\text{tree-level}} + \underbrace{4\zeta(2)y^{-1/2}}_{\text{one-loop}}}^{\text{perturbative terms}} + \overbrace{2\pi\sum_{m\neq 0}\sqrt{|m|}\sigma_{-2}(m)e^{-S_{\text{inst}}(z)}\left[1+O\left(\frac{1}{y}\right)\right]}^{\text{non-perturbative terms}}, \tag{14.7}$$

amplitudes in the presence of instantons

where we have defined the *instanton action*

$$S_{\text{inst}}(z) := 2\pi\,|m|\,y - 2\pi i m x. \tag{14.8}$$

It is clear from this expression that the infinite series is exponentially suppressed by e^{-y} in the limit $y = g_s^{-1} \to \infty$ corresponding to the weak coupling limit of the theory, and this exponential suppression is characteristic for non-perturbative effects as we saw in Section 13.6.

In string theory, these corrections come from so-called D-*instantons* [306, 308], where 'D' stands for 'Dirichlet' as in the *Dirichlet boundary conditions* that are imposed on strings attached to them (see Section 13.5). D-instantons are special cases of D-branes which are localised to a single point in space-time, and were discussed in more detail in Section 13.6.

In this setting the divisor sum $\sigma_{-2}(m) = \sum_{d|m} d^{-2}$ is called the *instanton measure* which weights the contributions of different modes m called *instanton charges*. As proposed in [308] and elaborated upon in Section 13.6.3, the summation over divisors counts the degeneracy of D-instanton states of charge m. When m is a negative integer it corresponds to an *anti-instanton*.

The higher-order corrections in $g_s = y^{-1}$ in the non-perturbative terms in (14.7) are higher-genus corrections to the scattering amplitude in the presence of instantons.

In summary, the discrete symmetry leading us to the study of automorphic forms gives a lot of information about string theory. For instance, it tells us that there are no perturbative corrections to the R^4 term from world-sheets of genus larger than one, and gives an indirect way of computing scattering amplitudes in the presence of instantons in string theory, which is otherwise very difficult to do. The absence of higher-genus corrections to R^4 was confirmed independently through *non-renormalisation theorems* in [59].

The above example provides the first hint of an intriguing relation between quantum corrections in string theory and automorphic forms. There is by now a vast literature on this subject; see [41, 39, 42, 43, 44, 45, 322, 318, 323, 331, 336, 393, 438, 437, 503, 504, 517, 524, 528, 530, 518, 525] for a sample. In recent years the representation-theoretic aspects have also proven to play an increasingly important rôle [511, 313, 315, 520, 212, 213, 214], thus providing ample motivation from physics for the emphasis on automorphic representations in this book.

Chapter 15 continues the discussions of this chapter with the topic of automorphic representations for the coefficients $\mathcal{E}^{(D)}_{(p,q)}$ and the $\nabla^6 R^4$ correction, which differs from the lower-order terms by also requiring a non-constant source term in the differential Equation (14.1c).

Remark 14.4 The specific Eisenstein series that appear in (14.2) are the spherical vectors of small representations in the sense of Section 6.4.3 (see Table 6.2). The function $\mathcal{E}^{(D)}_{(0,0)}$ of the R^4 derivative correction belongs to the minimal representation and the function $\mathcal{E}^{(D)}_{(1,0)}$ of the $\nabla^4 R^4$ correction belongs to the next-to-minimal representation [313, 520, 315]. The first term $\mathcal{E}^{(D)}_{(0,-1)}$ in (13.39) is associated with the standard Einstein–Hilbert term given by the Ricci scalar R and is a constant and thus belongs to the trivial automorphic representation. We will discuss these representation-theoretic aspects further in Section 14.3.3.

14.2.2 Different Cusps and Their Physical Interpretation

In Section 14.2.1, we explained in a ten-dimensional example how the constant terms of the functions $\mathcal{E}^{(D)}_{(p,q)}$ appearing in the low-energy α'-expansion of the scattering amplitude are related to terms in string perturbation theory. For type IIB string theory in ten space-time dimensions with the coordinate $z = \chi + ig_s^{-1}$ parametrising the upper half-plane, the constant terms are polynomials in g_s and correspond to contributions in (14.7) coming from Riemann surfaces of different genus.

The *weak coupling limit* $g_s \to 0$ is the only cusp of the moduli space $SL(2,\mathbb{Z})\backslash SL(2,\mathbb{R})/SO(2,\mathbb{R})$. There is one type of Fourier expansion in the case of $SL(2)$ and it is associated with the maximal unipotent subgroup N. For string theory compactified on a torus T^d from ten dimensions to $10-d$ space-time dimensions the relevant group is E_{d+1}, as displayed in Table 13.2, and there are many different ways of computing Fourier expansions associated with various unipotent subgroups U of E_{d+1}. The constant terms with respect to various unipotents U have different physical interpretations.

There are three preferred choices of unipotent subgroup U that have been mainly studied in string theory. These are displayed in Figure 14.1 and are the unipotent subgroups of maximal parabolics in E_{d+1}. For computing the constant terms with respect to these unipotent subgroups one can use Theorem 8.9. For maximal parabolic subgroups, there is always a $GL(1)$ factor in the Levi subgroup and approaching the cusp in the corresponding variable is the generalisation of the weak coupling limit discussed above. The corresponding unipotent subgroups are not necessarily abelian and therefore one is in principle faced with the non-abelian Fourier expansions discussed in Section 6.2.3. We will restrict the discussion below to the abelian Fourier coefficients.

The three preferred choices have the following physical interpretation for string theory in $(10-d)$ space-time dimensions with an internal torus T^d [313, 520, 315]:

(a) *M-theory limit*: The parabolic subgroup is P_2 with Levi subgroup $GL(d+1) = GL(1) \times SL(d+1)$. The group $SL(d+1)$ is associated with a $(d+1)$-dimensional torus whose overall size is controlled by the $GL(1)$ factor. Approaching the cusp in this parabolic is tantamount to making this $(d+1)$-dimensional torus large and thus approaching a macroscopic space-time of $10-d+(d+1) = 11$ dimensions. Physics in

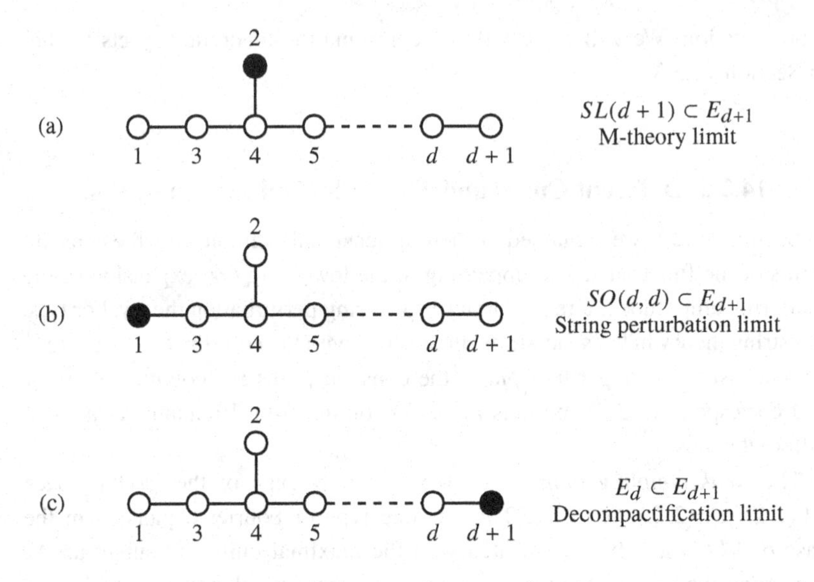

Figure 14.1 Three distinguished maximal parabolic subgroups of E_{d+1}. The filled black node indicates the simple root i_* defining the maximal parabolic P_{i_*} in the sense of (3.43). The semi-simple parts of the Levi subgroups are displayed together with their physical significance.

11 space-time dimensions in the context of string theory is often referred to as M-theory, giving rise to the name of this limit.

The constant terms in this limit correspond to terms coming from the semi-classical approximation by eleven-dimensional supergravity on a $(d+1)$-dimensional torus. The non-zero Fourier modes correspond to non-perturbative states of M-theory, such as the M2- and M5-brane.

(b) *String perturbation limit*: The parabolic subgroup is P_1 with Levi subgroup $GL(1) \times SO(d, d)$. The group $SO(d, d)$ is associated with the T-duality group of string theory on T^d whereas the $GL(1)$ factor controls the strength of the string coupling in $10 - d$ space-time dimensions. The T-duality symmetry is a *perturbative* symmetry, which is preserved order by order in string perturbation theory [554].

The constant terms in this limit correspond to the contributions from world-sheets of different genus just as in (14.7). The non-zero Fourier modes correspond to the various D-instantons that can arise for string theory on a d-torus.

(c) (Single) *decompactification limit*: The parabolic subgroup is P_{d+1} with Levi subgroup $GL(1) \times E_d$. The group E_d is associated with moduli space of string compactifications on a torus T^{d-1}, i.e., of one dimension less than the original torus. The $GL(1)$ factor controls the size of the extra circle and the cusp associated with this parabolic corresponds to the limit when this circle becomes large, whence the name decompactification limit.

The constant terms in this limit are the terms that can be obtained from direct compactification of the theory on T^{d-1} along with certain threshold terms. The non-zero Fourier modes are associated with black holes in the theory on T^{d-1} whose world-line wraps the extra circle. (There is one exception to this in the case $d = 8$ [315].)

Remark 14.5 One could of course also consider other unipotent subgroups, for instance those associated with maximal parabolic for nodes in the middle of the E_{d+1} Dynkin diagram. These correspond to the simultaneous decompactification of several dimensions but not all. As the essential physical effects of interest are typically covered by the three limits above, such intermediate Fourier expansions are rarely considered in string theory.

14.2.3 Constant Terms and Perturbative Contributions

The constant terms in the three distinguished limits of Figure 14.1 can be determined with the help of Theorem 8.9 and by using the techniques introduced in Chapter 10. The results for these calculations are given in detail for all string duality groups E_d in [313, 520].

As an illustration of the physical significance of the constant terms, we shall now analyse the string perturbation limit (limit (b)) of the R^4 correction $\mathcal{E}^{(D)}_{(0,0)}$ from (14.2) in the case of E_7. The function $\mathcal{E}^{(4)}_{(0,0)}$ is given by a maximal parabolic Eisenstein series on node 1 of the Dynkin diagram and we write an arbitrary Eisenstein series in this degenerate principal series as

$$E_{\left[\substack{\quad 0 \\ s\,0\,0\,0\,0\,0}\right]}(g) \tag{14.9}$$

by putting the weight $\lambda = 2s\Lambda_1 - \rho$ on the E_7 Dynkin diagram.

For the string perturbation limit (b) one obtains the constant term

$$\int\limits_{U_1(\mathbb{Q})\backslash U_1(\mathbb{A})} E_{\left[\begin{smallmatrix} & 0 & & & & \\ s & 0 & 0 & 0 & 0 & 0\end{smallmatrix}\right]}(ug)du = r^{-4s} + \frac{\xi(2s-1)}{\xi(2s)} r^{-(2s+1)} E_{\left[0\,0\,0\,0\,{}^{s-1/2}_{\;0}\right]}$$
$$+ \frac{\xi(2s-4)\xi(2s-7)}{\xi(2s)\xi(2s-3)} r^{-8} E_{\left[0\;s-2\;0\;0\;{}^{0}_{0}\right]}$$
$$+ \frac{\xi(2s-6)\xi(2s-8)\xi(2s-10)\xi(4s-17)}{\xi(2s)\xi(2s-3)\xi(2s-5)\xi(4s-16)} r^{2s-18} E_{\left[0\,0\,0\,0\,{}^{s-3}_{\;0}\right]}$$
$$+ \frac{\xi(2s-11)\xi(2s-13)\xi(2s-16)\xi(4s-17)}{\xi(2s)\xi(2s-3)\xi(2s-5)\xi(4s-16)} r^{4s-34}, \tag{14.10}$$

where we have suppressed the argument of the Eisenstein series on $SO(6,6)$ on the right-hand side. The variable r parametrises the $GL(1)$ factor of the Levi $GL(1) \times SO(6,6) \subset E_7$ in a convenient normalisation.

According to (14.2), the functions for the R^4 and $\nabla^4 R^4$ derivative corrections can be obtained from this by setting $s = \frac{3}{2}$ and $s = \frac{5}{2}$, respectively. For the R^4 term we obtain

$$\int\limits_{U_1(\mathbb{Q})\backslash U_1(\mathbb{A})} \mathcal{E}^{(D)}_{(0,0)}(ug)du = 2\zeta(3)r^{-6} + 2\zeta(3)\frac{\xi(2)}{\xi(3)} r^{-4} E_{\left[0\,0\,0\,0\,{}^{1}_{0}\right]}$$
$$= 2\zeta(3)r^{-6} + 2\zeta(3)\frac{\xi(4)}{\xi(3)} r^{-4} E_{\left[2\,0\,0\,0\,{}^{0}_{0}\right]}$$
$$= 2\zeta(3)r^{-6} + \frac{4\zeta(4)}{\pi} r^{-4} E_{\left[2\,0\,0\,0\,{}^{0}_{0}\right]}, \tag{14.11}$$

where we have inserted the normalisation of $\mathcal{E}^{(D)}_{(0,0)}$ from (14.2) and have used a functional relation on the $SO(6,6)$ Eisenstein series. The two terms in this expansion are related to the tree-level and one-loop contribution to the four-graviton scattering amplitude in $D = 10 - d$ space-time dimensions in Einstein frame, where r should be identified as the inverse of the D-dimensional string coupling constant. The first term $2\zeta(3)r^{-6}$ is just the reduction of the corresponding tree-level term in (14.7). The second term can be understood through the standard (Narain) one-loop theta function $\Theta_{\Gamma_{6,6}}$ on the 12-dimensional split, even and self-dual lattice $\Gamma_{6,6}$ that was introduced in Section 13.4.3:

$$\Theta_{\Gamma_{6,6}}(m,\tau) = \tau_2^3 \sum_{(v^i,w^i)\in\mathbb{Z}^{12}} \exp\left(-\frac{\pi}{\tau_2}(v^i - \tau w^i)m_{ij}(v^j - \bar{\tau}w^j)\right), \tag{14.12}$$

where $\tau = \tau_1 + i\tau_2 \in \mathbb{H}$ is the modulus of the world-sheet torus at one-loop and $m_{ij} = g_{ij} + b_{ij}$ contains the metric and the B-field on T^6 with volume τ_2^3, such that m_{ij} is a coordinate on the symmetric space of the simple Levi factor $SO(6,6)/S(O(6)\times O(6))$. Repeated indices are summed in the above expression in accord with the *Einstein summation convention*. The lattice function (14.12) describes the sum over momenta and winding of closed string states on T^6 that can be present at one-loop order; see also Section 13.4. Integrating over all possible genus-one surfaces (= one-loop graphs) then produces the following contribution to the scattering amplitude at R^4 order:

$$I^{(4)}_{(0,0)}(m) = \int\limits_{SL(2,\mathbb{Z})\backslash\mathbb{H}} \frac{d^2\tau}{\tau_2^2}\left(\Theta_{\Gamma_{6,6}}(m,\tau) - \tau_2^3\right). \tag{14.13}$$

(The extra τ_2^3 serves as regulator for the divergence coming from the $\tau_2 \to \infty$ behaviour of $\Theta_{\Gamma_{6,6}}$ [318].) The theta lift $I^4_{(0,0)}$ is a spherical function on $SO(6,6)$, which is invariant under $SO(6,6,\mathbb{Z})$, which that in fact is proportional to an Eisenstein series [503, 318]

$$I^{(4)}_{(0,0)}(m) = \frac{4\zeta(4)}{\pi^3} E_{\left[2\,0\,0\,0\,{}^{0}_{0}\right]}; \tag{14.14}$$

see also Example 13.13. Thus the two terms in (14.11) precisely match the expected contributions from tree-level and one-loop in $D = 4$ space-time dimensions. That there are no higher-loop contributions is due to the supersymmetry protection of the R^4 coupling as discussed in Section 14.2.1.

A similar analysis can be done for the $\nabla^4 R^4$ derivative correction by setting $s = \frac{5}{2}$ in (14.10); this is discussed for instance in [313, 520].

14.2.4 Fourier Coefficients and Instantons

We shall now discuss the physical interpretation of the Fourier coefficients in the limits of Figure 14.1 in more detail. Associated with any of the limits is a unipotent group U for which one can define Fourier coefficients of an automorphic form φ on a group G, as in Section 6.2. These (global) Fourier coefficients

$$F_\psi(\varphi, g) = \int\limits_{U(\mathbb{Q})\backslash U(\mathbb{A})} \varphi(ug)\overline{\psi(u)}du \tag{14.15}$$

depend on the choice of a unitary character $\psi : U(\mathbb{Q})\backslash U(\mathbb{A}) \to U(1)$ on the unipotent. As discussed in detail in Section 6.2, such characters are trivial on the commutator subgroup $[U, U]$ and can be parametrised by a set of charges $m_\alpha \in \mathbb{Q}$ where $\alpha \in \Delta^{(1)}(\mathfrak{u})$ are the roots of the abelianisation $[U, U]\backslash U$; see Proposition 6.9. Restricting the argument $g \in G(\mathbb{A})$ to the archimedean place implies that the set of possible charges m_α is an integer charge lattice $\Gamma \cong \mathbb{Z}^d$ with $d = \dim \Delta^{(1)}(\mathfrak{u})$.

In physical terms, this lattice Γ is interpreted as the lattice of *instanton* charges. Recall that instantons are localised objects in space-time; thus they occur at a some spatial point and at some temporal instant. This was discussed in the simplest case of $SL(2)$ in Section 14.2.1. Recall further that instantons are also called *non-perturbative effects*. A physical object or effect is called non-perturbative in a coupling constant r if it does not admit a Taylor expansion around weak coupling $r = 0$. In other words, 'non-perturbative' corresponds to smooth but not analytic at $r = 0$, with the prototypical example being a function of the form $e^{-1/r}$. This is exactly the kind of behaviour seen in Section 14.2.1 for the Fourier coefficients of $SL(2)$ Eisenstein series with $r = g_s = y^{-1}$ in terms of the coordinate $z = x + iy$ on the upper half-plane. We shall now prove in general that Fourier coefficients $F_\psi(\varphi, g)$ for maximal parabolics with r parametrising the $GL(1)$ factor in the Levi subgroup $L = GL(1) \times M$ always produce this kind of non-perturbative behaviour (even when continued to the non-abelian Fourier coefficients).

We formalise this discussion as follows:

Proposition 14.6 (Non-perturbative nature of Fourier coefficients) *Let Γ be the charge lattice of the unipotent U of a maximal parabolic subgroup $P \subset G$ with Levi subgroup $L = GL(1) \times M$. The Fourier coefficient* (14.15) *of a (spherical) automorphic form φ with respect to a character ψ on U has the asymptotic behaviour*

$$F_\psi(\varphi, g) \sim c_\psi(m) e^{-a_\psi/r} e^{2\pi i \sum_{\alpha \in \Delta^{(1)}(\mathfrak{u})} m_\alpha u_\alpha} \tag{14.16}$$

when approaching the cusp $r \to 0$ associated with the maximal parabolic. Here, r is a suitably normalised coordinate on the $GL(1)$ factor and $m \in M$. The variables u_α are coordinates on the abelianisation $[U, U]\backslash U$ as in (6.30), *$a_\psi \in \mathbb{Q}$ is a number determined by the character ψ, and $c_\psi(m)$ are the functions on the semi-simple M that determine the Fourier coefficients. The Fourier coefficients are non-perturbative when approaching the cusp $r \to 0$.*

We will actually prove the following more general proposition that proves a conjecture made in [253, App. A.6] about the asymptotic behaviour of Fourier coefficients in a non-abelian Fourier expansion of the type considered in Section 6.2.3.

Proposition 14.7 (Asymptotics of Fourier coefficients) *Let φ be a spherical automorphic form on a group $G(\mathbb{R})$ that is an eigenfunction of the Laplace operator on the symmetric space $G(\mathbb{R})/K(\mathbb{R})$ with real eigenvalue μ: $\Delta_{G/K}\varphi(g) = \mu\varphi(g)$. Let U be the unipotent of a standard maximal parabolic $P \subset G$ and let d be the generator of the $GL(1)$ factor in the Levi component $L = GL(1) \times M$ of P, normalised such that it takes integer values $1, \ldots$ on the unipotent U and with associated group element $e^{\phi d}$. The Lie algebra $\mathfrak{u}$ decomposes as $\mathfrak{u} = \bigoplus_\ell \mathfrak{u}_\ell$. A Fourier coefficient for the unipotent $U_\ell(\mathbb{R})$ with* *character* $\psi(x_\alpha(\chi)) = e^{2\pi i n\chi}$, *where* α *belongs to* $\Delta^{(1)}(\mathfrak{u}_\ell)$, *then behaves asymptotically as* $\phi \to -\infty$ *as*

$$\int_{U_\ell(\mathbb{Z})\backslash U_\ell(\mathbb{R})} \varphi(ue^{\phi d}mx_\alpha(\chi))\overline{\psi(u)}du \sim c_\psi(m)e^{-S_{\mathrm{E}}^\ell} + \cdots, \tag{14.17}$$

with

$$S_E^\ell = 2\pi|n|e^{-\ell\phi} + 2\pi i n\chi + \textit{sub-leading in } e^\phi\,. \tag{14.18}$$

Proof The element $d \in \mathfrak{h}(\mathbb{R})$ induces a grading of $\mathfrak{p}(\mathbb{R}) = \mathfrak{m}(\mathbb{R}) \oplus d\mathbb{R} \oplus \mathfrak{u}(\mathbb{R})$ with the properties

$$[d, m] = 0 \qquad \text{for } m \in \mathfrak{m} = \mathrm{Lie}(M), \tag{14.19a}$$

$$[d, u_\ell] = \ell u_\ell \qquad \text{for } u_\ell \in \mathfrak{u}_\ell \text{ with } \mathfrak{u} = \bigoplus_\ell \mathfrak{u}_\ell\,. \tag{14.19b}$$

(The decomposition of $\mathfrak{u}$ is the same one that arose in (11.151).) The relevant part of the Laplacian $\Delta_{G/K}$ for the directions ϕ and χ is found to be

$$\Delta_{G/K} \propto \partial_\phi^2 + \beta\partial_\phi + e^{-2\ell\phi}\partial_\chi^2 + \cdots, \tag{14.20}$$

where $\beta = \sum_\ell \ell\dim(\mathfrak{u}_\ell)$ and we have not fixed the overall normalisation of the Laplace operator as it can be absorbed into the eigenvalue μ for this discussion. The method for finding the Laplacian is given in Appendix B. For a character with charge n as given, we now make the ansatz for the automorphic function that, asymptotically for $\phi \to -\infty$,

$$\varphi_n(\phi, \chi) = e^{-ae^{-b\phi}+2\pi i n\chi}\left(1 + O(e^\phi)\right), \tag{14.21}$$

suppressing the dependence on the remaining coordinates of G/K. This form of φ_n corresponds to a Fourier coefficient for a character $\psi(x_\alpha(\chi)) = e^{2\pi i n\chi}$.

Acting with the relevant part (14.20) of the Laplace operator on this ansatz shows that it can only be an eigenfunction (asymptotically) if

$$a = 2\pi|n| \quad \text{and} \quad b = \ell. \tag{14.22}$$

Note that this reasoning is independent of the eigenvalue μ and of whether φ is an Eisenstein series or any other automorphic function. The important point about (14.22) is that it shows that the leading part of the *instanton action* is

$$S_{\mathrm{E}}(\phi, \chi) = \log\varphi_n(\phi, \chi) = 2\pi|n|e^{-\ell\phi} + 2\pi i n\chi + \text{sub-leading in } e^\phi, \tag{14.23}$$

as claimed. □

Remark 14.8 The form (14.18) is typical of what we call type-ℓ instantons, where 'type' here refers to the degree in $\mathfrak{u}(\mathbb{R})$. Treating the expectation value of e^ϕ as the coupling constant r, the *weak coupling limit* $r \to 0$ then corresponds to $\phi \to -\infty$. The rôle of the 'axion' to which the instanton couples is played by χ associated with the unipotent generator $E_\alpha \in \mathfrak{u}_\ell$. Making the link to the non-abelian Fourier expansion of Section 6.2 shows that the more non-abelian a Fourier coefficient is the faster it falls off in the corresponding weak coupling expansion. The typical cases encountered in string theory are when $e^\phi = g_s$ is the *string coupling*. Instantons with $\ell = 1$ are then D-instantons and those with $\ell = 2$ are NS-instantons [308, 309, 51, 525, 42, 43]. In low space-time dimensions one also expects instantons with $\ell > 2$ [502, 203].

Case Study: Decompactification Limit for E_7

We consider the case of the R^4 coefficient function $\mathcal{E}^{(4)}_{(0,0)}$ for the case E_7. In Section 14.2.3, we studied the constant terms of this function in the string perturbation limit. We shall now look at the decompactification limit (c) mentioned in Section 14.2.2 that uses the Levi $L = GL(1) \times E_6$. The abelian unipotent is a 27-dimensional representation of E_6. Recall from (14.2a) that the coefficient function is given by

$$\mathcal{E}^{(4)}_{(0,0)} = 2\zeta(3) E_{\left[\begin{smallmatrix} & & 0 & & & \\ s & 0 & 0 & 0 & 0 & 0 \end{smallmatrix}\right]}(g) \tag{14.24}$$

for $s = \frac{3}{2}$.

The Fourier coefficients in this unipotent can be studied by using degenerate Whittaker coefficients and their covariantisation similar to the Piatetski-Shapiro method of Section 6.5. In this particular case, an alternative derivation in terms of Poisson resummation is available as the automorphic form can be represented as a constrained Epstein series [82]. We require the degenerate Whittaker coefficient for charge $(0, 0, 0, 0, 0, 0, k)$ given in Example 10.27, from which we deduce that

$$\int\limits_{N(\mathbb{Z})\backslash N(\mathbb{R})} \mathcal{E}^{(4)}_{(0,0)}(na)\overline{\psi(n)}dn = \frac{4\zeta(3)}{\xi(4)} v_6^{3/2} v_7 |k|^{-3/2} \sigma_3(k) K_{3/2}(2\pi|k|v_7^2 v_6^{-1}) . \tag{14.25}$$

Using the property that in the minimal representation the Whittaker models agree with the Fourier–Whittaker models [246], this result can be translated into one for the Fourier coefficient of a charge in the intersection of the A_1-type nilpotent orbit with the 27-dimensional unipotent. We will call such charges rank-one and write them as kN_1. Under the $E_6(\mathbb{Z})$ action any rank-one charge on the 27-dimensional unipotent can be brought to the form where the associated character ψ_{kN_1} agrees with the restriction of the Whittaker character for charge $(0, 0, 0, 0, 0, 0, k)$ for some $k \in \mathbb{Z}$. Our parametrisation kN_1 thus means that N_1 is a primitive vector in the 27-dimensional charge lattice.

To write the Fourier coefficient, we parametrise the Levi element $l \in GL(1, \mathbb{R}) \times E_6(\mathbb{R})$ by a coordinate r on the $GL(1)$ whose generator is proportional to the fundamental weight Λ_7 of $\mathfrak{e}_7$ and an element $m \in E_6$. With this we can replace the torus variables v_6 and v_7 by combinations of r and $||m^{-1}N_1||$, where N_1 is the primitive rank-one charge, as follows:

$$v_7 = r^{3/2}, \quad v_6 = r^2||m^{-1}N_1||^{-1}. \tag{14.26}$$

The norm appearing in the above equation is the $K(E_6) = USp(8)$-invariant norm. These relations follow from considering N_1 in position $(0, \ldots, k)$ and using that the r is generated by Λ_7. Putting things together we then find the Fourier coefficient

$$\begin{aligned} F_{kN_1}(g) &= \int\limits_{U(\mathbb{Z})\backslash U(\mathbb{R})} \mathcal{E}^{(4)}_{(0,0)}(ug)\overline{\psi_{kN_1}(u)}du \\ &= \frac{4\zeta(3)}{\xi(4)} r^{9/2}|k|^{-3/2}\sigma_3(k)\frac{K_{3/2}(2\pi|k|r||m^{-1}N_1||)}{||m^{-1}N_1||^{3/2}} e^{2\pi i k\langle N_1, a\rangle}. \end{aligned} \tag{14.27}$$

Here, the dependence on g on the left-hand side is decomposed on the right-hand side as follows: a parametrises the 27 'axionic' directions of the abelian unipotent dual to the charge kN_1 and r and m parametrise the Levi.

In physical terms, the variable r is related to the radius of the decompactifying circle when going from $D = 4$ to $D = 5$ space-time dimensions. We can see this more explicitly by considering the asymptotics of the Fourier coefficient in the limit $r \to \infty$ by expanding the Bessel function. This leads to

$$F_{kN_1}(g) = \frac{180\zeta(3)}{\pi^2} \frac{r^4}{|k|^2||m^{-1}N_1||^2}\sigma_3(k)e^{-2\pi|k|r||m^{-1}N_1||+2\pi i k\langle N_1, a\rangle} + \cdots. \tag{14.28}$$

The non-perturbative terms arising in this Fourier coefficient are BPS black holes in $D = 5$ and with BPS mass (or central charge) $Z(kN_1) = |k|||m^{-1}N_1||$ and action

$$S_{\mathrm{E}} = 2\pi Z(kN_1)r - 2\pi i k\langle N_1, a\rangle. \tag{14.29}$$

We see the typical exponential suppression of instanton corrections for decompactifying $r \to \infty$ by non-analytic terms in $e^{-S_{\mathrm{E}}}$. Moreover, we note that all characters ψ_{Nk_1} contributing to the Fourier expansion are of rank one as is expected from the wavefront set of this function. The rank-one condition corresponds physically to the BPS particles in $D = 5$ being 1/2-BPS, in agreement with the fact that the function $\mathcal{E}^{(4)}_{(0,0)}$ for the R^4 correction is 1/2-BPS, a viewpoint that will be developed further in Section 14.3.

There are two aspects of the Fourier coefficient (14.27) that are not understood from string theory, to the best of our knowledge. The first concerns the divisor sum σ_3 that arises. There is no understanding of the order 3 from combinatorics of BPS states similar to what was discussed for σ_2 and D-instantons in Section 13.6.3. Second, the Bessel function that arises has order 3/2 which means that the asymptotic expansion is exact and consists of two terms. This should entail a certain non-renormalisation theorem that rules out corrections in the instanton background beyond the first non-trivial order in the large radius expansion.

14.3 Automorphic Representations and BPS Orbits

As we have argued above, the coefficient functions $\mathcal{E}^{(D)}_{(0,0)}$ and $\mathcal{E}^{(D)}_{(0,1)}$ are associated with Eisenstein series on the symmetry group E_{11-D}. In this section, we will analyse which automorphic representation they belong to and how the associated constraints on the possible non-trivial Fourier coefficients are related to the structure of so-called BPS orbits.

14.3.1 Levi Orbits on the Charge Lattice and Characters

We recall from Section 6.4.1 that the discrete Levi subgroup $L(\mathbb{Q})$ acts on the Fourier coefficients $F_\psi(\varphi, g)$ by grouping them into character variety orbits: for $\gamma \in L(\mathbb{Q})$ one has that the Fourier coefficients for the characters $\psi(u)$ and $\psi^\gamma(u) = \psi(\gamma u \gamma^{-1})$ are related. One therefore has to decompose the charge lattice Γ into orbits under the action of the Levi subgroup and it is sufficient to compute the Fourier coefficient for a representative of each orbit. Even though the description of the decomposition of the charge lattice over $L(\mathbb{Z})$ can be tricky, there are some cases where the orbits are known explicitly in terms of algebraic conditions on the charges.

Interpreting the charges in the lattice Γ as characteristics of physical objects, such that they correspond for instance to generalised electro-magnetic charges, these algebraic conditions can often be interpreted in terms of physical constraints. More precisely, these physical constraints can be related to so-called *Bogomol'nyi–Prasad–Sommerfield* (BPS) conditions which derive from the supersymmetry algebra of the theory [502] (see also Remark 13.19). These BPS conditions provide restrictions on a set of states, in our case instantons. However, the presence of these BPS states also breaks some of the supersymmetry present. The physical spectrum of string theory states consists of states that preserve some supersymmetry and states that do not preserve any supersymmetry. States that preserve some supersymmetry are called *BPS states* and are often labelled by the fraction of the maximal possible supersymmetry that they preserve. Other states are called non-BPS states.

Remark 14.9 In the context of supergravity, the BPS conditions can be understood in terms of the number of *Killing spinor equations* that determine the

Killing spinors of a space-time and matter configuration. Flat space without any matter is always a solution of the supergravity equations of motion and admits the maximal number of Killing spinors. For so-called maximal supergravity theories as they arise for type II superstrings on tori T^d this maximal number is 32. The complete classification of the possible number of independent Killing spinors that can be realised by other space-times for various maximal supergravity theories is open. See [257, 209, 210, 301, 302, 303, 304, 305] for recent work on this topic.

Some higher-derivative corrections enjoy so-called partial supersymmetry or *BPS protection*. This means that their computation only relies on understanding the contributions of certain BPS states. Examples are the R^4, $\nabla^4 R^4$ and $\nabla^6 R^4$ terms discussed above. They are $1/2$-, $1/4$- and $1/8$-BPS, respectively. This means for instance that the R^4 higher-derivative correction is sensitive only to string theory states that preserve at least $1/2$ of the maximal supersymmetry.

Let us now translate this discussion to Fourier coefficients of automorphic forms. If a certain higher-derivative correction is BPS-protected, this means that its coefficient function $\mathcal{E}^{(D)}_{(p,q)}$ can only have a restricted set of Fourier coefficients (14.15). For a given unipotent U with associated charge lattice Γ, only those Levi orbits in Γ can contribute that satisfy the relevant BPS conditions. This is reminiscent of the vanishing theorems by Mœglin–Waldspurger [486, 484] and Matumoto [470] already mentioned in Section 6.4 on wavefront sets and we will further elaborate on this connection in Section 14.3.3.

14.3.2 Case Study: BPS Orbits in Four Dimensions

To illustrate further the way BPS instantons are related to orbits in the charge lattice and to Fourier expansions, we consider in some detail the instantons in four space-time dimensions viewed in the decompactification limit.

In the decompactification limit from four to five space-time dimensions one can describe BPS instantons in three different ways. The first way is to consider BPS instantons in five space-time dimensions that descend to instantons in four space-time dimensions upon compactification of one dimension. The second way is to consider BPS point particles in five space-time dimensions. Unlike instantons that occur at a given instant, point particles trace out a one-dimensional world-line in space-time during their evolution. If one considers special classes of stationary BPS point particles then one can use their stationarity to compactify them along the time direction to four dimensions where they appear as instantons. BPS particles in five space-time dimensions appear as charged BPS black holes. They carry generalised electro-magnetic charges that transform in the 27-dimensional representation of $E_6(\mathbb{R})$ that is

Table 14.1 Orbits of BPS point particles in five space-time dimensions. The particles carry 27 generalised electro-magnetic charges q^i that transform under the group $GL(1) \times E_6$. There are three different orbits that can be described in terms of the cubic invariant I_3 of E_6 in its 27-dimensional representation. Table adapted from [208, 315].

BPS conditions	BPS type	Orbit	Dimension
$\begin{cases} I_3 = \frac{\partial I_3}{\partial q^i} = 0 \\ \frac{\partial^2 I_3}{\partial q^i \partial q^j} \neq 0 \end{cases}$	1/2	$\frac{E_6}{Spin(5,5)\ltimes\mathbb{R}^{16}}$	17
$I_3 = \frac{\partial I_3}{\partial q^i} \neq 0$	1/4	$\frac{E_6}{Spin(4,5)\ltimes\mathbb{R}^{16}}$	26
$I_3 \neq 0$	1/8	$\frac{GL(1)\times E_6}{F_4}$	27

the Cremmer–Julia symmetry group of five-dimensional maximal supergravity. Due to charge quantisation they lie in a rank-27 lattice in this 27-dimensional representation and this relates to the case study in Section 14.2.4 above. The third way consists of all remaining instantons that do not arise in either of the other two ways, and they will play a rôle in our analysis; see [315] for additional comments on this case.

The BPS point particle states in five dimensions have been classified for example in [207, 208, 455]. There are three types of BPS point particles that are called 1/2-, 1/4- and 1/8-BPS according to the fraction of supersymmetry generators they preserve, and which are shown in Table 14.1. They transform in orbits of the group $GL(1) \times E_6$ and carry quantised charges q^i, where $i = 1, \dots, 27$, where the $GL(1)$ is related to the so-called on-shell trombone symmetry [455]. The three types of BPS particles can be distinguished in terms of their cubic invariant

$$I_3 = \sum_{i,j,k} J_{ijk} q^i q^j q^k, \tag{14.30}$$

where J_{ijk} is the cubic invariant tensor of E_6 in its 27-dimensional representation. The three orbits are then characterised in terms of the vanishing of I_3 and its derivatives, as shown in Table 14.1. Alternatively, they can be described in terms of $GL(1) \times E_6$ quotiented by the stabiliser of a representative state.

Both the BPS instantons and BPS point particles in five space-time dimensions make their appearance in the Fourier expansion of E_7 automorphic forms in the decompactification limit. The decompactification limit (c) of Figure 14.1 has Levi factor $L = GL(1) \times E_6$, where the coordinate r of the $GL(1)$ can be identified with the radius of the compact dimension relating the four- and five-dimensional descriptions. The abelian unipotent U in this case is the 27-dimensional representation of E_6. The Fourier expansion of an E_7 automorphic form along this unipotent produces constant terms that contain powers of r and automorphic forms on E_6 according to Theorem 8.9. These automorphic forms on E_6 have Fourier expansions that will contain the BPS instantons that already exist in five space-time dimensions. The non-constant modes in the Fourier expansion of the E_7 automorphic form will contain instantons with charges in the 27-dimensional lattice in the Lie algebra of the unipotent U according to Proposition 14.6. These are the BPS point particle states with generalised electro-magnetic charges in five space-time dimensions that are listed in Table 14.1. The BPS orbits correspond to the character variety orbits relating F_ψ to F_{ψ^γ} where γ is an element of the discrete Levi subgroup.

This gives the following prediction for the decompactification limit of the E_7 automorphic forms $\mathcal{E}^{(D)}_{(p,q)}$ that appear in the low-energy α'-expansion of the four-graviton scattering amplitude. The function $\mathcal{E}^{(D)}_{(0,0)}$ associated with the 1/2-BPS coupling R^4 can only have Fourier coefficients for characters ψ that are associated with the 1/2-BPS orbit in the Lie algebra of the unipotent. Similarly, the function $\mathcal{E}^{(D)}_{(1,0)}$ associated with the 1/4-BPS coupling $\nabla^4 R^4$ can only have Fourier coefficients for characters ψ that associated with the 1/2- or 1/4-BPS orbit. Finally, the function $\mathcal{E}^{(D)}_{(0,1)}$ that multiplies the 1/8-BPS coefficient $\nabla^6 R^4$ has all possible types of Fourier coefficients. This last statement is specific to E_6.

Remark 14.10 The function multiplying the derivative correction $\nabla^6 R^4$ has non-vanishing Fourier coefficients for all possible orbits in the decompactification limit but is still 1/8-BPS and thus much more special than a generic non-BPS coupling. The relative simplicity that the remaining 1/8-BPS supersymmetry should imply for $\mathcal{E}^{(D)}_{(0,1)}$ is not visible from the Fourier support that was analysed in the above discussion, but should be visible in the precise form of the Fourier coefficients $c_\psi(m)$ (with $m \in E_6$) in (14.16). The exact form of these coefficients has not been worked out, to the best of our knowledge. This is partly due to the fact that $\mathcal{E}^{(D)}_{(0,1)}$ is not strictly an automorphic form in the sense of Definition 4.13. This theme will be further explored in Sections 15.1 and 14.3.3.

Similar predictions, *mutatis mutandis*, for the vanishing (or not) of the Fourier coefficients in the three limits can be made for all groups E_{d+1}. These predictions can then be compared to the character variety orbits tabulated in [476]. The BPS orbits and character variety orbits coincide perfectly as discussed in detail in [315].

14.3.3 Wavefront Sets

In this section, we collect and systematise some of the remarks on wavefront sets and curvature corrections that have been made in the preceding discussion. We will do this for the case $G = E_7(\mathbb{R})$ that is relevant for $D = 4$ space-time dimensions and maximal supersymmetry. The *closure diagram* of nilpotent orbits of $\mathfrak{e}_7(\mathbb{C})$ can be found for example in [581] and that of $\mathfrak{e}_7(\mathbb{R})$ in [185]. We display the closure (or Hasse) diagram of the smallest nilpotent orbits of $\mathfrak{e}_7(\mathbb{C})$ in Figure 14.2.

In the figure, we have also shown the maximal orbits of the various types of curvature corrections $\nabla^{2k}R^4$ following [520, 315, 83, 80, 81]. What is noticeable is that the maximal orbits of the wavefront sets appear to be associated only with *special* orbits [80]. Preliminary investigations of higher-derivative terms in [81] suggest that correction terms with more than six derivatives acting on R^4 will generically have contributions from the orbit $(A_3+A_1)''$. The expansion in increasing orders of derivatives seems to be related to an expansion in terms of size of the associated wavefront set (with only special orbits as maximal orbits). In general, one expects that the maximal orbits of the wavefront set are admissible orbits [295].

We note also that there can be several maximal nilpotent orbits contributing to a given curvature correction, as in the case of ∇^6R^4. This is related to the fact that the U-duality invariant functions $\mathcal{E}^{(D)}_{(p,q)}$ that arise are not necessarily automorphic forms of the standard type but more general, as discussed in Section 15.1. The branching of the diagram is associated in physics with the existence of independent (linearised) supersymmetry invariants [83, 84, 85].

Let us also relate this discussion back to the analysis of small representations of Sections 6.4.3, 10.4 and 14.2.4. In the case of $E_7(\mathbb{R})$ one has a degenerate principal series representation of functional dimension 33 that can be realised with a maximal parabolic Eisenstein series by choosing the weight $\lambda = 2s\Lambda_1-\rho$. The wavefront set in the case of generic s is of type A_2 of dimension 66. This is twice the dimension of the coset $P_1\backslash E_7$, where P_1 is the maximal parabolic subgroup associated with node 1 in the Dynkin diagram of Figure 13.5. Reductions occur in this case for the values $s = \frac{5}{2}$ and $s = \frac{3}{2}$ when the wavefront set collapses to $2A_1$ and A_1, respectively. The reduction can be analysed using Theorem 9.6 as discussed in Section 10.4. The two cases where the wavefront set reduces correspond to the R^4 and ∇^4R^4 curvature correction. There is a contribution of this function to the ∇^6R^4 correction for a non-special value of s. Similar results hold for other groups E_d.

Remark 14.11 The best-studied case of curvature corrections is that of $D = 10$ type IIB superstring theory where the symmetry group is $SL(2,\mathbb{R})$ with U-duality group $SL(2,\mathbb{Z})$. The set of nilpotent orbits of $\mathfrak{sl}(2,\mathbb{C})$ is very

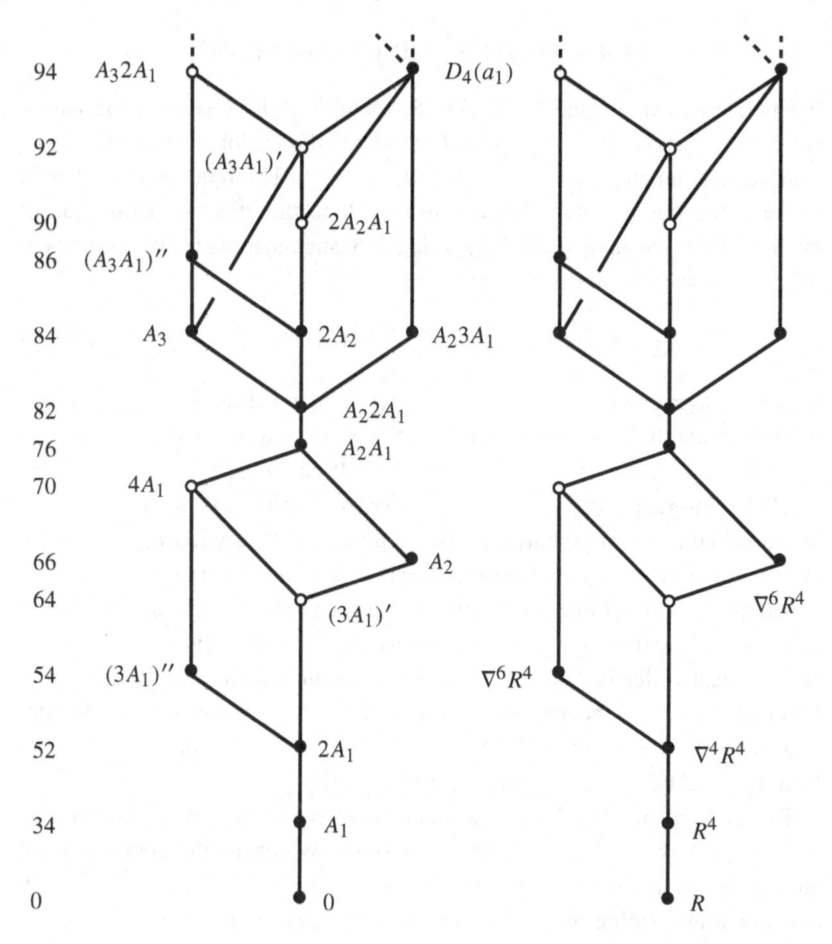

Figure 14.2 The smallest nilpotent orbits of $\mathfrak{e}_7(\mathbb{C})$ and their closure ordering. The vertical axis is the dimension of the orbit and on the left they are labelled according to the Bala–Carter classification, where we have denoted $2A_2 + A_1 \equiv 2A_2A_1$ etc. for brevity. The open circles indicate orbits that are not special. The figure is adapted from [581, 83, 81]. On the right, the wavefront sets of the various curvature terms appearing in the four-graviton scattering amplitude are shown on the same kind of diagram.

degenerate and consists only of either the trivial or the regular (A_1-type) orbit. Nevertheless, the various curvature corrections of type $\nabla^{2k}R^4$ come with very specific orders s of the non-holomorphic Eisenstein series $E(s, z)$. It is an open problem to understand the specific values that appear, notably $s = \frac{3}{2}$ and $s = \frac{5}{2}$, from a mathematical point of view, as there seems to be nothing special happening for the automorphic representation for these values.

14.4 Supersymmetry Constraints*

We will now explain the arguments for obtaining the eigenvalue Equation (14.1a) for the coefficient function $\mathcal{E}^{(10)}_{(0,0)}$ in ten dimensions using constraints from supersymmetry following [321, 310, 307, 579]. This requires us to include not only the four-graviton corrections (13.42) to the effective action but all interactions at the same order in α' related by supersymmetry. We denote the different corrections to the action as

$$S = S^{(0)} + (\alpha')^3 S^{(3)} + (\alpha')^4 S^{(4)} + \cdots, \tag{14.31}$$

where the four-graviton term R^4 of (13.42) is included in $S^{(3)}$ and where we have used that there are no corrections of order α' and $(\alpha')^2$. In fact, the term $S^{(4)}$ also vanishes but this does not affect our discussion.

All the different interaction terms at order $(\alpha')^3$ can be obtained by studying the so-called rigid (or global) limit of the linearised theory [310]. This means that we are, at first, only considering *global supersymmetry* transformations for which the transformation parameters are constant spinors ϵ, as opposed to the local case where they are space-time dependent. Moreover, the transformations are of linear order in the field fluctuations around some fixed background. From the linearised theory we will also find relations between the coefficient functions of different interactions. The sought-after eigenvalue equation can then be found using the full non-linear symmetry.

The physical fields of the ten-dimensional type IIB supergravity theory are conveniently packaged into a generating function as the coefficients of an expansion in formal, complex Grassmann variables θ^α, $\alpha = 1, \ldots, 16$ together transforming as a Weyl spinor under $Spin(9,1)$. Their complexity means that they also transform with non-trivial weight under the $U(1) = SO(2)$ *R-symmetry* of ten-dimensional type IIB symmetry. This $SO(2)$ symmetry is the denominator of the moduli space $SL(2,\mathbb{R})/SO(2)$. Together with the space-time coordinates, the variables θ^α parametrise a *superspace* and the generating function is called a *superfield*. This construction only makes sense here at the linearised level. Supersymmetry transformations can then be described as coordinate transformations on this space. The corresponding differential operators acting on the superfields are

$$Q_\alpha = \frac{\partial}{\partial \theta^\alpha}, \qquad Q^*_\alpha = -\frac{\partial}{\partial \theta^{*\,\alpha}} + 2i(\bar{\theta}\gamma^\mu)_\alpha \frac{\partial}{\partial x^\mu}, \tag{14.32}$$

which (by virtue of acting from the left and the right, respectively) anticommute with the covariant derivatives

$$D_\alpha = \frac{\partial}{\partial \theta^\alpha} + 2i(\gamma^\mu \theta^*)_\alpha \frac{\partial}{\partial x^\mu}, \qquad D^*_\alpha = -\frac{\partial}{\partial \theta^{*\,\alpha}}, \tag{14.33}$$

where $\overline{\theta} = \theta^\dagger \gamma^0$ is the Dirac conjugate, with $\theta^\dagger$ being the Hermitian conjugate and θ^* the complex conjugate. We use the Einstein summation convention where repeated indices are summed over. The γ-matrices are real matrices acting on spinors and satisfy the *Clifford algebra* $\{\gamma^\mu, \gamma^\nu\} = 2\eta^{\mu\nu}$ with the constant Minkowski background metric $\eta^{\mu\nu} = \mathrm{diag}(- + + \ldots +)$, where $\mu, \nu = 0, \ldots, 9$ are Lorentz indices for the ten directions of space-time. We also use the standard notation that a γ-matrix with multiple space-time indices $\gamma^{\mu_1 \cdots \mu_n}$ is the anti-symmetrised product of n γ-matrices with the normalisation $\gamma^{\mu_1 \cdots \mu_n} = \gamma^{\mu_1} \cdots \gamma^{\mu_n}$ for $\mu_1 \neq \ldots \neq \mu_n$.

Comparing the coefficients in the superfield expansion before and after an infinitesimal coordinate transformation in superspace gives the supersymmetry transformations for the physical fields. We see from (14.32) (acting on, for example, (14.34) below) that the supersymmetry transformation relates bosons (with an even number of free spinor indices) and fermions (with an odd number of spinor indices).

The superfield Φ of ten-dimensional type IIB supergravity satisfies the holomorphic constraint $D^*\Phi = 0$ and the on-shell condition $D^4\Phi = D^{*4}\Phi^*$ [310, 359] which narrows down the expansion of Φ into its component fields – the physical fields – into the following [310, 307]:

$$
\begin{aligned}
\Phi &= z_0 + \delta\Phi \\
&= z_0 + \delta z + \frac{1}{g_s}\Big(i\overline{\theta}^* \delta\lambda + \delta\hat{G}_{\mu\nu\rho}\overline{\theta}^* \gamma^{\mu\nu\rho}\theta + \cdots \\
&\quad + \mathcal{R}_{\mu\sigma\nu\tau}\overline{\theta}^* \gamma^{\mu\nu\rho}\theta\overline{\theta}^* \gamma^{\sigma\tau}{}_{\rho}\theta + \cdots + \theta^8 \partial^4 \overline{\delta z} \Big),
\end{aligned}
\tag{14.34}
$$

where $\delta\Phi$ is the linearised fluctuation around a flat background with constant axio-dilaton $z_0 = \chi_0 + ie^{-\phi_0} = \chi_0 + ig_s^{-1}$ and where $\delta\lambda$ is the fluctuation of the dilatino, $\mathcal{R}_{\mu\sigma\nu\tau}$ is the linearised curvature tensor of the metric fluctuation and $\hat{G}$ is a supercovariant combination of fermion bilinears and field strengths of Ramond and Neveu–Schwarz two-form potentials. The exact expression for $\hat{G}$ can be found in [307, App. C]. We will continue to use the notation $z = x + iy = z_0 + \delta z$ for the full axio-dilaton.

Remark 14.12 If we let the metric fluctuation be a plane wave with polarisation $\epsilon_{\mu\nu}$ and momentum k_ρ in the linearised curvature, we recover the expression $\mathcal{R}_{\mu\nu\rho\sigma} \propto k_\mu \epsilon_{\nu\rho} k_\sigma$ (with permutations) from Section 13.2 together with an exponential that later gives the momentum conservation in the amplitude.

The linearised action at order $(\alpha')^3$ can be obtained as an integral over half of superspace (that is, no integrations over θ^*) of some function $F[\Phi]$ [310, 321]:

$$
S^{(3)}_{\text{linear}} = \mathrm{Re} \int d^{10}x \, d^{16}\theta \, F[\Phi],
\tag{14.35}
$$

where the Grassmann integration amounts to taking the θ^{16} term of the expansion of $F[\Phi]$.

From the expansion (14.34), we see that the θ^4 term of Φ contains the linearised curvature $\mathcal{R}_{\mu\sigma\nu\tau}$ and the θ^{16} term of $S^{(3)}$ will therefore contain a contraction of four such curvature tensors. In fact, they form the $\mathcal{R}^4$ interaction of Section 13.2, which can be expressed as the superspace integral [501, 307]

$$\mathcal{R}^4 = \int d^{16}\theta\, (\mathcal{R}_{\mu\sigma\nu\tau}\overline{\theta}^* \gamma^{\mu\nu\rho}\theta\overline{\theta}^* \gamma^{\sigma\tau}{}_{\rho}\theta)^4 . \tag{14.36}$$

In the expansion, this interaction comes with a coefficient $g_s^{-4}\partial_z^4 F$, ignoring numerical prefactors. Inspecting (14.34), we see that another term in the linearised action $S^{(3)}_{\text{linear}}$ is the $\delta\lambda^{16}$ interaction with coefficient $g_s^{-16}\partial_z^{16}F$, where

$$(\lambda^n)_{\alpha_{n+1}\cdots\alpha_{16}} = \frac{1}{n!}\epsilon_{\alpha_1\cdots\alpha_{16}}\lambda^{\alpha_1}\cdots\lambda^{\alpha_n}, \tag{14.37}$$

with α_i being spinor indices. Similar terms can be constructed from the other pieces of the expansion (14.34).

The full (no longer linearised) supersymmetric action $S^{(3)}$ can, in this manner, be obtained as [321]

$$\begin{aligned} S^{(3)} = \int d^{10}x\sqrt{g}\Big(& f^{(12,-12)}(z)\lambda^{16} + f^{(11,-11)}(z)\hat{G}\lambda^{14} + \cdots \\ &+ f^{(0,0)}(z)R^4 + \cdots + f^{(-12,12)}(z)\lambda^{*\,16}\Big), \end{aligned} \tag{14.38}$$

where the coefficients $f^{(w,-w)}(z)$ are to be determined shortly. Note that we have renamed the R^4 coefficient which was labelled as $\mathcal{E}^{(10)}_{(0,0)}$ in Section 14.2.1, and that, unlike the indices for $\mathcal{E}^{(D)}_{(p,q)}$, the indices $(w,-w)$ denote the coefficient's transformation properties under $SL(2,\mathbb{Z})$ as described below. We let (14.38) define the relative normalisations between the coefficients.

Taking the weak coupling limit, we can relate different terms in the full action using the linearised theory. By reading off the coefficients in the expansion of $F[\Phi]$ in (14.35), letting $y^{-1} \sim g_s \to 0$ we see that [310, 307]

$$f^{(12,-12)}(z) \sim y^{12}\partial_z^{12} f^{(0,0)}(z), \tag{14.39}$$

up to a numerical factor.

However, the coefficient $f^{(0,0)}(z) = \mathcal{E}^{(10)}_{(0,0)}(z)$ is invariant under $SL(2,\mathbb{Z})$ transformations, and the same U-duality arguments give that each of the terms in (14.38) is $SL(2,\mathbb{Z})$-invariant. Due to the transformation properties of λ^{16}, $\hat{G}\lambda^{14}$ and the remaining interaction terms, the coefficient functions $f^{(w,\hat{w})}(z)$ have to be modular forms with holomorphic weight w and anti-holormorphic weight $\hat{w}$ transforming as

$$f^{(w,\hat{w})}\left(\frac{az+b}{cz+d}\right) = (cz+d)^{w}(c\bar{z}+d)^{\hat{w}} f^{(w,\hat{w})}(z), \qquad \begin{pmatrix} a & b \\ c & d \end{pmatrix} \in SL(2,\mathbb{Z}). \tag{14.40}$$

This generalises the discussion in Section 4.3.3.

This means that (14.39) cannot be the complete relation since $y^{12}\partial_z^{12} f^{(0,0)}(z)$ does not transform as a modular form of holomorphic and anti-holormorphic weights $(12,-12)$. In physics parlance, the derivative is not covariant with respect to R-symmetry. Instead, we need the modular covariant derivative

$$\mathcal{D}_{(w)} = \frac{\partial}{\partial z} - i\frac{w}{2y}, \tag{14.41}$$

which maps a modular form of weights $(w, \hat{w})$ to a modular form of weights $(w+2, \hat{w})$. Multiplying with $y = \operatorname{Im} z$, which transforms as

$$\operatorname{Im}\frac{az+b}{cz+d} = \frac{\operatorname{Im} z}{|cz+d|^2}, \tag{14.42}$$

we have that $y\mathcal{D}_{(w)} f^{(w,\hat{w})}(z)$ transforms with weights $(w+1, \hat{w}-1)$.

Then, we may covariantise (14.39) to [310]

$$f^{(12,-12)}(z) \propto y^{12}\mathcal{D}_{(22)}\mathcal{D}_{(20)}\cdots\mathcal{D}_{(0)} f^{(0,0)}(z), \tag{14.43}$$

where we do not need the numerical proportionality constant.

Remark 14.13 We note that the covariant derivative $\mathcal{D}_{(w)}$ reduces to the ordinary derivative for the terms in $f^{(w,-w)}$ which satisfy

$$\frac{\partial}{\partial z} f(z) \gg \frac{w}{y} f(z) \qquad \text{as } y \to \infty\,. \tag{14.44}$$

This is not satisfied for the perturbative terms in $f^{(w,-w)}$, which are highlighted in (14.7) for $f^{(0,0)}$. It is, however, satisfied for the non-perturbative terms in (14.7), meaning that the linearised theory (14.35) can only capture the leading instanton contributions to the effective action [307].

To simplify the notation of (14.43) we also use that $\mathcal{D}_{(w)}(y^n f^{(w+n,\hat{w})}) = y^n \mathcal{D}_{(w+n)} f^{(w+n,\hat{w})}$, which allows us to write (14.39) as

$$\begin{aligned} f^{(12,-12)}(z) &\propto y\mathcal{D}_{(11)}\, y\mathcal{D}_{(10)}\cdots y\mathcal{D}_{(0)} f^{(0,0)}(z) \\ &= D_{(11)}D_{(10)}\cdots D_{(0)} f^{(0,0)}(z), \end{aligned} \tag{14.45}$$

with

$$D_{(w)} = i\Big(y\frac{\partial}{\partial z} - i\frac{w}{2}\Big) = iy\mathcal{D}_{(w)}, \tag{14.46}$$

where we have introduced an extra factor of i to use the same normalisation as in the existing literature. This covariant derivative takes a modular form of weight $(w, \hat{w})$ to a modular form of weight $(w+1, \hat{w}-1)$.

The coefficient functions with negative holomorphic weight (such as $f^{(-12,12)}$) can be related to $f^{(0,0)}$ using the conjugated covariant derivative

$$\overline{D}_{(\hat{w})} := \overline{D_{(\hat{w})}} = -i\Big(y\frac{\partial}{\partial \bar{z}} + i\frac{\hat{w}}{2}\Big), \tag{14.47}$$

which maps a modular form of weights $(w, \hat{w})$ to a modular form of weight $(w-1, \hat{w}+1)$.

Remark 14.14 The modular forms above, transforming with holomorphic and anti-holomorphic weights $(w, -w)$, are so-called *Maaß forms* of weight $2w$, which were discussed in Sections 4.3.3 and 5.6, where the modular covariant derivatives are closely related to the raising and lowering operators of (5.66).

We will now use the full non-linear, local supersymmetry of the theory to determine the coefficient functions. When adding the $(\alpha')^3$ terms (14.38) to our effective action we also need to correct the supersymmetry transformation δ_ϵ, with local spinor parameter ϵ, acting on the different fields. Schematically, for some generic field Ψ we would then have the expansion

$$\delta_\epsilon \Psi = \Big(\delta^{(0)} + \alpha'\delta^{(1)} + (\alpha')^2\delta^{(2)} + \cdots\Big)\Psi. \tag{14.48}$$

Requiring that the action should be invariant under this supersymmetry transformation (up to total derivatives), we get the following first few conditions:

$$\delta^{(0)}S^{(0)} = 0, \qquad \delta^{(0)}S^{(3)} + \delta^{(3)}S^{(0)} = 0, \qquad \ldots, \tag{14.49}$$

where the first equation simply states that the classical supergravity action should be invariant under the ordinary local supersymmetry transformations.

To determine the coefficient functions $f^{(w,-w)}(z)$ it turns out that we do not have to analyse the supersymmetry transformations of every term in the action – we only need to focus on a few interaction terms with high powers of the dilatino λ. Expanding $\hat{G}\lambda^{14}$ in terms of the physical fields, we get, among others, a term $\lambda^{15}\gamma^\mu\psi^*_\mu$ [321]

$$S^{(3)} = \int d^{10}x\sqrt{g}\Big(f^{(12,-12)}(z)\lambda^{16} - 3\cdot 144 f^{(11,-11)}(z)\lambda^{15}\gamma^\mu\psi^*_\mu + \cdots\Big). \tag{14.50}$$

When collecting the terms of $\delta^{(0)}_{\epsilon_1}S^{(3)}$ proportional to $\lambda^{16}\overline{\epsilon}^*_1\gamma^\mu\psi^*_\mu$, the two highlighted terms in (14.50) are the only contributing interactions, giving [321]

$$\delta^{(0)}_{\epsilon_1}S^{(3)} = -i\int d^{10}x\sqrt{g}(\lambda^{16}\overline{\epsilon}^*_1\gamma^\mu\psi^*_\mu)\Big(8f^{(12,-12)}(z) + 6\cdot 144 D_{(11)}f^{(11,-11)}(z)\Big) + \cdots. \tag{14.51}$$

There are no possible $\delta^{(3)}$ variations which, when acting on $S^{(0)}$, would give a contribution proportional to $\lambda^{16}\overline{\epsilon}_1^*\gamma^\mu\psi_\mu^*$ [321]. Hence,

$$\begin{aligned}&\delta^{(0)}_{\epsilon_1} S^{(3)} + \delta^{(3)}_{\epsilon_1} S^{(0)} = 0 \\ \Longrightarrow \quad & D_{(11)} f^{(11,-11)}(z) = -\frac{4}{3\cdot 144} f^{(12,-12)}(z)\,.\end{aligned} \tag{14.52}$$

If we instead collect the terms of $\delta^{(0)}_{\epsilon_2^*} S^{(3)}$ proportional to $\lambda^{16}\overline{\epsilon}_2\lambda^*$ (again with contributions coming only from the two highlighted terms in (14.50)) we get [321]

$$\begin{aligned}\delta^{(0)}_{\epsilon_2^*} S^{(3)} = -2i\int d^{10}x\sqrt{g}(\lambda^{16}\overline{\epsilon}_2\lambda^*)\Big(&\overline{D}_{(12)} f^{(12,-12)}(z) \\ &- 3\cdot 144\cdot \frac{15}{2} f^{(11,-11)}(z)\Big) + \cdots\,.\end{aligned} \tag{14.53}$$

However, in this case we might also get a contribution from $\delta^{(3)}_{\epsilon_2^*} S^{(0)}$ if we assume that the $(\alpha')^3$ variation of λ^* is of the form

$$\delta^{(3)}_{\epsilon_2^*}\lambda^*_a = -\frac{i}{6} h(z)(\lambda^{14})_{cd}(\gamma^{\mu\nu\rho}\gamma^0)_{dc}(\gamma_{\mu\nu\rho}\epsilon_2^*)_a, \tag{14.54}$$

where $h(z)$ is an unkown function of the axio-dilaton. This gives a contribution [321]

$$\delta^{(3)}_{\epsilon_2^*} S^{(0)} = 180i\int d^{10}x\sqrt{g}(\lambda^{16}\overline{\epsilon}_2\lambda^*)h(z) + \cdots\,. \tag{14.55}$$

From the ϵ_2^* variation at order $(\alpha')^3$ we then get

$$\begin{aligned}&\delta^{(0)}_{\epsilon_2^*} S^{(3)} + \delta^{(3)}_{\epsilon_2^*} S^{(0)} = 0 \\ \Longrightarrow \quad & \overline{D}_{(-12)} f^{(12,-12)}(z) - 3\cdot 144\cdot\frac{15}{2} f^{(11,-11)}(z) - 90h(z) = 0.\end{aligned} \tag{14.56}$$

The two Equations (14.52) and (14.56) give us a system of ordinary differential equations for $f^{(12,-12)}$ and $f^{(11,-11)}$ together with the unknown $h(z)$, which we will now determine by requiring that the (α')-corrected supersymmetry algebra (14.48) closes. That is, we require that the commutator of two local supersymmetry transformations acting on some field Ψ gives a sum of local symmetry transformations that are present in the theory up to equations of motion:

$$[\delta_{\epsilon_1}, \delta_{\epsilon_2^*}]\Psi = \delta_{\text{local translation}}\Psi + \delta_{\text{local symmetries}}\Psi + (\text{equations of motion}). \tag{14.57}$$

Computing the commutator acting on λ^* to order $(\alpha')^3$ using our ansatz (14.54) for $\delta^{(3)}$ and comparing with the equations of motion for λ^* from the corrected action (14.38), one finds that [321, 579]

$$-32D_{(11)}h(z) = f^{(12,-12)}(z)\,. \tag{14.58}$$

Inserting this in (14.56) multiplied by $4D_{(11)}$ on the left and using (14.52) to eliminate $D_{(11)}f^{(11,-11)}(z)$, we obtain the following eigenvalue equation for $f^{(12,-12)}(z)$:

$$\begin{aligned} &4D_{(11)}\overline{D}_{(-12)}f^{(12,-12)}(z) \\ &\quad = -4\Big(-3\cdot 144\cdot\frac{15}{2}D_{(11)}f^{(11,-11)}(z) - 90D_{(11)}h(z)\Big) \\ &\quad = -4\left(-3\cdot 144\cdot\frac{15}{2}\Big(-\frac{4}{3\cdot 144}f^{(12,-12)}(z)\Big) - 90\Big(-\frac{1}{32}f^{(12,-12)}(z)\Big)\right) \\ &\quad = \Big(-132 + \frac{3}{4}\Big)f^{(12,-12)}(z). \end{aligned} \tag{14.59}$$

Then, using the relation (14.45) between $f^{(12,-12)}(z)$ and $f^{(0,0)}(z)$ we get that

$$4y^2\partial_z\partial_{\bar{z}}f^{(0,0)}(z) = \Delta_{\mathbb{H}}f^{(0,0)}(z) = \frac{3}{4}f^{(0,0)}(z), \tag{14.60}$$

which is exactly the eigenvalue Equation (14.5) for $\mathcal{E}^{(10)}_{(0,0)}$ with $\Delta_{\mathbb{H}}$ being the Laplace–Beltrami operator (1.4) on the upper half-plane. Note that we can now obtain eigenvalue equations for all coefficient functions $f^{(w,-w)}$ at order $(\alpha')^3$ by using the modular covariant derivatives $D_{(w)}$ and $\overline{D}_{(-w)}$.

With similar arguments requiring invariance of the action and supersymmetry algebra closure at order $(\alpha')^5$ one can find the eigenvalue Equation (14.5) for the $\nabla^4 R^4$ term in ten dimensions as was done in [579]. The eigenvalue Equations (14.1) for the R^4, $\nabla^4 R^4$ and $\nabla^6 R^4$ terms in arbitrary dimensions were given in [318]. In [83, 84, 85], an approach using linearised harmonic superspace for maximal supersymmetry was used to investigate possible supersymmetry invariants in arbitrary dimensions. The resulting equations for the coefficient functions are expressed in terms of tensorial differential operators on moduli space and specialise to the scalar Laplace equations presented here in Equation (14.1). The tensorial equations are more constraining than the scalar ones and are closely related to the associated varieties of small automorphic representations, as discussed in Section 14.1.

15

Further Occurrences of Automorphic Forms in String Theory

In this chapter, we discuss various other ways automorphic forms arise in string theory and also where string theory points to possible generalisations of the concepts of PART ONE. This will be most prominent when looking at $\nabla^6 R^4$ corrections to the four-graviton scattering as predicted by string theory where the standard $\mathcal{Z}$-finiteness condition appears to be violated. String scattering amplitudes are not the only instance where automorphic forms appear in theoretical physics. Other topics discussed in this chapter include modular graph functions, a structure that is closely associated with perturbative string one-loop amplitudes. We also discuss possible reformulations of Eisenstein series in terms of lattice sums. The final two sections of this chapter are devoted to black hole counting functions and generalisations of monstrous moonshine.

This whole chapter can be viewed as starred, but it contains some of the most interesting current research questions.

15.1 $\nabla^6 R^4$-amplitudes and Generalised Automorphic Forms

We consider the low-energy effective action (13.42) of the four-graviton scattering process in string theory with automorphic coefficients in the series of curvature correction terms. The inhomogeneous Laplace Equation (14.1c), written in Chapter 14 for the coupling function $\mathcal{E}^{(D)}_{(0,1)}$ of the $\nabla^6 R^4$ term, reproduced here for convenience, is

$$\left(\Delta - \frac{6(14-D)(D-6)}{D-2}\right)\mathcal{E}^{(D)}_{(0,1)} = -\left(\mathcal{E}^{(D)}_{(0,0)}\right)^2 + 40\zeta(3)\delta_{D,6} + \frac{55}{3}\mathcal{E}^{(5)}_{(0,0)}\delta_{D,5} + \frac{85}{2\pi}\mathcal{E}^{(4)}_{(1,0)}\delta_{D,4}\,. \quad (15.1)$$

Because of the source term, this does not represent a typical $\mathcal{Z}(\mathfrak{g})$-finiteness condition and therefore its coefficient function $\mathcal{E}^{(D)}_{(0,1)}(g)$ cannot be expected to be an automorphic form in the strict sense of Definition 4.13. Its solutions have nevertheless been investigated in detail recently by Green, Miller and Vanhove in [314] (see [323, 83, 84, 171] for earlier and related work). For $D = 10$, an $SL(2,\mathbb{Z})$-invariant solution was found and its Fourier expansion has been studied.

Green, Miller and Vanhove have also succeeded in expressing the solution as a sum over $G(\mathbb{Z})$-orbits similar to the standard form of Langlands–Eisenstein series [314]:

$$\mathcal{E}^{(10)}_{(0,1)}(g) = \sum_{\gamma \in B(\mathbb{Z})\backslash G(\mathbb{Z})} \Phi(\gamma g), \tag{15.2}$$

where $\Phi\colon G \to \mathbb{R}$ is a right $K = SO(2,\mathbb{R})$-invariant function and hence can be interpreted as a function on $B(\mathbb{R})$. It is furthermore invariant under $B(\mathbb{Z})$. However, unlike the case of Eisenstein series, the function Φ is *not* a character on the Borel subgroup $B(\mathbb{R})$ but rather a highly non-trivial function. More precisely, it is given by

$$\Phi(z) = \frac{2}{3}\zeta(3)^2 y^3 + \frac{1}{9}\pi^2\zeta(3)y + \sum_{n\neq 0} c_n(y)e^{2\pi i n x}, \tag{15.3}$$

where the $c_n(y)$ are complicated functions involving Bessel functions and rational functions. The Fourier expansion of such a function is not straightforward due to complications involving Kloosterman sums and nested integrals over Bessel functions. (See Appendix D for some details on Fourier expansions of Poincaré series and Kloosterman sums.)

In string theory a proper framework for $G(\mathbb{Z})$-invariant functions that satisfy differential equations of the type (14.1c) appears to be required. The class of functions extends the notion of an automorphic form as discussed in the preceding chapters. The analysis in [314] points in the direction of a relation to *automorphic distributions* [478, 477, 479, 549, 603].

For $D = 10$, the function $\mathcal{E}^{(10)}_{(0,1)}(g)$ has the following constant terms [314, Eq. (2.25)]:

$$\int_{N(\mathbb{Z})\backslash N(\mathbb{R})} \mathcal{E}^{(10)}_{(0,1)}(ng)dn = \frac{2\zeta(3)^2}{3}y^3 + \frac{4\zeta(2)\zeta(3)}{3}y + \frac{4\zeta(4)}{y} + \frac{4\zeta(6)}{27y^3}$$
$$+ \text{non-polynomial terms in } y. \tag{15.4}$$

Here, we have used the usual coordinates from Section 4.1 on $SL(2,\mathbb{R})$. The non-polynomial terms are of the form $\sum_{n>0} a_n e^{-4\pi n y}/y^2 + \cdots$ and do not have an expansion around weak coupling $y \to \infty$. These terms have an interpretation as instanton/anti-instanton bound states. We see that the structure of the constant terms is quite different from that of Eisenstein series where, according to the Langlands constant term formula of Theorem 8.1, one has a sum of polynomial terms in y only and the number of terms is bound from above by the order of the Weyl group $\mathcal{W}$, which would be $|\mathcal{W}(SL(2,\mathbb{Z}))| = 2$ here.

In terms of string perturbation theory, the four polynomial terms in (15.4) correspond to contributions from string world-sheets of genus $h = 0, 1, 2, 3$. We recognise the genus $h = 0$ contribution from (13.53). The genus $h = 2$ contribution predicted here was recently compared to a first-principles string theory calculation and found to agree [169, 171], where, remarkably, a connection was also found to the so-called *Kawazumi–Zhang invariant* on the moduli space of genus $h = 2$ Riemann surfaces [633, 392]. As a consequence of Equation (14.1c), [171] discovered that the Kawazumi–Zhang invariant must satisfy a simple Laplace eigenvalue equation on the moduli space of Riemann surfaces and [522] has found a theta lift for the Kawazumi–Zhang invariant. The genus $h = 3$ term in (15.4) has been verified directly from a string perturbation calculation in [294] in the pure spinor formalism.

In terms of wavefront sets and automorphic representations it seems natural to associate the $\nabla^6 R^4$ coupling for E_7 to the (special) nilpotent orbits of type $3A_1$ and A_2 [83]; see Figure 14.2. A proper interpretation of these wavefront sets for $SL(2,\mathbb{R})$ is missing since the largest nilpotent orbit is the regular A_1-type orbit. $\nabla^6 R^4$ correction terms have been analysed recently in various dimensions by different methods [46, 317, 316, 318, 171, 48, 521, 85, 81, 526]. From the analysis of [521, 81], a particular solution to the differential equation for the $\nabla^6 R^4$ coefficient function after compactifying on the torus T^d can be written as

$$\frac{2\pi^{6-d}}{9} \sum_{\substack{\omega_1,\omega_2\in\Lambda_d \\ \omega_i\times\omega_j=0}} \int_{\mathbb{R}^3_+} \frac{dL_1 dL_2 dL_3}{(L_1L_2 + L_2L_3 + L_3L_1)^{\frac{8-d}{2}}} \times \left(L_1 + L_2 + L_3 - 5\frac{L_1L_2L_3}{L_1L_2 + L_2L_3 + L_3L_1}\right) e^{-\omega_i M^{ij}\omega_j}, \tag{15.5}$$

where the sum is over a lattice Λ_d in the E_d-representation with highest weight Λ_d and the constraint has an interpretation in terms of $1/2$-BPS constraints on the charges. The matrix M^{ij} in the exponential depends on L_1, L_2 and L_3 as well as the moduli $g \in E_d/K(E_d)$. Similar lattice forms for the lower-order terms in α' are discussed in Remark 15.5.

15.2 Modular Graph Functions

In Chapter 13, we discussed the meaning of the low-energy expansion of string theory, as an expansion in α'. In Section 13.3, the string tree-level contribution to the four-graviton superstring scattering amplitude in ten dimensions was derived explicitly. *Modular graph functions*, which we want to introduce in this section, derive from the analysis of the one-loop contribution to the low-energy expansion. One-loop amplitudes are also discussed in Section 13.4 in connection with theta lifts.

Here, we will be interested in considering not only the four-graviton superstring scattering, but more generally the scattering of N superstrings. Scattering of an arbitrary number of open or closed strings has been studied at tree level, displaying a remarkably clear structure connected to the theory of multiple polylogarithms and zeta values [99, 587, 588, 585, 464, 465, 550, 96, 98, 97, 586, 589, 463], and to some extent at one-loop level hinting at an extension to their elliptic generalisations [462, 94, 95, 632]. Our main reference on modular graph functions, which our presentation follows closely, is the article [170].

The amplitude of four-graviton scattering at one-loop order can be obtained from an object denoted by $\mathcal{B}_4$ by integrating it together with the Narain genus-one partition function [496] over the fundamental domain of the genus-one Riemann surface; see Section 13.4. The quantity $\mathcal{B}_4$ is called the *partial amplitude* at a fixed moduli value and is also known as the *Koba–Nielsen factor*. Generalising from the four-graviton case, let us consider the form of the partial amplitude for an arbitrary number of gravitons N. The partial amplitude is then of the form

$$\mathcal{B}_N(s,\tau) = \prod_{n=1}^{N} \int_\Sigma \frac{d^2 z_n}{\tau_2} \exp\left(\sum_{1\le i<j\le N} s_{ij} G(z_i - z_j|\tau) \right), \tag{15.6}$$

where $\tau = \tau_1 + i\tau_2$ with $\tau_2 > 0$ is the complex structure of the genus-one torus, the s_{ij} are given by the Mandelstam variables $-\alpha'(k_i + k_j)^2/4$, where k_i are null momenta in ten-dimensional Minkowski space (see (13.34)), and z_i are the places of insertion of the vertex operators on the torus.

Defining the coordinate on the torus Σ as $z = u + \tau v$, where u, v are real in $[0, 1]$, the scalar *Green's function* $G(z|\tau)$ on the torus Σ satisfies

$$-\Delta_{z\bar{z}} G = -4\pi\delta^{(2)}(z) + \frac{4\pi}{\tau_2}, \tag{15.7}$$

where $\Delta_{z\bar{z}} = -4\partial_{\bar{z}}\partial_z$ and the integral of $G(z|\tau)$ over the compact world-sheet Σ vanishes. We use $\delta^{(2)}(z) = \tau_2^{-1}\delta(u)\delta(v)$ such that $\int_\Sigma d^2 z\delta^{(2)}(z) = 1$. The Green's function can be explicitly written in the form of the series

$$G(z|\tau) = \sum_{(m,n)\neq(0,0)} \frac{\tau_2}{\pi|m\tau + n|^2} e^{2\pi i(mu - nv)}. \tag{15.8}$$

Remark 15.1 The eigenfunctions $\psi_{m,n}(z)$ of the Laplacian $\Delta_{z\bar{z}}$ on the one-loop torus Σ were given in Section 13.4.1, where we calculated the one-loop partition function of the bosonic string. From these we can readily construct the Green's function as $\frac{4\pi}{\tau_2}\sum'_{(m,n)\in\mathbb{Z}^2}\lambda^{-1}_{m,n}\psi_{m,n}(z)$, where the eigenvalue $\lambda_{m,n}$ was given in (13.66) and the $\frac{4\pi}{\tau_2}$ is a combination of the τ_2^{-1} of the δ-function with the defining normalisation in (15.7), to give formula (15.8) above. An alternative view on the scalar Green's function (15.8) is that it is the Fourier transform (up to volume and normalisation factors) of the momentum space propagator $1/|p|^2$, where $p = m\tau + n$ represents a discrete momentum on the torus. The scalar propagator is related to theta functions through the *(second) Kronecker limit formula*, but we shall not require this connection here.

We stress that, in order to avoid any confusion, the variable $\tau = \tau_1 + i\tau_2$ appearing in the expression (15.8) is the modulus of the torus world-sheet, whereas the variable $z = x + iy$ that we used in previous discussions of ten-dimensional type IIB higher-derivative couplings $\mathcal{E}^{(10)}_{(p,q)}$ was associated with the moduli space (13.32) of inequivalent string theory compactifications. Although the variables z and τ are from the same mathematical space, their physical meaning is very different. An expression similar to the one-loop Green's function already appeared in Section 13.4.1 when we discussed one-loop partition functions.

The partial amplitude (15.6) represents (part of) the integrand for calculating the string scattering amplitude. $\mathcal{B}_N(s,\tau)$ still depends on the world-sheet modulus, and for getting the one-loop scattering amplitude one still needs to integrate over τ; see Sections 13.1.4 and 13.4. A closed expression for the integral is typically not available. As the Mandelstam variables involve α', one can consider performing the low-energy α'-expansion already at the level of the integrand $\mathcal{B}_N(s,\tau)$. This leads to integrals over vertices z_n on the world-sheet that are connected by (multiple) copies of the Green's function (15.8). These integrals can be thought of as world-sheet Feynman diagrams and represent the modular graph functions. They exhibit interesting properties, as we will now discuss in more detail.

A first step is to introduce so-called *elliptic polylogarithms* [628]. These are generalisations of (multiple) polylogarithms and are functions on an *elliptic curve* $\mathbb{C}^\times/(\mathbb{Z}+\tau\mathbb{Z})$. In the special case when the parameters a, b are integers, the elliptic polylogarithms are given by

$$D_{a,b}(q;\zeta) = \frac{(2i\tau_2)^{a+b-1}}{2\pi i}\sum_{(m,n)\neq(0,0)}\frac{e^{2\pi i(nu-mv)}}{(m\tau+n)^a(m\bar{\tau}+n)^b}, \tag{15.9}$$

where the change of variables $q = \exp(2\pi i\tau)$ and $\zeta = \exp(2\pi i z) = \exp(2\pi i(u\tau + v))$ has been used. These functions are single-valued and

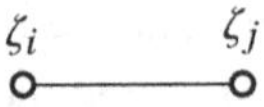

ζ_i ——— ζ_j $= D_{1,1}(q, \zeta_i/\zeta_j) = G(z_i - z_j|\tau)$

(a)

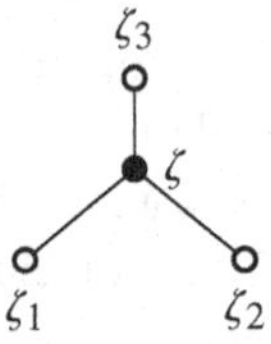

ζ_3, ζ, ζ_1, ζ_2 $= \int_\Sigma \frac{d^2 \log \zeta}{4\pi^2\tau_2} \prod_{i=1}^3 D_{1,1}(q, \zeta/\zeta_i)$

(b)

Figure 15.1 Two simple examples. (a) Green's function. (b) Integrated versus non-integrated vertices.

transform with a modular weight under $SL(2,\mathbb{Z})$ acting on τ. From this definition and the definition (15.8) of the Green's function we note the relation $G(z|\tau) = D_{1,1}(q;\zeta)$ or, with their application in the genus-one amplitude in mind,

$$G(z_i - z_j|\tau) = D_{1,1}(q;\zeta_i/\zeta_j)\,. \tag{15.10}$$

In the special case of $a = b = 1$, the function $D_{1,1}$ and hence also the Green's function is invariant under a modular transformation of τ. Consequently, the partial amplitude $\mathcal{B}_N$ from Equation (15.6) displays the same modular invariance.

The next step is to write $\mathcal{B}_N$ as an expansion in powers of the Mandelstam variables s_{ij}, which is equivalent to an expansion in α'. It is useful to do so, since there exists a diagrammatic way of representing each of the terms in the power series. The various contributing diagrams are *Feynman diagrams* of a conformal scalar field on the genus-one torus, with N vertices. The vertices represent the insertions of the vertex operators on the torus. At a given order w in s_{ij}, all possible graphs with N vertices, connected by up to w Green's functions, contribute to the coefficient. The Green's functions are represented by lines (see Figure 15.1(a)), while a vertex can be represented by either a closed or an open dot, depending on whether the vertex is integrated or not; see Figure 15.1(b). Physically, the integration of vertex ζ_i means that conservation of momentum is imposed on all Green's functions entering that vertex. This is a slight but useful generalisation of the immediate expansion of $\mathcal{B}_N$, where all vertices are integrated over.

When all vertices of a general graph $\Gamma = \Gamma_{N,w}$ are integrated over, we define a function by

$$I_\Gamma(q) \equiv I_{\Gamma_{N,w}}(q) = \prod_{k=1}^{N} \int_\Sigma \frac{d^2 \log \zeta_k}{4\pi^2\tau_2} \prod_{1\le i<j\le N} D_{1,1}(q,\zeta_i/\zeta_j)^{n_{ij}}\,, \tag{15.11}$$

where the n_{ij} are the components of the adjacency matrix defining the graph. When not all vertices are integrated over, the remaining positions ζ remain as arguments of $I_\Gamma(q, \zeta)$. The function $I_\Gamma(q)$ is a single-valued, non-holomorphic modular function. The property of single-valuedness is a consequence of the following non-trivial observation, phrased as a proposition.

Proposition 15.2 (Modular graph functions are single-valued) *The function $I_\Gamma(q)$ is given by a* single-valued elliptic multiple polylogarithm, *denoted by $I_\Gamma(q, \zeta)$, evaluated at the point $\zeta = 1$* [170]. *Single-valued elliptic multiple polylogarithms are defined in terms of products of* single-valued elliptic polylogarithms $D_{1,1}(q, \zeta_i)$, *where the argument ζ_i of one factor is left unintegrated. Consequently $I_\Gamma(q)$ is also single-valued.*

Due to its modular properties and its connection with Feynman graphs, $I_\Gamma(q)$ is called a *modular graph function*. We stress that in the following we shall use 'loop' to refer to properties of the world-sheet Feynman diagrams. From the point of view of string perturbation theory in g_s, all graphs appear in the string one-loop amplitude associated with a toroidal world-sheet.

Following [170], we will now provide two examples that illustrate Proposition 15.2. The first example is given by the chain graph with one loop, which is constructed from the graph of a linear chain. The linear graph is shown in Figure 15.2(a), where the box with label a represents $a - 1$ nodes which are integrated. Its associated function is given by

$$(-4\pi\tau_2)^{1-a} D_{a,a}(q; \zeta_{a+1}/\zeta_1) = \prod_{k=2}^{a} \int_\Sigma \frac{d^2 \log \zeta_k}{4\pi^2\tau_2} \prod_{j=1}^{a} D_{1,1}(q, \zeta_{j+1}/\zeta_j)\,; \tag{15.12}$$

see also [628]. The *one-loop* graph is shown in Figure 15.2(b) and is obtained from the linear graph, by letting ζ_{a+1} and ζ_1 coincide.

Another example is the *two-loop* modular graph function shown in Figure 15.3(b), obtained from 15.3(a) upon identification of ζ_1 and ζ_1'. The expression $C_{a,b,c}(q; \zeta)$ is given by

$$\begin{aligned} &C_{a,b,c}(q; \zeta) \\ &\quad = (-4\pi\tau_2)^{3-a-b-c} \int_\Sigma \frac{d^2 \log \zeta_2}{\tau_2} D_{a,a}(q; \zeta/\zeta_2) D_{b,b}(q; \zeta_2) D_{c,c}(q; \zeta_2), \end{aligned} \tag{15.13}$$

with $\zeta = \zeta_1/\zeta_1'$.

These two examples illustrate Proposition 15.2. It is worth pointing out that the modular graph function of the one-loop graph example is in fact proportional

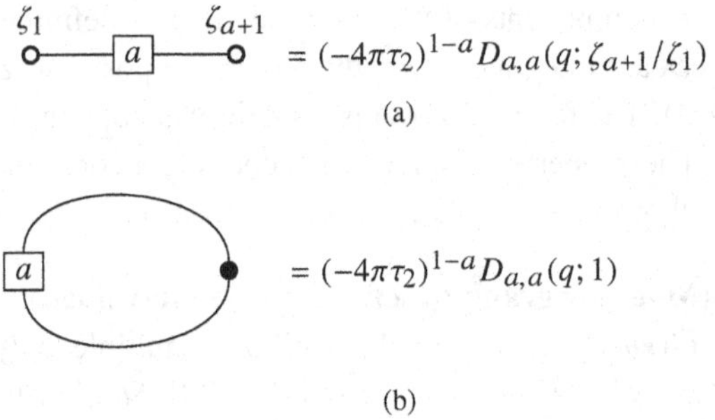

$$\zeta_1 \; \boxed{a} \; \zeta_{a+1} = (-4\pi\tau_2)^{1-a} D_{a,a}(q; \zeta_{a+1}/\zeta_1)$$

(a)

$$\boxed{a} = (-4\pi\tau_2)^{1-a} D_{a,a}(q; 1)$$

(b)

Figure 15.2 Linear graph and one-loop graph. (a) Linear chain graph. (b) One-loop graph.

to a non-holomorphic Eisenstein series

$$\mathbf{E}_a(\tau) = \sum_{(m,n)\neq(0,0)} \frac{\tau_2^a}{\pi^a |m\tau + n|^{2a}}, \tag{15.14}$$

where $a \in \mathbb{Z}_{\geq 1}$. Compared to the standard definition employed for example in (1.12), we see that the modular one-loop graph function satisfies

$$\mathbf{E}_a(\tau) = \frac{2\zeta(2a)}{\pi^a} E(\chi_a, \tau). \tag{15.15}$$

Finally, in [170], an interesting although yet unproven conjecture concerning the modular graph functions and their associated single-valued elliptic multiple polylogarithms was put forward. We shall briefly sketch the conjecture and its

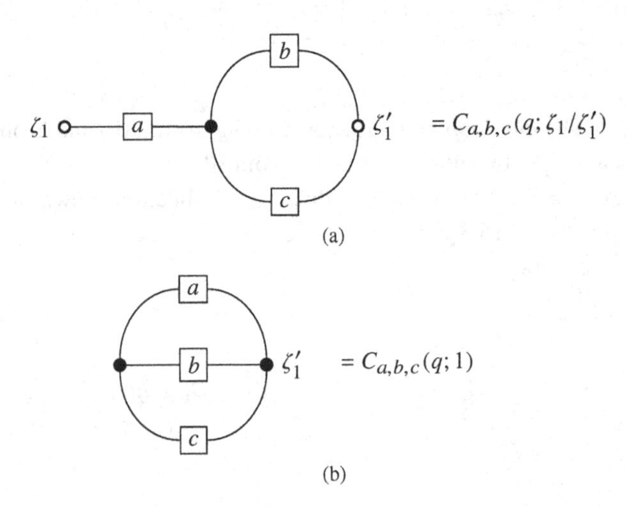

Figure 15.3 Two-loop example. (a) One-loop graph with tail. (b) Two-loop modular graph function.

implications for the two examples of modular graph functions that we have just introduced. The conjecture applies to the class of so-called single-component graphs.

Conjecture 15.3 (Modular graph functions and elliptic multiple polylogs) *The single-valued elliptic multiple polylogarithm $I_\Gamma(q;\zeta)$ associated with a single-component graph $\Gamma = \Gamma_{N+1,w}$, with w Green's functions, $N+1$ vertices and $L = w-N$ loops, is a linear combination of elliptic multiple polylogarithms of depth at most L and weight at most $w+N-1$ [170].*

Reasoning for the validity of the conjecture has been provided in [170] for a certain subclass of the single-component graphs. From Conjecture 15.3 follow a number of corollaries regarding the form of the zeroth Fourier mode $I^0_\Gamma(\tau_2)$ of a modular graph function defined by

$$I^0_\Gamma(\tau_2) = \int_0^1 d\tau_1 I_\Gamma(q), \tag{15.16}$$

assuming that Conjecture 15.3 holds true for a general single-component graph $\Gamma = \Gamma_{N,w}$. As done in [170], we shall also state these corollaries and illustrate them for the examples of modular graph functions provided above.

Corollary 15.4 (On the zeroth Fourier mode of a modular graph function)

(1) The constant Fourier mode $I^0_\Gamma(\tau_2)$ has an expansion for large τ_2 consisting of a Laurent polynomial with a term of highest degree τ_2^w and a term of lowest degree τ_2^{1-w} with exponentially suppressed corrections of the order $O(e^{-2\pi\tau_2})$.

(2) The coefficient of the $(\pi\tau_2)^w$ term in the Laurent polynomial is a rational number.

(3) The coefficients of the $(\pi\tau_2)^k$ terms in the Laurent polynomial are single-valued (multiple) zeta values for $1-w \le k \le w-1$.

It should be noted that single-valued (multiple) zeta values (from string tree level) are defined as values of single-valued polylogarithms (not to be confused with the elliptic polylogarithms (15.9) from above). In particular, single-valued zeta values ζ_{sv} have the property that

$$\zeta_{\text{sv}}(2n) = 0, \qquad \zeta_{\text{sv}}(2n+1) = 2\zeta(2n+1), \tag{15.17}$$

i.e., only odd zeta values are non-vanishing [98]. The first example for which we will illustrate the above corollaries is the one-loop modular graph function 15.2(b), given by the non-holomorphic Eisenstein series (15.14). From (15.15) we conclude that the zeroth Fourier mode of that function is given by

$$\int_0^1 d\tau_1 \mathbf{E}_a(q) = (-1)^{a-1}\frac{B_{2a}}{(2a)!}(4y)^a + \frac{4(2a-4)!}{(a-2)!(a-1)!}\zeta(2a-1)(4y)^{1-a}, \tag{15.18}$$

where $y = \pi\tau_2$ and B_{2a} are the even *Bernoulli numbers*. We note that in order to make a connection with the general expression for the constant term of the non-holomorphic Eisenstein series (see (1.17)) evaluated at values of $s \in \mathbb{N}$, the relation

$$\zeta(2n) = \frac{(-1)^{n+1}B_{2n}(2\pi)^{2n}}{2(2n)!} \tag{15.19}$$

is useful. The points of Corollary 15.4 are easily seen to be confirmed by the form of the Fourier mode.

Modular graph functions are often studied by considering the modular invariant differential equations that they satisfy. Turning to the case of the two-loop modular graph functions $C_{a,b,c}$, the form of the Laurent polynomial of the corresponding zeroth Fourier mode may be deduced from the inhomogeneous Laplace eigenvalue equations satisfied by these functions. Let us note first that the number of Green's functions, $w = a + b + c$, is also referred to as the weight of the modular graph function. One considers a linear combination $\mathfrak{C}_{w,s}$ of $C_{a,b,c}$'s with rational coefficients. From [172] it is known that such a linear combination of weight w and eigenvalue s satisfies an inhomogeneous eigenvalue equation which is of the schematic form

$$(\Delta - s(s-1))\,\mathfrak{C}_{w,s} \in \mathbb{Q}\mathbf{E}_w + \sum_{\substack{w_1,w_2\geq 2\\ w_1+w_2=w}} \mathbb{Q}\mathbf{E}_{w_1}\mathbf{E}_{w_2}\,. \tag{15.20}$$

From our previous discussion it is clear that the constant term on the right-hand side involves power terms y^k, with $1 - w \leq k \leq w$. Furthermore, only the coefficient of the term with y^w is rational, while the coefficients of all other terms are given in terms of single-valued zeta values.

The kernel of the differential operator on the left-hand side contains terms with powers y^s and y^{1-s}. It is then clear from the inhomogeneous form of the differential equation that only the term y^w in the Laurent polynomial of $\mathfrak{C}_{w,s}$ has a rational coefficient, while all other terms are single-valued zeta values. The eigenvalues s take values $s = 1, 2, \ldots, w-2$.

Corollary 15.4 and the above reasoning are confirmed by explicitly calculating the zeroth Fourier mode for a few examples of $C_{a,b,c}$. Note that, since no general expression for $C_{a,b,c}$ exists, the Laurent polynomials have to be computed case-by-case. The Laurent polynomials for various cases have been given in [172, 632] and we content ourselves here with providing just two examples:

$$\int_0^1 d\tau_1 C_{2,1,1}(q) = \frac{2y^4}{14175} + \frac{\zeta(3)y}{45} + \frac{5\zeta(5)}{12y} - \frac{\zeta(3)^2}{4y^2} + \frac{9\zeta(7)}{16y^3} + O(e^{-2y}), \tag{15.21}$$

$$\begin{aligned}\int_0^1 d\tau_1 C_{3,1,1}(q) = {}& \frac{2y^5}{155925} + \frac{2\zeta(3)y^2}{945} - \frac{\zeta(5)}{180} + \frac{7\zeta(7)}{16y^2} \\ & - \frac{\zeta(3)\zeta(5)}{2y^3} + \frac{43\zeta(9)}{64y^4} + O(e^{-2y}).\end{aligned} \tag{15.22}$$

A certain family of modular graph functions with three loops and their associated Laplace eigenvalue equations were studied in [50, 411]. In other recent work, some theorems regarding the type of allowed coefficients (rational/multiple zeta values) of the zeroth Fourier mode of two-loop modular graph functions were established [168].

The interest in the expansion of modular graph functions comes both from the theory of (elliptic) zeta values and from string theory when one attempts to perform the integral over the world-sheet modulus τ, for instance using *Rankin–Selberg techniques* [117, 19, 20, 170, 172]. To date, this integral has only been carried out for a few low-weight cases [170, 172, 49, 50].

15.3 Automorphic Functions and Lattice Sums

As discussed in the introduction to this book, the non-holomorphic Eisenstein series $E(\chi_s, z)$ of $SL(2,\mathbb{R})$ (see (1.12)) can be equivalently written in terms of a sum over an integral lattice:

$$E(\chi_s, z) = \sum_{\gamma\in B(\mathbb{Z})\backslash SL(2,\mathbb{Z})} \chi_s(\gamma\cdot z) = \frac{1}{\zeta(2s)} \sum_{\substack{(c,d)\in\mathbb{Z}^2 \\ (c,d)\neq(0,0)}} \frac{y^2}{|cz+d|^{2s}}. \tag{15.23}$$

In physics, this *lattice sum* for the Eisenstein series can sometimes be interpreted as the sum over the lattice of all possible charges that define the U-duality group (see Chapter 13). The sum over the cosets $B(\mathbb{Z})\backslash G(\mathbb{Z})$, on the other hand, can be interpreted as the contribution from a single U-duality orbit, if $G(\mathbb{Z})$ is the U-duality group of Table 13.2 in Chapter 13. From the latter point of view, each function discussed in (14.2) simply represents the U-duality orbit of the perturbative tree-level scattering amplitude, whereas the function in (15.2) is the U-duality orbit of a finite number of perturbative terms and an infinite number of non-perturbative terms.

Having a representation of an automorphic function as a lattice sum can be physically intuitive and it certainly opens up the possibility of employing

Poisson resummation for performing the Fourier expansion of the function, as is done in Appendix A for $SL(2, \mathbb{R})$.

Lattice sums for more general groups G were considered by Obers and Pioline in [503]. They write the group element $g \in G(\mathbb{R})$ in some linear finite-dimensional representation $\mathcal{R}$. In the same representation, a lattice $\Lambda_{\mathcal{R}}$ is embedded that is preserved by the action of $G(\mathbb{Z})$. This can be constructed for example by starting from the highest weight vector in the representation $\mathcal{R}$. One can form a scalar invariant by considering the object

$$||g^{-1}\omega||^2, \tag{15.24}$$

where $\omega \in \Lambda_{\mathcal{R}}$ and the norm is computed using the $K(\mathbb{R})$-invariant inner product on $\mathcal{R}$. In the example (15.23) above, this is realised by working in the two-dimensional representation, letting $\omega = \begin{pmatrix} -d \\ c \end{pmatrix} \in \mathbb{Z}^2$ and using the Euclidean norm. Then

$$\sum_{\substack{(c,d)\in\mathbb{Z}^2 \\ (c,d)\neq(0,0)}} \frac{y^2}{|cz+d|^{2s}} = \sum_{0\neq\omega\in\Lambda_{\mathcal{R}}} ||g^{-1}\omega||^{-2s}. \tag{15.25}$$

The quantity (15.24) is well-defined by construction on G/K for any $\omega \in \Lambda_{\mathcal{R}}$, and one can form a $G(\mathbb{Z})$-invariant function very generally by letting

$$\tilde{\mathcal{E}}_{\mathcal{R},s}(g) = \sum_{0\neq\omega\in\Lambda_{\mathcal{R}}} ||g^{-1}\omega||^{-2s}. \tag{15.26}$$

For $\mathrm{Re}(s)$ large enough, the sum converges absolutely and this function is K-finite, of moderate growth and $G(\mathbb{Z})$-invariant. Moreover, it is directly amenable to Poisson resummation on the lattice $\Lambda_{\mathcal{R}}$, and this has been exploited widely to obtain results about the constant terms and also partly the non-constant terms of $\tilde{\mathcal{E}}_{\mathcal{R},s}(g)$ [438, 437, 331, 40].

For some groups $G(\mathbb{R})$ and some representations $\mathcal{R}$ it can happen that the function $\tilde{\mathcal{E}}_{\mathcal{R},s}(g)$ is proportional to a (maximal parabolic) Eisenstein series as defined in Sections 4.5 and 5.7, and as is the case in the $SL(2, \mathbb{R})$ example in (15.23) above. However, as was already emphasised in [503], the function $\tilde{\mathcal{E}}_{\mathcal{R},s}(g)$ will in general *not* be an eigenfunction of the ring of invariant differential operators, i.e., it will not be $\mathcal{Z}(\mathfrak{g})$-finite and hence not a proper automorphic form.

The failure to be automorphic can be remedied by restricting the lattice sum over $\mathcal{R}$ to an appropriate $G(\mathbb{Z})$-invariant subset. Such a subset can be found for example by considering the symmetric tensor product $\mathcal{R} \otimes \mathcal{R}$ and then projecting onto the largest invariant subspace in this product [503]. The symmetric tensor product arises because (15.24) is computing a symmetric quantity in the $\mathcal{R}$-valued ω. An automorphic form is then given by

$$\mathcal{E}_{\mathcal{R},s}(g) = \sum_{0\neq\omega\in\Lambda_{\mathcal{R}}} \delta(\omega\otimes\omega)||g^{-1}\omega||^{-2s}, \tag{15.27}$$

where $\delta(\omega\otimes\omega)$ projects on the invariant subspace in $\mathcal{R}\otimes\mathcal{R}$ defined above. In physical applications, this projection has the interpretation of implementing certain conditions that are called *BPS conditions* and correspond to considering contributions only from a subset of all instantonic states. The presence of the projection $\delta(\omega\otimes\omega)$ in the sum often makes the direct application of Poisson resummation impossible and renders the Fourier expansion much more difficult. Examples where the full Fourier expansion of a constrained sum was carried out can be found in [42, 43]. Functions of the form (15.26) or (15.27) are sometimes referred to as *(constrained) Epstein series*.

Another way of turning $\tilde{\mathcal{E}}_{\mathcal{R},s}(g)$ into an automorphic form is by restricting to a single $G(\mathbb{Z})$-orbit within $\Lambda_{\mathcal{R}}$, and this leads back to Langlands' definition.

Remark 15.5 A specific variant of (15.27) was recently found when studying loop amplitudes in *exceptional field theory* [81]. Building on earlier work in [425, 84], it was shown for $d \leq 7$ that the maximal parabolic Eisenstein series on the exceptional group E_d induced from a character on the parabolic subgroup P_d can be written as

$$E(2s\Lambda_d - \rho, g) = \frac{1}{2\zeta(2s)} \sum_{\substack{\omega\in\Lambda_d\\ \omega\times\omega=0}} ||g^{-1}\omega||^{-2s}. \tag{15.28}$$

The lattice Λ_d on the right-hand side is the integer lattice in the E_d representation with highest weight Λ_d preserved by $E_d(\mathbb{Z})$ and the condition $\omega\times\omega = 0$ is a shorthand for the 1/2-BPS condition on the charge ω. This formula is completely in line with (15.27), even though the maximal parabolic Eisenstein series is non-BPS for generic values of s. This expression arises from a one-loop calculation in exceptional field theory and ω is then a BPS charge that circulates in the loop.

One can also perform a similar two-loop calculation and finds the following more unusual formula with two BPS charges circulating in the loops:

$$\begin{aligned}\frac{2}{3}\pi^{5-d} &\sum_{\substack{\omega_1,\omega_2\in\Lambda_d\\ \omega_i\times\omega_j=0,\ \epsilon^{ij}\omega_i\omega_j\neq 0}} \int \frac{d^3\Omega}{(\det\Omega)^{\frac{7-d}{2}}} e^{-\Omega^{ij}B(\omega_i,\omega_j)} \\ &= 8\pi\xi(d-5)\xi(d-4)E\left((d-4)\Lambda_{d-1} - \rho, g\right).\end{aligned} \tag{15.29}$$

Here, the integral is over the space of real symmetric (2×2) matrices and $B(\omega_i,\omega_j)$ is the inner product between the charges, which is schematically of the form $\omega_i^T(gg^T)^{-1}\omega_j$ with $g\in E_d$. Using a functional relation this function (when convergent) equals the coefficient function of the ∇^4R^4 derivative correction [81].

15.4 Black Hole Counting and Automorphic Representations

In this section, we will discuss relations between automorphic forms and quantum properties of black holes in string theory. In particular, we will show that the quantum entropy of a special class of black holes (BPS black holes) can be reproduced by a microscopic counting of microstates, using automorphic forms. We begin by discussing some generalities; after that we focus on some specific string theories of varying degrees of supersymmetry.

15.4.1 Index Versus Degeneracy

As explained in Chapter 13, string theory compactified on a compact six-dimensional manifold X gives rise to an effective supersymmetric gravitational theory in four space-time dimensions. The number of preserved supersymmetries, usually denoted by $\mathcal{N}$, depends on the properties of X. Previously we have discussed the case of $X = T^6$ (corresponding to $\mathcal{N} = 8$) but there are other interesting and relevant manifolds, such as $X = K3 \times T^2$ (yielding $\mathcal{N} = 4$) and when X is a Calabi–Yau threefold ($\mathcal{N} = 2$). The resulting theories have black hole solutions carrying electro-magnetic charges γ taking values in a lattice Γ.

A natural question is whether we can *count* all states with fixed charge γ; this is the total *degeneracy* $d(\gamma)$. This is an important observable since according to *Boltzmann's equation the quantum entropy* is given by

$$S_{\text{quantum}} = \log d(\gamma) \tag{15.30}$$

and it is related to the classical entropy $S(\gamma)$ of a black hole with charge γ by

$$S_{\text{quantum}} = S(\gamma) + \cdots, \tag{15.31}$$

where the ellipsis corresponds to quantum corrections that encode deviations from Einstein's general relativity. In a quantum theory of gravity one would like to be able to compute these quantum corrections and deduce the exact relation for which the above is an approximation. According to the the Bekenstein–Hawking *area law* the classical entropy is given by

$$S(\gamma) = \frac{A(\gamma)}{4}, \tag{15.32}$$

where $A(\gamma)$ is the area of the horizon of the black hole. In a quantum theory of gravity this is the classical limit of a *quantum entropy function* of the form [556, 155, 293]

$$S_{\text{quantum}}(\gamma) = \frac{A(\gamma)}{4} + c_0 \log A(\gamma) + c_1 A(\gamma)^{-1} + c_2 A(\gamma)^{-2} + \cdots O(e^{-A(\gamma)}), \tag{15.33}$$

where the last term represents the non-perturbative contributions. Hence, by (15.30), if we could find the total degeneracy $d(\gamma) = \exp(S_{\text{quantum}}(\gamma))$ we could in principle compute all the coefficients $c_0, c_1, c_2, \ldots$ and find the quantum entropy.

At present this is inaccessible since the absolute degeneracy is notoriously difficult to compute. However, in string theory it is possible to make progress when restricting to BPS states (see Remark 13.19 and Section 14.3).

To this end we consider the *BPS index* $\Omega(\gamma)$ which is a function

$$\Omega\colon \Gamma \to \mathbb{Z} \tag{15.34}$$

that counts the (signed) degeneracies of a certain class of black holes (called BPS black holes) with charge vector $\gamma \in \Gamma$. The simplest example is the *Witten index* $\text{Tr}(-1)^F$ (F being the fermion number), which counts the supersymmetric ground states in supersymmetric quantum mechanics [618] (see also Section 13.6.3).

More generally, for quantum field theories with extended supersymmetry the relevant object is the so-called *helicity supertrace*, whose precise form depends on the particular theory and degree of supersymmetry (see [406], App. E for details). Although we are really interested in the degeneracy $d(\gamma)$, under certain circumstances, namely for so-called *extremal black holes*, the index and degeneracy are actually equal [558]:

$$d(\gamma) = \Omega(\gamma). \tag{15.35}$$

Since the index is often computable, this represents a very interesting testing ground for quantum gravity effects. The first successful microscopic derivation of black hole entropy in string theory was carried out by Strominger and Vafa [590].

The BPS index $\Omega(\gamma)$ holds the key to many interesting connections between string theory and mathematics. The charge lattice Γ is the cohomology lattice $H^*(X, \mathbb{Z})$ of the compact manifold X, and the index $\Omega(\gamma)$ can roughly be thought of as counting certain submanifolds of X in the cohomology class $[\gamma] \subset H^*(X, \mathbb{Z})$. It is therefore naturally related to the enumerative geometry of X. Remarkably, the index also provides a link to automorphic forms. The BPS index $\Omega(\gamma)$ is also U-duality invariant and in many cases it turns out to be related to Fourier coefficients of modular forms, as we shall illustrate below.

15.4.2 $\mathcal{N} = 8$ Supersymmetry

Let us first consider the case when $X = T^6$, the real six-dimensional torus. This leads to $\mathcal{N} = 8$ supersymmetry in four dimensions with electric and magnetic charges taking values in a lattice $\Gamma \cong \mathbb{Z}^{56}$, and is the case discussed in Chapter 14. As reviewed in Chapter 13 and Section 13.1.6, this theory exhibits a classical $E_7(\mathbb{R})$ symmetry which is broken in the quantum theory to $E_7(\mathbb{Z}) = \{g \in E_7(\mathbb{R}) \mid g\Gamma = \Gamma\}$. This implies that the weighted degeneracy $\Omega(\gamma)$ of BPS black holes of charge $\gamma \in \Gamma$ must be invariant under $E_7(\mathbb{Z})$.

However, not all black holes have charges supported on the entire lattice Γ. For example, the 1/2-BPS black holes preserve half of the supersymmetries of the theory and can only have charges supported on a 28-dimensional (Lagrangian) subspace $C_{1/2} \subset \Gamma$. Similarly, 1/4-BPS black holes have support on a 45-dimensional subspace $C_{1/4} \subset \Gamma$.

Now denote by $\Omega_{1/A}(\gamma)$ the index counting $1/A$-BPS black holes ($A = 2, 4$). Due to the $E_7(\mathbb{Z})$-invariance it is natural to suspect that the index arises as the Fourier coefficient of an automorphic form, constrained so that $\Omega_{1/A}(\gamma)$ is non-vanishing only when $\gamma \in C_A$.

As discussed in Section 14.3.2, BPS black holes in D space-time dimensions are point particles that compactify on a circle to instantons in $D-1$ dimensions. Therefore, in order to study the BPS black holes in $D = 4$, we should study automorphic forms on E_8, which is the relevant symmetry group in $D = 3$, and analyse the Fourier expansion in the unipotent associated with decompactification. This limit was called limit (c) in Section 14.2.2.

Let us consider the $A = 2$ case for illustration. It turns out that all the expected properties are fulfilled by an automorphic form φ_{min} on $E_8(\mathbb{Z})\backslash E_8(\mathbb{R})$ attached to the minimal representation π_{min} of $E_8(\mathbb{R})$ [518, 336, 519]. This representation has Gelfand–Kirillov dimension 29 and can thus be realised as the unitary action of E_8 on a space of functions of $29 = 28 + 1$ variables. These variables should be thought of as associated with the $(56 + 1)$-dimensional Heisenberg unipotent radical $U_{\text{Heis}} \subset E_8$ that arises in the decompactification limit. Using the *Stone–von Neumann theorem*, half of the 56-dimensional representation of E_7 and the central direction can be taken as coordinates of a function space on which U_{Heis} acts, and this action can be extended to all of E_8 [334]. The centre $Z = [U_{\text{Heis}}, U_{\text{Heis}}]$ is one-dimensional and, according to the non-abelian Fourier expansion of Section 6.2.3, there are unitary characters $\psi_Z\colon Z(\mathbb{Z})\backslash Z(\mathbb{R}) \to U(1)$, trivial on the abelianisation $Z\backslash U_{\text{Heis}}$, that can be parametrised by a single integer k. On the other hand the electric and magnetic charges $\gamma = (q, p) \in \Gamma \cong \mathbb{Z}^{56}$ parametrise characters $\psi\colon U_{\text{Heis}}(\mathbb{Z})\backslash U_{\text{Heis}}(\mathbb{R}) \to U(1)$, trivial on $Z(\mathbb{R})$. Consider the constant term of φ_{min} with respect to Z:

$$\varphi_{Z,\text{min}} = \int_{Z(\mathbb{Z})\backslash Z(\mathbb{R})} \varphi_{\text{min}}(zg)dz\,. \tag{15.36}$$

This is a function $\varphi_{Z,\text{min}}\colon E_7(\mathbb{R}) \to \mathbb{C}$ invariant under $E_7(\mathbb{Z})$. By taking the constant term with respect to Z we have effectively removed the dependence on the variable k. The function $\varphi_{Z,\min}$ can be expanded further (see Section 6.2.3):

$$\varphi_{Z,\min}(g) = \varphi_{U_{\text{Heis}}} + \sum_{\psi \neq 1} F_{\psi}(\varphi_{\min}, g), \tag{15.37}$$

where

$$F_{\psi}(\varphi_{\min}, g) = \int_{U_{\text{Heis}}(\mathbb{Z})\backslash U_{\text{Heis}}(\mathbb{R})} \varphi(ug)\overline{\psi(u)}du\,. \tag{15.38}$$

In general, such a Fourier coefficient might not be *Eulerian* (i.e., have an Euler product factorisation); however, as we explained in Section 10.4.4, for the minimal representation this turns out to be the case:

$$F_{\psi}(\varphi_{\min}, g) = F_{\psi_\infty}(\varphi_{\min}, g_\infty) \times \prod_{p<\infty} F_{\psi_p}(\varphi_{\min}, g_p)\,. \tag{15.39}$$

It was shown in [393, 394] that these Fourier coefficients indeed have support on the Lagrangian subspace $C_{1/2}$. We can now state the relation to $1/2$-BPS black holes:

Conjecture 15.6 ($1/2$-BPS index) *The index $\Omega_{1/2}(q,p)$ counting charged $1/2$-BPS black holes in four-dimensional, $\mathcal{N} = 8$ supergravity is given by* [518, 336, 519]

$$\Omega_{1/2}(q,p) = \prod_{p<\infty} F_{\psi_p}(\varphi_{\min}, 1), \tag{15.40}$$

where $F_{\psi_p}(\varphi_{\min}, 1)$ is the p-adic spherical vector in the minimal representation $\pi_{\min}$ of E_8 (obtained in [394]*) and the electro-magnetic charges (q,p) parametrise the character ψ_p. Note that the p in ψ_p is unrelated to the magnetic charge p.*

Similarly, for the $1/4$-BPS black holes we have:

Conjecture 15.7 ($1/4$-BPS index) *The index $\Omega_{1/4}(q,p)$ counting charged $1/4$-BPS black holes is given by* [518, 336, 519]

$$\Omega_{1/4}(q,p) = \prod_{p<\infty} F_{\psi_p}(\varphi_{\text{ntm}}, 1), \tag{15.41}$$

where φ_{ntm} is an automorphic form in the next-to-minimal representation π_{ntm} of E_8.

Remark 15.8 For a certain class of $1/8$-BPS black holes in $\mathcal{N} = 8$ theories there exists a proposal for the degeneracies. More precisely, for black holes which are dual to bound states of D0-D2-D4-D6 branes with charges $(p^\Lambda, q_\Lambda) = (p^0, p^a, q_a, q_0)$, $a = 1, \ldots, 27$, the degeneracies are conjecturally given by [467, 572]

$$\Omega_{1/8}(q,p) = \hat{c}(I_4(q,p))\,, \tag{15.42}$$

where I_4 is the quartic E_7-invariant,

$$I_4(q,p) = q^2p^2 - (p \cdot q)^2, \tag{15.43}$$

and $\hat{c}$ are the Fourier coefficients of the *Jacobi form*

$$-\frac{\theta_1^2(\tau,z)}{\eta(\tau)^6} = \sum_{n,\ell} \hat{c}(4n-\ell^2)q^n y^\ell, \qquad q := e^{2\pi i\tau},\ z := e^{2\pi i z}. \tag{15.44}$$

(We recall that Jacobi forms and their Fourier expansions were discussed in Section 4.2.3.) The classical black hole entropy is in this case given by

$$S(q,p) = \pi\sqrt{I_4(q,p)}, \tag{15.45}$$

and by computing the asymptotics of the Fourier coefficients $\hat{c}$ for large charges one finds

$$\hat{c}(I_4(q,p)) \sim e^{\pi\sqrt{I_4(q,p)}} + \cdots, \tag{15.46}$$

in complete agreement with Boltzmann's formula (15.30).

It is natural to expect that the index $\Omega_{1/8}(q,p)$ can be extracted as a Fourier coefficient of the automorphic form $\mathcal{E}^{D=3}_{(0,1)}$ in Section 15.1, which controls the $\nabla^6 R^4$ couplings in type II string theory on T^7. As discussed above this is not an automorphic form in the strict sense, but should nevertheless in some appropriate sense be attached to the next-to-next-to minimal nilpotent orbit of E_8.

Remark 15.9 For the 1/8-BPS black holes, the index behaves like

$$\Omega_{1/8}(\gamma) \sim e^{\pi\sqrt{I_4(\gamma)}} \tag{15.47}$$

in the large-charge limit $\gamma \to k\gamma$ with $k \to \infty$. This is indeed reproduced by (15.42) since the Fourier coefficients of the weak Jacobi form (15.44) exhibit precisely the desired growth. It is therefore a prediction from string theory that the same type of exponential growth should be observed in the Fourier–Whittaker coefficients of the automorphic form associated to the next-to-next-to minimal nilpotent orbit(s) of E_8, relevant for the $\nabla^6 R^4$ couplings. See Section 15.1 for a discussion of these objects.

15.4.3 $\mathcal{N} = 4$ Supersymmetry

Let us now take $X = K3\times T^2$, where the first factor is a compact $K3$ surface [552, 30]. This yields $\mathcal{N} = 4$ supersymmetry in four space-time dimensions, which admits 1/2- and 1/4-BPS black holes with electro-magnetic charges $\gamma = (q,p)$ taking values in $\Gamma = H^*(X,\mathbb{Z})$. The quantum symmetry of this theory is $SL(2,\mathbb{Z}) \times SO(6,22;\mathbb{Z})$ [551, 554, 620] and we are interestested in finding invariant BPS indices $\Omega_{1/2}(q,p)$ and $\Omega_{1/4}(q,p)$. Mathematically, these indices are counting special Lagrangian submanifolds of X in the class $[\gamma] \subset H^*(X, \mathbb{Z})$. As we shall see, the counting works quite differently in this case compared to the $\mathcal{N} = 8$ theory considered above.

The 1/2-BPS states are purely electric $\gamma = (q, 0)$ or purely magnetic $\gamma = (0, p)$ and they are known to be exactly counted by [156, 153]

$$\Omega_{1/2}(q, 0) = d(q^2/2), \tag{15.48}$$

where $d(n)$ are the Fourier coefficients of the *discriminant modular form* (see Section 4.2.1) with $\tau \in \mathbb{H}$:

$$\Delta(\tau) = \frac{1}{\eta(\tau)^{24}} = \sum_{n=-1}^{\infty} d(n) e^{2\pi i n\tau}, \tag{15.49}$$

which is a cusp form of weight 12 for $SL(2, \mathbb{Z})$. Note that the index is automatically invariant under $SO(6, 22; \mathbb{Z})$ since it only depends on the invariant square $q^2 = q \cdot q$ of the charge vector q. On the other hand the $SL(2, \mathbb{Z})$ part of the quantum symmetry is broken since (q, p) transforms in a doublet. The 1/4-BPS index is preserved under the full symmetry group which means that we need to consider both electric and magnetic charges: $\Omega_{1/4}(q, p)$. Moreover, in order to preserve $SO(6, 22; \mathbb{Z})$ it can only depend on the invariant combinations $q^2, p^2, p \cdot q$ [555, 38]. The answer is that $\Omega_{1/4}(q, p)$ is the Fourier coefficient of the inverse of the unique weight 10 cusp form for $Sp(4; \mathbb{Z})$, known as the *Igusa cusp form* and usually denoted by Φ_{10}. The statement is [180, 573]

$$\Omega_{1/4}(q, p) = D(q^2/2, p^2/2, p \cdot q), \tag{15.50}$$

where the numbers $D(m, n, \ell)$ are extracted from the expansion of the inverse of the Igusa cusp form:

$$\frac{1}{\Phi_{10}(\rho, \sigma, \tau)} = \sum_{m,n,\ell} D(m, n, \ell) e^{2\pi i m\sigma} e^{2\pi i n\tau} e^{2\pi i \ell\rho}, \tag{15.51}$$

where (ρ, σ, τ) are complex variables parametrising the *Siegel upper half-plane* $Sp(4; \mathbb{R})/U(2)$ that we encountered in Remark 4.10. We note that, as it stands, the relation (15.50) is not really precise, as it depends on a choice of contour in extracting the Fourier coefficients $D(m, n, \ell)$ [135, 136]. The relation to the entropy of the black hole is given by

$$\log \Omega_{1/4}(q, p) \sim \pi\sqrt{q^2 p^2 - (q \cdot p)^2} + \cdots, \tag{15.52}$$

where the leading term is the classical entropy.

This can be generalised to orbifolds of $X = K3 \times T^2$ by some discrete subgroup $\mathbb{Z}_N$, in which case the counting is given by *Siegel modular forms* for paramodular groups (see [555] for a review and further references).

Remark 15.10 It was recently pointed out in [79] that the relation (15.50) can be understood from a similar perspective as the connection between BPS indices in $D = 4$ and BPS-saturated couplings in $D = 3$, discussed for $\mathcal{N} = 8$ above. The index $\Omega_{1/4}(q, p)$ counts dyons in four-dimensional $\mathcal{N} = 4$ string theory, where the moduli space is

$$SL(2,\mathbb{Z})\backslash SL(2,\mathbb{R})/SO(2) \times SO(6,22;\mathbb{Z})\backslash SO(6,22)/(SO(6) \times SO(22)). \tag{15.53}$$

From this point of view it is not clear why a Siegel modular form would be the relevant object to encode the black hole degeneracies. It was, however, shown in [79] that $\Omega_{1/4}(q, p)$ arises as a Fourier coefficient in the decompactification limit of an automorphic form on

$$SO(8,24;\mathbb{Z})\backslash SO(8,24)/(SO(8) \times SO(24)), \tag{15.54}$$

which is the moduli space on the same theory compactified on a circle to $D = 3$. Physically, this automorphic form controls a BPS-saturated higher-derivative coupling of the form $D^2(D\Phi)^4$, where Φ is a scalar field.

A generalisation to orbifolds by $\mathbb{Z}_N$ symmetries of $K3 \times T^2$ (so-called *CHL models*) was also obtained in [78], verifying previous results on the existence of Fricke-type symmetries in such models [513].

15.4.4 $\mathcal{N} = 2$ Supersymmetry

Finally, we consider the case when X is a *Calabi–Yau threefold* [121, 591, 357]. This gives rise to $\mathcal{N} = 2$ supersymmetry in four space-time dimensions. The lattice Γ of electric and magnetic charges is either $H^{\text{even}}(X,\mathbb{Z})$ or $H^3(X,\mathbb{Z})$ depending on whether we consider type IIA or type IIB string theory. According to Kontsevich's *homological mirror symmetry conjecture* [419] (see also Section 16.3.1) a BPS black hole with charge $\gamma \in H^{\text{even}}(X,\mathbb{Z})$ can be viewed as a (semi-)stable object in the (bounded) *derived category of coherent sheaves* $\mathrm{D^bCoh}(X)$, while black holes with charges $\gamma \in H^3(X,\mathbb{X})$ correspond to (semi-)stable objects (special Lagrangians) in the *derived Fukaya category* $\mathrm{D^bFuk}(X)$. The BPS index $\Omega\colon \Gamma \to \mathbb{Z}$ should then be identified with the *generalised Donaldson–Thomas invariants* of X [420, 237, 384]. String theory predicts that there should be an action of a discrete Lie group $G(\mathbb{Z})$ on the categories $\mathrm{D^bCoh}(X)$ and $\mathrm{D^bFuk}(X)$, which is unexpected from a purely mathematical viewpoint. In general it is not known what the group $G(\mathbb{Z})$ should be but it must at least contain the 'S-duality' group $SL(2,\mathbb{Z})$ (see, e.g., [154, 166]). For certain choices of X there are, however, precise conjectures regarding the nature of $G(\mathbb{Z})$.

Let X be a rigid Calabi–Yau threefold, i.e., $h_{2,1}(X) = 0$. In this case the classical moduli space is $SU(2,1)/U(2)$. We also assume X to be of CM-type, i.e., admitting *complex multiplication* by the ring of algebraic integers O_d in the quadratic number field $\mathbb{Q}(\sqrt{-d})$. In this case the intermediate Jacobian of X is an elliptic curve:

$$H^3(X,\mathbb{R})/H^3(X,\mathbb{Z}) \cong \mathbb{C}/O_d\,. \tag{15.55}$$

We then have the following conjecture:

Conjecture 15.11 (Picard modular group and rigid Calabi–Yau threefolds) *For type IIB string theory compactified on a rigid Calabi–Yau threefold X of CM-type, the 'U-duality group' $G(\mathbb{Z})$ is the* Picard modular group [42, 43]

$$PU(2,1;O_d) := U(2,1) \cap PGL(3,O_d). \tag{15.56}$$

In particular, this group acts on the charge lattice $H^3(X,\mathbb{Z})$ and consequently on $\mathrm{D^bFuk(X)}$.

If correct, this suggests that the BPS index $\Omega(\gamma)$ should arise as the Fourier coefficient of an automorphic form on $SU(2,1)$, in a similar vein as for $\mathcal{N} = 8$ and $\mathcal{N} = 4$ supergravity discussed above. (We discuss some aspects of automorphic forms on $SU(2,1)$ in Chapter 18.) Constraints from supersymmetry further imply that there exists a class of 1/2-BPS states that have support only on charges γ such that $Q_4(\gamma) \geq 0$, where $Q(\gamma)$ is a quartic polynomial in the charge vector γ. In other words, the BPS index is constrained such that

$$\Omega(\gamma) = \begin{cases} n \neq 0, & Q_4(\gamma) \geq 0 \\ 0, & Q_4(\gamma) < 0\,. \end{cases} \tag{15.57}$$

It turns out that this constraint is precisely satisfied for Fourier coefficients of automorphic forms attached to the quaternionic discrete series of Lie groups G in their quaternionic real form [327, 328, 608]. This leads to the following conjecture:

Conjecture 15.12 (DT-invariants of rigid Calabi–Yau threefolds) *The generalised Donaldson–Thomas invariants $\Omega(\gamma)$ of a CM-type rigid Calabi–Yau threefold X are captured by the Fourier coefficients of an automorphic form attached to the quaternionic discrete series of $SU(2,1)$* [42, 43, 511].

Another interesting case is when the hypermultiplet moduli space is of the form $G_2(\mathbb{R})/SO(4)$. This is realised as the hypermultiplet moduli space of type IIB string theory on a Calabi–Yau threefold X with $h_{1,1}(X) = 1$, or IIA string theory on X with $h_{2,1}(X) = 1$ [64]. One then expects that the U-duality group is an arithmetic subgroup $G_2(\mathbb{Z})$ of the split real form $G_2(\mathbb{R})$. Automorphic forms on $G_2(\mathbb{R})$ associated with the quaternionic discrete series have been analysed in detail by Gan, Gross and Savin [244], and one has the following conjecture:

Conjecture 15.13 (DT-invariants and automorphic forms on G_2) *There exist Calabi–Yau threefolds X with $h_{1,1} = 1$ whose Donaldson–Thomas invariants $\Omega(\gamma)$ are captured by automorphic forms attached to the quaternionic discrete series of G_2, as analysed by Gan–Gross–Savin* [525, 511].

Remark 15.14 We note that 1/2-BPS states in $\mathcal{N} = 2$ correspond to 1/8-BPS states in $\mathcal{N} = 8$ theories. As in Remark 15.9, we therefore expect that for large values of the charges the index should reproduce the macroscopic entropy of the black hole, which is known to be given by $S(\gamma) = \pi\sqrt{Q_4(\gamma)}$. Translated into mathematics this implies that the Fourier coefficient should have an asymptotic growth given by

$$\Omega(\gamma) \sim e^{\pi\sqrt{Q_4(\gamma)}} \quad \text{as} \quad \gamma \to \infty. \tag{15.58}$$

This gives rise to the following interesting puzzle. In general, Hecke eigenforms always give rise to Fourier coefficients that grow polynomially, and hence the growth in (15.58) does not seem to be compatible with the fact that the automorphic forms of Gan–Gross–Savin are indeed Hecke eigenforms. One possible resolution to this problem is that one should not consider honest automorphic forms in the quaternionic discrete series but rather some analogue of mock modular forms for G_2, a possibility proposed by Stephen D. Miller. This might also be consistent with the fact that the BPS index $\Omega(\gamma)$ jumps discontinuously at certain co-dimension one walls in parameter space (known as *wall-crossing*) and this phenomenon is closely related to mock modularity (see, e.g., [469, 468, 157]). Another possibility is that the relevant automorphic object is of the same type as the $\nabla^6 R^4$ coupling, discussed in Remark 15.9. These, in particular, fail to be $\mathcal{Z}$-finite and so are not automorphic forms in the strict sense (see also Section 15.1).

15.5 Moonshine

In mathematics and theoretical physics, the term 'moonshine' refers to webs of connections between different areas, including number theory, group theory and string theory. The most famous example is known as *monstrous moonshine*, and its impact can hardly be overstated; it led to numerous developments in mathematics which crucially relied on results in string theory. In many respects, moonshine is similar to the Langlands program in the sense that it connects many a priori distinct areas under a single framework, as illustrated in Figure 15.4. Moreover, automorphic forms play a pivotal rôle in moonshine. However, as we will see, the automorphic forms that appear in moonshine are of a different kind and the relation to automorphic representations is not clear. We shall comment on this in a little more detail at the end of this section.

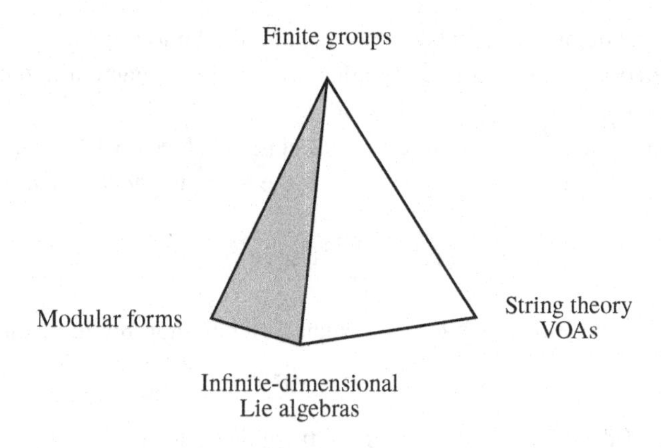

Figure 15.4 Pictorial overview of moonshine.

Remark 15.15 In order to stay closer to the original literature and to avoid confusion when talking about Jacobi forms, we change some of our standard notation in this section. The typical element of the upper half-plane $\mathbb{H}$ will be called τ (not z!) and typically the dependence on it will be holomorphic. This implies that $SL(2,\mathbb{Z})$-invariant modular forms $f(\tau)$ will admit a *q-expansion* with $q = e^{2\pi i\tau}$. The variable z will be used for the elliptic variable when we discuss Jacobi forms $\phi(\tau, z)$. These will then have a Fourier expansion in two variables which will be $q = e^{2\pi i\tau}$ and $y = e^{2\pi iz}$.

15.5.1 Monstrous Moonshine

Monstrous moonshine pertains to a relation between the largest of the 26 finite sporadic simple groups known as the *monster* $\mathbb{M}$, and a certain class of modular functions. The starting point of monstrous moonshine was McKay and Thompson's observation that the first few Fourier coefficients in the q-expansion of the unique (up to the constant term; see Section 4.2.1) meromorphic $SL(2,\mathbb{Z})$-invariant $J(\tau)$-function

$$J(\tau) = \sum_{n=-1}^{\infty} c(n)q^n = q^{-1} + 196884q + 21493760q^2 + \cdots \tag{15.59}$$

correspond to sums of dimensions of irreducible representations of $\mathbb{M}$:

$$196\,884 = 1 + 196\,883\,, \qquad 21\,493\,760 = 1 + 196\,883 + 1\,2196\,876\,. \tag{15.60}$$

This suggests the existence of an infinite-dimensional graded representation $V = \bigoplus_{n=-1}^{\infty} V_n$ of $\mathbb{M}$, such that each graded piece V_n decomposes into direct sums of irreducible representations of $\mathbb{M}$, and the Fourier coefficients can be reinterpreted as $c(n) = \dim V_n$; in other words, the J-function is the *graded dimension* of V.

One can then consider a more general class of objects where the $c(n)$ are replaced by characters $\text{Tr}_{V_n}(g)$; this yields the *McKay–Thompson series*,

$$T_g(\tau) = \sum_{n=-1}^{\infty} \text{Tr}_{V_n}(g) q^n. \tag{15.61}$$

That is, to each element $g \in \mathbb{M}$ one associates a holomorphic function

$$T_g : \mathbb{H} \longrightarrow \mathbb{C}. \tag{15.62}$$

By virtue of the cyclic properties of the trace these are really *class functions*, meaning that they only depend on the conjugacy class of g:

$$T_g(\tau) = T_{xgx^{-1}}(\tau), \qquad \forall x \in \mathbb{M}. \tag{15.63}$$

The J-function corresponds to the identity element $e \in \mathbb{M}$ for which $\text{Tr}_{V_n}(e) = \dim V_n$. Conway and Norton [148] proposed that all McKay–Thompson series are special types of modular forms known as *Hauptmodule* for special *genus zero* subgroups Γ_g of $SL(2,\mathbb{R})$. A discrete subgroup $\Gamma \subset SL(2,\mathbb{R})$ is genus zero if the closure of the quotient $\Gamma\backslash\mathbb{H}$ has the topology of a sphere when viewed as a Riemann surface. A Hauptmodul for Γ then generates the field of rational functions on $\Gamma\backslash\mathbb{H}$. The conjecture by Conway and Norton is indeed true for the J-function, which is the Hauptmodul for $\Gamma_e = SL(2,\mathbb{Z})$. Frenkel–Lepowsky–Meurman [225] constructed a *(conformal) vertex algebra* $V^\natural$, with $\text{Aut}(V^\natural) = \mathbb{M}$, central charge $c = 24$, and with graded dimension (L_0 = Virasoro generator)

$$J(\tau) = \text{Tr}_{V^\natural}(q^{L_0 - 1}), \tag{15.64}$$

providing an explanation for monstrous moonshine.

Remark 15.16 We recall that a *vertex algebra* is a vector space with some additional structure, such as a product obeying certain axioms; a *conformal* vertex algebra also carries an action of the *Virasoro algebra*, whose zeroth generator L_0 gives rise to the grading. Conformal vertex algebras are also known as *chiral algebras* or *holomorphic vertex operator algebras* (VOAs).

Remark 15.17 The monster vertex algebra $V^\natural$ has a natural interpretation in physics: it corresponds to a holomorphic conformal field theory describing bosonic string theory on the orbifold $(\mathbb{R}^{24}/\Lambda_{\text{Leech}})/\mathbb{Z}_2$, where Λ_{Leech} is the *Leech lattice*. From this point of view the modularity of the McKay–Thompson series becomes completely natural: it originates from the symmetry of the path integral over maps from the world-sheet torus, with modified boundary conditions along the spatial direction, into space-time. One missing piece is the *genus zero property*, namely the fact that the invariance groups Γ_g of the McKay–Thompson series are all genus zero subgroups of $SL(2, \mathbb{R})$, meaning that $\overline{\Gamma_g \backslash \mathbb{H}}$ has the topology of a sphere. This cannot be explained by conformal field theory methods alone since Γ_g contains elements, so-called *Fricke elements*, which act as $\tau \to -1/(N\tau)$, $N = \text{order}(g)$, and these lie outside of $SL(2, \mathbb{Z})$. Recently, a new type of string theory compactification was proposed [506], in which these Fricke symmetries appear naturally as space-time dualities, thus providing a complete physical explanation of monstrous moonshine. This construction also relates monstrous moonshine to BPS state counting, since in the proposal of [506], the McKay–Thompson series are reinterpreted as supersymmetric indices counting BPS states.

The complete Conway–Norton conjecture was subsequently proven by Borcherds [67], who made use of $V^\natural$ combined with an earlier 'no-ghost' theorem in string theory [287]. In the course of his proof, Borcherds also introduced new ingredients, notably the *monster Lie algebra* $\mathfrak{m}$, carrying an action of $\mathbb{M}$ by Lie algebra automorphisms. The construction of $\mathfrak{m}$ was inspired by physics: it arises as the so-called (degree one) BRST cohomology of a certain extension of $V^\natural$. The monster Lie algebra is an example of a *Borcherds–Kac–Moody algebra*, which is an infinite-dimensional Lie algebra constructed from a Cartan matrix by generators and relations. The main difference from ordinary Kac–Moody algebras is that Borcherds–Kac–Moody algebras allow for *imaginary simple roots* (i.e., with non-positive norm).

A few years after the Conway–Norton conjectures were put forward, Norton proposed a vast generalisation that he called *generalised moonshine*. This pertains to a wider class of modular functions $T_{g,h}(\tau)$ associated to *commuting pairs* of elements $g, h \in \mathbb{M}$. He gave a precise list of properties that these functions should satisfy; in particular that for the identity element $g = e$ they reproduce the McKay–Thompson series.

Norton's generalised moonshine conjecture has recently been proven in a series of papers by Carnahan [125, 126, 127], in particular by constructing a class of Borcherds–Kac–Moody-algebras $\mathfrak{m}_g$, labelled by classes $[g] \subset \mathbb{M}$, admitting an action of the centraliser $C_{\mathbb{M}}(g)$ of g in $\mathbb{M}$. Physically, the generalised moonshine functions can conjecturally be understood as [183]

$$T_{g,h}(\tau) = \text{Tr}_{V_g^\natural}(hq^{L_0-1}), \tag{15.65}$$

where $V_g^\natural$ denotes the so-called g-twisted sector in the orbifold of $V^\natural$ by the element $g \in \mathbb{M}$. For this formula to make sense, h must be a symmetry of $V_g^\natural$, which has automorphism group the centraliser $C_{\mathbb{M}}(g)$, and so indeed h and g must commute.

15.5.2 The Singular Theta Correspondence

We now want to focus on a particular aspect of Borcherds' proof of monstrous moonshine. The starting point is the following famous product formula for the modular invariant J-function:

$$J(p) - J(q) = p^{-1} \prod_{m>0, n\in\mathbb{Z}} (1 - p^m q^n)^{c(mn)}, \tag{15.66}$$

where $q = e^{2\pi i\tau}$, $p = e^{2\pi i\sigma}$ and $\tau, \sigma \in \mathbb{H}$. The exponent on the right-hand side involves the Fourier coefficients $c(n)$ of the J-function itself (15.59). Borcherds' key insight was to interpret this formula as the denominator formula (see (3.27)) of the monster Lie algebra $\mathfrak{m}$ from Section 15.5.1 above. The denominator formula of such a Borcherds–Kac–Moody algebra has the same structure as that of an ordinary Lie algebra (3.27), but with a correction factor due to the imaginary simple roots:

$$\sum_{w\in\mathcal{W}} \epsilon(w) w(e^\rho \Sigma) = e^\rho \prod_{\alpha>0} (1 - e^{-\alpha})^{\mathrm{mult}(\alpha)} . \tag{15.67}$$

The factor Σ contains a certain combination of the imaginary simple roots, including information about their multiplicities. The monster Lie algebra has only a single real simple root and therefore the Weyl group (which is defined with respect to reflections in the real simple roots only) is $\mathcal{W} = \{\pm 1\}$. The left-hand side of the denominator formula (15.67) for $\mathfrak{m}$ then has only two terms, and one can show that this gives precisely the difference of J-functions on the left-hand side of (15.66). Similarly, the product over positive roots in the denominator formula is identified with the product over $m > 0, n \in \mathbb{Z}$ in (15.66). The infinite product then corresponds to a product over all the positive roots of $\mathfrak{m}$.

The formula may also be viewed as a lift from a modular form on $\mathbb{H}$, namely the J-function, to an automorphic form on $SO(2,2)/(SO(2)\times SO(2)) \cong \mathbb{H}\times\mathbb{H}$. It is an example of a more general process known as a *Borcherds lift*, or *multiplicative lift*. This, in turn, can be understood in the context of the *theta correspondence* that was discussed in Chapter 12.

Consider the following integral:

$$\Theta_J(\sigma, \rho) = \int_{SL(2,\mathbb{Z})\backslash\mathbb{H}} \Theta_{\Gamma^{2,2}}(\tau; \sigma, \rho) J(\tau) \frac{d\tau_1 d\tau_2}{\tau_2^2}, \tag{15.68}$$

where $\Theta_{\Gamma^{2,2}}$ is the weight 0 Siegel theta series for the four-dimensional indefinite lattice $\Gamma^{2,2}$ of signature $(2,2)$, and $(\sigma, \rho) \in SO(2,2;\mathbb{R})/(SO(2)\times SO(2)) \cong \mathbb{H}\times\mathbb{H}$; see Section 13.4.4. The measure $d\tau_1 d\tau_2/\tau_2^2$ is invariant so the integrand is indeed $SL(2,\mathbb{Z})$-invariant for $\tau = \tau_1 + i\tau_2$. This integral has exactly the form of the theta integrals discussed in Chapter 12. However, there is a crucial difference. The seed function is now the modular invariant J-function $J(\tau) = q^{-1} + \cdots$, which has a pole at the cusp $\tau \to i\infty$. Hence, the integral is divergent and does not fit the requirements for the theta correspondence. It can however be evaluated after proper regularisation, as was first done by Harvey and Moore in [347]. The result is

$$\int_{SL(2,\mathbb{Z})\backslash\mathbb{H}} \Theta_{\Gamma^{2,2}}(\tau; g) J(\tau) \frac{d\tau_1 d\tau_2}{\tau_2^2} = -2 \log \left| e^{-2\pi i\sigma} \prod_{\substack{m>0 \\ n\in\mathbb{Z}}} (1 - e^{2\pi i(m\sigma+n\rho)})^{c(mn)} \right| . \tag{15.69}$$

By comparing with (15.66) we see that the result of the integral involves the product side of the denominator formula for the J-function. By replacing the seed function with the McKay–Thompson series $T_g(\tau)$ one can obtain similar product formulas for elements $g \in \mathbb{M}$. Borcherds later generalised the Harvey–Moore construction into the so-called *singular theta correspondence* [68], which deals with arbitrary integrals of the form

$$\Theta_\Lambda(F) := \int_{SL(2,\mathbb{Z})\backslash\mathbb{H}} \frac{d\tau_1 d\tau_2}{\tau_2^2} (\overline{\theta}_\Lambda(\tau), F(\tau)) . \tag{15.70}$$

Here Λ is a lattice, $\tau = \tau_1 + i\tau_2$, F is a vector-valued modular form of weight k (valued in the group ring $\mathbb{C}[\Lambda^\vee/\Lambda]$), $\overline{\theta}_\Lambda$ is a vector-valued Siegel theta series of weight $-k$ and $(\,,\,)$ denotes the scalar product in the vector space $\Lambda^\vee/\Lambda$. Let e_γ, $\gamma \in \Lambda^\vee/\Lambda$, be a basis for the group algebra $\mathbb{C}[\Lambda^\vee/\Lambda]$ so that

$$e_\gamma e_{\gamma'} = e_{\gamma+\gamma'} , \tag{15.71}$$

with inner product

$$(e_\gamma, e_{\gamma'}) = \delta_{\gamma+\gamma',0} . \tag{15.72}$$

In what follows we shall restrict to the case of weight $k = 0$ for simplicity. A vector-valued modular function F for a congruence subgroup $\Gamma \subset SL(2,\mathbb{Z})$ is then defined as

$$F(\tau) = \sum_{\gamma\in\Lambda^\vee/\Lambda} F_\gamma(\tau) e_\gamma , \tag{15.73}$$

where the components $F_\gamma(\tau)$ are modular functions for Γ, transforming in the metaplectic representation of (the double cover of) $SL(2,\mathbb{Z})$ on $\mathbb{C}[\Lambda^\vee/\Lambda]$:

$$\begin{aligned} F_\gamma(\tau+1) &= e^{2\pi i Q(\gamma)} F_\gamma(\tau) , \\ F_\gamma(-1/\tau) &= \frac{1}{\sqrt{|\Lambda^\vee/\Lambda|}} \sum_{\gamma'\in\Lambda^\vee/\Lambda} e^{-2\pi i(\gamma,\gamma')} F_{\gamma'}(\tau) , \end{aligned} \tag{15.74}$$

where $Q\colon \Lambda \to \mathbb{Z}$ is the quadratic form on $\Lambda^\vee/\Lambda$ and we have also defined the even bilinear form

$$(\gamma, \gamma') := Q(\gamma+\gamma') - Q(\gamma) - Q(\gamma') . \tag{15.75}$$

The vector-valued theta series is defined similarly:

$$\Theta_\Lambda(\tau) = \sum_{\gamma\in\Lambda^\vee/\Lambda} \theta_{\Lambda+\gamma} e_\gamma\,, \tag{15.76}$$

where $\theta_{\Lambda+\gamma}$ is the ordinary Siegel theta series for the shifted lattice $\Lambda + \gamma$.

In the course of his proof of generalised monstrous moonshine, Carnahan has computed the integral form (15.70) for Θ_Λ for the following choice of data. Let $\Lambda = \Pi^{1,1}(N) \times \Pi^{1,1}$, where $\Pi^{1,1}$ is the standard two-dimensional Lorentzian lattice, and $\Pi^{1,1}(N) \subset \Pi^{1,1}$ is the sublattice where the quadratic form is rescaled by $N > 0$. Thus, as a free abelian group of rank four we can view the lattice as

$$\Lambda \cong \mathbb{Z}\times N\mathbb{Z}\times\mathbb{Z}\times\mathbb{Z}, \tag{15.77}$$

and for any vector $u = (a, b, c, d) \in \Lambda$ the quadratic form is

$$Q(u) = Q(a, b, c, d) = ab + cd\,. \tag{15.78}$$

The quotient $\Lambda^\vee/\Lambda$ is then identified with $\mathbb{Z}/N\mathbb{Z}\times\mathbb{Z}/N\mathbb{Z}$, and for any vector $v = (a, b) \in \Lambda^\vee/\Lambda$ the quadratic form descends to $Q(v) = Q(a, b) = ab/N$. One can construct a vector-valued modular form F_g by starting from the generalised moonshine functions T_{g^i,g^j} in the following way. Let $\gamma = (i, k) \in \Lambda^\vee/\Lambda$ and construct the components $F_\gamma = F_{i,k}$ by discrete Fourier transform:

$$F_\gamma(\tau) = F_{i,k}(\tau) = \frac{1}{N}\sum_{j\in\mathbb{Z}/N\mathbb{Z}} e(-jk/N)T_{g^i,g^j}(\tau). \tag{15.79}$$

One can show that under modular transformations this satisfies

$$\begin{aligned}
F_\gamma(\tau+1) &= F_{i,k}(\tau+1) = e(ik/N)F_{i,k}(z) = e(Q(\gamma))F_\gamma(\tau),\\
F_\gamma(-1/\tau) &= F_{i,k}(-1/\tau) = \frac{1}{N}\sum_{j,l\in\mathbb{Z}/N\mathbb{Z}} e\left((-jk-il)/N\right) F_{j,l}(\tau)\\
&= \frac{1}{N}\sum_{\delta\in M^\vee/M} e(-(\gamma,\delta))F_\delta(\tau),
\end{aligned} \tag{15.80}$$

where $\delta = (j, l)$ so that $(\gamma, \delta) = Q(i+j, k+l) - Q(i, k) - Q(j, l) = (il + jk)/N$. Thus

$$F_g(\tau) = \sum_{i,k\in\mathbb{Z}/N\mathbb{Z}} F_{i,k}(\tau)e_{i,k} \tag{15.81}$$

is a vector-valued modular form of weight 0. Given this choice of data we have the following theorem due to Carnahan:

Theorem 15.18 (Singular theta lift of McKay–Thompson series) *The singular theta lift (15.70) of the vector-valued modular function $F_g(z)$ defined in (15.79) is given by* [126]

$$\Theta_\Lambda(F_g) = -4\log|T_{1,g}(\rho) - T_{g,1}(\sigma)|, \tag{15.82}$$

where $T_{1,g}$ and $T_{g,1}$ are specialisations of the generalised moonshine functions $T_{g,h}$.

Remark 15.19 (Product formula for McKay–Thompson series) After evaluating the integral $\Theta_\Lambda(F_g)$ one also obtains an infinite product formula generalising the one for the J-function:

$$T_{1,g}(\rho) - T_{g,1}(\sigma) = e^{-2\pi i\rho} \prod_{m>0,\, n\in\frac{1}{N}\mathbb{Z}} (1 - e^{2\pi i(m\rho+n\sigma)})^{c_g(m,n)}, \tag{15.83}$$

where $c_g(m,n)$ is the q^{mn}th coefficient of the vector-valued function $F_{m,Nn}$. For each element $g \in \mathbb{M}$ this is the denominator formula for a Borcherds–Kac–Moody algebra $\mathfrak{m}_g$ [125, 126].

15.5.3 Mathieu Moonshine

In 2010, a completely new moonshine phenomenon was conjectured by physicists Eguchi, Ooguri and Tachikawa [200], which involves another of the finite sporadic simple groups, namely the *Mathieu group* M_{24}. *Mathieu moonshine* was subsequently proved by Gannon [248].

In Mathieu moonshine, the rôle of the J-function is played by an object known as the *elliptic genus of K*3 *surfaces*, denoted ϕ_{K3}. This is a topological invariant of a family of two-dimensional complex surfaces, and it is also an example of a modular form, or, more precisely, a *Jacobi form* of the type introduced in Section 4.2.3. We recall that a Jacobi form is a two-variable generalisation of a modular form. The key to revealing the presence of M_{24} is to notice that ϕ_{K3} affords the following decomposition [197]:

$$\phi_{\mathrm{K3}}(\tau, z) = \frac{\theta_1(\tau,z)^2}{\eta(\tau)^3}\big(24\mu(\tau,z) + H(\tau)\big), \quad \tau \in \mathbb{H},\ z \in \mathbb{C}, \tag{15.84}$$

where θ_1 is the Jacobi theta function, η is the Dedekind eta function and μ is an Appell–Lerch sum. The key object for us is the function $H(\tau)$, which turns out to be a so-called *mock modular form* of weight 1/2. See [157] for details on mock modular forms and the decomposition (15.84) of ϕ_{K3}. The function $H(\tau)$ can be represented explicitly in terms of a Fourier expansion:

$$H(\tau) = q^{-1/8}(-2 + 2\sum_{n=1}^{\infty} A_n q^n), \qquad q = e^{2\pi i\tau}. \tag{15.85}$$

One notices now that the first few coefficients A_n are given by

$$A_1 = 45, \qquad A_2 = 231, \qquad A_3 = 770, \qquad A_4 = 2277, \ldots, \tag{15.86}$$

where $45, 231, 770, \ldots$ are all dimensions of representations of M_{24}. This observation thus shares many similarities with monstrous moonshine, although the modular object is of a different type. Based on these observations, Eguchi, Ooguri and Tachikawa conjectured that there exists an infinite-dimensional graded module $\mathcal{H} = \bigotimes_n \mathcal{H}_n$ whose automorphism group is M_{24} and such that $\dim \mathcal{H}_n = A_n$. Subsequent work [131, 234, 233, 198, 199, 235, 248, 236, 512] provided further evidence for this. In particular, the analogues of the McKay–Thompson series $\phi_g(\tau, z)$, $g \in M_{24}$ (called *twining genera*) were all computed and shown to decompose into irreducible representations of M_{24}. Gannon showed [248] that all the twining genera $\phi_g(\tau, z)$ are true characters of M_{24}. In particular, beyond the first negative term, their Fourier coefficients are all positive integers. This proves the Mathieu moonshine conjecture.

Remark 15.20 Mathieu moonshine was subsequently shown by Cheng, Duncan and Harvey to be a special case of a larger class of new moonshine phenomena called *umbral moonshine* [133]. There are 24 instances of umbral moonshine, classified by finite groups associated with the even self-dual Niemeier lattices [134]. The Mathieu moonshine case corresponds to the lattice A_1^{24}. Gannon's proof of Mathieu moonshine has been extended by Duncan, Griffin and Ono [192] to give a full proof of the umbral moonshine conjecture.

However, despite this amazing progress, the main question remains unanswered: *why does the elliptic genus carry information about the Mathieu group?* This is analogous to McKay's original question that initiated monstrous moonshine. There are several indications that the resolution to this problem can be found in the context of BPS black holes in string theory. Consider string theory compactified on a manifold X which is a product of a K3 surface S and a torus T^2. Recall from Section 15.4.3 that this yields a theory in four dimensions with $\mathcal{N} = 4$ supersymmetry. In (type IIA) string theory on $X = S \times T^2$ the charge lattice is $\Gamma = \Gamma^{6,22} \oplus \Gamma^{6,22}$ and generic BPS states have charge vectors $\gamma = (q, p)$, with electro-magnetic charges $q, p \in \Gamma^{6,22}$. A generic BPS state is a *dyon* and preserves 1/4 of the full $\mathcal{N} = 4$ supersymmetry. As mentioned in Section 15.4.3, the number of such states is counted (with signs) by the *sixth helicity supertrace*, which is an index defined as follows:

$$\Omega_{1/4}(q, p) = \mathrm{Tr}_{\mathcal{H}_{\mathrm{BPS}}(q,p)}\left((-1)^{2J}(2J)^6\right), \tag{15.87}$$

where J is the helicity. This index is invariant under $SL(2, \mathbb{Z}) \times SO(6, 22; \mathbb{Z})$, where the electro-magnetic charge vector $\gamma = (q, p)$ transforms as a doublet under $SL(2, \mathbb{Z})$, and can only enter through the $SO(6, 22; \mathbb{Z})$-invariant combinations q^2, p^2 and $q \cdot p$. Remarkably, the generating function that counts these states is known, and is given by the reciprocal of a Siegel modular form, known as the *Igusa cusp form* Φ_{10} [180, 573]:

$$\frac{1}{\Phi_{10}(\sigma,\tau,z)} = \sum_{m,n,\ell} D(m,n,\ell) e^{2\pi i m\sigma} e^{2\pi i n\tau} e^{2\pi i \ell z}, \tag{15.88}$$

with the identification

$$\Omega_{1/4}(q,p) = D\left(\frac{q^2}{2}, \frac{p^2}{2}, q\cdot p\right). \tag{15.89}$$

The Igusa cusp form also appears in Section 15.4.3 for 1/4-BPS black holes in $\mathcal{N}=4$ string theory.

The Igusa cusp form furthermore arises from a singular theta correspondence. One has the following result due to Kawai [391]:

Theorem 15.21 (Singular theta representation of the Igusa cusp form)

$$\int_{SL(2,\mathbb{Z})\backslash\mathbb{H}} \mathrm{Tr}_{\mathcal{H}_{\mathrm{BPS}}}\left((-1)^{2J}(2J)^6 q^{L_0}\bar{q}^{\bar{L}_0}\right)\frac{dxdy}{y} = \log(Y^{10}|\Phi_{10}(\sigma,\tau,z)|^2), \tag{15.90}$$

where $Y = \det\begin{pmatrix}\sigma & z\\ z & \tau\end{pmatrix}$ *is the determinant of the genus-two period matrix.*

On the left-hand side we use as an exception the notation $q = e^{2\pi i(x+iy)}$ to keep the standard variables on the right-hand side. Evaluating the integral directly yields the product representation

$$\Phi_{10}(\sigma,\tau,z) = e^{2\pi i(\sigma+\tau-z)} \prod_{\substack{n,m\geq 0,\,\ell\in\mathbb{Z}\\ \ell>0 \text{ when } m=n=0}} \left(1-e^{2\pi i(n\sigma+m\tau+\ell z)}\right)^{c(4mn-\ell^2)}, \tag{15.91}$$

where the exponents $c(4mn-\ell^2)$ are the Fourier coefficients of the K3 elliptic genus

$$\phi_{\mathrm{K3}}(\tau,z) = \sum_{m\geq 0,\,\ell\in\mathbb{Z}} c(m,\ell) q^m y^\ell. \tag{15.92}$$

The infinite product (15.91) corresponds to the right-hand side of the denominator formula (15.67) of a Borcherds–Kac–Moody algebra with Cartan matrix (associated with the real simple roots) [324]:

$$\begin{pmatrix} 2 & -2 & -2\\ -2 & 2 & -2\\ -2 & -2 & 2\end{pmatrix}. \tag{15.93}$$

The product form of Φ_{10} corresponds to an automorphic (Borcherds) lift of the K3 elliptic genus, analogously to the Borcherds lift of the J-function which yields the product in (15.66). The full denominator formula is given by the following expression:

$$\sum_{m>0} e^{2\pi i m\tau} T_m \phi_{10,1}(\sigma, z) = e^{2\pi i(\sigma+\tau-z)} \prod_{\substack{n,m\geq 0,\ell\in\mathbb{Z} \\ \ell>0 \text{ when } m=n=0}} \left(1 - e^{2\pi i(n\sigma+m\tau+\ell z)}\right)^{c(4mn-\ell^2)}, \tag{15.94}$$

where $\phi_{10,1}$ is the weak Jacobi form of weight 10, index 1, and T_m is the mth Hecke operator (to be precise, the denominator formula for the BKM with Cartan matrix (15.93) is actually the *square* of (15.94)). See [202] for the precise definition of how the Hecke operator T_m acts on Jacobi forms. Just as for the monster Lie algebra, the left-hand side encodes the structure of the imaginary simple roots, whose multiplicities are in this case captured by the Fourier coefficients of the Jacobi form $\phi_{10,1}$. The left-hand side is known as the *additive automorphic lift* (or *Saito–Kurokawa–Maaß lift* from its inventors), in contrast to the right-hand side which is a multiplicative lift.

Remark 15.22 We have seen that the multiplicative lift can be understood within the singular theta correspondence. It is an interesting open question whether there is some alternative version of the (singular) theta lift that encodes also the additive automorphic lift.

It has been speculated (starting with [347, 348]) that the space of BPS states in $\mathcal{N} = 4$ string theory carries an action of the Mathieu group M_{24}, thereby providing a possible explanation for the Mathieu moonshine phenomenon. Further support for this was given in [131, 512, 513], where a class of functions

$$\Phi_{g,h} : \mathbb{H}^{(2)} \longrightarrow \mathbb{C} \tag{15.95}$$

on the Siegel upper half-plane $\mathbb{H}^{(2)}$ was defined for each commuting pair of elements $g, h \in M_{24}$ (see [512] for the precise definition of $\Phi_{g,h}$). It was proven that these functions are Siegel modular forms with respect to discrete subgroups $\Gamma^{(2)}_{g,h} \subset Sp(4;\mathbb{R})$, and they play a key rôle in generalised Mathieu moonshine [236]. When restricted to $(g, h) = (1, 1)$, where 1 is the identity element of M_{24}, this reproduces the Igusa cusp form $\Phi_{10} = \Phi_{1,1}$. A central observation was that $\Phi_{g,h}$ satisfies the surprising property

$$\Phi_{g,h}(\sigma, \tau, z) = \Phi_{g,h}(\tfrac{\tau}{N}, \sigma, z), \tag{15.96}$$

where N is the order of the element g. What could be the physical interpretation of the functions $\Phi_{g,h}$? In the follow-up paper [513] it was demonstrated that the symmetry (15.96) corresponds precisely to a novel type of electro-magnetic duality in certain orbifold compactifications of heterotic string theory on T^6.

In these theories the most general index, counting h-twisted black hole states in the g-orbifold theory [557], is given by

$$\Omega_{1/4}^{g,h}(q,p) = \mathrm{Tr}_{\mathcal{H}_{\mathrm{BPS},g}(q,p)}((-1)^{2J}(2J)^6 h), \tag{15.97}$$

where the trace is taken over the g-orbifold sector $\mathcal{H}_{\mathrm{BPS},g}(q,p)$. This suggests a very natural interpretation of all the functions $\Phi_{g,h}$. Fourier expanding its reciprocal, one obtains

$$\frac{1}{\Phi_{g,h}(\sigma,\tau,z)} = \sum_{m,n,\ell} D_{g,h}(m,n,\ell) e^{2\pi i m\sigma} e^{2\pi i n\tau} e^{2\pi i \ell z}, \tag{15.98}$$

generalising (15.88). This leads to the following conjecture:

Conjecture 15.23 (Counting of twisted dyons) *For each $g \in M_{24}$ and $h \in C_{M_{24}}(g)$ the twisted black hole states in (generalised) g-orbifold CHL models are counted by the Fourier coefficients of* $1/\Phi_{g,h}$ [512, 513], *i.e.,*

$$\Omega_{1/4}^{g,h}(q,p) = D_{g,h}\left(\frac{q^2}{2}, \frac{p^2}{2}, q\cdot p\right). \tag{15.99}$$

These results certainly indicate that BPS black holes will ultimately play a rôle in explaining Mathieu moonshine.

Remark 15.24 A possible connection between BPS states and Mathieu moonshine was also observed recently in [390, 132] where certain stable pair invariants (Gopakumar–Vafa invariants) of type II string theory on a K3 surface were studied. The authors found a numerical coincidence between the stable invariants and dimensions of M_{24} representations. This also prompted recent ongoing work, aimed at proving (or disproving) the assertion that M_{24} is the automorphism group of the algebra of BPS states of string theory on K3 [492].

15.5.4 Representation-Theoretic Interpretation?

It is natural to ask whether the automorphic forms which appear in monstrous moonshine can be understood in the context of automorphic representations. Put differently, can parts of monstrous moonshine be understood in the context of the Langlands program? These questions have in particular been raised by Borcherds on several occasions. The first thing to note is that the automorphic infinite products that arise in the singular theta correspondence have poles at cusps and are therefore not Hecke eigenforms in general. This means that the Hecke algebra acts freely and Flath's theorem (see Theorem 5.13) does not apply. Hence, if there was an underlying automorphic representation it is most likely not factorisable. Another way to see this is to try to construct L-functions associated with automorphic infinite products. If one naively defines these in the standard way via Dirichlet series formed from their Fourier coefficients, one ends up with divergent series. Borcherds has proposed that instead one might define L-functions via some regularised version of the Mellin transform. However, to the best of our knowledge, this has not yet been realised. We are therefore forced to admit that a satisfactory interpretation of the singular theta correspondence in general, and moonshine in particular, is still an open problem.

PART THREE

ADVANCED TOPICS

16

Connections to the Langlands Program

It is a deeper subject than I appreciated and, I begin to suspect, deeper than anyone yet appreciates.
To see it whole is certainly a daunting, for the moment even impossible, task.

— Robert P. Langlands[1]

This chapter represents the beginning of the final part of this book, in which we survey a variety of topics that are opened up by the ideas developed in the previous chapters. These topics are mostly in the focus of current research and many questions do not have final answers. For this reason, we will be more descriptive, with an emphasis on exposing the central ideas and their interconnections. As we merely scratch the surface of very wide and deep fields, we provide many references for the reader who wishes to delve deeper into a given question.

This first chapter, containing advanced topics, deals with the classical and geometric Langlands program and connections between certain L-functions and elliptic curves through the modularity theorem. Some of Langlands' ideas have a counterpart in physics, which we briefly describe.

16.1 The Classical Langlands Program

Any survey of automorphic forms would be incomplete without at least mentioning some of the key ideas involved in the *classical Langlands program*, the collective name given to the visionary conjectures outlined by Langlands

[1] Langlands, R. P. 2006. A review of Haruzo Hida's book 'p-adic Automorphic Forms on Shimura Varieties'. *Bull. Amer. Math. Soc.*, **44**, 291–308.

in his letter to Weil in 1967 [444], and later expanded upon in the lecture notes 'Problems in the theory of automorphic forms' [446]. To give a complete account of these conjectures goes far beyond the scope of this survey. However, we would like to give a heuristic discussion of some of the ingredients and their implications. This section leans on the discussions in Sections 11.6 to 11.8.

The context of Langlands' letter to Weil was reductive groups G defined over an arbitrary field $\mathbb{F}$ that can be either local (like $\mathbb{Q}_p$) or global (like $\mathbb{Q}$). Let us focus on the global situation. As usual we restrict our treatment to $\mathbb{F} = \mathbb{Q}$, and we let G be a split group over $\mathbb{Q}$, for example $GL(n, \mathbb{Q})$. Recall that being split over $\mathbb{Q}$ means that there exists a maximal torus which is a product of $GL(1, \mathbb{Q})$s. However, Langlands also considered groups G that were *quasi-split*, meaning that they contain a Borel subgroup which is defined over $\mathbb{Q}$. Equivalently, a quasi-split group is split over an *unramified finite extension* $\mathbb{E}/\mathbb{F}$. We recall that a finite extension of a field $\mathbb{F}$ is another field $\mathbb{E}$ that contains $\mathbb{F}$ and which has finite dimension as a vector space over $\mathbb{F}$, so in this case it is a finite-dimensional vector space over $\mathbb{Q}$. 'Unramified' here means that the finite index of the residue field of $\mathbb{F}$ in the residue field of $\mathbb{E}$ is the same as that of $\mathbb{F}$ in $\mathbb{E}$ and implies that the prime ideals behave nicely under the field extension.

The group of automorphisms of the extension $\mathbb{E}$ is called the *Galois group* and denoted by $\text{Gal}(\mathbb{E}/\mathbb{F})$. In this more general context the L-group of $G(\mathbb{Q})$ is really defined as the semi-direct product

$$^LG = \widehat{G}(\mathbb{C}) \rtimes \text{Gal}(\mathbb{E}/\mathbb{F}), \tag{16.1}$$

where the first factor is the complex group that we discussed in Section 11.6, denoted ^LG there since we were in a more restricted setting: in the case when G is split over $\mathbb{F} = \mathbb{Q}$, as for $GL(n, \mathbb{Q})$, the Galois group acts trivially and the L-group becomes a direct product $^LG = \widehat{G}(\mathbb{C}) \times \text{Gal}(\mathbb{E}/\mathbb{Q})$. In this situation one can take the representation $\rho : {}^LG \to GL(n, \mathbb{C})$ that enters in the construction of L-functions $L(\pi, s, \rho)$, defined in (11.139), to have a trivial projection on the second Galois factor in ^LG of (16.1) and only consider the complex group. We therefore recover the description of L-functions in Section 11.8, where we had simply assumed $^LG = \widehat{G}(\mathbb{C})$; see also Remark 11.11.

One of the main parts of Langlands' conjectures is the *principle of functoriality*. To state it, let G and G' be reductive groups over $\mathbb{Q}$. The principle of functoriality asserts that whenever we have a group homomorphism between the associated L-groups

$$\Psi : {}^LG \longrightarrow {}^LG', \tag{16.2}$$

there should be a relation between the associated automorphic forms on $G(\mathbb{Q})\backslash G(\mathbb{A})$ and $G'(\mathbb{Q})\backslash G'(\mathbb{A})$, as follows. Suppose π is an automorphic representation of G associated with a Satake class $[A_\pi]$ in the dual group ^LG.

Functoriality implies that there exists an automorphic representation π' of G' with Satake class $[A_{\pi'}] \subset {}^L G'$, such that

$$[A_{\pi'}] \cong [\Psi(A_\pi)]. \tag{16.3}$$

Moreover, the functoriality extends to the L-functions associated with the automorphic representations.

It turns out that this has far-reaching consequences even for the case when the first group is taken to be trivial. Suppose for example that $G = \{1\}$ and $G' = GL(n)$. In this situation the dual group of G is simply the *(absolute) Galois group* $\mathrm{Gal}(\bar{\mathbb{Q}}/\mathbb{Q})$, where the extension is $\mathbb{E} = \bar{\mathbb{Q}}$, the algebraic closure of $\mathbb{Q}$ that can also be viewed as the union of all Galois number fields. (Recall that the algebraic closure $\bar{\mathbb{F}}$ of a number field $\mathbb{F}$ is obtained by adjoining to $\mathbb{F}$ all roots of all polynomials over $\mathbb{F}$. This not a finite extension and so generalises the discussion above.) The L-dual group of G' is the direct product ${}^L G' = GL(n, \mathbb{C}) \times \mathrm{Gal}(\bar{\mathbb{Q}}/\mathbb{Q})$. The map Ψ then yields a homomorphism

$$\Psi : \mathrm{Gal}(\bar{\mathbb{Q}}/\mathbb{Q}) \longrightarrow GL(n, \mathbb{C}). \tag{16.4}$$

This has the remarkable consequence that to each automorphic representation π of $GL(n, \mathbb{A})$ there should exist an associated n-dimensional representation R of $\mathrm{Gal}(\bar{\mathbb{Q}}/\mathbb{Q})$:

$$\left[\begin{array}{c} n\text{-dimensional Galois} \\ \text{representation} \\ R\colon \mathrm{Gal}(\bar{\mathbb{Q}}/\mathbb{Q}) \to GL(n, \mathbb{C}) \end{array}\right] \longleftrightarrow \left[\begin{array}{c} \text{Cuspidal automorphic} \\ \text{representation } \pi \text{ of} \\ GL(n, \mathbb{Q})\backslash GL(n, \mathbb{A})/K_{\mathbb{A}} \end{array}\right], \tag{16.5}$$

such that

$$L_A(s, R) = L(s, \pi), \tag{16.6}$$

where the object on the right is the standard L-function of π discussed in Section 11.8 (i.e., corresponding to ρ in Section 11.8 being the fundamental representation of $GL(n, \mathbb{C})$) and the object on the left is the so-called *Artin L-function* of the Galois representation R. We shall not go into the details of Artin L-functions but rather refer to [448] for a nice discussion of the two sides of (16.6), and also to Remark 11.17.

There are numerous sources which give overviews of various aspects of the Langlands program; we would like to especially mention [259, 448, 129, 28, 412, 413, 414] for further references.

16.2 The Geometric Langlands Program

There exists a version of the Langlands program which does not have its roots in number theory, but rather in the geometry of Riemann surfaces. This is commonly referred to as the *geometric Langlands program* and was proposed by Beilinson, Deligne, Drinfeld and Laumon [450, 449, 190, 52] (for a nice survey see the lectures notes by Frenkel [219], and for a recent update see the note by Gaitsgory [240]). To each object in the original classical Langlands program discussed in the previous section there exist geometric counterparts; for instance, the rôle of the Galois group is played by the fundamental group of the Riemann surface, while automorphic forms are replaced by certain 'automorphic sheaves' on the moduli space of principal bundles on the Riemann surface. A complete review of the geometric Langlands program goes far beyond the scope of this book, but in this section we will highlight some interesting structures that can be viewed as geometric analogues to Eisenstein series, Whittaker coefficients and the Casselman–Shalika formula.

16.2.1 The Geometric Langlands Correspondence

Recall from (16.4) in the previous section that one incarnation of the classical Langlands correspondence involves a homomorphism Ψ from the Galois group $\mathrm{Gal}(\bar{\mathbb{Q}}/\mathbb{Q})$ into $GL(n,\mathbb{C})$. The latter group is essentially the Langlands dual group ${}^LG$ in the case when $G = GL(n,\mathbb{Q})$. One can approach the geometric story by first moving from the case of number fields to *function fields* that we define for a smooth algebraic variety.

Let X be a curve over a field $\mathbb{F}$. We can then consider the field $F = \mathbb{F}(X)$ of rational functions on X, and the associated Galois group $\mathrm{Gal}(\bar{F}/F)$. This yields an analogue of the Galois representation (16.4) in terms of a homomorphism from $\mathrm{Gal}(\bar{F}/F)$ to ${}^LG$.

Remark 16.1 One advantage of function fields over number fields is that any completion of a function field is non-archimedean, whereas the number field $\mathbb{Q}$ has also the archimedean completion $\mathbb{R}$, which behaves very differently from the non-archimedean completions $\mathbb{Q}_p$. The classical Langlands correspondence has been proven in the function field case for $GL(n)$ by Drinfeld and Lafforgue [189, 190, 431].

We now take one step further and consider a curve X defined over $\mathbb{C}$, i.e., a Riemann surface. The analogue of the Galois group $\mathrm{Gal}(\bar{F}/F)$ is given by the fundamental group $\pi_1(X)$. An intuitive way of understanding this relation is by thinking of $\mathrm{Gal}(\bar{F}/F)$ as the *deck transformations* of the maximal cover of X, as follows. For any (unramified) cover $Y \to X$, one has a field extension $\mathbb{F}(Y)/\mathbb{F}(X)$ of the rational functions, with associated Galois group $\mathrm{Gal}(\mathbb{F}(Y)/\mathbb{F}(X))$ being a quotient of the fundamental group $\pi_1(X)$. For the maximal cover one then obtains $\mathrm{Gal}(\bar{F}/F) \leftrightarrow \pi_1(X)$. Thus, an essential ingredient in the geometric version of the Langlands program is a homomorphism

$$\Psi : \pi_1(X) \longrightarrow {}^LG \tag{16.7}$$

that replaces the Galois representations in (16.4). In order to get a more geometric picture of this homomorphism we use the *Riemann–Hilbert correspondence*, which associates to every Ψ a rank-n holomorphic vector bundle $\mathcal{E} \to X$ with a flat holomorphic connection ∇; see for example [219].

In the classical Langlands program, the correspondence (16.5) associated an n-dimensional representation of the Galois group $\mathrm{Gal}(\bar{\mathbb{F}}/\mathbb{F})$ with an automorphic representation of $G(\mathbb{A}_\mathbb{F})$, i.e., certain functions on $G(\mathbb{F})\backslash G(\mathbb{A}_\mathbb{F})/K_{\mathbb{A}_\mathbb{F}}$. The Galois representations are replaced in the geometric Langlands program by rank-n holomorphic vector bundles with flat connections. We next address the problem of finding the geometric analogue of the automorphic representations.

To formulate the geometric structure, it is useful to first go back to the case of function fields $F = \mathbb{F}(X)$. Weil's *uniformisation theorem* states that the quotient space

$$G(F)\backslash G(\mathbb{A}_F)/K_{\mathbb{A}_F} \tag{16.8}$$

is in bijection with the moduli space (actually, moduli *stack*) of principal G-bundles on X, denoted $\mathrm{Bun}_G(X)$. This suggests that in the geometric setting of Riemann surfaces (i.e., curves X over $\mathbb{C}$) the space $G(\mathbb{F})\backslash G(\mathbb{A}_\mathbb{F})/K_{\mathbb{A}_\mathbb{F}}$ should be replaced by the moduli space Bun_G. We are thus looking for some analogue of an automorphic form on $\mathrm{Bun}_G(X)$. It turns out that the correct notion is not a *function*, but rather a *sheaf* on $\mathrm{Bun}_G(X)$. Automorphic representations are in particular characterised by the fact that they yield eigenfunctions of the spherical Hecke algebra; see Chapter 11. Similarly, in the geometric setting, the analogous structure is captured by so-called *Hecke eigensheaves*. These behave similarly to Hecke eigenfunctions with respect to a certain integral transform on $\mathrm{Bun}_G(X)$ known as a *Hecke correspondence* [219]. To summarise, the geometric Langlands correspondence can be captured by the following equivalence of mathematical structures:

$$\begin{bmatrix} \text{Rank-}n \text{ holomorphic } G\text{-bundles} \\ \mathcal{E} \to X \\ \text{with a holomorphic connection } \nabla \end{bmatrix} \longleftrightarrow \begin{bmatrix} \text{Hecke eigensheaves on} \\ \mathrm{Bun}_G(X) \end{bmatrix}. \tag{16.9}$$

The geometric Langlands correspondence was proven by Laumon [449, 450] (following Deligne) in the case $G = GL(1)$, by Drinfeld [190] for $G = GL(2)$

and by Frenkel, Gaitsgory and Villonen [223, 238] for $G = GL(n)$. There is also a conjectured version for arbitrary G that has been proposed by Beilinson and Drinfeld [52].

16.2.2 Geometric Eisenstein Series

We will now discuss the analogue of Eisenstein series in the geometric setting. First let us recall the set-up from Chapter 5 using a slightly different perspective. Let $\mathbb{F}$ be a number field, $\mathbb{A}_{\mathbb{F}}$ the associated ring of adeles, G an algebraic group defined over $\mathbb{F}$, $B \subset G$ a choice of Borel subgroup and $K \subset G$ its maximal compact subgroup. Let $\chi\colon B(\mathbb{F})\backslash B(\mathbb{A}_{\mathbb{F}}) \to \mathbb{C}^{\times}$ be a global character. In Section 5.4, we saw that an Eisenstein series on $G(\mathbb{F})\backslash G(\mathbb{A}_{\mathbb{F}})/K_{\mathbb{A}_{\mathbb{F}}}$ can be viewed as a map

$$\begin{aligned} \text{Eis} : \text{Ind}_{B(\mathbb{A}_{\mathbb{F}})}^{G(\mathbb{A}_{\mathbb{F}})}\chi \quad &\longrightarrow \quad \mathcal{A}(G(\mathbb{F})\backslash G(\mathbb{A}_{\mathbb{F}})/K_{\mathbb{A}_{\mathbb{F}}}), \\ \mathsf{f}_{\chi} \quad &\longmapsto \quad \big[\text{Eis}(f_{\chi})\big](g) = \sum_{\gamma\in B(\mathbb{F})\backslash G(\mathbb{F})} \mathsf{f}_{\chi}(\gamma g). \end{aligned} \tag{16.10}$$

We shall now look at this construction under a slightly different light. The character χ is a function on $B(\mathbb{F})\backslash B(\mathbb{A}_{\mathbb{F}})$ which is determined by its restriction to the maximal torus $A(\mathbb{A}_{\mathbb{F}}) \subset B(\mathbb{A}_{\mathbb{F}}) = N(\mathbb{A}_{\mathbb{F}})A(\mathbb{A}_{\mathbb{F}})$. Equivalently, it can be viewed as a function on $N(\mathbb{A}_{\mathbb{F}})B(\mathbb{F})\backslash G(\mathbb{A}_{\mathbb{F}})/K_{\mathbb{A}_{\mathbb{F}}}$. From χ we construct the representation space $\text{Ind}_{B(\mathbb{A}_{\mathbb{F}})}^{G(\mathbb{A}_{\mathbb{F}})}\chi$, which is the space of ($B(\mathbb{A}_{\mathbb{F}})$-equivariant) functions on $B(\mathbb{F})\backslash G(\mathbb{A}_{\mathbb{F}})/K_{\mathbb{A}_{\mathbb{F}}}$. Consider the diagram

$$\begin{array}{ccc} B(\mathbb{F})\backslash G(\mathbb{A}_{\mathbb{F}})/K_{\mathbb{A}_{\mathbb{F}}} & \xrightarrow{\rho} & N(\mathbb{A}_{\mathbb{F}})B(\mathbb{F})\backslash G(\mathbb{A}_{\mathbb{F}})/K_{\mathbb{A}_{\mathbb{F}}} \\ \pi\downarrow & & \\ G(\mathbb{F})\backslash G(\mathbb{A}_{\mathbb{F}})/K_{\mathbb{A}_{\mathbb{F}}}. & & \end{array} \tag{16.11}$$

Here, ρ is the restriction to functions on the maximal torus $A(\mathbb{A}_{\mathbb{F}})$, the fibers of which are compact spaces $N(\mathbb{F})\backslash N(\mathbb{A}_{\mathbb{F}})$. Hence, the pull-back of χ yields $\rho^*(\chi) = \mathsf{f}_{\chi} \in \text{Ind}_{B(\mathbb{A}_{\mathbb{F}})}^{G(\mathbb{A}_{\mathbb{F}})}\chi$. The map π, on the other hand, denotes the lift to $G(\mathbb{F})$-invariant functions on $G(\mathbb{A}_{\mathbb{F}})/K_{\mathbb{A}_{\mathbb{F}}}$, the fibers of which are the infinite-dimensional discrete quotients $B(\mathbb{F})\backslash G(\mathbb{F})$. The push-forward π_* implies integration along the fibers of π, which in this case corresponds to summation over $B(\mathbb{F})\backslash G(\mathbb{F})$. We conclude that the map (16.10) can be written as the composition of maps $\text{Eis} = \pi_* \circ \rho^*$, and, for any character χ, we obtain

$$\pi_*(\rho^*(\chi))(g) = \sum_{\gamma\in B(\mathbb{F})\backslash G(\mathbb{F})} \mathsf{f}_{\chi}(\gamma g). \tag{16.12}$$

This is the first step towards geometric Eisenstein series. Let us further note that due to the Iwasawa decomposition $G = NAK$ we have the following isomorphisms:

$$\begin{aligned} B(\mathbb{F})\backslash G(\mathbb{A}_{\mathbb{F}})/K_{\mathbb{A}_{\mathbb{F}}} &\cong B(\mathbb{F})\backslash B(\mathbb{A}_{\mathbb{F}})/B_K, \\ N(\mathbb{A}_{\mathbb{F}})B(\mathbb{F})\backslash G(\mathbb{A}_{\mathbb{F}})/K_{\mathbb{A}_{\mathbb{F}}} &\cong T(\mathbb{F})\backslash T(\mathbb{A}_{\mathbb{F}})/T_K, \end{aligned} \tag{16.13}$$

where $B_K = B(\mathbb{A}_{\mathbb{F}}) \cap K_{\mathbb{A}_{\mathbb{F}}}$ and $T_K = A(\mathbb{A}_{\mathbb{F}}) \cap K_{\mathbb{A}_{\mathbb{F}}}$. This implies that we can equivalently write the diagram (16.11) as

$$\begin{array}{ccc} B(\mathbb{F})\backslash B(\mathbb{A}_{\mathbb{F}})/B_K & \stackrel{\rho}{\to} & T(\mathbb{F})\backslash T(\mathbb{A}_{\mathbb{F}})/T_K \\ \pi\downarrow & & \\ G(\mathbb{F})\backslash G(\mathbb{A}_{\mathbb{F}})/K_{\mathbb{A}_{\mathbb{F}}}. & & \end{array} \tag{16.14}$$

This is the structure we have been seeking to understand the geometric underpinnings of Eisenstein series. By the general discussion in the beginning of this section, the analogues of each of the spaces above are given by the associated moduli spaces (stacks) of bundles on the Riemann surface X. Thus we have a diagram

$$\begin{array}{ccc} \mathrm{Bun}_B(X) & \stackrel{\rho}{\to} & \mathrm{Bun}_A(X) \\ \pi\downarrow & & \\ \mathrm{Bun}_G(X). & & \end{array} \tag{16.15}$$

Note that in the number-theoretic context (16.14) we are really considering the maps (ρ, π) between the associated C^∞-spaces of functions on $G(\mathbb{F})\backslash G(\mathbb{A}_{\mathbb{F}})/K_{\mathbb{A}_{\mathbb{F}}}$, $B(\mathbb{F})\backslash B(\mathbb{A}_{\mathbb{F}})/B_K$ and $T(\mathbb{F})\backslash T(\mathbb{A}_{\mathbb{F}})/T_K$. As we have mentioned, in the geometric context, functions on $G(\mathbb{A}_{\mathbb{F}})$ should be replaced by sheaves on $\mathrm{Bun}_G(X)$, i.e., objects in a category over $\mathrm{Bun}_G(X)$. The relevant category turns out to be the *derived category of perverse sheaves* [219], denoted $\mathcal{D}(\mathrm{Bun}_G(X))$.

We are now equipped to define a geometric Eisenstein series following Braverman and Gaitsgory [86]. More precisely, we define an *Eisenstein functor* Eis to be the map

$$\mathrm{Eis} : \mathcal{D}(\mathrm{Bun}_A(X)) \longrightarrow \mathcal{D}(\mathrm{Bun}_G(X)), \tag{16.16}$$

which sends a sheaf $\mathfrak{F} \in \mathcal{D}(\mathrm{Bun}_A(X))$ to

$$\mathrm{Eis} : \mathfrak{F} \longmapsto \pi_*(\rho^*(\mathfrak{F})), \tag{16.17}$$

which is then a 'geometric Eisenstein sheaf'.

One can prove many properties of these geometric Eisenstein series which are analogous to the number-theoretic setting. Let us list a few properties below:

- Geometric Eisenstein series satisfy a version of a functional relation with respect to the Weyl group of G and they are Hecke eigensheaves [86].
- One can define the constant term of Eis by the inverse map (see, e.g., [239])

$$\mathrm{CT}: \mathcal{D}(\mathrm{Bun}_G(X)) \longrightarrow \mathcal{D}(\mathrm{Bun}_A(X)), \tag{16.18}$$

  defined by

$$\mathrm{CT} = \rho_* \circ \pi^{-1}, \tag{16.19}$$

  where π^{-1} denotes the inverse image.
- One can also define the analogue of a Whittaker coefficient. To this this end, one has to define the analogue of the local fields and groups in the geometric setting, which we describe roughly following [221]. For curves X over finite fields $\mathbb{F} = \mathbb{F}_q$, one can complete the field of rational functions $F = \mathbb{F}(X)$ on X at any closed point $x \in X$, and we denote this field by F_x, which is the analogue of $\mathbb{Q}_p$ in the number-theoretic setting. There is also a notion of integers in F_x and one can define the adeles $\mathbb{A}_X = \prod'_{x\in X} F_x$, where the product only runs over closed points (see also Section 16.3.2 below). For any local F_x one can then define Whittaker functions much like (11.130) using characters of representations of $GL(n, \bar{\mathbb{Q}}_\ell)$, where ℓ is relatively prime to q, and where $\mathrm{Gal}(\bar{F}/F)$ from the L-group embeds in $GL(n, \bar{\mathbb{Q}}_\ell)$. These local Whittaker functions can be strung together in a global object through an Euler product. More precisely, one has a certain *Whittaker category* [221, 222, 239]

$$\mathrm{Whit}_G(X) \subset \mathcal{D}(\mathrm{Bun}_G(X)), \tag{16.20}$$

  consisting of Hecke eigensheaves which satisfy an appropriate equivariance condition with respect to the action of the unipotent radical N.
- Viewed as tensor categories, one has the following isomorphism [221, 222]:

$$\mathrm{Whit}_G(X) \cong \mathrm{Rep}(^LG), \tag{16.21}$$

  where $\mathrm{Rep}(^LG)$ denotes the category of finite-dimensional representations of the dual group ^LG. This may be viewed as the geometric counterpart of the Casselman–Shalika formula [221]. In fact, this was used in [339] to give a new proof of the Casselman–Shalika formula which does not rely on the uniqueness of Whittaker models.
- One can extend the construction to consider 'twisted' Whittaker categories $\mathrm{Whit}_{\widetilde{G}}(X)$, where $\widetilde{G}$ is a metaplectic group (see Section 12.2) [459, 461]. This may be viewed as a geometric counterpart of the metaplectic Eisenstein series [110] and the metaplectic Casselman–Shalika formula [508]. See also

Chapter 17 for more details on metaplectic Eisenstein series and the relation with multiple Dirichlet series.

- Lafforgue and Lysenko have constructed 'theta sheaves', which are objects in the category that is the geometric analogue of the Weil representation of the metaplectic group [432, 458]. More generally, they have also developed the notion of a 'geometric minimal automorphic representation' in the case of orthogonal groups [434]. This was further used to construct geometric analogues of theta correspondences [433, 460]. Theta correspondences in the number-theoretic setting were already discussed in Chapter 12.

16.3 The Langlands Program and Physics

In this section, we give some pointers to places where the Langlands program has appeared in physics.

16.3.1 The Geometric Langlands Program and Yang–Mills Theory

Kapustin and Witten have shown [388] that the geometric Langlands program can be naturally understood in the context of quantum field theory (more precisely, a twisted version of $\mathcal{N} = 4$ supersymmetric quantum field theory in four space-time dimensions). In this context the analogue of the 'Langlands duality' (16.4) corresponds to a variant of *(homological) mirror symmetry*. We shall now provide a brief overview of the salient features of this story.

Homological mirror symmetry was proposed by Kontsevich [419] as the mathematical incarnation of mirror symmetry. To state it, let X be a Calabi–Yau threefold. String theory on X then depends on a choice of complex structure I of X as well as a complexified Kähler class $B + iJ \in H^2(X, \mathbb{C})$, where B is the B-field (see Section 13.1.3). Let $\mathcal{M}_{\mathbb{C}}(X)$ be the moduli space of complex structure deformations of X and let $\mathcal{M}_K(X)$ be the analogue for complexified Kähler deformations (sometimes called the 'stringy Kähler moduli space').

For a mirror pair of Calabi–Yau threefolds $(X, \widehat{X})$ the *mirror symmetry conjecture* implies that there are isomorphisms

$$\begin{aligned} \mathcal{M}_{\mathbb{C}}(X) &\cong \mathcal{M}_K(\widehat{X}), \\ \mathcal{M}_{\mathbb{C}}(\widehat{X}) &\cong \mathcal{M}_K(X). \end{aligned} \tag{16.22}$$

Kontsevich suggested considering the (bounded derived) *category of coherent sheaves* on X, denoted $\mathrm{D^bCoh}(X)$. This is a triangulated category whose objects are complexes of sheaves on X. Physically, these objects should be viewed as certain classes of D-branes, called *B-branes*. On the mirror manifold $\widehat{X}$ we consider instead the (derived) *Fukaya category* $\mathrm{D^bFuk}(\widehat{X})$, whose objects are *special Lagrangian submanifolds* of $\widehat{X}$ (or *A-branes*). The homological mirror symmetry conjecture then asserts that there are equivalences

$$\begin{aligned} \mathrm{D^bCoh}(X) &\cong \mathrm{D^bFuk}(\widehat{X}), \\ \mathrm{D^bCoh}(\widehat{X}) &\cong \mathrm{D^bFuk}(X). \end{aligned} \tag{16.23}$$

To see the relation with the geometric Langlands program, recall from Equation (16.9) that this relates holomorphic G-bundles to Hecke eigensheaves on Bun_G. Holomorphic G-bundles on a Riemann surface X are a *local system* and hence on the left-hand side we are studying the space of local systems $\mathrm{Loc}_G(X)$. According to Beilinson and Drinfeld [52], the most general formulation of the geometric Langlands correspondence is then an equivalence between the derived category of $\mathcal{O}$-modules on $\mathrm{Loc}_G(X)$ and the derived category of $\mathcal{D}$-modules on $\mathrm{Bun}_G(X)$.

Kapustin and Witten were able to relate this categorical version of the geometric Langlands program to a version of homological mirror symmetry. They considered a certain topological supersymmetric gauge theory (twisted version of $\mathcal{N} = 4$ super Yang–Mills), with gauge group G, formulated on a 4-manifold M_4 of the form

$$M_4 = \Sigma \times X, \tag{16.24}$$

where Σ is a Riemann surface (with a boundary) and X is an algebraic curve (closed Riemann surface). Compactifying the theory on X (i.e., taking X to be very small in an appropriate sense) yields an effective two-dimensional topological sigma model describing maps

$$\Sigma \longrightarrow \mathcal{M}_H(G), \tag{16.25}$$

where the target space is the *Hitchin moduli space* $\mathcal{M}_H(G)$ of so-called *Higgs G-bundles* on X. In the original gauge theory there is also an *S-duality* (*Montonen–Olive duality*) which yields an equivalence of physical theories related by the map

$$G \longleftrightarrow {}^L G, \tag{16.26}$$

where ${}^L G$ is the Langlands dual group, combined with the following action on the coupling:

$$\tau \longleftrightarrow -1/\tau. \tag{16.27}$$

In the effective two-dimensional sigma model on X this yields an isomorphism

$$\mathcal{M}_H(G) \cong \mathcal{M}_H({}^L G). \tag{16.28}$$

Kapustin and Witten then go on to show that this extends to an equivalence of branes on these moduli spaces. Roughly, the category of $\mathcal{O}$-modules on

$\mathrm{Loc}_G(X)$ corresponds to the category of so-called B-branes, while the category of $\mathcal{D}$-modules corresponds to A-branes. Hence, one deduces that the geometric Langlands correspondence can be interpreted as homological mirror symmetry for the Hitchin moduli space, in the following sense:

$$\left[\text{A-branes on } \mathcal{M}_H(G)\right] \longleftrightarrow \left[\text{B-branes on } \mathcal{M}_H({}^L G)\right]. \tag{16.29}$$

Since the original paper of Kapustin and Witten, there have been several developments, including the papers [332, 623, 224, 624, 625], to which we refer the interested reader. For a nice overview, emphasising the comparison with other forms of the geometric Langlands program, see Frenkel's Bourbaki seminar [220].

16.3.2 The Geometric Langlands Program and CFT

In an influential paper [619], Witten made some intriguing observations regarding relations between two-dimensional conformal field theories and automorphic representations. In standard formulations of conformal field theory one considers a quantum field theory on a Riemann surface X invariant under the Virasoro algebra. Witten suggested that it might be beneficial to generalise this and consider X to be an algebraic curve over some algebraically closed field $\mathbb{F}$ and define a conformal field theory on $X(\mathbb{F})$; denote this theory by $\mathrm{CFT}[X(\mathbb{F})]$. Let $\mathcal{L}^{1/2}$ be a choice of square root of the canonical bundle $\mathcal{L} \to X$ and let Y be the space of (rational) sections of $\mathcal{L}^{1/2}$. Roughly, these sections should be viewed as fermions in the theory. For any point $P \in X$, denote by Y_P the completion of Y at P; these are the sections of Y that can be defined by formal power series in the neighbourhood of P. Witten showed that one can define a quadratic form on Y_P by taking residues of differential forms at P, and this allows for the construction of a Clifford algebra structure CY_P on Y_P. The space of physical observables V_P in the conformal field theory at the point $P \in X$ then forms a representation of CY_P. An arbitrary observable $\mathcal{O} \in \mathrm{CFT}[X(\mathbb{F})]$ is a product $\mathcal{O} = \prod_{i=1}^{N} \mathcal{O}_{P_i}$ of local observables $\mathcal{O}_{P_i} \in V_{P_i}$ at the points P_i. Witten observed that an alternative way to view this is to instead represent $\mathcal{O}$ as an infinite product over all points $P \in X$, but with the restriction that $\mathcal{O}_P = 1$ for all but finitely many points P. In other words, an arbitrary observable in $\mathrm{CFT}[X(\mathbb{F})]$ is a restricted product

$$\mathcal{O} = \prod_{P\in X}{}' \mathcal{O}_P, \tag{16.30}$$

in complete analogy with the definition of the adeles of the field of functions on X. The observable $\mathcal{O}$ should be viewed as an element of the vector space $V = \otimes_P V_P$, representing the space of all observables. The identity vector $1 = \otimes 1_P \in V$ represents the vacuum state (using the state–operator correspondence). This is indeed very reminiscent of standard constructions in automorphic representations.

Subsequently, Frenkel has extensively developed the connection between certain special types of holomorphic conformal field theories and automorphic representations in the context of the geometric Langlands program; see [219] for a survey. It is quite striking, however, that Witten's observations were made long before the geometric Langlands correspondence was actually developed.

16.3.3 Connections with String Theory?

Automorphic forms occur in abundance in string theory, as discussed in PART TWO of this book. Despite this fact, the physical rôle of the classical Langlands program remains unclear. We have seen that automorphic representations play a rôle in understanding BPS states and instantons in string theory, but we have no clue as to the physical interpretation of the dual side in (16.5), involving representations of the Galois group. It would be very interesting to find out whether such an interpretation exists. Given that automorphic L-functions lie at the heart of the Langlands program (see (16.6)) a very natural question, posed by Moore in [491], is the following:

Open question: *Is there a natural rôle for automorphic L-functions in BPS state counting problems?*

For some speculation on this and related issues, see [490, 475], and for a conjectured connection between BPS states in string theory and Galois representations, see [607]. Interesting ideas on the connection between the geometry of Calabi–Yau manifolds and modular forms constructed from string world-sheet conformal field theories can be found in [547].

Recently, Kim has also made the intriguing suggestion that L-functions should be viewed as *wave functions*, in the sense that they correspond to sections of determinant line bundles over arithmetic schemes [402].

16.4 Modular Forms and Elliptic Curves

The aim of this section is to explain in broad terms the *modularity theorem* that relates modular forms to elliptic curves. Through a further connection to Galois representations it can be seen as a special proven case of the (classical) Langlands conjectures discussed in Section 16.1 above. General references for the material of this section include [145, 24, 179, 140, 141, 560, 540].

16.4.1 Elliptic Functions and Complex Elliptic Curves

The starting point for our discussion is *doubly periodic complex functions* $F: \mathbb{C} \to \mathbb{C}$. 'Doubly periodic' means that there are two complex numbers ω_1 and ω_2 with $\omega_1/\omega_2 \notin \mathbb{R}$ such that $F(z + m\omega_1 + n\omega_2) = F(z)$ for all $z \in \mathbb{C}$ and $m, n \in \mathbb{Z}$. In other words, the function f is defined on the *complex torus* $\mathbb{C}/\Lambda$, where

$$\Lambda = \{m\omega_1 + n\omega_2 \in \mathbb{C} \mid m, n \in \mathbb{Z}\} \tag{16.31}$$

is the period lattice in $\mathbb{C}$ generated by the two independent (over $\mathbb{R}$) periods ω_1 and ω_2. If the function F is moreover meromorphic, it is called an *elliptic function*. (The reason for this name is the connection of such functions to elliptic integrals as studied by Abel and Jacobi, a connection that will not be used in the sequel.)

For a given lattice Λ as in (16.31), an example of an elliptic function is given by the *Weierstraß function*

$$\wp(z) = \frac{1}{z^2} + \sum_{\omega\in\Lambda}{}' \left(\frac{1}{(z+\omega)^2} - \frac{1}{\omega^2} \right) \tag{16.32}$$

for $z \in \mathbb{C} \setminus \Lambda$; the primed summation excludes $\omega = 0$ as elsewhere in this book. The expression above is absolutely convergent and periodicity is most easily seen by considering its derivative

$$\wp'(z) = -2 \sum_{\omega\in\Lambda} \frac{1}{(z+\omega)^3}, \tag{16.33}$$

which is clearly doubly periodic and odd, such that the even primitive $\wp(z)$ also turns out doubly periodic. An important observation is that any elliptic function can be written as a rational function of $\wp(z)$ and $\wp'(z)$. In fact, $\wp(z)$ and $\wp'(z)$ are not even algebraically independent but satisfy the cubic equation

$$(\wp'(z))^2 = 4\,(\wp(z))^3 - g_2\wp(z) - g_3, \tag{16.34}$$

where

$$g_2 \equiv g_2(\Lambda) = 60 \sum_{\omega\in\Lambda}{}' \frac{1}{\omega^4} \quad \text{and} \quad g_3 \equiv g_3(\Lambda) = 140 \sum_{\omega\in\Lambda}{}' \frac{1}{\omega^6} \tag{16.35}$$

are numerical invariants associated with the lattice Λ. One may therefore quotient the ring of rational functions in $\wp$ and $\wp'$ by the relation (16.34). Relation (16.34) can be shown for instance by considering the Laurent expansions of $\wp(z)$ and $\wp'(z)$ [24].

Including a point at infinity, (16.34) therefore implies that we have a bijection

$$z \mapsto \begin{cases} (\wp(z), \wp'(z), 1), & z \notin \Lambda \\ (0, 1, 0), & z \in \Lambda \end{cases} \tag{16.36}$$

between the complex torus $\mathbb{C}/\Lambda$ and points on the projective algebraic variety with equation $y^2t = 4x^3 - g_2xt^2 - g_3t^3$. This algebraic variety is a curve that is called an *elliptic curve over* $\mathbb{C}$ since it is in bijection with the generators of elliptic functions. Going to inhomogeneous coordinates one could write the elliptic curve as the cubic

$$E: y^2 = 4x^3 - g_2x - g_3 \tag{16.37}$$

over $\mathbb{C}$, sometimes called the Weierstraß form. We will denote by $E(\mathbb{C})$ the $\mathbb{C}$-points of the elliptic curve given by this equation, together with the point at infinity. We collect some further properties of complex elliptic curves and notions in the form of remarks.

Remark 16.2 (Isomorphisms of complex tori and the upper half-plane) Two complex tori for lattices Λ and Λ' are isomorphic if and only if there is a non-zero complex number ρ such that $\rho\Lambda = \Lambda'$. Considering complex tori up to this multiplication we can therefore choose the periods to be $\omega_1 = 1$ and $\omega_2 = \tau$ with $\text{Im}(\tau) > 0$. Moreover, we know from Chapter 4 that the modular group $SL(2,\mathbb{Z})$ permutes the lattice points on such a standard lattice so that we have to identify τ-values that are related by the $SL(2,\mathbb{Z})$ action (4.7). In total, complex tori are parametrised by $SL(2,\mathbb{Z})\backslash\mathbb{H}$ and we write the corresponding lattice as Λ_τ. We here follow the convention that the modulus of the space of complex tori is denoted by $\tau \in \mathbb{H}$ rather than by the variable z we have been using in PART ONE. The reason is to avoid confusion with the variable z appearing for example in the elliptic function $\wp(z)$ and also because τ here is closer in spirit to world-sheet modulus for one-loop amplitudes in string theory; see Section 13.4.

Remark 16.3 (Modular invariants and discriminant of an elliptic curve) In light of the previous remark, we can view the invariants (16.35) as functions of the modulus τ of the lattice Λ_τ such that

$$g_2(\tau) = 60 \sum_{(m,n)\in\mathbb{Z}^2}{}' \frac{1}{(m+n\tau)^4}, \qquad g_3(\tau) = 140 \sum_{(m,n)\in\mathbb{Z}^2}{}' \frac{1}{(m+n\tau)^6}. \tag{16.38}$$

Under the modular group $SL(2,\mathbb{Z})$, these transform with weights 4 and 6, respectively. Comparing to (4.21) we see that these functions are identical, up to an overall normalisation, to the holomorphic Eisenstein series E_4 and E_6. This is our first indication of a relation of elliptic curves to modular forms. Since the torus $\mathbb{C}/\Lambda_\tau$ is non-singular, so also is the elliptic curve associated to it. In other words the *discriminant*

$$\Delta(\tau) = 16\left(g_2(\tau)^3 - 27g_3(\tau)^2\right) \tag{16.39}$$

is non-vanishing and actually agrees with the discriminant cusp form seen in (4.26). It transforms under the modular group with weight 12. Moreover, the multiplicative scaling symmetry of the lattice means that

$$j(\tau) = \frac{1728 g_2(\tau)^3}{g_2(\tau)^3 - 27 g_3(\tau)^2} \tag{16.40}$$

is a function only of the isomorphism class of the lattice, i.e., it is constructed such that it is invariant under scaling and modular invariant: $j(\gamma\tau) = j(\tau)$ for all $\gamma \in SL(2, \mathbb{Z})$. However, as we know from (4.27), it has a pole at infinity.

Remark 16.4 (Elliptic curves as abelian groups) Using the above bijection (16.36), one can transfer the abelian group law of addition from the complex torus to the elliptic curve. There is a simple geometric interpretation of this addition law for two points on the curve $E(\mathbb{C})$ by drawing a straight line through the two points. Generically, this line will intersect the curve at a third distinct point that is then the negative of the sum of the two points. (The negative of a point can be constructed similarly by taking the point and the point at infinity.) Algebraic expressions for this group operation on the elliptic curve can be found for example in [145], but here we require only that the group law exists and is abelian. The existence of the algebraic group law makes the elliptic curve an *abelian variety*.

Remark 16.5 (Complex multiplication) Viewed as an abelian algebraic group, an elliptic curve $E(\mathbb{C})$ can have endomorphisms. For any elliptic curve, these endomorphisms include multiplication by an integer, as this amounts simply to multiple additions. If there are more endomorphisms of $E(\mathbb{C})$ one says that the elliptic curve possesses *complex multiplication*, a concept we will encounter again for rigid Calabi–Yau threefolds in Section 18.6. It can be shown that for the complex torus $\mathbb{C}/\Lambda \cong E(\mathbb{C})$ complex multiplication means that there is a non-integral complex number α such that $\alpha\Lambda \subset \Lambda$, whence the name. One can prove [145] that if $E(\mathbb{C})$ has complex multiplication, then the endomorphism ring of the elliptic curve is isomorphic to $\mathbb{Z} + \tau\mathbb{Z}$ with τ an algebraic integer of degree two in a quadratic number field $\mathbb{Q}(\tau)$. The endomorphism ring is then isomorphic to the order of integers in $\mathbb{Q}(\tau)$. Moreover, the function value $j(\tau)$ in such a case is then also an algebraic integer.

16.4.2 Rational Elliptic Curves and L-Functions

There are many different ways of writing an elliptic curve in terms of an algebraic equation. Over $\mathbb{C}$ a minimal form was given above in (16.37). We will now be interested in elliptic curves over the rationals $\mathbb{Q}$ and over finite fields. In this case it is not always possible to achieve the simple form (16.37), but the best one can do in general is

$$E\colon y^2 + a_1 xy + a_3 y = x^3 + a_2 x^2 + a_4 x + a_6, \tag{16.41}$$

where we take the coefficients to be integers. The Mordell–Weil theorem [494, 610] implies that the points $E(\mathbb{Q})$ on E with rational coordinates is again a finitely generated abelian group (written additively):

$$E(\mathbb{Q}) \cong E(\mathbb{Q})_{\text{tors}} \oplus \mathbb{Z}^r , \tag{16.42}$$

where $r \in \mathbb{Z}_{\geq 0}$ is called the *rank* of the elliptic curve and $E(\mathbb{Q})_{\text{tors}}$ its so-called *torsion group*. The torsion group is a finite group of order at most 16 and the possibilities have been completely classified by Mazur [471]. It is known how to compute the torsion group for a given rational elliptic curve, but determining its rank r is an open problem that we will come back to in Section 16.4.3.

To study the rational elliptic curve $E(\mathbb{Q})$ further, we can consider the reduction of its defining Equation (16.41) modulo p for primes p. This is in the spirit of the local-global principle and we are thus considering the elliptic curve over the finite field $\mathbb{F}_p$. The $\mathbb{F}_p$-points of this curve are finitely many and will be denoted by $E(\mathbb{F}_p)$. As already shown by Hasse [350], one has for the number of these that

$$\left|E(\mathbb{F}_p)\right| = p + 1 - a_p(E), \tag{16.43}$$

with $|a_p| < 2\sqrt{p}$. The number $a_p(E)$ is a good parametrisation of the number of points in that it turns out to be related to modular forms, as we will discuss below. One has to be careful that an originally smooth rational elliptic curve may be singular over $\mathbb{F}_p$; if this happens, we call p a prime of bad reduction.

Hasse's theorem, or a special case of a theorem by Weil and Deligne [611, 165], shows that one can associate local zeta factors to the elliptic curve (16.41) at primes p of good reduction through

$$\zeta_p(E, s) = \exp\left[\sum_{n\geq 1} \frac{N_n(p)}{n} p^{-ns}\right] = \frac{1 - a_p(E)p^{-s} + p^{1-2s}}{(1 - p^{-s})(1 - p^{1-s})}, \tag{16.44}$$

where $N_n(p) = \left|E(\mathbb{F}_{p^n})\right|$ is the number of points on the smooth elliptic curve over $\mathbb{F}_{p^n}$ and a_p the number that appears in (16.43). Augmenting this local zeta factor by appropriate factors for the bad primes and removing the normalising local Riemann zeta factors in the denominator in (16.44), one can define the following global *Hasse–Weil L-function* for the elliptic curve E:

$$L(E, s) = \prod_{p\in S} \frac{1}{1 - \epsilon(p)p^{-s}} \prod_{p\notin S} \frac{1}{1 - a_p(E)p^{-s} + p^{1-2s}}, \tag{16.45}$$

where S denotes the finite set of bad primes, similarly to the discussion in Section 11.8. The parameter $\epsilon(p) \in \{-1, 0, 1\}$ depends on the type of non-smooth reduction of the elliptic curve at p [145]. The definition above is valid initially for $\mathrm{Re}(s) > \frac{3}{2}$. The bad primes can be characterised in terms of the discriminant of the rational elliptic curve E when written in the minimal form (16.41). The (potentially) bad primes are just those that divide this discriminant. An associated quantity is the (algebraic) *conductor* N_E of the elliptic curve E, which is given as the product of the bad primes raised to a small power that depends on the reduction type at the divisor p; see for instance [179] for more details.

As with any L-function, its functional and analytic properties are of central importance here, in order to study the point counts of elliptic curves. We will address these issues by connecting the Hasse–Weil L-function to automorphic L-functions.

16.4.3 The Modularity Theorem

In Section 11.2.4, we studied L-functions associated with modular forms f of weight k for the modular group $SL(2, \mathbb{Z})$, in particular cusp forms. As mentioned in Remark 11.7, the definition given in (11.34) can be generalised to modular forms for congruence subgroups of the modular group of the type discussed in Section 4.2.2. For the congruence subgroup $\Gamma_0(N) \subset SL(2, \mathbb{Z})$ and a weight 2 cusp form f that is also a Hecke eigenform (see Chapter 11), the definition becomes [613, 614]

$$L(f, s) = \prod_{p \in S} \frac{1}{1 - \epsilon(p)p^{-s}} \prod_{p \notin S} \frac{1}{1 - a_p(f)p^{-s} + p^{1-2s}}, \tag{16.46}$$

where $a_p(f)$ are the Fourier coefficients of the modular form $f = \sum_{n \geq 0} a_n(f) e^{2\pi i n \tau}$ and the set of bad primes S is given by those p that divide N. The factor $\epsilon(p)$ turns out to be identical to the one in (16.45). The Hecke L-function (16.46) above has a known completion, analytic continuation and functional relation.

Clearly, the L-functions in (16.45) and (16.46) are very similar in form but constructed from quite different objects. The first definition (16.45) makes use of rational elliptic curves, whereas the second L-function (16.46) is constructed from a cusp form for a congruence subgroup $\Gamma_0(N)$. One can prove [201, 574] that, given a weight 2 cusp form for $\Gamma_0(N)$, one can construct a rational elliptic curve E with conductor N_E such that the Hasse–Weil L-function (16.45) equals the Hecke L-function (16.46):

$$L(E, s) = L(f, s). \tag{16.47}$$

The rational elliptic curve E associated with the modular form f is then called a *modular elliptic curve*. According to (16.43) we can count its points using the Fourier expansion of f. The Hasse–Weil L-function $L(E, s)$ inherits the analytic properties from the Hecke L-functions and this can be used to study the elliptic curve. The functional relation in this case involves s going to $2 - s$ and the critical line for the generalised Riemann conjecture is $\text{Re}(s) = 1$. The conductor N_E is equal to N only if f is a so-called *newform*; otherwise N_E is a divisor of N such that f is a newform for this divisor. For a discussion of newforms and oldforms see Remark 4.4.

The inverse problem has long been a conjecture, known as the *Taniyama–Shimura–Weil conjecture* [576, 575, 613]: every rational elliptic curve is modular, such that one can find a weight 2 cusp and Hecke eigenform on $\Gamma_0(N_E)$ whose Hecke L-function agrees with the Hasse–Weil L-function of the rational elliptic curve. In other words, the conjecture was that every rational elliptic curve is actually modular. This is only one way of phrasing the conjecture and this particular form is sometimes also known as the *Hasse–Weil conjecture*. We also note that one can define analogues of the Hecke operators T_p for the elliptic curve, and these play a crucial rôle in the so-called *Eichler–Shimura relation* [201, 574].

The Taniyama–Shimura–Weil conjecture was proven by Wiles, Taylor, Diamond, Conrad and Breuil (in various combinations [616, 595, 178, 147, 93]) and the result is now known as the *modularity theorem*. The version (16.47) is known as the arithmetic version and there are many different formulations. We note the similarity of (16.47) with (16.6), and the similarity to the general Langlands correspondence will become even more transparent when we consider the reformulation of the modularity theorem in terms of Galois representations.

Connection with Galois Representations

The proofs of the modularity theorem typically involve a formulation in terms of Galois representations, and this also brings out connections to the classical Langlands conjectures. We will state briefly how these Galois representations arise from elliptic curves; for more details see [179, 540]. Let $E(\mathbb{Q})$ be a rational elliptic curve. We can also view the same equation over $\mathbb{C}$ and obtain a complex elliptic curve $E(\mathbb{C}) \cong \mathbb{C}/\Lambda$ of the type discussed in Section 16.4.1, and $E(\mathbb{C})$ is also an algebraic group. For a prime ℓ we can consider the *ℓ-division points* $E[\ell]$ of $E(\mathbb{C})$ as those elements of the group that have order ℓ. Since $E(\mathbb{C}) \cong \mathbb{C}/\Lambda$, we can view this $E[\ell]$ as $\frac{1}{\ell}\Lambda/\Lambda$ and hence as isomorphic (non-canonically) to $(\mathbb{Z}/\ell\mathbb{Z})^2$. (Sometimes $E[\ell]$ is called the *ℓ-power torsion* subgroup of $E(\mathbb{Q})$.) Since the points in $E[\ell]$ satisfy algebraic equations, one has that $E[\ell] \subset E(\bar{\mathbb{Q}})$, where $\bar{\mathbb{Q}}$ is the algebraic closure of $\mathbb{Q}$. The Galois group $\mathrm{Gal}(\bar{\mathbb{Q}}/\mathbb{Q})$ leaves $E[\ell]$ invariant. In fact, $E[\ell]$ is a module for $\mathrm{Gal}(\bar{\mathbb{Q}}/\mathbb{Q})$, such that we have a representation

$$\rho_{E,\ell} \colon \mathrm{Gal}(\bar{\mathbb{Q}}/\mathbb{Q}) \to \mathrm{Aut}(E[\ell]) \cong GL(2, \mathbb{Z}/\ell\mathbb{Z}), \qquad (16.48)$$

where we have made a choice of basis for $E[\ell]$ in the last isomorphism that is not canonical. However, this choice of basis is irrelevant when one considers traces and determinants, as they are invariant under inner automorphisms corresponding to a choice of basis. If $p \neq \ell$ is now another prime for which the elliptic curve E has good reduction, we call $\rho_{E,\ell}$ unramified at p. Let σ_p be the *Frobenius element* (up to inner automorphisms) of the Galois group $\mathrm{Gal}(K_\ell/\mathbb{Q})$, where K_ℓ is the (Galois) extension of $\mathbb{Q}$ by including the coordinates of the points on $E[\ell]$. This Frobenius element then has the property

$$\mathrm{Tr}\rho_{E,\ell}(\sigma_p) = a_p(E) \mod \ell, \quad \det \rho_{E,\ell}(\sigma_p) = p, \tag{16.49}$$

associating Galois representations with point counts of elliptic curves. It is sufficient to consider the Frobenius elements as their conjugacy classes exhaust $\mathrm{Gal}(K_\ell/\mathbb{Q})$ and the Frobenius element σ_p lifts to an absolute Frobenius element $\mathrm{Frob}_p \in \mathrm{Gal}(\bar{\mathbb{Q}}/\mathbb{Q})$ with the same properties. For ramified p and some special ℓ, the above definitions have to be refined. Repeating the same analysis for all powers ℓ^n of a prime ℓ and combining them gives a Galois representation using ℓ-adic integers, i.e., the representation matrices belong to $GL(2, \mathbb{Z}_\ell)$. The Galois representation $\rho_{E,\ell}$ is irreducible. The Galois representation (16.48) is a special case of what appears in the Langlands correspondence (16.4).

There is a similar, slightly more involved construction of a Galois representation $\rho_{f,\ell}$ starting with a modular form f instead of a rational elliptic curve E. A technical point in the definition of $\rho_{f,\ell}$ is that the traces of the Frobenius elements then give rise to the Fourier coefficients $a_p(f)$ mod ℓ through a connection with Hecke operators. The details can be found in [179, 540]. Such a representation of the Galois group $\rho_{f,\ell}$ is called *modular*. The modularity theorem in this language is that any Galois representation that arises for an elliptic curve is modular in the sense that it comes from a representation $\rho_{f,\ell}$.

A more general version of the Taniyama–Shimura–Weil conjecture is the *Serre conjecture* [560] that every (irreducible, odd and continuous) representation $\rho\colon \mathrm{Gal}(\bar{\mathbb{Q}}/\mathbb{Q}) \to GL(2, \overline{\mathbb{F}_\ell})$ is modular. Unlike the Taniyama–Shimura–Weil conjecture, the Serre conjecture has not been proven.

Fermat's Last Theorem

The modularity theorem is at the heart of the proof of *Fermat's last theorem* by Wiles [616], and Taylor and Wiles [595]. It was observed earlier by Frey [226] and by Ribet [539] that it is possible to connect integer solutions of the equation $a^p + b^p = c^p$ for primes $p > 3$ to a rational elliptic curve $y^2 = x(x - a^p)(x + b^p)$, called the *Frey–Hellegouarch curve*, with discriminant $\Delta = (abc)^{2p}/256$ and, importantly, that this curve could not be modular; see also [560]. Therefore, proving that all rational elliptic curves of this type are modular would imply Fermat's last theorem. The curve belongs to the class of semi-stable elliptic curves, meaning that every prime factor appears once in its conductor N_E, and Wiles and Taylor proved the modularity theorem in these cases, obtaining Fermat's last theorem as a corollary. The modularity theorem was subsequently proven in full generality for any rational elliptic curve [178, 147, 93].

Birch and Swinnerton-Dyer Conjecture

The group structure of a rational elliptic curve E was shown in (16.42) to decompose into a finite abelian group, called the torsion group, and a rank-r lattice $\mathbb{Z}^r$. The determination of the rank r is an open question but empirical data has led to the *Birch and Swinnerton-Dyer conjecture* [61] that an elliptic curve has rank r if and only if its Hasse–Weil L-function has a zero of order exactly r at $s = 1$:

$$\lim_{s\to 1}(s-1)^{-r}L(E,s) \neq 0\,. \tag{16.50}$$

There is a more precise conjecture that also predicts the value on the right-hand side in terms of the elliptic curve data, but to state it would require notions beyond the scope of this book. They can be found for example in [629, 179, 617].

Combining the modularity theorem with the results of Gross–Zagier [329] and Kolyvagin [418] establishes the conjecture for $r = 0$ and $r = 1$. The modularity theorem is also crucial for defining the Hasse–Weil L-function at $s = 1$ since one can use the functional equation of an associated Hecke L-function $L(f, s)$. The Birch and Swinnerton-Dyer conjecture has applications for example to the class number problem [627] and is one of the millennium problems [617].

Remark 16.6 Connections between the Birch and Swinnerton-Dyer conjecture and moonshine have recently been discovered in [194, 193]. This may provide the first hint of the type of representation-theoretic interpretation of moonshine that was discussed in Section 15.5.4.

17

Whittaker Functions, Crystals and Multiple Dirichlet Series

In this chapter, we discuss some issues related to a fascinating connection between Whittaker functions and statistical mechanics. Starting from a rewriting of the Casselman–Shalika formula, generalisations of Whittaker functions to metaplectic groups will be given. Their relation to Weyl group multiple Dirichlet series will be discussed and an alternative interpretation in terms of lattice models mentioned. This is an active area of research that has received a lot of momentum through the work of Brubaker, Bump, Chinta, Friedberg, Gunnells, Hoffstein and many others. We rely in our exposition mainly on the collection [118] and on [107], and refer the reader also to [105] for an overview.

17.1 Generalisations of the Weyl Character Formula

The Casselman–Shalika formula for the spherical Whittaker function $W^\circ(\lambda, a)$ on a group $G(\mathbb{Q}_p)$ was discussed in detail in Chapter 9 and given an interpretation in terms of characters ch_Λ of the Langlands dual group ${}^LG$ in Equation (9.81). This formula can actually be inverted to give an alternative formula for highest weight characters of ${}^LG$ through

$$\mathrm{ch}_\Lambda(a_\lambda) = \frac{W^\circ(\lambda, a_\Lambda)\delta^{-1/2}(a_\Lambda)}{\prod_{\alpha>0}(1 - p^{-1}a_\lambda^\alpha)}. \tag{17.1}$$

Here, λ is a weight of the original group G parametrising the principal series representation and a_λ and a_Λ are distinguished elements of A and ${}^LA$, respectively. These distinguished elements were defined in Section 9.6.

Formula (17.1) resembles the standard Weyl character formula (3.30), in particular the denominator. Independent of Whittaker functions, Tokuyama [600] considered a one-parameter family of deformations of the Weyl character formula that can be written as

$$\mathrm{ch}_\Lambda(a_\lambda) = \frac{\sum_{v\in\mathcal{B}_{\Lambda+\rho}} G(v,t) a_\lambda^{\mathrm{wt}(v)+\rho}}{\prod_{\alpha>0}(1+ta_\lambda^\alpha)}, \tag{17.2}$$

where $t \in \mathbb{C}$ and the function $G(v,t)$ has to be such that the left-hand side is independent of t. In this expression, all quantities refer to the Langlands dual group ${}^LG$. The sum here is over all v in the *crystal* $\mathcal{B}_{\Lambda+\rho}$. The crystal $\mathcal{B}_{\Lambda+\rho}$ is a directed graph with vertices v given by all the weights (with multiplicity) of the irreducible highest weight representation $V_{\Lambda+\rho}$ of ${}^LG$, where the ρ shift is important, and the edges labelled by simple roots. The map wt: $\mathcal{B}_{\Lambda+\rho} \to \mathfrak{h}^*$ identifies the vertices with points in the weight lattice of ${}^LG$. Crystals were introduced by Kashiwara [389] in his study of the *quantum deformed universal enveloping algebra* $U_q({}^L\mathfrak{g})$ (closely related to *quantum groups* [457]) and possess a *canonical basis* in the sense of Kashiwara and Lusztig [389, 456]. The operators corresponding to the edges are the simple step operators f_i in the limit $q \to 0$. Kashiwara also introduced the crystal $\mathcal{B}_\infty$, which is modelled on the canonical (free) *Verma module* of $U({}^L\mathfrak{n}_-)$.

Following [100], we will call the complex function $G(v,t)$ a *Tokuyama function* and it is the main object of interest in expression (17.2). In the original paper [600], the numerator was not written in terms of the crystal $\mathcal{B}_{\Lambda+\rho}$ but in terms of *Gelfand–Tsetlin patterns* [268] with top row $\Lambda+\rho$ and the analysis restricted to the special linear group. We will comment later on the status for other groups.

Interesting special cases of the deformed character formula (17.2) are:

- $t = -1$. This is the value for the standard character formula (3.30). In this case the sum over the crystal collapses to a sum over the Weyl orbit of the shifted highest weight $\Lambda+\rho$. In other words, $G(v,-1) = 0$ unless $\mathrm{wt}(v) = w(\Lambda+\rho)$ for some $w \in \mathcal{W}$, and in that case $G(v,-1) = \epsilon(ww_{\mathrm{long}})$.
- $t = 1$. In this case one obtains a relation to the original formulation of Gelfand–Tsetlin patterns.
- $t = 0$. The denominator trivialises and Tokuyama used this case to recover a relation of Stanley's [583] between Gelfand–Tsetlin patterns and 'most singular' values of Hall–Littlewood polynomials. In the crystal formulation, the only contributing terms arise from the embedding of $\mathcal{B}_\Lambda \to \mathcal{B}_{\Lambda+\rho}$ [118] and the sum then becomes the character in the form (3.25).
- $t = -p^{-1}$. This is the case relevant for the Casselman–Shalika formula and will be discussed in more detail below.

The first three cases were originally studied by Tokuyama [600]. The relation of the last case to Whittaker functions was first explored by Brubaker, Bump, Friedberg and Hoffstein [109]; see also [341] for combinatorial aspects.

17.2 Whittaker Functions and Crystals

In the case $t = -p^{-1}$, we can compare (17.2) and (17.1) to deduce that we have an alternative description of Whittaker functions in terms of a sum over a crystal with a Tokuyama function $G(v, -p^{-1})$:

$$W^\circ(\lambda, a_\Lambda) = \delta^{1/2}(a_\Lambda) \sum_{v \in \mathcal{B}_{\Lambda+\rho}} G(v, -p^{-1}) a_\lambda^{\mathrm{wt}(v)+\rho}. \tag{17.3}$$

In order to ease notation, from here on we will suppress the t-value in the Tokuyama function and will simply write $G(v)$ instead of $G(v, -p^{-1})$.

The identity (17.3) in some sense defines the Tokuyama function $G(v)$ given the spherical Whittaker function. But it is desirable to have an independent description of the function $G(v)$. This was achieved in crystal form in [107] and can be given in terms of so-called decorated *Berenstein–Zelevinsky–Littelmann paths* (BZL paths) in the crystal [58, 453]. A BZL path of a vertex $v \in \mathcal{B}_{\Lambda+\rho}$ is given by first fixing a choice of a reduced expression of the longest Weyl word w_{long}:

$$w_{\mathrm{long}} = w_{i_1} \cdots w_{i_\ell}, \tag{17.4}$$

where $\ell = \ell(w_{\mathrm{long}})$ is the length of the longest Weyl word and w_i is the ith fundamental reflection. The BZL path $\mathrm{BZL}(v)$ of a crystal vertex $v \in \mathcal{B}_{\Lambda+\rho}$ is then obtained by following the simple lowering operators f_i as far as possible through the crystal in the order given in the reduced expression of w_{long}. Let b_1 be the largest integer such that $f_{i_1}^{b_1} v \neq 0$, that is, b_1 is the maximum number of steps in the direction of f_{i_1} one can take in the crystal without leaving it. Starting from the point obtained in this way one then constructs b_2 as the maximum number of steps in the f_{i_2} direction and so on. This yields a sequence of non-negative integers

$$\mathrm{BZL}(v) = (b_1, b_2, \ldots, b_\ell), \tag{17.5}$$

and the endpoint of the crystal always corresponds to the 'lowest weight' $v_- \in \mathcal{B}_{\Lambda+\rho}$ with $\mathrm{wt}(v_-) = w_{\mathrm{long}}(\Lambda + \rho)$. A vertex v is uniquely characterised by its string $\mathrm{BZL}(v)$ (for a fixed choice of reduced expression (17.4)).

For determining the function $G(v)$, the BZL string $(b_1, \ldots, b_\ell)$ needs to be decorated further. In the case of $G = GL(r+1)$ of rank r this is described in [107] for two choices of reduced w_{long} words. We will here give the version for

$$w_{\mathrm{long}} = w_1 w_2 w_1 w_3 w_2 w_1 \cdots w_r w_{r-1} \cdots w_2 w_1 \tag{17.6}$$

and note that $\ell = r(r+1)/2$.

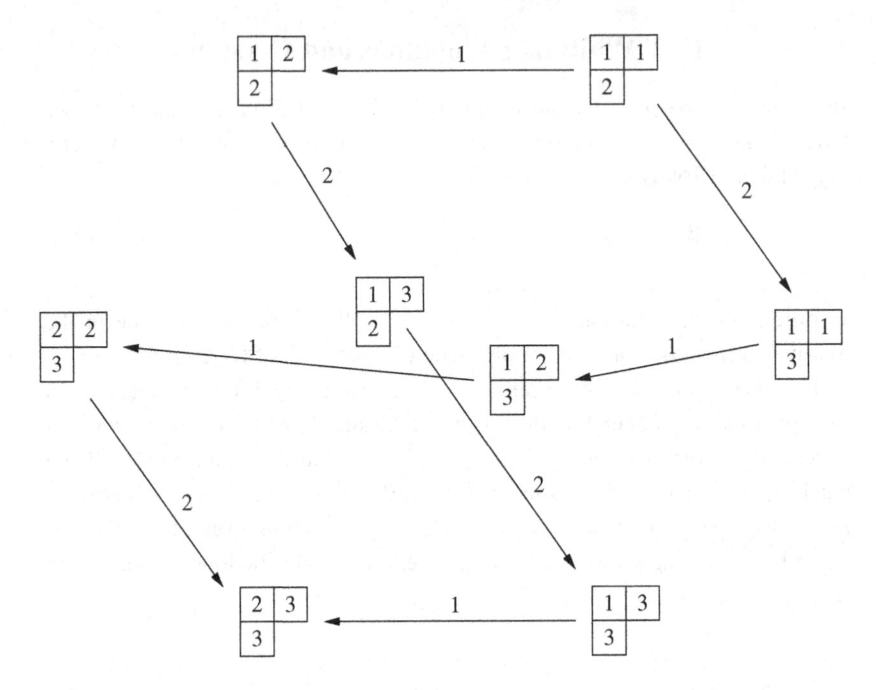

Figure 17.1 The crystal $\mathcal{B}_{\Lambda+\rho}$ for $SL(3)$ and $\Lambda = 0$. The ρ-shift turns this into the weight diagram of the adjoint representation and we label the different vertices of the crystal by filled Young tableaux. The arrows with numbers indicate the action of the operators f_1 and f_2. The two tableaux in the centre correspond to the multiplicity two weight space of the adjoint representation associated with the two-dimensional Cartan subalgebra. For the choice of vertex $v = \begin{smallmatrix}\boxed{1}\boxed{1}\\ \boxed{3}\end{smallmatrix}$, and the reduced expression $w_{\text{long}} = w_1 w_2 w_1$, the BZL path is $\text{BZL}(v) = (2, 1, 0) = \left\{\begin{matrix}\textcircled{0} & \boxed{1}\\ \boxed{2} & \end{matrix}\right\}$, where we have circled and boxed the entries according to the rules described in the text.

Remark 17.1 There is another common choice of reduced expression $w_{\text{long}} = w_r w_{r-1} w_r w_{r-1} w_{r-2} \cdots w_r \cdots w_2 w_1$ that is obtained by starting at the other end of the Dynkin diagram [108]. We will not use it here – it corresponds to Δ-ice in a statistical mechanics interpretation whereas the choice here corresponds to Γ-ice.

The numbers in the BZL string $\text{BZL}(v)$ of (17.5) are then arranged in a triangular (Gelfand–Tsetlin-like) pattern according to

$$\text{BZL}(v) = \left\{\begin{matrix} \cdots & \cdots & \cdots \\ b_3 & b_2 & \\ b_1 & & \end{matrix}\right\}, \tag{17.7}$$

such that the ith column contains all numbers associated with the w_i fundamental reflection. Littelmann proved that the numbers along a fixed row are weakly increasing [453]. Entries in this tableaux now get circle or box decorations according to the following rules: (i) if an entry b_k is equal to its left neighbour (or equal to 0 if it does not have one) it is circled; (ii) if the crystal point $f_{i_{k-1}}^{b_{k-1}} \cdots f_{i_1}^{b_1} v$ does not have a neighbour in the e_{i_k} direction, i.e., it sits on the boundary, then b_k is boxed. An example of this description is given in Figure 17.1 for the case $r = 3$. The boxing and circling rules can be given a geometrical interpretation in terms of the embedding of $\mathcal{B}_{\Lambda+\rho}$ into $\mathcal{B}_\infty$ [119].

The decorated BZL string can then be used to define the Tokuyama function $G(v)$ via [118]

$$G(v) = \prod_{b_k \in \mathrm{BZL}(v)} \begin{cases} 1 & \text{if } b_k \text{ is circled but not boxed} \\ -p^{-1} & \text{if } b_k \text{ is boxed but not circled} \\ 1 - p^{-1} & \text{if } b_k \text{ is neither boxed nor circled} \\ 0 & \text{if } b_k \text{ is boxed and circled.} \end{cases} \tag{17.8}$$

More complicated versions of this rule exist for other values of the Tokuyama deformation parameter t [107]. An equivalent description of $G(v, t)$ directly in terms of Gelfand–Tsetlin patterns was given in [104, 108, 109].

Tokuyama's formula for the Tokuyama function $G(v)$ of (17.8) gives a purely combinatorial description of the spherical Whittaker function $W^\circ(\lambda, a_\Lambda)$ for the case $G = GL(r + 1, \mathbb{Q}_p)$. One may wonder whether other choices of Tokuyama function $G(v)$ also correspond to objects related to automorphic forms. An affirmative answer to this was provided by Bump, Brubaker and Friedberg, and we will discuss this next, in a broader context [118]. But first we illustrate the crystal picture for the p-adic spherical Whittaker function of the $SL(2, \mathbb{A})$ Eisenstein series.

Example 17.2: Crystal description of $SL(2, \mathbb{Q}_p)$ Whittaker function

We consider the case $G = SL(2, \mathbb{Q}_p)$ and verify formulas (17.3) and (17.8). The Whittaker function $W^\circ(\lambda, a_\Lambda)$ is (see (9.89))

$$W^\circ(\lambda, a_\Lambda) = (1 - p^{-2s}) \frac{p^{sN-N} - p^{-2s+1-sN}}{1 - p^{-2s+1}}. \tag{17.9}$$

To work out the crystal sum we fix the longest Weyl word as $w_{\text{long}} = w_1$ and the highest weight as $\Lambda + \rho = (N + 1)\rho$, where we recall that everything refers to Langlands dual group of $SL(2, \mathbb{Q}_p)$. Then the crystal $\mathcal{B}_{\Lambda+\rho}$ consists of the vertices $v \in \{(N + 1)\rho, (N - 1)\rho, \ldots, -(N + 1)\rho\}$, which we label $v_k = (N + 1 - 2k)\rho$ for $k = 0, \ldots, N + 1$. The highest weight representation $V_{\Lambda+\rho}$ is of dimension $N + 2$. The BZL path of a vertex v_k is

$$\mathrm{BZL}(v_k) = (N + 1 - k) \tag{17.10}$$

and its single entry is circled for $k = N + 1$ and boxed for $k = 0$; otherwise it is undecorated. Therefore

$$G(v_k) = \begin{cases} -p^{-1} & \text{for } k = 0 \\ 1 & \text{for } k = N + 1 \\ 1 - p^{-1} & \text{otherwise.} \end{cases} \tag{17.11}$$

The right-hand side of Equation (17.3) therefore becomes ($a_\lambda^\rho = p^{-(2s-1)/2}$)

$$\begin{aligned} p^{-N/2} \sum_{k=0}^{N+1} G(v_k) a_\lambda^{(N+2-2k)\rho} &= p^{-sN-2s+1}\Big(-p^{-1} + (1-p^{-1}) \sum_{k=1}^{N} p^{k(2s-1)} \\ &\qquad + p^{(N+1)(2s-1)} \Big) \\ &= p^{-sN-2s+1}\Big(-1 + (1-p^{-1}) \frac{1-p^{(N+1)(2s-1)}}{1-p^{2s-1}} \\ &\qquad + p^{(N+1)(2s-1)} \Big) \\ &= p^{-(N+2)s}(1-p^{2s}) \frac{1-p^{(N+1)(2s-1)}}{p^{2s-1}-1} \\ &= p^{-sN}(1-p^{-2s}) \frac{p^{N(2s-1)} - p^{-2s+1}}{1-p^{-2s+1}}, \end{aligned} \tag{17.12}$$

which agrees with (17.9).

17.3 Weyl Group Multiple Dirichlet Series

In Sections 11.2.4 and 11.8 we introduced Dirichlet series and automorphic L-functions. Both are meromorphic functions of a single complex variable s, satisfy functional equations for $s \leftrightarrow 1-s$ and have an Euler product form. They correspond to multiplicative sequences a_n of numbers; in the simplest case of an $SL(2,\mathbb{Z})$ cuspidal Hecke eigenform f these are just the Fourier coefficients of f (see (11.34)) so there is a close connection between Fourier expansions of automorphic forms and Dirichlet series. See Sections 11.8 and 11.9 for more details.

It is natural to wonder whether these concepts can be generalised to functions of *several* complex variables $s_1, \ldots, s_r$. This is a non-trivial problem and it turns out that multiplicativity of the coefficients cannot be maintained; see [118], which also discusses the history of the subject. One way of constructing such multiple Dirichlet series is as so-called *Weyl group multiple Dirichlet series* [100].

To introduce them we again restrict to $G = GL(r+1)$ of rank r and introduce the following additional definitions [102]. Let $\mathbb{F}$ be a number field that contains the group μ_{2n} of $2n$th roots of unity. Let S be a finite set of places of $\mathbb{F}$ such that S includes all archimedean places (e.g., $p = \infty$) and all divisors of n. We denote by $\mathbb{F}_p$ the completion of $\mathbb{F}$ at a place p, and by $\mathfrak{o}_p$ the corresponding integers for p non-archimedean. The ring of S-integers $\mathfrak{o}_S$ in $\mathbb{F}$ are those $x \in \mathbb{F}$ whose component x_p is in $\mathfrak{o}_p$ for all $p \notin S$. We can, if necessary, enlarge S such that the S-integers $\mathfrak{o}_S$ are a principal ideal domain. We denote $\mathbb{F}_S = \prod_{p\in S} \mathbb{F}_p$.

A general form for a multiple Dirichlet series is then given by

$$Z_\Psi(\mathbf{s}, \mathbf{m}) = \sum_{\text{ideals}(C_i)} \Psi(C_1, \ldots, C_r) H(C_1, \ldots, C_r; m_1, \ldots, m_r) |C_1|^{-2s_1} \cdots |C_r|^{-2s_r} \tag{17.13}$$

for $\mathbf{s} = (s_1, \ldots, s_r) \in \mathbb{C}^r$ and $\mathbf{m} = (m_1, \ldots, m_r) \in (\mathfrak{o}_S)^r$. Here, $\Psi\colon (\mathbb{F}_S^\times)^r \to \mathbb{C}$ and $H\colon (\mathbb{F}_S^\times)^r \times (\mathfrak{o}_S)^r \to \mathbb{C}$ are functions with multiplicativity properties that ensure that the sum over ideals in the principal ideal domain $\mathfrak{o}_S$ is well-defined. These properties rely on the properties of the *nth-order Hilbert symbol* and on nth-order reciprocity. We will make no further use of the precise conditions here and refer the reader to [102] for the details. We note, however, that the conditions make Ψ a member of a finite-dimensional vector space $\mathcal{M}$ and that this space carries an action of the Weyl group $\mathcal{W}$ [102].

The function H satisfies an additional multiplicative property, called *twisted multiplicativity*, that ensures that it is completely determined by its values on prime powers $H(p^{k_1}, \ldots, p^{k_r}; p^{l_1}, \ldots, p^{l_r})$. The parameters $\mathbf{m}$ appearing in (17.13) are called the twisting parameters and they enter crucially in the twisted multiplicativity relation. The problem of finding interesting multiple Dirichlet series is then reduced to specifying the $H(p^{k_1}, \ldots, p^{k_r}; p^{l_1}, \ldots, p^{l_r})$. One requirement that one would naturally impose on them is that, when viewed as a function of one s_i alone, one obtains sums of single Dirichlet functions with standard functional relations. (Since we are working over a field that contains μ_{2n} these will actually be so-called *Kubota Dirichlet series* [426], and we refer again to [102] for the details.)

This requirement will ensure that the multiple Dirichlet series $Z_\Psi(\mathbf{s}, \mathbf{m})$ will satisfy a functional relation under the fundamental Weyl reflection w_i of the form

$$Z_\Psi(w_i \mathbf{s}, \mathbf{m}) = Z_{\Psi'}(\mathbf{s}, \mathbf{m}), \tag{17.14}$$

where Ψ' is some other element of the finite-dimensional space $\mathcal{M}$. By choosing a suitable normalisation

$$Z^\star_\Psi(\mathbf{s}, \mathbf{m}) = \left(\prod_{\alpha>0} \zeta_\alpha(\mathbf{s}) G_\alpha(\mathbf{s}) \right) Z_\Psi(\mathbf{s}, \mathbf{m}) \tag{17.15}$$

in terms of factors of the Dedekind zeta function of $\mathbb{F}$ and appropriate Γ-function factors evaluated at places parametrised by the positive roots α and $\mathbf{s}$, one can bring this functional relation into the nicer form

$$Z^{\star}_{w\Psi}(w\mathbf{s}, \mathbf{m}) = Z^{\star}_{\Psi}(\mathbf{s}, \mathbf{m}) \tag{17.16}$$

by using the action of $\mathcal{W}$ on $\mathcal{M}$. A functional equation of this type allows meromorphic continuation of the multiple Dirichlet series from the domain of convergence of (17.13) to $\mathbb{C}^r$ by means of a variant of Bochner's theorem about complex functions in tube domains [100].

Finding $H(p^{k_1}, \dots, p^{k_r}; p^{l_1}, \dots, p^{l_r})$ that satisfy this requirement is a non-trivial combinatorial problem. Essential information on the multiple Dirichlet series is contained in the so-called *p-part* of Z_Ψ, which is defined by suppressing Ψ:

$$\sum_{k_i=0}^{\infty} H(p^{k_1}, \dots, p^{k_r}; p^{l_1}, \dots, p^{l_r}) |p|^{-2k_1 s_1 - \dots - 2k_r s_r}. \tag{17.17}$$

The p-part depends on the s_i and on the twisting parameters that are now given in terms of the integers l_i. We would like to interpret the p-part as an expression on a crystal of ${}^LG$ evaluated at a special point a_λ as in (17.3). To this end, we consider the case when the non-negative integers $(k_1, \dots, k_r)$ correspond to a vector κ linking a (strongly dominant) highest weight $\Lambda + \rho$ to one of its Weyl images, i.e.,

$$\kappa = \sum_i k_i \alpha_i = \Lambda + \rho - w(\Lambda + \rho) \tag{17.18}$$

for some $w \in \mathcal{W}$, such that the second term in (17.17) is basically $\delta^{1/2}(a_\Lambda) a_\lambda^{\kappa+\rho}$. The highest weight Λ here is determined by the integers l_i through $\Lambda = \sum_i l_i \varpi_i$, where ϖ_i are the fundamental weights of ${}^LG$. Suppose now that the twisting parameters l_i (and hence also Λ) are fixed. Then one can evaluate $H(p^{k_1}, \dots, p^{k_r}; p^{l_1}, \dots, p^{l_r})$ for those k_i for which (17.18) is satisfied as a finite product of Gauss sums [103]

$$H(p^{k_1}, \dots, p^{k_r}; p^{l_1}, \dots, p^{l_r}) = \prod_{\substack{\alpha > 0 \\ w\alpha < 0}} g_2(p^{\langle \Lambda+\rho | \alpha \rangle - 1}, p^{\langle \Lambda+\rho | \alpha \rangle}), \tag{17.19}$$

where all elements now refer to the Langlands dual group and g_2 is a certain Gauss sum [103]. The points k_i, for which one thus has a relatively simple formula for the value of H, lie on the Weyl orbit of the highest weight $\Lambda + \rho$. It is an important observation that the only other values of k_i for which $H(p^{k_1}, \dots, p^{k_r}; p^{l_1}, \dots, p^{l_r})$ can be non-zero are the other points of the crystal $\mathcal{B}_{\Lambda+\rho}$. We are thus in a very similar situation to the discussion of the Weyl character formula and its generalisations above. For the strict Weyl character of the highest weight representation $V_{\Lambda+\rho}$ the crystal sum (17.2) only had support on the Weyl images of $\Lambda+\rho$, but for the spherical Whittaker function one needed to consider also the other points of $\mathcal{B}_{\Lambda+\rho}$, in particular those in its interior.

Determining $H(p^{k_1},\dots,p^{k_r};p^{l_1},\dots,p^{l_r})$ at the other points of $\mathcal{B}_{\Lambda+\rho}$ is non-trivial and a number of approaches to this problem exist:

- One approach is called the *averaging method* of Chinta–Gunnells [137, 138, 118], which employs a deformed character constructed using the averaged Weyl group action on the field of rational functions in several variables. This approach works uniformly for any type of root system. It has also been extended to the affine case in [451].
- The approach by Bump, Brubaker and Friedberg [103, 107] starts from the just mentioned analogy with the crystal sum and finds rules for computing the Tokuyama function $G(v)$ on the crystal. These resemble the rules (17.8) above but instead one gives different weights to the various parts of the BZL path. These weights are not simple powers of p but instead involve nth-order Gauss sums. In this approach one has to treat each type of root system separately. For the various classical and some exceptional types we refer the reader to [101, 54, 53, 230, 139, 228, 229].
- Given a Tokuyama function $G(v)$ on the crystal one might wonder, in view of (17.3), whether the crystal sum can be interpreted as the Whittaker coefficient of some Eisenstein series. It turns out that in order for this to be true one needs to consider *metaplectic Eisenstein series*. These are functions defined over the group $G(\mathbb{F})$ where $\mathbb{F}$ is the number field that contains the roots of unity μ_{2n} (see also Section 12.2). One can define Eisenstein series over $G(\mathbb{F})$ and consider their Whittaker coefficients in the same way as for $G(\mathbb{Q}_p)$. However, it turns out that for $n>1$ one no longer has uniqueness of the Whittaker functional and that over a global field the Eulerian property is similarly no longer guaranteed. The multiplicative property of standard (non-metaplectic) Whittaker coefficients is replaced by the *twisted multiplicativity* that we have encountered above. The *metaplectic Whittaker functions* are sources for the coefficients of multiple Dirichlet series as shown by Brubaker, Bump and Friedberg [107, 118, 472, 473, 110]. This approach works for all types of root system.
- Finally, one can interpret the crystal sum as the partition function of a statistical mechanical model, as done by Brubaker, Bump and Friedberg in [106, 118]. In the case $n=1$, this opens up new tools from the theory of integrable systems, most notably the *Yang–Baxter equation*. This goes back to work of Kostant on the (quantum) Toda lattice and representation theory [422]. In this approach one has to treat different types of root systems differently.

The equivalence of these different approaches has been shown in many cases and we refer to [118] for an overview.

18

Automorphic Forms on Non-split Real Forms

So far in this book we have mainly assumed that $G(\mathbb{R})$ is a Lie group in its split real form. The representation theory differs considerably if this assumption is relaxed. Instead of attempting a general discussion, we shall consider the detailed example of $G = SU(2,1)$, a non-split real form of $SL(3,\mathbb{C})$, in this chapter. This case is interesting for several reasons. The maximal compact subgroup of $SU(2,1)$ is $U(2) = SU(2) \times U(1)$ and this is therefore one of the few examples that allows for both holomorphic discrete series and quaternionic discrete series representations. We have also seen in Section 15.4.4 that automorphic forms on $SU(2,1)$ have interesting applications in string theory, following [42, 43, 511]. In this chapter, we will revisit some of the results of [42] and put them more firmly in the mathematical context of the present treatment.

18.1 Eisenstein Series on $SU(2,1)$

In this section we will introduce the basic theory of the group $SU(2,1)$ and the associated symmetric space $SU(2,1)/U(2)$. We will further recall some properties of the number field $\mathbb{Q}(i)$ and the associated Gaussian integers $\mathbb{Z}[i]$, required to define the Picard modular group. Finally we use this data to construct an Eisenstein series on $SU(2,1)/U(2)$. Some of the relevant structure theory of the Lie algebra $\mathfrak{su}(2,1)$ is recalled in Appendix C.

18.1.1 The Symmetric Space $SU(2,1)/U(2)$

Let

$$SU(2,1) = \left\{ g \in GL(3,\mathbb{C}) \;\middle|\; g^\dagger \eta g = \eta \quad \text{and} \quad \det(g) = 1 \right\}, \tag{18.1}$$

where $(\cdot)^\dagger$ denotes the conjugate transpose and the invariant metric is given by

$$\eta = \begin{pmatrix} 0 & 0 & -i \\ 0 & 1 & 0 \\ i & 0 & 0 \end{pmatrix}. \tag{18.2}$$

$SU(2,1)$ is an algebraic group that can be defined over $\mathbb{Q}$ or any algebraic number field.

Let $B = AN$ be a Borel subgroup of $SU(2,1)$. The maximal non-compact torus $A \cong GL(1)$ is generated by the non-compact Cartan generator H and the unipotent N is a three-dimensional Heisenberg group, corresponding to the subspace $\mathfrak{g}_1 \oplus \mathfrak{g}_2$ in the five-grading (C.10).

We parametrise a representative $g \in SU(2,1)/U(2)$ in Iwasawa form by

$$\begin{aligned} g = na &= e^{\chi X_{(1)} + \tilde{\chi} X'_{(1)} + 2\psi X_{(2)}} e^{-\phi H} \\ &= \begin{pmatrix} e^{-\phi} & \tilde{\chi} - \chi + i(\chi + \tilde{\chi}) & e^{\phi}(2\psi + i(\chi^2 + i\tilde{\chi}^2)) \\ 0 & 1 & e^{\phi}(\chi + \tilde{\chi} + i(\tilde{\chi} - \chi)) \\ 0 & 0 & e^{\phi} \end{pmatrix}. \end{aligned} \tag{18.3}$$

In these coordinates the invariant metric on $SU(2,1)/U(2)$ takes the form

$$ds^2 = d\phi^2 + e^{2\phi}(d\chi^2 + d\tilde{\chi}^2) + e^{4\phi}(d\psi + \chi d\tilde{\chi} - \tilde{\chi} d\chi)^2. \tag{18.4}$$

The space $SU(2,1)/SU(2) \times U(1)$ is an example of a so-called *quaternionic Kähler manifold*. These are real manifolds of dimension $4d$ admitting a triplet of almost complex structures but generically no integrable complex structure, as we will see in more detail in Section 18.6. Among general quaternionic Kähler spaces, $SU(2,1)/SU(2) \times U(1)$ is special in that it is also really a complex Kähler manifold, and we will display the Kähler potential underlying the metric (18.4) below. (In general, quaternionic Kähler spaces are not Kähler manifolds, which is a confusing but standard terminology.)

Constructing the following complex coordinates

$$\begin{aligned} z_1 &= 2\psi + i\left(e^{-2\phi} + \tfrac{1}{2}|z_2|^2\right) = 2\psi + i\left(e^{-2\phi} + \chi^2 + \tilde{\chi}^2\right), \\ z_2 &= \chi + \tilde{\chi} + i(\tilde{\chi} - \chi), \end{aligned} \tag{18.5}$$

we also have an isomorphism with the complex hyperbolic space

$$\begin{aligned} \mathbb{CH}^1 &= \left\{ Z = (z_1, z_2) \in \mathbb{C}^2 \;\middle|\; \mathcal{F}(Z) = \mathrm{Im}(z_1) - \tfrac{1}{2}|z_2|^2 > 0 \right\} \\ &\cong SU(2,1)/U(2). \end{aligned} \tag{18.6}$$

In this picture the metric (18.4) can be obtained from the *Kähler potential*

$$K(Z) = \log \mathcal{F}(Z), \tag{18.7}$$

and $SU(2,1)$ acts via fractional transformations

$$g \cdot Z = \frac{AZ + B}{CZ + D} \qquad \text{for} \quad g = \begin{pmatrix} A & B \\ C & D \end{pmatrix}, \tag{18.8}$$

where the blocks A, B, C and D have the sizes (2×2), (2×1), (1×2) and (1×1), respectively, so that $CZ + D \in \mathbb{C}$. The height function $\mathcal{F}(Z)$ transforms as

$$\mathcal{F}(g \cdot Z) = \frac{\mathcal{F}(Z)}{|CZ + D|^2}, \tag{18.9}$$

implying that the condition $\mathcal{F}(Z) > 0$ is preserved and the action is indeed isometric. Note that this transformation is very similar to that of the imaginary part on the Poincaré upper half-plane, given for example in (14.42), and thus $\mathcal{F}(Z)$ can also be thought of as a character on the Borel subgroup.

18.1.2 The Picard Modular Group

We are interested in the following special types of discrete transformations:

- *Heisenberg shifts* for $a, b, c \in \mathbb{Z}$

$$\begin{aligned} \phi &\longmapsto \phi, \\ \chi &\longmapsto \chi + a, \\ \tilde{\chi} &\longmapsto \tilde{\chi} + b, \\ \psi &\longmapsto \psi + \frac{1}{2}c - a\tilde{\chi} + b\chi; \end{aligned} \tag{18.10}$$

- *Symplectic transformation*

$$(\chi, \tilde{\chi}) \longmapsto (-\tilde{\chi}, \chi); \tag{18.11}$$

- *Involution*

$$\begin{aligned} \phi &\longmapsto -\frac{1}{2} \ln \left[\frac{e^{-2\phi}}{4\psi^2 + [e^{-2\phi} + \chi^2 + \tilde{\chi}^2]^2} \right], \\ \chi &\longmapsto \frac{2\psi\tilde{\chi} - (e^{-2\phi} + \chi^2 + \tilde{\chi}^2)\chi}{4\psi^2 + [e^{-2\phi} + \chi^2 + \tilde{\chi}^2]^2}, \\ \tilde{\chi} &\longmapsto -\frac{2\psi\chi + (e^{-2\phi} + \chi^2 + \tilde{\chi}^2)\tilde{\chi}}{4\psi^2 + [e^{-2\phi} + \chi^2 + \tilde{\chi}^2]^2}, \\ \psi &\longmapsto -\frac{\psi}{4\psi^2 + [e^{-2\phi} + \chi^2 + \tilde{\chi}^2]^2}. \end{aligned} \tag{18.12}$$

These transformations (with $a = b = c = 1$) generate the *Picard modular group* (non-minimally) [205], defined by

$$SU(2,1;\mathbb{Z}[i]) := SU(2,1) \cap SL(3,\mathbb{Z}[i]), \tag{18.13}$$

where $\mathbb{Z}[i]$ denotes the *Gaussian integers* further discussed in Remark 18.1.3. The matrix representations of the generators of the Picard modular group are

$$T_1 = \begin{pmatrix} 1 & -1+i & i \\ 0 & 1 & 1-i \\ 0 & 0 & 1 \end{pmatrix}, \tilde{T}_1 = \begin{pmatrix} 1 & 1+i & i \\ 0 & 1 & 1+i \\ 0 & 0 & 1 \end{pmatrix}, T_2 = \begin{pmatrix} 1 & 0 & 1 \\ 0 & 1 & 0 \\ 0 & 0 & 1 \end{pmatrix},$$
$$R = \begin{pmatrix} i & 0 & 0 \\ 0 & -1 & 0 \\ 0 & 0 & i \end{pmatrix}, \quad S = \begin{pmatrix} 0 & 0 & i \\ 0 & -1 & 0 \\ -i & 0 & 0 \end{pmatrix}. \tag{18.14}$$

18.1.3 A Digression on the Number Field $\mathbb{Q}(i)$

The integers in the *imaginary quadratic number field*

$$\mathbb{Q}(i) = \{p + iq \mid p, q \in \mathbb{Q}\} \tag{18.15}$$

are called the *Gaussian integers* $\mathbb{Z}[i] = \{m + in \mid m, n \in \mathbb{Z}\}$. The field $\mathbb{Q}(i)$ is an extension of $\mathbb{Q}$ of degree two, and the *principal ideal domain* $\mathbb{Z}[i]$ has unique prime factorisation with the following behaviour of primes compared to the primes of $\mathbb{Z}$. The rational primes $p \equiv 3 \bmod 4$ are called *inert primes* and also primes of $\mathbb{Z}[i]$. The rational primes $p \equiv 1 \bmod 4$ are called *split primes* and can be written as $p = g\bar{g}$ in terms of Gaussian primes, e.g. $5 = (1 + 2i)(1 - 2i)$. Finally, the rational prime $p = 2$ *ramifies* in $\mathbb{Z}[i]$ which means that $2 = -i(1+i)^2$, such that the Gaussian prime $(1+i)$ appears in the factorisation of $p = 2$ with a power greater than one. There are four units $\{\pm 1, \pm i\} \subset \mathbb{Z}[i]$ and any Gaussian prime g defines a *prime ideal* $g\mathbb{Z}[i] \subset \mathbb{Z}[i]$.

The *Dedekind zeta function* of the imaginary quadratic number field $\mathbb{Q}(i)$ is given by

$$\zeta_{\mathbb{Q}(i)}(s) = \frac{1}{4} \sum_{\omega \in \mathbb{Z}[i]}{}' |\omega|^{-2s}. \tag{18.16}$$

The $1/4$ comes from the number of units of $\mathbb{Z}[i]$. As for all principal ideal domains, the Dedekind zeta function can also be expressed as an Euler product [500]:

$$\zeta_{\mathbb{Q}(i)}(s) = \prod_{g \text{ prime in } \mathbb{Z}[i]} \frac{1}{1 - N(g)^{-s}}, \tag{18.17}$$

generalising the Euler product of the Riemann zeta function (1.22). The *absolute norm* in this simple case is $N(g) = |g|^2$; thus the product, when separated into the three types of primes, becomes

$$\zeta_{\mathbb{Q}(i)}(s) = \frac{1}{1-2^{-s}} \prod_{p\equiv 1 \bmod 4} \frac{1}{(1-p^{-s})^2} \prod_{p\equiv 3 \bmod 4} \frac{1}{1-p^{-2s}}$$
$$= \prod_{p<\infty} \frac{1}{1-p^{-s}} \prod_{p\equiv 1 \bmod 4} \frac{1}{1-p^{-s}} \prod_{p\equiv 3 \bmod 4} \frac{1}{1+p^{-s}} . \tag{18.18}$$

We recognise the Euler product (1.22) of the Riemann zeta function, and the other two factors constitute the *Dirichlet beta function*

$$\beta(s) = \prod_{p\equiv 1 \bmod 4} \frac{1}{1-p^{-s}} \prod_{p\equiv 3 \bmod 4} \frac{1}{1+p^{-s}} , \tag{18.19}$$

such that

$$\zeta_{\mathbb{Q}(i)}(s) = \zeta(s)\beta(s). \tag{18.20}$$

The Dirichlet beta function can be completed by

$$\beta^\star(s) = (\pi/4)^{-(s+1)/2}\Gamma((s+1)/2)\beta(s), \tag{18.21}$$

where the first factor should be thought of as the archimedean contribution in an adelic formulation. The completed Dirichlet beta function satisfies

$$\beta^\star(s) = \beta^\star(1-s), \tag{18.22}$$

and thus the completed Dedekind zeta function $\xi_{\mathbb{Q}(i)}(s) = \xi(s)\beta^\star(s)$ obeys

$$\xi_{\mathbb{Q}(i)}(s) = \xi_{\mathbb{Q}(i)}(1-s) . \tag{18.23}$$

18.1.4 Constructing the Eisenstein Series

We are now finally equipped to construct an $SU(2,1;\mathbb{Z}[i])$-invariant Eisenstein series on the space $SU(2,1)/U(2)$ by

$$E(s,Z) = \sum_{\gamma\in B(\mathbb{Z}[i])\backslash SU(2,1;\mathbb{Z}[i])} \mathcal{F}(\gamma\cdot Z)^s. \tag{18.24}$$

As we will see below, this is attached to the principal series $\mathrm{Ind}_B^{SU(2,1)}\chi$, and converges absolutely for $\mathrm{Re}(s) > 2$. We also note from [42] that this function can be written as a constrained lattice sum of the type discussed in Section 15.3:

$$E(s,Z) = \frac{1}{4\zeta_{\mathbb{Q}(i)}(s)} \sum_{\substack{\omega\in\mathbb{Z}[i]^3\\ \omega^\dagger\eta\omega=0}}{}' \left[\omega^\dagger g g^\dagger \omega\right]^{-s} , \tag{18.25}$$

where the primed sum indicates the exclusion of $\omega = 0$ and the conjugate transpose $\omega^\dagger$ appears naturally for the embedding in a complex group. η denotes the invariant metric (C.8) defining $SU(2,1)$ and the constraint is needed for the sum to be an eigenfunction of the Laplacian. The relation above is similar to (1.11) in the $SL(2,\mathbb{R})$ case. For other work on $SU(2,1)$ automorphic forms see [261, 260, 266, 356, 371, 435, 505, 578].

18.2 Constant Term and L-functions

In [42] the following was proven by Poisson resummation of the lattice form:

Theorem 18.1 (Constant term and functional relation) *The constant term of $E(s,Z)$ is given by*

$$C(s,Z) = \int_{N(\mathbb{Z}[i])\backslash N} E(s, n\cdot Z)dn = y^s + \frac{\mathfrak{Z}(2-s)}{\mathfrak{Z}(s)} y^{2-s}, \tag{18.26}$$

where $y = e^{-2\phi} = \mathcal{F}(Z)$ is the coordinate along the torus A, and $\mathfrak{Z}(s)$ is the 'zeta function'

$$\mathfrak{Z}(s) := \xi(s)\beta^\star(s)\beta^\star(2s-1). \tag{18.27}$$

The constant term furthermore satisfies the functional relation

$$\mathfrak{Z}(s)C(s,Z) = C(2-s,Z)\mathfrak{Z}(2-s). \tag{18.28}$$

Remark 18.2 Note from its definition that $\mathfrak{Z}(s)$ does not satisfy a simple functional equation of its own, but rather has properties similar to (11.154).

In what follows we will reinterpret the explicit constant term expression (18.26) and compare it to earlier works on L-functions in this case, in particular [266]. We begin by verifying that (18.26) is compatible with a suitably modified form of Langlands' general constant term formula for Eisenstein series.

To this end, let G be a reductive group over $\mathbb{R}$ (not necessarily split) and let $B = AN$ be a Borel subgroup. Let $\Gamma \subset G$ be an arithmetic subgroup. A general Eisenstein series on $G(\mathbb{R})$ can then be defined as follows:

$$E(\lambda, g) = \sum_{\gamma\in(B\cap\Gamma)\backslash\Gamma} e^{\langle\lambda+\rho|H(\gamma g)\rangle}, \tag{18.29}$$

where λ is a complex weight of G and ρ is the Weyl vector. Note that ρ and λ are defined only with respect to the *restricted root system* associated with the maximal split torus. In the case of $\mathfrak{su}(2,1)$, this *split* torus is one-dimensional; ρ is twice the fundamental weight Λ_H dual to the split Cartan generator H and can also be written as $\rho = 2\alpha$ in terms of the non-divisible root α of the restricted root system $(BC)_1$; see Appendix C.

The constant term of $E(\lambda, g)$ with respect to the unipotent N is given by [443, 569]

$$\int_{(N\cap\Gamma)\backslash N} E(\lambda, na)dn = \sum_{w\in\mathcal{W}} e^{\langle w\lambda+\rho|H(a)\rangle} \prod_{\substack{\alpha\in\Sigma_+^{\text{nd}}\\ w\alpha<0}} \frac{L_\alpha(\langle\lambda|\alpha\rangle)}{L_\alpha(\langle\lambda|\alpha\rangle+1)}. \tag{18.30}$$

It is important to note that the sum in (18.30) is taken over the *restricted* Weyl group $\mathcal{W}$. For split real forms this is just the ordinary Weyl group, but for non-split forms the Weyl group is defined with respect to only the restricted root system (see Appendix C for some details on this). The product in (18.30) is, moreover, taken over *only the non-divisible positive roots* Σ_+^{nd} in the terminology of [130], which can also be called primitive roots. In (18.30), L_α is a certain L-function that depends on the choice of group G and the root α. In the case of a root of standard A_1-type in the restricted root system, the function L_α is simply the completed Riemann zeta function $\xi(s)$ (see Theorem 8.1).

Remark 18.3 The way we have written the formula (18.30) is slightly unconventional. If a root α is not of A_1-type, this means that α and an integer multiple of α are both roots, potentially with multiplicity, and the action of the corresponding split torus generator on the corresponding root space is then a representation of dimension greater than one. For the L-function L_α one therefore has to consider a graded decomposition similar to (11.151) in Section 11.9, and work out L-functions in a way similar to (11.145). We will come back to this point below in Section 18.3 (see in particular Example 18.5) when we discuss in more detail the connection to the Langlands–Shahidi method.

Let us now try to match the explicit constant term in (18.26) to the general formula (18.30). For $SU(2,1)$ the (restricted) Weyl group $\mathcal{W}$ is $\mathbb{Z}_2$ and the restricted root system is non-reduced of type $(BC)_1$, which implies that both $\alpha/2$ and α are roots (see Appendix C for details on this). We denote the non-divisible root by α; it has multiplicity two, corresponding to the generators $X_{(1)}$ and $X'_{(1)}$ in (18.3), while 2α has multiplicity one, corresponding to the generator $X_{(2)}$ in (18.3). The action of the split Cartan generator H associated with α is then on this $(2+1)$-dimensional space, and we will see below that the L-function L_α in (18.30) gets different contributions from the associated root spaces.

To begin with, the Eisenstein series $E(s, Z)$ on $SU(2,1)$ defined in (18.24) fits into the general framework (18.29) if we fix $\lambda+\rho = 2s\Lambda_H$, where $s\in\mathbb{C}$ and Λ_H is the fundamental weight for H satisfying $\Lambda_H(H) = 1$. Since $[H, X_{(1)}] = X_{(1)}$, one has $\Lambda_H = \alpha$ and also from (C.18) that $\rho = 2\alpha$. The only non-trivial Weyl element then acts on s as follows:

$$w(s) = 2 - s. \tag{18.31}$$

The trivial Weyl element will simply contribute to the constant term by a factor

$$e^{\langle \lambda+\rho|H(a)\rangle} = e^{-2s\phi} = y^s, \tag{18.32}$$

with unit coefficient, consistent with the first term in (18.26).

For the non-trivial Weyl element w we have the non-divisible positive root α that is mapped to a negative root, and hence from (18.30) we obtain

$$e^{\langle w\lambda+\rho|H(a)\rangle} \frac{L_\alpha(\langle\lambda|\alpha\rangle)}{L_\alpha(\langle\lambda|\alpha\rangle+1)}. \tag{18.33}$$

As we mentioned above, the function L_α for non-divisible α receives contributions from *all* the root spaces that H acts on. This means that we should properly think of the intertwiner as

$$\frac{L_\alpha(\langle\lambda|\alpha\rangle)}{L_\alpha(\langle\lambda|\alpha\rangle+1)} = \frac{\widetilde{L}_\alpha(\langle\lambda|\alpha\rangle)}{\widetilde{L}_\alpha(\langle\lambda|\alpha\rangle+1)} \frac{\widetilde{L}_{2\alpha}(\langle\lambda|2\alpha\rangle)}{\widetilde{L}_{2\alpha}(\langle\lambda|2\alpha\rangle+1)} \tag{18.34}$$

by introducing independent $\widetilde{L}$-functions for the two root spaces. For the weight $\lambda = (2s-2)\Lambda_H = (2s-2)\alpha$ as above one obtains

$$\begin{aligned} \langle\lambda|\alpha\rangle &= s-1, \\ \langle\lambda|2\alpha\rangle &= 2s-2, \end{aligned} \tag{18.35}$$

since $\langle\alpha|\alpha\rangle = 1/2$ for the non-divisible root α as induced from the bilinear form of $\mathfrak{sl}(3,\mathbb{C})$. See Appendix C for additional details. Inserting this together with (18.31) into the expanded (18.33) yields

$$y^{2-s} \frac{\widetilde{L}_\alpha(s-1)}{\widetilde{L}_\alpha(s)} \frac{\widetilde{L}_{2\alpha}(2s-2)}{\widetilde{L}_{2\alpha}(2s-1)}. \tag{18.36}$$

Comparing this expression with (18.26) we find perfect agreement if we set

$$\begin{aligned} \widetilde{L}_\alpha(s) &= \xi^\star_{\mathbb{Q}(i)}(s) = \xi(s)\beta^\star(s), \\ \widetilde{L}_{2\alpha}(s) &= \beta^\star(s). \end{aligned} \tag{18.37}$$

Indeed, using (18.37) the formula (18.36) becomes

$$y^{2-s} \frac{\xi(s-1)\beta^\star(s-1)\beta^\star(2s-2)}{\xi(s)\beta^\star(s)\beta^\star(2s-1)}, \tag{18.38}$$

and by using the functional Equations (18.22) of $\beta^\star(s)$ and $\xi(s)$ we can rewrite this as follows:

$$y^{2-s} \frac{\xi(2-s)\beta^\star(2-s)\beta^\star(3-2s)}{\xi(s)\beta^\star(s)\beta^\star(2s-1)}, \tag{18.39}$$

which precisely equals $y^{2-s}\mathfrak{Z}(2-s)/\mathfrak{Z}(s)$ in (18.26).

Remark 18.4 In the literature, the constant term factor is often written differently in terms of an intertwining coefficient. For instance, Gelbart and Rogawski state in [266, p. 462] the relative factor between the two contributions to the constant term more generally as

$$R(s-1) = \frac{L(2(s-1), \xi_F \omega_{E/F}) L(s-1, \xi)}{L(2(s-1)+1, \xi_F \omega_{E/F}) L((s-1)+1, \xi)}, \tag{18.40}$$

where we have translated the conventions of the parameter s. Here, F is a number field and E/F denotes a quadratic extension of F. ξ is a unitary Hecke character of E and ξ_F is its restriction to F. Similarly, $\omega_{E/F}$ denotes the restriction to the idele group I_F of a certain Hecke character on the idele group I_E. In our case, we should take $F = \mathbb{Q}$ and $E = \mathbb{Q}(i)$. The coefficient (18.40) then agrees with our formula (18.38) upon identification of the L-functions similar to (18.37) as

$$\begin{aligned} L(s, \xi) &= \xi(s)\beta^\star(s), \\ L(s, \xi_{\mathbb{Q}} \omega_{\mathbb{Q}(i)/\mathbb{Q}}) &= \beta^\star(s). \end{aligned} \tag{18.41}$$

18.3 Connection with the Langlands–Shahidi Method

According to the Langlands–Shahidi method discussed in Section 11.9, the local L-factors appearing in the constant term should correspond to the inverse of Casselman–Shalika's unramified Whittaker function discussed in Chapter 9 for the split case. In this section, we wish to verify this for the constant term of the $SU(2,1)$ Eisenstein series $E(s, Z)$ defined in (18.24).

To this end we generalise some terminology on Whittaker coefficients from Section 5.3 and Chapter 9. Let F be a non-archimedean local field (e.g., $F = \mathbb{Q}_p$ or a completion of $\mathbb{Q}(i)$ that we discuss below) and O its ring of integers (e.g., $\mathbb{Z}_p$). Let χ be a character on the Borel subgroup $B(F) \subset G(F)$, extended to all of G by using the Iwasawa decomposition and demanding that χ is right-invariant under $K = G(O)$. We denote the principal series by $I(\chi) = \mathrm{Ind}_{B(F)}^{G(F)} \chi$. Fix an unramified character ψ on the unipotent $N(F)$ that is trivial on $N(O)$. Recall from Definition 6.12 that an unramified character can be defined by

$$\psi(n) = \psi\left(\exp \sum_{\alpha>0} x_\alpha E_\alpha\right) = \exp\left(2\pi i \sum_{\alpha \in \Pi} x_\alpha\right), \qquad x_\alpha \in F, \tag{18.42}$$

where Π denotes the simple roots. In the context of non-split groups, 'simple root' refers to all non-divisible simple roots of the restricted root system Σ. If there are non-trivial root multiplicities, there are as many x_α as mult(α) demands. For $SU(2,1)$ this means two directions, associated with χ and $\tilde{\chi}$ in the parametrisation (18.3).

Recall from Chapter 9 that the unramified Whittaker function is given by the Jacquet–Whittaker integral (9.13):

$$W_\psi(\chi, a) = \int_{N(F)} \chi(w_{\text{long}} n a)\overline{\psi(n)} dn, \qquad a \in A(F), \tag{18.43}$$

where w_{long} is the longest Weyl word in the restricted Weyl group $\mathcal{W}$. The Langlands–Shahidi method ensures that the unramified local Whittaker function at the identity $a = \mathbb{1}$ equals the inverse of the local L-factor associated with the longest Weyl word w_{long} in the constant term. For $F = \mathbb{Q}_p$ (see Equation (11.150)):

$$W_{\psi,p}(\chi, \mathbb{1}) = \frac{1}{\prod\limits_{\substack{\alpha\in\Sigma_+^{\text{nd}} \\ w_{\text{long}}\alpha<0}} L_{\alpha,p}(\langle\lambda|\alpha\rangle)}. \tag{18.44}$$

In the present case, this formula should reproduce the local L-factors in the Euler product of (18.39). To see this we need to discuss in more detail the possible non-archimedean fields that can arise in this case.

The algebraic number field $\mathbb{Q}(i)$ can be completed at any of its primes $g \in \mathbb{Z}[i]$, where the corresponding norm is given by [578, 42]

$$|z|_g = |g|^{-k} \qquad z \in \mathbb{Q}(i)\,. \tag{18.45}$$

Here, k denotes the maximum power of g in the prime factorisation of the Gaussian rational z. This generalises the p-adic norm (2.15). Note that (18.45) is not invariant under complex conjugation of z.

The completion of $\mathbb{Q}(i)$ at g now depends strongly on whether g sits over a splitting rational prime $p \in \mathbb{Q}$ or not. Recall from Remark 18.1.3 that the splitting rational primes are $p \equiv 1 \bmod 4$ in the field extension $\mathbb{Q} \hookrightarrow \mathbb{Q}(i)$ of degree two. We will treat the ramified case $p = 2$ separately later. The complex numbers $\mathbb{C}$ appearing in the definition of $SU(2, 1)$ in (18.1) arise from the completion at the archimedean place using the standard norm on $\mathbb{Q}(i)$.

Let $p \equiv 1 \bmod 4$ be a splitting prime of $\mathbb{Z}$. Then, by *Hensel's lemma* [440], we can solve the equation $x^2 = -1$ over the p-adic integers $\mathbb{Z}_p \subset \mathbb{Q}_p$. For example, for $p = 5$, we can choose

$$-1 = (2 + 1\cdot 5^1 + 2\cdot 5^2 + 1\cdot 5^3 + 3\cdot 5^4 + \cdots)^2, \tag{18.46}$$

where the expansion of the solution in $\mathbb{Z}_p$ can be computed recursively, solving modulo p^k. (There is another solution to the quadratic equation starting with $3 + 3\cdot 5 + 2\cdot 5^2 + \cdots$.) For inert primes $p \equiv 3 \bmod 4$ there is no solution to $x^2 = -1$ in $\mathbb{Q}_p$. The existence of a solution to $x^2 = -1$ in $\mathbb{Q}_p$ for split primes means that the completion of $\mathbb{Q}(i)$ at a Gaussian prime g such that $g\bar{g} = p$ is in fact isomorphic to $\mathbb{Q}_p$ [441, 440]:

$$(\mathbb{Q}(i))_g \cong \mathbb{Q}_p \tag{18.47}$$

since -1 is already part of $\mathbb{Q}_p$. In other words, the degree of the local extension $\mathbb{Q}_p \hookrightarrow (\mathbb{Q}(i))_g$ is equal to one for split primes while the degree of the global extension $\mathbb{Q} \hookrightarrow \mathbb{Q}(i)$ is always equal to two. The two Gaussian primes g and $\bar{g}$ give inequivalent isomorphisms of the above form and one has $\overline{(\mathbb{Q}(i))_g} = (\mathbb{Q}(i))_{\bar{g}}$. For inert primes the degree of the local extension is the same as that of the global extension.

The different behaviour of the completions of $\mathbb{Q}(i)$ for split and inert primes means that for inert primes p one has a genuine unitary group $SU(2,1;(\mathbb{Q}(i))_p)$ with a restricted $(BC)_1$ type root system. In the case of split primes $p = g\bar{g}$ the situation for the group $SU(2,1;(\mathbb{Q}(i))_g)$ is quite different. The (Galois) conjugation $(\cdot)^\dagger$ entering the definition (18.1) relates the two primes g and $\bar{g}$ of the global field extension and thus the two local field extensions must be treated together. However, since $i = \sqrt{-1}$ is now in the base field by (18.46), the analysis of the restricted root system proceeds very differently from the archimedean case in Appendix C; combining the contribution from the two primes g and $\bar{g}$ one deals effectively with the group $SL(3,\mathbb{Q}_p)$ for split primes $p = g\bar{g}$.

With this preliminary, we can now express the local Whittaker coefficients (18.44) using the Casselman–Shalika formula. For a split prime p, we can use the formula (9.23) that was already evaluated for $SL(3,\mathbb{Q}_p)$ at the identity in (9.99a). A generic character on $SL(3,\mathbb{Q}_p)$ maps

$$\begin{pmatrix} v_1 & & \\ & v_2/v_1 & \\ & & 1/v_2 \end{pmatrix} \longmapsto |v_1|_p^{2s_1}|v_2|_p^{2s_2}. \tag{18.48}$$

Here, we have from (18.3) and (18.32) the character $\operatorname{diag}(v,1,1/v) \mapsto |v|_p^{2s}$, such that $s_1 = s_2 = s/2$. Substituting this into (9.99a), we get the local Whittaker coefficient for a split prime p as

$$W_{\psi,p}(\chi,\mathbb{1}) = (1-p^{-s})(1-p^{-s})(1-p^{1-2s}). \tag{18.49}$$

For inert primes p, one has to use a modification of the Casselman–Shalika formula for $SU(2,1;(\mathbb{Q}(i))_p)$ proved in the original paper [130, Thm. 5.4]. The result shown there is, in our notation,

$$\begin{aligned} W_{\psi,p}(\chi,\mathbb{1}) &= \left(1+q^{-1}e^{\langle\lambda|H(a)\rangle}\right)\left(1-q^{-2}e^{\langle\lambda|H(a)\rangle}\right) \\ &= (1+p^{1-2s})(1-p^{-2s}) = (1+p^{1-2s})(1+p^{-s})(1-p^{-s}), \end{aligned} \tag{18.50}$$

where in the first line q is the cardinality of the residue field of the maximal ideal (generated by the uniformiser) within the integers O in $(\mathbb{Q}(i))_p$ and so equals p here. From (18.32) we also get $e^{\langle\lambda|H(a)\rangle} = p^{2-2s}$, leading to the second line in the above equation. The origin of (18.50) is discussed more in Example 18.5 below.

The local Whittaker coefficients should be compared with the local factors in the global L-function $\mathfrak{Z}(s)$, defined in (18.27) and appearing as the denominator in the constant term (18.26). Writing $\mathfrak{Z}(s)$ as an Euler product over the rational primes p as

$$\mathfrak{Z}(s) = \prod_{p\leq\infty} \mathfrak{Z}_p(s), \tag{18.51}$$

we find from (18.18) and (18.19) that

$$\begin{aligned} p \text{ split:} \qquad & \frac{1}{\mathfrak{Z}_p} = (1-p^{-s})(1-p^{-s})(1-p^{1-2s}), \\ p \text{ inert:} \qquad & \frac{1}{\mathfrak{Z}_p} = (1-p^{-s})(1+p^{-s})(1+p^{1-2s}), \end{aligned} \tag{18.52}$$

in precise agreement with (18.49) and (18.50) above. That the ramified prime $p = 2$ and the archimedean place $p = \infty$ in $\mathfrak{Z}(s)$ also work out between the constant term and the Fourier coefficients will follow from the global results presented in the following Section 18.4.

Example 18.5: Local L-functions and reducible actions

Rather than proving the $SU(2,1)$ unramified Whittaker formula (18.50) by a p-adic integration as we did in the split $SL(2)$ case in Chapter 9, we provide here a different motivation using the relation to L-functions of the type discussed in Section 11.9, in particular the discussion around (11.148). We wish to evaluate for *inert* primes p the expression

$$L_p(1,\lambda,\rho) = \frac{1}{\det\left(\mathbb{1} - \rho(A_{\pi_p})p^{-1}\right)}, \tag{18.53}$$

where we have to work with representations of the L-group $\mathrm{Gal}(\mathbb{Q}(i)_p/\mathbb{Q}_p) \ltimes PGL(3,\mathbb{C})$ (see (16.1)). The Galois group of the degree-two extension is $\mathrm{Gal}(\mathbb{Q}(i)_p/\mathbb{Q}(i)) = \{\mathbb{1},\sigma\}$, where σ is related to complex conjugation. The Satake parameter $A_{\pi_p} \in {}^LA$ corresponding to λ is given by

$$A_{\pi_p} = \begin{pmatrix} p^{1-s} & & \\ & 1 & \\ & & p^{s-1} \end{pmatrix}. \tag{18.54}$$

The action $\rho(A_{\pi_p})$ of the dual torus ${}^LA$ on the $(2+1)$-dimensional complex representation ${}^L\mathfrak{n}$ includes the Galois action of σ, which is complex conjugation in this case. Taking the basis

$$X^\vee_{(1)} = \begin{pmatrix} 0 & (1+i)/2 & 0 \\ 0 & 0 & (1-i)/2 \\ 0 & 0 & 0 \end{pmatrix}, \quad X^\vee_{(1')} = \overline{X^\vee_{(1)}} = \begin{pmatrix} 0 & (1-i)/2 & 0 \\ 0 & 0 & (1+i)/2 \\ 0 & 0 & 0 \end{pmatrix},$$

$$X^\vee_{(2)} = \left[X^\vee_{(1)}, X^\vee_{(1')}\right] = \begin{pmatrix} 0 & 0 & i \\ 0 & 0 & 0 \\ 0 & 0 & 0 \end{pmatrix} \tag{18.55}$$

for ${}^L\mathfrak{n}$, the action of A_{π_p} is

$$\rho(A_{\pi_p})(X^\vee_{(1)}) = p^{1-s} X^\vee_{(1')}, \quad \rho(A_{\pi_p})(X^\vee_{(1')}) = p^{1-st} X^\vee_{(1)},$$
$$\rho(A_{\pi_p})(X^\vee_{(2)}) = -p^{2-2s} X^\vee_{(2)}. \tag{18.56}$$

Writing the above action as a matrix, we can evaluate the local L-function (18.53) as

$$\det(\mathbb{1} - \rho_L(A_{\pi_p})p^{-1}) = \det\begin{pmatrix} 1 & -p^{-s} & 0 \\ -p^{-s} & 1 & 0 \\ 0 & 0 & 1+p^{1-2s} \end{pmatrix}$$
$$= (1-p^{-2s})(1+p^{1-2s}). \tag{18.57}$$

This equals (18.50), such that the central feature of the Langlands–Shahidi method, that the unramified Whittaker coefficient at the identity equals the inverse of the local L-function, is verified.

Since the Borel subgroup is both maximal and minimal for $SU(2,1)$, we can easily include a non-trivial character on the parabolic in the sense of (11.146). This just amounts to a shift or reinterpretation of the meaning of s as not being associated with a 'cuspidal' representation λ but with the modulus character of B. Then the above result can also be written as

$$L_p(1,\lambda,\rho) = L_p(1+(s-1),\lambda_0,\rho)$$
$$= L_p(1+(s-1),\lambda_0,\rho_1)L_p(1+2(s-1),\lambda_0,\rho_2), \tag{18.58}$$

where λ_0 indicates that the s-dependence was shifted to the argument of the L-function, making λ_0 trivial. In the last equality, we have also shown explicitly the factorisation of the local L-function into the factors corresponding to the decomposition of reducible representation of the (Levi) torus on ${}^L\mathfrak{n}$, analogous to (11.151). ρ_1 denotes the two-dimensional 'level-one' piece, spanned by $X^\vee_{(1)}$ and $X^\vee_{(1')}$, and ρ_2 the one-dimensional 'level-two' piece, spanned by $X^\vee_{(2)}$. The factorisation above is an instance of the general behaviour exhibited in (11.152).

Finally, comparing to (18.40), we recognise that the two factors in that intertwining coefficient are due to the reducibility of the action of the Levi of the L-group on the nilpotent radical.

18.4 Global Whittaker Coefficients

Let us also consider the global Whittaker coefficients of the Eisenstein series $E(s, Z) = E(s, g)$, with $g \in SU(2, 1)$. Let $\psi\colon N(\mathbb{Z}[i])\backslash N \to \mathbb{U}(1)$ be a unitary character on the Heisenberg unipotent radical N of $SU(2, 1)$. The Whittaker coefficients are given by

$$W_\psi(s, g) = \int_{N(\mathbb{Z}[i])\backslash N} E(s, ng)\overline{\psi(n)}dn = W_\infty(s, g)\left[\prod_{p<\infty} W_p(s, 1)\right], \quad (18.59)$$

where $N(\mathbb{Z}[i])$ denotes the unipotent subgroup generated by integer powers of T_1, $\tilde{T}_1$ and T_2 of (18.14). In [42] the following was proven:

Theorem 18.6 (Whittaker coefficients) *The non-archimedean contribution to the Whittaker coefficients in (18.59) is given by*

$$\prod_{p<\infty} W_p(s, 1) = \frac{2\zeta_{\mathbb{Q}(i)}(s)}{\mathcal{Z}(s)}|\Lambda|^{2(s-1)} \sum_{\omega|\Lambda} |\omega|^{2(1-s)} \sum_{z|\frac{\Lambda}{\omega}} |z|^{4(1-s)}, \quad (18.60)$$

where $\sum_{\omega|\Lambda}$ *is a sum over Gaussian divisors of* $\Lambda \in \mathbb{Z}[i]$ *and* $\omega \in \mathbb{Z}[i]$ *is subject to the constraint* $gcd(\mathrm{Re}(\omega), \mathrm{Im}(\omega)) = 1$. *This means that* ω *is a primitive vector on the lattice* $\mathbb{Z}[i]$. *The Dedekind zeta function* $\zeta_{\mathbb{Q}(i)}(s)$ *was defined in* (18.16).

In addition, the archimedean contribution to (18.59) takes the form

$$W_\infty(s, g) = yK_{2s-2}(2\pi\sqrt{y}|\Lambda|)e^{2\pi i(\mathrm{Re}(\Lambda)\chi + \mathrm{Im}(\Lambda)\tilde{\chi})}, \quad (18.61)$$

where K_t *is the modified Bessel function.*

The global Whittaker coefficient furthermore satisfies the functional relation

$$\mathcal{Z}(s)W_\psi(s, g) = W_\psi(2 - s, g)\mathcal{Z}(2 - s). \quad (18.62)$$

As was explained in Section 6.2.3, when the unipotent N is non-abelian, the Whittaker coefficients are not sufficient to reconstruct the full automorphic form, unless one can invoke some form of the Piatetski-Shapiro–Shalika formula. For $SU(2, 1)$ such a formula is not known, as far as we know. However, since N is a Heisenberg group, the only coefficients that are missing are those that are non-trivial along the one-dimensional centre $Z = [N, N]$ of N. To describe these, we introduce a character along the centre:

$$\psi_Z\colon Z(\mathbb{Z})\backslash Z \to \mathbb{C}^\times. \quad (18.63)$$

We can consider the following Z-Fourier coefficient:

$$F_{\psi_Z}(s, g) = \int_{Z(\mathbb{Z})\backslash Z} E(s, zg)\overline{\psi_Z(z)}dz. \quad (18.64)$$

We then have the following result from [42]:

Theorem 18.7 (Non-abelian Fourier coefficients) *The Z-Fourier coefficient of the Eisenstein series $E(s, g)$ with $g \in SU(2,1)$ can be written as*

$$F_{\psi Z}(s,g) = \sqrt{y} \sum_{k\in\mathbb{Z}}{}' \sum_{\ell=0}^{4|k|-1} \sum_{n\in\mathbb{Z}+\frac{\ell}{4|k|}} \sum_{r=0}^{\infty} C_{r,k,\ell}(s)\, |k|^{1/2-s} e^{-4\pi|k|(\tilde{\chi}-n)^2}$$
$$\times H_r\left(\sqrt{8\pi|k|}(\tilde{\chi}-n)\right) W_{-r-1/2,s-1}\left(4\pi|k|y\right) e^{8\pi ikn\chi - 4\pi ik(\psi+\chi\tilde{\chi})}, \tag{18.65}$$

where the primed sum excludes $k = 0$, and H_r and $W_{-r-1/2,s-1}$ denote the Hermite polynomial and Whittaker function, respectively. The coefficients $C_{r,k,\ell}(s)$ are numerical Fourier coefficients that are, as of yet, unknown.

Remark 18.8 It would be interesting to derive an explicit formula for the numerical coefficients $C_{r,k,\ell}(s)$, analogously to (18.60). This should be possible by an adelic analysis of (18.64). An integral form of the non-abelian Fourier coefficient was given in [42].

18.5 More General Number Fields

It is straightforward to generalise the previous discussion to quadratic extensions $\mathbb{Q}(\sqrt{-d})$, where $d > 0$ is a square-free positive integer. The associated ring of integers $\mathcal{O}_d \subset \mathbb{Q}(\sqrt{-d})$ is the lattice

$$\mathcal{O}_d = \mathbb{Z} + \omega_d\mathbb{Z}, \tag{18.66}$$

where $\omega_d = (-1 + i\sqrt{d})/2$. The Gaussian case considered above corresponds to $d = 1$. Let us for simplicity restrict to the *Stark–Heegner sequence*

$$\{1, 2, 3, 7, 11, 19, 43, 67, 163\}. \tag{18.67}$$

These are the cases for which $\mathcal{O}_d$ has *class number* 1, i.e., is a *principal ideal domain* and hence exhibits unique prime factorisation.

We define the associated *Picard modular group* as follows:

$$PU(2,1;\mathcal{O}_d) := U(2,1) \cap PGL(3,\mathcal{O}_d). \tag{18.68}$$

An Eisenstein series can be constructed by analogy with the Gaussian case above:

$$E_d(s,Z) = \sum_{\gamma\in B(\mathcal{O}_d)\backslash PU(2,1;\mathcal{O}_d)} \mathcal{F}(\gamma\cdot Z)^s. \tag{18.69}$$

We then have from [43]:

Theorem 18.9 (Constant term and functional relation) *The constant term of the Eisenstein series $E_d(s, Z)$ is given by*

$$\mathcal{C}_d(s, Z) = \int_{N(\mathcal{O}_d)\backslash N} E_d(s, n \cdot Z) dn = y^s + \frac{\mathfrak{Z}_d(2-s)}{\mathfrak{Z}_d(s)} y^{2-s}, \tag{18.70}$$

and satisfies the functional relation

$$\mathfrak{Z}_d(s) \mathcal{C}_d(s, Z) = \mathfrak{Z}_d(2-s) \mathcal{C}_d(2-s, Z), \tag{18.71}$$

where $\mathfrak{Z}_d$ is the Picard zeta function, defined below.

The Picard zeta function is given by

$$\mathfrak{Z}_d(s) = \zeta^{\star}_{\mathbb{Q}(\sqrt{-d})}(s) \beta^{\star}_d(2s-1), \tag{18.72}$$

where $\zeta^{\star}_{\mathbb{Q}(\sqrt{-d})}(s)$ is the completed Dedekind zeta function and $\beta^{\star}_d(s)$ is the completed Dirichlet L-function, i.e., $\beta^{\star}_d(s) = L\left(\left(\frac{\cdot}{d}\right), s\right)$, with $\left(\frac{\cdot}{d}\right)$ the *Legendre symbol.*

Also, the Whittaker coefficients were obtained in [43]:

Theorem 18.10 (Whittaker coefficients) *The Whittaker coefficients of the Eisenstein series $E_d(s, Z)$ are given by*

$$\begin{aligned} W^{(d)}_{\psi}(s, Z) &= \int_{N(\mathcal{O}_d)\backslash N} E_d(s, n \cdot Z) \overline{\psi(n)} dn \\ &= \frac{\zeta_{\mathbb{Q}(\sqrt{-d})}(s)}{\mathfrak{Z}(s)} C_s(\Lambda) K_{2s-2}(2\pi\sqrt{y}|\Lambda|) e^{2\pi i (m\chi + n\tilde{\chi})}, \end{aligned} \tag{18.73}$$

where $\Lambda = -i(m + \omega_d n)/\operatorname{Im}\omega_d$, with $m, n \in \mathbb{Z}$. The numerical coefficients take the form

$$C_s(\Lambda) = \sum_{\substack{\Omega \in \mathcal{O}_d \\ \frac{\Lambda}{\Omega} \in \mathcal{O}^*_d}} \left|\frac{\Lambda}{\Omega}\right|^{2s-2} \sum_{\substack{z \in \mathcal{O}_d \\ \frac{\Lambda}{z\Omega} \in \mathcal{O}^*_d}} |z|^{4-4s}, \tag{18.74}$$

where $\mathcal{O}^*_d$ is the dual lattice and $\Omega = m_1 + \omega_d m_2 \in \mathcal{O}_d$ is subject to the constraint $\gcd(m_1, m_2) = 1$.

The global Whittaker coefficient furthermore satisfies the functional relation

$$\mathfrak{Z}_d(s) W^{(d)}_{\psi}(s, Z) = \mathfrak{Z}_d(2-s) W^{(d)}_{\psi}(2-s, Z). \tag{18.75}$$

18.6 String Theory and Enumerative Geometry

In Section 15.4.4, it was conjectured that the Picard modular group $PU(2, 1; \mathcal{O}_d)$ should be the quantum symmetry of (type IIB) string theory compactified on a rigid Calabi–Yau variety X (i.e., such that $h_{2,1}(X) = 0$). Some precise conjectures were put forward on the connection between the enumerative geometry of X and automorphic forms for $PU(2,1;\mathcal{O}_d)$. Let us now elaborate further on these conjectures and connect the discussion to the results above.

It was argued in Section 15.4.4 that a certain automorphic form on $SU(2,1)$ should be relevant for counting BPS black hole states arising from D3-branes in type IIB string theory wrapping special Lagrangian three-cycles in a Calabi–Yau threefold. D3-branes are $(3+1)$-dimensional extended objects in spacetime and the three spatial dimensions are wrapped so that the remaining one-dimensional time-like line is interpreted as the world-line of the BPS black hole.

There is another variant of this story in type IIA string theory, corresponding to D2-brane instantons fully wrapping a special Lagrangian three-cycle in a Calabi–Yau threefold X. These instantons contribute with non-perturbative corrections to the moduli space metric of the general form (see also Section 13.6.4 for a discussion of general D-instanton effects)

$$\sum_{[\gamma]\in H_3(X,\mathbb{Z})} \Omega(\gamma) \exp\left[2\pi\frac{1}{g_s}\left|\int_\gamma \omega\right| + 2\pi i \int_\gamma C_{(3)}\right], \tag{18.76}$$

where $\omega \in H^{3,0}(X,\mathbb{C})$ is the holomorphic 3-form on X, while $C_{(3)} \in H^3(X,\mathbb{R})/H^3(X,\mathbb{Z})$ is the so-called Ramond–Ramond 3-form and g_s is the string coupling, which is given in terms of the dilaton scalar field ϕ by $g_s = e^{\phi}$. The numerical coefficient $\Omega(\gamma)$ is a function

$$\Omega:\ H_3(X,\mathbb{Z}) \longrightarrow \mathbb{Z}, \tag{18.77}$$

the *BPS index* (a.k.a. instanton measure) which counts D2-branes in the homology class $[\gamma] \in H_3(X,\mathbb{Z})$ (see, e.g., [511, 10]). As mentioned in Section 15.4.4, for a rigid Calabi–Yau threefold one has $h_{2,1} = 0$ and hence only two inequivalent three-cycles. Let $(\mathcal{A},\mathcal{B})$ denote a choice of symplectic basis of $H_3(X,\mathbb{Z})$. Then we can write $\gamma = \ell_1\mathcal{A} + \ell_2\mathcal{B}$, where $\ell_1,\ell_2 \in \mathbb{Z}$ are the (electromagnetic) instanton charges. The exponential in (18.76) is the instanton action S_γ. For a rigid Calabi–Yau threefold, we can write

$$\int_\gamma \omega = \frac{|\ell_1 + \tau\ell_2|}{\operatorname{Im}\tau}, \tag{18.78}$$

where τ is the *period matrix*

$$\tau = \frac{\int_{\mathcal{A}}\omega}{\int_{\mathcal{B}}\omega}. \tag{18.79}$$

Let us also define the periods

$$\chi := \int_{\mathcal{A}} C_{(3)}, \qquad \tilde{\chi} := \int_{\mathcal{B}} C_{(3)}, \tag{18.80}$$

which can be viewed as coordinates along the intermediate Jacobian torus

$$H^3(X,\mathbb{R})/H^3(X,\mathbb{Z}) = \mathbb{C}/(\mathbb{Z}+\tau\mathbb{Z}). \tag{18.81}$$

The instanton action S_γ can now be written more explicitly as

$$S_\gamma = \frac{2\pi}{g_s}\frac{|\ell_1+\tau\ell_2|}{\operatorname{Im}\tau} + 2\pi i(\ell_1\chi+\ell_2\tilde{\chi}), \tag{18.82}$$

and the instanton sum (18.76) becomes (see Examples 13.23 and 13.24)

$$\sum_{(\ell_1,\ell_2)\in\mathbb{Z}^2} \Omega(\ell_1,\ell_2)\exp\left[2\pi e^{-\phi}\frac{|\ell_1+\tau\ell_2|}{\operatorname{Im}\tau} + 2\pi i(\ell_1\chi+\ell_2\tilde{\chi})\right]. \tag{18.83}$$

The conjecture in Section 15.4.4 claims that for rigid Calabi–Yau threefolds of CM-type, i.e., when the intermediate Jacobian is the elliptic curve $\mathbb{C}/\mathcal{O}_d$ admitting complex multiplication [66] and where $\mathcal{O}_d$ is the ring of integers in $\mathbb{Q}(\sqrt{-d})$, we have that $PU(2,1;\mathcal{O}_d)$ is a symmetry of the full string theory. This implies, in particular, that the instanton corrections to the metric on the moduli space must be $PU(2,1;\mathcal{O}_d)$-invariant. From this perspective, the instanton sum (18.83) should correspond to a Fourier coefficient of an automorphic form on $SU(2,1)$.

To this end, let us examine the Whittaker coefficient (18.73) a little more closely. From a string theory perspective we have $\sqrt{y} = e^{-\phi} = 1/g_s$, and the weak-coupling limit $g_s \to 0$ corresponds to $y \to \infty$. In this limit we can expand the Bessel function and write the partial Fourier sum as

$$\begin{aligned}\sum_{\psi:\,N(\mathcal{O}_d)\backslash N\to U(1)} \int_{N(\mathcal{O}_d)\backslash N} E_d(s,n\cdot Z)\overline{\psi(n)}dn &\sim_{y\to\infty} \frac{\zeta_{\mathbb{Q}(\sqrt{-d})}(s)}{\mathfrak{Z}(s)}e^{\phi/2}\\ \times \sum_{\substack{(m,n)\in\mathbb{Z}^2\\(m,n)\neq(0,0)}} C_s(\Lambda)|\Lambda|^{-1/2}&\exp\left[2\pi e^{-\phi}|\Lambda| + 2\pi i(m\chi+n\tilde{\chi})\right].\end{aligned} \tag{18.84}$$

If we now note that $\Lambda = -i(m+\omega_d n)/\operatorname{Im}\omega_d$, this is exactly of the same form as the instanton sum in (18.83). The key difference is that the Fourier–Whittaker coefficients depend on a parameter $s\in\mathbb{C}$ which is absent in (18.83). Is there a particular choice of s that allows us to identify the two expressions? Put differently, can the instanton measure $\Omega(\gamma)$ be extracted from the numerical Fourier coefficients (18.74)? If possible, this would give the first explicit generating function of special Lagrangian submanifolds of compact Calabi–Yau varieties. To make this question more precise we have to analyse what type of automorphic object is expected from a string theory perspective. This will be the topic of the following section.

18.7 Twistors and Quaternionic Discrete Series

In Conjecture 15.12, it was argued that the relevant automorphic form that counts BPS states in string theory on rigid Calabi–Yau threefolds should be attached to a particular representation of $SU(2,1)$ called the quaternionic discrete series. This is a class of unitary representations of quaternionic real forms that was constructed by Gross and Wallach [327, 328]. One of the main reasons for this conjecture is that it follows from Wallach's result [608] that the Fourier coefficients attached to the quaternionic discrete series have support on the same set of charges as the BPS index $\Omega(\gamma)$. We shall now review some basic properties of quaternionic real forms and their discrete series, after which we restrict to $SU(2,1)$ and the connection with BPS state counting.

18.7.1 Some Structure Theory

Here, we introduce a little bit of structure theory for the Lie group G that will be needed later. We also refer the reader to Section 3.1.1 and Example 5.32 for split real forms of exceptional Lie algebras.

Let $\mathfrak{g}$ be the rank-r Lie algebra of a complex simple Lie group $G(\mathbb{C})$, and $\Pi = \{\alpha_1, \ldots, \alpha_r\}$ a system of simple roots spanning the dual $\mathfrak{h}^*$ of the Cartan subalgebra $\mathfrak{h} \subset \mathfrak{g}$. We denote by θ the highest root. There is a canonical five-grading of $\mathfrak{g}$ associated with θ [327, 328], namely

$$\mathfrak{g} = \mathfrak{g}_{-2} \oplus \mathfrak{g}_{-1} \oplus \mathfrak{g}_0 \oplus \mathfrak{g}_1 \oplus \mathfrak{g}_2, \tag{18.85}$$

where the subscript denotes the eigenvalue under the Cartan generator H_θ. The eigenspaces $\mathfrak{g}_{\pm 2}$ are one-dimensional and are given by

$$\mathfrak{g}_2 = \mathbb{C}E_\theta\,, \qquad \mathfrak{g}_{-2} = \mathbb{C}F_\theta\,, \tag{18.86}$$

where E_θ and F_θ are the Chevalley generators associated the $\mathfrak{sl}(2,\mathbb{R})_\theta$ subalgebra:

$$[H_\theta, E_\theta] = 2E_\theta\,, \qquad [H_\theta, F_\theta] = -2F_\theta\,, \qquad [E_\theta, F_\theta] = H_\theta. \tag{18.87}$$

The zeroth subspace $\mathfrak{g}_0$ is the 'Levi factor' given explicitly by

$$\mathfrak{g}_0 = \mathbb{C}H_\theta \oplus \mathfrak{m} \equiv \mathfrak{l}, \tag{18.88}$$

where $\mathfrak{m}$ is the commutant of $\mathfrak{sl}(2,\mathbb{C})_\theta$ in $\mathfrak{g}$. In general $\mathfrak{m}$ is a reductive Lie algebra. The subspace $\mathfrak{u} = \mathfrak{g}_1 \oplus \mathfrak{g}_2$ is a Heisenberg subalgebra of dimension $\dim_{\mathbb{C}} \mathfrak{g}_1 + 1 = 2d + 1$, with centre $\mathfrak{g}_2$. We thus call

$$\mathfrak{p} \equiv \mathfrak{l} \oplus \mathfrak{u} \tag{18.89}$$

the *Heisenberg parabolic*. The value of d depends on the particular group under consideration; its values for the finite simple Lie algebras are listed in Table 18.1.

Table 18.1 The values of the parameter d for all finite-dimensional simple Lie algebras. The table is reproduced from [328].

Lie algebra	d
A_n $(n \geq 1)$	$n - 1$
B_n $(n \geq 2)$	$2n - 3$
C_n $(n \geq 2)$	$n - 1$
D_n $(n \geq 3)$	$2n - 4$
G_2	2
F_4	7
E_6	10
E_7	16
E_8	28

Following the terminology of the previous sections we also denote the opposite Heisenberg parabolic and unipotent radical by $\overline{\mathfrak{p}}, \overline{\mathfrak{u}}$, respectively.

The quotient $V = \mathfrak{u}/\mathfrak{g}_2$ is an abelian Lie algebra of dimension $2d$. As a vector space it carries a non-degenerate symplectic structure $\Lambda^2 V \to \mathfrak{g}_2$, afforded by the Lie bracket $[\mathfrak{g}_1, \mathfrak{g}_1] \subseteq \mathfrak{g}_2$ ([328], Prop. 2.4).

The Lie groups associated with the Lie algebras $\mathfrak{p}, \mathfrak{u}, \mathfrak{l}$ are denoted by P, U, L, respectively. The centre $\mathfrak{g}_2$ of $\mathfrak{u}$ exponentiates to the centre Z of U, which coincides with the commutator subgroup: $Z = [U, U]$ (group commutator). The quotient U/Z is an abelian group, whose tangent space is isomorphic to the symplectic space V.

18.7.2 Quaternionic Discrete Series

Let $G = G(\mathbb{R})$ be a real, semi-simple Lie group in its *quaternionic real form* with Lie algebra $\mathfrak{g}$. This implies that the maximal compact subgroup $K \subset G$ decomposes according to $K = SU(2) \times M$ and the $SU(2)$ sits inside the $SL(2, \mathbb{C})$ of the five-grading (18.85) above. The associated symmetric space $\mathcal{M} = G/K$ is of real dimension $4d$ and does not admit an integrable complex structure, but is rather a real manifold of quaternionic Kähler type, sometimes also known as a *Wolf space*. This is in marked contrast with, for instance, the case of $SL(2, \mathbb{R})/U(1)$, which is a complex manifold of special Kähler type.

Quaternionic real forms of Lie groups do not carry unitary representations analogous to the holomorphic discrete series of $SL(2, \mathbb{R})$. This is roughly because one cannot realise the action of G on holomorphic functions on the space G/K. There is nevertheless a very similar representation, known as the *quaternionic discrete series* (QDS), constructed by Gross and Wallach [327, 328]. Their insight was to instead construct a unitary action of G on holomorphic functions on a slightly larger space, namely the *twistor space*

$\mathcal{Z} = \mathbb{C}P^1 \times \mathcal{M}$. More specifically, Gross and Wallach identify the module $\pi_k, k \in \mathbb{Z}$, of the quaternionic discrete series with the cohomology group $H^1(\mathcal{Z}, \mathcal{L}_{-k})$, where $\mathcal{L}_k$ is a G-equivariant line bundle over $\mathcal{Z}$. In fact, $\mathcal{L}_{-k}$ is trivial on $\mathcal{M}$ and its restriction to the twistor fiber is equivalent to the line bundle $\mathcal{O}(-k) \to \mathbb{C}P^1$ whose sections are holomorphic functions on $\mathbb{C}P^1$ homogeneous of degree $-k$. So, we may identify

$$\pi_k = H^1(\mathcal{Z}, \mathcal{O}(-k)). \tag{18.90}$$

The quaternionic discrete series thus acts on sections $f \in H^1(\mathcal{Z}, \mathcal{O}(-k))$ via the standard right-regular representation π.

The complexified Lie algebra $\mathfrak{g}_\mathbb{C} = \text{Lie } G_\mathbb{C}$ decomposes under $\mathfrak{k}_\mathbb{C} = \text{Lie } K_\mathbb{C}$ as (Cartan decomposition)

$$\mathfrak{g}_\mathbb{C} = \mathfrak{k}_\mathbb{C} \oplus \mathfrak{p}_\mathbb{C} = \mathfrak{sl}(2) \oplus \mathfrak{m} \oplus (\mathbb{C}^2 \otimes V), \tag{18.91}$$

where $\mathbb{C}^2$ is a two-dimensional representation of $\mathfrak{sl}(2)$ and V is an irreducible symplectic representation of $\mathfrak{m}$. Gross and Wallach show that the quaternionic discrete series affords the following K-type decomposition (see Definition 5.12) under $SU(2) \times M$:

$$\pi_k = \bigoplus_{n \geq 0} S^{k-2+n}(\mathbb{C}^2) \otimes S^n(V), \tag{18.92}$$

where $S^n(X)$ denotes the nth symmetric algebra of X. In contrast to the principal series $\text{Ind}_B^G \chi$, the module π_k does not admit a spherical vector, but only a 'minimal K-type' transforming under $SU(2) \times M$ according to the representation $S^{k-2}(\mathbb{C}^2) \otimes \mathbb{C}$. Taking the limit $n \to \infty$ of (18.92) shows that π_k has Gelfand–Kirillov dimension $2d + 1$ [328].

18.7.3 The Twistor Space

As already mentioned, the symmetric space $\mathcal{M} = G/K = G/(SU(2) \times M)$ is a real manifold of quaternionic Kähler type. The twistor space over $\mathcal{M}$ is a $\mathbb{C}P^1$-fibration $\mathcal{Z} \to \mathcal{M}$, which is topologically a product $\mathcal{Z} = \mathcal{M} \times \mathbb{C}P^1$. If we view $\mathbb{C}P^1$ as the 2-sphere $SU(2)/U(1)$, the twistor space can be realised as the following coset:

$$\mathcal{Z} = G/(U(1) \times M). \tag{18.93}$$

The twistor space is a complex manifold of complex dimension $\dim_\mathbb{C} \mathcal{Z} = 2d + 1$. This fact yields an alternative description that will be very useful in what follows. Consider the complexified Lie group $G_\mathbb{C}$ with associated complex Lie algebra $\mathfrak{g}_\mathbb{C}$. The complex Lie algebra affords the same five-grading as in (18.85), where the unipotent radical $\mathfrak{u}_\mathbb{C}$ is a complex vector space of complex dimension $2d + 1$. This may be viewed as the tangent space $T_{\mathbb{C}}\mathscr{Z}$, and using the Borel embedding

$$G/K \hookrightarrow G_{\mathbb{C}}/\overline{P}_{\mathbb{C}}, \tag{18.94}$$

where $\overline{P}_{\mathbb{C}}$ is the lower-triangular Heisenberg parabolic in $G_{\mathbb{C}}$, we obtain an isomorphism [328, 335]

$$G/(U(1) \times M) \cong G_{\mathbb{C}}/\overline{P}_{\mathbb{C}}. \tag{18.95}$$

This description provides us with a natural set of local complex coordinates on $\mathscr{Z}$ denoted by $(\xi^\Lambda, \tilde{\xi}_\Lambda, \alpha)$, $\Lambda = 0, 1, \dots, d-1$ [498]. These coordinates are adapted to the Heisenberg group U, which simply acts by translations:

$$\begin{aligned} \xi^\Lambda &\longrightarrow \xi^\Lambda + m^\Lambda, \\ \tilde{\xi}_\Lambda &\longrightarrow \tilde{\xi}_\Lambda + n_\Lambda \\ \alpha &\longrightarrow \alpha + 2\kappa - n_\Lambda \xi^\Lambda + m^\Lambda \tilde{\xi}_\Lambda, \end{aligned} \tag{18.96}$$

with $(m^\Lambda, n_\Lambda, \kappa) \in \mathbb{R}^{2d} \times \mathbb{R}$.

18.7.4 Automorphic Realisation of the Quaternionic Discrete Series

Let $\chi_P \colon P(\mathbb{Z})\backslash P(\mathbb{R}) \to \mathbb{C}^\times$ be a quasi-character on the Heisenberg parabolic $P(\mathbb{R})$. As discussed in Sections 5.7.1 and 5.7.2, such a character can be written in two different ways,

$$\chi_P(p) = e^{\langle \lambda_P + \rho_P | H_P(p) \rangle} = e^{\langle \lambda + \rho | H(p) \rangle}, \tag{18.97}$$

using either a weight $\lambda_P \in \mathfrak{a}_P^*$ and ρ_P of (5.85) determining the modulus character of P or by referring everything to the standard Borel subgroup and using Proposition 5.30. For the case of the Heisenberg parabolic we note from Example 5.32 that $\mathfrak{a}_P^*$ is spanned by the highest root θ and $\rho_P = \frac{d+1}{2}\theta$. We recall that $2d = \dim \mathfrak{g}_1$.

We thus obtain for the character χ_P a one-parameter family

$$\chi_s(p) = e^{\left\langle \frac{1}{2} s\theta | H_P(p) \right\rangle} = e^{\langle \lambda_s + \rho | H(p) \rangle}, \tag{18.98}$$

with $\lambda_s = \frac{1}{2}s\theta - \rho$ and where $s \in \mathbb{C}$ parametrises the space of quasi-characters of P. This parametrises a degenerate principal series representation, and we have chosen a slightly unconventional normalisation of s to match [328].

The quaternionic discrete series π_k can be embedded as a submodule into this degenerate principal series when $s = k$ and $k \geq 2d + 1 \in \mathbb{Z}$ [328]:

$$i : \pi_k \hookrightarrow \mathrm{Ind}_{P(\mathbb{R})}^{G(\mathbb{R})} \chi_k. \tag{18.99}$$

Additional 'limit quaternionic discrete series representations' exist for $k \geq d+1$.

Remark 18.11 We note that the literature on the quaternionic discrete series uses different conventions for parametrising the quasi-character χ_k. Our parameter $k = k_{\text{here}}$ is the same as Gross–Wallach [328], here denoted k_{GW}, while it differs from that of Gan–Gross–Savin [244], denoted k_{GGS}, i.e.,

$$k = k_{\text{here}} = k_{\text{GW}} = 2k_{\text{GGS}} + 2. \tag{18.100}$$

Moreover, our parameter s in the degenerate principal series is related to the parameter in Jiang–Rallis [378], denoted s_{JR}, via $s = s_{\text{here}} = 6s_{\text{JR}}$.

Following [244], we can view the character χ_k as the archimedean component $\chi_{k,\infty}$ of a global character:

$$\chi_k : P(\mathbb{Q})\backslash P(\mathbb{A}) \longrightarrow \mathbb{C}^\times. \tag{18.101}$$

We thus have an associated global, degenerate principal series $\text{Ind}_{P(\mathbb{A})}^{G(\mathbb{A})}\chi_k$. At the finite places $p < \infty$, $\chi_{k,p}$ is a character in the p-adic degenerate principal series $\text{Ind}_{P(\mathbb{Q}_p)}^{G(\mathbb{Q}_p)}\chi_{k,p}$. The global representation decomposes as

$$\text{Ind}_{P(\mathbb{A})}^{G(\mathbb{A})}\chi_k = \text{Ind}_{P(\mathbb{R})}^{G(\mathbb{R})}\chi_{k,\infty} \otimes {\bigotimes_{p<\infty}}' \text{Ind}_{P(\mathbb{Q}_p)}^{G(\mathbb{Q}_p)}\chi_{k,p}. \tag{18.102}$$

Now let $\mathsf{f}_{k,p}$ be the unique, spherical (i.e., $G(\mathbb{Z}_p)$-invariant) vector in $\text{Ind}_{P(\mathbb{Q}_p)}^{G(\mathbb{Q}_p)}\chi_{k,p}$ which satisfies $\mathsf{f}_{k,p}(1) = 1$ (see Definition 5.14). For $\varphi_k \in \pi_k$ we have $i(\varphi_k) \in \text{Ind}_{P(\mathbb{R})}^{G(\mathbb{R})}\chi_{k,\infty}$, and we construct a global standard section $\widehat{\varphi}_k \in \text{Ind}_{P(\mathbb{A})}^{G(\mathbb{A})}\chi_k$ as follows:

$$\widehat{\varphi}_k = i(\varphi_k) \otimes \bigotimes_{p<\infty} \mathsf{f}_{k,p}. \tag{18.103}$$

Remark 18.12 The section $i(\varphi_k)$ is a *non-spherical* standard section of $\text{Ind}_{P(\mathbb{R})}^{G(\mathbb{R})}\chi_k$. It corresponds to a one-dimensional representation of the Levi $L(\mathbb{R}) = P(\mathbb{R})/U(\mathbb{R})$, parametrised by the integer k. This is analogous to the construction of the non-spherical section used in Example 5.6 to embed the holomorphic discrete series of $SL(2,\mathbb{R})$ in the continuous principal series.

For each global section $\widehat{\varphi}_k \in \text{Ind}_{P(\mathbb{A})}^{G(\mathbb{A})}\chi_k$ we form the adelic Eisenstein series as in [244]:

$$E(\widehat{\varphi}_k, g) = \sum_{\gamma\in P(\mathbb{Q})\backslash G(\mathbb{Q})} \widehat{\varphi}_k(\gamma g). \tag{18.104}$$

In the special case when $G = G_2$, Gan–Gross–Savin show [244] when translated to our conventions:

Theorem 18.13 (Eisenstein series on $G_2(\mathbb{A})$ attached to the QDS) *When $G = G_2(\mathbb{A})$ the Eisenstein series $E(\widehat{\varphi}_k, g)$ converges absolutely for even integers*

$k > 4$ and defines a non-trivial element of the space M_{k-2} (defined in [244]*) of weight $(k-2)$ modular forms on $G_2(\mathbb{R})/SO(4)$.*

Remark 18.14 When restricted to a function on $G_2(\mathbb{R})/SO(4)$ the Eisenstein series $E(\widehat{\varphi}_k, g)$ transforms with weight $k-2$ under $G_2(\mathbb{Z})$ because of the non-spherical character $i(\varphi_k)$. This is equivalent to the way the weight of the non-holomorphic Eisenstein series arises in Equation (5.62).

Remark 18.15 The construction of the Eisenstein series $E(\widehat{\varphi}, g)$ generating the quaternionic discrete series differs from the ordinary degenerate Eisenstein series only at the archimedean place. Thus, the non-archimedean contribution to Fourier–Whittaker coefficients $E(\widehat{\varphi}, g)$ will coincide with those for the degenerate Eisenstein series [244], similar to the construction in Section 5.6.

The construction of Eisenstein series for the quaternionic discrete series given above can in particular be applied to $G = SU(2,1)$. The Fourier–Whittaker coefficients attached to the degenerate principal series of $SU(2,1)$ were obtained in (18.73) and the non-archimedean contribution is the double divisor sum $C_s(\Lambda)$, $\Lambda \in \mathcal{O}_d$, in (18.74). For $s = k/2$ these coefficients should also arise from an Eisenstein series realisation of π_k. Combining all the arguments given previously in this section then motivates the following conjecture:

Conjecture 18.16 (Counting cycles in rigid Calabi–Yau threefolds) *For some integer value of $k \geq 2$, the BPS index (or instanton measure) $\Omega(\gamma)$ counting special Lagrangian three-cycles in homology class $[\gamma] \in H_3(X,\mathbb{Z})$, with X a rigid Calabi–Yau threefold of CM-type, is given by*

$$\Omega(\gamma) = |\Lambda|^{-1/2} \sum_{\substack{\Omega \in \mathcal{O}_d \\ \frac{\Lambda}{\Omega} \in \mathcal{O}_d^*}} \left|\frac{\Lambda}{\Omega}\right|^{k-2} \sum_{\substack{z \in \mathcal{O}_d \\ \frac{\Lambda}{z\Omega} \in \mathcal{O}_d^*}} |z|^{4-2k}, \tag{18.105}$$

where $\gamma = \ell_1 \mathcal{A} + \ell_2 \mathcal{B} \in H_3(X,\mathbb{Z})$ and $\Lambda = -i(\ell_1 + \omega_d \ell_2)/\operatorname{Im}\omega_d \in \mathcal{O}_d$.

Remark 18.17 Which of the quaternionic discrete series π_k (i.e., the relevant value of k) that arises in the above conjecture must be determined by constraints from string theory. Some preliminary attempts were made in [42, 43, 511, 16] but a more careful analysis is required before this question can be settled.

18.7.5 Automorphic Forms on Twistor Space

Although the construction of the Eisenstein series $E(\widehat{\varphi}, g)$ given above provides an automorphic realisation of the quaternionic discrete series, it does not immediately connect with the twistor space picture. Recall that Gross and Wallach realise π_k as holomorphic sections on twistor space $\mathscr{Z}$ over G/K, i.e., elements of π_k can be viewed as sheaf cohomology classes in $H^1(\mathscr{Z}, \mathcal{O}(-k))$. This leads to the following:

Open question: *Is it possible to construct an automorphic realisation of π_k in terms of holomorphic sections in $H^1(\mathscr{Z}, \mathcal{O}(-k))$?*

This is in fact very natural from the point of view of string theory. The (hypermultiplet) moduli space in string theory on a Calabi–Yau threefold X is a quaternionic Kähler manifold $\mathscr{M}(X)$ of real dimension $4(h_{2,1}+1)$ in type IIA and $4(h_{1,1}+1)$ in type IIB. The metric on $\mathscr{M}(X)$ receives quantum corrections due to D-brane instantons. These deformations can be efficiently encoded in deformations of the complex (contact) structure on the twistor space (see [10] for a review):

$$\mathbb{P}^1 \to \mathscr{Z} \to \mathscr{M}(X). \tag{18.106}$$

It is by now very well understood that the $SL(2,\mathbb{Z})$ S-duality symmetry of $\mathscr{M}(X)$ lifts to a holomorphic isometry of $\mathscr{Z}$, which is compatible with the contact structure as well as mirror symmetry [542, 15, 3, 17, 13, 12, 4, 14, 11, 5, 7, 6, 8, 9].

For rigid Calabi–Yau threefolds the exact quaternionic Kähler metric on $\mathscr{M}(X)$, including all perturbative and non-perturbative quantum corrections, is conjectured to be $PU(2,1;\mathcal{O}_d)$-invariant. To illustrate what this entails, let us give the action of the Picard modular group $SU(2,1;\mathbb{Z}[i])$ on the twistor space over the classical moduli space $SU(2,1)/U(2)$. In practice this means that we must lift the action of the generators in (18.10), (18.11) and (18.12) to a holomorphic action on the twistor space.

Let us first explain some basic properties of $\mathscr{Z}$, mainly following [335]. If we identify the twistor fiber with the Riemann sphere $\mathbb{P}^1 \cong SU(2)/U(1)$, the twistor space $\mathscr{Z}$ over $\mathcal{M} = SU(2,1)/U(2)$ is locally given by

$$\mathscr{Z} = \mathcal{M} \times \mathbb{P}^1 \cong SU(2,1)/(U(1)\times U(1)). \tag{18.107}$$

This is a complex contact manifold of complex dimension 2 + 1 with local Darboux coordinates $(\xi, \tilde{\xi}, \alpha)$ and holomorphic contact one-form

$$\mathcal{X} = d\alpha + \xi d\tilde{\xi} - \tilde{\chi} d\xi. \tag{18.108}$$

The Kähler potential is

$$\begin{aligned} K_{\mathscr{Z}} &= \frac{1}{2}\log\left[\left((\xi-\bar{\xi})^2 + (\tilde{\xi}-\bar{\tilde{\xi}})^2\right)^2 + 4\left(\alpha - \bar{\alpha} + \bar{\xi}\tilde{\xi} - \xi\bar{\tilde{\xi}}\right)^2\right] \\ &= \log\frac{1+z\bar{z}}{|z|} - 2\phi. \end{aligned} \tag{18.109}$$

In the second line we have related the Darboux coordinates $(\xi, \tilde{\xi}, \alpha)$ to the coordinates $(\phi, \chi, \tilde{\chi}, \psi)$ on the base $SU(2,1)/U(2)$ (see Section 18.1 for our parametrisation of $\mathcal{M}$) together with a stereographic coordinate z on the $\mathbb{P}^1$-fiber. This is known as 'parametrising the twistor lines'. In terms of the coordinates on $\mathcal{M}$ given by (18.3) we find

$$\begin{aligned} \xi &= -\sqrt{2}\chi + \frac{1}{\sqrt{2}}e^{-\phi}(z - z^{-1}), \\ \tilde{\xi} &= -\sqrt{2}\tilde{\chi} - \frac{i}{\sqrt{2}}e^{-\phi}(z + z^{-1}), \\ \alpha &= 2\psi - e^{-\phi}\left[z(\tilde{\chi} + i\chi) - z^{-1}(\tilde{\chi} - i\chi)\right]. \end{aligned} \tag{18.110}$$

We now want to lift the action of $SU(2,1;\mathbb{Z}[i])$ to a holomorphic action on $\mathscr{Z}$. To this end we note the isomorphism (see [335])

$$SU(2,1)/(U(1) \times U(1)) \cong SL(3,\mathbb{C})/\bar{B}(\mathbb{C}), \tag{18.111}$$

where $SL(3,\mathbb{C})$ is the complexification $SU(2,1) \otimes \mathbb{C}$ and $\bar{B}(\mathbb{C})$ is the lower-triangular Borel subgroup. We can parametrise an element $g_{\mathbb{C}} \in SL(3,\mathbb{C})/\bar{B}(\mathbb{C})$ by

$$g_{\mathbb{C}} = \begin{pmatrix} 1 & \tilde{\xi} - \xi + i(\xi + \tilde{\xi}) & \alpha + i(\xi^2 + i\tilde{\xi}^2) \\ 0 & 1 & \xi + \tilde{\xi} + i(\tilde{\xi} - \xi) \\ 0 & 0 & 1 \end{pmatrix}. \tag{18.112}$$

This shows that the Heisenberg action lifts trivially to a holomorphic action on the twistor space

$$\begin{aligned} \xi &\longmapsto \xi + a, \\ \tilde{\xi} &\longmapsto \tilde{\xi} + b, \\ \alpha &\longmapsto \alpha + c - 2a\tilde{\xi} + 2b\xi, \end{aligned} \tag{18.113}$$

as we have already seen in general in (18.96).

Similarly, the symplectic transformation (18.11) acts in precisely the same way as on the base:

$$(\xi, \tilde{\xi}) \longmapsto (-\tilde{\xi}, \xi), \tag{18.114}$$

accompanied by a simple action on the fiber

$$z \longmapsto iz. \tag{18.115}$$

It remains to determine the lifted action of the involution (18.12). The matrix which realises the involution is

$$S = \begin{pmatrix} 0 & 0 & i \\ 0 & -1 & 0 \\ -i & 0 & 0 \end{pmatrix} \in SU(2,1). \tag{18.116}$$

The action (18.12) on $SU(2,1)/U(2)$ is obtained by acting with S on $g = na$ from the left and then compensating by an element $k \in U(2)$ from the right in order to bring the result back into Iwasawa form, i.e., $g' = n'a'$. We can now implement the same strategy on $\mathscr{Z}$: act with S on the left of the representative $g_{\mathbb{C}} \in SL(3,\mathbb{C})/\bar{B}(\mathbb{C})$ and then compensate by a transformation of $\bar{b}_{\mathbb{C}} \in \bar{B}(\mathbb{C})$ on the right in order to bring the result back into upper-triangular form. This yields

$$Sg_{\mathbb{C}}\bar{b}_{\mathbb{C}} = \begin{pmatrix} 0 & 0 & i \\ 0 & -1 & 0 \\ -i & 0 & 0 \end{pmatrix}\begin{pmatrix} 1 & \tilde{\xi} - \xi + i(\xi + \tilde{\xi}) & \alpha + i(\xi^2 + i\tilde{\xi}^2) \\ 0 & 1 & \xi + \tilde{\xi} + i(\tilde{\xi} - \xi) \\ 0 & 0 & 1 \end{pmatrix}\bar{b}_{\mathbb{C}}$$

$$= \begin{pmatrix} 1 & \tilde{\xi}' - \xi' + i(\xi' + \tilde{\xi}') & \alpha' + i(\xi'^2 + i\tilde{\xi}'^2) \\ 0 & 1 & \xi' + \tilde{\xi}' + i(\tilde{\xi}' - \xi') \\ 0 & 0 & 1 \end{pmatrix}, \tag{18.117}$$

where the transformed coordinates $(\xi', \tilde{\xi}', \alpha')$ are given by

$$\begin{aligned} \xi &\longmapsto -\frac{-2\alpha\tilde{\xi} + (\xi^2 + \tilde{\xi}^2)\xi}{4\alpha^2 + (\xi^2 + \tilde{\xi}^2)^2}, \\ \tilde{\xi} &\longmapsto -\frac{2\alpha\xi + (\xi^2 + \tilde{\xi}^2)\tilde{\xi}}{4\alpha^2 + (\xi^2 + \tilde{\xi}^2)^2}, \\ \alpha &\longmapsto -\frac{\alpha}{4\alpha^2 + (\xi^2 + \tilde{\xi}^2)^2}, \end{aligned} \tag{18.118}$$

with an accompanying transformation on the fiber z that is too cumbersome to write out. The associated transformation of the contact one-form is a holomorphic rescaling

$$\mathcal{X} \longmapsto \frac{\mathcal{X}}{4\alpha^2 + (\xi^2 + \tilde{\xi}^2)^2}, \tag{18.119}$$

thus leaving the complex contact structure on $\mathscr{Z}$ invariant. One may also verify that the Kähler potential $K_{\mathscr{Z}}$ transforms by a Kähler transformation,

$$K_{\mathscr{Z}} \longmapsto K_{\mathscr{Z}} + \log f(\xi, \tilde{\xi}, \alpha) + \log \bar{f}(\bar{\xi}, \bar{\tilde{\xi}}, \bar{\alpha}), \tag{18.120}$$

implying that this is an isometry of the twistor space. We have thus found the desired lift of $SU(2,1;\mathbb{Z}[i])$ to a holomorphic action on $\mathscr{Z}$.

We conclude this section with the following:

Open question: *Is it possible to construct an automorphic form attached to π_k of $SU(2,1)$ in terms of a holomorphic section $\varphi(\xi, \tilde{\xi}, \alpha)$ on $\mathscr{Z}$, whose Penrose transform (i.e., integration along the $\mathbb{P}^1$-fibers of $\mathscr{Z} \to \mathcal{M}$) reproduces the Eisenstein series $E(\widehat{\varphi}, g)$ in (18.104)?*

19

Extension to Kac–Moody Groups

In this book, we have concentrated on the study of automorphic forms and in particular Eisenstein series defined on mainly split real finite-dimensional Lie groups, as categorised in the Cartan classification. In this chapter, we discuss the extension to infinite-dimensional *Kac–Moody groups*, which are generated by an infinite number of raising and lowering operators. A complete classification of Kac–Moody groups is at present not known and we will restrict our attention here mainly to Eisenstein series defined on *affine, hyperbolic and Lorentzian Kac–Moody groups.*

General indefinite Kac–Moody algebras are only known in terms of generators and relations. In the particular case of affine Kac–Moody algebras one can use a reformulation based on loop algebras. Full accounts of Kac–Moody algebras can be found in the books [386, 489, 606, 430, 286]. The definition of Kac–Moody groups proceeds similarly from generating one-parameter subgroups with relations. The construction is not dissimilar to that of the Steinberg presentation in Section 3.1.4 and this is the language we will mainly use. For details on Kac–Moody groups we refer the reader to [514, 430, 250, 163].

The main emphasis in this chapter will be on Eisenstein series on loop groups, which is the setting that is most developed. The main result in this chapter is a detailed proof of the analogue of the Langlands constant term formula for affine Kac–Moody groups. We will also discuss some recent results and conjectures concerning automorphic representations on more general Kac–Moody groups, exemplified by the exceptional cases E_9, E_{10}, and E_{11}.

Before we embark on this, however, we shall provide some motivation for the extension of the theory of automorphic forms to Kac–Moody groups, from both a mathematical and a physical viewpoint.

19.1 Motivation

The motivation for studying Eisenstein series defined on infinite-dimensional Kac–Moody groups is (at least) threefold.

19.1.1 String Theory Motivation: Infinite-Dimensional U-duality

In string theory, Kac–Moody groups appear, for example, conjecturally in the list of U-duality groups encoding discrete symmetries of type II string theory compactified on a $(10 - D)$-dimensional torus from ten down to D space-time dimensions. The list of these groups was given in Table 13.2 in Chapter 13 and consists of the groups in the exceptional series of the Cartan classification, where in $D \geq 3$ dimensions the respective U-duality group is given by the finite-dimensional and discrete group $E_{11-D}(\mathbb{Z})$. In $D = 2, 1$ and 0 dimensions, however, the corresponding U-duality groups are infinite-dimensional and are conjectured to be given by the affine, hyperbolic and Lorentzian Kac–Moody groups $E_9(\mathbb{Z})$, $E_{10}(\mathbb{Z})$ and $E_{11}(\mathbb{Z})$, respectively [365]. In particular, the groups E_{10} and E_{11} are of special relevance [160, 615], since they have been conjectured as fundamental symmetries of *M-theory* [620, 535], a theory whose low-energy limit is eleven-dimensional supergravity and from which the five different known types of string theories can be derived as particular limits. Automorphic forms on these groups capture all types of non-perturbative effects in string theory on tori, and can be viewed as important book-keeping devices that also organise the non-perturbative effects in higher dimensions [212, 213, 214, 40].

19.1.2 Gravity Motivation: Quantum Cosmological Billiards

There is also an intriguing connection of the theory of Kac–Moody automorphic forms to *quantum cosmology*. Classical cosmology, as the study of the evolution of the Universe in the form of a dynamical space-time subject to Einstein's field equations, often assumes the existence of a space-like singularity like the *big bang*. Similar space-like singularities arise in the interior of Schwarzschild and Kerr black holes. In seminal work devoted to the study of Einstein's partial differential equations, Belinskii, Khalatnikov and Lifshitz (BKL) discovered that (*i*) there is most probably a regime in the vicinity of the singularity where one can effectively use ordinary differential equations and (*ii*) the behaviour of the four-dimensional empty (matterless) Universe in this set-up becomes chaotic [56, 57]; see also [481].

This was later generalised to the context of maximal supergravity by Damour and Henneaux [158, 159], where it was found that similar phenomena persist and moreover a Weyl group structure related to the hyperbolic Kac–Moody group E_{10} was found; see also [161, 354] for reviews. The Weyl group determines the shape of a certain effective billiard system, giving rise to the name *cosmological billiards* for this field of theoretical cosmology. Moreover, in [160] it was shown that there are traces of a full $E_{10}(\mathbb{R})$ structure for part of the ordinary differential equations arising in this analysis, lending strong support to the conjecture that Kac–Moody symmetries are fundamental for string theory and M-theory.

At the quantum level, it is interesting to contemplate how much of the Kac–Moody structure remains relevant. In a first step, one can try to quantise the cosmological billiard system associated with the Weyl group acting on some generalised upper half-plane. This was undertaken in [408, 409, 410], where it was shown that the spectrum of the Laplace operator in this context is such that one obtains a quantum-mechanical resolution of the space-like singularity; see also [300, 372, 403, 529, 216]. This *quantum cosmological billiards* analysis, however, should properly be embedded in a full quantum gravitational analysis that goes beyond the crude approximations used in these first investigations. At any rate, it is to be expected that automorphic forms will play a central rôle in the final formulation of this model.

19.1.3 Mathematical Motivation: New Automorphic L-functions

From a mathematics perspective an important motivation to study Kac–Moody Eisenstein series is to consider them as a potential source for deriving new L-functions through a Fourier expansion of the series. Using finite-dimensional algebras, it has been possible to consider L-functions up to the symmetric cube on $GL(2)$ using the Langlands–Shahidi method for the largest finite-dimensional exceptional E_8 and converse theorems [400]; see also Remark 11.17. To achieve higher symmetric tensor powers, one should speculatively use Dynkin diagrams, not for finite-dimensional but for infinite-dimensional Kac–Moody groups [568]. It is, however, not precisely clear if the direct extension of the theory of Eisenstein series will necessarily yield to new types of L-functions, and the focus of the discussion has so far been on series defined on affine groups. In fact, in [568] an argument was provided that no new functions will be found, while in [252] a new method for obtaining such functions was devised. This new method relies on an expansion of the series with respect to 'lower triangular parabolics', instead of only 'upper triangular parabolics'. See Section 19.2.4 below, and in particular Remark 19.5, for more details.

In recent years, there has been some work developing the theory of Eisenstein series for Kac–Moody groups. The most well-developed part is that of Eisenstein series defined on affine Kac–Moody groups which was started by Garland [250, 251]. While for finite-dimensional groups convergence with respect to the (complex) defining weight λ was proven over almost all of the complex plane (see (4.134)), convergence for infinite-dimensional Kac–Moody groups is restricted and the defining weight has to lie inside the Tits cone [386]. Furthermore, a restriction on the group element forming the argument of the series has to be imposed [251]. First steps towards a definition of Eisenstein series on hyperbolic Kac–Moody groups have been made in [122], where the case of rank-two hyperbolics was considered. Furthermore, in [212, 214] Eisenstein series defined on the hyperbolic E_{10} group (along with E_9 and E_{11}) were discussed. A general proof of convergence of Eisenstein series on general hyperbolics remains to be developed.

19.2 Eisenstein Series on Affine Kac–Moody Groups

A theory of Eisenstein series on loop groups was initiated in a series of seminal works by Garland [250, 251, 252], and further developed by many other authors (see, e.g., [454, 253, 87]). First we introduce some basic ingredients of adelic loop groups, after which we introduce Eisenstein series and derive their constant terms.

Remark 19.1 We note that in this section we will follow the standard convention in the literature on affine groups to denote the affine extension of a finite-dimensional Lie group G by $\widetilde{G}$. This tilde should not be confused with the same notation used commonly for metaplectic covers that we saw in Chapters 12, 16 and 17. In the present section, there will be no metaplectic covers and thus no confusion should arise.

19.2.1 Preliminaries on Adelic Loop Groups

We now introduce some basic material on loop groups, mainly following [250, 454]. Let $\mathfrak{g}$ be a complex rank-n finite simple Lie algebra, $\overline{\mathfrak{g}}$ the associated *loop algebra*, $\widehat{\mathfrak{g}} = \overline{\mathfrak{g}} \oplus \mathbb{C}c$ the centrally extended loop algebra and $\widetilde{\mathfrak{g}}$ the full non-twisted affine extension. As a complex vector space the affine Kac–Moody algebra $\widetilde{\mathfrak{g}}$ is given by

$$\widetilde{\mathfrak{g}} = \widehat{\mathfrak{g}} \oplus \mathbb{C}d = (\mathbb{C}[t, t^{-1}] \otimes_{\mathbb{C}} \mathfrak{g}) \oplus \mathbb{C}c \oplus \mathbb{C}d, \tag{19.1}$$

where c is the *central element* and d is the *derivation* (see [386]). The loop algebra $\overline{\mathfrak{g}}$ is realised by (finite) Laurent polynomials in $\mathfrak{g}$. We note that $\overline{\mathfrak{g}}$, $\widehat{\mathfrak{g}}$ and $\widetilde{\mathfrak{g}}$ differ only by the elements in the Cartan subalgebra but have identical non-semi-simple generators.

We denote by $\Pi = \{\alpha_1, \dots, \alpha_n\}$ the set of simple roots of $\mathfrak{g}$ and by $\widetilde{\Pi} = \{\alpha_0, \alpha_1, \dots, \alpha_n\}$ those of $\widetilde{\mathfrak{g}}$. Associated to this choice of simple roots we have corresponding root systems

$$\Delta = \Delta_+ \cup \Delta_-, \qquad \widetilde{\Delta} = \widetilde{\Delta}_+ \cup \widetilde{\Delta}_-. \tag{19.2}$$

A novel feature of affine Kac–Moody algebras compared to finite-dimensional algebras is that they possess *imaginary roots* whose associated generators are not locally nilpotent.

In what follows we will restrict to split real forms, and work over $\mathbb{Q}$ and the adeles $\mathbb{A}$ of $\mathbb{Q}$. To define affine Kac–Moody groups we form the formal Laurent series $\mathbb{Q}_p((t))$, such that an element is of the form

$$\sum_{n\in\mathbb{Z}} a_n t^n, \quad \text{with } a_n \in \mathbb{Q}_p\,. \tag{19.3}$$

Similarly the Laurent series over $\mathbb{A}$ is given by the restricted direct product of all $\mathbb{Q}_p((t))$, and we will denote it by $\mathbb{A}\langle t\rangle$. Over $\mathbb{Q}$ the affine group is defined as

$$\widetilde{G}(\mathbb{Q}((t))) = \widehat{G}(\mathbb{Q}((t))) \rtimes \sigma(\mathbb{Q}^\times), \tag{19.4}$$

where the factor $\sigma(\mathfrak{q})$, with $\mathfrak{q} \in \mathbb{Q}^\times$, in the semi-direct product acts on $\widehat{G}(\mathbb{A}\langle t\rangle)$ as a rescaling, $a\langle t\rangle \mapsto a\langle t\mathfrak{q}\rangle$, where $a\langle t\rangle \in \mathbb{A}\langle t\rangle$. This action is referred to as the 'rotation of the loop'. It can be understood as the formal exponentiation of the derivation operator of the underlying affine algebra.

Now denote by $\mathbb{Q}\langle t\rangle$ the subset of elements $x \in \mathbb{Q}((t))$ such that $x \in \mathbb{Z}_p((t))$ for all but finitely many finite primes p. In other words,

$$\mathbb{Q}\langle t\rangle = \mathbb{Q}((t)) \cap \mathbb{A}\langle t\rangle, \tag{19.5}$$

with $\mathbb{Q}((t))$ embedded diagonally. Then we define

$$\mathbb{Q}\langle t\rangle_+ = \mathbb{Q}\langle t\rangle \cap \mathbb{Q}[[t]], \qquad \mathbb{A}\langle t\rangle_+ = \mathbb{A}\langle t\rangle \cap \mathbb{A}[[t]], \tag{19.6}$$

where $\mathbb{A}[[t]]$ denotes Taylor series in t. Let $\widehat{B}$ be the Borel subgroup of $\widehat{G}$ and $\widehat{K}$ be the maximal compact subgroup. We then have the Iwasawa decomposition [163]

$$\widehat{G}(\mathbb{A}\langle t\rangle) = \widehat{B}\widehat{K}, \tag{19.7}$$

and, similarly for $\widetilde{G}$,

$$\widetilde{G}(\mathbb{A}\langle t\rangle) = \widetilde{B}\widehat{K}. \tag{19.8}$$

We note that $\widetilde{K} = \widehat{K}$ since $\widetilde{K} \subset \widehat{G}$. An arbitrary element $g \in \widetilde{G}$ then splits as follows according to the Iwasawa decomposition:

$$g = b\sigma(\mathfrak{q})k, \tag{19.9}$$

where $b \in \widehat{B}$, $k \in \widehat{K}$ and $\sigma(\mathfrak{q}) = v^d$ for $v \in \mathbb{A}\langle t\rangle$ is the derivation direction in (19.4).

We further let $\widehat{A}(\mathbb{A})$ denote the maximal Cartan torus in $\widehat{B}(\mathbb{A})$, and $\widehat{N}(\mathbb{A})$ the associated unipotent radical. Thus an element $b \in \widehat{B}(\mathbb{A})$ splits according to

$$b = na \in \widehat{N}(\mathbb{A})\widehat{A}(\mathbb{A}). \tag{19.10}$$

The affine group (over $\mathbb{Q}$) also enjoys a Bruhat decomposition similar to (8.5) of the form

$$\widehat{G} = \bigcup_{w \in \widetilde{\mathcal{W}}} \widehat{B}w\widehat{B} = \bigcup_{w \in \widetilde{W}} \widehat{B}wN_w, \tag{19.11}$$

where $\widetilde{\mathcal{W}}$ is the affine Weyl group and we have defined

$$N_w = \prod_{\substack{\alpha>0 \\ w\alpha<0}} N_\alpha. \tag{19.12}$$

Note that N_w is $\ell(w)$-dimensional, where $\ell(w) < \infty$ is the length of the reduced affine Weyl word w.

Group elements in $\widehat{G}$ can be most conveniently written in terms of the Steinberg presentation described in Section 3.1.4. For example, we will denote the exponential of the step operator for a positive real root multiplied by u as $x_\alpha(u)$. Note that imaginary roots α will never show up in N_w since they cannot be mapped to negative roots using the Weyl group [386]. Moreover, the definition of the group we are using only relies on being able to exponentiate real root generators [514].

In what follows we will shorten the notation, writing simply $\widetilde{G}(\mathbb{Q})$ and $\widetilde{G}(\mathbb{A})$ and not displaying explicitly the spectral parameter t since its presence is implicit by the hats and the tildes on the groups.

19.2.2 Eisenstein Series on Loop Groups

Let $\chi_{\widehat{B}}$ be a (quasi-)character on $\widehat{B}(\mathbb{A})$ trivial on $\widehat{B}(\mathbb{Q})$:

$$\chi_{\widehat{B}} : \widehat{B}(\mathbb{Q})\backslash\widehat{B}(\mathbb{A}) \longrightarrow \mathbb{C}^\times. \tag{19.13}$$

We assume the character splits as usual over the adeles

$$\chi_{\widehat{B}} = \bigotimes_p \chi_{\widehat{B}_p}, \tag{19.14}$$

and we extend it to all of $\widetilde{B} = \widehat{B} \rtimes \sigma(\mathfrak{q})$ by taking it to be trivial on σ. This means that in terms of weight $\lambda \in \widetilde{\mathfrak{h}}$ parametrising the character expanded on the fundamental weights $\Lambda_0, \ldots, \Lambda_n$ and δ we assume vanishing δ-component. The extension of the character to $\widetilde{B}$ is denoted by $\widetilde{\chi}_{\widehat{B}}$. We further extend it to all of $\widetilde{G}$ by making it right-invariant under K:

$$\widetilde{\chi}_{\widehat{B}}(b\sigma(\mathfrak{q})k) = \widetilde{\chi}_{\widehat{B}}(b\sigma(\mathfrak{q})) = \chi_{\widehat{B}}(b), \tag{19.15}$$

where $b \in \widehat{B}(\mathbb{A})$ and we recall that $\widetilde{\chi}_{\widehat{B}}$ was taken trivial on σ.

We now construct an Eisenstein series on $\widetilde{G}(\mathbb{A}(\langle t \rangle)$ attached to the degenerate principal series $\mathrm{Ind}_{\widetilde{G}(\mathbb{A})}^{\widehat{B}(\mathbb{A})} \widetilde{\chi}_{\widehat{B}}$ in the following way for $g \in \widetilde{G}(\mathbb{A})$:

$$E(\chi_{\widehat{B}}, g) = \sum_{\gamma \in \widehat{B}(\mathbb{Q}) \backslash \widehat{G}(\mathbb{Q})} \widetilde{\chi}_{\widehat{B}}(\gamma g). \tag{19.16}$$

Let $\widetilde{G}^{\tau}(\mathbb{A})$ be the slice of constant value for the derivation in the adelic loop group $\widetilde{G}(\mathbb{A})$ defined by

$$\widetilde{G}^{\tau}(\mathbb{A}) = \widehat{G}(\mathbb{A})\sigma(\tau). \tag{19.17}$$

For $g \in \widetilde{G}^{\tau}(\mathbb{A}\langle t \rangle)$ with $|\tau| < 1$ the Eisenstein series $E(\chi_{\widehat{B}}, g)$ converges absolutely for a certain range of $\widetilde{\chi}_{\widehat{B}}$ similar to the Godement range (4.134) and satisfies a functional equation [250, 251]. We stress that for $|\tau| \geq 1$ there is *no convergence* of the Eisenstein series (see also Remark 19.5).

Remark 19.2 One could consider a mild generalisation of the definition (19.16) above by also allowing a non-trivial dependence on the derivation direction $\sigma(\mathfrak{q}) = e^{vd}$ in the character: $\widetilde{\chi}_{\widehat{B}}(av^d) = \chi_{\widehat{B}}(a)v^{s_d}$ for some $s_d \in \mathbb{C}$ and all $a \in \widehat{A}$, generalising (19.15). Since v is not changed by the action of γ in (19.16) this amounts to simply multiplying (19.16) by the function v^{s_d}. In fact, v^{s_d} is formally an automorphic function on the loop group itself.

Let ψ be a character on the unipotent radical $\widehat{N}(\mathbb{A})$ of $\widehat{B}(\mathbb{A})$:

$$\psi : \widehat{N}(\mathbb{Q}) \backslash \widehat{N}(\mathbb{A}) \longrightarrow \mathbb{C}^{\times}. \tag{19.18}$$

Then $\psi = \prod_{\alpha \in \widehat{\Pi}} \psi_{\alpha}$ where ψ_{α} is a character on the abelian group $N_{\alpha}(\mathbb{Q}) \backslash N_{\alpha}(\mathbb{A})$. We distinguish two types of characters in a slight generalisation of Definition 6.10:

- ψ is N-generic if ψ_{α} is non-trivial for all $\alpha \in \Pi$
- ψ is generic if in addition ψ_{α_0} is also non-trivial.

The constant term with respect to $\widehat{B}$ is defined by (see Definition 6.17)

$$C_{\widehat{B}}(\chi_{\widehat{B}}, g) = \int_{\widehat{N}(\mathbb{Q}) \backslash \widehat{N}(\mathbb{A})} E(\chi_{\widehat{B}}, ng) dn, \tag{19.19}$$

while the Whittaker–Fourier coefficients with respect to $\widehat{B}$ are given by (see Definition 6.15)

$$W_{\widehat{B}, \psi}(\chi_{\widehat{B}}, g) = \int_{\widehat{N}(\mathbb{Q}) \backslash \widehat{N}(\mathbb{A})} E(\chi_{\widehat{B}}, ng) \overline{\psi(n)} dn. \tag{19.20}$$

19.2.3 Constant Term

The constant term formula for affine Eisenstein series was established in [250], generalising Theorem 8.1.

Theorem 19.3 (Constant term of the loop Eisenstein series) *Let $E(\chi_{\widehat{B}}, g)$ be an Eisenstein series defined as above in* (19.16). *Then we have*

$$\int_{\widehat{N}(\mathbb{Q})\backslash\widehat{N}(\mathbb{A})} E(\chi_{\widehat{B}}, nb\sigma(q))dn$$
$$= \sum_{w\in\widetilde{W}} (a\sigma(q))^{w\lambda+\rho} \prod_{\alpha\in\widetilde{\Delta}_+ | w\alpha\in\widetilde{\Delta}_-} \frac{\xi(\langle\lambda|\alpha\rangle)}{\xi(\langle\lambda|\alpha\rangle+1)}, \qquad (19.21)$$

where λ is the weight parametrising the character $\chi_{\widehat{B}}$, such that $\chi_{\widehat{B}}(a) = a^{\lambda+\rho}$ for $a \in \hat{A}$. ρ is the Weyl vector of the affine algebra and is given by the sum of the fundamental weights $\rho = \sum_{i=0}^{n} \Lambda_i$.

Proof The steps of the proof completely parallel those of Chapter 8. The starting point is the Bruhat decomposition (19.11). Since Weyl words $w \in \widetilde{W}$ are of finite length, the space N_w of (19.12) appearing in the Bruhat decomposition can be parametrised in exactly the same way as in Section 8.3 in terms of $\ell(w)$ Chevalley factors $x_\alpha(u)$. Iwasawa decomposing these factors successively and performing appropriate changes of integration variables, one generates the terms appearing in the constant term formula, where the completed Riemann zeta functions are due to the archimedean and non-archimedean Gindikin–Karpelevich formulas of Section 8.6. Since all roots α appearing in N_w are real, one only has the same $SL(2, \mathbb{A})$ calculations as in the finite-dimensional case. A nice detailed exposition of the proof for $\widetilde{G} = \widetilde{SL(2)}$ can be found in [454]. □

19.2.4 Fourier–Whittaker Coefficients

Now let us discuss the non-constant Whittaker coefficients of (19.20). Using the Bruhat decomposition of the loop group (19.11) we can rewrite this in the form

$$W_{\widehat{B},\psi}(\chi_{\widehat{B}}, g) = \sum_{w\in\widetilde{W}} \int_{w^{-1}\widehat{B}(\mathbb{Q})w\cap\widehat{N}(\mathbb{Q})\backslash\widehat{N}(\mathbb{A})} \widetilde{\chi}_{\widehat{B}}(wng)\overline{\psi(n)}dn. \qquad (19.22)$$

In the case of finite-dimensional groups it was shown in Chapter 9 that for generic characters ψ all the terms in the above sum vanish except the contribution from the longest Weyl word $w = w_0$ (see (9.13)). This fact is the source of a major difference between finite-dimensional Lie groups and Kac–Moody groups, namely that for Kac–Moody groups there is *no longest Weyl word*. This implies that the entire generic Whittaker coefficient vanishes:

$$\int_{\widehat{N}(\mathbb{Q})\backslash\widehat{N}(\mathbb{A})} E(\chi_{\widehat{B}}, ng)\overline{\psi(n)}dn = 0, \qquad \text{for } \psi \text{ generic.} \tag{19.23}$$

This is a general phenomenon that holds for all Kac–Moody groups. It does not mean, however, that only the constant terms contribute to the Fourier expansion, since we can still consider the Whittaker coefficients for degenerate characters ψ, which do not vanish in general. Indeed, it was shown in [214] that the reduction formula in Theorem 9.6 holds also for Eisenstein series on Kac–Moody groups (assuming convergence). This implies that for degenerate characters the Whittaker coefficients on a Kac–Moody group G can be written as finite sums of generic Whittaker coefficients on finite-dimensional subgroups $G' \subset G$.

Affine Casselman–Shalika Formula

Because of the absence of a longest Weyl word, the generic Whittaker coefficients do not receive non-zero contributions from the sum in (19.22). The characteristic feature of the longest Weyl word w_0 in the finite-dimensional case is that it maps *all* positive roots $\alpha > 0$ to negative ones $\alpha < 0$, and vice versa. In particular, under w_0 the unipotent radical N of the Borel subgroup B is mapped to the *opposite* unipotent N^- associated to the negative nilpotent root space $\mathfrak{n}_-$ in the triangular decomposition of the Lie algebra, $\mathfrak{g} = \mathfrak{n}_- \oplus \mathfrak{h} \oplus \mathfrak{n}_+$. This implies that the Jacquet–Whittaker integral (9.13) can be equivalently written as an integral over N^-, without the explicit appearance of w_0, i.e.,

$$W_\psi^\circ(\chi, a) = \int\limits_{N^-(\mathbb{A})} \chi(n^- a)\overline{\psi(n^-)}dn^-. \tag{19.24}$$

In the affine case, the opposite unipotent $\widetilde{N}^-(\mathbb{A})$ still exists, even though w_0 does not. Hence, we can still formally construct the analogue of the Jacquet integral also for Kac–Moody groups:

$$\begin{aligned} W_{\widehat{B}^-,\psi}(\chi_{\widehat{B}}, g) &= \int_{\widetilde{N}^-(\mathbb{A})} \widetilde{\chi}_{\widehat{B}}(n^- g)\overline{\psi(n^-)}dn^- \\ &= \left[\prod_{p<\infty} \int\limits_{N^-(\mathbb{Q}_p)} \widetilde{\chi}_{\widehat{B}}(n^- g)\overline{\psi(n^-)}dn^-\right] \int\limits_{N^-(\mathbb{R})} \widetilde{\chi}_{\widehat{B}}(n^- g)\overline{\psi(n^-)}dn^-. \end{aligned} \tag{19.25}$$

Remark 19.4 Braverman and Kazhdan have argued that even though there is no longest Weyl word w_0 the integral (19.25) should be added by hand to the Fourier expansion of the Eisenstein series $E(\chi_{\widehat{B}}, g)$ [507].

Remark 19.5 In the finite-dimensional case, there is no real difference between integrating over N and over N^-; after all, in the constant term formula we do integrate over the positive unipotent while for the Jacquet integral the opposite unipotent appears. However, this is not so in the affine case. It is not at all clear how to perform the integral over N^-, one obstacle being that there is no natural choice of coordinates over which to integrate. In the finite case, such a set of coordinates was provided by the reduced decomposition of the longest Weyl word w_0, which is not available for affine Kac–Moody groups. The reason for this asymmetry is the restriction on τ which is needed for convergence. Recall that convergence of $E(\chi_{\widehat{B}}, g)$ required us to take $g \in \widetilde{G}^\tau(\mathbb{A})$, with $|\tau| < 1$. This is a constraint on the allowed values of the derivation. This restriction is related to the fact that in defining the Borel subgroup for affine Kac–Moody groups we allowed for formal Taylor series in t but not in t^{-1}. If we were to consider series in t^{-1} instead, then convergence of the Eisenstein series would require the opposite regime $|\tau| > 1$ [507].

For non-archimedean local fields (e.g., p-adic fields, function fields) the problems with integrating over N^- have now been resolved, thanks to work by Braverman, Garland, Kazhdan and Patnaik [88, 90, 87, 509]. In particular, the integral over $\mathbb{Q}_p$ in (19.25) has recently been evaluated by Patnaik, leading to an affine generalisation of the Casselman–Shalika formula [509] (see also [397]):

Theorem 19.6 (Affine Casselman–Shalika formula) *Let $\widetilde{G}(\mathbb{Q}_p)$ be an untwisted affine Kac–Moody group and let ${}^L\widetilde{G}$ be the Langlands dual group. Then we have*

$$\int_{N^-(\mathbb{Q}_p)} \widetilde{\chi}_{\widehat{B}}(n^-)\overline{\psi(n^-)}dn^- = p^{-\langle\rho,\lambda\rangle}\mathfrak{m} \prod_{\alpha>0} \left(1 - p^{-(\langle\lambda|\alpha\rangle+1)}\right) \mathrm{ch}_\lambda, \tag{19.26}$$

where ch_λ is the Weyl–Kac character of a unitary highest weight representation of ${}^L\widetilde{G}$ with highest weight λ. The factor $\mathfrak{m}$ is furthermore given by

$$\mathfrak{m} = \prod_{i=1}^{\ell}\prod_{j=1}^{\infty} \frac{1 - p^{-(\langle\lambda|j\mathbf{c}\rangle+1+m_i)}}{1 - p^{-(\langle\lambda|j\mathbf{c}\rangle+m_i)}}. \tag{19.27}$$

Here m_i, $i = 1, \ldots, \ell$ are the exponents of the root system ${}^L\Delta$ of the underlying finite-dimensional Lie algebra ${}^L\mathfrak{g}$ and $\mathbf{c}$ is the minimal imaginary root of ${}^L\widetilde{\mathfrak{g}}$.

Proof For the proof, see [509]. □

19.3 Extension to General Kac–Moody Groups

In this section, we denote by G an arbitrary Kac–Moody group with $\mathfrak{g}$ its underlying Lie algebra. The general definition of a Borel–Eisenstein series, given in (4.133) for the finite-dimensional case, can be formally extended to the case when G is any Kac–Moody group:

$$E(\lambda, g) = \sum_{\gamma \in B(\mathbb{Q})\backslash G(\mathbb{Q})} e^{\langle \lambda+\rho | H(\gamma g)\rangle}. \tag{19.28}$$

Several remarks on this are in order:

- As we already mentioned, Garland has proven absolute convergence and analytic continuation in the special case when G is an affine Kac–Moody group.
- One can also formally define Kac–Moody Eisenstein series induced from cusp forms on finite-dimensional subgroups of G, generalising (5.104). Entirety of such cuspidal Eisenstein series was established in the affine case by Garland, Miller and Patnaik [253].
- Absolute convergence in the rank-two hyperbolic case was established by Carbone, Lee and Liu [122].

Remark 19.7 When restricting to the affine case, the expression (19.28) must be slightly modified due to the presence of the derivation d. See Section 19.2.2 above for a detailed discussion of this case.

19.3.1 Collapse of the Constant Term

Despite the absence of a mathematically rigorous definition of Eisenstein series on general Kac–Moody groups, quite a bit can be said about the Fourier expansion of these series, as we have already illustrated above. The foundation for this work was laid in [250], where the analogue of Langlands' formula (8.41) for the constant term was developed for the case of affine Kac–Moody Eisenstein series. Assuming convergence of the Eisenstein series $E(\lambda, g)$ one can formally derive a generalisation of Langlands constant term formula (8.3) for any Kac–Moody group:

$$\int_{N(\mathbb{Q})\backslash N(\mathbb{A})} E(\lambda, ng)dn = \sum_{w\in\mathcal{W}} a^{w\lambda+\rho} \prod_{\alpha>0\,|\,w\alpha<0} \frac{\xi(\langle\lambda|\alpha\rangle)}{\xi(1+\langle\lambda|\alpha\rangle)}. \tag{19.29}$$

While for Eisenstein series on finite-dimensional groups we have explained how to evaluate Langlands' formula in Section 10.3, it is not clear how to evaluate this formula in the case of Kac–Moody Eisenstein series. The reason for this is that the sum over Weyl words appearing in the formula is an infinite sum due to the infinite-dimensional nature of the associated Weyl groups $\mathcal{W}$.

This problem was addressed in [212] (see also [213] for a summary), where it was shown that, for special types of Kac–Moody Eisenstein series, the naively infinite sum 'collapses' to a finite sum and can be explicitly *computed*. On a more technical level, to evaluate Langlands' formula one proceeds just as in the case of a finite-dimensional group, and one successively constructs Weyl words in the set $C(\lambda)$ by the orbit method; see Section 10.3. It can then be shown that for particular types of Eisenstein series, which we will discuss in a moment, only the coefficients $M(w, \lambda)$ associated with the first few Weyl words w in the carefully constructed orbit are non-zero. All other coefficients associated with the infinite number of Weyl words that follow in the orbit are, however, zero and therefore do not contribute to the constant term. Let us now offer some details on this phenomenon.

As in Section 10.3.1 we define the Harish-Chandra c-function

$$c(\langle\lambda|\alpha\rangle) := \frac{\xi(\langle\lambda|\alpha\rangle)}{\xi(1+\langle\lambda|\alpha\rangle)}. \tag{19.30}$$

Recall that the only simple zeroes and poles of this function are

$$c(\langle\lambda|\alpha\rangle) = \begin{cases} 0, & \langle\lambda|\alpha\rangle = -1, \\ \infty, & \langle\lambda|\alpha\rangle = +1, \end{cases} \tag{19.31}$$

and it satisfies (for a suitable limit)

$$c(-1)c(+1) = 1. \tag{19.32}$$

This has the crucial implication that if $M(w, \lambda)$ contains an excess of $c(-1)$-factors in its product it must vanish. Moreover, from the mutiplicative property (see Lemma 8.6)

$$M(w_1 w_2, \lambda) = M(w_1, w_2\lambda)M(w_2, \lambda), \qquad \forall\, w_1, w_2 \in \mathcal{W}, \tag{19.33}$$

one can deduce that if $M(\tilde{w}, \lambda)$ vanishes for some $\tilde{w} \in \mathcal{W}$, then any other Weyl word w' containing $\tilde{w}$ will also yield $M(w', \lambda) = 0$.

Let us now consider the implications of this for special choices of λ. We shall restrict to the special values considered in Section 10.2:

$$\lambda_{i_*}(s) = 2s\Lambda_{i_*} - \rho, \qquad s \in \mathbb{C}, \tag{19.34}$$

where Λ_{i_*} is the fundamental weight associated with the simple root α_{i_*}. We then have the following lemma [212]:

Lemma 19.8 (Reduction of the constant term for special choices of λ) *For the particular choice of weight $\lambda = \lambda_{i_*}$ the constant term reduces to*

$$\int_{N(\mathbb{Q})\backslash N(\mathbb{A})} E(\lambda_{i_*}(s), ng)dn = \sum_{w \in \mathcal{S}_{i_*}} a^{w\lambda_{i_*}(s)+\rho} \prod_{\alpha>0\,|\,w\alpha<0} \frac{\xi(\langle\lambda_{i_*}(s)|\alpha\rangle)}{\xi(1+\langle\lambda_{i_*}(s)|\alpha\rangle)}, \tag{19.35}$$

where the sum runs over

$$\mathcal{S}_{i_*} := \{w \in \mathcal{W} \mid w\alpha_i > 0,\ \forall i \neq i_*\}, \tag{19.36}$$

which can be identified with the quotient group $\mathcal{W}/\mathcal{W}(\mathrm{stab}(\lambda))$.

Even though this considerably reduces the number of terms in the sum, there are still generically an infinite number of Weyl elements in the set $\mathcal{S}_{i_*}$. However, for very special choices of the parameter s, the infinite sum over the Weyl group actually collapses to a finite sum. This was analysed in detail in [212] for the exceptional Kac–Moody groups E_9, E_{10}, E_{11}, where it was shown that, for the choice of weight $\lambda_1(s) = 2s\Lambda_1 - \rho$ and fixing the parameter s to the special values $s = 3/2$ and $s = 5/2$, the associated Eisenstein series

$$E(s, g) := E(2s\Lambda_1 - \rho, g) \tag{19.37}$$

has only a finite number of terms in the constant term. We summarise this in Table 19.1.

Table 19.1 The number of terms contributing to the constant term of the Kac–Moody Eisenstein series $E(s, g)$ on E_9, E_{10}, E_{11} for the special values $s = 3/2$ and $s = 5/2$.

Kac–Moody group	$s = 3/2$	$s = 5/2$
E_9	10	54
E_{10}	11	65
E_{11}	12	77

Remark 19.9 The fact that the constant terms collapse to a finite number of terms has an important interpretation in string theory, which was in fact the main motivation for the analysis in [212]. In Chapter 14, we learned that the Eisenstein series $E(3/2, g)$ and $E(5/2, g)$ for $g \in E_8$ determine the couplings of the R^4 and $\nabla^4 R^4$ terms, respectively, in the effective action of type II string theory on T^7. These terms preserve a large amount of supersymmetry, ensuring that the perturbative expansion of their couplings should terminate after a small number of loop corrections. As we compactify further on T^8, T^9 and T^{10}, the U-duality groups are conjecturally enhanced to the Kac–Moody groups $E_9(\mathbb{Z}), E_{10}(\mathbb{Z})$

and $E_{11}(\mathbb{Z})$. Still, the constraints from supersymmetry ensure that there should only be a finite number of perturbative contributions, in perfect agreement with Table 19.1.

19.3.2 Collapse of the Fourier–Whittaker Coefficients

By an argument completely analogous to for the generic Whittaker coefficients of affine Kac–Moody groups in (19.23), the generic Whittaker coefficients of any Kac–Moody group vanishes due to the absence of a longest Weyl word:

$$W_\psi(\lambda, a) = \int_{N(\mathbb{Q})\backslash N(\mathbb{A})} E(\lambda, ng)\overline{\psi(n)}dn = 0, \qquad \psi \text{ generic.} \tag{19.38}$$

Therefore, only degenerate characters ψ on $N(\mathbb{A})$ yield non-vanishing Whittaker coefficients. Formally, Theorem 9.6 still holds in the Kac–Moody case, with the understanding that ψ is a degenerate character of N such that the support supp(ψ) is determined by a finite-dimensional subgroup $G' \subset G$. In other words, ψ is a *generic* character on the unipotent radical N' of the Borel subgroup $B' \subset G'$. We then have the formula as in Theorem 9.6:

$$W_\psi(\lambda, a) = \sum_{w_c w'_{\text{long}} \in \mathcal{W}/\mathcal{W}'} a^{(w_c w'_{\text{long}})^{-1}\lambda+\rho} M(w_c^{-1}, \lambda) W'_{\psi^a}(w_c^{-1}\lambda, \mathbb{1}), \tag{19.39}$$

where w'_{long} is the longest Weyl word in the finite-dimensional Weyl group $\mathcal{W}' = \mathcal{W}(\mathfrak{g}')$, and W'_ψ denotes a Whittaker function on $G'(\mathbb{A})$.

When trying to evaluate this formula one faces the same problems as for the constant term: generically, this is an infinite linear combination of Whittaker coefficients W'_ψ. It was, however, demonstrated in [214] that a similar collapse as in the constant term occurs also for degenerate Whittaker coefficients of E_9, E_{10} and E_{11}. Introduce the degenerate characters

$$\begin{aligned} \psi_\alpha &: N(\mathbb{Q})\backslash N(\mathbb{A}) \longrightarrow U(1), \\ \psi_{\alpha,\beta} &: N(\mathbb{Q})\backslash N(\mathbb{A}) \longrightarrow U(1), \end{aligned} \tag{19.40}$$

such that ψ_α is non-trivial only on the one-parameter subgroup $N_\alpha \subset N$ corresponding to the simple root α, and $\psi_{\alpha,\beta}$ is similarly only non-trivial on the pairs of disconnected simple roots (α, β), i.e., roots for which the associated root generators commute: $[E_\alpha, E_\beta] = 0$. The following result was proven in [214]:

Theorem 19.10 (Collapse of the Kac–Moody Whittaker coefficients) *Let G be E_9, E_{10} or E_{11} and $\mathfrak{g}$ the corresponding Lie algebra. Further, let $E(s, g)$ be the associated Kac–Moody Eisenstein series, defined as in (19.37), and ψ a* *non-trivial character on $N(\mathbb{A})$, trivial on $N(\mathbb{Q})$. We then have*

$$\sum_{\psi} W_{\psi}(3/2, na) = \sum_{\alpha\in\Pi}\sum_{\psi_\alpha} c_\alpha(3/2, a) W'_{\psi_\alpha^a}(3/2, \mathbb{1})\psi_\alpha(n),$$

$$\sum_{\psi} W_{\psi}(5/2, na) = \sum_{\alpha\in\Pi}\sum_{\psi_\alpha} c_\alpha(5/2, a) W'_{\psi_\alpha^a}(5/2, \mathbb{1})\psi_\alpha(n) + \sum_{\substack{\alpha,\beta\in\Pi\\ [E_\alpha,E_\beta]=0}}\sum_{\psi_{\alpha,\beta}} c_{\alpha,\beta}(5/2, a) W'_{\psi_{\alpha,\beta}^a}(5/2, \mathbb{1})\psi_{\alpha,\beta}(n), \tag{19.41}$$

where the coefficients $c_\alpha(s,a)$ and $c_{\alpha,\beta}(s,a)$ are simple polynomial functions on the torus A that can be found in [214].

The above theorem is quite remarkable. It states that the only non-vanishing Fourier–Whittaker coefficients on N are those associated with the degenerate characters ψ_α and $\psi_{\alpha,\beta}$. This is an enormous simplication compared to the generic behaviour of Kac–Moody Eisenstein series. The function $W'_{\psi_\alpha}(3/2, \mathbb{1})$ is a simple Whittaker coefficient of an $SL(2)$-subgroup, while $W'_{\psi_{\alpha,\beta}}(5/2, \mathbb{1})$ is the product of two such $SL(2)$-Whittaker coefficients. To provide some more details, first define the function (see the examples in Section 10.4.3)

$$B(s_\alpha, m_\alpha) := \frac{2}{\xi(2s_\alpha)}|m_\alpha|^{1/2-s_\alpha}\sigma_{2s_\alpha-1}(m_\alpha)K_{s_\alpha-1/2}(2\pi|m_\alpha|), \tag{19.42}$$

where $m_\alpha \in \mathbb{Q}$ parametrises the character ψ_α and the s_α parametrises the projection of the weight $\lambda_1(s) = 2s\Lambda_1 - \rho$ of $\mathfrak{g}$ onto the $\mathfrak{sl}(2)_\alpha$-subalgebra, such that $\lambda_1(s)$ projects onto $\lambda_\alpha = 2s_\alpha\Lambda_\alpha - \rho_\alpha$, where Λ_α is the unique fundamental weight of $\mathfrak{sl}(2)_\alpha$ and ρ_α is the associated Weyl vector.

In terms of the function $B(s_\alpha, m_\alpha)$ we have

$$\begin{aligned} W'_{\psi_\alpha}(3/2, \mathbb{1}) &= B(s_\alpha, m_\alpha), \\ W'_{\psi_{\alpha,\beta}}(5/2, \mathbb{1}) &= B(s_\alpha, m_\alpha)B(s_\beta, m_\beta). \end{aligned} \tag{19.43}$$

The function $W'_{\psi_\alpha}(5/2, \mathbb{1})$, on the other hand, is slightly more involved, being given by a linear combination of $B(s_\alpha, m_\alpha)$ and products of the form $B(s_\alpha, m_\alpha)B(s_\beta, m_\beta)$. The precise details can be found in [214].

Remark 19.11 It follows from these results that the Whittaker coefficients of the Eisenstein series $E(3/2, g)$ are Eulerian, i.e.,

$$W_\psi(3/2, g) = \prod_{p\leq\infty} W_{p,\psi}(3/2, g), \tag{19.44}$$

where the local factors $W_{p,\psi}(3/2, g)$ can be read off from the first equation in (19.41). On the other hand, the Whittaker coefficients of the Eisenstein series $E(5/2, g)$ are *not* Eulerian, as is clear from the second equation in (19.41). However, because of the factorisation (19.43) the Whittaker coefficient corresponding to the character $\psi_{\alpha,\beta}$ is Eulerian:

$$W_{\psi_{\alpha,\beta}}(3/2, \mathbb{1}) := c_{\alpha,\beta}(5/2, \mathbb{1}) W'_{\psi_{\alpha,\beta}}(5/2, \mathbb{1}) = \prod_{p \le \infty} W_{p,\psi_{\alpha,\beta}}(5/2, \mathbb{1}). \quad (19.45)$$

In general, we expect the 'maximal' (in terms of support) non-vanishing degenerate Whittaker coefficients to be Eulerian. We will discuss the representation-theoretic interpretation of these observations in the next section.

19.3.3 Small Representations of Kac–Moody Groups

As we have indicated repeatedly in this chapter, the theory of automorphic forms on Kac–Moody groups is still in its infancy, and beyond the affine case almost nothing is known except for a few low-rank hyperbolic groups. In spite of this, we have seen above that it is possible to obtain very explicit results for the Fourier coefficients of Kac–Moody Eisenstein series in some non-trivial cases. Although these results are only formal, since the convergence of the associated Eisenstein series are yet to be proven, it is nevertheless interesting to speculate as to what they might indicate from the point of view of the representation theory of $G(\mathbb{A})$.

To this end we note that, for finite-dimensional Lie groups, the structure of the Whittaker coefficients given in Theorem 19.10 is the hallmark of spherical vectors of small automorphic representations [476, 214, 340]. Indeed, for G one of the finite-dimensional Lie groups in Table 13.2 the Eisenstein series $E(3/2, g) = E(3\Lambda_1 - \rho, g)$ and $E(5/2, g) = E(5\Lambda_1 - \rho, g)$ are spherical vectors in the minimal π_{min} and next-to-minimal π_{ntm} automorphic representations, respectively [281, 520, 315].

Based on these observations and the results of [212, 214] presented above we are led to the following:

Conjecture 19.12 (Small representations of Kac–Moody groups) *Let G be either of the exceptional Kac–Moody groups E_9, E_{10}, E_{11}.*

- *Then there exist a minimal π_{min} and a next-to-minimal π_{ntm} unitary representation of $G(\mathbb{A})$, realised as submodules*

$$\pi_{\text{min}} \subset \text{Ind}_{B(\mathbb{A})}^{G(\mathbb{A})} e^{2s\Lambda_1 - \rho}\Big|_{s=3/2},$$

$$\pi_{\text{ntm}} \subset \text{Ind}_{B(\mathbb{A})}^{G(\mathbb{A})} e^{2s\Lambda_1 - \rho}\Big|_{s=5/2}. \quad (19.46)$$

The degenerate principal series can equivalently be constructed by induction from the maximal parabolic subgroup P_n of E_n, $n = 9, 10, 11$, with *semi-simple Levi group of type D_{n-1}, which is obtained by deleting the first node of the E_n Dynkin diagram (using Bourbaki labelling; see Figure 13.5).*

- *The canonically associated Eisenstein series $E(3\Lambda_1-\rho,g)$ and $E(5\Lambda_1-\rho,g)$ (obtained by analytic continuation) are the unique spherical vectors in π_{min} and π_{ntm}, respectively. The wavefront sets $WF(\pi_{\text{min}})$ and $WF(\pi_{\text{ntm}})$ have Bala–Carter labels A_1 and $2A_2$.*
- *We further have the factorisations*

$$\pi_{\text{min}} = \bigotimes_{p\le\infty} \pi_{\text{min},p},$$

$$\pi_{\text{ntm}} = \bigotimes_{p\le\infty} \pi_{\text{ntm},p}, \tag{19.47}$$

and the local factors $W_{p,\psi_\alpha}(3/2)$ and $W_{p,\psi_{\alpha,\beta}}(5/2)$ in Equations (19.44) and (19.45) are (abelian limits of) unique spherical vectors in $\pi_{\text{min},p}$ and $\pi_{\text{min},p}$, respectively.

Remark 19.13 In the finite-dimensional setting the wavefront sets of the minimal and next-to-minimal representations are defined as the closures of the smallest nilpotent orbits $\mathcal{O}_{\text{min}}$ and $\mathcal{O}_{\text{ntm}}$; see Section 6.4.2. The theory of nilpotent orbits is not developed for Kac–Moody groups, but one can guess a definition of these orbits by analogy with the finite-dimensional cases. Indeed, the minimal nilpotent can always be obtained as the G-orbit of a single simple root generator E_α, $\alpha \in \Pi$, while the next-to-minimal nilpotent orbit is obtained as the orbit of $E_\alpha + E_\beta$, where α and β are disconnected simple roots. Thus, we expect that also in the Kac–Moody case the following definitions make sense:

$$\mathcal{O}_{\text{min}} := \mathcal{O}_G(E_\alpha), \qquad \mathcal{O}_{\text{ntm}} := \mathcal{O}_G(E_\alpha + E_\beta). \tag{19.48}$$

There are many open questions concerning automorphic representations of Kac–Moody groups that are not addressed by the conjectures above. Below we discuss some of these.

Gelfand–Kirillov Dimension

What is the analogous notion to Gelfand–Kirillov dimension in functional representations of Kac–Moody groups? In the finite-dimensional case, the Gelfand–Kirillov dimension is half the dimension of the highest nilpotent orbit $\mathcal{O}$ in the wavefront set. This does not seem like a good notion in the Kac–Moody case since all nilpotent orbits are infinite-dimensional. Perhaps a better notion is to use the co-dimension of $\mathcal{O}$?

Joseph Ideal

Recall from Section 6.4.3 that for real finite-dimensional Lie groups $G(\mathbb{R})$ the minimal representation $\pi_{\min}$ can be characterised by the fact that their annihilator ideal is the *Joseph ideal*, which is a two-sided ideal in the universal enveloping algebra $\mathcal{U}(\mathfrak{g}_{\mathbb{C}})$. Can one extend the notion of a Joseph ideal to the case of universal enveloping algebras of Kac–Moody algebras? Some steps in this direction for affine Kac–Moody algebras have been taken in [25].

Character Distribution

In the non-archimedean setting, a representation π of $G(\mathbb{Q}_p)$ is minimal if its Harish-Chandra character distribution takes the form

$$\operatorname{tr}(\pi) = \widehat{\mu}_{\mathcal{O}_{\min}} + c_0 , \tag{19.49}$$

where $\widehat{\mu}_{\mathcal{O}_{\min}}$ is the Fourier transform of a distribution $\mu_{\mathcal{O}_{\min}}$ on the unique minimal nilpotent orbit $\mathcal{O}_{\min}$; see [246] for details. As was pointed out around (6.100) for finite-dimensional Lie groups (following the earlier local results of Matumoto [470] and Mœglin-Waldspurger [486]), the structure of the Whittaker coefficients of the Eisenstein series $E(3/2, g)$ can be viewed as a global analogue of the character distribution (19.49). Could we similarly interpret the structure of the Whittaker coefficients displayed in Theorem 19.10 as automorphic analogoues of the character distribution of the minimal representation of Kac–Moody groups?

Reflection Representation of Double Affine Hecke Algebras

For non-archimedean groups there is a different realisation of minimal representations via the spherical Hecke algebra. Recall from Section 11.3 that the spherical Hecke algebra $\mathcal{H}(G(\mathbb{Q}_p))$ is the convolution algebra of $K(\mathbb{Q}_p)$-bi-invariant functions on $G(\mathbb{Q}_p)$; that is, $\mathbb{C}$-valued functions on $K(\mathbb{Q}_p)\backslash G(\mathbb{Q}_p)/K(\mathbb{Q}_p)$.

The spherical Hecke algebra has a special representation, the *reflection representation* $\mathcal{E}$, which can be viewed as a deformation of the ordinary reflection representation of the Weyl group $\mathcal{W}(\mathfrak{g})$. The minimal representation $\pi_{\min,p}$ can then equivalently be defined as the unique irreducible representation of $G(\mathbb{Q}_p)$ such that its space of Iwahori-fixed vectors is isomorphic to $\mathcal{E}$, viewed as an $\mathcal{H}(G(\mathbb{Q}_p))$-module [545]. Savin has suggested that this might be a good approach to understanding minimal representations of p-adic Kac–Moody groups, at least in the affine case.

For p-adic loop groups, the analogue of the spherical Hecke algebra is the *double affine Hecke algebra* (or DAHA for short) [387, 88, 90, 258]. Some properties of the DAHA resembling the reflection representation have also been analysed in [369]. It would be very interesting if one could define the analogue of Iwahori-fixed vectors in this context and possibly verify that $W_{p,\psi_\alpha}(3/2)$

indeed corresponds to the unique spherical vector of $\pi_{\min,p}$ for E_9, E_{10} and E_{11}, thereby generalising the results of Kazhdan and Polishchuk on spherical vectors of minimal representations of $E_6(\mathbb{Q}_p)$, $E_7(\mathbb{Q}_p)$ and $E_8(\mathbb{Q}_p)$ [394].

Langlands Program for Kac–Moody Groups?

The developments discussed in this chapter obviously beg the question of whether it is possible to develop a complete theory of automorphic representations of Kac–Moody groups. Or, even more ambitiously: *is there a generalisation of the Langlands program to arbitrary Kac–Moody groups?* This is 'the dream' as nicely formulated by Braverman and Kazhdan in [89]:

> Our dream would be to develop an analog of the above representation theories and the Langlands correspondence for the group $\widetilde{G}$ (or, more generally, for any symmetrizable Kac–Moody group). This is a fascinating task by itself but we also believe that a fully developed theory of automorphic forms for $\widetilde{G}$ will have powerful applications to automorphic forms on G.

Although at present this remains a dream, the recent developments reviewed above certainly provide hope that such a theory is within reach.

APPENDICES

Appendix A

$SL(2,\mathbb{R})$ Eisenstein Series and Poisson Resummation

In this appendix, we perform the Fourier expansion of the series (1.2),

$$f_s(z) = \sum_{\substack{(c,d)\in\mathbb{Z}^2 \\ (c,d)\neq(0,0)}} \frac{y^s}{|cz+d|^{2s}}, \tag{A.1}$$

which is related to the standard $SL(2,\mathbb{R})$ Eisenstein series through $f_s(z) = 2\zeta(2s)E(s,z)$; see (1.11). Here, $z = x + iy$ lies on the upper half-plane $\mathbb{H}$ as defined in Section 4.1.

The invariance of $f_s(z)$ under shifts $z \to z+1$ implies that it should have a Fourier expansion

$$f_s(z) = C(y) + \sum_{m\neq 0} a_m(y) e^{2\pi i m x} . \tag{A.2}$$

The 'constant term(s)' $C(y)$ and the non-zero Fourier coefficients $a_m(y)$ are determined in the following. We suppress the label s on the constant terms and Fourier coefficients for ease of notation.

The technique to be used rests on *Poisson resummation*, whose fundamental equation here is (see [535, Eq. (8.2.10])

$$\sum_{m\in\mathbb{Z}} \exp(-\pi a m^2 + 2\pi i b m) = a^{-1/2} \sum_{\tilde{m}\in\mathbb{Z}} \exp\left(-\frac{\pi(\tilde{m}-b)^2}{a}\right) . \tag{A.3}$$

Another useful form of this same formula is

$$\sum_{m\in\mathbb{Z}} \exp\left(-\frac{\pi}{t}(m+nx)^2\right) = t^{1/2} \sum_{\tilde{m}\in\mathbb{Z}} \exp\left(-\pi t \tilde{m}^2 - 2\pi i \tilde{m} n x\right) . \tag{A.4}$$

Note that the sums are over all integers and not constrained to a single $SL(2,\mathbb{Z})$-orbit.

We will also use the following representation of powers for $\mathrm{Re}(s) > 0$ and $\mathrm{Re}(M) > 0$:

$$M^{-s} = \frac{\pi^s}{\Gamma(s)} \int_0^\infty \frac{dt}{t^{s+1}} e^{-\frac{\pi}{t}M} . \tag{A.5}$$

Finally, we require the following integral representation of the modified Bessel function for real $a, b \neq 0$:

$$\int_0^\infty \frac{dt}{t^{s+1}} e^{-\pi t a^2 - \frac{\pi}{t} b^2} = 2\left|\frac{a}{b}\right|^s K_s(2\pi|ab|) . \tag{A.6}$$

A.1 Constant Term(s)

First extract the term $c = 0$ from (A.1). Then $d \neq 0$ and

$$f_s(z) = y^s \sum_{d\neq 0} |d|^{-2s} + \underbrace{y^s \sum_{c\neq 0} \sum_{d\in\mathbb{Z}} |cz+d|^{-2s}}_{f_s^{(1)}(z)} = 2\zeta(2s) y^s + f_s^{(1)}(z) . \tag{A.7}$$

The power $|cz+d|^{-2s}$ appearing in the second term can be rewritten as an integral using (A.5). Then one can Poisson resum over $d \in \mathbb{Z}$ using (A.4):

$$\begin{aligned} f_s^{(1)}(z) &= \frac{\pi^s}{\Gamma(s)} y^s \sum_{c\neq 0} \sum_{d\in\mathbb{Z}} \int_0^\infty \frac{dt}{t^{s+1}} \exp\left(-\frac{\pi}{t}|cz+d|^2\right) \\ &= \frac{\pi^s}{\Gamma(s)} y^s \sum_{c\neq 0} \sum_{d\in\mathbb{Z}} \int_0^\infty \frac{dt}{t^{s+1}} \exp\left(-\frac{\pi}{t}((cx+d)^2 + (cy)^2)\right) \\ &= \frac{\pi^s}{\Gamma(s)} y^s \sum_{c\neq 0} \sum_{\tilde{d}\in\mathbb{Z}} \int_0^\infty \frac{dt}{t^{s+1}} t^{1/2} \exp\left(-\pi t \tilde{d}^2 - 2\pi i \tilde{d} c x - \frac{\pi}{t}(cy)^2\right) . \end{aligned} \tag{A.8}$$

In the final line of (A.8) one can separate out the term with $\tilde{d} = 0$ by

$$f_s^{(1)}(z) = \frac{\pi^s}{\Gamma(s)} y^s \sum_{c\neq 0} \int_0^\infty \frac{dt}{t^{s+1/2}} \exp\left(-\frac{\pi}{t}(cy)^2\right) + f_s^{(2)}(z), \tag{A.9}$$

since it does not have any x dependence and where $f_s^{(2)}$ are the terms with $\tilde{d} \neq 0$:

$$f_s^{(2)}(z) = \frac{\pi^s}{\Gamma(s)} y^s \sum_{c\neq 0} \sum_{\tilde{d}\neq 0} \int_0^\infty \frac{dt}{t^{s+1/2}} \exp\left(-\pi t \tilde{d}^2 - 2\pi i \tilde{d} c x - \frac{\pi}{t}(cy)^2\right) . \tag{A.10}$$

The integral in the term with $\tilde{d} = 0$ can be undone using (A.5) and the sum over $c \neq 0$ can be carried out afterwards. Hence the first term in (A.9) becomes

$$\begin{aligned}
&\frac{\pi^s}{\Gamma(s)} y^s \sum_{c\neq 0} \int_0^\infty \frac{dt}{t^{s+1/2}} \exp\left(-\frac{\pi}{t}(cy)^2\right) \\
&= \frac{\pi^s}{\Gamma(s)} \frac{\Gamma(s-1/2)}{\pi^{s-1/2}} y^{s-2(s-1/2)} \sum_{c\neq 0} c^{-2(s-1/2)} \\
&= 2\zeta(2s) \frac{\pi^{-(s-1/2)}\Gamma(s-1/2)\zeta(2s-1)}{\pi^{-s}\Gamma(s)\zeta(2s)} y^{1-s} \\
&= 2\zeta(2s) \frac{\xi(2s-1)}{\xi(2s)} y^{1-s}, \qquad\qquad (A.11)
\end{aligned}$$

where we have pulled out the same overall factor as in (A.7) and regrouped the π-factors to use the definition of the completed Riemann zeta function $\xi(k) = \pi^{-k/2}\Gamma(k/2)\zeta(k)$.

A.2 Non-zero Fourier Modes

The current status of the Fourier expansion is then

$$f_s(z) = 2\zeta(2s)\left(y^s + \frac{\xi(2s-1)}{\xi(2s)} y^{1-s}\right) + f_s^{(2)}(z), \qquad (A.12)$$

with the non-zero mode part $f_s^{(2)}$ given by (A.10). The t-integral appearing in that expression is a Bessel integral and can be evaluated using (A.6) as

$$\begin{aligned}
f_s^{(2)}(z) &= \frac{2\pi^s}{\Gamma(s)} y^s \sum_{c\neq 0} \sum_{\tilde{d}\neq 0} \left|\frac{\tilde{d}}{cy}\right|^{s-1/2} K_{s-1/2}(2\pi|\tilde{d}c|y) e^{-2\pi i \tilde{d} c x} \\
&= \frac{2\pi^s}{\Gamma(s)} y^{1/2} \sum_{c\neq 0} \sum_{\tilde{d}\neq 0} \left|\frac{\tilde{d}}{c}\right|^{s-1/2} K_{s-1/2}(2\pi|\tilde{d}c|y) e^{-2\pi i \tilde{d} c x}. \qquad (A.13)
\end{aligned}$$

To find the Fourier coefficient $a_m(y)$ of a mode $e^{2\pi i m x}$ we transform the double summation to one over $m \neq 0$ and the (positive) divisors $d|m$. Then

$$\begin{aligned}
f_s^{(2)}(z) &= \frac{4\pi^s}{\Gamma(s)} y^{1/2} \sum_{m\neq 0} \sum_{d|m} d^{1-2s} |m|^{s-1/2} K_{s-1/2}(2\pi|m|y) e^{2\pi i m x} \\
&= 2\zeta(2s) \frac{2y^{1/2}}{\xi(2s)} \sum_{m\neq 0} |m|^{s-1/2} \sigma_{1-2s}(m) K_{s-1/2}(2\pi|m|y) e^{2\pi i m x},
\end{aligned}$$

(A.14)

again pulling out the same overall factor $2\zeta(2s)$ and using the divisor sum

$$\sigma_s(m) = \sum_{d|m} d^s, \tag{A.15}$$

where only positive divisors are included.

The full Fourier expansion is therefore given by

$$f_s(z) = 2\zeta(2s)\Bigg[y^s + \frac{\xi(2s-1)}{\xi(2s)}y^{1-s} + \frac{2y^{1/2}}{\xi(2s)}\sum_{m\neq 0}|m|^{s-1/2}\sigma_{1-2s}(m)K_{s-1/2}(2\pi|m|y)e^{2\pi imx}\Bigg]. \tag{A.16}$$

The term in the square brackets is the full expansion of the Eisenstein series $E(s,\tau)$ for $SL(2,\mathbb{R})$. This agrees with the adelic derivation of Theorem 7.1.

Appendix B

Laplace Operators on G/K and Automorphic Forms

In this appendix, we briefly review the connection between the *scalar Laplace operator* on the symmetric space $G(\mathbb{R})/K(\mathbb{R})$ and the quadratic Casimir. We do this first for a general simple, simply-laced split group $G(\mathbb{R})$ and then give a very explicit analysis for $G = SL(2, \mathbb{R})$.

B.1 Scalar Laplace Operator and Quadratic Casimir

For a simple, simply-laced split $G(\mathbb{R})$ we denote by $\mathfrak{h}$ a fixed Cartan subalgebra of the Lie algebra $\mathfrak{g}(\mathbb{R})$ of $G(\mathbb{R})$. With respect to $\mathfrak{h}$ and a choice of simple roots α_i ($i = 1, \ldots, r$ with $r = \dim_{\mathbb{R}}(\mathfrak{h})$) the remaining generators arrange into positive and negative step operators; see (3.16). We denote by E_α the step operator of a given root α. In *Iwasawa gauge* we choose to write an arbitrary element $g \in G(\mathbb{R})/K(\mathbb{R})$ as

$$g = na = \exp\left(\sum_{\alpha>0} u_\alpha E_\alpha\right) \prod_{i=1}^{r} v_i^{h_i}, \tag{B.1}$$

where h_i are the Cartan generators associated with the choice of simple roots (see (3.18)). The variables v_i (for $= 1, \ldots, r$) and u_α (for $\alpha \in \Delta_+$) are coordinates on the symmetric space $G(\mathbb{R})/K(\mathbb{R})$.

The $G(\mathbb{R})$-invariant metric on the symmetric space can be constructed from

$$ds^2_{G/K} = 2\langle \mathcal{P} | \mathcal{P} \rangle, \tag{B.2}$$

where we choose a convenient normalisation and

$$\mathcal{P} = \frac{1}{2}\left(g^{-1}dg - \theta(g^{-1}dg)\right) \tag{B.3}$$

is the coset projection of the *Maurer–Cartan form* $g^{-1}dg$ associated with the vector space decomposition $\mathfrak{g} = \mathfrak{p} \oplus \mathfrak{k}$. Here, $\mathfrak{k}$ is the Lie algebra of K. The

(Cartan) involution θ leaving $\mathfrak{k}$ fixed can be defined by

$$\theta(E_\alpha) = -E_{-\alpha}, \quad \theta(h_i) = -h_i. \tag{B.4}$$

With this convention, $\mathfrak{k}$ and $\mathfrak{p}$ have the bases

$$\begin{aligned}\mathfrak{k} &= \langle E_\alpha - E_{-\alpha} \mid \alpha > 0\rangle, \\ \mathfrak{p} &= \langle E_\alpha + E_{-\alpha} \mid \alpha > 0\rangle \oplus \langle h_i \mid i = 1, \dots, r\rangle.\end{aligned} \tag{B.5}$$

We further choose the normalisation (A_{ij} is the Cartan matrix (3.19) of the simply-laced $\mathfrak{g}(\mathbb{R})$)

$$\langle E_\alpha | E_{-\beta}\rangle = \delta_{\alpha,\beta}, \quad \langle h_i | h_j\rangle = A_{ij}. \tag{B.6}$$

Working out the Maurer–Cartan form for the element (B.1) one finds

$$\begin{aligned}g^{-1}dg &= \sum_{i=1}^{r} v_i^{-1} dv_i h_i + a^{-1}\left(\sum_{\alpha>0} Du_\alpha E_\alpha\right) a \\ &= \sum_{i=1}^{r} v_i^{-1} dv_i h_i + \sum_{\alpha>0} a^{-\alpha} Du_\alpha E_\alpha,\end{aligned} \tag{B.7}$$

where $Du_\alpha = du_\alpha + \cdots$, and the dots represent finitely many terms coming from commutator terms when expanding out the *Baker–Campbell–Hausdorff identity*

$$e^{-X} d(e^X) = dX - \frac{1}{2!}[X, dX] + \frac{1}{3!}[X,[X,dX]] + \cdots \tag{B.8}$$

for the nilpotent E_α. The expression (B.7) together with (B.6) leads to a block-diagonal metric of the form

$$ds^2_{G/K} = g_{\mu\nu} dx^\mu dx^\nu = 2\sum_{i,j=1}^{r} v_i^{-1} v_j^{-1} dv_i dv_j A_{ij} + \sum_{\alpha>0} a^{-2\alpha}(Du_\alpha)^2. \tag{B.9}$$

The scalar Laplacian associated with this metric is ($\partial_\mu \equiv \frac{\partial}{\partial x^\mu}$ and $\sqrt{g} = \sqrt{\det(g_{\mu\nu})}$)

$$\begin{aligned}\Delta_{G/K} &= \frac{1}{\sqrt{g}} \partial_\mu \left(\sqrt{g} g^{\mu\nu} \partial_\nu\right) \\ &= \frac{1}{2}\sum_{i,j=1}^{r} (A^{-1})^{ij} a^{2\rho} v_i \partial_i \left(a^{-2\rho} v_j \partial_j\right) + \sum_{\alpha>0} a^{2\alpha} \partial_\alpha^2 + \cdots,\end{aligned} \tag{B.10}$$

where the dots come from inverting the metric in the $du_\alpha du_\beta$ sector and $(A^{-1})^{ij}$ is the inverse of the Cartan matrix. We have used the relation $\sum_{\alpha>0} \alpha = 2\rho$ for the Weyl vector; see (3.6). The Laplace operator (B.10) is $G(\mathbb{R})$-invariant since the Maurer–Cartan form trivially is: the transformation of $g \in G(\mathbb{R})/K(\mathbb{R})$ is given by $g \to g_0 g k^{-1}$ with constant $g_0 \in G(\mathbb{R})$ and $k \in K(\mathbb{R})$ such that $g^{-1}dg \to k(g^{-1}dg)k - dkk^{-1}$ is independent of g_0.

We can evaluate the eigenvalue of the Laplacian (B.10) when acting on an Eisenstein series $E(\lambda, g)$ as defined in (4.133). Due to the invariance of the Laplacian, it suffices to evaluate it on the summand $\chi(g) = \chi(a) = a^{\lambda+\rho}$, corresponding to $\gamma = \mathbb{1}$. For this term, the derivatives ∂_α with respect to the coordinates u^α vanish and one finds

$$\Delta_{G/K} a^{\lambda+\rho} = \Delta_{G/K} \prod_{i=1}^{r} v_i^{2s_i} = \frac{1}{2} \sum_{i,j=1}^{r} (A^{-1})^{ij} 2s_i(2s_j - 2)a^{\lambda+\rho}$$
$$= \frac{1}{2} \left(\langle\lambda|\lambda\rangle - \langle\rho|\rho\rangle\right) a^{\lambda+\rho}, \tag{B.11}$$

where we stress that we have assumed $\mathfrak{g}$ to be simply-laced. As already mentioned, $G(\mathbb{R})$-invariance implies that this is also the eigenvalue for the full Eisenstein series:

$$\Delta_{G/K} E(\lambda, g) = \frac{1}{2} \left(\langle\lambda|\lambda\rangle - \langle\rho|\rho\rangle\right) E(\lambda, g). \tag{B.12}$$

This agrees up to a factor with the standard quadratic Casimir evaluated on a lowest weight representation with lowest weight $\Lambda = \lambda + \rho$ [386].

B.2 $SL(2,\mathbb{R})$ Automorphic Forms as Laplace Eigenfunctions

For the case of $SL(2,\mathbb{R})$ we can give fully explicit expressions. Using

$$g = na = \exp(ue)v^h \tag{B.13}$$

one finds from (B.2)

$$ds^2_{G/K} = 4v^{-2}dv^2 + e^{-4v}du^2 \quad \Rightarrow \quad \Delta_{G/K} = \frac{1}{4}e^{2v}v\partial_v\left(e^{-2v}v\partial_v\right) + e^{4v}\partial_u^2. \tag{B.14}$$

This can be brought into a more familiar form by using $v = y^{1/2}$ and $u = x$; see (4.5). This leads to

$$\Delta_{G/K} = y^2\left(\partial_x^2 + \partial_y^2\right), \tag{B.15}$$

which agrees with the Laplacian on the upper half-plane $\mathbb{H}$ given in (4.56).

Consider now a real eigenfunction $\varphi(z)$, where $z = x + iy$, of the Laplace operator with eigenvalue $s(s-1)$:

$$\Delta_{G/K}\varphi(z) = s(s-1)\varphi(z). \tag{B.16}$$

If the function is furthermore invariant under $SL(2,\mathbb{Z})$, this implies $\varphi(z) = \varphi(z+1)$ and one has a Fourier expansion of the form

$$\varphi(z) = \sum_{m\in\mathbb{Z}} a_m(y)e^{2\pi imx}, \tag{B.17}$$

where $m \in \mathbb{Z}$ denotes the 'instanton charge' of the character, in the terminology of Section 6.2, and $a_0(y)$ is the constant term. This Fourier expansion is due to the translations $x \to x+1$ contained in the action of $SL(2,\mathbb{Z})$ on $SL(2,\mathbb{R})$. Reality of $\varphi(z)$ implies that $a_m(y) = a_{-m}(y)$ for all $m > 0$. We therefore restrict to $m \geq 0$.

Plugging the Fourier expansion (B.17) into the Laplace Equation (B.16) and analysing each mode individually leads to the following equations:

$$m = 0: \qquad y^2\partial_y^2 a_0(y) = s(s-1)a_0(y), \tag{B.18a}$$

$$m \neq 0: \qquad y^2\left(\partial_y^2 - 4\pi^2 m^2\right)a_m(y) = s(s-1)a_m(y). \tag{B.18b}$$

The Equation (B.18a) for the constant term has two linearly independent solutions:

$$s \neq 1/2: \qquad a_0(y) = y^s \qquad \text{or} \qquad a_0(y) = y^{1-s}, \tag{B.19a}$$

$$s = 1/2: \qquad a_0(y) = y^{1/2} \qquad \text{or} \qquad a_0(y) = y^{1/2}\log y. \tag{B.19b}$$

All these solutions are at most power laws when y approaches any cusp, e.g., $y \to \infty$.

Equation (B.18b) for the non-zero modes becomes more familiar when one uses $a_m(y) = y^{1/2}b_m(y)$, which leads to

$$y^2\partial_y^2 b_m(y) + y\partial_y b_m(y) - \left(4\pi m^2 y^2 + (s-1/2)^2\right)b_m(y) = 0. \tag{B.20}$$

After a rescaling of the y coordinate this becomes the modified Bessel equation with the two modified Bessel functions $K_{s-1/2}$ and $I_{s-1/2}$ as linearly independent solutions. Translated back to $a_m(y)$ these are

$$a_m(y) = y^{1/2}K_{s-1/2}(2\pi|m|y) \quad \text{or} \quad a_m(y) = y^{1/2}I_{s-1/2}(2\pi|m|y). \tag{B.21}$$

If one insists on at most power law growth near the cusp $y \to \infty$, the solution involving the function $I_{s-1/2}$ is disallowed. This is an instance of the uniqueness of Whittaker functions mentioned in Remark 6.22.

Putting everything together, we see that any real function $\varphi(g)$ on $SL(2,\mathbb{R})$ that is right-invariant under $SO(2,\mathbb{R})$ and satisfies the three conditions stated for automorphic forms in the introduction can be expanded as

$$\varphi(z) = a_0^{(s)}y^s + a_0^{(1-s)}y^{1-s} + y^{1/2}\sum_{m\neq 0} a_m K_{s-1/2}(2\pi|m|y)e^{2\pi imx} \tag{B.22}$$

with $a_m = a_{-m}$, and these are purely numerical coefficients. For cusp forms one has that the numerical coefficients $a_0^{(s)}$ and $a_0^{(1-s)}$ vanish identically. The above expansion is valid for $s \neq 1/2$; for $s = 1/2$ one has to replace the constant terms by the solutions of (B.19b).

As shown in Section 11.2, the coefficients a_m can also be determined for cusp forms from the values a_p for primes p if one demands in addition to the Laplace condition that $\varphi(z)$ is a simultaneous eigenfunction of all the Hecke operators T_p. These can be thought of as the analogues of the Laplace operator for finite $p < \infty$ and therefore an automorphic function that obeys simple equations for all $p \leq \infty$ is uniquely fixed (up to an overall normalisation); see Remark 11.8 and Example 11.10.

Appendix C

Structure Theory of $\mathfrak{su}(2,1)$

In this appendix, we give some details on the structure of real forms of $\mathfrak{sl}(3,\mathbb{C})$, with emphasis on the non-split real form $\mathfrak{su}(2,1)$. This material is needed in Chapter 18. We recall very briefly some relevant properties of the structure theory of the Lie algebra $\mathfrak{su}(2,1)$.

As a vector space, the complex Lie algebra $\mathfrak{sl}(3,\mathbb{C})$ can be represented as

$$\mathfrak{sl}(3,\mathbb{C}) = \bigoplus_{k=1}^{3} \mathbb{C}F_k \oplus \bigoplus_{k=1}^{2} \mathbb{C}H_k \oplus \bigoplus_{k=1}^{3} \mathbb{C}E_k, \tag{C.1}$$

where the eight generators $\{H_i, E_i, F_i\}$ have the following matrix realisations in the fundamental representation:

$$E_1 = \begin{pmatrix} 0 & 1 & 0 \\ 0 & 0 & 0 \\ 0 & 0 & 0 \end{pmatrix}, \quad E_2 = \begin{pmatrix} 0 & 0 & 0 \\ 0 & 0 & 1 \\ 0 & 0 & 0 \end{pmatrix}, \quad E_3 = \begin{pmatrix} 0 & 0 & 1 \\ 0 & 0 & 0 \\ 0 & 0 & 0 \end{pmatrix},$$
$$H_1 = \begin{pmatrix} 1 & 0 & 0 \\ 0 & -1 & 0 \\ 0 & 0 & 0 \end{pmatrix}, \quad H_2 = \begin{pmatrix} 0 & 0 & 0 \\ 0 & 1 & 0 \\ 0 & 0 & -1 \end{pmatrix}, \tag{C.2}$$

as well as

$$F_i = (E_i)^T. \tag{C.3}$$

One may readily check that these matrices satisfy the standard Chevalley–Serre relations, as implied by the Dynkin diagram •—•.

The split real form $\mathfrak{sl}(3,\mathbb{R})$ is then simply obtained by restricting the base field to the real numbers:

$$\mathfrak{sl}(3,\mathbb{R}) = \bigoplus_{k=1}^{3} \mathbb{R}F_k \oplus \bigoplus_{k=1}^{2} \mathbb{R}H_k \oplus \bigoplus_{k=1}^{3} \mathbb{R}E_k. \tag{C.4}$$

The compact real form $\mathfrak{su}(3)$ is the fixed point subalgebra with respect to the Cartan involution τ. For the case at hand, this is given by

$$\tau(X) = -X^{\dagger}. \tag{C.5}$$

The combination of generators of $\mathfrak{sl}(3,\mathbb{C})$ which are fixed pointwise by τ defines $\mathfrak{su}(3)$, and we find

$$\mathfrak{su}(3) = \bigoplus_{k=1}^{2} \mathbb{R}(iH_k) \oplus \bigoplus_{k=1}^{3} \mathbb{R}(E_k - F_k) \oplus \bigoplus_{k=1}^{3} \mathbb{R}i(E_k + F_k). \tag{C.6}$$

The Killing form is negative definite on $\mathfrak{su}(3)$. Going from the split to the compact real form can be done in general using the *Weyl unitary trick* to multiply the non-compact generators H_k and $E_k + F_k$ by the imaginary unit i.

The (intermediate) real form $\mathfrak{su}(2,1)$ is defined as

$$\mathfrak{su}(2,1) = \{X \in \mathfrak{sl}(3,\mathbb{C}) \mid X^{\dagger}\eta + \eta X = 0\}, \tag{C.7}$$

where the invariant bilinear form is chosen as

$$\eta = \begin{pmatrix} 0 & 0 & -i \\ 0 & 1 & 0 \\ i & 0 & 0 \end{pmatrix}. \tag{C.8}$$

To analyse the structure theory of $\mathfrak{su}(2,1)$ it is natural to change basis and define the following new Cartan generators:

$$H = \begin{pmatrix} 1 & 0 & 0 \\ 0 & 0 & 0 \\ 0 & 0 & -1 \end{pmatrix}, \quad J = \begin{pmatrix} i & 0 & 0 \\ 0 & -2i & 0 \\ 0 & 0 & i \end{pmatrix}. \tag{C.9}$$

These have the property that H has only real eigenvalues while J has only imaginary eigenvalues on the nilpotent subspaces $\mathfrak{n}_{\pm}$ of $\mathfrak{sl}(3,\mathbb{C})$. For this reason, J is referred to as *compact* while H is *non-compact*. In particular, H generates a non-compact five-grading

$$\mathfrak{su}(2,1) = \mathfrak{g}_{-2} \oplus \mathfrak{g}_{-1} \oplus \mathfrak{g}_0 \oplus \mathfrak{g}_1 \oplus \mathfrak{g}_2, \tag{C.10}$$

where

$$\begin{aligned} \mathfrak{g}_{\pm 2} &= \mathbb{R}X_{(\pm 2)}, \\ \mathfrak{g}_{\pm 1} &= \mathbb{R}X_{(\pm 1)} \oplus \mathbb{R}X'_{(\pm 1)}, \\ \mathfrak{g}_0 &= \mathbb{R}H \oplus \mathbb{R}J. \end{aligned} \tag{C.11}$$

The matrix realisations of the positive and negative root generators are

$$X_{(1)} = \begin{pmatrix} 0 & -1+i & 0 \\ 0 & 0 & 1-i \\ 0 & 0 & 0 \end{pmatrix}, \quad X_{(-1)} = \begin{pmatrix} 0 & 0 & 0 \\ 1+i & 0 & 0 \\ 0 & -1-i & 0 \end{pmatrix},$$

$$X'_{(1)} = \begin{pmatrix} 0 & 1+i & 0 \\ 0 & 0 & 1+i \\ 0 & 0 & 0 \end{pmatrix}, \quad X'_{(-1)} = \begin{pmatrix} 0 & 0 & 0 \\ -1+i & 0 & 0 \\ 0 & -1+i & 0 \end{pmatrix},$$

$$X_{(2)} = \begin{pmatrix} 0 & 0 & 1 \\ 0 & 0 & 0 \\ 0 & 0 & 0 \end{pmatrix}, \quad X_{(-2)} = \begin{pmatrix} 0 & 0 & 0 \\ 0 & 0 & 0 \\ -1 & 0 & 0 \end{pmatrix}. \tag{C.12}$$

The nilpotent subspace $\mathfrak{g}_1 \oplus \mathfrak{g}_2$ is a Heisenberg subalgebra

$$[\mathfrak{g}_1, \mathfrak{g}_1] \subseteq \mathfrak{g}_2. \tag{C.13}$$

Let us denote by $\mathfrak{a}$ the non-compact part of the Cartan subalgebra:

$$\mathfrak{a} = \mathbb{R}H \subset \mathfrak{g}_0. \tag{C.14}$$

It is now easy to determine the *restricted root system* Σ. Recall that the restricted root spaces are defined as

$$\mathfrak{g}_\lambda = \{X \in \mathfrak{g} \mid \forall h \in \mathfrak{a} : \; [h, X] = \lambda(h)X\}. \tag{C.15}$$

The restricted roots therefore belong to the dual space $\mathfrak{a}^*$ of the non-compact part of the Cartan subalgebra. Indeed, one can easily check that on $\mathfrak{a}$ the roots are all real. As is clear from the five-grading (C.10), the two generators $X_{(1)}$ and $X'_{(1)}$ restrict to the same root space of a root α on $\mathfrak{a}$ that has eigenvalue $\alpha(H) = +1$, while the generator $X_{(2)}$ restricts to the root 2α. Hence, the restricted root system consists of the simple root α, which has multiplicity two, and the highest root 2α, with multiplicity one. This can be identified with the non-reduced root system $(BC)_1$:

$$\Sigma = (BC)_1. \tag{C.16}$$

The Weyl vector of a restricted root system is defined in general as a sum over the positive restricted roots by

$$\rho = \frac{1}{2} \sum_{\alpha \in \Sigma_+} \mathrm{mult}(\alpha)\alpha\,. \tag{C.17}$$

For the case of the $(BC)_1$ root system with α and 2α being positive roots in terms of the non-divisible α, this leads to

$$\rho = 2\alpha. \tag{C.18}$$

The bilinear form on the space of restricted roots is given by restriction from the one on the complex Lie algebra. For $\mathfrak{su}(2,1)$ with a single non-divisible root α this means $\langle \alpha | \alpha \rangle = 1/2$.

For more information on real forms and restricted root systems we refer the reader to [353].

Appendix D

Poincaré Series and Kloosterman Sums

In this appendix, we briefly review some classic material related to the Fourier expansion of Poincaré series on the upper half-plane that are invariant under $G(\mathbb{Z}) = SL(2, \mathbb{Z})$. References for this exposition include [373].

D.1 Poincaré Series

The primary ingredient for a Poincaré series is the *seed function* $\sigma\colon \mathbb{H} \to \mathbb{C}$, which is invariant under discrete shifts:

$$\sigma(z+1) = \sigma(z), \tag{D.1}$$

or, more group-theoretically,

$$\sigma(\gamma z) = \sigma(z) \quad \text{for all} \quad \gamma \in B(\mathbb{Z}) = \left\{ \begin{pmatrix} \pm 1 & n \\ & \pm 1 \end{pmatrix} \middle|\, n \in \mathbb{Z} \right\} \subset SL(2, \mathbb{Z}). \tag{D.2}$$

Note that the seed function can depend on both the real and imaginary parts of z and it is therefore more general than the character that enters in the construction of non-holomorphic Eisenstein series in (4.76). The periodicity assumption, however, entails that the seed $\sigma(z)$ has a Fourier expansion

$$\sigma(z) = \sigma_0(y) + \sum_{n \neq 0} \sigma_n(y) e^{2\pi i n x}. \tag{D.3}$$

The general form of a *Poincaré series* is

$$f(z) \equiv P(\sigma, z) = \sum_{\gamma \in B(\mathbb{Z}) \backslash G(\mathbb{Z})} \sigma(\gamma z). \tag{D.4}$$

This sum makes sense for suitably fast-decaying seeds. The Poincaré series is invariant under $G(\mathbb{Z}) = SL(2, \mathbb{Z})$ by construction, and we would like to say as much as possible about its Fourier expansion

$$f(z) = \sum_{n\in\mathbb{Z}} f_n(y)e^{2\pi inx}. \tag{D.5}$$

D.2 Fourier Expansion

We will tackle the Fourier expansion of $f(z)$ by starting from the Poincaré series representation and performing yet another right quotient to study

$$B(\mathbb{Z})\backslash SL(2,\mathbb{Z})/B(\mathbb{Z}). \tag{D.6}$$

Choosing a unique representative then includes a sum over $k \in \mathbb{Z}$ for the right quotient that will be very useful.

The left quotient can be parametrised in the standard way in terms of coprime c and d such that

$$\begin{pmatrix} a & b \\ c & d \end{pmatrix} \tag{D.7}$$

is any fixed matrix representing the single coset class (i.e., one fixes arbitrarily any a and b that satisfy $ad - bc = 1$ over $\mathbb{Z}$). Performing the right quotient then allows us to bring d in the range between 0 and $c-1$ since it includes translates by c in the lower-right component. This is true for $c \neq 0$; for $c = 0$ one has to have $d = 1$ and the coset is represented by the identity. Let $(\mathbb{Z}/c\mathbb{Z})^\times$ denote the invertible integers mod c; these are all the numbers d coprime with c.

Therefore

$$\begin{aligned} f(z) &= \sum_{\gamma\in B(\mathbb{Z})\backslash SL(2,\mathbb{Z})} \sigma(\gamma z) \\ &= \sigma(z) + \sum_{c>0}\sum_{d\in(\mathbb{Z}/c\mathbb{Z})^\times}\sum_{k\in\mathbb{Z}} \sigma\left(\frac{a(z+k)+b}{c(z+k)+d}\right) \\ &= \sigma(z) + \sum_{c>0}\sum_{d\in(\mathbb{Z}/c\mathbb{Z})^\times}\sum_{k\in\mathbb{Z}} \sigma\left(\frac{a}{c} - \frac{1}{c(c(z+k)+d)}\right), \end{aligned} \tag{D.8}$$

where the first term comes from the $c = 0$ coset. The k-sum represents the right quotient. a here is a chosen upper-left entry of the matrix in the left quotient.

Next we use the Poisson summation formula

$$\sum_{k\in\mathbb{Z}} \varphi(z+k) = \sum_{n\in\mathbb{Z}} \int_{\mathbb{R}} \varphi(z+\omega)e^{-2\pi i\omega n}d\omega. \tag{D.9}$$

Applied to $f(z)$, we get

$$f(z) = \sigma(z) + \sum_{c>0} \sum_{d\in(\mathbb{Z}/c\mathbb{Z})^\times} \sum_{n\in\mathbb{Z}} \int_{\mathbb{R}} e^{-2\pi i\omega n} \sigma\left(\frac{a}{c} - \frac{1}{c^2((z+\omega)+d/c))}\right) d\omega. \tag{D.10}$$

Now letting $z = x + iy$ and introducing $\tilde{\omega} = \omega + x + d/c$, one gets

$$f(z) = \sigma(z) + \sum_{c>0} \sum_{d\in(\mathbb{Z}/c\mathbb{Z})^\times} \sum_{n\in\mathbb{Z}} e^{2\pi in(x+d/c)} \int_{\mathbb{R}} e^{-2\pi i\tilde{\omega}n} \sigma\left(\frac{a}{c} - \frac{1}{c^2(\tilde{\omega}+iy)}\right) d\tilde{\omega}. \tag{D.11}$$

This rearrangement has completely brought out the Fourier mode $e^{2\pi inx}$ in the second term and so one finds the Fourier modes of the Poincaré sum to be

$$f_n(y) = \sigma_n(y) + \sum_{c>0} \sum_{d\in(\mathbb{Z}/c\mathbb{Z})^\times} e^{2\pi in\frac{d}{c}} \int_{\mathbb{R}} e^{-2\pi i\tilde{\omega}n} \sigma\left(\frac{a}{c} - \frac{1}{c^2(\tilde{\omega}+iy)}\right) d\tilde{\omega}. \tag{D.12}$$

In the next step we use the Fourier expansion of the seed $\sigma(z)$ given in (D.3), which here leads to

$$\sigma\left(\frac{a}{c} - \frac{1}{c^2(\tilde{\omega}+iy)}\right) = \sum_{m\in\mathbb{Z}} \sigma_m\left(\frac{y}{c^2(\tilde{\omega}^2+y^2)}\right) e^{2\pi im\frac{a}{c}} e^{-2\pi im\frac{\tilde{\omega}}{c^2(\tilde{\omega}^2+y^2)}}. \tag{D.13}$$

Substituting this back into (D.12) yields

$$\begin{aligned} f_n(y) &= \sigma_n(y) + \sum_{c>0} \sum_{m\in\mathbb{Z}} \sum_{d\in(\mathbb{Z}/c\mathbb{Z})^\times} \Bigg[e^{2\pi in\frac{d}{c}+2\pi im\frac{a}{c}} \\ &\quad \times \int_{\mathbb{R}} e^{-2\pi i\tilde{\omega}n - 2\pi im\frac{\tilde{\omega}}{c^2(\tilde{\omega}^2+y^2)}} \sigma_m\left(\frac{y}{c^2(\tilde{\omega}^2+y^2)}\right) d\tilde{\omega}\Bigg] \\ &= \sigma_n(y) \\ &\quad + \sum_{c>0} \sum_{m\in\mathbb{Z}} S(m,n;c) \int_{\mathbb{R}} e^{-2\pi i\tilde{\omega}n - 2\pi im\frac{\tilde{\omega}}{c^2(\tilde{\omega}^2+y^2)}} \sigma_m\left(\frac{y}{c^2(\tilde{\omega}^2+y^2)}\right) d\tilde{\omega}, \end{aligned} \tag{D.14}$$

where we have carried out the sum over $d\in(\mathbb{Z}/c\mathbb{Z})^\times$ in terms of the *Kloosterman sum*

$$S(m,n;c) = \sum_{d\in(\mathbb{Z}/c\mathbb{Z})^\times} e^{2\pi i(md+nd^{-1})/c}. \tag{D.15}$$

Here, we have used that $ad - bc = 1$, which means that $ad \equiv 1 \mod c$ and therefore $a \equiv d^{-1} \mod c$. (This is the meaning of x^{-1} in the definition above.) Note that $S(m,n;c) = S(n,m;c)$.

For the zero mode $f_{n=0}(y)$ of $f(z)$ we therefore obtain

$$f_0(y) = \sigma_0(y) + \sum_{c>0}\sum_{m\in\mathbb{Z}} S(m,0;c) \int_{\mathbb{R}} e^{-2\pi i m \frac{\tilde{\omega}}{c^2(\tilde{\omega}^2+y^2)}} \sigma_m\left(\frac{y}{c^2(\tilde{\omega}^2+y^2)}\right) d\tilde{\omega}. \tag{D.16}$$

For the zero mode, the Kloosterman sum $S(m,0;c)$ simplifies to the *Ramanujan sum*.

D.3 The Case of Eisenstein Series

Let us evaluate the Fourier modes for the case of a standard non-holomorphic Eisenstein series. In this case, the seed is a character on the Borel subgroup and has the Fourier expansion

$$\sigma(z) = y^s \qquad \Rightarrow \qquad \sigma_m(y) = \begin{cases} y^s & \text{if } m = 0, \\ 0 & \text{if } m \neq 0. \end{cases} \tag{D.17}$$

The constant term (D.16) then collapses to the $m = 0$ term and reads

$$\begin{aligned} f_0(y) &= y^s + \sum_{c>0} S(0,0;c) c^{-2s} \int_{\mathbb{R}} \left(\frac{y}{\tilde{\omega}^2 + y^2}\right)^s d\tilde{\omega} \\ &= y^s + \sqrt{\pi}\frac{\Gamma(s-1/2)}{\Gamma(s)} y^{1-s} \sum_{c>0} \varphi(c) c^{-2s}. \end{aligned} \tag{D.18}$$

The integral converges for $\mathrm{Re}(s) > 1/2$. The Euler totient function $\varphi(c) = S(0,0;c)$ gives the cardinality of $(\mathbb{Z}/c\mathbb{Z})^\times$, i.e., the number of non-zero integers less than c that are also coprime with c. It is a standard result for the Dirichlet series that [373, Eq. (2.28)]

$$\sum_{c>0} \varphi(c) c^{-2s} = \frac{\zeta(2s-1)}{\zeta(2s)}. \tag{D.19}$$

In this way one recovers the standard constant terms. The sum converges for $\mathrm{Re}(s) > 1$. For $s = 1$ it diverges. Combining the constant terms leads to

$$f_0(y) = y^s + \frac{\xi(2s-1)}{\xi(2s)} y^{1-s}, \tag{D.20}$$

with the completed Riemann zeta function $\xi(s) = \pi^{-s/2}\Gamma(s/2)\zeta(s)$.

To see the y-dependence of the constant term, the following change of variables is sufficient ($t = \tilde{\omega}/y$):

$$\int_{\mathbb{R}} \left(\frac{y}{\tilde{\omega}^2 + y^2} \right)^s d\tilde{\omega} = y^{1-s} \int_{\mathbb{R}} (1 + t^2)^{-s} dt. \tag{D.21}$$

(Note that the integrand coincides with the spherical vector of the principal series representation that the Eisenstein series belongs to.)

For the non-zero mode one also quickly recognises the Bessel integral. The number-theoretic sum in this case is more complicated but of course yields the standard divisor sum [373, Eq. (2.27)]. In some more detail,

$$\begin{aligned}
f_n(y) &= \sum_{c>0} S(0, n; c) c^{-2s} \int_{\mathbb{R}} e^{-2\pi i n \tilde{\omega}} \left(\frac{y}{\tilde{\omega}^2 + y^2} \right)^s d\tilde{\omega} \\
&= \sum_{c>0} S(0, n; c) c^{-2s} y^{1-s} \int_{\mathbb{R}} e^{-2\pi i n y t} \left(1 + t^2\right)^{-s} dt \\
&= \sum_{c>0} S(0, n; c) c^{-2s} y^{1-s} \frac{2\pi^s}{\Gamma(s)} |ny|^{s-1/2} K_{s-1/2}(2\pi |n| y) \\
&= \frac{2\pi^s}{\Gamma(s)} |n|^{s-1/2} y^{1/2} K_{s-1/2}(2\pi |n| y) \sum_{c>0} S(0, n; c) c^{-2s} \\
&= \frac{2\pi^s}{\Gamma(s)} |n|^{s-1/2} y^{1/2} K_{s-1/2}(2\pi |n| y) \frac{\sigma_{1-2s}(n)}{\zeta(2s)} \\
&= \frac{2}{\xi(2s)} \sigma_{1-2s}(n) |n|^{s-1/2} y^{1/2} K_{s-1/2}(2\pi |n| y),
\end{aligned} \tag{D.22}$$

the familiar result for $SL(2)$ Eisenstein series.

References

[1] Ahlén, O. 2016. Global Iwasawa-decomposition of $SL(n, \mathbb{A}_{\mathbb{Q}})$. arXiv:1609.06621 [math.NT].

[2] Ahlén, O., Gustafsson, H. P. A., Kleinschmidt, A., Liu, B., and Persson, D. 2017. Fourier coefficients attached to small automorphic representations of $SL_n(\mathbb{A})$. arXiv:1707.08937 [math.RT].

[3] Alexandrov, S. 2009. D-instantons and twistors: Some exact results. *J. Phys.*, **A42**, 335402.

[4] Alexandrov, S. 2013. Twistor approach to string compactifications: A review. *Phys. Rep.*, **522**, 1–57.

[5] Alexandrov, S., and Banerjee, S. 2013. Modularity, quaternion-Kähler spaces, and mirror symmetry. *J. Math. Phys.*, **54**, 102301.

[6] Alexandrov, S., and Banerjee, S. 2014a. Dualities and fivebrane instantons. *JHEP*, **11**, 040.

[7] Alexandrov, S., and Banerjee, S. 2014b. Fivebrane instantons in Calabi-Yau compactifications. *Phys. Rev.*, **D90**(4), 041902.

[8] Alexandrov, S., and Banerjee, S. 2015. Hypermultiplet metric and D-instantons. *JHEP*, **02**, 176.

[9] Alexandrov, S., Banerjee, S., Manschot, J., and Pioline, B. 2017. Multiple D3-instantons and mock modular forms I. *Commun. Math. Phys.*, **353**(1), 379–411.

[10] Alexandrov, S., Manschot, J., Persson, D., and Pioline, B. 2015. Quantum hypermultiplet moduli spaces in N=2 string vacua: A review. *Proceedings of Symposia in Pure Mathematics*, **90**, 181–212.

[11] Alexandrov, S., Manschot, J., and Pioline, B. 2013. D3-instantons, mock theta series and twistors. *JHEP*, **04**, 002.

[12] Alexandrov, S., Persson, D., and Pioline, B. 2011a. Fivebrane instantons, topological wave functions and hypermultiplet moduli spaces. *JHEP*, **03**, 111.

[13] Alexandrov, S., Persson, D., and Pioline, B. 2011b. On the topology of the hypermultiplet moduli space in type II/CY string vacua. *Phys. Rev.*, **D83**, 026001.

[14] Alexandrov, S., and Pioline, B. 2012. S-duality in twistor space. *JHEP*, **08**, 112.

[15] Alexandrov, S., Pioline, B., Saueressig, F., and Vandoren, S. 2009. D-instantons and twistors. *JHEP*, **03**, 044.

[16] Alexandrov, S., Pioline, B., and Vandoren, S. 2010. Self-dual Einstein spaces, heavenly metrics and twistors. *J. Math. Phys.*, **51**, 073510.

[17] Alexandrov, S., and Saueressig, F. 2009. Quantum mirror symmetry and twistors. *JHEP*, **09**, 108.

[18] Alvarez-Gaumé, L., Moore, G. W., and Vafa, C. 1986. Theta functions, modular invariance and strings. *Commun. Math. Phys.*, **106**, 1–40.

[19] Angelantonj, C., Florakis, I., and Pioline, B. 2012a. A new look at one-loop integrals in string theory. *Commun. Num. Theor. Phys.*, **6**, 159–201.

[20] Angelantonj, C., Florakis, I., and Pioline, B. 2012b. One-loop BPS amplitudes as BPS-state sums. *JHEP*, **06**, 070.

[21] Angelantonj, C., Florakis, I., and Pioline, B. 2013. Rankin-Selberg methods for closed strings on orbifolds. *JHEP*, **07**, 181.

[22] Angelantonj, C., Florakis, I., and Pioline, B. 2015. Threshold corrections, generalised prepotentials and Eichler integrals. *Nucl. Phys.*, **B897**, 781–820.

[23] Apostol, T. M. 1976. *Introduction to Analytic Number Theory*. Springer, New York, London.

[24] Apostol, T. M. 1997. *Modular Functions and Dirichlet Series in Number Theory*. Graduate Texts in Mathematics, vol. 41. Springer, London.

[25] Arakawa, T., and Moreau, A. 2016. Joseph ideals and lisse minimal W-algebras. *J. Inst. Math. Jussieu*, 1–21.

[26] Argurio, R. 1998. *Brane Physics in M-Theory*. Ph.D. thesis, Free University of Brussels. `arXiv:hep-th/9807171 [hep-th]`.

[27] Arthur, J. 1989. Orbites unipotentes et représentations, II (Unipotent automorphic representations: Conjectures). *Astérisque*, **171–172**, 13–71.

[28] Arthur, J. 2003. The principle of functoriality. *Bull. Amer. Math. Soc. (N.S.)*, **40**(1), 39–53.

[29] Arthur, J. 2005. An introduction to the trace formula. Pages 1–263 of Arthur, J., Ellwood, D., and Kottwitz, R. (eds), *Harmonic Analysis, the Trace Formula, and Shimura Varieties*. Clay Mathematics Proceedings, vol. 4. American Mathematical Society, Providence, RI.

[30] Aspinwall, P. S. 1997. $K3$ surfaces and string duality. Pages 421–540 of Efthimiou, C., and Greene, B. (eds), *Fields, Strings and Duality (Boulder, CO, 1996)*. World Scientific Publishing, River Edge, NJ.

[31] Aspinwall, P. S. 2001. Compactification, geometry and duality: $N = 2$. Pages 723–805 of Harvey, J. A., Kachru, S., and Silverstein, E. (eds), *Strings, Branes and Gravity, TASI 99 (Boulder, CO)*. World Scientific Publishing, River Edge, NJ.

[32] Atiyah, M. 1987. The logarithm of the Dedekind η-function. *Math. Ann.*, **278** (1–4), 335–380.

[33] Aurich, R., Lustig, S., Steiner, F., and Then, H. 2005. Can one hear the shape of the Universe? *Phys. Rev. Lett.*, **94**, 021301.

[34] Bachas, C. 1996. D-brane dynamics. *Phys. Lett.*, **B374**, 37–42.

[35] Bachas, C., and Porrati, M. 1992. Pair creation of open strings in an electric field. *Phys. Lett.*, **B296**, 77–84.

[36] Bala, P., and Carter, R. W. 1976a. Classes of unipotent elements in simple algebraic groups, I. *Math. Proc. Cambridge Philos. Soc.*, **79**(3), 401–425.

[37] Bala, P., and Carter, R. W. 1976b. Classes of unipotent elements in simple algebraic groups, II. *Math. Proc. Cambridge Philos. Soc.*, **80**(1), 1–17.

[38] Banerjee, S., and Sen, A. 2008. Duality orbits, dyon spectrum and gauge theory limit of heterotic string theory on T**6. *JHEP*, **03**, 022.

[39] Bao, L., Bielecki, J., Cederwall, M., Nilsson, B. E. W., and Persson, D. 2008a. U-duality and the compactified Gauss-Bonnet term. *JHEP*, **0807**, 048.

[40] Bao, L., and Carbone, L. 2013. Integral forms of Kac-Moody groups and Eisenstein series in low dimensional supergravity theories. `arXiv:1308.6194 [hep-th]`.

[41] Bao, L., Cederwall, M., and Nilsson, B. E. W. 2008b. Aspects of higher curvature terms and U-duality. *Class. Quant. Grav.*, **25**, 095001.

[42] Bao, L., Kleinschmidt, A., Nilsson, B. E. W., Persson, D., and Pioline, B. 2010. Instanton corrections to the universal hypermultiplet and automorphic forms on SU(2,1). *Commun. Num. Theor. Phys.*, **4**, 187–266.

[43] Bao, L., Kleinschmidt, A., Nilsson, B. E. W., Persson, D., and Pioline, B. 2013. Rigid Calabi-Yau threefolds, Picard Eisenstein series and instantons. *J. Phys. Conf. Ser.*, **462**(1), 012026.

[44] Basu, A. 2007. The D**10 R**4 term in type IIB string theory. *Phys. Lett.*, **B648**, 378–382.

[45] Basu, A. 2008a. The D**4 R**4 term in type IIB string theory on T**2 and U-duality. *Phys. Rev.*, **D77**, 106003.

[46] Basu, A. 2008b. The D**6 R**4 term in type IIB string theory on T**2 and U-duality. *Phys. Rev.*, **D77**, 106004.

[47] Basu, A. 2011. Supersymmetry constraints on the R^4 multiplet in type IIB on T^2. *Class. Quant. Grav.*, **28**, 225018.

[48] Basu, A. 2014. The $D^6 R^4$ term from three loop maximal supergravity. *Class. Quant. Grav.*, **31**(24), 245002.

[49] Basu, A. 2016a. Poisson equation for the three loop ladder diagram in string theory at genus one. *Int. J. Mod. Phys.*, **A31**(32), 1650169.

[50] Basu, A. 2016b. Proving relations between modular graph functions. *Class. Quant. Grav.*, **33**(23), 235011.

[51] Becker, K., Becker, M., and Strominger, A. 1995. Five-branes, membranes and nonperturbative string theory. *Nucl. Phys.*, **B456**, 130–152.

[52] Beilinson, A., and Drinfeld, V. 1991. Quantization of Hitchin's integrable system and Hecke eigensheaves. `www.math.uchicago.edu/~mitya/langlands/hitchin/BD-hitchin.pdf`.

[53] Beineke, J., Brubaker, B., and Frechette, S. 2011. Weyl group multiple Dirichlet series of type *C*. *Pacific J. Math.*, **254**(1), 11–46.

[54] Beineke, J., Brubaker, B., and Frechette, S. 2012. A crystal definition for symplectic multiple Dirichlet series. Pages 37–63 of Bump, D., Friedberg, S., and Goldfeld, D. (eds), *Multiple Dirichlet Series, L-functions and Automorphic Forms*. Progress in Mathematics, vol. 300. Birkhäuser/Springer, New York.

[55] Bekaert, X. 2011. Singletons and their maximal symmetry algebras. Pages 71–89 of Dragovich, B., and Rakic, Z. (eds), *Modern Mathematical Physics: Proceedings, 6th Summer School, Belgrade Serbia, September 14–23, 2010*. `arXiv:1111.4554 [math-ph]`

[56] Belinsky, V. A., Khalatnikov, I. M., and Lifshitz, E. M. 1970. Oscillatory approach to a singular point in the relativistic cosmology. *Adv. Phys.*, **19**, 525–573.

[57] Belinsky, V. A., Khalatnikov, I. M., and Lifshitz, E. M. 1982. A general solution of the Einstein equations with a time singularity. *Adv. Phys.*, **31**, 639–667.

[58] Berenstein, A., and Zelevinsky, A. 2001. Tensor product multiplicities, canonical bases and totally positive varieties. *Invent. Math.*, **143**(1), 77–128.

[59] Berkovits, N. 1998. Construction of R(4) terms in N=2 D = 8 superspace. *Nucl. Phys.*, **B514**, 191–203.
[60] Bhargava, M. 2004. Higher composition laws. I: A new view on Gauss composition, and quadratic generalizations. *Ann. of Math. (2)*, **159**(1), 217–250.
[61] Birch, B. J., and Swinnerton-Dyer, H. P. F. 1965. Notes on elliptic curves, II. *J. Reine Angew. Math.*, **218**, 79–108.
[62] Blasius, D. 1994. On multiplicities for SL(n). *Israel J. Math.*, **88**(1-3), 237–251.
[63] Blumenhagen, R., Lüst, D., and Theisen, S. 2013. *Basic Concepts of String Theory*. Theoretical and Mathematical Physics. Springer, Heidelberg.
[64] Bodner, M., and Cadavid, A. C. 1990. Dimensional reduction of type IIB supergravity and exceptional quaternionic manifolds. Class. Quant. Grav., **7**, 829pp.
[65] Bogomolny, E. B., Georgeot, B., Giannoni, M. J., and Schmit, C. 1997. Arithmetical chaos. *Phys. Rep.*, **291**, 219–324.
[66] Borcea, C. 1992. Calabi-Yau threefolds and complex multiplication. Pages 489–502 of Yau, S. T. (ed.), *Essays on Mirror Manifolds*. International Press, Hong Kong.
[67] Borcherds, R. E. 1992. Monstrous moonshine and monstrous Lie superalgebras. *Invent. Math.*, **109**(2), 405–444.
[68] Borcherds, R. E. 1998. Automorphic forms with singularities on Grassmannians. *Invent. Math.*, **132**(3), 491–562.
[69] Borel, A. 1966. Introduction to automorphic forms. Pages 199–210 of Borel, A., and Mostow, G. D. (eds), *Algebraic Groups and Discontinuous Subgroups*. Proceedings of Symposia in Pure Mathematics, vol. 9. American Mathematical Society, Providence, RI.
[70] Borel, A. 1972. *Représentations de Groupes Localement Compacts*. Lecture Notes in Mathematics, vol. 276. Springer, Berlin, New York.
[71] Borel, A. 1979. Automorphic L-functions. Pages 27–61 of Borel, A., and Casselman, W. (eds), *Automorphic Forms, Representations and L-functions*. Proceedings of Symposia in Pure Mathematics, vol. 33. American Mathematical Society, Providence, RI.
[72] Borel, A. 1991. *Linear Algebraic Groups*. Graduate Texts in Mathematics, vol. 126. Springer, Berlin, New York.
[73] Borel, A. 1997. *Automorphic Forms on* $SL(2,\mathbb{R})$. Cambridge Tracts in Mathematics, vol. 130. Cambridge University Press, Cambridge.
[74] Borel, A., and Tits, J. 1972. Compléments à l'article: 'Groupes réductifs'. *Inst. Hautes Études Sci. Publ. Math.*, **41**, 253–276.
[75] Borho, W. 1977. Berechnung der Gelfand-Kirillov-Dimension bei induzierten Darstellungen. *Math. Ann.*, **225**(2), 177–194.
[76] Borho, W., and Brylinski, J.-L. 1982. Differential operators on homogeneous spaces, I: Irreducibility of the associated variety for annihilators of induced modules. *Invent. Math.*, **69**(3), 437–476.
[77] Born, M., and Infeld, L. 1934. Foundations of the new field theory. *Proc. Roy. Soc. Lond.*, **A144**, 425–451.
[78] Bossard, G., Cosnier-Horeau, C., and Pioline, B. 2017a. Four-derivative couplings and BPS dyons in heterotic CHL orbifolds. *SciPost Phys.*, **3**, 008.
[79] Bossard, G., Cosnier-Horeau, C., and Pioline, B. 2017b. Protected couplings and BPS dyons in half-maximal supersymmetric string vacua. *Phys. Lett.*, **B765**, 377–381.

[80] Bossard, G., and Kleinschmidt, A. 2015. Supergravity divergences, supersymmetry and automorphic forms. *JHEP*, **08**, 102.

[81] Bossard, G., and Kleinschmidt, A. 2016. Loops in exceptional field theory. *JHEP*, **01**, 1.

[82] Bossard, G., and Pioline, B. 2017. Exact $\nabla^4 R^4$ couplings and helicity supertraces. *JHEP*, **01**, 050.

[83] Bossard, G., and Verschinin, V. 2014. Minimal unitary representations from supersymmetry. *JHEP*, **1410**, 008.

[84] Bossard, G., and Verschinin, V. 2015a. $\mathcal{E}\nabla^4 R^4$ type invariants and their gradient expansion. *JHEP*, **03**, 089.

[85] Bossard, G., and Verschinin, V. 2015b. The two $\nabla^6 R^4$ type invariants and their higher order generalisation. *JHEP*, **07**, 154.

[86] Braverman, A., and Gaitsgory, D. 2002. Geometric Eisenstein series. *Invent. Math.*, **150**(2), 287–384.

[87] Braverman, A., Garland, H., Kazhdan, D., and Patnaik, M. 2014. An affine Gindikin-Karpelevich formula. Pages 43–64 of Etingof, P., Khovanov, M., and Savage, A. (eds), *Perspectives in Representation Theory*. Contemporary Mathematics, vol. 610. American Mathematical Society, Providence, RI.

[88] Braverman, A., and Kazhdan, D. 2011. The spherical Hecke algebra for affine Kac-Moody groups I. *Ann. of Math. (2)*, **174**(3), 1603–1642.

[89] Braverman, A., and Kazhdan, D. 2013. Representations of affine Kac-Moody groups over local and global fields: A survey of some recent results. Pages 91–117 of Latata, R., Ruciński, A., Strzelecki, P., Swiatkowski, J., Wrzosek, D., and Zakrzewski, P. (eds), *European Congress of Mathematics: Proceedings of the 6th Congress (6 ECM), Held at the Jagiellonian University, Kraków, July 2–7, 2012*. European Mathematical Society, Zürich.

[90] Braverman, A., Kazhdan, D., and Patnaik, M. M. 2016. Iwahori-Hecke algebras for p-adic loop groups. *Invent. Math.*, **204**(2), 347–442.

[91] Breckenridge, J. C., Michaud, G., and Myers, R. C. 1997. More D-brane bound states. *Phys. Rev.*, **D55**, 6438–6446.

[92] Brekke, L., and Freund, P. 1993. p-adic numbers in physics. *Phys.Rept.*, **233**, 1–66.

[93] Breuil, C., Conrad, B., Diamond, F., and Taylor, R. 2001. On the modularity of elliptic curves over $\mathbf{Q}$: Wild 3-adic exercises. *J. Amer. Math. Soc.*, **14**(4), 843–939.

[94] Broedel, J., Mafra, C. R., Matthes, N., and Schlotterer, O. 2015. Elliptic multiple zeta values and one-loop superstring amplitudes. *JHEP*, **07**, 112.

[95] Broedel, J., Matthes, N., and Schlotterer, O. 2016. Relations between elliptic multiple zeta values and a special derivation algebra. *J. Phys.*, **A49**(15), 155203.

[96] Broedel, J., Schlotterer, O., and Stieberger, S. 2013. Polylogarithms, multiple zeta values and superstring amplitudes. *Fortsch. Phys.*, **61**, 812–870.

[97] Broedel, J., Schlotterer, O., Stieberger, S., and Terasoma, T. 2014. All order α'-expansion of superstring trees from the Drinfeld associator. *Phys. Rev.*, **D89**(6), 066014.

[98] Brown, F. 2014. Single-valued motivic periods and multiple zeta values. *SIGMA*, **2**, e25.

[99] Brown, F. C. S. 2006. Multiple zeta values and periods of moduli spaces $\mathfrak{M}_{0,n}$. `arXiv:math/0606419 [math.AG]`.

[100] Brubaker, B., Bump, D., Chinta, G., Friedberg, S., and Hoffstein, J. 2006a. Weyl group multiple Dirichlet series, I. Pages 91–114 of Friedberg, S., Bump, D., Goldfeld, D., and Hoffstein, J. (eds), *Multiple Dirichlet Series, Automorphic Forms, and Analytic Number Theory*. Proceedings of Symposia in Pure Mathematics, vol. 75. American Mathematical Society, Providence, RI.

[101] Brubaker, B., Bump, D., Chinta, G., and Gunnells, P. E. 2012. Metaplectic Whittaker functions and crystals of type B. Pages 93–118 of Bump, D., Friedberg, S., and Goldfeld, D. (eds), *Multiple Dirichlet Series, L-functions and Automorphic Forms*. Progress in Mathematics, vol. 300. Birkhäuser/Springer, New York.

[102] Brubaker, B., Bump, D., and Friedberg, S. 2006b. Weyl group multiple Dirichlet series, II: The stable case. *Invent. Math.*, **165**(2), 325–355.

[103] Brubaker, B., Bump, D., and Friedberg, S. 2008. Twisted Weyl group multiple Dirichlet series: The stable case. Pages 1–26 of Gan, W. T., Kudla, S. S., and Tschinkel, Y. (eds), *Eisenstein Series and Applications*. Progress in Mathematics, vol. 258. Birkhäuser Boston, Boston, MA.

[104] Brubaker, B., Bump, D., and Friedberg, S. 2009. Gauss sum combinatorics and metaplectic Eisenstein series. Pages 61–81 of Ginzburg, D., Lapid, E., and Soudry, D. (eds), *Automorphic Forms and L-functions I: Global Aspects*. Contemporary Mathematics, vol. 488. American Mathematical Society, Providence, RI.

[105] Brubaker, B., Bump, D., and Friedberg, S. 2011a. Eisenstein series, crystals, and ice. *Notices Amer. Math. Soc.*, **58**(11), 1563–1571.

[106] Brubaker, B., Bump, D., and Friedberg, S. 2011b. Schur polynomials and the Yang-Baxter equation. *Comm. Math. Phys.*, **308**(2), 281–301.

[107] Brubaker, B., Bump, D., and Friedberg, S. 2011c. Weyl group multiple Dirichlet series, Eisenstein series and crystal bases. *Ann. of Math. (2)*, **173**(2), 1081–1120.

[108] Brubaker, B., Bump, D., and Friedberg, S. 2011d. *Weyl Group Multiple Dirichlet Series: Type A Combinatorial Theory*. Annals of Mathematics Studies, vol. 175. Princeton University Press, Princeton, NJ.

[109] Brubaker, B., Bump, D., Friedberg, S., and Hoffstein, J. 2007. Weyl group multiple Dirichlet series, III: Eisenstein series and twisted unstable A_r. *Ann. of Math. (2)*, **166**(1), 293–316.

[110] Brubaker, B., and Friedberg, S. 2015. Whittaker coefficients of metaplectic Eisenstein series. *Geom. Funct. Anal.*, **25**(4), 1180–1239.

[111] Bruhat, F., and Tits, J. 1967. Groupes algébriques simples sur un corps local. Pages 23–36 of Springer, T. A. (ed.), *Proceedings of a Conference on Local Fields (Driebergen, 1966)*. Springer, Berlin.

[112] Bruhat, F., and Tits, J. 1972. Groupes réductifs sur un corps local. *Inst. Hautes Études Sci. Publ. Math.*, **41**, 5–251.

[113] van der Geer, G. 2008. Siegel modular forms and their applications. Pages 181–245 of Ranestad, K. (ed.), *The 1-2-3 of Modular Forms: Lectures from the Summer School on Modular Forms and their Applications, Held at Nordfjordeid, Norway, June 2004*. Universitext. Springer, Berlin.

[114] Brylinski, R., and Kostant, B. 1994. Minimal representations, geometric quantization, and unitarity. *Proc. Nat. Acad. Sci. U.S.A.*, **91**(13), 6026–6029.

[115] Bump, D. Hecke algebras. `http://sporadic.stanford.edu/bump/math263/hecke.pdf`.

[116] Bump, D. 1998. *Automorphic Forms and Representations*. Cambridge Studies in Advanced Mathematics, vol. 55. Cambridge University Press, Cambridge.

[117] Bump, D. 2005. The Rankin-Selberg method: An introduction and survey. Pages 41–73 of Cogdell. J. W., Jiang, D., Kudla, S. S., Soudry, D., and Stanton, R. J. (eds), *Automorphic Representations, L-functions and Applications: Progress and Prospects*. Ohio State University Mathematics Research Institute Publications, vol. 11. de Gruyter, Berlin.

[118] Bump, D., Friedberg, S., and Goldfeld, D. (eds) 2012. *Multiple Dirichlet Series, L-functions and Automorphic Forms*. Progress in Mathematics, vol. 300. Birkhäuser/Springer, New York.

[119] Bump, D., and Nakasuji, M. 2010. Integration on p-adic groups and crystal bases. *Proc. Amer. Math. Soc.*, **138**(5), 1595–1605.

[120] Bump, D. W. 1982. *Automorphic Forms on GL(3,R)*. Lecture Notes in Mathematics, vol. 1083. Springer, Berlin, Heidelberg.

[121] Candelas, P., Horowitz, G. T., Strominger, A., and Witten, E. 1985. Vacuum configurations for superstrings. *Nucl. Phys.*, **B258**, 46–74.

[122] Carbone, L., Lee, K.-H., and Liu, D. 2017. Eisenstein series on rank 2 hyperbolic Kac-Moody groups. *Math. Ann.*, **367**(3-4), 1173–1197.

[123] Cardy, J. L. 1989. Boundary conditions, fusion rules and the Verlinde formula. *Nucl. Phys.*, **B324**, 581–596.

[124] Cardy, J. L. 2004. Boundary conformal field theory. `arXiv:hep-th/0411189 [hep-th]`.

[125] Carnahan, S. 2010. Generalized moonshine, I: Genus-zero functions. *Algebra Number Theory*, **4**(6), 649–679.

[126] Carnahan, S. 2012. Generalized moonshine, II: Borcherds products. *Duke Math. J.*, **161**(5), 893–950.

[127] Carnahan, S. 2012. Generalized moonshine, IV: Monstrous Lie algebras. `arXiv:1208.6254 [math.RT]`.

[128] Carter, R. W. 1993. *Finite Groups of Lie Type: Conjugacy Classes and Complex Characters*. Wiley Classics Library. John Wiley & Sons, Chichester. Reprint of the 1985 original.

[129] Casselman, B. The L-group. `www.math.ubc.ca/~cass/research/pdf/miyake.pdf`.

[130] Casselman, W., and Shalika, J. 1980. The unramified principal series of p-adic groups, II. The Whittaker function. *Compos. Math.*, **41**(2), 207–231.

[131] Cheng, M. C. N. 2010. K3 surfaces, N=4 dyons, and the Mathieu group M24. *Commun. Num. Theor. Phys.*, **4**, 623–658.

[132] Cheng, M. C. N., Duncan, J. F. R., Harrison, S. M., and Kachru, S. 2017. Equivariant K3 Invariants. *Commun. Num. Theor. Phys.*, **11**, 41–72.

[133] Cheng, M. C. N., Duncan, J. F. R., and Harvey, J. A. 2014a. Umbral moonshine. *Commun. Num. Theor. Phys.*, **08**, 101–242.

[134] Cheng, M. C. N., Duncan, J. F. R., and Harvey, J. A. 2014b. Umbral moonshine and the Niemeier lattices. *Res. Math. Sci.*, **1**, Art. 3, 81.

[135] Cheng, M. C. N., and Verlinde, E. 2007. Dying dyons don't count. *JHEP*, **09**, 070.

[136] Cheng, M. C. N., and Verlinde, E. P. 2008. Wall crossing, discrete attractor flow, and Borcherds algebra. *SIGMA*, **4**, 068.

[137] Chinta, G., and Gunnells, P. E. 2007. Weyl group multiple Dirichlet series constructed from quadratic characters. *Invent. Math.*, **167**(2), 327–353.

[138] Chinta, G., and Gunnells, P. E. 2010. Constructing Weyl group multiple Dirichlet series. *J. Amer. Math. Soc.*, **23**(1), 189–215.

[139] Chinta, G., and Gunnells, P. E. 2012. Littelmann patterns and Weyl group multiple Dirichlet series of type D. Pages 119–130 of Bump, D., Friedberg, S., and Goldfeld, D. (eds), *Multiple Dirichlet Series, L-Functions and Automorphic Forms*. Progress in Mathematics, vol. 300. Birkhäuser/Springer, New York.

[140] Cogdell, J. W. 2004. Lectures on L-functions, converse theorems, and functoriality for GL_n. Pages 1–96 of Cogdell, J. W., Kim, H. H., and Ram Murty, M. (eds), *Lectures on Automorphic L-Functions*. Fields Institute Monographs, vol. 20. American Mathematical Society, Providence, RI.

[141] Cogdell, J. W. 2005. Converse theorems, functoriality, and applications. *Pure Appl. Math. Q.*, **1**(2, Special Issue: In memory of Armand Borel. Part 1), 341–367.

[142] Cogdell, J. W., Kim, H. H., Piatetski-Shapiro, I. I., and Shahidi, F. 2001. On lifting from classical groups to GL_N. *Inst. Hautes Études Sci. Publ. Math.*, **93**, 5–30.

[143] Cogdell, J. W., and Piatetski-Shapiro, I. I. 1994. Converse theorems for GL_n. *Inst. Hautes Études Sci. Publ. Math.*, **79**, 157–214.

[144] Cogdell, J. W., and Piatetski-Shapiro, I. I. 1999. Converse theorems for GL_n, II. *J. Reine Angew. Math.*, **507**, 165–188.

[145] Cohen, H. 1992. Elliptic curves. Pages 212–237 of Waldschmidt, M., Moussa, P., Luck, J. M., and Itzykson, C. (eds), *From Number Theory to Physics*. Springer, Berlin.

[146] Collingwood, D. H., and McGovern, W. M. 1993. *Nilpotent Orbits in Semisimple Lie Algebras*. Van Nostrand Reinhold Mathematics Series. Van Nostrand Reinhold, New York.

[147] Conrad, B., Diamond, F., and Taylor, R. 1999. Modularity of certain potentially Barsotti-Tate Galois representations. *J. Amer. Math. Soc.*, **12**(2), 521–567.

[148] Conway, J. H., and Norton, S. P. 1979. Monstrous moonshine. *Bull. London Math. Soc.*, **11**(3), 308–339.

[149] Conway, J. H., and Sloane, N. J. A. 1999. *Sphere Packings, Lattices and Groups*. Springer, New York.

[150] Cremmer, E., and Julia, B. 1978. The N=8 supergravity theory, 1: The Lagrangian. *Phys. Lett.*, **B80**, 48.

[151] Cremmer, E., and Julia, B. 1979. The SO(8) supergravity. *Nucl. Phys.*, **B159**, 141–212.

[152] Cremmer, E., Julia, B., Lu, H., and Pope, C. 1998. Dualization of dualities, 1. *Nucl. Phys.*, **B523**, 73–144.

[153] Dabholkar, A. 2005. Exact counting of black hole microstates. *Phys. Rev. Lett.*, **94**, 241301.

[154] Dabholkar, A., Denef, F., Moore, G. W., and Pioline, B. 2005. Precision counting of small black holes. *JHEP*, **0510**, 096.

[155] Dabholkar, A., Gomes, J., Murthy, S., and Sen, A. 2011. Supersymmetric index from black hole entropy. *JHEP*, **04**, 034.

[156] Dabholkar, A., and Harvey, J. A. 1989. Nonrenormalization of the superstring tension. *Phys. Rev. Lett.*, **63**, 478.

[157] Dabholkar, A., Murthy, S., and Zagier, D. 2012. Quantum black holes, wall crossing, and mock modular forms. `arXiv:1208.4074 [hep-th]`.

[158] Damour, T., and Henneaux, M. 2000. Chaos in superstring cosmology. *Phys. Rev. Lett.*, **85**, 920–923.

[159] Damour, T., and Henneaux, M. 2001. E(10), BE(10) and arithmetical chaos in superstring cosmology. *Phys. Rev. Lett.*, **86**, 4749–4752.

[160] Damour, T., Henneaux, M., and Nicolai, H. 2002. E(10) and a 'small tension expansion' of M theory. *Phys. Rev. Lett.*, **89**, 221601.

[161] Damour, T., Henneaux, M., and Nicolai, H. 2003. Cosmological billiards. *Class. Quant. Grav.*, **20**, R145–R200.

[162] de Graaf, W. A. 2011. Computing representatives of nilpotent orbits of θ-groups. *J. Symbolic Comput.*, **46**(4), 438–458.

[163] De Medts, T., Gramlich, R., and Horn, M. 2009. Iwasawa decompositions of split Kac-Moody groups. *J. Lie Theory*, **19**(2), 311–337.

[164] Deitmar, A. 2012. *Automorphic Forms*. Universitext. Springer, London.

[165] Deligne, P. 1974. La conjecture de Weil, I: *Inst. Hautes Études Sci. Publ. Math.*, **43**, 273–307.

[166] Denef, F., and Moore, G. W. 2011. Split states, entropy enigmas, holes and halos. *JHEP*, **1111**, 129.

[167] D'Hoker, E. 1999. Perturbative string theory. Pages 807–1011 of Deligne, P., Kazhdan, D., Etingof, P., Morgan, J. W., Free, D. S., Morrison, D. R., and Witten, E. (eds), *Quantum Fields and Strings : A Course for Mathematicians*. American Mathematical Society, Providence, RI.

[168] D'Hoker, E., and Duke, W. 2017. Fourier series of modular graph functions. `arXiv:1708.07998 [math.NT]`.

[169] D'Hoker, E., and Green, M. B. 2014. Zhang-Kawazumi invariants and superstring amplitudes. *Journal of Number Theory*, **144**, 111–150.

[170] D'Hoker, E., Green, M. B., Gurdogan, O., and Vanhove, P. 2017. Modular graph functions. *Commun. Num. Theor. Phys.*, **11**, 165–218.

[171] D'Hoker, E., Green, M. B., Pioline, B., and Russo, R. 2015a. Matching the D^6R^4 interaction at two-loops. *JHEP*, **01**, 031.

[172] D'Hoker, E., Green, M. B., and Vanhove, P. 2015b. On the modular structure of the genus-one type II superstring low energy expansion. *JHEP*, **08**, 041.

[173] D'Hoker, E., Gutperle, M., and Phong, D. H. 2005. Two-loop superstrings and S-duality. *Nucl. Phys.*, **B722**, 81–118.

[174] D'Hoker, E., and Phong, D. H. 2002. Lectures on two loop superstrings. *Conf. Proc.*, **C0208124**, 85–123. (Proceedings of International Conference on Superstring Theory, Hangzhou, China, August 12-15, 2002.)

[175] D'Hoker, E., and Phong, D. H. 1986. Multiloop amplitudes for the bosonic Polyakov string. *Nucl. Phys.*, **B269**, 205–234.

[176] D'Hoker, E., and Phong, D. H. 2005. Two-loop superstrings, VI: Non-renormalization theorems and the 4-point function. *Nucl. Phys.*, **B715**, 3–90.

[177] Di Francesco, P., Mathieu, P., and Senechal, D. 1997. *Conformal Field Theory*. Graduate Texts in Contemporary Physics. Springer, New York.

[178] Diamond, F. 1996. On deformation rings and Hecke rings. *Ann. of Math. (2)*, **144**(1), 137–166.

[179] Diamond, F., and Shurman, J. 2005. *A First Course in Modular Forms*. Graduate Texts in Mathematics, vol. 228. Springer, New York.

[180] Dijkgraaf, R., Verlinde, E. P., and Verlinde, H. L. 1997. Counting dyons in N=4 string theory. *Nucl. Phys.*, **B484**, 543–561.

[181] Dirac, P. A. M. 1931. Quantized singularities in the electromagnetic field. *Proc. Roy. Soc. Lond.*, **A133**, 60–72.

[182] Dirac, P. A. M. 1962. An extensible model of the electron. *Proc. Roy. Soc. Lond.*, **A268**, 57–67.

[183] Dixon, L. J., Ginsparg, P. H., and Harvey, J. A. 1988. Beauty and the beast: Superconformal symmetry in a monster module. *Commun. Math. Phys.*, **119**, 221–241.

[184] Dixon, L. J., Kaplunovsky, V., and Louis, J. 1991. Moduli dependence of string loop corrections to gauge coupling constants. *Nucl. Phys.*, **B355**, 649–688.

[185] Đoković, D. Ž. 2001. The closure diagram for nilpotent orbits of the split real form of E_7. *Represent. Theory*, **5**, 284–316.

[186] Đoković, D. Ž. 2003. The closure diagram for nilpotent orbits of the split real form of E_8. *Cent. Eur. J. Math.*, **1**(4), 573–643.

[187] Donagi, R., and Witten, E. 2013. Supermoduli space is not projected. Pages 19–72 of Donagi, R., Katz. S., Klemm, A., and Morrison, D. R. (eds), *String-Math 2012*. Proceedings of Symposia in Pure Mathematics, vol. 90. American Mathematical Society, Providence, RI.

[188] Douglas, M. R. 1995. Branes within branes. Pages 267–275 of Baulieu, L., Kazakov, V., Picco, M., Windey, P., Di Francesco, P., and Douglas, M. R. (eds), *Strings, Branes and Dualities: Proceedings, NATO Advanced Study Institute, Cargèse, France, May 26-June 14, 1997*. NATO ASI Series C, vol. 520. Springer, Dordrecht.

[189] Drinfeld, V. G. 1980. Langlands' conjecture for GL(2) over functional fields. Pages 565–574 of Lehto, O. (ed.), *Proceedings of the International Congress of Mathematicians (Helsinki, 1978)*. Academia Scientiarum Fennica, Helsinki.

[190] Drinfeld, V. G. 1983. Two-dimensional l-adic representations of the fundamental group of a curve over a finite field and automorphic forms on GL(2). *Amer. J. Math.*, **105**(1), 85–114.

[191] Duff, M. J., Lu, H., and Pope, C. N. 1996. The black branes of M theory. *Phys. Lett.*, **B382**, 73–80.

[192] Duncan, J. F. R., Griffin, M. J., and Ono, K. 2015. Proof of the umbral moonshine conjecture. *Res. Math. Sci.*, **2**, Art. 26, 47.

[193] Duncan, J. F. R., Mertens, M. H., and Ono, K. 2017a. O'Nan moonshine and arithmetic. `arXiv:1702:03516 [math. NT]`.

[194] Duncan, J. F. R., Mertens, M. H., and Ono, K. 2017b. Pariah moonshine. *Nat. Commun.* **8**(1). `arXiv:1709.08867 [math.RT]`.

[195] Dvorsky, A., and Sahi, S. 1999. Explicit Hilbert spaces for certain unipotent representations, II. *Invent. Math.*, **138**(1), 203–224.

[196] Dynkin, E. B. 1952. Semisimple subalgebras of semisimple Lie algebras. *Mat. Sbornik N.S.*, **30**(72), 349–462.

[197] Eguchi, T., and Hikami, K. 2009. Superconformal algebras and mock theta functions 2: Rademacher expansion for K3 surface. *Commun. Num. Theor. Phys.*, **3**, 531–554.

[198] Eguchi, T., and Hikami, K. 2011. Note on twisted elliptic genus of $K3$ surface. *Phys. Lett.*, **B694**, 446–455.

[199] Eguchi, T., and Hikami, K. 2012. Twisted elliptic genus for K3 and Borcherds product. *Lett. Math. Phys.*, **102**, 203–222.

[200] Eguchi, T., Ooguri, H., and Tachikawa, Y. 2011. Notes on the K3 surface and the Mathieu group M_{24}. *Exper. Math.*, **20**, 91–96.

[201] Eichler, M. 1954. Quaternäre quadratische Formen und die Riemannsche Vermutung für die Kongruenzzetafunktion. *Arch. Math. (Basel)*, **5**, 355–366.

[202] Eichler, M., and Zagier, D. 1985. *The theory of Jacobi forms*. Progress in Mathematics, vol. 55. Birkhäuser, Basel.

[203] Englert, F., Houart, L., Kleinschmidt, A., Nicolai, H., and Tabti, N. 2007. An E(9) multiplet of BPS states. *JHEP*, **0705**, 065.

[204] Enriquez, B. 2014. Elliptic associators. *Selecta Math. (N.S.)*, **20**(2), 491–584.

[205] Falbel, E., Francsics, G., Lax, P. D., and Parker, J. R. 2011. Generators of a Picard modular group in two complex dimensions. *Proc. Amer. Math. Soc.*, **139**(7), 2439–2447.

[206] Fernando, S., and Günaydin, M. 2016. Massless conformal fields, AdS_{d+1}/CFT_d higher spin algebras and their deformations. *Nucl. Phys.*, **B904**, 494–526.

[207] Ferrara, S., and Günaydin, M. 1998. Orbits of exceptional groups, duality and BPS states in string theory. *Int. J. Mod. Phys.*, **A13**, 2075–2088.

[208] Ferrara, S., and Maldacena, J. M. 1998. Branes, central charges and U duality invariant BPS conditions. *Class. Quant. Grav.*, **15**, 749–758.

[209] Figueroa-O'Farrill, J. M., and Papadopoulos, G. 2003. Maximally supersymmetric solutions of ten-dimensional and eleven-dimensional supergravities. *JHEP*, **03**, 048.

[210] Figueroa-O'Farrill, J. M., and Simon, J. 2004. Supersymmetric Kaluza-Klein reductions of AdS backgrounds. *Adv. Theor. Math. Phys.*, **8**(2), 217–317.

[211] Flath, D. 1979. Decomposition of representations into tensor products. Pages 179–183 of Borel, A., and Casselman, W. (eds), *Automorphic Forms, Representations and L-functions*: Proceedings of Symposia in Pure Mathematics, vol. 33. American Mathematical Society, Providence, RI.

[212] Fleig, P., and Kleinschmidt, A. 2012a. Eisenstein series for infinite-dimensional U-duality groups. *JHEP*, **1206**, 054.

[213] Fleig, P., and Kleinschmidt, A. 2012b. Perturbative terms of Kac-Moody-Eisenstein series. Pages 265–275 of Donagi, R., Katz, S., Klemm, A., and Morrison, D. R. (eds), *Proceedings, String-Math 2012, Bonn, Germany, July 16-21, 2012*. Proceedings of Symposia in Pure Mathematics, vol. 90. American Mathematical Society, Providence, RI.

[214] Fleig, P., Kleinschmidt, A., and Persson, D. 2014. Fourier expansions of Kac-Moody Eisenstein series and degenerate Whittaker vectors. *Commun. Num. Theor. Phys.*, **08**, 41–100.

[215] Florakis, I., and Pioline, B. 2017. On the Rankin-Selberg method for higher genus string amplitudes. *Commun. Num. Theor. Phys.*, **11**, 337–404.

[216] Forte, L. A. 2009. Arithmetical chaos and quantum cosmology. *Class. Quant. Grav.*, **26**, 045001.

[217] Freed, D. S. 1986. Determinants, torsion, and strings. *Commun. Math. Phys.*, **107**, 483–513.

[218] Freedman, D. Z., and Van Proeyen, A. 2012. *Supergravity*. Cambridge University Press, Cambridge.

[219] Frenkel, E. 2007. Lectures on the Langlands program and conformal field theory. Pages 387–533 of Cartier, P., Moussa, P., Julia, B., and Vanhove, P. (eds), *Frontiers in Number Theory, Physics and Geometry, II: On Conformal Field Theories, Discrete Groups and Renormalization*. Springer, Berlin, Heidelberg.

[220] Frenkel, E. 2009. Gauge theory and Langlands duality. `arXiv:0906.2747 [math.RT]`.

[221] Frenkel, E., Gaitsgory, D., Kazhdan, D., and Vilonen, K. 1998. Geometric realization of Whittaker functions and the Langlands conjecture. *J. Amer. Math. Soc.*, **11**(2), 451–484.

[222] Frenkel, E., Gaitsgory, D., and Vilonen, K. 2001. Whittaker patterns in the geometry of moduli spaces of bundles on curves. *Ann. of Math. (2)*, **153**(3), 699–748.

[223] Frenkel, E., Gaitsgory, D., and Vilonen, K. 2002. On the geometric Langlands conjecture. *J. Amer. Math. Soc.*, **15**(2), 367–417.

[224] Frenkel, E., and Witten, E. 2008. Geometric endoscopy and mirror symmetry. *Commun. Num. Theor. Phys.*, **2**, 113–283.

[225] Frenkel, I., Lepowsky, J., and Meurman, A. 1988. *Vertex Operator Algebras and the Monster*. Pure and Applied Mathematics, vol. 134. Academic Press, Boston, MA.

[226] Frey, G. 1986. Links between stable elliptic curves and certain Diophantine equations. *Ann. Univ. Sarav. Ser. Math.*, **1**(1), iv+40.

[227] Friedan, D., Martinec, E. J., and Shenker, S. H. 1986. Conformal invariance, supersymmetry and string theory. *Nucl. Phys.*, **B271**, 93.

[228] Friedberg, S., and Zhang, L. 2015. Eisenstein series on covers of odd orthogonal groups. *Amer. J. Math.*, **137**(4), 953–1011.

[229] Friedberg, S., and Zhang, L. 2016. Tokuyama-type formulas for characters of type B. *Israel J. Math.*, **216**(2), 617–655.

[230] Friedlander, H., Gaudet, L., and Gunnells, P. E. 2015. Crystal graphs, Tokuyama's theorem, and the Gindikin-Karpelevič formula for G_2. *J. Algebraic Combin.*, **41**(4), 1089–1102.

[231] Fulton, W. 1997. *Young Tableaux*. London Mathematical Society Student Texts, vol. 35. Cambridge University Press, Cambridge.

[232] Fulton, W., and Harris, J. 2008. *Representation Theory: A First Course*. Graduate Texts in Mathematics, vol. 129. Springer, London.

[233] Gaberdiel, M. R., Hohenegger, S., and Volpato, R. 2010a. Mathieu moonshine in the elliptic genus of K3. *JHEP*, **10**, 062.

[234] Gaberdiel, M. R., Hohenegger, S., and Volpato, R. 2010b. Mathieu twining characters for K3. *JHEP*, **09**, 058.

[235] Gaberdiel, M. R., Hohenegger, S., and Volpato, R. 2012. Symmetries of K3 sigma models. *Commun. Num. Theor. Phys.*, **6**, 1–50.

[236] Gaberdiel, M. R., Persson, D., Ronellenfitsch, H., and Volpato, R. 2013. Generalized Mathieu moonshine. *Commun. Num. Theor Phys.*, **07**, 145–223.

[237] Gaiotto, D., Moore, G. W., and Neitzke, A. 2010. Four-dimensional wall-crossing via three-dimensional field theory. *Commun. Math. Phys.*, **299**, 163–224.

[238] Gaitsgory, D. 2004. On a vanishing conjecture appearing in the geometric Langlands correspondence. *Ann. of Math. (2)*, **160**(2), 617–682.

[239] Gaitsgory, D. 2010. Notes on geometric Langlands: The extended Whittaker category. `www.math.harvard.edu/~gaitsgde/GL/extWhit.pdf`.

[240] Gaitsgory, D. 2016. Recent progress in geometric Langlands theory. `arXiv:1606.09462 [math.AG]`.

[241] Gan, W. T. Lecture slides on automorphic forms and representations. `www.math.nus.edu.sg/~matgwt/`.

[242] Gan, W. T. 2000a. An automorphic theta module for quaternionic exceptional groups. *Canad. J. Math.*, **52**(4), 737–756.

[243] Gan, W. T. 2000b. A Siegel-Weil formula for exceptional groups. *J. Reine Angew. Math.*, **528**, 149–181.

[244] Gan, W. T., Gross, B., and Savin, G. 2002. Fourier coefficients of modular forms on G_2. *Duke Math. J.*, **115**(1), 105–169.

[245] Gan, W. T., Qiu, Y., and Takeda, S. 2014. The regularized Siegel-Weil formula (the second term identity) and the Rallis inner product formula. *Invent. Math.*, **198**(3), 739–831.

[246] Gan, W. T., and Savin, G. 2005. On minimal representations: Definitions and properties. *Represent. Theory*, **9**, 46–93 (electronic).

[247] Gannon, T. 2002. Boundary conformal field theory and fusion ring representations. *Nucl. Phys.*, **B627**, 506–564.

[248] Gannon, T. 2016. Much ado about Mathieu. *Adv. Math.*, **301**, 322–358.

[249] Gannon, T., and Lam, C. S. 1992. Lattices and theta function identities, 2: Theta series. *J. Math. Phys.*, **33**, 871–887.

[250] Garland, H. 2001. Certain Eisenstein series on loop groups: Convergence and the constant term. Pages 275–319 of Dani, S. G., and Prasad, G. (eds), *Proceedings of the International Conference on Algebraic Groups and Arithmetic, Mumbai, 2001*. Narosa Publishing House, New Delhi, for Tata Institute of Fundamental Research, Mumbai.

[251] Garland, H. 2006. Absolute convergence of Eisenstein series on loop groups. *Duke Math. J.*, **135**, 203–260.

[252] Garland, H. 2011. On extending the Langlands-Shahidi method to arithmetic quotients of loop groups. Pages 151–167 of Adams, J., Lian, B., and Sahi, S. (eds), *Representation Theory and Mathematical Physics*. Contemporary Mathematics, vol. 557. American Mathematical Society, Providence, RI.

[253] Garland, H., Miller, S. D., and Patnaik, M. M. 2017. Entirety of cuspidal Eisenstein series on loop groups. *Amer. J. Math.*, **139**(2), 461–512.

[254] Garrett, P. 1999. Satake parameters versus unramified principal series. www.math.umn.edu/~garrett/m/v/satake_urps.pdf.

[255] Garrett, P. 2014. Transition: Eisenstein series on adele groups. www.math.umn.edu/~garrett/m/mfms/notes_2013-14/12_2_transition_Eis.pdf.

[256] Gatti, V., and Viniberghi, E. 1978. Spinors of 13-dimensional space. *Adv. Math.*, **30**(2), 137–155.

[257] Gauntlett, J. P., Gutowski, J. B., Hull, C. M., Pakis, S., and Reall, H. S. 2003. All supersymmetric solutions of minimal supergravity in five dimensions. *Class. Quant. Grav.*, **20**, 4587–4634.

[258] Gaussent, S., and Rousseau, G. 2014. Spherical Hecke algebras for Kac-Moody groups over local fields. *Ann. of Math. (2)*, **180**(3), 1051–1087.

[259] Gelbart, S. 1984. An elementary introduction to the Langlands program. *Bull. Amer. Math. Soc.*, **10**(2), 177–219.

[260] Gelbart, S., Jacquet, H., and Rogawski, J. 2001. Generic representations for the unitary group in three variables. *Israel J. Math.*, **126**, 173–237.

[261] Gelbart, S., and Piatetski-Shapiro, I. 1984. Automorphic forms and L-functions for the unitary group. Pages 141–184 of Herb, R., Kudla, S., Lipsman, R., and Rosenberg, J. (eds), *Lie Group Representations, II (College Park, MD, 1982/1983)*. Lecture Notes in Mathematics, vol. 1041. Springer, Berlin, New York.

[262] Gelbart, S., and Shahidi, F. 1988. *Analytic Properties of Automorphic L-functions*. Perspectives in Mathematics, vol. 6. Academic Press, Boston, MA.

[263] Gelbart, S. S. 1975. *Automorphic Forms on Adele Groups*. Annals of Mathematics Studies, vol. 83. Princeton University Press, Princeton, NJ.

[264] Gelbart, S. S. 1976. *Weil's Representation and the Spectrum of the Metaplectic Group*. Lecture Notes in Mathematics, vol. 530. Springer, Berlin, New York.

[265] Gelbart, S. S., and Miller, S. D. 2003. Riemann's zeta function and beyond. *Bull. Amer. Math. Soc.*, **41**(1), 59–112.

[266] Gelbart, S. S., and Rogawski, J. D. 1991. L-functions and Fourier-Jacobi coefficients for the unitary group U(3). *Invent. Math.*, **105**(3), 445–472.

[267] Gelfand, I., Graev, M., and Piatetski-Shapiro, I. 1968. *Representation Theory and Automorphic Functions*. Saunders Mathematics Books. Saunders, Philadelphia, PA.

[268] Gelfand, I. M., and Cetlin, M. L. 1950. Finite-dimensional representations of the group of unimodular matrices. *Doklady Akad. Nauk SSSR (N.S.)*, **71**, 825–828.

[269] Gerasimov, A., Lebedev, D., and Oblezin, S. 2006. Givental integral representation for classical groups. arXiv:math/0608152.

[270] Gerasimov, A., Lebedev, D., and Oblezin, S. 2009. On Baxter Q-operators and their arithmetic implications. *Lett. Math. Phys.*, **88**(1-3), 3–30.

[271] Gerasimov, A., Lebedev, D., and Oblezin, S. 2010a. On q-deformed $\mathfrak{gl}_{l+1}$-Whittaker function, I. *Comm. Math. Phys.*, **294**(1), 97–119.

[272] Gerasimov, A., Lebedev, D., and Oblezin, S. 2010b. On q-deformed $\mathfrak{gl}_{l+1}$-Whittaker function, II. *Comm. Math. Phys.*, **294**(1), 121–143.

[273] Gerasimov, A., Lebedev, D., and Oblezin, S. 2011a. On q-deformed $\mathfrak{gl}_{\ell+1}$-Whittaker function, III. *Lett. Math. Phys.*, **97**(1), 1–24.

[274] Gerasimov, A., Lebedev, D., and Oblezin, S. 2011b. Parabolic Whittaker functions and topological field theories, I. *Commun. Num. Theor. Phys.*, **5**, 135–202.

[275] Gerasimov, A., Lebedev, D., and Oblezin, S. 2014. Baxter Operator Formalism for Macdonald Polynomials. *Lett. Math. Phys.*, **104**, 115–139.

[276] Gerasimov, A. A., Lebedev, D. R., and Oblezin, S. V. 2012. New integral representations of Whittaker functions for classical Lie groups. *Uspekhi Mat. Nauk*, **67**(1(403)), 3–96.

[277] Gibbons, G. W., Green, M. B., and Perry, M. J. 1996. Instantons and seven-branes in type IIB superstring theory. *Phys. Lett.*, **B370**, 37–44.

[278] Ginzburg, D. 2006. Certain conjectures relating unipotent orbits to automorphic representations. *Israel J. Math.*, **151**, 323–355.

[279] Ginzburg, D. 2014. Towards a classification of global integral constructions and functorial liftings using the small representations method. *Adv. Math.*, **254**, 157–186.

[280] Ginzburg, D., and Hundley, J. 2013. Constructions of global integrals in the exceptional group F_4. *Kyushu J. Math.*, **67**(2), 389–417.

[281] Ginzburg, D., Rallis, S., and Soudry, D. 1997a. On the automorphic theta representation for simply laced groups. *Israel J. Math.*, **100**, 61–116.

[282] Ginzburg, D., Rallis, S., and Soudry, D. 1997b. A tower of theta correspondences for G_2. *Duke Math. J.*, **88**(3), 537–624.

[283] Ginzburg, D., Rallis, S., and Soudry, D. 2003. On Fourier coefficients of automorphic forms of symplectic groups. *Manuscripta Math.*, **111**(1), 1–16.

[284] Givental, A. 1997. Stationary phase integrals, quantum Toda lattices, flag manifolds and the mirror conjecture. Pages 103–115 of Khovanski, A., Varclienko, A., and Vassiliev, V. (eds), *Topics in Singularity Theory*. American Mathematical Society Translations, Series 2, vol. 180. American Mathematical Society, Providence, RI.

[285] Goddard, P., Nuyts, J., and Olive, D. I. 1977. Gauge theories and magnetic charge. *Nucl. Phys.*, **B125**, 1.

[286] Goddard, P., and Olive, D. 1986. Kac-Moody and Virasoro algebras in relation to quantum physics. *Internat. J. Modern Phys. A*, **1**(2), 303–414.

[287] Goddard, P., and Thorn, C. B. 1972. Compatibility of the dual pomeron with unitarity and the absence of ghosts in the dual resonance model. *Phys. Lett.*, **B40**, 235–238.

[288] Godement, R. 1962. Domaines fondamentaux des groupes arithmétiques. Pages 201–225 of *Séminaire Bourbaki, Vol. 8, Exp. No. 257*. Société Mathématique de France, Paris.

[289] Godement, R. 1995. Introduction à la théorie de Langlands. Pages 115–144 of *Séminaire Bourbaki, Vol. 10*, Exp. No. 321. Société Mathématique de France, Paris.

[290] Goldfeld, D. 2006. *Automorphic Forms and L-functions for the Group* $GL(n,\mathbb{R})$. Cambridge Studies in Advanced Mathematics, vol. 99. Cambridge University Press, Cambridge.

[291] Goldfeld, D., and Hundley, J. 2011a. *Automorphic Representations and L-Functions for the General Linear Group. Volume I.* Cambridge Studies in Advanced Mathematics, vol. 129. Cambridge University Press, Cambridge.

[292] Goldfeld, D., and Hundley, J. 2011b. *Automorphic Representations and L-Functions for the General Linear Group. Volume II.* Cambridge Studies in Advanced Mathematics, vol. 130. Cambridge University Press, Cambridge.

[293] Gomes, J. 2017. U-duality invariant quantum entropy from sums of Kloosterman sums. `arXiv:1709.06579 [hep-th]`.

[294] Gomez, H., and Mafra, C. R. 2013. The closed-string 3-loop amplitude and S-duality. *JHEP*, **1310**, 217.

[295] Gomez, R., Gourevitch, D., and Sahi, S. 2016. Whittaker supports for representations of reductive groups. `arXiv:1610.00284 [math.RT]`.

[296] Gomez, R., Gourevitch, D., and Sahi, S. 2017. Generalized and degenerate Whittaker models. *Compos. Math.*, **153**(2), 223–256.

[297] Goncharov, A., and Shen, L. 2015. Geometry of canonical bases and mirror symmetry. *Invent. Math.*, **202**(2), 487–633.

[298] Gourevitch, D., and Sahi, S. 2013. Annihilator varieties, adduced representations, Whittaker functionals, and rank for unitary representations of GL(n). *Selecta Math. (N.S.)*, **19**(1), 141–172.

[299] Gourevitch, D., and Sahi, S. 2015. Degenerate Whittaker functionals for real reductive groups. *Amer. J. Math.*, **137**(2), 439–472.

[300] Graham, R., and Szepfalusy, P. 1990. Quantum creation of a generic universe. *Phys. Rev.*, **D42**, 2483–2490.

[301] Gran, U., Gutowski, J., and Papadopoulos, G. 2010a. Classification of IIB backgrounds with 28 supersymmetries. *JHEP*, **01**, 044.

[302] Gran, U., Gutowski, J., and Papadopoulos, G. 2010b. M-theory backgrounds with 30 Killing spinors are maximally supersymmetric. *JHEP*, **03**, 112.

[303] Gran, U., Papadopoulos, G., and von Schultz, C. 2014. Supersymmetric geometries of IIA supergravity, I. *JHEP*, **05**, 024.
[304] Gran, U., Papadopoulos, G., and von Schultz, C. 2015. Supersymmetric geometries of IIA supergravity, II. *JHEP*, **12**, 113.
[305] Gran, U., Papadopoulos, G., and von Schultz, C. 2016. Supersymmetric geometries of IIA supergravity, III. *JHEP*, **06**, 045.
[306] Green, M. B. 1995. A gas of D instantons. *Phys. Lett.*, **B354**, 271–278.
[307] Green, M. B. 1999. Interconnections between type II superstrings, M theory and N=4 supersymmetric Yang-Mills. Pages 22–96 of Cevesole, A., Kounnas, C., Lüst, D., and Theisen, S. (eds), *Quantum Aspects of Gauge Theories, Supersymmetry and Unification: Proceedings, 2nd International Conference, Corfu, Greece, September 20-26, 1998*. Lecture Notes in Physics, vol. 525. Springer, Berlin, New York.
[308] Green, M. B., and Gutperle, M. 1997. Effects of D instantons. *Nucl. Phys.*, **B498**, 195–227.
[309] Green, M. B., and Gutperle, M. 1998. D particle bound states and the D instanton measure. *JHEP*, **01**, 005.
[310] Green, M. B., Gutperle, M., and Kwon, H.-H. 1998. Sixteen fermion and related terms in M theory on T**2. *Phys. Lett.*, **B421**, 149–161.
[311] Green, M. B., Gutperle, M., and Vanhove, P. 1997. One loop in eleven-dimensions. *Phys. Lett.*, **B409**, 177–184.
[312] Green, M. B., Kwon, H.-H., and Vanhove, P. 2000. Two loops in eleven-dimensions. *Phys. Rev.*, **D61**, 104010.
[313] Green, M. B., Miller, S. D., Russo, J. G., and Vanhove, P. 2010. Eisenstein series for higher-rank groups and string theory amplitudes. *Commun. Num. Theor. Phys.*, **4**, 551–596.
[314] Green, M. B., Miller, S. D., and Vanhove, P. 2015a. $SL(2,\mathbb{Z})$-invariance and D-instanton contributions to the D^6R^4 interaction. *Commun. Num. Theor. Phys.*, **09**, 307–344.
[315] Green, M. B., Miller, S. D., and Vanhove, P. 2015b. Small representations, string instantons, and Fourier modes of Eisenstein series. *J. Number Theory*, **146**, 187–309. With an appendix by D. Ciubotaru and P. Trapa.
[316] Green, M. B., Russo, J. G., and Vanhove, P. 2008a. Low energy expansion of the four-particle genus-one amplitude in type II superstring theory. *JHEP*, **02**, 020.
[317] Green, M. B., Russo, J. G., and Vanhove, P. 2008b. Modular properties of two-loop maximal supergravity and connections with string theory. *JHEP*, **07**, 126.
[318] Green, M. B., Russo, J. G., and Vanhove, P. 2010a. Automorphic properties of low energy string amplitudes in various dimensions. *Phys. Rev.*, **81**(8), 086008.
[319] Green, M. B., Russo, J. G., and Vanhove, P. 2010b. String theory dualities and supergravity divergences. *JHEP*, **1006**, 075.
[320] Green, M. B., Schwarz, J., and Witten, E. 1987. *Superstring Theory. Vols I & II.* Cambridge University Press, Cambridge.
[321] Green, M. B., and Sethi, S. 1999. Supersymmetry constraints on type IIB supergravity. *Phys. Rev. D*, **59**(4), 046006.
[322] Green, M. B., and Vanhove, P. 1997. D instantons, strings and M theory. *Phys. Lett.*, **B408**, 122–134.
[323] Green, M. B., and Vanhove, P. 2006. Duality and higher derivative terms in M theory. *JHEP*, **0601**, 093.

[324] Gritsenko, V. A., and Nikulin, V. V. 1997. Siegel automorphic form corrections of some Lorentzian Kac–Moody Lie algebras. *Amer. J. Math.*, **119**(1), 181–224.

[325] Gross, B. H. On the Satake isomorphism. `www.math.harvard.edu/~gross/preprints/sat.pdf`.

[326] Gross, B. H., and Kudla, S. S. 1992. Heights and the central critical values of triple product L-functions. *Compos. Math.*, **81**(2), 143–209.

[327] Gross, B. H., and Wallach, N. R. 1994. A distinguished family of unitary representations for the exceptional groups of real rank = 4. Pages 289–304 of Kostant, B., and Brylinski, J.-L. (eds), *Lie Theory and Geometry*. Progress in Mathematics, vol. 123. Birkhäuser Boston, Boston, MA.

[328] Gross, B. H., and Wallach, N. R. 1996. On quaternionic discrete series representations, and their continuations. *J. Reine Angew. Math.*, **481**, 73–123.

[329] Gross, B. H., and Zagier, D. B. 1986. Heegner points and derivatives of L-series. *Invent. Math.*, **84**(2), 225–320.

[330] Gross, D. J., and Witten, E. 1986. Superstring modifications of Einstein's equations. *Nucl. Phys.*, **B277**, 1.

[331] Gubay, F., Lambert, N., and West, P. 2010. Constraints on automorphic forms of higher derivative terms from compactification. *JHEP*, **1008**, 028.

[332] Gukov, S., and Witten, E. 2006. Gauge theory, ramification, and the geometric Langlands program. `arXiv:hep-th/0612073 [hep-th]`.

[333] Günaydin, M., Koepsell, K., and Nicolai, H. 2001a. Conformal and quasiconformal realizations of exceptional Lie groups. *Comm. Math. Phys.*, **221**(1), 57–76.

[334] Günaydin, M., Koepsell, K., and Nicolai, H. 2001b. The minimal unitary representation of $E_{8(8)}$. *Adv. Theor. Math. Phys.*, **5**(5), 923–946.

[335] Günaydin, M., Neitzke, A., Pavlyk, O., and Pioline, B. 2008. Quasi-conformal actions, quaternionic discrete series and twistors: $SU(2,1)$ and $G_{2(2)}$. *Commun. Math. Phys.*, **283**, 169–226.

[336] Günaydin, M., Neitzke, A., Pioline, B., and Waldron, A. 2006. BPS black holes, quantum attractor flows, and automorphic forms. *Phys. Rev. D*, **73**(8), 084019.

[337] Günaydin, M., and Pavlyk, O. 2005. Minimal unitary realizations of exceptional U-duality groups and their subgroups as quasiconformal groups. *JHEP*, **01**, 019.

[338] Günaydin, M., and Pavlyk, O. 2006. A unified approach to the minimal unitary realizations of noncompact groups and supergroups. *JHEP*, **09**, 050.

[339] Gurevich, N. 2013. The twisted Satake isomorphism and Casselman-Shalika formula. `arXiv:1307.7510 [math.RT]`.

[340] Gustafsson, H. P. A., Kleinschmidt, A., and Persson, D. 2016. Small automorphic representations and degenerate Whittaker vectors. *J. Number Theory*, **166**, 344–399.

[341] Hamel, A. M., and King, R. C. 2002. Symplectic shifted tableaux and deformations of Weyl's denominator formula for sp($2n$). *J. Algebraic Combin.*, **16**(3), 269–300 (2003).

[342] Harish-Chandra. 1968. *Automorphic Forms on Semisimple Lie Groups*. Notes by J. G. M. Mars. Lecture Notes in Mathematics, vol. 62. Springer, Berlin, New York.

[343] Harish-Chandra. 1978. Admissible Invariant Distributions on Reductive p-adic Groups. Pages 281–347 of Rossmann, W. (ed.), *Lie Theories and Their Applications: Proceedings of the 1977 Annual Seminar of the Canadian Mathematical Congress, Queen's University, Kingston, Ontario, 1977*. Queen's Papers in Pure and Applied Mathematics, vol. 48. Queen's University, Kingston, ON.

[344] Harris, M. 2002. On the local Langlands correspondence. Pages 583–597 of Tatsien, L. (ed.), *Proceedings of the International Congress of Mathematicians, Beijing, 2002, vol. II*. Higher Education Press, Beijing.

[345] Harris, M., and Kudla, S. S. 1991. The central critical value of a triple product L-function. *Ann. of Math. (2)*, **133**(3), 605–672.

[346] Harris, M., and Taylor, R. 2001. *The Geometry and Cohomology of Some Simple Shimura Varieties*. Annals of Mathematics Studies, vol. 151. Princeton University Press, Princeton, NJ. With an appendix by V. G. Berkovich.

[347] Harvey, J. A., and Moore, G. W. 1996. Algebras, BPS states, and strings. *Nucl. Phys.*, **B463**, 315–368.

[348] Harvey, J. A., and Moore, G. W. 1998. On the algebras of BPS states. *Commun. Math. Phys.*, **197**, 489–519.

[349] Hashizume, M. 1982. Whittaker functions on semisimple Lie groups. *Hiroshima Math. J.*, **12**(2), 259–293.

[350] Hasse, H. 1933. Beweis des Analogons der Riemannschen Vermutung für die Artinschen und F. K. Schmidtschen Kongruenzzetafunktionen in gewissen elliptischen Fällen. *Nachr. Gesell. Wissen. Göttingen*, **42**, 253–262.

[351] Hecke, E. 1937a. Über Modulfunktionen und die Dirichletschen Reihen mit Eulerscher Produktentwicklung, I. *Math. Ann.*, **114**(1), 1–28.

[352] Hecke, E. 1937b. Über Modulfunktionen und die Dirichletschen Reihen mit Eulerscher Produktentwicklung, II. *Math. Ann.*, **114**(1), 316–351.

[353] Helgason, S. 2001. *Differential Geometry, Lie Groups and Symmetric Spaces*. Graduate Studies in Mathematics, vol. 34. American Mathematical Society, Providence, RI.

[354] Henneaux, M., Persson, D., and Spindel, P. 2008. Spacelike singularities and hidden symmetries of gravity. *Living Rev. Rel.*, **11**, 1.

[355] Henniart, G. 2000. Une preuve simple des conjectures de Langlands pour $\mathrm{GL}(n)$ sur un corps p-adique. *Invent. Math.*, **139**(2), 439–455.

[356] Hickey, T. J. 1986. *On the Fourier-Jacobi Coefficients of Certain Eisenstein Series on a Unitary Group*. Ph.D. thesis, University of Chicago.

[357] Hori, K., Katz, S., Klemm, A., Pandharipande, R., Thomas, R., Vafa, C., Vakil, R., and Zaslow, E. 2003. *Mirror Symmetry*. Clay Mathematics Monographs, vol. 1. American Mathematical Society, Providence, RI.

[358] Horowitz, G. T., and Strominger, A. 1991. Black strings and P-branes. *Nucl. Phys.*, **B360**, 197–209.

[359] Howe, P. S., and West, P. C. 1984. The complete N=2, D=10 supergravity. *Nucl. Phys.*, **B238**, 181–220.

[360] Howe, R. 1974. The Fourier transform and germs of characters (case of Gl_n over a p-adic field). *Math. Ann.*, **208**, 305–322.

[361] Howe, R. 1979. θ-series and invariant theory. Pages 275–285 of Borel, A., and Casselman, W. (eds), *Automorphic Forms, Representations and L-Functions*. Proceedings of Symposia in Pure Mathematics, vol. 33. American Mathematical Society, Providence, RI.

[362] Howe, R. 1985. Dual pairs in physics: Harmonic oscillators, photons, electrons, and singletons. Pages 179–207 of Flato, M., Sally, P., and Zukerman, G. (eds), *Applications of Group Theory in Physics and Mathematical Physics (Chicago, 1982)*. Lectures in Applied Mathematics, vol. 21. American Mathematical Society, Providence, RI.

[363] Howe, R. 1990. Another look at the local θ-correspondence for an unramified dual pair. Pages 93–124 of Gelbart, S., Howe, R., and Sarnak, P. (eds), *Festschrift in Honor of I. I. Piatetski-Shapiro on the Occasion of His Sixtieth Birthday, Part I (Ramat Aviv, 1989)*. Israel Mathematical Conference Proceedings, vol. 2. Weizmann Science Press, Jerusalem.

[364] Huang, J.-S., Pandžić, P., and Savin, G. 1996. New dual pair correspondences. *Duke Math. J.*, **82**(2), 447–471.

[365] Hull, C., and Townsend, P. 1995. Unity of superstring dualities. *Nucl. Phys.*, **B438**, 109–137.

[366] Humphreys, J. E. 1975. *Linear Algebraic Groups*. Graduate Texts in Mathematics, vol. 21. Springer, Heidelberg.

[367] Humphreys, J. E. 1980. *Arithmetic Groups*. Lecture Notes in Mathematics, vol. 789. Springer, Berlin, New York.

[368] Humphreys, J. E. 1997. *Introduction to Lie Algebras and Representation Theory*. Graduate Texts in Mathematics, vol. 9. Springer, London.

[369] Ion, B., and Sahi, S. 2015. Double affine Hecke algebras and congruence groups. arXiv:1506.06417 [math.QA].

[370] Ishii, T., and Stade, E. 2007. New formulas for Whittaker functions on $\mathrm{GL}(n, \mathbb{R})$. *J. Funct. Anal.*, **244**(1), 289–314.

[371] Ishikawa, Y.-H. 1999. The generalized Whittaker functions for SU(2, 1) and the Fourier expansion of automorphic forms. *J. Math. Sci. Univ. Tokyo*, **6**, 477–526.

[372] Ivashchuk, V. D., and Melnikov, V. N. 1995. Billiard representation for multidimensional cosmology with multicomponent perfect fluid near the singularity. *Class. Quant. Grav.*, **12**, 809–826.

[373] Iwaniec, H. 2002. *Spectral Methods of Automorphic Forms*. Second edn. Graduate Studies in Mathematics, vol. 53. American Mathematical Society, Providence, RI.

[374] Jacquet, H. 1967. Fonctions de Whittaker associées aux groupes de Chevalley. *Bull. Soc. Math. France*, **95**, 243–309.

[375] Jacquet, H. 1984. On the residual spectrum of $\mathrm{GL}(n)$. Pages 185–208 of Herb, R., Kudla, S., Lipsman, R., and Rosenberg, J. (eds), *Lie Group Representations, II (College Park, MD, 1982/1983)*. Lecture Notes in Mathematics, vol. 1041. Springer, Berlin, New York.

[376] Jacquet, H., and Langlands, R. P. 1970. *Automorphic Forms on* GL(2). Lecture Notes in Mathematics, vol. 114. Springer, Berlin, New York.

[377] Jiang, D., Liu, B., and Savin, G. 2016. Raising nilpotent orbits in wave-front sets. *Represent. Theory*, **20**, 419–450.

[378] Jiang, D., and Rallis, S. 1997. Fourier coefficients of Eisenstein series of the exceptional group of type G_2. *Pacific J. Math.*, **181**(2), 281–314.

[379] Jiang, D., and Soudry, D. 2004. Generic representations and local Langlands reciprocity law for p-adic SO_{2n+1}. Pages 457–519 of Hida, H., Ramakrishnan, D., and Shahidi, F. (eds), *Contributions to Automorphic Forms, Geometry, and Number Theory*. Johns Hopkins University Press, Baltimore, MD.

[380] Jones, G., and Jones, J. 1998. *Elementary Number Theory*. Springer Undergraduate Mathematics Series. Springer, London.

[381] Joseph, A. 1974. Minimal realizations and spectrum generating algebras. *Comm. Math. Phys.*, **36**, 325–338.

[382] Joseph, A. 1976. The minimal orbit in a simple Lie algebra and its associated maximal ideal. *Ann. Sci. École Norm. Sup. (4)*, **9**(1), 1–29.

[383] Joseph, A. 1985. On the associated variety of a primitive ideal. *J. Algebra*, **93**(2), 509–523.

[384] Joyce, D., and Song, Y. 2012. A theory of generalized Donaldson-Thomas invariants. *Mem. Amer. Math. Soc.*, **217**(1020), iv+199.

[385] Julia, B. 1981. Group disintegrations. Pages 331–350 of Hawking, S. W., and Roček, M. (eds), *Superspace and Supergravity (Proceedings of Nuffield Workshop, Cambridge, June 16 to July 12, 1980)*. Cambridge University Press, Cambridge.

[386] Kac, V. G. 1990. *Infinite Dimensional Lie Algebras*. Third edn. Cambridge University Press, Cambridge.

[387] Kapranov, M. 1998. Double affine Hecke algebras and 2-dimensional local fields. `arXiv:math/9812021`.

[388] Kapustin, A., and Witten, E. 2007. Electric-magnetic duality and the geometric Langlands program. *Commun. Num. Theor. Phys.*, **1**, 1–236.

[389] Kashiwara, M. 1990. Crystalizing the q-analogue of universal enveloping algebras. *Comm. Math. Phys.*, **133**(2), 249–260.

[390] Katz, S., Klemm, A., Pandharipande, R., and Thomas, R. P. 2016. On the motivic stable pairs invariants of K3 surfaces. Pages 111–146 of Faber, C., Farkas, G., and van der Geer, G. (eds), *K3 Surfaces and Their Moduli*. Progress in Mathematics, vol. 315. Birkhäuser Boston, Boston, MA.

[391] Kawai, T. 1996. N=2 heterotic string threshold correction, K3 surface and generalized Kac-Moody superalgebra. *Phys. Lett.*, **B372**, 59–64.

[392] Kawazumi, N. 2008. Johnson's homomorphisms and the Arakelov-Green function. `arXiv:0801.4218 [math.GT]`.

[393] Kazhdan, D., Pioline, B., and Waldron, A. 2002. Minimal representations, spherical vectors, and exceptional theta series. *Commun. Math. Phys.*, **226**, 1–40.

[394] Kazhdan, D., and Polishchuk, A. 2004. Minimal representations: Spherical vectors and automorphic functionals. Pages 127–198 of Dani, S. G., and Prasad, G. (eds), *Algebraic Groups and Arithmetic*. Narosa Publishing House, New Delhi, for Tata Institute of Fundamental Research, Mumbai.

[395] Kazhdan, D., and Savin, G. 1990. The smallest representation of simply laced groups. Pages 209–223 of Gelbart, S., Howe, R., and Sarnak, P. (eds), *Festschrift in honor of II Piatetski-Shapiro on the Occasion of His Sixtieth Birthday, Part I (Ramat Aviv, 1989)*. Israel Mathematical Conference Proceedings, vol. 2. Weizmann Science Press, Jerusalem.

[396] Kim, H. H. 1996. The residual spectrum of G_2. *Canad. J. Math.*, **48**(6), 1245–1272.

[397] Kim, H. H., and Lee, K.-H. 2012. Quantum affine algebras, canonical bases, and q-deformation of arithmetical functions. *Pacific J. Math.*, **255**(2), 393–415.

[398] Kim, H. H., and Shahidi, F. 1999. Symmetric cube L-functions for GL_2 are entire. *Ann. of Math. (2)*, **150**(2), 645–662.

[399] Kim, H. H., and Shahidi, F. 2000. Holomorphy of Rankin triple L-functions: Special values and root numbers for symmetric cube L-functions. *Israel J. Math.*, **120**, 449–466.

[400] Kim, H. H., and Shahidi, F. 2002. Functorial products for $\mathrm{GL}_2 \times \mathrm{GL}_3$ and the symmetric cube for GL_2. *Ann. of Math. (2)*, **155**(3), 837–893. With an appendix by C. J. Bushnell and G. Henniart.

[401] Kim, H. H., and Shahidi, F. 2004. On simplicity of poles of automorphic L-functions. *J. Ramanujan Math. Soc.*, **19**(4), 267–280.

[402] Kim, M. 2015. Arithmetic Chern-Simons theory I. `arXiv:1510.05818 [math.NT]`.

[403] Kirillov, A. A. 1995. Reduction of additional dimensions in nonuniform quantum Kaluza-Klein cosmological models. *JETP Lett.*, **62**, 89–94. [Pisma Zh. Eksp. Teor. Fiz.62,81(1995)].

[404] Kirillov, A. A. 1999. Merits and demerits of the orbit method. *Bull. Am. Math. Soc., New Ser.*, **36**(4), 433–488.

[405] Kirillov, A. N., and Berenstein, A. D. 1995. Groups generated by involutions, Gel'fand-Tsetlin patterns, and combinatorics of Young tableaux. *Algebra i Analiz*, **7**(1), 92–152.

[406] Kiritsis, E. 1998. *Introduction to Superstring Theory*. Leuven Notes in Mathematical and Theoretical Physics, vol. B9. Leuven University Press, Leuven. `arXiv:hep-th/9709062 [hep-th]`.

[407] Kiritsis, E., and Pioline, B. 1997. On R**4 threshold corrections in IIb string theory and (p, q) string instantons. *Nucl. Phys.*, **B508**, 509–534.

[408] Kleinschmidt, A., Koehn, M., and Nicolai, H. 2009. Supersymmetric quantum cosmological billiards. *Phys. Rev.*, **D80**, 061701.

[409] Kleinschmidt, A., and Nicolai, H. 2009. Cosmological quantum billiards. Pages 106–124 of Murugan, J., Weltman, A., and Ellis, G. F. R. (eds), *Proceedings, Foundations of Space and Time: Reflections on Quantum Gravity*. Cambridge University Press, Cambridge.

[410] Kleinschmidt, A., Nicolai, H., and Palmkvist, J. 2012. Modular realizations of hyperbolic Weyl groups. *Adv. Theor. Math. Phys.*, **16**(1), 97–148.

[411] Kleinschmidt, A., and Verschinin, V. 2017. Tetrahedral modular graph functions. *JHEP*, **09**, 155.

[412] Knapp, A. W. 1997. Introduction to the Langlands program. Pages 245–302 of Baitey, T. N., and Knapp, A. W. (eds), *Representation Theory and Automorphic Forms*. Proceedings in Symposia in Pure Mathematics, vol. 61, American Mathematical Society, Providence, RI.

[413] Knapp, A. W. 2009a. First steps with the Langlands program. Pages 10–20 of Ji, L., Liu, K., Yau, S.-T., and Zheng, Z.-J. (eds), *Automorphic Forms and the Langlands Program*. Advanced Lectures in Mathematics, vol. 9. International Press of Boston, Somerville, MA.

[414] Knapp, A. W. 2009b. Prerequisites for the Langlands program. Pages 1–9 of Ji, L., Liu, K., Yau, S.-T., and Zheng, Z.-J. (eds), *Automorphic Forms and the Langlands Program*. Advanced Lectures in Mathematics vol. 9. International Press of Boston, Somerville, MA.

[415] Kobayashi, T., and Savin, G. 2015. Global uniqueness of small representations. *Math. Z.*, **281**(1-2), 215–239.

[416] Koblitz, N. 1984. *p-adic Numbers, p-adic Analysis, and Zeta-Functions*. Springer, New York.

[417] Koblitz, N. 1993. *Introduction to Elliptic Curves and Modular Forms*. Second edn. Graduate Texts in Mathematics, vol. 97. Springer, New York.

[418] Kolyvagin, V. A. 1988. Finiteness of $E(\mathbf{Q})$ and $Ш(E, \mathbf{Q})$ for a subclass of Weil curves. *Izv. Akad. Nauk SSSR Ser. Mat.*, **52**(3), 522–540, 670–671.

[419] Kontsevich, M. 1995. Homological algebra of mirror symmetry. Pages 120–139 of Chatterji, S. D. (ed.), *Proceedings of the International Congress of Mathematicians, Vol. 1, 2 (Zürich, 1994)*. Birkhäuser, Basel.

[420] Kontsevich, M., and Soibelman, Y. 2008. Stability structures, motivic Donaldson-Thomas invariants and cluster transformations. arXiv:0811.2435 [math.AG].

[421] Kostant, B. 1963. Lie group representations on polynomial rings. *Amer. J. Math.*, **85**, 327–404.

[422] Kostant, B. 1978. On Whittaker vectors and representation theory. *Invent. Math.*, **48**(2), 101–184.

[423] Kostant, B., and Rallis, S. 1971. Orbits and representations associated with symmetric spaces. *Amer. J. Math.*, **93**, 753–809.

[424] Kostov, I. K., and Vanhove, P. 1998. Matrix string partition functions. *Phys. Lett.*, **B444**, 196–203.

[425] Krutelevich, S. 2007. Jordan algebras, exceptional groups, and Bhargava composition. *J. Algebra*, **314**(2), 924–977.

[426] Kubota, T. 1969. *On Automorphic Functions and the Reciprocity Law in a Number Field.* Lectures in Mathematics, Department of Mathematics, Kyoto University, No. 2. Kinokuniya Book-Store Co., Tokyo.

[427] Kudla, S. S., and Rallis, S. 1988a. On the Weil-Siegel formula. *J. Reine Angew. Math.*, **387**, 1–68.

[428] Kudla, S. S., and Rallis, S. 1988b. On the Weil-Siegel formula, II: The isotropic convergent case. *J. Reine Angew. Math.*, **391**, 65–84.

[429] Kudla, S. S., and Rallis, S. 1994. A regularized Siegel-Weil formula: The first term identity. *Ann. of Math. (2)*, **140**(1), 1–80.

[430] Kumar, S. 2002. *Kac-Moody Groups, Their Flag Varieties and Representation Theory*. Progress in Mathematics, vol. 204. Birkhäuser Boston, Boston, MA.

[431] Lafforgue, L. 2002. Chtoucas de Drinfeld et correspondance de Langlands. *Invent. Math.*, **147**(1), 1–241.

[432] Lafforgue, V., and Lysenko, S. 2009. Geometric Weil representation: Local field case. *Compos. Math.*, **145**(1), 56–88.

[433] Lafforgue, V., and Lysenko, S. 2011. Compatibility of the theta correspondence with the Whittaker functors. *Bull. Soc. Math. France*, **139**(1), 75–88.

[434] Lafforgue, V., and Lysenko, S. 2013. Geometrizing the minimal representations of even orthogonal groups. *Represent. Theory*, **17**, 263–325.

[435] Lai, K. F. 1974. *On the Tamagawa Number of Quasi-split Groups*. Ph.D. thesis, Yale University.

[436] Lai, K. F. 1976. On the Tamagawa number of quasi-split groups. *Bull. Amer. Math. Soc.*, **82**(2), 300–302.

[437] Lambert, N., and West, P. 2010. Perturbation theory from automorphic forms. *JHEP*, **1005**, 098.

[438] Lambert, N., and West, P. C. 2007. Duality groups, automorphic forms and higher derivative corrections. *Phys. Rev.*, **D75**, 066002.

[439] Lang, S. 1975. $SL(2,\mathbb{R})$. Graduate Texts in Mathematics, vol. 105. Springer, New York.

[440] Lang, S. 1994. *Algebraic Number Theory*. Second edn. Graduate Texts in Mathematics, vol. 110. Springer, New York.

[441] Lang, S. 2002. *Algebra*. Third edn. Graduate Texts in Mathematics, vol. 211. Springer, New York.

[442] Langlands, R. P. 1966. Eisenstein Series. Pages 235–252 of Borel, A., and Mostow, G. D. (eds), *Algebraic Groups and Discontinuous Subgroups*. Proceedings of Symposia in Pure Mathematics, vol. 9. American Mathematical Society, Providence, RI.

[443] Langlands, R. 1967a. *Euler Products*. James K. Whittemore Lectures in Mathematics, Yale Univesity, April 1967. Published with corrections by the author as Yale Mathematical Monographs, vol. 1, Yale Univesity Press, New Haven, CT (1971).

[444] Langlands, R. P. 1967b. Letter to André Weil. `http://publications.ias.edu/sites/default/files/ltw_1.pdf`.

[445] Langlands, R. P. 1967c. Letter to Godement. `http://publications.ias.edu/sites/default/files/godement-ps.pdf`.

[446] Langlands, R. P. 1970. Problems in the theory of automorphic forms. *Lectures in Modern Analysis and Applications III, Lecture Notes in Mathematics*, **170**.

[447] Langlands, R. P. 1976. *On the Functional Equations Satisfied by Eisenstein Series*. Lecture Notes in Mathematics, vol. 544. Springer, Berlin, New York.

[448] Langlands, R. P. 1978. L-functions and automorphic representations. Pages 165–175 of Lehto, O. (ed.), *Proceedings of the International Congress of Mathematicians (Helsinki, 1978)*. Academia Scientiarium Fennica, Helsinki (1980).

[449] Laumon, G. 1987. Correspondance de Langlands géométrique pour les corps de fonctions. *Duke Math. J.*, **54**(2), 309–359.

[450] Laumon, G. 1990. Faisceaux automorphes liés aux séries d'Eisenstein. Pages 227–281 of Clozel, L., and Milne, J. S. (eds), *Automorphic Forms, Shimura Varieties, and L-functions, Vol. I (Ann Arbor, MI, 1988)*. Perspectives in Mathematics, vol. 10. Academic Press, Boston, MA.

[451] Lee, K.-H., and Zhang, Y. 2015. Weyl group multiple Dirichlet series for symmetrizable Kac-Moody root systems. *Trans. Amer. Math. Soc.*, **367**(1), 597–625.

[452] Littelmann, P. 1996. An effective method to classify nilpotent orbits. Pages 255–269 of González-Vega, L., and Recio, T. (eds), *Algorithms in Algebraic Geometry and Applications (Santander, 1994)*. Progress in Mathematics, vol. 143. Birkhäuser, Basel.

[453] Littelmann, P. 1998. Cones, crystals, and patterns. *Transform. Groups*, **3**(2), 145–179.

[454] Liu, D. 2015. Eisenstein series on loop groups. *Trans. Amer. Math. Soc.*, **367**(3), 2079–2135.

[455] Lu, H., Pope, C. N., and Stelle, K. S. 1998. Multiplet structures of BPS solitons. *Class. Quant. Grav.*, **15**, 537–561.

[456] Lusztig, G. 1990. Canonical bases arising from quantized enveloping algebras. *J. Amer. Math. Soc.*, **3**(2), 447–498.

[457] Lusztig, G. 2010. *Introduction to Quantum Groups*. Modern Birkhäuser Classics. Birkhäuser/Springer, New York. Reprint of the 1994 edition.

[458] Lysenko, S. 2006a. Moduli of metaplectic bundles on curves and theta-sheaves. *Ann. Sci. École Norm. Sup. (4)*, **39**(3), 415–466.

[459] Lysenko, S. 2006b. Whittaker and Bessel functors for GSP4. *Ann. Inst. Fourier (Grenoble)*, **56**(5), 1505–1565.

[460] Lysenko, S. 2011. Geometric theta-lifting for the dual pair. *Ann. Sci. École Norm. Sup. (4)*, **44**(3), 427–493.

[461] Lysenko, S. 2017. Twisted Whittaker models for metaplectic groups. *Geom. Funct. Anal.*, **27**(2), 289–372.

[462] Mafra, C. R., and Schlotterer, O. 2014. The structure of n-point one-loop open superstring amplitudes. *JHEP*, **08**, 099.

[463] Mafra, C. R., and Schlotterer, O. 2017. Non-abelian Z-theory: Berends-Giele recursion for the α'-expansion of disk integrals. *JHEP*, **01**, 031.

[464] Mafra, C. R., Schlotterer, O., and Stieberger, S. 2013a. Complete N-point superstring disk amplitude, I: Pure spinor computation. *Nucl. Phys.*, **B873**, 419–460.

[465] Mafra, C. R., Schlotterer, O., and Stieberger, S. 2013b. Complete N-point superstring disk amplitude, II: Amplitude and hypergeometric function structure. *Nucl. Phys.*, **B873**, 461–513.

[466] Magaard, K., and Savin, G. 1997. Exceptional Θ-correspondences, I. *Compos. Math.*, **107**(1), 89–123.

[467] Maldacena, J. M., Moore, G. W., and Strominger, A. 1999. Counting BPS black holes in toroidal type II string theory. `arXiv:hep-th/9903163 [hep-th]`.

[468] Manschot, J. 2010. Stability and duality in N=2 supergravity. *Commun. Math. Phys.*, **299**, 651–676.

[469] Manschot, J., and Moore, G. W. 2010. A modern farey tail. *Commun. Num. Theor. Phys.*, **4**, 103–159.

[470] Matumoto, H. 1987. Whittaker vectors and associated varieties. *Invent. Math.*, **89**(1), 219–224.

[471] Mazur, B. 1978. Rational isogenies of prime degree. *Invent. Math.*, **44**(2), 129–162. With an appendix by D. Goldfeld.

[472] McNamara, P. J. 2011. Metaplectic Whittaker functions and crystal bases. *Duke Math. J.*, **156**(1), 1–31.

[473] McNamara, P. J. 2012. Principal series representations of metaplectic groups over local fields. Pages 299–327 of Bump, D., Friedberg, S., and Goldfeld, D. (eds), *Multiple Dirichlet Series, L-functions and Automorphic Forms*. Progress in Mathematics, vol. 300. Birkhäuser/Springer, New York.

[474] Miller, S. D. 2013. Residual automorphic forms and spherical unitary representations of exceptional groups. *Ann. of Math. (2)*, **177**(3), 1169–1179.

[475] Miller, S. D., and Moore, G. 2000. Landau-Siegel zeroes and black hole entropy. *Asian J. Math.*, **4**(1), 183–211.

[476] Miller, S. D., and Sahi, S. 2012. Fourier coefficients of automorphic forms, character variety orbits, and small representations. *J. Number Theory*, **132**(12), 3070–3108.

[477] Miller, S. D., and Schmid, W. 2004a. The highly oscillatory behavior of automorphic distributions for SL(2). *Lett. Math. Phys.*, **69**, 265–286.

[478] Miller, S. D., and Schmid, W. 2004b. Summation formulas, from Poisson and Voronoi to the present. Pages 419–440 of Delorme, P., and Vergne, M. (eds), *Noncommutative Harmonic Analysis*. Progress in Mathematics, vol. 220. Birkhäuser Boston, Boston, MA.

[479] Miller, S. D., and Schmid, W. 2008. The Rankin-Selberg method for automorphic distributions. Pages 111–150 of Kobayashi, T., Schmid, W., and Yang, J.-H. (eds), *Representation Theory and Automorphic Forms*. Progress in Mathematics, vol. 255. Birkhäuser Boston, Boston, MA.

[480] Milne, J. S. 2015. Algebraic groups (v2.00). `www.jmilne.org/math/`.

[481] Misner, C. W. 1969. Mixmaster universe. *Phys. Rev. Lett.*, **22**, 1071–1074.

[482] Mizoguchi, S., and Schroeder, G. 2000. On discrete U duality in M theory. *Class. Quant. Grav.*, **17**, 835–870.

[483] Mœglin, C. 1994. Représentations unipotentes et formes automorphes de carré intégrable. *Forum Math.*, **6**(6), 651–744.

[484] Mœglin, C. 1996. Front d'onde des représentations des groupes classiques p-adiques. *Amer. J. Math.*, **118**(6), 1313–1346.

[485] Mœglin, C. 1998. Correspondance de Howe et front d'onde. *Adv. Math.*, **133**(2), 224–285.

[486] Mœglin, C., and Waldspurger, J.-L. 1987. Modèles de Whittaker dégénérés pour des groupes p-adiques. *Math. Z.*, **196**(3), 427–452.

[487] Mœglin, C., and Waldspurger, J.-L. 1989. Le spectre résiduel de $\mathrm{GL}(n)$. *Ann. Sci. École Norm. Sup. (4)*, **22**(4), 605–674.

[488] Mœglin, C., and Waldspurger, J.-L. 1995. *Spectral Decomposition and Eisenstein Series*. Cambridge University Press, Cambridge.

[489] Moody, R. V., and Pianzola, A. 1995. *Lie Algebras with Triangular Decompositions*. Canadian Mathematical Society Series of Monographs and Advanced Texts. John Wiley & Sons, New York.

[490] Moore, G. W. 1998. Arithmetic and attractors. arXiv:hep-th/9807087 [hep-th].

[491] Moore, G. W. 2014. Physical mathematics and the future. www.physics.rutgers.edu/~gmoore/PhysicalMathematicsAndFuture.pdf.

[492] Moore, G. W. 2016. Desperately seeking moonshine. www.physics.rutgers.edu/~gmoore/DaveDayFinal.pdf.

[493] Moore, G. W., Nekrasov, N., and Shatashvili, S. 2000. D particle bound states and generalized instantons. *Commun. Math. Phys.*, **209**, 77–95.

[494] Mordell, L. J. 1922. On the rational solutions of the indeterminate equations of the third and fourth degree. *Proc. Cam. Phil. Soc.*, **21**, 179–192.

[495] Myers, R. C. 1999. Dielectric branes. *JHEP*, **12**, 022.

[496] Narain, K. S. 1986. New heterotic string theories in uncompactified dimensions <10. *Phys. Lett.*, **169B**, 41–46.

[497] Narita, H. 2006. Fourier-Jacobi expansion of automorphic forms on Sp(1, q) generating quaternionic discrete series. *J. Funct. Anal.*, **239**, 638–682.

[498] Neitzke, A., Pioline, B., and Vandoren, S. 2007. Twistors and black holes. *JHEP*, **04**, 038.

[499] Nepomechie, R. I. 1985. Magnetic monopoles from antisymmetric tensor gauge fields. *Phys. Rev.*, **D31**, 1921.

[500] Neukirch, J. 2006. *Algebraische Zahlentheorie*. Springer, Berlin.

[501] Nilsson, B. E. W., and Tollsten, A. K. 1986. Supersymmetrization of zeta (3) (R $\mu\nu\rho\sigma$)**4 in superstring theories. *Phys. Lett.*, **B181**, 63–66.

[502] Obers, N., and Pioline, B. 1999. U duality and M theory. *Phys. Rep.*, **318**, 113–225.

[503] Obers, N. A., and Pioline, B. 2000a. Eisenstein series and string thresholds. *Comm. Math. Phys.*, **209**(2), 275–324.

[504] Obers, N. A., and Pioline, B. 2000b. Eisenstein series in string theory. *Class. Quant. Grav.*, **17**, 1215–1224.

[505] Orloff, T. 1985. Dirichlet series and automorphic forms on unitary groups. *Trans. Amer. Math. Soc.*, **290**(2), 431–456.

[506] Paquette, N. M., Persson, D., and Volpato, R. 2016. Monstrous BPS-algebras and the superstring origin of moonshine. *Commun. Num. Theor. Phys.*, **10**, 433–526.

[507] Patnaik, M. September, 2017. Automorphic forms on loop groups. Talk at the workshop 'Automorphic Forms, Mock Modular Forms and String Theory', Simons Center for Geometry and Physics, Stony Brook, NY, 31 August 2016.

[508] Patnaik, M., and Puskás, A. 2017. On Iwahori-Whittaker functions for metaplectic groups. *Adv. Math.*, **313**, 875–914.

[509] Patnaik, M. M. 2017. Unramified Whittaker functions on p-adic loop groups. *Amer. J. Math.*, **139**(1), 175–213.

[510] Persson, D. 2010. *Arithmetic and Hyperbolic Structures in String Theory*. Ph.D. thesis, Free University of Brussels. `arXiv:1001.3154 [hep-th]`.

[511] Persson, D. 2012. Automorphic instanton partition functions on Calabi-Yau threefolds. *J. Phys. Conf. Ser.*, **346**.

[512] Persson, D., and Volpato, R. 2014. Second quantized Mathieu moonshine. *Commun. Num. Theor. Phys.*, **08**, 403–509.

[513] Persson, D., and Volpato, R. 2015. Fricke S-duality in CHL models. *JHEP*, **12**, 156.

[514] Peterson, D. H., and Kac, V. G. a. 1983. Infinite flag varieties and conjugacy theorems. *Proc. Nat. Acad. Sci. U.S.A.*, **80**(6 i.), 1778–1782.

[515] Petropoulos, P. M., and Vanhove, P. 2012. Gravity, strings, modular and quasimodular forms. *Ann. Math. Blaise Pascal*, **19**(2), 379–430.

[516] Piatetski-Shapiro, I. I. 1979. Multiplicity one theorems. Pages 209–212 of Borel, A., and Casselman, W. (eds), *Automorphic Forms, Representations and L-functions*. Proceedings of Symposia in Pure Mathematics, vol. 33. American Mathematical Society, Providence, RI.

[517] Pioline, B. 1998. A note on nonperturbative R**4 couplings. *Phys. Lett.*, **B431**, 73–76.

[518] Pioline, B. 2005. BPS black hole degeneracies and minimal automorphic representations. *JHEP*, **08**, 071.

[519] Pioline, B. 2006. Lectures on black holes, topological strings and quantum attractors. *Class. Quant. Grav.*, **23**, S981.

[520] Pioline, B. 2010. R**4 couplings and automorphic unipotent representations. *JHEP*, **03**, 116.

[521] Pioline, B. 2015. D^6R^4 amplitudes in various dimensions. *JHEP*, **04**, 057.

[522] Pioline, B. 2016. A theta lift representation for the Kawazumi-Zhang and Faltings invariants of genus-two Riemann surfaces. *J. Number Theory*, **163**, 520–541.

[523] Pioline, B., and Kiritsis, E. 1998. U duality and D-brane combinatorics. *Phys. Lett.*, **B418**, 61–69.

[524] Pioline, B., Nicolai, H., Plefka, J., and Waldron, A. 2001. R**4 couplings, the fundamental membrane and exceptional theta correspondences. *JHEP*, **0103**, 036.

[525] Pioline, B., and Persson, D. 2009. The automorphic NS5-brane. *Commun. Num. Theor. Phys.*, **3**, 697–754.

[526] Pioline, B., and Russo, R. 2015. Infrared divergences and harmonic anomalies in the two-loop superstring effective action. *JHEP*, **12**, 102.

[527] Pioline, B., and Vandoren, S. 2009. Large D-instanton effects in string theory. *JHEP*, **0907**, 008.

[528] Pioline, B., and Waldron, A. 2003a. Automorphic forms: A physicist's survey. Pages 277–302 of Cartier, P. E., Julia, B., Moussa, P., and Vanhove, P. (eds), *Frontiers in Number Theory, Physics, and Geometry 1: On Random Matrices,*

Zeta Functions and Dynamical Systems. Les Houches, France, March 9-21, 2003. Springer, Berlin.

[529] Pioline, B., and Waldron, A. 2003b. Quantum cosmology and conformal invariance. *Phys. Rev. Lett.*, **90**, 031302.

[530] Pioline, B., and Waldron, A. 2004. The automorphic membrane. *JHEP*, **0406**, 009.

[531] Platonov, V., and Rapinchuk, A. 1994. *Algebraic Groups and Number Theory*. Pure and Applied Mathematics, vol. 139. Academic Press, Boston, MA. Translated from the 1991 Russian original by Rachel Rowen.

[532] Polchinski, J. 1994. Combinatorics of boundaries in string theory. *Phys. Rev.*, **D50**, 6041–6045.

[533] Polchinski, J. 1995. Dirichlet branes and Ramond-Ramond charges. *Phys. Rev. Lett.*, **75**, 4724–4727.

[534] Polchinski, J. 1996. Tasi lectures on D-branes. 293–356. `arXiv:hep-th/9611050 [hep-th]`.

[535] Polchinski, J. 2007. *String Theory. Vol. 1: An Introduction to the Bosonic String; Vol. 2: Superstring Theory and Beyond*. Cambridge Monographs on Mathematical Physics. Cambridge University Press, Cambridge.

[536] Polchinski, J., and Cai, Y. 1988. Consistency of open superstring theories. *Nucl. Phys.*, **B296**, 91–128.

[537] Prasad, G. 1977. Strong approximation for semi-simple groups over function fields. *Ann. of Math. (2)*, **105**(3), 553–572.

[538] Proskurin, N. V. 1984. Expansion of automorphic functions. *J. Sov. Math.*, **26**(3), 1908–1921.

[539] Ribet, K. A. 1990. On modular representations of Gal($\overline{\mathbf{Q}}/\mathbf{Q}$) arising from modular forms. *Invent. Math.*, **100**(2), 431–476.

[540] Ribet, K. A. 1995. Galois representations and modular forms. *Bull. Amer. Math. Soc. (N.S.)*, **32**(4), 375–402.

[541] Riemann, B. 1859. Über die Anzahl der Primzahlen unter einer gegebenen Größe. *Monatsber. Berliner Akademie*, 671–680.

[542] Robles-Llana, D., Rocek, M., Saueressig, F., Theis, U., and Vandoren, S. 2007. Nonperturbative corrections to 4D string theory effective actions from SL(2,Z) duality and supersymmetry. *Phys. Rev. Lett.*, **98**, 211602.

[543] Rodier, F. 1975. Modèle de Whittaker et caractères de représentations. Pages 151–171 of Carmona, J., Dixmier, J., and Vergne, M. (eds), *Non-commutative Harmonic Analysis*. Lecture Notes in Mathematics, vol. 466. Springer, Berlin, New York.

[544] Satake, I. 1963. Theory of spherical functions on reductive algebraic groups over p-adic fields. *Inst. Hautes Études Sci. Publ. Math.*, **18**, 1–69.

[545] Savin, G. K-types of minimal representations (p-adic case). `www.math.utah.edu/~savin/k-tipovi.pdf`.

[546] Savin, G., and Woodbury, M. 2007. Structure of internal modules and a formula for the spherical vector of minimal representations. *J. Algebra*, **312**(2), 755–772.

[547] Schimmrigk, R. 2011. Emergent spacetime from modular motives. *Commun. Math. Phys.*, **303**, 1–30.

[548] Schlotterer, O., and Stieberger, S. 2013. Motivic multiple zeta values and superstring amplitudes. *J. Phys.*, **A46**, 475401.

[549] Schmid, W. 2000. Automorphic distributions for $SL(2, \mathbb{R})$. Pages 345–387 of Dito, G., and Sternheimer, D. (eds), *Conférence Moshé Flato 1999, Vol. I (Dijon)*. Mathematical Physics Studies, vol. 21. Kluwer Academic Publishers, Dordrecht.

[550] Schulze-Pillot, R. 1998. Theta liftings: A comparison between classical and representation-theoretic results. Pages 142–153 of Sugano, T. (ed.), *Automorphic Forms and Number Theory*. RIMS Kôkyûroku, vol. 1052. Research Institute for Mathematical Sciences, Kyoto University, Kyoto.

[551] Schwarz, J. H., and Sen, A. 1993. Duality symmetries of 4-D heterotic strings. *Phys. Lett.*, **B312**, 105–114.

[552] Seiberg, N. 1988. Observations on the moduli space of superconformal field theories. *Nucl. Phys.*, **B303**, 286–304.

[553] Sekiguchi, J. 1987. Remarks on real nilpotent orbits of a symmetric pair. *J. Math. Soc. Japan*, **39**(1), 127–138.

[554] Sen, A. 1994. Strong–weak coupling duality in four-dimensional string theory. *Int. J. Mod. Phys.*, **A9**, 3707–3750.

[555] Sen, A. 2008. Black hole entropy function, attractors and precision counting of microstates. *Gen. Relativity Gravitation*, **40**, 2249–2431.

[556] Sen, A. 2009. Arithmetic of quantum entropy function. *JHEP*, **08**, 068.

[557] Sen, A. 2010. Discrete information from CHL black holes. *JHEP*, **11**, 138.

[558] Sen, A. 2014. Microscopic and macroscopic entropy of extremal black holes in string theory. *Gen. Relativity Gravitation*, **46**, 1711.

[559] Serre, J.-P. 1973. *A Course in Arithmetic*. Graduate Texts in Mathematics, vol. 7. Springer, New York, Heidelberg.

[560] Serre, J.-P. 1987. Sur les représentations modulaires de degré 2 de $\mathrm{Gal}(\overline{\mathbf{Q}}/\mathbf{Q})$. *Duke Math. J.*, **54**(1), 179–230.

[561] Sethi, S., and Stern, M. 1998. D-brane bound states redux. *Commun. Math. Phys.*, **194**, 675–705.

[562] Shahidi, F. 1978. Functional equation satisfied by certain L-functions. *Compos. Math.*, **37**(2), 171–207.

[563] Shahidi, F. 1981. On certain L-functions. *Amer. J. Math.*, **103**(2), 297–355.

[564] Shahidi, F. 1985. Local coefficients as Artin factors for real groups. *Duke Math. J.*, **52**(4), 973–1007.

[565] Shahidi, F. 1990. A proof of Langlands' conjecture on Plancherel measures: Complementary series for p-adic groups. *Ann. of Math. (2)*, **132**(2), 273–330.

[566] Shahidi, F. 1996. *Intertwining Operators, L-functions and Representation Theory*. Lecture Notes of the Eleventh KAIST Mathematics Worskshop, vol. 11. Korea Advanced Institute of Science and Technology, Daejeon, South Korea.

[567] Shahidi, F. 2002. Automorphic L-functions and functoriality. Pages 655–666 of Tatsien, L. (ed.), *Proceedings of the International Congress of Mathematicians, Beijing, 2002. vol. II*. Higher Education Press, Beijing.

[568] Shahidi, F. 2005. Infinite dimensional groups and automorphic L-functions. *Pure Appl. Math. Q.*, **1**(3, part 2), 683–699.

[569] Shahidi, F. 2010. *Eisenstein Series and Automorphic L-functions*. American Mathematical Society Colloquium Publications, vol. 58. American Mathematical Society, Providence, RI.

[570] Shalika, J. A. 1974. The multiplicity one theorem for GL_n. *Ann. of Math. (2)*, **100**, 171–193.

[571] Shenker, S. H. 1990. The strength of nonperturbative effects in string theory. Pages 191–200 of Alvarez, O., Marinari, E., and Windey, P. (eds), *Random Surfaces and Quantum Gravity: Proceedings, NATO Advanced Study Institute, Cargèse, France, May 27-June 2, 1990*. NATO ASI Series B, vol. 262. Plenum Press, New York.

[572] Shih, D., Strominger, A., and Yin, X. 2006a. Counting dyons in N=8 string theory. *JHEP*, **06**, 037.

[573] Shih, D., Strominger, A., and Yin, X. 2006b. Recounting dyons in N=4 string theory. *JHEP*, **0610**, 087.

[574] Shimura, G. 1958. Correspondances modulaires et les fonctions ζ de courbes algébriques. *J. Math. Soc. Japan*, **10**, 1–28.

[575] Shimura, G. 1971. On elliptic curves with complex multiplication as factors of the Jacobians of modular function fields. *Nagoya Math. J.*, **43**, 199–208.

[576] Shimura, G. 1989. Yutaka Taniyama and his time: Very personal recollections. *Bull. London Math. Soc.*, **21**(2), 186–196.

[577] Shintani, T. 1976. On an explicit formula for Class-1 'Whittaker Functions' on $GL(n)$ over p-adic fields. *Proc. Japan. Acad.*, **52**, 180–182.

[578] Shintani, T. 1979. On automorphic forms on unitary groups of order 3. Unpublished manuscript.

[579] Sinha, A. 2002. The $\hat{G}^4\lambda^{16}$ term in IIB supergravity. *JHEP*, **08**, 017.

[580] Soulé, C. 2007. An introduction to arithmetic groups. Pages 247–276 of Cartier, P., Moussa, P., Julia, B., and Vanhove, P. (eds), *Frontiers in Number Theory, Physics, and Geometry II*. Springer, Berlin.

[581] Spaltenstein, N. 1982. *Classes Unipotentes et Sous-groupes de Borel*. Lecture Notes in Mathematics, vol. 946. Springer, Berlin, New York.

[582] Stade, E. 1990. On explicit integral formulas for GL(n, **R**)-Whittaker functions. *Duke Math. J.*, **60**(2), 313–362. With an appendix by D. Bump, S. Friedberg and J. Hoffstein.

[583] Stanley, R. P. 1986. A baker's dozen of conjectures concerning plane partitions. Pages 285–293 of Labelle, G., and Leroux, P. (eds), *Combinatoire Énumérative (Montreal, Que., 1985/Quebec, Que., 1985)*. Lecture Notes in Mathematics, vol. 1234. Springer, Berlin, New York.

[584] Stelle, K. S. 1996. Lectures on supergravity p-branes. Pages 287–339 of Gava, E., Masiero, A., Narain, K. S., Randjbar-Daemi, S., and Shafi, Q. (eds), *High Energy Physics and Cosmology. Proceedings, Summer School, Trieste, Italy, June 10-July 26, 1996*. World Scientific, Singapore.

[585] Stieberger, S. 2011. Constraints on tree-level higher order gravitational couplings in superstring theory. *Phys. Rev. Lett.*, **106**, 111601.

[586] Stieberger, S. 2014. Closed superstring amplitudes, single-valued multiple zeta values and the Deligne associator. *J. Phys.*, **A47**, 155401.

[587] Stieberger, S., and Taylor, T. R. 2006a. Amplitude for N-gluon superstring scattering. *Phys. Rev. Lett.*, **97**, 211601.

[588] Stieberger, S., and Taylor, T. R. 2006b. Multi-gluon scattering in open superstring theory. *Phys. Rev.*, **D74**, 126007.

[589] Stieberger, S., and Taylor, T. R. 2014. Closed string amplitudes as single-valued open string amplitudes. *Nucl. Phys.*, **B881**, 269–287.

[590] Strominger, A., and Vafa, C. 1996. Microscopic origin of the Bekenstein-Hawking entropy. *Phys. Lett.*, **B379**, 99–104.

[591] Strominger, A., and Witten, E. 1985. New manifolds for superstring compactification. *Commun. Math. Phys.*, **101**, 341.

[592] Takhtajan, L. A. 1992. A simple example of modular forms as tau-functions for integrable equations. *Theor. Math. Phys.*, **93**, 1308–1317.

[593] Tamagawa, T. 1963. On the ζ-functions of a division algebra. *Ann. of Math. (2)*, **77**, 387–405.

[594] Tate, J. T. 1967. Fourier analysis in number fields, and Hecke's zeta-functions. Pages 305–347 of Cassels, J. W. S., and Fröhlich, A. (eds), *Algebraic Number Theory: Proceedings of an Instructional Conference Organized by the London Mathematical Society, Brighton, September 1-17, 1965*. London Mathematical Society, London.

[595] Taylor, R., and Wiles, A. 1995. Ring-theoretic properties of certain Hecke algebras. *Ann. of Math. (2)*, **141**(3), 553–572.

[596] Teitelboim, C. 1986a. Gauge invariance for extended objects. *Phys. Lett.*, **B167**, 63–68.

[597] Teitelboim, C. 1986b. Monopoles of higher rank. *Phys. Lett.*, **B167**, 69–72.

[598] Terras, A. 1985. *Harmonic Analysis on Symmetric Spaces and Applications, I*. Springer, New York.

[599] Terras, A. 1988. *Harmonic Analysis on Symmetric Spaces and Applications, II*. Springer, Berlin.

[600] Tokuyama, T. 1988. A generating function of strict Gel′fand patterns and some formulas on characters of general linear groups. *J. Math. Soc. Japan*, **40**(4), 671–685.

[601] Tong, D. 2009. String theory. `arXiv:0908.0333 [hep-th]`.

[602] Tseytlin, A. A. 1997. On non-abelian generalisation of the Born-Infeld action in string theory. *Nucl. Phys.*, **B501**, 41–52.

[603] Unterberger, A. 2011. *Pseudodifferential Analysis, Automorphic Distributions in the Plane and Modular Forms*. Pseudo-Differential Operators: Theory and Applications, vol. 8. Birkhäuser/Springer, Basel.

[604] Vinberg, È. B. 1975. The classification of nilpotent elements of graded Lie algebras. *Dokl. Akad. Nauk SSSR*, **225**(4), 745–748.

[605] Vinogradov, A., and Takhtadžjan, L. 1978. Theory of the Eisenstein series for the group SL(3, R) and its application to a binary problem, I: Fourier expansion of the highest Eisenstein series. *Zap. Nauchn. Sem. Leningrad. Otdel. Mat. Inst. Steklov.*, **76**, 5–52.

[606] Wakimoto, M. 2001. *Lectures on Infinite-Dimensional Lie Algebra*. World Scientific Publishing, River Edge, NJ.

[607] Walcher, J. 2012. On the arithmetic of D-brane superpotentials: Lines and conics on the mirror quintic. *Commun. Num. Theor. Phys.*, **6**, 279–337.

[608] Wallach, N. R. 2003. Generalized Whittaker vectors for holomorphic and quaternionic representations. *Comment. Math. Helv.*, **78**(2), 266–307.

[609] Wang, Y., and Yin, X. 2015. Supervertices and non-renormalization conditions in maximal supergravity theories. `arXiv:1505.05861 [hep-th]`.

[610] Weil, A. 1929. L'arithmétique sur les courbes algébriques. *Acta Math.*, **52**(1), 281–315.

[611] Weil, A. 1949. Numbers of solutions of equations in finite fields. *Bull. Amer. Math. Soc.*, **55**, 497–508.

[612] Weil, A. 1965. Sur la formule de Siegel dans la théorie des groupes classiques. *Acta Math.*, **113**, 1–87.

[613] Weil, A. 1967. Über die Bestimmung Dirichletscher Reihen durch Funktionalgleichungen. *Math. Ann.*, **168**, 149–156.

[614] Weil, A. 1995. Séries de Dirichlet et fonctions automorphes. Pages 547–552 of *Séminaire Bourbaki, Vol. 10, Exp. No. 346*. Société Mathématique de France, Paris.

[615] West, P. C. 2001. E(11) and M theory. *Class. Quant. Grav.*, **18**, 4443–4460.

[616] Wiles, A. 1995. Modular elliptic curves and Fermat's last theorem. *Ann. of Math. (2)*, **141**(3), 443–551.

[617] Wiles, A. 2006. The Birch and Swinnerton-Dyer conjecture. Pages 31–41 of Carlson, J., Jaffe, A., and Wiles, A. (eds), *The Millennium Prize Problems*. Clay Mathematics Institute, Cambridge, MA.

[618] Witten, E. 1982. Constraints on supersymmetry breaking. *Nucl. Phys.*, **B202**, 253.

[619] Witten, E. 1988. Quantum field theory, Grassmannians, and algebraic curves. *Commun. Math. Phys.*, **113**, 529.

[620] Witten, E. 1995. String theory dynamics in various dimensions. *Nucl. Phys.*, **B443**, 85–126.

[621] Witten, E. 1996. Bound states of strings and p-branes. *Nucl. Phys.*, **B460**, 335–350.

[622] Witten, E. 2000. World sheet corrections via D instantons. *JHEP*, **02**, 030.

[623] Witten, E. 2007. Gauge theory and wild ramification. `arXiv:0710.0631 [hep-th]`.

[624] Witten, E. 2010. Mirror Symmetry, Hitchin's Equations, and Langlands Duality. Pages 113–128 of Garcia-Prada, O., Bourguignon, J.-P., and Salamon, S. (eds), *The Many Facets of Geometry: A Tribute to Nigel Hitchin*. Oxford Science Publications. Oxford University Press, Oxford.

[625] Witten, E. 2015. More on gauge theory and geometric Langlands. `arXiv:1506.04293 [hep-th]`.

[626] Yi, P. 1997. Witten index and threshold bound states of D-branes. *Nucl. Phys.*, **B505**, 307–318.

[627] Zagier, D. 1984. *L*-series of elliptic curves, the Birch-Swinnerton-Dyer conjecture, and the class number problem of Gauss. *Notices Amer. Math. Soc.*, **31**(7), 739–743.

[628] Zagier, D. 1990. The Bloch-Wigner-Ramakrishnan polylogarithm function. *Math. Ann.*, **286**(1-3), 613–624.

[629] Zagier, D. 1991. The Birch-Swinnerton-Dyer conjecture from a naive point of view. Pages 377–389 of van der Geer, G., Oort, F., and Steenbrink, J. H. M. (eds), *Arithmetic Algebraic Geometry (Texel, 1989)*. Progress in Mathematics, vol. 89. Birkhäuser Boston, Boston, MA.

[630] Zagier, D. 2008. Elliptic modular forms and their applications. Pages 1–103 of Ranestad, K. (ed.), *The 1-2-3 of Modular Forms: Lectures from the Summer School on Modular Forms and Their Applications, Held at Nordfjordied, Norway, June 2004*. Universitext. Springer, Berlin.

[631] Zelevinsky, A. V. 1980. Induced representations of reductive $\mathfrak{p}$-adic groups, II: On irreducible representations of GL(n). *Ann. Sci. École Norm. Sup. (4)*, **13**(2), 165–210.

[632] Zerbini, F. 2016. Single-valued multiple zeta values in genus 1 superstring amplitudes. *Commun. Num. Theor. Phys.*, **10**, 703–737.

[633] Zhang, S.-W. 2010. Gross-Schoen cycles and dualising sheaves. *Invent. Math.*, **179**(1), 1–73.

[634] Zwiebach, B. 2004. *A First Course in String Theory*. Cambridge University Press, Cambridge.

Index

Printed in the United States
by Baker & Taylor Publisher Services